NEUROSCIENCE IN MEDICINE

NEUROSCIENCE IN MEDICINE

EDITOR

P. MICHAEL CONN, Ph.D.

Associate Provost for Extended Campus
and Biotechnology Development
Professor of Pharmacology
Oregon Health Sciences University
Portland, Oregon
Associate Director for Research and Development
Oregon Regional Primate Research Center
Beaverton, Oregon

J. B. LIPPINCOTT COMPANY
Philadelphia

Acquisitions Editor: Richard Winters
Assistant Editors: Jody Schott, Melissa James
Project Editor: Ellen M. Campbell
Indexer: Sandra King
Design Coordinator: Doug Smock
Interior Designer: Bill Donnelly
Cover Designer: Louis Fuiano
Production Manager: Caren P. Erlichman
Senior Production Coordinator: Kevin P. Johnson
Compositor: Compset Inc.
Printer/Binder: Courier Book Company/Westford

6   5   4   3   2   1

Library of Congress Cataloging-in-Publication Data

Neuroscience in medicine/[edited by] P. Michael Conn.
   p.   cm.
  Includes bibliographical references and index.
  ISBN 0-397-51279-1
  1. Neurobiology. 2. Neurophysiology. I. Conn, P. Michael.
  [DNLM: 1. Nervous System—physiology. WL 102 N50594 1994]
QP355.2.N53 1994
612.8—dc20
DNLM/DLC
for Library of Congress   94-22364
                                      CIP

∞ This Paper Meets the Requirements of ANSI/NISO Z39.48-1992
(Permanence of Paper).

The authors and publishers have exerted every effort to ensure that drug
selection and dosage set forth in this text are in accord with current
recommendations and practice at the time of publication. However, in view of
ongoing research, changes in government regulations, and the constant flow
of information relating to drug therapy and drug reactions, the reader is
urged to check the package insert for each drug for any change in indications
and dosage and for added warnings and precautions. This is particularly
important when the recommended agent is a new or infrequently employed
drug.

Contributors

Nancy C. Andreasen, M.D., Ph.D.
Andrew H. Woods Professor of Psychiatry
The University of Iowa College of Medicine
Director, Mental Health Clinical Research Center
The University of Iowa Hospitals and Clinics
Iowa City, Iowa

Mark F. Bear, Ph.D.
Associate Professor
Department of Neuroscience
Brown University
Associate Investigator
Howard Hughes Medical Institute
Providence, Rhode Island

Margery C. Beinfeld, Ph.D.
Professor
Department of Pharmacological and Physiological Science
St. Louis University Medical School
St. Louis, Missouri

Michael J. Brownstein, M.D., Ph.D.
Laboratory of Cell Biology
National Institute of Mental Health
Bethesda, Maryland

Rochelle S. Cohen, Ph.D.
Professor of Anatomy and Cell Biology
Department of Anatomy and Cell Biology
University of Illinois at Chicago
College of Medicine
Chicago, Illinois

Marc E. Freeman, Ph.D.
Professor
Department of Biological Science
The Florida State University
Tallahassee, Florida

Julio H. Garcia, M.D., FCAP
Professor of Pathology
Case Western Reserve University
Director, Division of Neuropathology
Henry Ford Hospital
Detroit, Michigan

Daniel Gardner, Ph.D.
Professor of Physiology and Biophysics in Neuroscience
Cornell University Medical College
New York, New York

G. F. Gebhart, Ph.D.
Professor
Department of Pharmacology
The University of Iowa College of Medicine
Iowa City, Iowa

J. Fielding Hejtmancik, M.D., Ph.D.
Medical Officer
National Eye Institute
Bethesda, Maryland

Igor A. Ilinsky, M.D., Ph.D.
Research Scientist
Department of Anatomy
The University of Iowa College of Medicine
Iowa City, Iowa

Kristy Kultas-Ilinsky, Ph.D.
Professor of Anatomy
Department of Anatomy
The University of Iowa College of Medicine
Iowa City, Iowa

Conrad E. Johanson, Ph.D.
Professor of Clinical Neuroscience and Physiology
Brown University
Director of Cerebrospinal Fluid Laboratory
Rhode Island Hospital
Providence, Rhode Island

Jean M. Lauder, Ph.D.
Professor of Cell Biology and Anatomy
University of North Carolina
School of Medicine
Chapel Hill, North Carolina

Michael D. Lumpkin, Ph.D.
Professor and Acting Chairman
Department of Physiology and Biophysics
Georgetown University School of Medicine
Washington, DC

Bruce E. Maley, Ph.D.
Associate Professor
Department of Anatomy and Neurobiology
University of Kentucky Medical Center
Lexington, Kentucky

Robert W. McCarley, M.D.
Professor and Director
Laboratory of Neuroscience
Department of Psychiatry
Harvard Medical School
Medical Investigator and Associate Chief, Psychiatry
Brockton/West Roxbury VA Medical Center
Brockton, Massachusetts

Michael W. Miller, Ph.D.
Professor, Department of Psychiatry and Pharmacology
The University of Iowa College of Medicine
Iowa City, Iowa

Marion Murray, Ph.D.
Professor of Anatomy and Neurobiology
Department of Anatomy and Neurobiology
Medical College of Pennsylvania
Philadelphia, Pennsylvania

Gerry S. Oxford, Ph.D.
Professor of Physiology
Department of Physiology
University of North Carolina at Chapel Hill
Chapel Hill, North Carolina

Suresh C. Patel, M.D.
Director, Section of Neuroradiology
Department of Diagnostic Radiology
Henry Ford Hospital
Detroit, Michigan

Donald W. Pfaff, Ph.D.
Professor of Neurobiology and Behavior
The Rockefeller University
New York, New York

Paul J. Reier, Ph.D.
Mark F. Overstreet Professor
Department of Neurological Surgery and Neuroscience
University of Florida College of Medicine
Gainesville, Florida

Robert L. Rodnitzky, M.D.
Professor and Vice Chairman
Department of Neurology
The University of Iowa College of Medicine
Attending Physician
University of Iowa Hospitals and Clinics
Iowa City, Iowa

William Z. Rymer, M.D., Ph.D.
John G. Searle Professor
Departments of Physical Medicine and
    Rehabilitation and Physiology
Northwestern University Medical School
Director of Research
Rehabilitation Institute of Chicago
Chicago, Illinois

Michael T. Shipley, Ph.D.
Professor and Chairman
Department of Anatomy
University of Maryland School of Medicine
Baltimore, Maryland

David V. Smith, Ph.D.
Professor
Department of Anatomy
University of Maryland School of Medicine
Baltimore, Maryland

Robert F. Spencer, Ph.D.
Professor of Anatomy
School of Medicine
Medical College of Virginia
Virginia Commonwealth University
Richmond, Virginia

David K. Sundberg, Ph.D.
Associate Professor
Department of Physiology and Pharmacology
Bowman Gray School of Medicine
Wake Forest University
Winston-Salem, North Carolina

Daniel Tranel, Ph.D.
Professor of Neurology and Psychology
University of Iowa
Director, Benton Neuropsychology Laboratory
The University of Iowa Hospitals and Clinics
Iowa City, Iowa

Harold II. Traurig, B.S., Ph.D.
Professor and Vice Chairman
Department of Anatomy and Neurobiology
University of Kentucky College of Medicine
Lexington, Kentucky

Thomas van Groen, Ph.D.
Research Assistant Professor
Department of Cell Biology
School of Medicine
The University of Alabama at Birmingham
Birmingham, Alabama

Brent A. Vogt, Ph.D.
Associate Professor of Physiology and Pharmacology
Bowman Gray School of Medicine
Wake Forest University
Winston-Salem, North Carolina

Simone Wagner, M.D.
Resident Physician
Department of Neurology
University of Heidelberg
Heidelberg, Germany

James R. West, Ph.D.
Professor and Head
Department of Anatomy and Neurobiology
College of Medicine
Texas A&M University Health Science Center
College Station, Texas

J. Michael Wyss, Ph.D.
Professor
Department of Cell Biology
School of Medicine
The University of Alabama at Birmingham
Birmingham, Alabama

Tom C.T. Yin, Ph.D.
Professor
Department of Neurophysiology
University of Wisconsin Medical School
Madison, Wisconsin

Preface

Neuroscience is a fascinating discipline. It encompasses a broad range of fields: gene expression and protein synthesis at the molecular level, cell biology, anatomy, biochemistry, physiology, and behavior, from learning and memory to thought processes. Accordingly, neuroscientists have been poised to take advantage of the growth of knowledge in these fields and the recent technological advances that have promoted this growth. In recognition of this thriving field, the 1990s were proclaimed by then-President Bush as the "Decade of the Brain." The National Institutes of Health has a separate Institute for Neuroscience (NINCDS). This is a heady time indeed for the field.

Neuroscience has become a course in its own right in most medical schools worldwide. Because medical students today have enormous demands upon their time, a major challenge in the development of *Neuroscience in Medicine* was to identify the "core" material associated with a course in the neurosciences. "Core" was interpreted to mean not only the essential material, but also enough subject diversity to allow the student to sample from the range of topics represented in this discipline and get a sense of the excitement in and direction of the field.

In order to meet this challenge, not only did the individual authors have to be leaders in the research end of their topics, but they also had to have credentials as excellent teachers. Such individuals are rare and, not surprisingly, are very careful with their time. Happily, the authors involved here recognized the significance of the project and generously made the necessary commitment to it. Once the manuscripts were in hand, it was the editor's job to make the writing uniform, remove duplicative material except where essential for ease of understanding, and incorporate additional critical material.

The text is designed to reveal the basic science underlying diseases and treatments for disorders of a neuroscience nature. While the chapters are intended to interdigitate, each chapter can be read as a "stand alone"—that is, it contains a complete discussion of the topic. Fourteen Clinical Correlations are presented to emphasize the scientific basis of disease. The book is thoroughly illustrated and uses color for emphasis. Tables are included to summarize and compare. The interested reader may also wish to consult the companion volume, *Atlas of the Human Brain,* by Jennes, Traurig, and Conn, which includes the traditional neuroanatomy of the central nervous system along with chapters on current imaging techniques.

I appreciate the help of many outstanding contributors who not only took time out from their busy research and teaching schedules to write, but also submitted their contributions in a timely fashion and responded favorably to the comments of outside readers. We have also been aided in our task by the art and editorial staff at Lippincott, whose help I gratefully acknowledge.

P. Michael Conn

Contents

Chapter 3
Cytology and Organization of Cell Types: Light and Electron Microscopy 21

Rochelle S. Cohen
Donald W. Pfaff

Chapter 11
Spinal Cord   197
Marion Murray

Clinical Correlation: DISORDERS OF THE SPINAL CORD   210
Robert L. Rodnitzky

Chapter 20
Spinal Mechanisms for Control of Muscle Length and Force 369
William Z. Rymer

Chapter 25
Audition 485
Tom C. T. Yin

Chapter 26
The Vestibular System 501
Robert F. Spencer

Chapter 27
The Gustatory System 511
David V. Smith
Michael T. Shipley

Chapter 30
Higher Brain Functions **555**
Daniel Tranel

NEUROSCIENCE IN MEDICINE

Section *I*

ORGANIZATIONAL LEVELS OF THE NERVOUS SYSTEM

Neuroscience in Medicine, edited by P. Michael Conn.
J. B. Lippincott Company, Philadelphia © 1995

Chapter *1*

Molecular Biology: An Overview

MICHAEL J. BROWNSTEIN

Compared with other organs, the human brain is extraordinarily complex. It is composed of about 10^{11} nerve cells, called neurons. As a class, these cells uniquely possess specialized processes for receiving (i.e., dendrites) and transmitting (i.e., axons) information, and they are excitable. Neurons are structurally and functionally diverse. There are thought to be several thousand types of neurons that differ in the signals they emit, the locations where they reside, and the cells from which they receive and to which they transmit information. The nature of each neuron is determined in the course of the brain's development and subsequently modified by experience.

Gene expression plays an important role in determining the structure and function of each neuron. In this chapter, I summarize how genes are activated, resulting in the production of specific proteins. That gene activation must occur in the correct cell at the correct time and in response to particular extracellular signals accounts for the complexity of the process of regulation.

The Genome

The genetic material inherited by each person is encoded in the form of double-stranded DNA. DNA is a polymer composed of four deoxyribonucleotides (Figs. 1-1 through 1-3). The sugar moiety forms the backbone of the DNA, and the bases vary. Two of the bases are purines, adenine (A) and guanine (G), and two are pyrimidines, thymine (T) and cytosine (C). Although the genome may look like a randomly arranged pack of deoxyribonucleotide cards, the sequence of the bases and the structure of the DNA are highly ordered, and this is of great consequence.

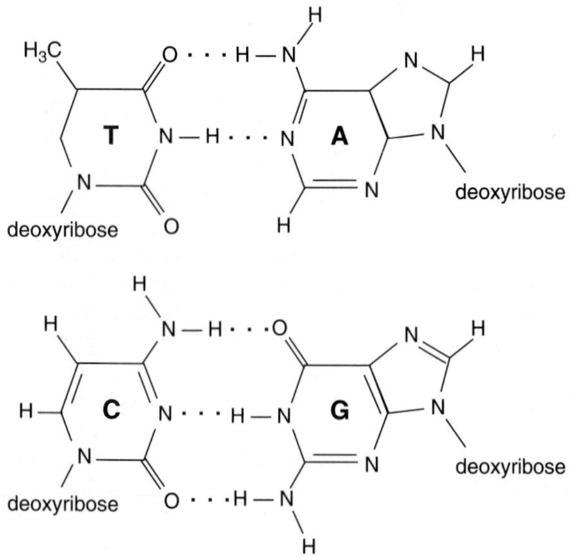

Figure 1-1
The purine-pyrimidine base pairs are held together by hydrogen bonds. Thymine (T) always pairs with adenine (A), and cytosine (C) always pairs with guanine (G).

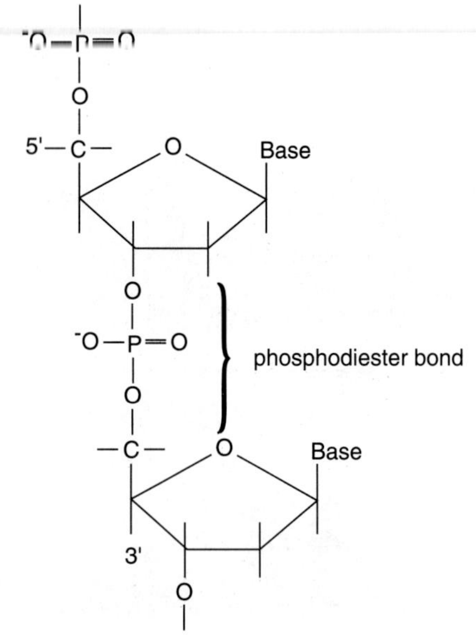

Figure 1-2
The backbone of the DNA polymer is formed by deoxyriboses attached by phosphodiester bridges. Notice that the chain is polar.

The human haploid genome contains about 3×10^9 nucleotides or base pairs (bp). In eukaryotes, most of the DNA is found in the nucleus. (We are ignoring mitochondrial DNA for the sake of simplicity.) The DNA does not exist as long, extended chains, but it is instead associated with proteins (Fig. 1-4). This DNA-protein complex is known as *chromatin*. The basic unit of chromatin is the *nucleosome*, which is a double-stranded (i.e., duplex) DNA wrapped around a protein octamer known as the core histone. DNA in this form—a coiled coil—has about one sixth of the length of the extended duplex. During metaphase, when the DNA condenses into chromosomes, its length is less than one thousandth of the extended length. This condensed DNA in which the nucleosomes are packed together to form higher-order structures does not function as a template for protein production.

The genome is a type of memory that works forward. It encodes polypeptides, linear polymers composed of as many as 20 different amino acids. Each amino acid residue in a protein is encoded by a triplet of nucleotides, a group called a *codon*. Sixty-four codons can be produced by the permutation of the four bases ($4 \times 4 \times 4$), but only 20 different amino acids are used to make proteins. The reason for this apparent discrepancy is that some amino acids can be coded for by more than one codon. The code is therefore unambiguous but degenerate, because more than one three-letter "word" or codon can specify certain amino acids, but each codon has only one meaning and specifies only one amino acid. DNA mutations are caused by substitutions of one base for another or by deletions or insertions of bases. Mutations can change single words or entire phrases of genetic information, and these alterations can result in diseases.

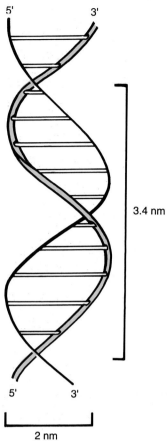

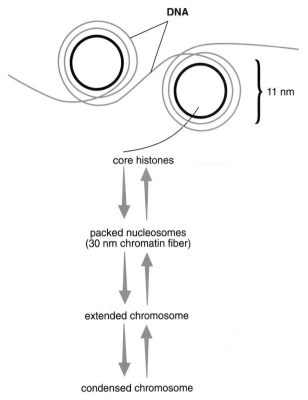

Figure 1-3
In the double helix configuration, the complimentary strands of the DNA duplex are held together by A-T and G-C base pairings, which are depicted as the rungs of a spiral staircase. Each chain is a mirror image of the other. The two antiparallel chains twist around one another and run in opposite directions. Ten residues form each complete turn (3.4 nm) of the coil around the central axis.

Genes

The genome is divided into genes, each one of which is copied (i.e., transcribed) by enzymes, yielding RNA copies. The sugar backbone of RNA is composed of ribose rather than deoxyribose found in DNA; the bases are of RNA are A, G, C, and uracil (U), instead of T. After the RNA transcript is converted into messenger RNA (mRNA), the product is exported to the cytoplasm. It is the RNA copy of the DNA blueprint that serves as a template for protein biosynthesis by the ribosome.

If the entire genome were transcribed, assuming that the average polypeptide or protein is about 500 amino acids long, the genome could direct the synthesis of 2×10^6 proteins. However, this is not the case; more than 95% of the DNA in the genome is noncoding. This leaves enough DNA for 100,000 proteins, almost half of which are thought to be present in the brain. The number of genes transcribed in the brain is much higher than in any other tissue.

Figure 1-4
Decondensed chromatin has the appearance of beads on a string. The beads, which are spools of core histone protein with two 80-nucleoside pairs of double stranded DNA wound about them, are connected by segments of linker DNA, which is depicted as the string. The bead-like nucleosomes are packed on one another to form 30-nm fibers. These fibers can form a series of loops that coil around a central axis, producing the superhelical form that must further condense to generate the metaphase chromosomes.

☐ One gene, one polypeptide?

The notion that each gene encodes a single polypeptide is no longer tenable. More than one primary transcript can be produced from a single gene, and these primary transcripts can give rise to multiple mRNA molecules that encode different proteins.

All mRNAs have two features in common: a methylated G residue added to the 5′ (i.e., upstream) end and a tail of 200 to 250 adenylic acid residues added to the 3′ (i.e., downstream) end of the message. The latter is typically found about 20 bases downstream of the sequence AAUAAA. Polyadenylation can occur at more than one site in the same primary transcript, and the choice of the polyadenylation site is cell specific. Polyadenylation is thought to be important protein synthesis. Messenger RNAs with A tails less than 30 to 40 bases long fail to form polysomes. Rather than returning to the beginning of the message and translating it again, ribosomes fall off such templates. Upstream, noncoding sequences also participate in reinitiation, and downstream noncoding sequences sometimes affect message stability.

In mRNA, the coding region is uninterrupted. The ribosome initiates synthesis at an ATG initiation codon and adds the amino acid dictated by each subsequent codon until it encounters a stop signal (i.e., UAA, UAG, or UGA). The coding region in the primary transcript may not be continuous; it may be interrupted by segments called *introns*, which divide the coding region into segments called *exons*. Before the mRNA leaves the nucleus, the introns have to be removed or spliced out. The pattern of splicing can differ in different cells, and the consequences of this are discussed later.

☐ Transcription: the core promoter establishes the direction

In principal, either strand of the genomic DNA duplex can be transcribed. The *core promoter* is a DNA sequence that defines which strand of the duplex is transcribed and where transcription begins (Fig. 1-5). It is found immediately upstream of the transcriptional start site, also known as the *cap site* because it has the methyl G cap added to it immediately after transcription. The most important component of the core promoter is the *TATA box*, a DNA segment with the following sequence:

$$5'\text{-GNGTATA}^{A}_{T}A^{A}_{T}\text{-}3'$$

in which N indicates any one of the four bases. It binds and orients *RNA polymerase II*, the enzyme responsible for synthesizing transcripts destined to be processed into mRNAs. There are two other RNA polymerases: RNA polymerases I and III. The former transcribes the large ribosomal RNAs; the latter transcribes small, stable RNAs such as transfer RNAs.

The promoter cannot be cut out and turned around and still work properly. Assuming that a particular gene is capable of being transcribed (i.e., it is not found in an inaccessible, condensed portion of the genome, and it has not been inactivated by methylation), the process of transcription begins with the binding of one of several transcription factors (e.g., TFIID) to the TATA box. Sub-

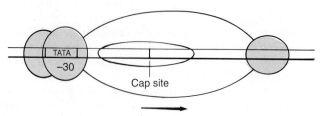

Figure 1-5
DNA transcription requires RNA polymerase and certain nucleotide sequences, called the core promoter, that define which DNA strand is transcribed and where transcription begins. The most important component of the promoter is the TATA box, a DNA segment that orients and binds the RNA polymerase. The initiation complex depicted consists of transcription initiation factors (*shaded areas*) and RNA polymerase II (*white area*). The arrow indicates the direction of transcription. Not all genes have a TATA box; the neural cell adhesion molecule gene is one example. The cap site is the transcriptional initiation site.

sequently, additional transcription factors (e.g., TFIIA, TFIIB, TFIIE, TFIIF) and RNA polymerase II also attach themselves to this domain, forming what is known as the *initiation complex*. At this point, the complex is ready to begin unwinding the DNA duplex and copying the "minus strand." The RNA copy grows from the 5' end to the 3' end; the DNA template is read from the 3' end to the 5' end. The RNA transcript is complementary to the template.

Transcription initiation depends on activation of the initiation complex, after which the chain is elongated at a brisk pace. Elongation ends when the polymerase encounters a termination sequence.

Transactivation

Transcription can be studied by introducing chimeric gene expression elements into eukaryotic cells. These elements consist of noncoding DNA from one gene that is presumed to contain regulatory domains and the coding region from a second, *reporter gene*. Examples of the latter include coding regions for β-galactosidase, chloramphenicol acetyltransferase, and luciferase. The protein products of these genes are easy to detect in cells (Fig. 1-6). Constructs that have only the core promoter upstream of the reporter are often transcribed poorly; additional (*cis*-acting) DNA sequences 8 to 15 nucleotides long are required for robust expression. These *enhancer elements* are typically found upstream of the core promoter, but they can be in introns or even downstream of the transcriptional termination site. They can be close to the core promoter (<30 bp) or far away (>10,000 bp). Unlike the core promoter sequence that has to face the minus strand it directs the transcription of, enhancer elements function in either orientation (i.e., 5' to 3' or 3' to 5').

Because enhancers can activate or inhibit transcription, the name given these elements is a somewhat unfortunate one. They are binding sites for gene regulatory proteins, which are often referred to as transactivators, although transregulator is more appropriate. These proteins often are capable of binding to DNA targets and to protein members of the initiation complex (Fig. 1-7). Binding to DNA may bring the factor into proximity to the initiation complex, facilitating or prolonging its interaction with the latter and activating, inhibiting, or suppressing it. The requirement that regulatory proteins must bind DNA to act guarantees that they affect specific genes. (Genomic DNA is highly structured, and an enhancer element hundreds or thousands of base pairs away from the core promoter may actually be near it because of DNA bending.) The precise mechanisms by which transactivators act are unknown, but they appear to stimulate the assembly and activity of the initiation complex. The interaction of an activator with the core promoter can occur only if the DNA separating the two forms a loop. Some activators may bind to nucleosomal DNA; others may only bind when DNA is not associated with histones.

Several classes of transactivator proteins have been isolated. Some of these, such as SP1 and OTF1, are consti-

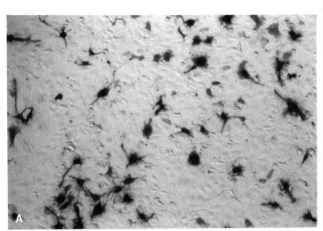

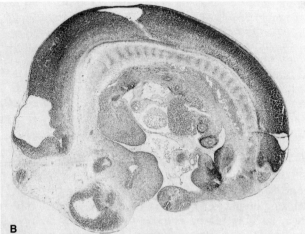

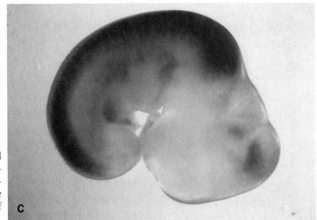

Figure 1-6
Coding regions, called reporters, can be introduced into cells and used to detect the activity of transcription promoters and their enhancers. The bacterial enzyme, β-galactosidase, was used as a reporter in (**A**) cultured cells, (**B**) tissues from a transgenic mouse embryo, and (**C**) an intact transgenic mouse embryo. (Courtesy of Andreas Zimmer, National Institutes of Mental Health)

tutively produced by all cells and guarantee that "housekeeping" proteins are made. Others, such as PIT1, are found in specific cells or are activated or inactivated by hormones or other chemical messengers.

It is worth describing how some gene regulatory proteins are thought to operate. The ATF or CREB protein is expressed in most cells. It is a substrate of protein kinase A, an enzyme that is activated by cyclic AMP, which is produced in response to activation of cell surface receptors by neurotransmitters. ATF binds to a specific DNA sequence:

$$5'\text{-}^{TT}_{GA}\text{CGTCA-}3'$$

When ATF is phosphorylated, it interacts with proteins of the core promoter.

The steroid-hormone receptor family of proteins have DNA-binding, effector, and hormone-binding domains. If the hormone-binding domains are removed, the receptors are constitutively active, indicating that the hormone-binding domain normally inhibits these proteins' actions. Binding of hormone appears to reverse this inhibition. Members of the helix-loop-helix family of transactivators, such as MYOD and E12/47, are only active as heterodimers. Introduction of MYOD into embryonal fibroblasts

induces them to differentiate into muscle cells, and myoD is thought to play a major role in the differentiation of myoblasts into mature muscle cells in vivo. It was surprising to find that myoblasts have high levels of MYOD, and they proved to have high levels of a related protein, Id. Id forms a heterodimer with myoD. Before differen-

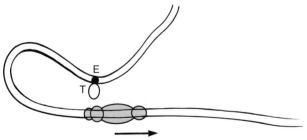

Figure 1-7
Enhancers can activate or inhibit transcription. A transcription factor (T) is bound to an enhancer element (E). Looping of the DNA strand positions the enhanced transcription factor near the proteins of the initiation complex (*shaded area*) and stimulates transcription. The initiation factors seem to stay bound to the core promoter site, facilitating its interaction with a series of RNA polymerase II molecules.

tiation, Id levels drop, and in the absence of its "poison partner," MYOD is active. This poison partner represents one class of repressor. Others bind to enhancer sites but fail to stimulate transcription or bind to the initiation complex without activating it and simultaneously prevent the binding of activators. V16 is a transactivator that is incapable of binding to DNA. Instead, it binds to another DNA-binding protein OCT1 and then activates the core promoter.

These variations on a theme illustrate a few of the ways in which regulatory proteins can turn transcription off; there are others. Multiple regulatory proteins acting on multiple enhancers can affect the transcription of single genes. Some of these activators are strong and can act over long distances, but others are weak and need to be nearby the core promoter to be effective. How multiple bound activators communicate with the promoter is a matter of speculation. Some may act simultaneously and cooperatively, or they may act independently, each one binding in its turn to some member of the promoter complex. The more activators there are, the more likely a productive interaction will occur or be sustained.

Splicing

The coding region in many primary transcripts is interrupted by intronic segments (Fig. 1-8). A 5' splice site or donor site is found on the upstream end of each intron; a 3' splice site or acceptor site is found on the downstream end. Roughly 30 bases upstream of the acceptor site is a branch point, a *cis*-acting splicing element in the intron. These elements interact with a complex composed of proteins and small nuclear RNAs (snRNAs). The ribonucleoprotein complex (i.e., snurp) together with the pre-RNA it acts on form a structure called the *spliceosome*. In the course of the splicing process, the upstream exon is transiently separated from the adjacent intron. Held in place by the spliceosome complex, the free 3' hydroxyl group on the last base of this exon subjects the acceptor site to nucleophilic attack. The resulting trans-

esterification reaction links the exons and liberates the intron.

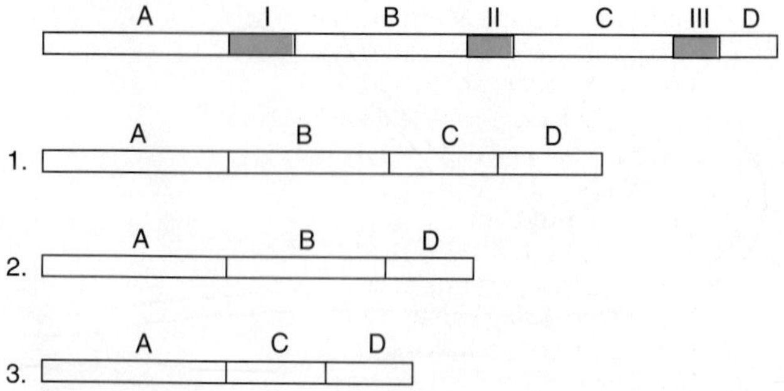

The pattern of splicing varies in different cells, and there are many potential outcomes. The intracellular distribution of the protein product can be altered (e.g., one species can be membrane bound, another secreted); a family of functionally similar but kinetically different proteins can be produced; or functionally different products can be made, as in the case of the precursors of calcitonin and CGRP, which derive from the same transcript. Alterations in the 5' or 3' untranslated domains of the mRNA could affect translational efficiency or message stability, both of which are also actively regulated. The mechanisms underlying alternative splicing by different cells are poorly understood. Removal of all introns is the default condition, and alternative modes of splicing require active intervention by the cell (i.e., production of putative splicing inhibitors).

A mode of splicing observed previously in lower eukaryotes has been demonstrated to occur in mammalian cells: *trans*-splicing. This involves splicing the 5' end of one transcript onto the 3' end of another. The physiologic importance of this phenomenon remains to be determined.

Conclusion

The description of genes and regulation of gene expression presented here is not comprehensive and not illustrated with experimental data. Because of a lack of space, several important topics were not discussed: the cell cycle, DNA replication, DNA repair, translation of RNA, intracellular protein targeting, and posttranslational modifications of proteins that affect their function. Some of these themes are dealt with elsewhere in this text, and others are discussed in textbooks of biochemistry and cell biology. Suggestions for further reading are provided in the Selected Readings section of this chapter. The student of neuroscience should become familiar with the fundamentals of each of these areas of study.

Figure 1-8

Splice choices. The primary transcript shown has four exons (A–D) and three introns (I–III). In case 1, all three introns are removed; in case 2, segments II, C, and III are treated as one large intron and spliced out, along with intron I; in case 3, segments I, B, and II are removed along with intron III.

SELECTED READINGS

Alberts B, Bray D, Lewis J, Raff M, Roberts K, Watson JD. Molecular biology of the cell, 2nd ed. New York: Garland Publishing, 1989.
Felsenfeld G. DNA. Sci Am 1985;253:58.
Hall Z. An introduction to molecular neurobiology. Sunderland, MA: Sinauer Associates, 1992.
Kriegler M. Assembly of enhancers, promoters, and splice signals. In Goeddel DV (ed). Gene expression technology, vol 195 of Methods in enzymology. New York: Academic Press, 1990.
Lewin B. Genes, 3rd ed. New York: Wiley, 1987.
Ptashne M. A genetic switch: phage and higher organisms, 2nd ed. Cambridge, MA: Cell Press and Blackwell Scientific Publications, 1992.
Steitz JA. Snurps. Sci Am 1987;258:56.
Watson JD, Hopkins NH, Roberts JW, Steitz JA, Weiner AM. Molecular biology of the gene, 4th ed. Menlo Park, CA: Benjamin-Cummings, 1987.

Neuroscience in Medicine, edited by P. Michael Conn.
J. B. Lippincott Company, Philadelphia © 1995

Chapter *2*

*P*rotein Synthesis and Distribution: Structural and Secreted Products

MARGERY C. BEINFELD

continued

The Golgi is a Carbohydrate Factory Responsible for Additional Protein Glycosylation Reactions, the Biosynthesis of Glycolipids, Proteoglycans, and Mucins
Proteins are Further Modified in the Golgi
Proteins Targeted for Regulated Secretion are Packaged in Condensing Vesicles in the Trans-Golgi Network
Hydrolytic Enzymes That Carry a Mannose-6-phosphate are Selected

for Transport to Lysosomes in the Trans-Golgi Network

Constitutive and Regulated Pathways of Secretion
Many Cell Surface Receptors With Their Attached Soluble Ligands are Internalized From Coated Pits by Clathrin-Coated Vesicles
Neurons Require an Additional Mechanism to Transport Proteins to the Synaptic Cleft

---

This chapter provides an overview of protein synthesis and traffic in cells, with an emphasis on the structural features of individual proteins that determine their fate.

Mechanism and Factors Required for Protein Synthesis

☐ Protein synthesis translates the genetic code

Protein synthesis is an energy-intensive process that represents an irreversible commitment to specific biologic function. In eukaryotes, cytosolic protein synthesis takes place on an 80S ribosome, which is composed of a 40S and 60S subunits. These subunits are composed of several proteins and nucleic acids. They are assembled in the nucleus and transported by transfer RNA (tRNA) into the cytoplasm. The 40S subunit contains a P site (i.e., peptidyl-tRNA site) and an A site (i.e., amino-acyl-tRNA site).

The overall mechanism of protein synthesis has been elucidated, but the identity of all the factors and their interaction at the molecular level is still under investigation. Protein synthesis involves a large number of components. In addition to the ribosomal subunits, the different tRNAs, and the amino-acyl-tRNA synthetases for each amino acid, there are at least 18 other protein factors, ranging from 17 to 550 kDa, that are involved in initiation, elongation, and termination. It is the amino-acyl-tRNAs that are the "bilingual" components that translate the information from nucleic acid to amino acid. Perhaps all this complexity is necessary to ensure the high rate of fidelity of this process.

☐ The rate of synthesis of specific proteins is controlled by several factors

The overall rate of protein synthesis is limited by the rate of initiation. The state of phosphorylation and dephosphorylation of certain initiation factors also exerts positive and negative control of the rate of protein synthesis, and this type of control has been correlated with cellular

growth. The rate of synthesis of specific proteins is regulated by the abundance of their mRNAs and several features of the structure of their mRNAs, including the presence of the 5'-m^{7}G cap and its accessibility, their secondary and tertiary structure, placement and context of their AUG initiation codon, presence of the poly A$^+$ tail, and the presence of protein repressor factors that inhibit translation.

Ribosomal scanning model of initiation

Figure 2-1 depicts a simplified version of the current model of protein synthesis. For the sake of clarity, all of the factors that act at each step are not shown in the diagram. Protein synthesis is divided into three stages: initiation, elongation, and termination. This model accounts for the initiation of most proteins. For some proteins, the initial binding of the preinitiation complex occurs downstream from the 5' cap, closer to the initiation codon.

Initiation

The dissociation of the 40S and 60S subunits is promoted by the binding of the initiation factors EIF1A and EIF3. The binding of Met-tRNA to the 40S subunit is stabilized by EIF2, the preinitiation complex. In mRNA binding, the preinitiation complex binds near the 5' terminus of mRNA. The mRNA is scanned to find the AUG initiation site. The initiation factors EIF4E, EIF4A, EIF4B, and EIF4F participate in a complex that displays ATP-dependent RNA "helicase" activity, which unwinds the 5' untranslated region up to the initiation codon while the 40S subunit follows. The 60S subunit binds, and the initiation factors dissociate.

Elongation

In the binding of amino-acyl-tRNA, an elongation factor, and GTP, the initiation factor EEF1A combined with GTP binds to the amino-acyl-tRNA before it binds to the 80S complex. The GTP on EEF1A is hydrolyzed, and the elongation factor dissociates. The peptidyltransferase activity in the 60S subunit catalyzes peptide bond formation. The peptidyl-tRNA then translocates from the A to the P site.

Termination

A release factor binds to the stop codon. The factor ERF promotes the cleavage of the protein from its peptidyl-tRNA site. Protein is released, and the factors and mRNA dissociate.

Two Major Pathways of Protein Synthesis: Cytosol and Endoplasmic Reticulum

☐ Divergent pathways produce proteins for all the cellular organelles

With the exception of a few proteins synthesized on ribosomes in mitochondria, the synthesis of most proteins begins on ribosomes in the cytosol. Although both pathways use the same cellular machinery for protein synthesis and share a common pool of ribosomes, they diverge immediately. Proteins destined to be retained in the cytosol or to be transported into mitochondria, the nucleus, or peroxisomes are synthesized in the cytosol. Proteins that will be retained in the ER or Golgi or proteins destined for secretion, lysosomes, or the cell surface have a different fate.

All of the proteins that are not made in the cytosol have a signal sequence (i.e., a stretch of 5 to 10 hydrophobic amino acids) that is recognized by several specific proteins and that results in the insertion of the growing peptide chain into the lumen of the ER. After extensive modification in the ER, they move in transport vesicles to the Golgi apparatus, where additional modifications are made and they are sorted for delivery to their final destinations. This pathway is illustrated in Figure 2-2.

☐ Proteins synthesized in the cytosol but destined for the nucleus, mitochondria, and peroxisomes contain targeting information in their primary sequence

Most of the mitochondrial proteins and all of the proteins of the nucleus and peroxisomes are imported from the cytosol. The targeting information resides in the primary sequence of these proteins.

For all three organelles, protein import probably involves specific receptors that recognize their respective targeting signals. In all cases, these proteins are thought to be imported through pores in the membrane by active, energy-dependent processes. Mitochondrial import also requires an electrochemical gradient produced by energy-dependent inward proton transport.

The nucleus must import histones, other chromosomal proteins, and the proteins contained in ribosomal subunits and in ribonucleoprotein complexes. In exchange, it exports ribosomal subunits, RNA, and ribonucleopro-

tein complexes. The nuclear membrane contains specialized nuclear pore complexes through which these materials move. In this 125-Md complex, the pore opening is about 150 nm in diameter.

Many proteins destined to be imported into the nucleus have a series of five adjacent positively charged amino acids (e.g., Lys-Lys-Lys-Arg-Lys). Cytosolic proteins factors may participate in nuclear protein import, and several putative nuclear localization signal receptors have been found. Because of the two-way nature of nuclear transport, different factors and receptors may be involved in controlling the traffic in each direction.

Mitochondrial proteins destined for import have a series of five or six positively charged amino acids spaced every three to five amino acids. In solution, these proteins assume an amphiphatic α-helix conformation with the positively charged amino acids on the same side of the helix. Mitochondrial proteins are unfolded before they are translocated into the matrix of the mitochondria at special membrane contact sites. Cytosolic proteins called chaperones help mitochondrial protein import by keeping proteins in their unfolded conformation. For proteins whose final destination is the inner mitochondrial membrane or inner mitochondrial space, an additional hydrophobic signal is unmasked when the initial transport signal is removed, and they are inserted into the inner mitochondrial membrane by a process similar to what occurs in the ER.

For import into the peroxisomes, there is a characteristic tripeptide sequence found in the carboxyl termini of proteins destined for import. In many cases, these targeting signals are removed during or immediately after the proteins are imported.

☐ Cytosolic proteins can be extensively modified after their synthesis

There are numerous posttranslational modifications that occur to proteins in the cytosol. Some of these are reversible, like acetylation, methylation, and phosphorylation. Many cytosol protein have amino termini that are acetylated. This modification increases their half-life by protecting them from proteolysis. Reversible phosphorylation of proteins by protein kinases is a widely used mechanism to reversibly regulate the activity of proteins. Other modifications, such as ubiquination, glycosylation, palmitylation, and myristylation, are irreversible. Covalent attachment of a ubiquitin group to a lysine side chain marks a protein for rapid destruction. This ensures that proteins that have a short half-life such as the products of cellular oncogenes and rate-limiting enzymes in metabolic pathways are appropriately degraded. Misfolded, damaged, or incorrectly made proteins are recognized and rapidly degraded. Addition of myristic or palmitic acid anchors cytosolic proteins to the cytoplasmic face of the plasma membrane. Other modifications, such as the addition of cofactors like biotin to enzymes, are essential for their activity. In each case, specific structural features

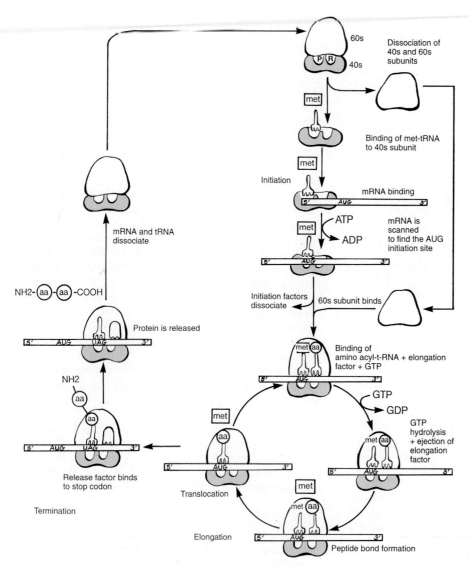

Figure 2-1

In eukaryotic protein synthesis, the major steps are initiation, elongation, and termination. After initiation, multiple cycles of elongation add amino acids to the growing protein chain. When the stop codon on the mRNA is encountered, termination takes place, and the peptide chain is released.

of the proteins target them for modification by cytosolic enzymes.

Protein Synthesis in the Endoplasmic Reticulum

☐ Signal sequences and factors involved in signal sequence recognition

It is the ribosomes that cover the surface of the rough ER and give it its characteristic rough appearance. This insertion process separates the two pathways of protein synthesis and is the site of several important modifications of proteins that have major functional consequences. One advantage of this complicated system of cotranslational insertion of the protein into the ER is that it circumvents the folding problem. If the synthesis of ER proteins is allowed to proceed beyond a certain point in the cytosol and they begin to assume their normal folded conformation, it is no longer possible to import them into the ER. Recognition, insertion, and translocation of proteins into the lumen of the ER involves many different proteins, some of which are composed of multiple subunits. There are different factors for each stage of the process; some are in the cytosol, such as the ribosomes, and some are integral membrane proteins in the ER.

The elucidation of the mechanism of protein translocation into the ER and of transport among the ER, Golgi, and the cell surface has been an area of very active investigation, and progress has been facilitated by the discovery of a large number of temperature-sensitive mutant strains of yeast that, when grown at the nonpermissive temperature, have defects in various stages of this path-

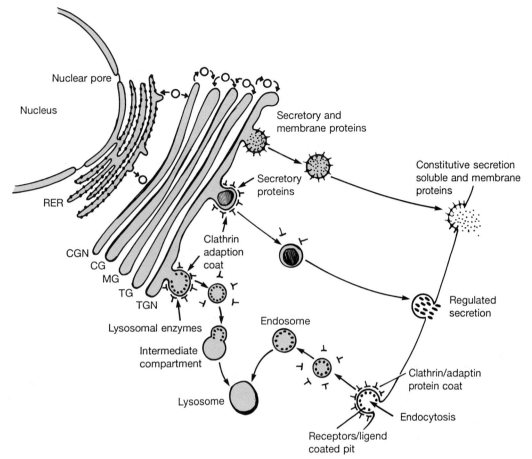

Figure 2-2
Overview of the protein traffic initiated in the endoplasmic reticulum in a cell with the regulated secretory pathway. Proteins pass in transport vesicles from the rough endoplasmic reticulum (RER) to the *cis* Golgi network (CGN), *cis* Golgi (CG), medial Golgi (MG), *trans* Golgi (TG), and *trans* Golgi network (TGN). From the TGN, vesicles move proteins to different destinations. Not shown in this simplified drawing is the recycling of receptors from the endosomes to the cell surface, the recovery of membrane from the cell surface to the Golgi, and the recovery of Golgi proteins like the mannose-6-phosphate receptor to the TGN.

way. Using the powerful technique of complementation, it has been possible to clone the proteins coded by the genes missing in these mutants. Although yeast are far removed from mammals on the evolutionary scale, some of these proteins are highly conserved, and the yeast proteins can substitute for mammalian proteins in in vitro functional assays.

The recognition and translocation process occurs in a series of steps. First, the signal sequence of the nascent peptide chain binds to the signal recognition particle (SRP). The SRP is composed of six polypeptides and 7S RNA. The SRP has an amino-terminal GTP-binding domain and a carboxyl-terminal, methionine-rich domain that has an amphiphatic α-helix, which contains a superficial hydrophobic domain that has high affinity for the signal sequence.

In the second step, the SRP-nascent chain binds to a ribosome, resulting in translation arrest. Third, the SRP-nascent chain, ribosome complex is targeted to the ER by the affinity of the SRP for the SRP receptor (i.e., docking protein). Fourth, the SRP dissociates from the complex and returns to the cytosol.

In the fifth step, the nascent chain-ribosome complex interacts with the signal sequence receptor in the ER membrane, and the nascent chain is translocated through the membrane into the lumen of the ER. Biophysical experiments with reconstituted systems suggest that a large aqueous channel opens when the ribosome binds, which allows the protein to pass. Additional integral membrane proteins of the ER are required for the translocational step. This macromolecular complex, which is involved in translocation, has been called the translocon.

Sixth, for most soluble proteins, the signal sequence is cleaved by the signal peptidase complex on the luminal side of the ER membrane. It has been proposed that the cotranslational modifications of proteins, such as folding, glycosylation, and signal peptide cleavage, are vital to the thermodynamics of translocation. These modifications make the translocation irreversible, preventing the proteins from moving backwards through the channel.

☐ Membrane proteins targeted to the cell surface are cotranslationally inserted into the membrane in the endoplasmic reticulum

The topology that is established in the ER is maintained during the subsequent transport steps, and cytoplasmic loops remain in the cytosol, while loops in the ER lumen are exposed on the cell surface. Proteins that have a single membrane-spanning domain frequently have a signal anchor sequence that can occur at either the amino or carboxyl end of the protein. This sequence is a series of 19 to 30 hydrophobic amino acids that is flanked on the amino or carboxyl side by regions with abundant positively charged basic residues. Stop and start transfer sequences are similar to signal anchor sequences. It is thought that this hydrophobic sequence in conjunction with the adjacent hydrophilic sequences stops translocation, causes a conformational change that disrupts or closes the translocation channel, and allows the hydrophobic domain to integrate into the lipid bilayer. The adjacent hydrophilic amino acids ensure its stable position in the membrane.

Three major types of topology are observed. Most soluble proteins have a single signal sequence that is cleaved, leaving them free in the lumen of the ER. Proteins with a single membrane-spanning domain have a signal sequence that is cleaved and a stop transfer or membrane anchor sequence that is not. Their sequence contains the information to specify whether the membrane-spanning domain occurs at their amino or carboxyl terminals. Proteins with multiple membrane-spanning domains have a series of start and stop transfer sequences that ensure that they assume their correct membrane topology. This is illustrated in Figure 2-3.

☐ The endoplasmic reticulum is a major site of membrane phospholipid synthesis

Phospholipid synthesis occurs on the cytoplasmic face of the ER. Redistribution of these lipids to the luminal face is promoted by phospholipid translocators. The ER provides membrane phospholipids to the Golgi, lysosomes, and the plasma membrane by means of transport vesicles, and the mitochondria and peroxisomes are supplied by water-soluble lipid-exchange proteins.

☐ Proteins are extensively modified in the endoplasmic reticulum

Complex carbohydrates are added to specific asparagine residues of some proteins (N-linked glycosylation). Glycosylation is a common cotranslational and posttranslational modification of proteins. Many proteins are N-glycosylated in the ER. The functional significance of glycosylation is not entirely clear, but it does influence protein folding, and highly glycosylated proteins are protected from proteolytic degradation.

N-glycosylation involves the addition of a complex carbohydrate to an asparagine side chain. The complex carbohydrate to be attached to the protein is assembled partly in the cytosol. It is assembled one residue at a time to a dolichol in the membrane by a high-energy pyrophosphate bond. The precursor is flipped to the luminal side, and additional glucose and mannose residues are added. The structure of this complex carbohydrate is identical for all N-linked sugars, and it contains two N-acetyl-glucosamines and three glucose and nine mannose residues. This reaction is catalyzed by a glycosyltransferase enzyme that reside on the luminal side of the ER membrane and occurs cotranslationally. Only proteins with the sequence Asn-X-Ser or Asn-X-Thr (X is any amino acid except proline) have this group added.

The specificity and diversity seen in complex carbohydrates attached to proteins arises from trimming and further modifications of the carbohydrate groups that occurs almost immediately in the ER and continues in the Golgi, in which O-glycosylation also takes place. The information for the final structure of the oligosaccharides to be added is dictated by the primary sequence of the protein.

Some proteins are anchored to the ER membrane by a phosphatidylinositol group. This reaction is similar to the fatty acid addition observed with certain proteins in cytosol. The amino group of a preassembled glycosyl-phosphatidylinositol intermediate forms a peptide bond with the free carboxyl termini of proteins just after they emerge from the ER membrane. Proteins with this modification, such as neural cell adhesion molecule, end up on the exterior surface of the cell.

Proteins assume their correct conformation in the ER. After the soluble proteins enter the lumen of the ER, they are free to assume their correct conformation. In the ER, there are proteins like binding protein (BiP), which is homologous to the cytosol heat shock protein HSP70, which facilitate this process. Bip is thought to bind to the hydrophobic regions of incorrectly folded proteins, retaining them in the ER until they assume the correct orientation. They behave like "reversible detergents" that use the energy of ATP hydrolysis to bind and release proteins.

The ER possesses an active system of protein degradation that is independent of lysosomes that degrades incorrectly folded proteins. This route may be the fate of the cystic fibrosis transmembrane conductance regulator, a plasma membrane chloride channel that has a triple base deletion seen in most patients with cystic fibrosis. When this mutant protein was expressed in cultured cells, it was retained in the ER. Cystic fibrosis may really be a protein-sorting defect caused by the inability of the mutant chloride channel protein to reach the plasma membrane.

Disulfide bond formation occurs in the ER. Disulfide bond formation is promoted by the nonreducing milieu

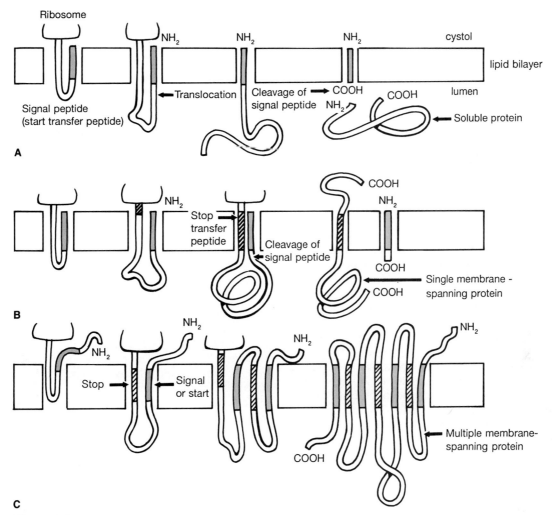

Figure 2-3
Protein translocation into endoplasmic reticulum (ER) membrane determines the topology for membrane proteins. (**A**) A soluble protein is released into the lumen of the ER after its signal sequence is removed. (**B**) A protein with a single membrane-spanning domain is anchored in the membrane with a single anchor or stop-transfer sequence. (**C**) A protein with seven membrane-spanning domains is inserted by a series of stop and start transfer sequences.

of the ER. A specific protein, protein disulfide isomerase, speeds the search for the correct and lowest-energy conformation. This enzyme is versatile and is also part of two other enzyme complexes that modify proteins: the prolyl-4-hydroxylase enzyme complex that converts proline to hydroxyproline in collagen and the oligosaccharide transferase enzyme, in which it is responsible for recognition of the consensus sequence for *N*-linked glycosylation.

amino acid (e.g., Lys-Asp-Glu-Leu or KDEL) retention signal near their carboxyl-terminal KDEL. Proteins with this retention signal do travel to a compartment that is transitional between the ER and Golgi, called the salvage compartment. There a KDEL receptor binds them and returns them to the ER in a transport vesicle. An additional ER retention signal that is independent of KDEL has been identified consisting of a stretch of about 12 amino acids at the carboxyl terminal of some proteins.

☐ Resident endoplasmic reticulum proteins are brought back to the endoplasmic reticulum from the salvage compartment

The enzymes that assist in protein folding, disulfide bond formation, and complex carbohydrate modification are resident soluble proteins in the ER. They contain a four-

☐ Transport vesicles ferry proteins and lipids among the endoplasmic reticulum, golgi, lysosomes, and the plasma membrane

Proteins that do not have a retention signal move by bulk flow to the Golgi where, unless they have a targeting signal for lysosomes or secretory granules, they are rapidly secreted by the nonselective or constitutive secretory pathway. Some viral proteins pass from the ER to the cell surface in as little as 15 minutes, but 2 hours is more typical for serum proteins. This difference is thought to be caused partially by differences in how long the proteins reside in the ER. Some proteins spend more time interacting with resident ER soluble proteins. After proteins enter the ER, their subsequent passage to the Golgi, lysosomes, and plasma membrane takes place in transport vesicles. This transport is not unidirectional, and vesicles move in both directions between the ER and Golgi and between the Golgi subcompartments. This bidirectional movement is particularly important for restoring membrane lipids to the ER membrane. These lipids are depleted with a half-time of about 15 minutes by the budding and exit of transfer vesicles to the Golgi. Most of the ER membrane lipids are thought to be returned from the salvage compartment. Whether these lipids are returned by vesicles or lipid transporting proteins is unknown. In contrast, most of the protein transport is thought to be unidirectional.

A detailed understanding of the mechanism of vesicle budding, transport, uncoating, docking, and release is lacking, although progress in this area has been rapid. Many protein factors are important for each stage of this process. Soluble factors in the cytosol; specific membrane proteins, including the small GTP-binding proteins; GTP; ATP; and calcium are required for many of the steps in this process.

☐ Transport between different cellular compartments uses different kinds of vesicle proteins

Our understanding of the biochemistry of this transport system has been greatly advanced by the discovery of many different secretory mutants of yeast and by the ability to reconstruct each step of these complex reactions in in vitro systems using permeabilized cells or isolated organelles. Additional information on the participation of small GTP-binding proteins and the composition of the protein coats of these vesicles have greatly accelerated the pace of discovery.

☐ Vesicles for regulated secretion, endocytosis, and transport from golgi to prelysosomes use a clathrin-adaptin protein coat

Vesicles that carry a selected cargo are coated with clathrin and adaptin, and vesicles that move by bulk flow between different Golgi compartments have a different coat protein called COP (for coat protein).

The basic subunit of clathrin-coated vesicles is the clathrin triskelion, which is composed of three clathrin heavy chains (180 kDa). These triskelions are capable of self-assembly in solution to form the characteristic polyhedral lattice observed microscopically. Clathrin light chains are thought to be involved in the disassembly and release of the coat from vesicles. An additional group of proteins called adaptins is involved that mediates the association of clathrin with the membrane.

At the site where vesicles emerge, adaptin molecules bind to the cytoplasmic face of the membrane, followed by clathrin. After maturation and formation of the vesicle, the coat proteins are removed and returned to the cytosol for reuse.

☐ Intra-Golgi transport vesicles use a different protein coat

Intra-Golgi transport vesicles have a protein coat that is composed of four different COP protein subunits, ranging from 61 to 160 kDa, and a number of other proteins, including ADP-ribosylation factor (ARF). ARF is a small GTP-binding protein. There are some similarities and differences between the two protein coats. Like the clathrin system, the COP proteins exist in soluble and membrane forms so they may be in dynamic equilibrium. β-COP also shares sequence homology with β-adaptin. The COP proteins are removed immediately before fusion, unlike clathrin, which is removed right after the vesicle is formed. Unlike clathrin, the uncoating of COP proteins is blocked by the nonhydrolyzable GTP analog GTPγS (i.e., guanosine-5′-O-thiophosphate), suggesting a role for GTP hydrolysis in this process.

Protein Modification and Sorting in the Golgi Apparatus

☐ The Golgi is composed of different subcompartments

Vesicles from the ER containing membrane proteins and proteins intended for lysosomes or secretion arrive at the cis face of the Golgi. They traverse the cis, medial, and trans Golgi by "bulk flow" and exit by the trans-Golgi network (TGN). Protein travel between the subcompartments is thought to be mainly unidirectional and not involve specific signals. The existence of resident Golgi proteins implies that a mechanism exists for their retention. This mechanism has not been identified. In the TGN, proteins targeted to lysosomes are separated from secretory and plasma membrane proteins. Secretory proteins targeted to the "regulated" secretory pathway are separated from plasma membrane proteins and secretory proteins traveling in the "constitutive" secretory pathway.

□ The Golgi is a carbohydrate factory responsible for additional protein glycosylation reactions, the biosynthesis of glycolipids, proteoglycans, and mucins

The Golgi apparatus performs several important biosynthetic reactions that are catalyzed by specific enzymes located in different Golgi subcompartments. The *N*-linked carbohydrates added in the ER can be extensively modified in the Golgi. These modifications include further trimming of mannose residues, followed by addition of *N*-acetyl glucosamine, galactose, and sialic acid.

□ Proteins are further modified in the Golgi

O-linked glycosylation

Certain proteins are *O*-glycosylated through the hydroxyl group of threonine and serine residues in the cis-Golgi. The first sugar to be added is *N*-acetylgalactosamine, and this reaction is the rate-limiting step. These relatively simple carbohydrates added include galactose, *N*-acetylgalactosamine, and terminal sugars like sialic acid. The carbohydrates added to the proteoglycans and mucins are also sulfated in the Golgi.

The functional significance of *O*-linked glycosylation is most apparent for the glycopeptide hormones like thyrotropin, in which correct *O*-glycosylation of the subunits is required for efficient assembly and secretion. Alpha subunits that are not glycosylated do not bind correctly to the beta subunits aggregate and are degraded intracellularly. They glycosylation state of proopiomelanocortin (i.e., precursor of adrenocorticotrophic hormone and β-endorphin) influences its processing.

Tyrosine sulfation and acetylation

Proteins are sulfated on the hydroxyl group of their tyrosines by tyrosylprotein sulfotransferase, a membrane-associated enzyme found in the TGN. It is only secretory proteins and plasma membrane proteins that receive this modification. This reaction is specific and occurs only on tyrosines that have adjacent acidic amino acids.

Tyrosine sulfation is important for protein-protein interactions among proteins involved in blood coagulation. Tyrosine sulfation of blood coagulation factor VIII is essential for its binding to von Willebrand factor. The importance of this modification is supported by the observation that a natural mutation of the essential tyrosine to a phenylalanine results in hemophilia.

Cholecystokinin (CCK) is a peptide thought to be a neurotransmitter in the brain and a hormone in the gastrointestinal tract. Pro-CCK has three tyrosine sulfation sites, one of which occurs in the processed secreted form of the peptide. Sulfation at this site is essential for its biologic activity at CCK-A receptors. Blockade of the sulfation of pro-CCK alters its processing and inhibits its secretion.

The acetylation of amino terminal serine residues in proteins is known to occur in the cytosol. Some biologically active peptides like α-melanocyte-stimulating hormone are acetylated in the Golgi or in secretory granules. This modification blocks the opiate activity of this peptide, while leaving the melanotropic activity intact. This acetylation reaction provide a mechanism to regulate the activity of the released peptide.

Proteolysis and amidation reactions

Proteolysis and amidation reactions begin in the Golgi and continue in the secretory granules to produce the final product. Many biologically active protein and peptides originate as prohormones, which are significantly larger than their final released product. In some cases, there are multiple copies of the identical active product or there are several active products that arise from the same prohormone. Some examples of proneuropeptides are illustrated in Figure 2-4. In many cases, the biologically active product is embedded in the prohormone, and the flanking sequences on both ends must be cleaved; with the isolation of a large number of bioactive peptides and the cloning of their cDNAs, certain patterns of cleavage are apparent. Most cleavages occur at specific pairs of basic residues; the most common pair is lysine and arginine. The actual bond that is cut depends on the prohormone, although the most common cleavage is between the lysine and arginine. Some cleavages occur at single basic residues, mainly at arginine residues.

The three-dimensional structure of the prohormone appears to determine the site and the efficiency of its cleavage. The sites that are cleaved are highly specific. Many other sites in the prohormones containing similar sequences are not cleaved. This pattern of selective cleavage differentiates these processing enzymes from the action of broad-spectrum degradative enzymes like trypsin, which can cleave on the carboxyl-terminal side of most arginine and lysine residues in proteins. These processing enzymes have a limited distribution and are found mainly in cells of neural or endocrine origin.

Several highly specific enzymes have the correct subcellular distribution and catalytic activity to perform these cleavages. Most of these enzymes cleave at dibasic pairs or single basic amino acids. Few cleave both types of sites. This implies that different enzymes are responsible for these cleavages in vivo. A family of mammalian subtilisin-like enzymes have been identified by homology with similar enzymes in yeast that cleave the pro-α-mating factor. The mammalian versions of this enzyme are capable of making several of the dibasic cleavages observed in proneuropeptides. A number of different enzymes that make the appropriate single basic cleavages have also been identified. Which of these enzymes is responsible for proneuropeptide processing in vivo is still unanswered.

The initial endoproteolytic cleavages frequently leave a single basic amino acid on the amino or carboxyl terminus of the active peptide. This extra amino acid is fre-

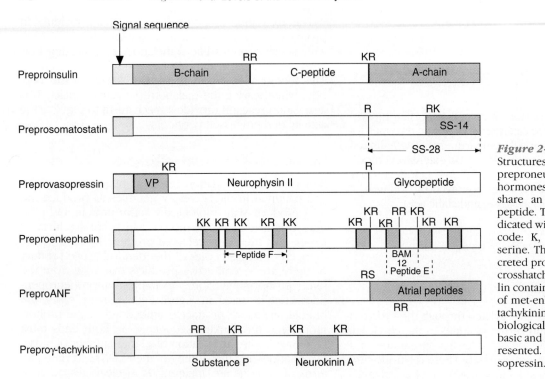

Figure 2-4
Structures of several mammalian preproneuropeptides and prepro-hormones. All the prohormones share an amino-terminal signal peptide. The cleavage sites are indicated with the single amino acid code: K, lysine; R, arginine; S, serine. The biologically active secreted product is indicated by the crosshatched bars. Pro-enkephalin contains seven identical copies of met-enkephalin, and prepro-γ-tachykinin contains two different biologically active peptides. Dibasic and monobasic sites are represented. SS, somatostatin; VP, vasopressin.

quently, although not exclusively, removed by an amino- or carboxypeptidase.

Amidation of the carboxyl-terminal site is a modification seen in most biologically active peptides. This modification is essential for biologic activity of these peptides and may prolong their biologic half-life by protecting them from attack by carboxypeptidases. In all cases, a glycine residue is on the carboxyl side of the amino acid that is amidated in the final product. The reaction takes place in two stages, both of which are catalyzed by the same enzyme, called peptidylglycine-amidating monooxygenase. The reaction involves an oxidative introduction of a hydroxyl group in the α carbon of glycine. This is followed by a disproportionation of the hydroxyl intermediate to the peptide amide and glyoxylate. The net result is that the carboxyl-terminal amino acid is amidated. This reaction is thought to be the last reaction before release, and it probably takes place in the vesicles. These reactions are summarized in Figure 2-5.

The cleavages begin in the Golgi and continue in the secretory granules. The prohormones and all the appropriate enzymes and cofactors are packaged into the same vesicles in the TGN.

☐ Proteins targeted for regulated secretion are packaged in condensing vesicles in the trans-Golgi network

Discrete primary amino acid sequences that target proteins to the regulated secretory pathway have not been identified. It is thought that the three-dimensional structure of the prohormones is recognized as a sorting signal. Two hypotheses have been proposed for the mechanism by which proteins destined for secretion are packaged into secretory granules: a receptor-mediated mechanism or selective aggregation of secretory products at sites in the TGN where adaptin- and clathrin-coated vesicle buds are being formed. The major argument against the receptor hypothesis is that the amount of secretory material carried by these vesicles is too large for each molecule to be associated with a membrane-bound receptor. The aggregation model is supported by the observation that secretory granules have cores of electron-dense material that may be aggregated secretory proteins. It is not known whether secretory proteins aggregate on their own or as a complex with carrier proteins. It has been suggested that the chromogranin-secretogranin family of proteins that are present in many secretory cells may perform this function. These proteins undergo self-aggregation when exposed to the low pH, high calcium environment thought to exist in the TGN. After aggregation, the vesicle formation process may resemble the phagocytosis at the cell surface in which particles are enclosed by a membrane coated by clathrins and adaptins.

How the information that directs the adaptin and clathrin molecules to this site traverses the membrane is unknown, but the involvement of G proteins in this form of signal transduction is a possibility. After the immature secretory vesicles are formed and uncoated, several of them fuse with each other and undergo condensation to become a mature secretory vesicle. This process involves acidification of the lumen, which is accomplished by an ATP-dependent proton pump. This condensation results in a concentration of proteins of about 200-fold over the Golgi lumen. These vesicles store large quantities of neu-

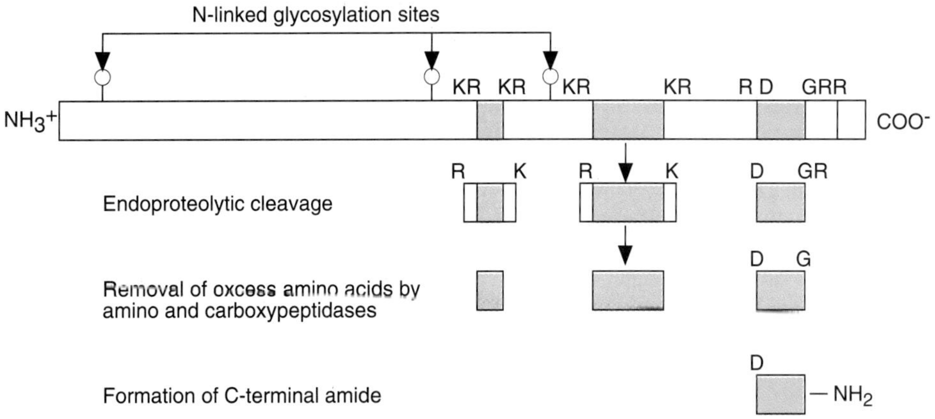

Figure 2-5
Cleavage of a hypothetical pro-neuropeptide containing *N*-linked glycosylation sites, multiple biologically active products, one of which has a carboxyl-terminal amide residue. D, aspartic acid; G, glycine; K, lysine; R, arginine.

ropeptides and other secretory material until they are released.

☐ Hydrolytic enzymes that carry a mannose-6-phosphate are selected for transport to lysosomes in the trans-Golgi network

Soluble hydrolytic enzymes destined for lysosomes contain a mannose-6-phosphate residue. Phosphorylation of this mannose residue takes place in the cis Golgi. A specific mannose-6-phosphate receptor in the TGN binds this residue, the cytoplasmic tail of the receptor interacts with adaptins and clathrin, and the complex is removed from the membrane in a vesicle and transported to a pre-lysosomal compartment. In this acidic compartment, the receptor has a lower affinity for the enzyme, and it dissociates. The receptor is carried back to the TGN to be reused while the enzyme is transferred to the lysosomes.

I-cell disease (i.e., a rare lysosomal storage disease) is caused by the loss of the Golgi enzyme that phosphorylates mannose. In this disease, the lysosomal hydrolases are not sorted to the lysosomes but are instead secreted. As a result, products that are normally degraded accumulate in the lysosomes with severe pathologic consequences.

Lysosomes have a pH of about 5, which is maintained by an ATP-dependent proton pump. It imports a large number of hydrolytic enzymes from the Golgi that operate at this pH but are largely inactive at physiologic pH. These include proteases, lipases, phosphatases, nucleases, sulfatases, phospholipases, and glycosidases. Lysosomal membranes are highly glycosylated to protect them from these proteases. The lysosomal membrane is also specialized to allow the final products of degradation to diffuse out to the cytosol, where they can be reused.

Lysosomes are responsible for degrading receptors and ligands captured from the endocytic pathway. They also degrade used cellular organelles such as the ER, mitochondria, and secretory vesicles. The organelles to be degraded are enveloped by a membrane after fusion with a lysosome.

Constitutive and Regulated Pathways of Secretion

All cells have a constitutive secretory pathway, but regulated secretion is a property of endocrine and exocrine secretory cells, epithelial cells, and neurons. Constitutive secretion is rapid, continuous, and independent of calcium and external stimuli. It delivers lipids, transmembrane receptors, serum proteins like albumin, and the proteins of the extracellular matrix. Regulated secretion involves electron-dense secretory granules in which the secretory products are highly concentrated, and their release depends on calcium and is triggered by external stimuli. Vesicles containing material for the regulated pathway have a clathrin coat when they leave the TGN, but the constitutive secretory vesicles do not. Both forms of secretion involve fusion of the vesicle with the plasma membrane.

☐ Many cell surface receptors with their attached soluble ligands are internalized from coated pits by clathrin-coated vesicles

Some cell surface receptors, including growth factors, cytokines, hormones, protein-clearing systems, and transferrin, are internalized in coated pits by a process called endocytosis. This is achieved by the same mechanism as TGN, sorting to lysosomes in which adaptins and clathrin recognize sorting signals in the cytoplasmic domain of these receptors and forming vesicles that deliver these receptors with their ligands to endosomes. Four different routes from the endosome have been identified for these internalized receptors and ligands.

In the first path, the receptor and ligand dissociate, receptors are recycled to plasma membrane, and ligands are transported to lysosomes. This is the path taken by the low-density lipoprotein (LDL) receptor that moves to the late endosome, where the receptor and ligand dissociate, the receptor is returned to the plasma membrane, and the ligand is transferred to the lysosome.

In the second path, the receptor and ligand are returned to the cell surface. This is the path taken by the transferrin receptor, which carries iron from outside the cell to the endosome. The acidic pH of the endosome causes the iron to dissociate from the transferrin. The transferrin, still bound to the transferrin receptor, returns to the plasma membrane.

In the third path, the receptor and ligand are transported to lysosomes. This is the route taken for short-lived receptors for which their target cell response is regulated by down-regulation. These ligands and their receptors move to late endosomes and then to lysosomes.

In the fourth path, the receptor and ligand are transported to the opposite side of polarized cells, where the ligand is released intact. In some polarized cells, insulin and epidermal growth factor ligands and receptors are delivered to the apical surface of the cell by transcytosis.

Endocytosis is essential for cell function. It delivers nutrients such as iron and cholesterol, clears a variety of enzymes from the extracellular space, and transports immunoglobulins across cells. By transporting receptors and ligands into the cell, it terminates the action of some hormones. Endocytosis may also deliver the ligands and growth factors like nerve growth factor to intracellular sites (e.g., nucleus), where they may influence RNA synthesis and gene regulation.

□ Neurons require an additional mechanism to transport proteins to the synaptic cleft

Because of their structure, nerve cells have a special problem distributing the products of protein synthesis to their sites of action. In the extreme case of a motor neuron, whose axon is a meter long and whose cellular contents are thousands of times larger than a single liver cell, the demands placed on its single nucleus and its transport system are enormous. To deliver proteins and organelles to its axon terminals, neurons have developed a special transport mechanism. This transport was studied in detail by observing transport of radiolabeled proteins in large central nervous system axons. Mechanistic studies were aided by the ability to observe microscopically the transport outside of axons in extruded squid axoplasm.

Although neurons contain several different filaments, it is the microtubules that conduct the vesicle traffic to the nerve terminal. These filaments, which are composed of tubulin, have a distinct polarity. They start at the centrosome near the nucleus, and the polarity is maintained all the way to the axon terminal.

Axoplasmic transport is bidirectional. Different materials move at different speeds. The most rapid anterograde transport (i.e., forward in the direction of the terminal) is about 400 mm/day, and the slowest is about 0.1 to 4 mm/day. Organelles like mitochondria and membrane-bound material like vesicles containing neuropeptides and other secretory material move by fast transport. Cytosolic enzymes, neurofilaments, clathrin, and cytoskeletal and membrane components travel by slow transport.

Fast retrograde transport (i.e., in the direction of the cell body) is about 300 mm/day and is used for moving mitochondria, small vesicles, and smooth ER. Some of this retrograde transport represents uptake of receptors with their bound ligands from coated pits. The extra membrane that is added to the nerve terminal when secretory granules fuse and deliver their contents to the synapse is recovered. Without this recovery, the surface area of the nerve terminal would increase. Substances such as horseradish peroxidase and other chemicals are taken up from the extracellular space by the terminals of nerve cells and transported to the cell body. This uptake has been used to map the projections of neurons in the brain.

Fast axoplasmic transport is thought to involve two proteins originally discovered to be "motor" molecules in ciliary and flagellar axonemes. These proteins are thought to coat reversibly the surface of vesicles and other organelles and to move them toward the positive end (i.e., anterograde) or the negative end (i.e., retrograde) of the polarized microtubule. A protein called kinesin is thought to play this role in the anterograde direction, and dynein works in the retrograde direction. Both of these proteins are composed of multiple subunits and have ATPase activity that is stimulated by the presence of microtubules. Vesicles coated with these proteins move in the appropriate direction as a result of ATP-dependent cycles of attachment and detachment from the microtubules.

SELECTED READINGS

Hershey JWB. Translational control in mammalian cells. Annu Rev Biochem 1991;60:717.

Loh YP (ed). Mechanism of intracellular trafficking and processing of pro-proteins. Boca Raton, FL: CRC Press, 1992.

Mellman I, Simons K. The Golgi complex: in vitro veritas? Cell 1992;68:829.

Pryer NK, Wuestehube LJ, Shekman R. Vesicle-mediated protein sorting. Annu Rev Biochem 1992;61:471.

Rapoport TA. Transport of proteins across the endoplasmic reticulum membrane. Science 1992;258:931.

Simon SM, Peskin CS, Oster GF. What drives the translocation of proteins? Proc Natl Acad Sci U S A 1992;89:3770.

Vallee RN, Bloom GS. Mechanisms of fast and slow axonal transport. Annu Rev Neurosci 1991;14:59.

Neuroscience in Medicine, edited by P. Michael Conn.
J. B. Lippincott Company, Philadelphia © 1995

Chapter 3

Cytology and Organization of Cell Types: Light and Electron Microscopy

ROCHELLE S. COHEN
DONALD W. PFAFF

Neuronal Response to a Changing Environment

Although neurons are cells that conform to fundamental cellular and molecular principles, they are differentiated from other cells in ways that reflect their unique ability to receive, integrate, store, and send information. Signals received from the internal and external environment are processed by neurons, resulting in the generation of a response that can be communicated to other neurons or tissues. In this way, the organism can successfully adapt to rapidly changing events, ensuring its survival.

All organisms possess stimulus-response systems that permit them to sense environmental fluctuations. In bacteria, for example, intracellular regulatory molecules couple the stimulus to the proper response. Multicellular organisms must communicate information to other cells, some of which may be local, but others are positioned some distance away. Communication may be accomplished by the release of chemical messengers that bind to specific complementary proteins called receptors, located on the surface of other cells. For local communication, diffusion can deliver the messenger to the receptive surface. Messengers such as hormones, secreted by endocrine cells in response to changes in the internal milieu, may have to travel long distances through the bloodstream to reach their targets. This voyage takes time—seconds, hours, or even days—and hormones are considered slow-acting agents. Because hormones are diluted in the bloodstream, they must be very potent and act at low concentrations to be effective. These properties are sufficient for endocrine functions necessary to keep the organism in a homeostatic state, but more immediate challenges require a rapid coupling of stimulus and response and a faster rate of communication among relevant cells. Rapid communication is achieved exquisitely by the neuron, whose form and function are designed to meet such demands.

The rapidity with which neurons can conduct signals (i.e., time is measured in less than 1 millisecond) is primarily a function of certain basic features common to all neurons: polarization of their form, unique associations with their neighboring neurons, and special properties of their plasma membranes. Information is conveyed within and between neurons in the form of electrical and chemical signals, respectively. Neurons are organized into complex networks, or functional circuits, which translate these signals into the myriad responses that constitute an organism's behavioral repertoire. Neural circuits develop in a predictable manner, achieving organizational specificity at functional sites of contact called *synapses*. At the synapses, neurons transmit signals with a great degree of fidelity, allowing some behaviors, particularly those necessary for survival, to be stereotyped. Neurons also possess a remarkable ability to modify the way messages are received, processed, and transmitted and may exert profound changes in behavioral patterns. This special property is called *plasticity* and depends on the molecular and structural properties of the neuron.

In addition to forming specific contacts with other nerve cells, neurons exist in relation to a group of cells collectively known as *glia*. Some glial cells envelop neurons and their processes and appear to provide them with mechanical and metabolic support. Others are arranged along specific neuronal processes so as to increase the rate of conduction of electrical signals. Glia are considered later in this chapter. We first focus on the neuron, which is the fundamental structural and functional unit of the nervous system.

☐ Synapses are the sites of directed communication between neurons

The polarization of neuronal shape permits its functioning within a simple or complex circuit. Like other cells, neurons possess a cell body, or perikaryon, which is the metabolic hub of the cell. However, cellular processes extend from this center and give the neuron its unique form and ability to receive and rapidly send signals often over long distances (Fig. 3-1).

Signals are communicated between neurons at synapses. One neuron forms the presynaptic element, and the subsequent neuron forms the postsynaptic element. Chemical signals, in the form of neurotransmitters or neuropeptides, are concentrated at the presynaptic site. The postsynaptic site contains a high concentration of receptor molecules that are specific for each messenger. Presynaptic and postsynaptic elements are separated by a space of only 20 to 40 μm, ensuring precise and directed transmission of signals.

☐ Neuronal polarity is a function of axons and dendrites

The two main types of cellular processes are called dendrites and axons (see Fig. 3-1). *Dendrites* are usually postsynaptic and form an enormous receptive surface, which branches extensively. In some neurons, such as cerebellar Purkinje cells, the dendritic branches form a characteristic elaborate arborization; others are less distinctive. In addition to their growth during normal development, some of these processes remain plastic and can change in length dramatically in the adult. For example, the extent of arborization of the dendritic tree of male rat motor neurons that innervate penile muscles and mediate copulatory behavior appears to be under steroid hormone (i.e., androgen) regulation. In some dendrites, the membranous surfaces are further elaborated to form protrusions called dendritic spines. Chemical signals received by dendrites and their spines are integrated and transduced into an electrical signal, known as the *synaptic potential*.

The signal triggers the action potential, which is propagated along the neuron's plasma membrane down an elongated process called the *axon*. Although axons are

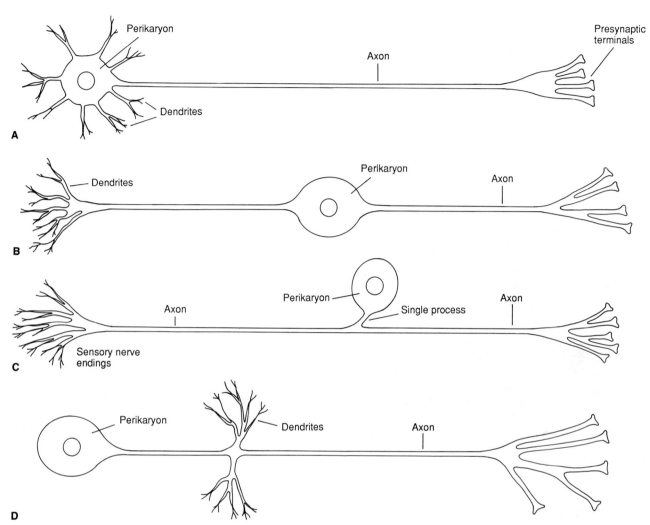

Figure 3-1

Polarization of the neuronal form. Nerve cells conduct impulses in a directed manner, although the position of axons and dendrites relative to the cell body and to each other may vary. (**A**) The multipolar neuron is the most prevalent type and shows extensive branching of the dendrites that emanate from the cell body. The axons emerge from the opposite end. (**B**) Bipolar neurons have sensory functions and transmit information received by dendrites along two processes that emerge from the cell body. (**C**) Pseudounipolar neurons are found in the dorsal root ganglion. These neurons are bipolar during early development. Later, the two processes fuse to form a stalk, which subsequently bifurcates into two axons. The action potentials are conducted from sensory nerve endings in skin and muscle to the spinal cord. The action potential usually bypasses the cell body, but in conditions such as a prolapsed vertebral disc, in which the dorsal nerve roots are put under pressure, the cell bodies may also generate impulses. (**D**) Unipolar neurons are found in invertebrates. Axons arise from the dendrites, which emerge from the cell body.

not as expansive as dendrites, they branch and may innervate more than one effector. The terminal end of an axon is modified to form a bulbous structure, the *presynaptic terminal* (see Fig. 3-1). The incoming action potentials cause the release of neurotransmitter molecules that bind to complementary postsynaptic receptors. This binding initiates the other type of electrical signal, the synaptic potential at the postsynaptic site.

The unidirectional or polarized flow of information consists of action potentials at the axonal level eliciting synaptic potentials in the postsynaptic cell, which triggers

an action potential in that cell and so on. Axons may be as long as 2 meters, permitting long-distance and rapid communication within the circuit.

□ Diversity in form is a distinctive property of neurons

Although neuronal form follows the basic plan described above, nerve cells show a tremendous diversity in size, shape, and function, which allows them to discriminate the multitude of different types of incoming signals. The detection of commands for muscle contraction is under the control of motor neurons. Information in the form of light, mechanical force, or chemical substances is distinguished by sensory neurons highly specialized for each particular type of sensation. It is perhaps in this group of neurons that structural and functional diversity is most apparent. The rigorous demands of sensory discrimination have imposed on these neurons the requirement to develop specific, highly sensitive detection systems that are able to perceive various degrees of stimulus intensities.

A classic example is the olfactory cell of the male gypsy moth, which can detect a molecule of the female's sex attractant or pheromone released a mile away. Equally impressive is the ability of the mammalian nasal epithelium to detect and discriminate more than 10,000 different odiferous substances. In terms of the ability to recognize diverse molecules, olfactory neurons are second only to the cells of the immune system. The olfactory neuron, however, is markedly different in form and receptor locale from, for example, the retinal photoreceptor cell. Receptive surfaces consisting of specialized cilia on olfactory neurons contain the receptors and are the primary sites of sensory transduction. In retinal photoreceptor neurons, visual transduction occurs within special cylindrical cellular domains containing stacks of about 1000 membranous discs in which the photon receptors called rhodopsin are embedded.

Sensory and other types of information enter the nervous system by means of a particular neuron, but those signals are rarely sent to the effector neuron directly. Rather, the initial signal is sent to a third class of nerve cell, the *interneurons*. Interneurons integrate various inputs from other neurons before the information, often in a modified form, reaches its final destination. In this way, various types of inputs (e.g., sensory, hormonal) can be integrated and relayed to other parts of the central nervous system or to motor or endocrine targets. Interneurons contribute to the formation of neural circuits and are to a great extent responsible for the relatively large size and extraordinary complexity of the mammalian nervous system.

☐ Neuronal polarity allows the directed flow of electrical and chemical signals

A single neuron may receive a multitude of different inputs from a variety of sources. A motor neuron, for example, receives thousands of presynaptic terminals from many different neurons on its dendritic surface and on its somal and, to some extent, axonal membranes, which may also bear postsynaptic receptors. This input has to be organized into a cohesive message that can be transmitted to its postsynaptic neighbor.

The neuronal plasma membrane plays a key role in integrating and relaying the information in a directed manner and with exceptional speed. Information is conducted within neurons in the form of electrical signals, which are actually changes in the distribution of electrical charges across the neuronal membrane. Charge distribution is highly regulated in neurons by specific and selective proteins called ion channels embedded in the plasma membrane. These transmembrane proteins control the flow of ions and, consequently, the distribution of positive and negative charges across the membrane. In resting neurons, the membrane potential is about -60 to -70 mV (i.e., an excess of positive charges outside and negative charges inside). When this potential becomes less negative or depolarized, electrical excitation occurs in the membrane. In axons, this excitation is known as the *action potential*. The action potential is generated when the membrane potential of the axonal membrane is decreased beyond a threshold value.

Mechanisms of Neuronal Function

Compared with other cells, neurons are unsurpassed in their complexity of form and ability to communicate with lightning speed over long distances. Nevertheless, as eukaryotic cells, they adhere to basic laws that govern cellular function. In many ways, neurons are not so different from other cells, and qualities once ascribed only to neurons may be found elsewhere. Egg membranes, for example, can depolarize during fertilization and release granules in response to Ca^{2+} entry through specific channels, although the process takes much longer than comparable signaling mechanisms in neurons. Sites of ribosomal RNA synthesis, called nucleoli, are found in all eukaryotic cells but are especially prominent in neurons, because there is a constant need for ribosomes for new protein synthesis. It appears that basic cellular mechanisms are present in nerve cells but that some of these are amplified to meet the rigorous demands of neuronal form and function.

☐ The nucleus is the command center of the neuron

The molecular and cellular diversity within and among neurons reflects a highly controlled differential expression of genes. Gene regulation also determines nerve cell connectivity, which dictates patterns of stereotyped behaviors. It is also becoming evident that more complex behaviors, such as memory and learning, may require the synthesis of new proteins, which ultimately depends on the expression of particular genes.

As eukaryotic cells, neurons sequester their genome in the nucleus, the largest and most conspicuous feature of the perikaryon (Fig. 3-2). The nucleus contains chromosomal DNA and the machinery for synthesizing and processing RNA, which is subsequently transported to the cytoplasm, where information encoded in the DNA is expressed as specific proteins. The nucleus is separated from the rest of the cytoplasm by a porous double mem-

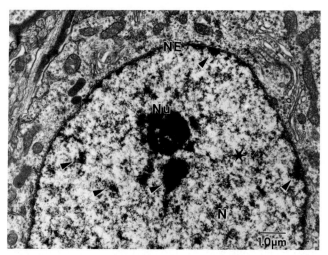

Figure 3-2
The nucleus of a hypothalamic neuron. The nucleus (N) is separated from the cytoplasm by a nuclear envelope (NE). A conspicuous nucleolus (Nu) signifies a great demand for ribosomal RNA by the neuron. Proteins called histones associate with DNA to form chromatin, which may appear extended (i.e., euchromatin) or condensed (i.e., heterochromatin), depending on the translational activity of specific regions of the genome. Most of the nucleus contains fine fibers of euchromatin (*asterisk*). Heterochromatin (*arrowheads*) is seen as clumps within the nucleus, on the inner surface of the nuclear envelope, or associated with the nucleolus.

brane, the nuclear envelope. The nuclear envelope protects the DNA molecules from mechanical perturbations caused by cytoplasmic filaments. Moreover, it separates the process of RNA synthesis (i.e., transcription) from that of protein synthesis (i.e., translation). This segregation of function has an important advantage over the situation in prokaryotes, in which transcription and translation occur simultaneously. In these organisms, protein synthesis begins before the completion of transcription, limiting the opportunity for modifying the RNA. Translation in eukaryotes does not begin until the RNA is transported into the cytoplasm. In the nucleus, the RNA may be modified in such a way that specific portions of the RNA molecule are removed (i.e., RNA splicing) or altered. These complex changes have important implications for cell function. Mechanisms that control variability at the level of transcribed RNA allow a single gene to code for several different proteins, resulting in the rich diversity seen, especially in neurons, in the form of neuropeptides, receptors, ion channels, and cytoskeletal proteins.

Chromatin structure in the regulation of gene activity

The degree of complexity of gene expression is further multiplied at another, even more fundamental level of gene regulation: the DNA. Genes are turned on by complexes of proteins called *transcription factors*. Each of these proteins possesses special DNA-binding domains and requires direct contact with the DNA to function.

However, the long stretch of DNA, which in humans measures about 3 cm long, is folded thousands of times to fit into a nucleus only a few micrometers in diameter. Such compaction creates a potential problem for factors that must gain free access to corresponding binding sites and other regulatory regions on the DNA strand. Another group of proteins, called *histones*, packs the DNA so that it is folded, coiled, and compressed many times over to form fibers called *chromatin*, visible as fine threads in the interphase nucleus (see Fig. 3-2) and, in an even more contracted form, as chromosomes in the dividing cell.

In the mature neuron, the nucleus remains in interphase. DNA is segregated into morphologically distinct areas, reflecting the degree of chromatin condensation, which is a function of nuclear activity (see Fig. 3-2). Ribosomal DNA genes and their products are separately packed into a structurally defined compartment called the *nucleolus* that is specialized for ribosomal RNA synthesis. Highly coiled regions for the genome appear as dense, irregularly shaped clumps known as *heterochromatin*. These areas of condensed chromatin are situated within the nucleus along the inner nuclear membrane, in association with the nucleolus or dispersed within the nucleus proper. Other regions of the genome readily available for transcription into messenger RNA appear as fine filaments and are known as *euchromatin*.

Nucleolus as the site of ribosomal RNA synthesis

The nucleolus is a prominent spherical region of the nucleus (see Fig. 3-2) containing that portion of the genome dedicated to the transcription of ribosomal DNA and the mechanisms for the assembly of ribosomal subunits, the precursors of mature ribosomes. Large precursor ribosomal RNA molecules are processed in the nucleus, resulting in the degradation of almost half of the nucleotide sequences. The remaining ribonucleoprotein molecules form two subunits that are independently transported into the cytoplasm, where the mature ribosomes are assembled. The nucleolus is evident only in the interphase nucleus; in other cells that undergo mitosis, it decondenses, ribosomal RNA synthesis stops, and ribosomal DNA genes associate with specific regions of the chromosomes called *nucleolar organizing regions*.

Nuclear pore complex controls traffic between the nucleus and cytoplasm

The double membrane comprising the nuclear envelope presents a formidable barrier between the nucleus and cytoplasm. Macromolecular traffic into and out of the nucleus is achieved by perforations in the nuclear envelope. At various points along the envelope, the inner and outer membranes are in continuity around the edges of each pore (Fig. 3-3). The *nuclear pore* is not a simple opening. A complex of eight large proteins surround the aperture on the external surface of the inner and outer membranes. These proteins form the nuclear pore complex (see Fig. 3-3), central to which is the actual orifice

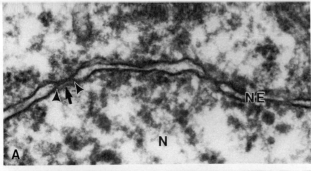
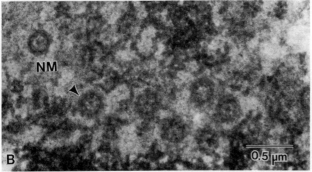

Figure 3-3
Electron micrographs of nuclear pores. (**A**) The perpendicular section of the nuclear envelope (NE) shows the continuity of the inner and outer nuclear membranes (*arrowheads*) around a nuclear pore (*arrow*) of the nucleus (N). (**B**) A surface view shows the arrangement of nuclear pores in the nuclear membrane (NM). One of the nuclear pores (*arrowhead*) shows the octagonal configuration of proteins comprising the nuclear pore complex.

through which particles are transported. Some particles, such as the assembled mature ribosomes, are too large to gain entrance through these passageways, ensuring that protein synthesis is restricted to the cytoplasm. Other particles, such as ribosomal subunits, may be larger than the channel but can be actively transported through the pores, although they may be distorted along the way.

Dynamic nuclear morphology reflects alterations in the genome

Although nuclear events occur on a molecular scale, they may be detected by gross adjustments in nuclear morphology. The overall size and shape of nuclei and nucleoli can change with the metabolic and physiologic demands of the neuron. Depending on transcriptional activity, various segments of the chromatin can condense or decondense, resulting in a relative change in the disposition of heterochromatin and euchromatin and altering the general appearance of the nucleus. For example, in the hypothalamus, a brain area controlling female reproductive behavior, the gonadal steroid hormone estrogen exerts a profound influence on nuclear morphology, altering its size, shape, and position of heterochromatic regions. Nucleolar size is also subject to the physiologic

conditions of the cell. Estrogen treatment has a pronounced effect on precursor ribosomal RNA levels, which are accompanied by a significant increase in nucleolar area in the hypothalamus of ovariectomized animals. Nucleolar hypertrophy is followed by a massive increase in rough endoplasmic reticulum in these neurons.

☐ Neurons are actively engaged in protein synthesis

Information contained within the genome is expressed as biologically active peptides in the cell body or perikaryon of the neuron. Some peptides are neuron specific, such as some of the neurosecretory peptides, cytoskeletal proteins, ion channels and receptors. Others are common to all cells and are involved in increasing the efficiency of transcriptional and translational events related to the production, transport, and release of these proteins.

Proteins synthesized by neurons for export

Appreciation of the tremendous protein synthetic activity of the nerve cell and its functional implications is relatively recent. Neuronal form and membrane properties were the main focus of earlier neurobiologists. However, the compelling discovery of glandular cells in the spinal cord of fish by Carl Speidel in 1919 and of neurosecretory cells in the hypothalamus by Ernst Scharrer in 1928 directed attention to the great degree of biosynthetic activity in the perikaryon. Ernst Scharrer noticed that the secretory activity of diencephalic neurons was comparable to that seen in endocrine cells. This observation stimulated further interest in finding structural counterparts of the secretory process in neurons. The mechanisms involved in peptide biosynthesis and posttranslational processing are now well known. The related molecular and biochemical events have been detailed in the previous chapters. In this chapter, we describe the structural correlates of these functions as they occur in various parts of the nerve cell.

Synthesis of exportable neuropeptides on the rough endoplasmic reticulum. Neurons must transmit chemical information and electrical signals over very long distances. Proteins are synthesized, packaged, processed, stored, and released in different domains of the neuron. The synthesis of exportable proteins begins in the perikaryon. Preribosomal subunits produced in the nucleolus enter the cytoplasm, where they are assembled and activated to form functional ribosomes. In some cases, they attach to membranous cisternae comprising the rough endoplasmic reticulum (Fig. 3-4). The outer nuclear membrane ramifies within the cytoplasm as it encircles the nucleus and, studded with ribosomes, also becomes part of the protein synthetic apparatus (Fig. 3-5).

Messenger RNAs associated with ribosomes are translated into precursor proteins. The precursors are larger than the biologically active peptides and must be enzy-

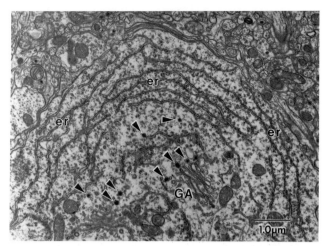

Figure 3-4
Electron micrograph of a hypothalamic neuron actively engaged in the synthesis and packaging of exportable proteins. Many cisternae of rough endoplasmic reticulum (er) are arranged in parallel stacks. The Golgi apparatus (GA) and associated dense-cored vesicles (*arrowheads*) are also visible. (Cohen RS, Pfaff DW. Ultrastructure of neurons in the ventromedial nucleus of the hypothalamus with or without estrogen treatment. Cell Tissue Res 1981; 217:463.)

matically cleaved and modified to attain their final form. Instructions about whether a given protein is destined for export are also encoded in the DNA of its precursor. A portion of the complementary messenger RNA is translated into a *signal peptide*, which directs ribosomes to the cisternae of the rough endoplasmic reticulum. Peptides lacking a signal sequence can not gain entry to the cisternae.

The disposition and extent of rough endoplasmic reticulum vary and appear to depend on the prevailing demands for secretory protein synthesis in a functional group of neurons. In nerve cells that are quiescent in regard to the production of exportable proteins, cisternae of the rough endoplasmic reticulum appear as discrete sacs and seem to occupy only a small fraction of the cellular space. Neurons actively engaged in secretory protein synthesis contain large stacks of elongated cisternae that fill a considerable portion of the perikaryon.

Packaging and modification of exportable neuropeptides in the Golgi apparatus. The Golgi apparatus is composed of a series of flattened, smooth-surfaced, membranous sacs and a variety of associated small vesicles, which encompass the Golgi apparatus on all sides (Fig. 3-6). The Golgi cisternae are arranged in a polarized fashion, reflecting their morphology and function. The forming or cis face approaches the rough endoplasmic reticulum; the opposite side is designated as the maturing or trans face or trans-Golgi network. Vesicles bud off from the smooth transitional portion of the rough endoplasmic reticulum and transport their contents to Golgi sacs on the cis face. Here, they deliver the newly synthesized precursor protein by fusion of the vesicle membrane with the Golgi membrane. The secretory material is later detected within the cisternae occupying the trans-Golgi network. Often, at the periphery of the sacs, the membranes appear to constrict around a dense granule that is the concentrated precursor protein (see Fig. 3-6). The granule-containing membranous compartment is pinched off, forming a dense-cored vesicle. Although the details of protein processing within the Golgi apparatus have not yet been elucidated, its overall function is known: the preparation and packaging of the secretory protein, still

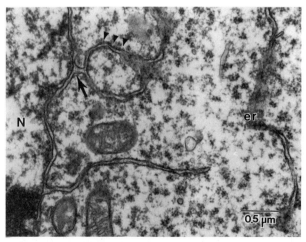

Figure 3-5
Electron micrograph of the outer nuclear membrane ramifying within the cytoplasm. Portions of the outer nuclear membrane (*arrow*) extend into the cytoplasm and, together with associated ribosomes (*arrowheads*), become part of the protein synthetic machinery of the neuron. A cisterna of the rough endoplasmic reticulum (er) is seen to the right of the micrograph. *N*, nucleus.

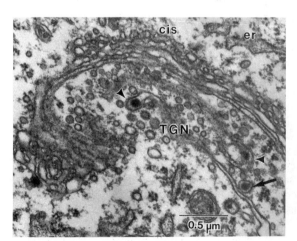

Figure 3-6
Electron micrograph of a Golgi apparatus of a hypothalamic neuron. Small vesicles surround the Golgi apparatus on all sides. The *cis* face is located near the rough endoplasmic reticulum (er); the opposite side is designated as the *trans* Golgi network (TGN). Secretory material is detected within the peripheral portions of the cisternae (*arrowheads*) of the TGN. A dense-cored vesicle (*arrow*) containing material of similar density is seen at the periphery of the TGN.

in its precursor form for transport to the axon terminal for storage or release.

Formation of the active neuropeptide within the neurosecretory vesicle. The actual enzymatic processing of the precursor and generation of individual biologically active peptides appears to occur distal to the Golgi apparatus as the secretory vesicle makes its way down the axon. Conditions, such as pH, necessary for the maximal activity of the processing enzymes are optimal. The relatively simple structure of the secretory vesicle—a dense core surrounded by a single membrane—belies its dynamic role in generating active neuropeptides. Various enzymatic activities have been located in neurosecretory vesicles, and some have been purified and characterized. The vesicle membrane contains proteins that regulate the internal vesicular environment. On the appropriate signal, active neuropeptides are released by exocytosis, which is fusion of the vesicle membrane with the plasma membrane at the presynaptic site.

Proteins synthesized for use within the neuron

The synthesis of exportable proteins represents only a portion of the total protein synthetic effort by the neuron. The elaboration of the cytoskeletal framework and the synthesis of other proteins including ion channels, receptors, second messenger systems, and other proteins destined for maintenance and renewal of the cytoplasm and its organelles, also depend on translation of messenger RNAs. These proteins are synthesized on free ribosomes that are plentiful in all nerve cells.

Axons possess a highly organized transport system to convey proteins to presynaptic terminals or back to the cell body. Dendrites, their postsynaptic specializations, and dendritic spines also require proteins for growth, maintenance, and function. Although there is an indication of molecular transport to these sites, the identities of the molecules and the nature of the transport mechanism has just begun to be explored. Moreover, evidence for local protein synthesis suggests that not all dendritic and postsynaptic proteins arrive from the cell body. The presence of polyribosomes beneath postsynaptic membrane specializations and at the base of dendritic spines provides the machinery for local protein synthesis. The existence of local protein synthetic mechanisms is further supported by the presence of some messenger RNAs in distant dendritic arbors, suggesting that growth-dependent and activity-dependent synaptic alterations, including changes in morphology, may be regulated partially by the local synthesis of key synaptic proteins.

☐ Smooth membrane compartments in neurons serve as reservoirs for calcium

In addition to the intramembranous compartments that directly participate in the synthesis, packaging, and transport of secretory peptides, other cisternal and vesicular structures are evident within the various domains of the neuron. The membranous sacs are visible as smooth-surfaced compartments in a variety of configurations. Some appear as relatively short sacs, but others are longer and anastomose within neuronal processes. In the cell body, smooth membrane profiles are arranged in stacks or emanate from cisternae of the rough endoplasmic reticulum. One of the most morphologically complex variations of these structures resides in the dendritic spine, where parallel, smooth-surfaced cisternae alternate with electron-dense bands of unknown composition.

All of the diverse structures are thought to function in the release and *sequestration of calcium* within the neuron. Calcium is central to most aspects of neuronal function, including membrane permeability; mediation of the effects of neurotransmitters, hormones, and growth factors; cytoskeletal function; and vesicle release. In neurons, calcium mobilization is achieved, at least in part, by the binding of inositol 1,4,5-triphosphate (IP_3) to intracellular receptors located on the aforementioned membranous cisternae. IP_3 is a second messenger generated on receptor-stimulated hydrolysis of phosphatidylinositol 4,5-biphosphate by phospholipase C, an enzyme activated by signal transduction mechanisms.

Cytoskeleton Determination of Neuronal Form

A singular feature of neurons is their overall extraordinary length, enabling them to transmit signals over great distances. This property is reflected in the polarity of neuronal form and function, which is governed by regional specialization of the plasma membrane and by differences in the cytoskeletal composition of dendritic and axonal processes emerging from the cell body. Although the neuronal cytoskeleton provides a structural framework on which various organelles and cellular events are organized, it is by no means a static configuration. Throughout the neuron, molecular alterations in cytoskeletal proteins reverberate as microscopically visible changes in movement of the cytoskeleton, its associated organelles, and the shape and extent of some of the processes. Although the cytoskeleton permits the general pattern of individual neurons to remain constant and identifiable, alterations in cytoskeletal dynamics enable the neuron to respond to environmental-dependent or activity-dependent fluctuations. This apparent contradiction is resolved by the inherent nature of cytoskeletal elements that exist in different structural and functional states of assembly and disassembly. Moreover, these structures may be stabilized and destabilized, providing yet another dimension to the number of possible conformations of cytoskeletal form.

The interactions of various cytoskeletal elements with their associated proteins or with each other contribute to the unique structural and functional identity of axons and dendrites and their associated dendritic spines. The cytoskeleton also interacts with the neuronal plasma membrane at specific sites, including the initial segment of the axon, special loci along the axon called nodes of Ranvier, and presynaptic and postsynaptic membranes, forming complex submembrane filamentous arrays. Such membranous-cytoskeletal associations may restrict the movement of important membrane proteins, such as receptors, at that site or communicate events occurring at the membrane to underlying areas. In this section, we describe components of the neuronal cytoskeleton and how they contribute to the architecture of the neuron and confer specificity to each structural domain.

☐ The neuronal cytoskeleton provides internal support

Axons and dendrites emerge from the perikaryon as delicate strands. Axons may be as much as a million times longer than wide. Consequently, these fragile processes require internal support. The rigidity of the cytoskeletal network is apparent after removal of the neuronal membrane with detergents that selectively extract membrane lipids and proteins. In experiments using detergent-treated cultured nerve cells, isolated neuronal processes, and isolated submembranous cytoskeletal patches, the cytoskeleton remains intact, and its shape is virtually identical to its original conformation. The cylindrical form of the axonal cytoskeleton is so cohesive that investigators, using the classic model of the squid giant axon to study axonal transport mechanisms, equate the extrusion of its contents, the axoplasm, to that of toothpaste being squeezed out of its tube. Even isolated submembrane filamentous arrays, such as those found beneath the postsynaptic membrane, appear to retain their curvature after the rigorous processes of homogenization of brain and centrifugation, lysis, and detergent treatment of isolated synaptic compartments; only after sonication or extremely acidic conditions do these tenacious structures dissociate into their component parts.

☐ The components of the neuronal cytoskeleton include microtubules, neurofilaments, and microfilaments and their associated proteins

The dual nature of the neuronal cytoskeleton, reflected in its rigidity and plasticity, is a function of three filament types: microtubules, neurofilaments, and microfilaments or actin filaments. Each cytoskeletal element acts in conjunction with a specific set of associated or binding proteins. Some of these cross-link the filaments to each other, the plasma membrane, and other intracellular organelles and are responsible for the gelatinous and relatively stiff consistency of the cytoskeleton. Other associated and binding proteins affect the rate and extent of filament polymerization, providing a mechanism for localized plastic changes.

Microfilaments consisting of the protein actin are 6 nm in diameter and are prominent in cortical regions, particularly in the highly specialized submembrane filamentous structures, such as the presynaptic and postsynaptic membrane specializations. Microtubules are long, tubular structures that are 25 nm in diameter and form tracks for the transport of various organelles and molecules, although the microtubules are themselves also capable of movement (Fig. 3-7). The microtubules and actin consist of globular subunits that can assemble and disassemble with relative ease. Neurofilaments that are 10 nm in diameter are a subdivision of the ubiquitous class of intermediate filaments found in all cells (see Fig. 3-7). Mammalian neurofilaments consist of three fibrous subunits that have a very high affinity for each other, and polymers composed of these subunits are very stable. Neurofilament subunits are synthesized and assembled in the cell body and then directed down the axon, where they contribute to its resiliency and its caliber. Neurofilaments are degraded at the entrance to the nerve terminal by Ca^{2+}-activated proteases located at that site.

Actin and tubulin polymers

Subunits of actin, a 43-kd globular protein, and microtubules, a heterodimer of two 50-kd globular proteins called α-tubulin and β-tubulin, assemble into polymers that bind to identical subunits at each end of a preexisting polymer. The lengths of the polymer are determined by

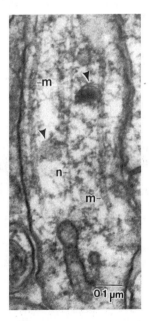

Figure 3-7
Electron micrograph of the axonal cytoskeleton. The two prominent cytoskeletal elements in this region are microtubules (m) and neurofilaments (n). The microtubules are 25 nm in diameter and form tracks for the transport of various organelles, such as vesicles (*arrowheads*). The neurofilaments belong to the ubiquitous class of intermediate filaments and are 10 nm in diameter. Neurofilaments are relatively stable polymers that may contribute to the resiliency and caliber of axons.

cellular mechanisms that control the rates of association and disassociation at the ends of each rod. In some polymers, there is a constant flux of monomers at each end. In more stable polymers, dissociation of the subunits at each end is slow or does not occur at all. Stability can be achieved by blocking the dissociation reaction at either end. Both tubulin and actin monomers are asymmetric so that they can only link up with each other in a specific orientation. Consequently, the resultant polymer is polarized and has plus and minus ends, a feature permitting polymers to grow in a directed manner.

Cytoskeletal-associated proteins

Cytoskeletal-associated proteins regulate cytoskeletal structure and function and characterize specific neuronal domains. Purified tubulin monomers can spontaneously assemble into microtubules in the presence of GTP. However, polymerization is greatly enhanced in impure preparations. The impurities are actually a group of accessory proteins that are subdivided into two categories: microtubule-associated proteins (MAPs) and tau proteins. These proteins induce the assembly and stabilization of microtubules by binding to them. The tau proteins facilitate polymerization by binding to more than one tubulin

dimer at the same time. MAPs have two domains, one of which binds to the microtubule and the other to an adjacent MAP molecule, filament, or cell organelle. MAPs provide the neuron with a mechanism for structural plasticity and variability. About 10 kinds of MAPs have been identified, and they appear to be differentially expressed during brain development. Specific MAPs appear to be restricted to different neuronal processes. MAP 2, for example, is expressed in dendrites but not axons (Fig. 3-8); conversely, MAP 3 is present in axons but not in dendrites.

Although microtubules appear in parallel array, actin filaments in neurons are usually visible as a tangled meshwork. The network sometimes appears as a dense submembranous array, as in the *postsynaptic density* (PSD) immediately beneath the postsynaptic membrane. However, the network may be less dense, as in the subsynaptic web immediately beneath the PSD and extending throughout the dendritic spine. Actin filaments are also associated with a group of accessory proteins, the actin-binding proteins, which bundle them or cross-link them to form a gel. Some binding proteins join the filament ends to obstruct further polymerization, and others link actin to the membrane. Actin-binding proteins are regulated by second messengers, such as calcium or cyclic nucleotides.

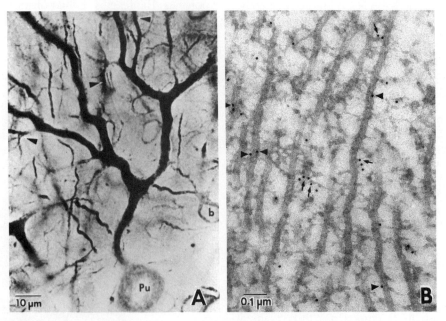

Figure 3-8
Dendrites labeled with antibodies to microtubule-associated protein (MAP) 2. The technique of immunocytochemistry uses antibodies to localize specific proteins in neurons. These antibodies are detected by secondary antibodies tagged with an enzyme or gold particle. (**A**) The enzyme, such as peroxidase, catalyzes a reaction that results in an electron-dense precipitate, as seen in the light micrograph, in which antibodies to MAP 2 label the dendritic tree (*arrowheads*) of a Purkinje cell. (**B**) Alternatively, the secondary antibody can be tagged with gold particles, seen as black dots in the electron micrograph, in which they localize MAP 2 to microtubules (*arrowheads*) and cross-bridges (*arrows*) between them. *Pu*, Purkinje cell; *b*, basket cell. (**A**) is reprinted from Bernhardt R, Matus A. J Comp Neurol 1984; 226:207, with permission of Wiley-Liss, a division of John Wiley and Sons, Inc., New York).

Molecular motors

Other proteins associated with the cytoskeleton are the molecular motors, which harness energy to propel themselves along filaments. These proteins are enzymes that hydrolyze ATP and GTP and use the liberated energy to move themselves along the polymer. Motion is achieved because each of the steps of nucleotide binding, hydrolysis, and release of ADP or GDP plus phosphate causes a concomitant change in the conformation of the motor protein such that it is directed forward. Myosin motors walk along actin filaments, which are pulled along in the process. This action is important in the motility of growth cones, the pioneering tip of developing nerve cell processes. Motor proteins are also associated with microtubules, where they are involved in organelle transport in neuronal processes.

Cytoskeletal-based transport system of neurons

Neuronal processes span great distances to reach their presynaptic and postsynaptic targets, which can be meters away from the cell body in large organisms. Neuropeptides synthesized and packaged in the perikaryon must embark on long journeys to far-removed nerve terminals. Although some synaptic components can be synthesized and recycled locally, others must be imported from the cell body or returned there for degradation by lysosomes, for example. Molecular and organelle traffic to and from these remote areas requires active mechanisms, because diffusion alone would take an inordinate amount of time.

Bidirectional traffic in axons delivers proteins and organelles to and from the nerve terminal by anterograde and retrograde transport, respectively. Anterograde transport delivers neuropeptide-containing vesicles and cytoskeletal proteins; retrograde transport returns endocytotic and other vesicles. Axonal transport has two components: a fast component, traveling at rates of 200 to 400 mm/day, and slow component, which consists of a slow component a (SC_a) and a slow component b (SC_b), moving at rates of 0.2 to 1 mm/day and 2 to 8 mm/day, respectively. The fast component carries membranous vesicles, and the slow compartment carries cytoskeletal proteins.

Fast transport involves movement of vesicles along tracks composed of microtubules. Microtubule motors generate the force required to propel organelles along the path. Two of these motors, kinesin and dynein, are ATPases that are activated on binding to the microtubule. Kinesin directs movement toward the plus end of the microtubule and dynein toward the minus end. In axons, most microtubules have their positive end toward the terminal and can guide movement in either direction. In dendrites, about half of the microtubules are oriented with their plus ends toward the dendritic tip, and the other half have their minus ends toward the tip. Differences in microtubule orientation in axons and dendrites may be one of the mechanisms underlying the selective transport of organelles into dendrites or axons.

Neuronal Synapses

Electrical signals can be conveyed directly from cell to cell at special sites called *gap junctions* (Fig. 3-9). Ions pass from one cell to another through relatively large channels that connect the cytoplasm of the two cells while isolating the flow from the intracellular space. In neurons, these electrical synapses permit rapid and direct electrical transmission and may also play a role in synchronizing neuronal activity. However, their invariant form and paucity of regulatory molecules preclude any major involvement in plastic events. Gap junctions are also found elsewhere in the nervous system between supportive cells, called astrocytes, where they participate in buffering the extracellular ionic milieu.

Chemical *synapses* are the main sites of interactions of nerve cells (Fig. 3-10). All synaptic junctions share common features that guarantee precise and directed transmission of signals. Although the general scheme remains relatively constant, variability in some of the molecular components, such as neurotransmitters, ion channels, receptors, and second messengers, and the dynamic properties of the supporting cytoskeleton enable each synapse to maintain its individuality, record past experiences, and vary responses to new signals.

☐ Synaptic structure follows a basic plan

Chemical synapses conform to a basic architectural plane despite their location along the neuron or within the nervous system itself, but during the past two decades, it has become apparent that the morphology of synaptic connections in the adult mammalian brain is not static. Synapses display structural plasticity, undergoing alterations in size, shape, and number. These changes have important implications in synaptic transmission, because they may modify the way in which incoming signals are received. Because morphologic alterations may reflect marked rearrangements of the molecular structure of synapses, these changes may be long lasting and signify the generation or maintenance of long-term processes, such as memory.

All synaptic junctions consist of presynaptic and postsynaptic elements (see Fig. 3-10). In synapses in the central nervous system, very fine filamentous material extends between the two processes in the cleft region. The most conspicuous feature of the presynaptic terminal are the small (40 nm in diameter), clear synaptic vesicles containing acetylcholine or amino acid neurotransmitters. Mitochondria are also visible, as are dense-cored vesicles, some containing neuropeptides imported from the perikaryon and others containing catecholamines synthesized in the terminal. The portion of presynaptic membrane directly apposing the postsynaptic membrane is called the *active zone*. Sometimes, this area is marked by a dense submembranous array. The presynaptic nerve terminal contacts a postsynaptic element. In the central

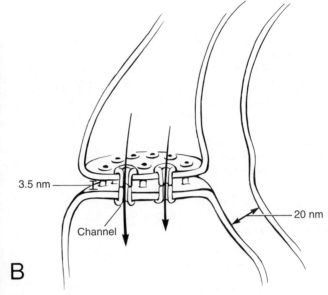

Figure 3-9
Structure of the gap junction. (**A**) In the electron micrograph, the membranes of two glial cell processes in close apposition form a gap junction (*arrows*). Notice the wider extrajunctional space (*arrowheads*) to the right of the gap junction. (**B**) The diagram shows a section of the half-channels in each membrane that join to form a pore, which allows communication between the cytoplasm of the two cells. The space between the two membranes of the gap junction is only 3.5 nm, much smaller than the extrajunctional space of 20 nm.

nervous system, this may be a cell body, dendrite (see Fig. 3-10*A*), dendritic spine (see Fig. 3-10*B*), another axon, or axon terminal.

The most striking postsynaptic feature is the dense, submembrane filamentous array beneath the postsynaptic membrane called the PSD (see Fig. 3-10). Extending from this area is a fine meshwork of actin filaments and their binding proteins. In dendrites, this network is contiguous with microtubules; in dendritic spines, the filamentous web fills the head and neck region of this process. Some postsynaptic membranes lack a well-developed PSD.

Synapses also form between nerves and muscles (Fig. 3-11). With the light microscope, axons can be seen to approach the muscle. At a specific site on the muscle, called the *end-plate region*, the axons give rise to several

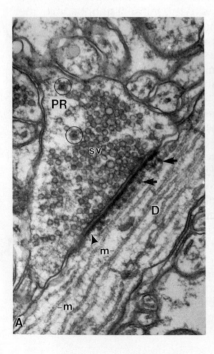
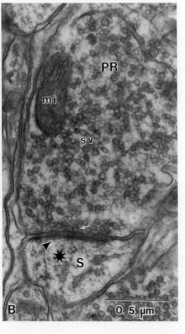

Figure 3-10
Basic features of synaptic junctions in the central nervous system. (**A**) The presynaptic terminal (PR) ends on a dendrite (D) characterized by microtubules (m), and (**B**) another that ends on a spine (S) characterized by an actin filament network (*asterisk*). In both cases, synaptic vesicles (sv) are seen in the presynaptic terminal, and a prominent postsynaptic density (*arrowhead*) is located behind the postsynaptic membrane. In **A**, the presynaptic terminal also contains dense-cored vesicles (*circled*), and in **B**, the terminal contains a mitochondrion (mi). Some of the synaptic vesicles (*white arrow*) in **B** are located or docked near the presynaptic membrane. In the dendrite in **A**, specializations called subsynaptic bodies (*black arrows*) are located on the cytoplasmic face of the postsynaptic density.

Figure 3-11
Neuromuscular junction seen by (**A**) light and (**B**) electron microscopy. In **A**, an axon (a) gives rise to a motor end plate (mep) on the muscle fiber (mf). The motor end plate exhibits many swellings (*arrowheads*). These represent the presynaptic terminals, one of which is seen at higher magnification in **B**. The presynaptic terminal (PR), filled with synaptic vesicles (sv), contacts the sarcolemma folds (*asterisks*) of the muscle cell (mc). A basal lamina (*arrowhead*) is found in between the terminal and sarcolemma. A portion of glial cell, known as a Schwann cell (SC), surrounds the terminal. (**B** Courtesy of Virginia Kriho, University of Illinois at Chicago, Chicago, IL)

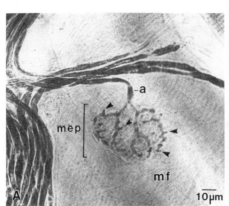
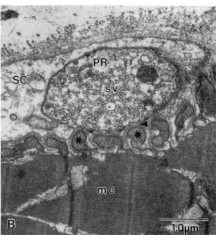

branches that display multiple swellings, each of which represents a presynaptic terminal. Ultrastructural examination reveals that the nerve terminal is closely apposed to the sarcolemma of skeletal muscle with a basal lamina lying in between the two components. The sarcolemma is thrown into distinctive folds. Acetylcholine receptors are located on the portions of the membrane directly opposing the active zones. A dense microfilamentous network extends from the membrane on the cytoplasmic face of the postsynaptic membrane and holds the receptors in place.

☐ The presynaptic nerve terminal is the site of transmitter release

Information transfer between neurons occurs in a matter of milliseconds. During this time, action potentials speed along the axonal membrane at rates between 1 and 100 m/second to the presynaptic nerve terminal, where the frequency of their firing is translated into specific quantities of neurotransmitter release. The rapidity of signal conduction over long distances demands that the nerve terminal be ready to respond to a barrage of incoming action potentials at all times. Although axonal transport replenishes the terminal with some needed molecules, organelles, and neuropeptide-containing vesicles, the rates of delivery are not fast enough to prepare the terminal with the small molecules (i.e., acetylcholine, amino acids, catecholamines) comprising the bulk of chemical messengers that mediate neurotransmission. Even fast anterograde transport can only convey membranous vesicles at a rate of 200 to 400 mm ($\approx$1 ft) each day. Consequently, the nerve terminal comes equipped with mechanisms for neurotransmitter synthesis, storage, and release and mechanisms for membrane recycling. Presynaptic terminals, however, lack the elaborate protein synthetic and packaging machinery found in the perikaryon. Even free polyribosomes are difficult to detect, although there is some evidence for presynaptic protein synthesis. Most of the synthetic enzymes and some membrane proteins required to construct synaptic vesicles must be imported from the cell body.

Neurotransmitters are primarily released from synaptic vesicles, although there is compelling evidence for additional nonvesicular release of acetylcholine from a cytoplasmic pool. Knowledge that synaptic vesicles mediate neurotransmitter release comes from the important discovery in 1952 by Paul Fatt and Bernard Katz that acetylcholine is released from terminals at the neuromuscular junction in quanta. A relatively constant number (about 10,000) of neurotransmitter molecules are released simultaneously. Since that time, neuroscientists have been actively engaged in experiments supporting this finding. Electron microscopic studies of nerve terminals revealed the presence of synaptic vesicles. When these vesicles were isolated, they were shown to contain acetylcholine. It was then proposed that the number of transmitter molecules in each quanta is equivalent to the number of acetylcholine molecules in each vesicle.

Synaptic vesicles occupy precise locales within the terminal; they are clustered and then docked near the active zone (see Fig. 3-10B). They appear to be held in place there by actin, which is connected to the vesicle by a neuron-specific protein, synapsin. However, vesicles that have already released transmitter must be replaced by those next in line, necessitating a transient depolymerization of actin filaments so that the vesicles are free to approach the membrane. Phosphorylation of synapsin releases vesicles from the cytoskeleton, permitting them to proceed to the docking site at the presynaptic membrane. Specific proteins within the vesicle membrane and presynaptic membrane interact to hold the vesicle in place. Vesicle fusion occurs so fast that it is thought to involve a conformational change in a specific protein—perhaps a change in a preassembled calcium-dependent pore from a closed to an open state. The extra vesicle membrane, now a part of the presynaptic membrane, is internalized by a clathrin-dependent mechanism and brought to an endosomal sac, where membrane proteins are sorted and new vesicles pinch off the cisternae.

It is thought that there are two release pathways: one for clear vesicles and another for neurosecretory vesicles.

The latter are not concentrated near the active zone and require a lower calcium concentration and a higher frequency of stimulation for release at other sites along the terminal membrane. Dense-cored vesicle release may represent a basal secretion, in contrast to the phasic release of clear vesicles.

☐ The postsynaptic element is the site of signal transduction

Neurotransmitters bind to specific sites called receptors on the postsynaptic membrane. This interaction is the initial step in a cascade of events that transduce the chemical message into an intracellular signal that affects the behavior of the postsynaptic neuron. Neuronal responses to incoming signals may include immediate alterations in membrane permeability or more long-lasting modifications in synaptic or neuronal architecture, which may modify the nature of the postsynaptic response. Signal transduction pathways in neurons attain an extraordinary level of complexity, increasing the number of adaptive responses by logarithmic proportions.

Postsynaptic density

Beneath the postsynaptic membrane is a dense filamentous array, the PSD (see Fig. 3-10). The intimacy of its association with the overlying membrane suggests that it restricts receptors at that site, similar to the way acetylcholine receptors are clustered at the sarcolemmal membrane of the neuromuscular junction by actin and its binding proteins. However, several properties of the PSD suggest a more dynamic role in nerve transmission. Although the PSD usually is a saucer-shaped structure, there are variations of this basic form, including differences in length, curvature, and the presence or absence of perforations. Moreover, quantitative electron microscopic analyses reveal that these parameters change in specific brain areas with various physiologic and behavioral inputs. The PSD contains actin that, together with its binding proteins also found here, may mediate dynamic changes in shape.

The major protein in cerebral cortex PSDs, for example, is the 51-kd, autophosphorylatable, Ca^{2+}/calmodulin-dependent protein kinase II, which comprises 30% to 50% of this structure. Mutant mice lacking one of the isoforms of this enzyme are also deficient in their ability to produce long-term potentiation (LTP). LTP is an electrophysiologic correlate of memory; after a given input, a synapse gets stronger and retains this new strength for a long period.

Dendritic spines

Approximately 100 years ago, Santiago Ramón y Cajal wrote that cortical dendrites seem to "bristle with teeth." He called these protuberances collateral spines, and it is only recently that we have gained some insight into their precise function. Dendritic spines are protrusions of the dendritic surface that receive synapses, almost all of which are excitatory. They consist of spine heads of various diameters that are connected to the parent dendrite by necks, which also vary in length and thickness (Fig. 3-12). The spine shape is often categorized as thin, stubby, or mushroom shaped. Cortical neurons have thousands of dendritic spines, each located every few micrometers along the dendritic shaft.

The cytoskeletal network comprising the interior of the spine includes actin and its binding proteins, myosin, Ca^{2+}-ATPase, calmodulin, and spectrin. Several reports suggest a role for actin and myosin in spine contractility that may lead to changes in the shape or number of spines. Among the inputs that appear to result in changes in spine shape or number are hormonal manipulations, long-term potentiation, dietary factors, and chronic ethanol consumption. The importance of proper dendritic spine structure is underscored by observed aberrations in the shape of dendritic spines in mental retardation. Recent studies implicate spines as calcium isolation compartments that decouple calcium changes in the spine head from those in the dendritic shaft and from neighboring spines. These experiments have important implications for explaining LTP. LTP may be synapse spe-

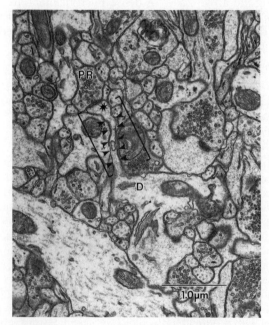

Figure 3-12
Electron micrograph of a dendritic spine of the hippocampus. A presynaptic terminal (PR) contacts a thin dendritic spine (*brackets*) that emanates from a dendrite (D). The dendritic spine is characterized by a head (*asterisk*) and neck (*arrowheads*) region. Notice the complexity of the surrounding nervous tissue, known as the neuropil. (Modified from Miller RJ. Neuronal Ca^{2+}: getting it up and keeping it up, Trends Neurosci 1992; 15:317, and from Harris KM, Jensen FF, Tsao B, Dendritic spines of CA1 pyramidal cells in the rat hippocampus: serial electron microscopy with reference to their biophysical characteristics, J Neurosci 1992; 12:2685. Courtesy of Dr. Kristen M. Harris, Children's Hospital and Harvard Medical School, Boston, MA, and with permission from Richard Miller, University of Chicago, Chicago, IL)

cific; its expression is a function of activation of calcium-dependent processes at the synapse, such as the activity of the Ca^{2+} and calmodulin-dependent protein kinase II described previously, which is also implicated in long-term processes. Calcium levels may be regulated by a unique membranous structure called the spine apparatus, because Ca^{2+}-ATPase, the IP_3 receptor, and calcium are localized there.

Glial Cells

Neuronal form is largely a function of the internal cytoskeletal framework. However, neurons are also supported externally by a second type of cell in the nervous system, the glial cell. Glial cells are present in the central and peripheral nervous systems. Glia fill in all spaces not occupied by neurons and blood vessels, surrounding and investing virtually all exposed surfaces in the central nervous system and axons in the peripheral nervous system. Glia vary in morphology, and their function is not restricted to mechanical support. In the central nervous system, glia are subdivided into four main types: astrocytes, oligodendrocytes, ependymal cells, and microglia. In the peripheral nervous system, Schwann cells function in a manner similar to that of oligodendrocytes, forming insulating myelin sheaths around axons that facilitate conduction.

□ Glia play a variety of roles in the nervous system

Astrocytes are stellate-shaped cells with a multitude of processes that radiate from the cell. These processes are supported internally by a glial-specific intermediate filament protein, glial fibrillary acidic protein (Fig. 3-13).

Some astrocytic processes may terminate as swellings called end-feet on neurons and blood vessels. Astrocytes may accumulate extracellular potassium resulting from the repeated firing of neurons. The potassium may then be released by astrocytic end-feet onto blood vessels to increase their diameter. In this way, increased neuronal activity may be supported by a concomitant increase in blood flow and oxygen consumption. Astrocytes may provide structural and metabolic support for neurons and regulate the ionic composition of the milieu around the nerve cell. Because of the emphasis on these supportive roles, astrocyte structure was historically viewed as being subservient to that of the neuron, with the form of the neuron dictating the shape of particular astrocytic processes. This view has evolved into a more dynamic one for the astrocyte, which may regulate neuronal shape, synaptic connectivity, and some aspects of neuronal function.

Morphologic analyses indicate that astrocytic processes preferentially contact neuronal surfaces over those of other glia, despite a ratio of glia to neurons of at least ten to one. Other evidence suggests that associations between neurons and glia are constantly in flux throughout

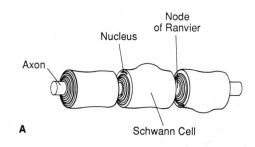

A

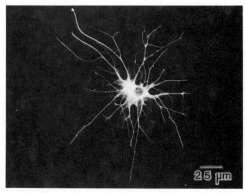

Figure 3-13
Light micrograph of an astrocyte in a primary culture prepared from neonatal rat brain. The cell was incubated with antibodies to glial fibrillary acidic protein, which was detected by a fluorescein-labeled secondary antibody. Many processes emerge from the cell body, giving the cell its star-shaped appearance. (Courtesy of Dr. Harry Yang, University of Illinois at Chicago, Chicago, IL)

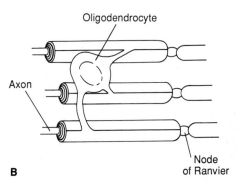

B

Figure 3-14
Diagrams of myelinated axons of the peripheral and central nervous systems. (**A**) In the peripheral axon, several Schwann cells wrap their plasma membrane concentrically around a single axon. The stretch of axonal membrane, called the axolemma, between adjacent Schwann cells is known as the node of Ranvier. (**B**) In the central axon, several glial processes emerge from one oligodendrocyte and ensheathe several axons of different origins.

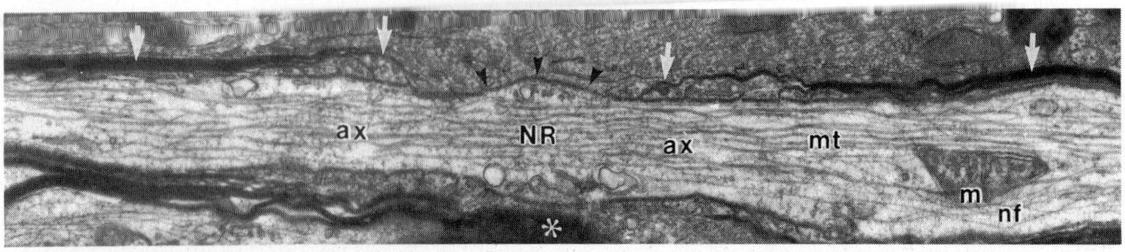

Figure 3-15
An electron micrograph of the myelin sheath of an axon (ax). The sheath is interrupted at regular intervals, called nodes of Ranvier (NR), where portions of the axonal membrane, called the axolemma (*arrowheads*), are exposed. Sodium ion channels are concentrated in the axolemma at the nodes. *Mt*, microtubles; m, mitochondron.

the life of the organism. Glia may promote or inhibit the outgrowth of neuronal processes during development by synthesizing and secreting various adhesion molecules. In some parts of the developing nervous system, such as the cerebellum and neural tube, radial glial cells form a transient scaffold that guides the migration of immature neurons to their final destinations. The migrating neurons wrap around these pole-shaped cells and crawl along them. After completion of the trip, the radial glia disappear and may be transformed into astrocytes.

In the central nervous system, astrocytes and macrophages, called microglia, remove the cellular debris resulting from degenerative processes. Ependymal cells form the ciliated lining of the central canal system of the brain and spinal cord; processes on the opposite surface may terminate on blood vessels.

☐ Glia form myelin sheaths that increase the speed and efficiency of conduction in axons

Oligodendrocytes and Schwann cells form *myelin sheaths* around axons. These sheaths are formed by the attenuation of the glial cytoplasm to such an extent that most of the sheath is composed of concentric layers of plasma membrane wrapping around the axon (Fig. 3-14). In oligodendrocytes, several processes extend out from the cell, tapering as they encounter an axon, and wrap around a portion of its length. One oligodendrocyte can ensheathe many axons, all of different neuronal origins. Individual Schwann cells dedicate themselves to a single axon. The exposed patch of axon in between adjacent segments of the myelin sheath is called the *node of Ranvier* (Fig. 3-15). Most of the Na⁺ ion channels of the axon are confined to this site. Because of the lack of channels between the nodes and great insulating action of the my-

elin sheath, there is virtually no current flow across these segments. The action potential bypasses these stretches of membrane by jumping from node to node. This type of rapid propagation is known as *saltatory conduction*. An added advantage is that energy is conserved; fewer ions enter and leave the axon so less energy is expended in returning the membrane to its original polarized state by active transport mechanisms.

SELECTED READINGS

Alberts B, Bray D, Lewis J, Raff M, Roberts K, Watson JD. Molecular biology of the cell, 2nd ed. New York: Garland Publishing, 1989.

Firestein S. A noseful of odor receptors. Trends Neurosci 1991; 14:270.

FitzGerald MJT. Neuroanatomy basic and applied. London: Bailliere Tindall, 1985.

Hall ZW. 1992. An introduction to molecular neurobiology. Sunderland, MA: Sinauer Associates, 1992.

Kandel ER, Schwartz JH, Jessell TM. Principles of neuroscience, 3rd ed. New York: Elsevier, 1991.

Kurs EM, Sengelaub DR, Arnold AP. Androgens regulate the dendritic length of mammalian motoneurons in adulthood. Science 1986; 232:395.

Linstedt AD, Kelly RB. Molecular architecture of the nerve terminal. Curr Opin Neurobiol 1991;1:382.

Pomeroy SL, Purves D. Neuron/glia relationships observed over intervals of several months in living mice. J Cell Biol 1988;107:1167.

Rasmussen AT. Some trends in neuroanatomy. Dubuque, IA: William C. Brown, 1947.

Rodriguez EM. Design and perspectives of peptide secreting neurons. In Nemeroff CB, Dunn AJ (eds). Peptides, hormones and behavior. New York: Spectrum Publications, 1984:1.

Silva AJ, Stevens CF, Tonegawa S, Wang Y. Deficient hippocampal long-term potentiation in α-calcium-calmodulin kinase II mutant mice. Science 1992;257:201.

Stevens CF. Bristling with teeth. Curr Biol 1991;1:369.

Steward O, Banker GA. Getting the message from the gene to the synapse: sorting and intracellular transport of RNA in neurons. Trends Neurosci 1992;15:180.

Wolff JR. Quantitative aspects of astroglia. Comptes Rendu du VIe Congres International de Neuropathologic 1970;31:327.

Section *II*

CELLULAR ELECTROPHYSIOLOGY AND SYNAPTIC TRANSMISSION

Neuroscience in Medicine, edited by P. Michael Conn.
J. B. Lippincott Company, Philadelphia © 1995

Chapter 4

Membranes and Ion Channels

GERRY S. OXFORD

Cell Membrane Regulation of Neuronal Ionic Environment

Nerve cells, like all other cells in the body, are bounded by a plasma membrane that characterizes the limits of each cell's identity. The plasma membrane serves several important functions. First, it provides structural integrity, isolating cell-specific elements from the extracellular environment. Second, it provides subcellular functional diversity by virtue of the distribution of various membrane components over different parts of some cells. Most importantly, the membrane regulates the intracellular and extracellular ionic environment crucial to the development and maintenance of the signaling properties of nerve cells. In this and the next chapter, we explore the composition of the neuronal cell membrane and how the constituent molecular components support communication and adaptation throughout the nervous system.

Structure and Characteristics of Neuronal Membranes and Compartments

☐ The lipid bilayer foundation

Cell membranes can be viewed in oversimplified terms as two-element structures. The basic element that comprises most of the membrane in terms of actual numbers of molecules is the lipid bilayer. Although various types of lipids contribute to this structure, phospholipids are among the most important. Phospholipid molecules are dipolar, exhibiting a marked difference in their ability to interact with charged or polar molecules such as water along their length. Water molecules themselves are dipolar, because the oxygen atom tends to be more electronegative than the hydrogen atoms. When water molecules interact with other polar substances, they orient so as to bring polar regions of unlike sign together, giving water molecules apparent "structure."

When exposed to water, phospholipid molecules also tend to orient with their hydrocarbon (i.e., hydrophobic) tails adjacent to one another and their polar or charged (i.e., hydrophilic) head groups adjacent to the water mol-

ecules. In free aqueous solution, this leads to the formation of micelles, clustered spheres of lipid with hydrocarbon cores and polar surfaces. In the case of cell membranes, the hydrocarbon tails orient "back to back" to form a type of bilayer (Fig. 4-1). The structural energy of association of the molecules is minimized in this configuration, and the hydrocarbon core of the bilayer is shielded from the water molecules of the intracellular or extracellular fluids by the hydrophilic head groups.

This bilayer structure serves four basic functions with regard to signaling in the nervous system. First, the hydrophobic nature of the interior of the bilayer serves as a barrier to the movement of charged or polar molecules (e.g., ions) between intracellular and extracellular compartments. The bilayer is central to the segregation of polar molecules in their respective functional compartments. Second, the exceptional hydrophobicity of the packed hydrocarbon tails in such a thin (1.5 to 3.0 nm) space imparts the property of electrical capacitance to the membrane. A capacitor is a device that can store electrical energy by separating mutually attractive charges (i.e., unlike sign) across a nonpolar region. In the same manner, the lipid bilayer permits the separation of ions, leading to the storage of electrochemical energy that drives the basic signaling process. Third, the bilayer provides a framework in which integral membrane proteins (e.g., ion channels, transporters) are oriented and grouped for their respective functions (see Fig. 4-1). Fourth, certain lipids comprising the bilayer are used as substrates for signal transducing enzymes activated during interneuronal chemical communication. We consider the first three functions in this chapter and the fourth in the Chapter 6.

☐ Membrane proteins contribute to neuronal identity

The second element of the neuronal cell membrane is a population of integral and associated proteins. The nature and distribution of membrane proteins imparts much of the uniqueness that discriminates one neuronal cell type from another. Neuronal membrane proteins include receptors for neurotransmitter molecules that stereospecifically recognize and bind these key signal messengers, receptors for growth and adhesion factors that

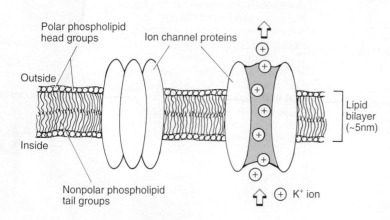

Figure 4-1
Two of the several fundamental components of neuronal cell membranes are the phospholipid bilayer and integral membrane proteins. The polar phospholipid molecules orient with their hydrocarbon tails in proximity and the more hydrophilic head groups adjacent to water. The ion channels are typical neuronal integral membrane proteins that facilitate the movement of ions across the otherwise energetically resistive lipid bilayer.

regulate differentiation and target specificity of dendritic and axonal processes, and cytoskeletal-associated proteins that regulate the movement and distribution of other proteins along the cell surface. Among the most important integral membrane proteins are ion channels and ion transporters that regulate the movement and distribution of ions such as Na^+, K^+, Cl^-, and Ca^{2+} across the cell membrane.

☐ Ionic concentration differences exist across cell membranes

The concentration of dissolved ions in a solution is a potential source of energy to drive cellular processes. The potential energy of a solution of any given ion is proportional to the concentration of that ion. For example, as the concentration of K^+ ions is increased in a potassium chloride (KCl) solution, the probability that a given K^+ ion will encounter another K^+ ion rather than a water molecule is increased. The resulting mutual repulsion between like-charged K^+ ions constitutes an energy source referred to as the *electrochemical energy* of K^+ ions in that solution. The ion-ion interactions that develop as the electrochemical energy increases can power a net diffusional movement of ions away from the area of their highest concentration. The electrochemical forces of concentrated ions in cytoplasmic or extracellular fluids provide an energy source that the neuronal cell membrane can use for signaling.

An important distinction between the intracellular and extracellular environments of nerve cells is the dramatic difference in the concentrations of various ions in the two compartments. Typical concentrations of several important ions are given in Table 4-1. The concentration differences have important functional consequences. For example, Ca^{2+} entry from the extracellular compartment serves as a trigger to activate many intracellular biochemical cascades in response to external stimuli. Ca^{2+} is therefore maintained at low basal levels inside nerve cells to prevent spontaneous activation of these processes under unstimulated conditions. Other ions (e.g., Na^+, K^+) are concentrated on one side of the membrane or the other to provide electrochemical energy to drive the basic signaling processes on which nervous system function depends.

☐ Electrical potential differences exist across cell membranes

There are electrical charge imbalances in the various cellular compartments as well as concentration imbalances. If it were possible to extract 1 mL of intracellular fluid from several million neurons and somehow count the number of cations (i.e., positive charges such as K^+) and all of the anions (i.e., negative charges such as Cl^-), the result would be an equal number of each charge type. Within the experimental error involved in such a heroic effort, the number of positive and negative ions in the intracellular compartment balance each other. The same would be true for an assessment of the charge distribution in the extracellular fluid.

Instead of going through this labor-intensive effort, the researcher could alternatively try to determine whether there were regions of the cytoplasm or of the extracellular fluid that had an excess of positive or negative charges by measuring the electrical potential difference, or voltage (V), between any two regions in the volume of fluid. This could be accomplished by placing fine glass microelectrodes (Display 4-1) at different places in the fluid and probing for regions of excess positive or negative charge with a voltmeter. If significant imbalances in the number of positive and negative ions were present, the researcher might expect to find regions of excess of one or the other charge and to detect this as a voltage. This expectation is analogous to using a voltmeter to detect the stored charges in different parts of an automobile battery. In reality, as the electrodes are moved to different positions in the cellular fluids, the researcher would find the voltage between the electrodes to be zero at any position (Fig. 4-2). This observation of negligible differences in the numbers of cations and anions in biologic fluids reflects the principle of *macroscopic electroneutrality*, which describes a balance of electric charge within any cellular fluid compartment.

The researcher would next compare the degree of charge separation between the cytoplasm and extracellular fluids. At first, this might seem a ridiculous question to ask, because the independent determinations of each compartment yielded no intracompartment charge differences. Remembering the great effort required to individually count cations and anions (not to mention the high probability of a miscount during such an enter-

Table 4-1
EXTRACELLULAR AND INTRACELLULAR CONCENTRATIONS OF SEVERAL ION SPECIES IN A TYPICAL MAMMALIAN NERVE CELL

Ion Species	Extracellular Concentration (mM/L)	Intracellular Concentration (mM/L)
Na^+	120	12
K^+	5	130
Cl^-	125	5
Ca^{2+}	2	0.0001*
Organic anions	<2	100

*Value for intracellular free calcium in a resting neuron.

Glass microelectrodes are important tools for understanding cellular electrical behavior

The use of fine glass microelectrodes for the measurement of electrical potentials across cell membranes is perhaps the most fundamental technical development contributing to our understanding of cellular excitability. Small glass capillaries can be heated to a partially molten state at a region along their length and the two cool ends pulled to cause the heated region to lengthen and narrow in diameter. After the two ends separate and are allowed to cool, each component tapers to an ultrafine tip of often less than 1 μm in diameter. The resulting glass microelectrode can be filled with a concentrated electrolyte solution, such as 3 M potassium chloride. One end of a silver wire is inserted into the electrolyte, and the other end is typically connected by another wire to an amplifier that measures voltage. Because the fine tip is connected by a conductive fluid to the amplifier and the glass wall electrically isolates the electrolyte from the outside, the experimenter can readily probe the voltage at the small open tip of the electrode. The microelectrode tip is maneuvered under microscopic control toward a single cell and rapidly forced through the membrane, which seals around the electrode shaft at the point of penetration. The amplifier then registers the intracellular voltage where the electrolyte makes contact with the cytoplasm. A separate electrode in the extracellular fluid acts as a voltage reference point. In this manner, the resting potentials and impulse activity associated with signal transmission can be accurately monitored in single cells and displayed on oscilloscopes or other record-logging devices. Our understanding and quantification of the electrical behavior of excitable cells, such as neurons, is a direct result of the application of this technique to cells throughout the nervous system.

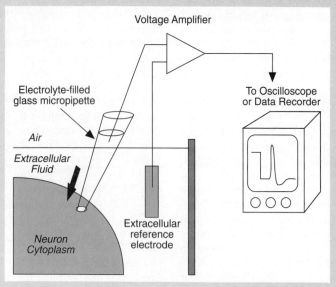

prise), the researcher wisely decides to use electrodes to measure voltage differences and estimate the balance in charge. One of the fine glass microelectrodes penetrates the neuronal membrane and makes contact with the cytoplasm, and the other is placed in the external fluid (see Fig. 4-2). In this case, a steady electrical potential difference is detected corresponding to a measured voltage of about −70 mV (0.07 V). By convention, the sign of the potential difference indicates the excess negative charge in the cytoplasm relative to the extracellular fluid. The measured voltage is always given as that of the intracellular compartment relative to the extracellular fluid.

☐ Membrane potentials are powerful cellular forces

Membrane potentials represent the force required to separate charged ion species. This force can be calculated for any separation of like or unlike charges by Coulomb's Law

$$F = k\frac{Q_1 Q_2}{d}$$

where F is the force, Q_1 and Q_2 are separated charges of like or unlike sign, d is the distance separating the two

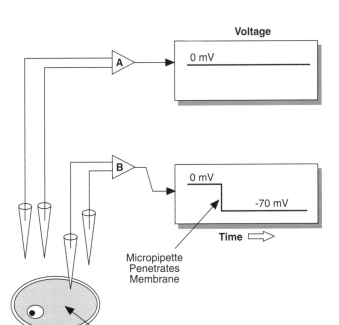

Figure 4-2

Probing the extracellular and intracellular environment of a neuron with microelectrodes reveals differences in the charges between the cytoplasmic and extracellular compartments but not within either compartment. Voltage amplifier A is recording the electrical potential difference (i.e., voltage) between two adjacent extracellular electrodes. Regardless of their relative position, no differences in voltage—hence, no major charge differences—are detected. Amplifier B measures the voltage between an extracellular electrode and one that has just penetrated the cell membrane to make contact with the neuronal cytoplasm. In this case, a steady membrane potential difference is recorded, indicating an excess of negative charge in the cytoplasm compared with the extracellular fluid. This resting potential is a fundamental property of nerve cells. In real neurons, this steady potential would probably be interrupted by brief changes in potential corresponding to the electrical signaling activity with which the nervous system functions.

charges, and k is a proportionality constant. The force is attractive for charges of unlike sign and repulsive for charges of like sign. Unlike charges separated across a very thin cell membrane (i.e., small d) constitute a strong attractive force, tending to unite the charges and dissipate the voltage.

Measured resting membrane potentials of neurons are on the order of -40 to -90 mV (0.04 to 0.09 V). These values are not impressive, especially considering that the voltage of a common AAA battery is about 20 times greater. Although this is a small voltage, it actually represents a very large electric field (i.e., voltage per unit distance), because the small charge separation giving rise to the membrane voltage occurs over a distance of only 3.0 to 5.0 nm, the thickness of the lipid bilayer. This corresponds to an electric field of 100,000 to 200,000 V/cm! In contrast, the electric field across a AAA battery is only 0.33 V/cm. The electric field of the membrane is thus nearly 1 million times greater than that of the AAA battery.

Such large membrane electric fields and the equivalently strong forces are attributable to the capacitance properties of the cell membrane. The lipid bilayer of neuronal cell membranes represents an electrical insulator between two highly conducting fluids. Because electrical capacitance is inversely related to the thickness of the insulating region separating two conductors, the thin cell membrane makes a very efficient capacitor. Capacitance values for cell membranes are on the order of 1 microfarad (μF)/cm^2 of membrane surface. The voltage across a capacitor is related to the amount of charge separated by the equation

$$V = \frac{Q}{C}$$

where V is the electrical potential or voltage, C is the capacitance, and Q is the charge. Using this equation, it is possible to calculate how many charges need to be separated across a typical neuronal cell membrane to yield normal resting potentials. We shall not go into the details of these calculations here, but the results reveal an important point. For a neuron of approximately spherical geometry that is 20 μm in diameter, a resting membrane potential of -70 mV corresponds to a charge separation of roughly 9×10^{-13} coulombs (a coulomb is a standard unit of charge). Assuming an intracellular concentration of potassium of 150 mM/L and calculating the volume of the neuron, the amount of charge available from only the pool of K$^+$ ions is roughly 6×10^{-5} coulombs. Assuming that the escape of some K$^+$ ions from the cytoplasm to the extracellular space produces the charge imbalance associated with the normal membrane potential, the amount required is only about 0.0000015% of the total cellular potassium! As we shall discuss in the next chapter, such movements of ions across cell membranes are the central processes in the generation and dissipation of membrane potentials. The amount of charge required to move across cell membranes to support physiologic changes in membrane potential produces only negligible changes in the concentrations of the ion species involved. Although this statement is generally true, this concept can be compromised for axons or neurons of very small size and volumes. For such neurons, accumulation or depletion of ions in the cytoplasm during electrical activity may have significant functional consequences.

Ion Channels

☐ Ion channels are membrane proteins

The lipid bilayer portion of the nerve cell membrane is hydrophobic and constitutes a high energy barrier to the movement of polar substances, including ions between intracellular and extracellular environments. Ions interact with polar water molecules to minimize the energy required to stay in free solution uncomplexed with a counter ion (i.e., phenomenon of hydration). The interaction energy between ions and the hydrophobic interior of the lipid bilayer is unfavorable and is not sufficiently strong to pull an ion from its energetically favorable hydration state in water. The probability that an ion will cross the lipid bilayer unassisted is vanishingly small. How then do ions cross this barrier? Two possible mechanisms can be suggested. Either an external source imparts sufficient energy to the ion to propel it across the barrier, or the energetic requirement to cross the bilayer is lowered. The latter mechanism usually is the case and is the function of ion channels.

Ion channels are integral membrane proteins that specifically facilitate the transfer of ions across the membrane. These movements are driven by electrochemical energy differences. It is thought that a channel accomplishes this facilitation by the formation of an aqueous pore through the membrane. This pore offers coordination sites (Fig. 4-3) with a sufficiently strong attraction for ions to replace energetically the water hydration shell surrounding an ion in free solution, allowing it to enter and eventually cross the membrane.

☐ Ion channels exist in open or closed conformations

The evidence that ion channels are actually pores was circumstantial until the recent development of experimental methods to detect ion flow through individual ion channels. Single-channel recording methods (Display 4-2) have revealed that ion channels can exist in one of two basic states: open (i.e., ion permeable) and closed (i.e., ion exclusive). Any given channel molecule may actually exhibit several protein conformations reflecting closed or open states. Regardless of the number of conformations, the primary functional difference among these states is the degree to which ions can permeate or pass through the channel.

Regulation of ion permeability by channels is achieved by two distinct but interacting properties: *gating* and *ion selectivity*. Transitions among the various conformations that regulate permeability are referred to as gating. When in their closed conformations, channel "gates" are presumed shut and prohibit the passage of ions. Little or no electrical current (in this case, current reflects the movement of ionic charges per second) is detected flowing through the membrane when channels are closed. When a channel is activated, it changes conformation and opens the functional gate, allowing ions to pass across the membrane (Fig. 4-4). Single-channel experiments have revealed an important feature of this gating. Channels switch among a finite number of functional gating states with nearly instantaneous speed. At the most basic level, channel gating transitions simulate the operation of a standard light switch; the channel is open or closed as a light is on or off. This mechanism is contrasted with a light on a dimmer switch in which variable amounts of light can be achieved. In general, ion channels do not exhibit conducting states of various magnitudes, although some can under certain circumstances. When in an open conformation, a channel can discriminate, or select, among various ion species for passage across the membrane, allowing some to cross more readily than others (e.g., sodium versus potassium). In later sections, we consider the possible mechanisms responsible for these two properties.

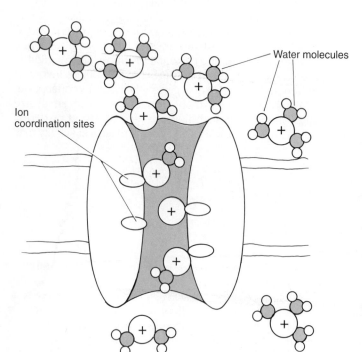

Ion coordination sites

Water molecules

Figure 4-3

Ion channels provide an energetically friendly environment for ions to pass through the cell membrane. It is thought that the aqueous pore region of an open ion channel presents a number of chemical groups that act as coordination sites along which ions pass during transit through the channel. These sites are energetically similar to the coordination afforded an ion by its normal waters of hydration in free solution. An ion can be passed along a sequence of water-ion, channel-ion, and back to water-ion interactions without paying a high energetic price. For some channels, as depicted here, it is thought that more than a single permeant ion can reside in the pore at one time. In some other channels, single ion occupancy may be the rule.

The patch clamp technique allows detection of the opening and closing of single ion channel molecules

For many decades after the Nobel Prize-winning studies of excitation in the squid giant axon by Hodgkin, Huxley, and Katz, it was assumed that transmembrane currents flowed through discrete ion channel proteins and that the permeability of an individual channel was consistent with the concept of ion movement through an aqueous pore with relatively minor energetic restrictions. The development of the patch clamp technique by Drs. Erwin Neher and Bert Sakmann in the early 1980s provided a way to examine these assumptions and, for the first time, to record the activity of an individual ion channel molecule in its native cell membrane. Drs. Neher and Sakmann won the 1991 Nobel Prize in physiology and medicine for their work with this method, three decades after the pioneering studies of Hodgkin, Huxley, and Katz.

The patch clamp method uses glass electrodes similar to those for cell membrane voltage recording, except that the tip openings are usually several micrometers in diameter and are heat polished to a smooth surface. When the electrode is placed against a cell membrane and gentle suction is applied, a small patch of membrane is drawn slightly into the electrode opening. After a few seconds, a remarkably strong bond is formed between the membrane and the glass surface such that ion flow between the two is negligible. Electrical resistances as high as 10^{11} ohms or 100 gigaohms occur between the inside of the electrode and the extracellular fluid. This bond is often referred to as a "gigaohm seal." The bond between the membrane and electrode is so tight that withdrawal of the electrode often rips the patch of membrane from the cell without reducing the adhesion to the electrode surface! This process of patch excision from a cell can be used experimentally to gain precise control of the composition of the solutions on either side of the membrane patch.

The voltage of the electrode interior can be effectively clamped to any level, because leakage of charge from the electrode tip is very small. If an individual ion channel residing in the patch of membrane has occasion to open, it affords the lowest resistance pathway between the interior and exterior of the electrode. Most of the current that passes from the electrode flows through the single open channel. With very sensitive current amplifiers, this molecularly focused current can be detected as brief pulsatile current steps on the order of several picoamperes (10^{12} A) in amplitude and several milliseconds in duration. The inset illustrates K^+ ion movement through a sporadically opening Ca^{2+}-activated K channel, where downward deflections correspond to an open, conducting channel. These transitions between two or more discrete current levels reflect the rapid switching of an individual channel protein among its most stable open or closed conformational states. The amazing sensitivity of this method can be appreciated when it is realized that it is the only biological measurement technique able to resolve the conformational changes of a single integral membrane protein, and in this sense may be regarded as the only true *molecular* biological technique.

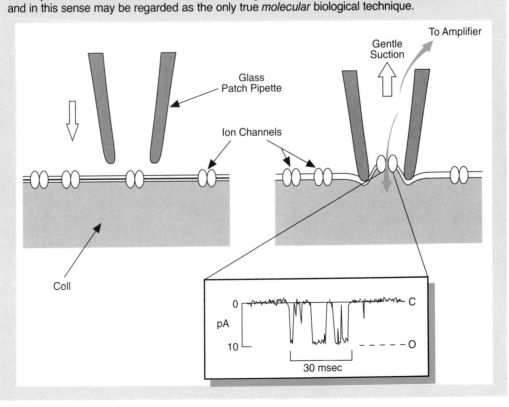

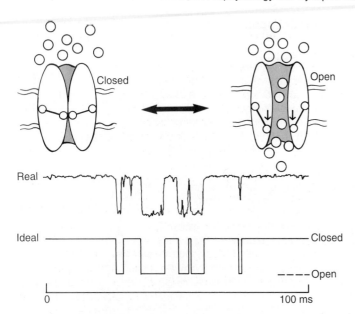

Figure 4-4
Ion channels can abruptly change between closed (i.e., ion-exclusive) and open (i.e., ion-permeable) conformations. Ion channel proteins exist in various conformations that can be grouped into two classes: closed or open. When an individual channel changes between these two classes of states, it does so very rapidly. The real tracing is an actual recording of K$^+$ ion current through a single potassium channel as it opens (i.e., downward deflections) and closes. The noisy and rounded character of the record reflects the normal limitations of the electronic recording equipment. The ideal tracing represents the actual conformational behavior of the channel estimated from the real record. When the channel appears to be open, it always conducts ions at the same rate, and the current is the same (8 pA in this case). This channel exhibits only two fundamental conformations during this sequence: a closed state and a single open state with a finite permeability to K$^+$ ions.

☐ Conformational changes can be triggered by several mechanisms

The gating function of an ion channel represents shifts of the protein conformation between various open and closed states. These conformational changes can be triggered or increased in probability by two basic types of stimuli: voltage or ligand binding. These stimuli are often used to define two classes of ion channel. Some ion channels are voltage gated. The stimulus for gating this class of channel is a change in the voltage across the membrane in which the channels are embedded. It is thought that the actual stimulus for activating these channels is the change in the intense electric field experienced by the channels. Considering the amazing forces represented by such small voltages, it should not be surprising that these forces can induce conformational changes in membrane proteins, such as ion channels. Voltage-gated channels are intimately involved in the initiation, shaping, and conduction of basic nerve impulses (i.e., action potentials) throughout the nervous system. Many voltage-gated channels are named after the ions that they primarily conduct under physiologic conditions (e.g., Na channels, Ca channels).

The other type of channel is gated by the stereospecific binding of a ligand to special receptor regions of the channel molecule. This type of channel is called ligand gated. Typically, an appropriate ligand such as a neurotransmitter molecule (e.g., acetylcholine or glutamate) or even cyclic nucleotides (e.g., cAMP), can directly activate some channels. The nicotinic acetylcholine receptor channel in skeletal muscle membranes at the neuromuscular junction is the classic example of a ligand-gated ion channel.

Some channels exhibit dual gating mechanisms, responding to ligand binding and to voltage changes. For example, a class of potassium channel (i.e., calcium-activated potassium channels) open in response to the binding of calcium ions to a region of the channel facing the cytoplasm (Fig. 4-5). In addition to calcium activation, the probability that these channels will open is increased by more positive cytoplasmic membrane voltages. In some channels, the influences of ligand binding on channel gating are not the result of direct activation, but they are mediated indirectly through an intracellular molecular signaling cascade. In these cases, ligands bind to specific membrane receptors that are distinct and separate proteins from the channels. For example, the binding of acetylcholine to some muscarinic receptor subtypes or of dopamine to some dopamine receptor subtypes triggers the activation of a membrane-bound GTP-binding protein, which promotes the opening of a specific class of K channels. Neurotransmitter receptors can trigger the activation of adenylyl cyclase, increasing cytoplasmic cAMP levels and activating specific protein kinases. Phosphorylation of channels or channel-associated proteins by these activated kinases can alter the probability that the channels will open (see Fig. 4-5). Certain types of voltage-gated calcium channels are subject to this type of regulation.

These are a few examples of the rich and complex mechanisms for the activation and modification of channel gating functions that have been discovered. The interactions between channel gating and cellular biochemical events are fundamental elements of nervous system signaling and add to the tremendous diversity and plasticity of communication in the nervous system.

☐ Ion channels can discriminate among ion species allowed to cross membranes

Channels are frequently named after the ions that they most often transport under physiologic conditions. This nomenclature reflects the fundamental property of ion

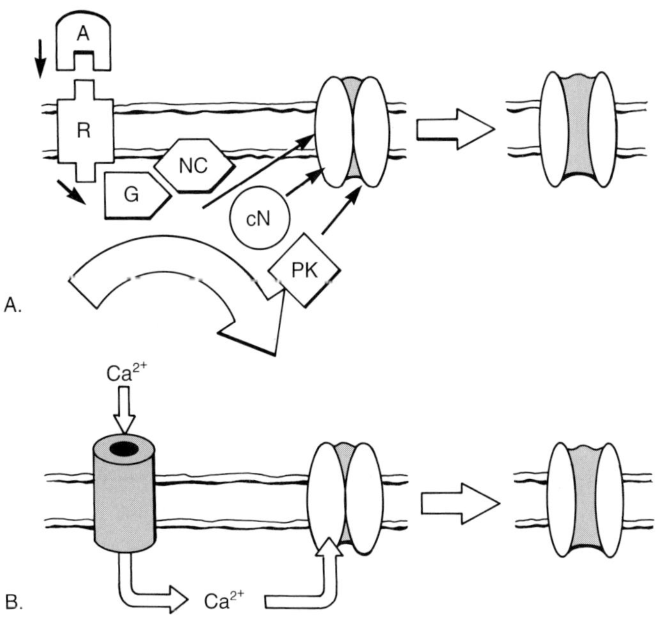

Figure 4-5

The probability that ion channels will open can be altered by biochemical signal transduction processes and second messenger molecules. (**A**) Several ion channel types can be biased toward the open state by interactions with various components of different signal transduction pathways. This illustration depicts a generic phosphorylation signal cascade, involving sequential activation of a membrane receptor (R) by an agonist (A), activation of a guanine nucleotide-binding G-protein (G), which activates (or inhibits) a nucleotide cyclase (NC), such as adenylyl cyclase. The resultant change in cyclic nucleotide (cN) concentration alters the activity of a protein kinase (PK) that may phosphorylate an ion channel target. Evidence exists for direct interactions of channels with G and cN and for phosphorylation by PK that alter gating behavior. The variety of pathways between a receptor and an independent channel target yield amazing diversity in the modulation of channel gating by agonists. (**B**) Ca^{2+}-activated K channels (K(Ca) channels) can be biased toward the open state by increases in intracellular Ca^{2+} concentrations. Ca^{2+} fluxes through voltage-gated calcium channels can produce such elevations in the cytoplasmic Ca^{2+} and activate K(Ca) channels. The opening of calcium channels depolarizes the cell, and the delayed opening of K(Ca) channels restores a resting potential and terminates the calcium response, a negative-feedback effect within a single cell.

selectivity. Discrimination among ion species by different channels reflects differences in the molecular architecture of the aqueous pore formed by an open channel. The fundamental mechanisms of ion selectivity are not well understood, but certain key properties of the channel and the transported ions appear important in defining this feature for a given channel type.

At the most basic level, ion channels can select among ions on the basis of sign. Channels that pass positively charged monovalent cations such as sodium and potassium do not generally facilitate passage of negatively charged anions such as chloride or sulfate. This level of discrimination may arise from differences in the arrangement of charged groups lining the opening of the chan-

nel. At yet another level, channels may exclude species of ions on the basis of size. For example, voltage-gated Na channels readily pass sodium and lithium ions (crystal radii of 0.095 and 0.06 nm, respectively), but the permeability to potassium and rubidium ions (0.133 and 0.148 nm, respectively) is 10 to 50 times lower.

Although exclusion of ions on the basis of size, much like a filter or sieve can exclude large particles, can account for selectivity for some ions, it cannot be the only mechanism. For example, all K channels studied are 50 to 100 times more permeable to potassium ions than to the smaller sodium ions. Under normal ionic conditions, calcium channels are nearly 1000 times less permeable to sodium than calcium despite the similarities in crystal radii (0.095 and 0.099 nm, respectively). It appears that additional factors determining selectivity are required. Among the properties thought to be involved are the interactions of ions with groups comprising the pore wall such that specific sites can accommodate particular ions better than others. This should not be difficult to envision considering that organic chemists can routinely synthesize simple molecules that associate specifically with ions such as Na^+ or Ca^{2+} to the exclusion of, for example, K^+ or Mg^{2+}. Such selectivity sites in the channel probably afford an energetically favorable environment that can more readily substitute for one of the inner hydration shells of one ion species than for another.

The differences in pore architecture among sodium-, calcium-, and potassium-preferring channels are probably subtle. This is suggested by determinations of the primary amino acid sequence of channels using molecular biologic techniques. High degrees of similarity exist among several types of ion channels in regions thought to form the ion pore. Single amino acid substitutions created by experimental mutagenesis can render Na channels quite permeable to calcium ions, suggesting that the molecular differences underlying functional Na and Ca channel selectivity are not great.

□ Functional properties of voltage-gated sodium and potassium channels

Responses of ion channels to the membrane potential

Our understanding of the behavior of voltage-gated ion channels is derived from the application of an electrical measurement technique called the *voltage clamp*. The voltage clamp technique allows the potential of a neuronal membrane to be controlled and rapidly changed to any desired value. The ionic currents flowing through channels so stimulated to open are readily measured electronically. In the late 1940s and early 1950s, a British research team of Alan Hodgkin, Andrew Huxley, and Bernard Katz used this technique to observe and describe the ionic currents through Na and K channels. Their experiments were performed on the membranes of unusually large axons found in the nervous system of the squid,

Loligo forbesi. The observations on this membrane preparation have served as the foundation for essentially all subsequent studies of voltage-gated channels.

When the membrane potential of an axon is quickly stepped from the resting potential of −60 mV to 0 mV, an inward cation current initially develops, reaching a peak in 1 or 2 milliseconds, and then decays. The declining inward current gives way to a developing outward current that reaches a steady value that is maintained for the duration of the voltage step (Fig. 4-6). In experiments in which sodium or potassium ions have been substituted by impermeant cations, it has been determined that the inward transient current is carried by sodium ions, while the outward current is carried by potassium.

When the potential of the neuronal membrane is stepped to different values, the same basic pattern of currents is observed, but the kinetics and magnitude of each component changes with each change in potential. These different currents result from the opening (i.e., activation) of separate populations of ion channels for sodium and potassium. From knowledge of the peak Na⁺ or steady K⁺ current at different potentials, the electrical conductance (i.e., permeability) of the membrane to these ions can be calculated and represented as a function of the membrane potential (Fig. 4-7). The membrane permeability increases as the membrane potential is made more positive and remains very low for more negative potentials. The increase in permeability with depolarization results from the increasing probability that individual channels will attain the open conformation.

If the membrane potential is stepped to levels more negative than the resting potential, the ion channels fail to activate. At large positive voltages, the permeability saturates, indicating that all channels have attained their maximal open probability. This property by which currents are induced by potential changes of only one direction is called *rectification*. The sigmoidal shapes of these curves representing the plot of P_{Na} or P_K versus membrane potential are similar to those that describe responses to ligand-receptor binding. In this informal analogy, the change in membrane electric field that occurs with voltage changes can be thought of as the "ligand" or "agonist" for voltage-gated channels.

The observation that Na channels always open before K channels when elicited by a voltage stimulus is of fundamental importance to their role in the membrane excitability underlying nerve impulse signaling (see Chap. 5). The pattern of currents also reveals another important functional difference between Na and K channels in axons. The decline of inward Na current after the peak while the voltage stimulus is maintained contrasts with the persistence of outward K⁺ current during the stimulus (see Fig. 4-6). This loss of Na current reflects a decline in Na permeability during the stimulus and is referred to as *inactivation*. This phenomenon represents a transition of the Na channels to a new closed conformational state (Fig. 4-8). The inactivated conformations differs from the "resting" closed states of the channels in that they are insensitive or refractory to depolarizing stimuli. Channels are not able to open in response to depolarization when in an inactivated state; they must recover the ability to activate on return of the membrane to a negative potential. As the membrane potential remains at a negative level, the channels slowly return to a resting state over a period of several milliseconds. Experimental evidence suggests that an important region of channel proteins involved in the inactivation process is probably exposed on the intracellular surface of inactivating channels. It is thought that for some channels a portion of the inactivation process represents an occlusion of the inner mouth of the ion pathway by a cluster of amino acid residues tethered to the remainder of the protein, much like a bathtub plug prevents water from draining.

The inactivation process plays a fundamental role in setting limits on the impulse frequency an axon can support (see Chap. 5). Carrying the ligand-receptor analogy further, inactivation is similar to the phenomenon of receptor desensitization, whereby a membrane receptor becomes refractory to activation in the continued presence of appropriate ligands. Receptors recover from desensitization in the absence of ligand as inactivated channels recover during prolonged repolarization.

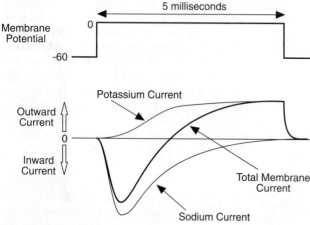

Figure 4-6
Depolarization of a nerve axon membrane induces a characteristic sequence of ion currents. The membrane potential of a squid giant axon can be controlled by an electronic circuit, known as the voltage clamp. When the membrane potential is abruptly changed from its normal resting value (−60 mV) to zero, the membrane rapidly responds by generating an inwardly directed positive current. This current reaches a peak within 1 millisecond and then reverses direction to become an outward current, which is maintained as long as the membrane potential is depolarized. This current pattern is the result of two overlapping ionic components. Replacement of the external Na⁺ by impermeant choline⁺ ions eliminates the inward current component (*upper thin line*). Independent replacement of the intracellular K⁺ by impermeant Cs⁺ ions eliminates the outward current component (*lower thin line*). The inward and outward currents reflect currents through the Na and K channels, respectively. The sodium current component is transient and decays despite the maintained voltage stimulus.

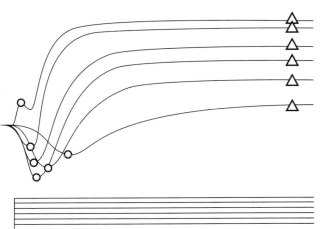

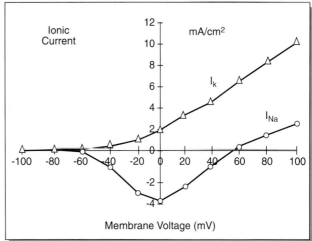

B

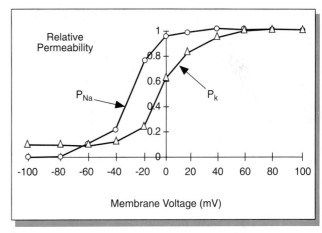

C

Figure 4-7

Analysis of sodium (Na) and potassium (K) currents at different membrane potentials reveals the voltage dependence of membrane ion permeability. (**A**) Different kinetic patterns of Na and K currents occur with different amplitudes of voltage steps. Because the early peak currents represent mainly sodium currents (*circles*) and the late steady currents represent potassium currents (*triangles*), they can be independently graphed as a function of the membrane potential. (**B**) The resulting current versus voltage relations illustrate that K currents increase monotonically with voltage, but inward Na currents increase and then become outward. The outward Na current occurs when the intracellular positive voltage is large enough to drive Na^+ ions out of the cell against the concentration gradient. (**C**) Analysis of the current versus voltage data yields an estimate of the permeabilities (P) of the membrane to Na^+ and K^+ at different voltages. Both P_{Na} and P_K increase as the membrane potential becomes more positive but not as it is made more negative. Both permeabilities saturate at very positive voltages, indicating the maximal open probabilities of each channel type. The shapes of the curves are reminiscent of the dose-response relations for ligand-receptor interactions.

Detection of the opening of individual ion channels with the patch clamp method indicates that channels open abruptly, remain open for some period, and then close abruptly. How then do these brief, pulsatile, single-channel current events give rise to the apparently smooth increases and decreases of membrane current seen in axonal membranes containing many channels? The answer lies in the probabilistic nature of single-channel events. The force exerted on the channels by a changing electric field does not invariably trigger an opening at a specific time for all channels, but the force increases the probability that any given channel will open. Each channel then responds in a random fashion, with some opening early, some opening late, and some not opening at all (Fig. 4-9). Any given channel will not respond identically to a series of voltage stimuli, but its average response (i.e., latency to open and number of successful openings) will be characteristic of the particular membrane potential. The summation of this behavior over all channels in the

membrane yields a characteristic smooth waveform reflecting the average gating behavior (see Fig. 4-9).

Diversity of voltage-gated ion channels

Within a given ion selectivity class (e.g., Na channels, K channels), ion channels can often be differentiated on the basis of their pharmacology or gating properties. These differences confer important functional properties on the membranes expressing a given channel subtype. A few subtypes from each of three major groups serve as examples.

Sodium channels. Voltage-dependent Na^+ channels are the principal membrane element underlying the nerve impulse (see Chap. 5). These channels are the primary target of local anesthetics often used to deaden pain sensation in peripheral nerves. Local anesthetic molecules block the ion pore from the intracellular side and

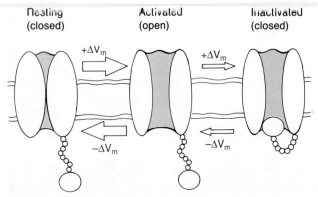

Resting   Activated   Inactivated
(closed)   (open)   (closed)

Figure 4-8

Voltage-gated sodium channels undergo sequential transitions among three basic functional states. This figure illustrates the major transitions that a sodium channel undergoes after depolarization or repolarization of the membrane potential. More positive membrane potentials shift the gating probabilities toward the right. On activation (resting → activated), the channel rapidly opens and, if the voltage is maintained, more slowly undergoes inactivation (activated → inactivated) to a different closed state. More negative potentials drive the gating sequence to the left, such that any inactivated channels undergo a slow reactivation (inactivated → activated), and any open channels rapidly deactivate (activated → resting). The relative rates of the different transitions can vary among sodium channels from different species, and other transitions among states occur much less often. Similar behavior is seen in inactivating K channels.

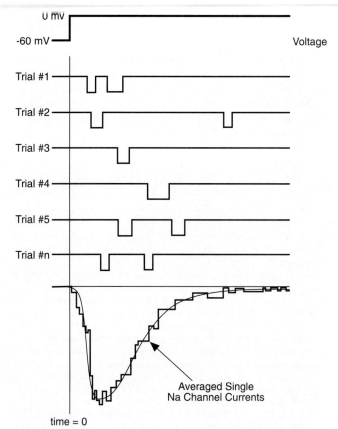

Figure 4-9

Ionic currents from populations of Na or K channels reflect the average behavior of stochastically operating individual channels. If the voltage of a membrane patch containing only a single sodium channel is repetitively stepped from −60 mV to 0 mV, the channel opens in each trial with a variable delay from the onset of the step and remains open for a variable period before closing. Occasionally, the channel may open more than once during a trial or may fail to open at all. Adding all of the individual current records and dividing by the number of trials yields a fluctuating estimate of the average behavior of that channel (*jagged heavy line*). This average time course closely matches the smooth current (*thin line*) resulting from a single step of the membrane voltage of an entire cell in which thousands of channels respond.

prevent the passage of Na ions on activation of the channels. In addition to these synthetic molecules, tetrodotoxin (TTX), a natural poison contained in the tissues of several species of pufferfish is a highly selective and potent blocker of Na$^+$ channels. TTX has proven an invaluable tool in experimental investigations of Na$^+$ channel function. Sensitivity to TTX is widely used to define the involvement of these channels in a given physiologic situation. Not all voltage-gated Na$^+$ channels are TTX sensitive. For example, a population of TTX-insensitive Na$^+$ channels supports the excitability of primary sensory neurons in dorsal root ganglia and functions otherwise identically to TTX-sensitive channels.

Calcium channels. Calcium channels provide a pathway for the stimulated entry of Ca^{2+} ions into neurons. Subsequent elevations in intracellular Ca^{2+} serve various functions, from promoting neurotransmitter release to the activation of various enzymes. Subtypes of calcium channels have been described and classified on the basis of gating properties and pharmacology. Because the nomenclature of these subtypes is controversial and not very important, we have chosen a widely used scheme employing letters.

The L-type channels are found in most muscle cells (i.e., skeletal, cardiac, smooth) and are involved in excitability and excitation-contraction coupling. These channels are blocked by a class of drugs called dihydropyridines that are used clinically for treating cardiovascular

disorders. This subtype is also present in some neuronal cell bodies and supports depolarizing electrical responses of the cell membrane. N-type channels appear to be expressed exclusively in neuronal or neurally related cells. Evidence suggests that they are the principal calcium channel subtype through which calcium enters nerve terminals and triggers the release of many chemical transmitter substances. A toxin from a fish-hunting marine cone snail, ω-conotoxin, selectively blocks and differentiates N-type calcium channels. Another subtype of calcium channel, the P-type channel, has been found to be abundant in Purkinje neurons of the cerebellum among other brain regions. A toxin extracted from certain species of spider, ω-agatoxin, blocks calcium flux exclusively through this channel subtype.

Potassium channels. Potassium channels exhibit an exceptional variety of functional properties. The K channels in the axon membrane that open with a delay on depolarization are called *delayed rectifiers*. This terminology reflects the lag in activation (see Fig. 4-6) and the probability that they will open at progressively depolarizing voltages but not at progressively negative voltages. These channels play a crucial role in action potential repolarization in a variety of neuronal cells.

Another class of voltage-gated K channel exhibits inactivation behavior similar to that seen in TTX-sensitive Na channels. On depolarization, the current through these channels rapidly increases and then declines at a rate depending on the type of neuron. Potassium currents associated with these channels are often called A currents. The inactivation behavior of these channels regulates the action potential frequency in some neurons during prolonged stimulation. These channels can also be involved in adaptation to sensory stimuli in other neurons.

Some K channels exhibit voltage-dependent gating properties that are the reverse of most channels. These K channels preferentially gate the inward movement of K^+ ions by opening when the membrane potential is made more negative rather than more positive. Such anomalous inward rectifiers are important regulators of the pacemaking activity in spontaneously active neurons. A subtype of these channels, the ATP-sensitive K channel, is activated by decreases in intracellular ATP and may play a protective role during periods of cellular hypoxia in the brain by suppressing excitability and preventing intracellular accumulation of Na^+ or Ca^{2+}.

Some K channels are activated by elevations in intracellular Ca^{2+} levels. These Ca^{2+}-dependent K channels provide a link between Ca^{2+} influx through calcium channels and the regulation of membrane potential. Experimental evidence suggests that they play a role in terminating action potential activity during bursting electrical behavior in some neurons. As intracellular calcium levels in neuronal cell bodies progressively increase during such repetitive firing, the activation of these channels hyperpolarizes the membrane below the level where action potential activity can be supported, terminating the burst. Other channels of this class provide the underlying permeability change for hyperpolarizing potentials that characteristically follow some action potentials.

The pharmacologic properties of K channels are often very similar. Well-documented blockers of delayed rectifier channels such as tetraethylammonium ions (TEA^+) or 4-aminopyridine have been shown to inhibit K^+ current in several different classes of K channel. Some toxins from scorpions and serpents appear to block and differentiate different types of K channels.

It is apparent that various K channel types support a wide variety of subtle but important signaling processes in neurons. The diversity of channel types serves in large part to define the functional role of specific neurons in complex nervous system behaviors.

☐ Many ion channels share common molecular structural characteristics

The application of molecular biologic and genetic engineering techniques to the study of ion channels has provided clues about the structure and composition of ion channel proteins and new methods for investigating function. The genes encoding several classes of ion channels have been cloned and sequenced. Physical chemical properties predicted from the primary amino acid structures encoded by the nucleotide sequences suggest various structural models for the channels. By analyzing the relative hydrophobicity of the amino acids, several stretches of approximately 18 to 22 amino acids long that are highly hydrophobic have been identified in most sequences. These "water-hating" regions are predicted to interact favorably with membrane lipids, forming a series of membrane spanning domains with α-helical structure. In the case of several voltage-gated channels, it has been suggested that six such transmembrane helical domains are connected by stretches of more hydrophilic amino acids alternating in the cytoplasmic and extracellular fluids and that these connected sequences form a basic subunit for channel function (Fig. 4-10). These subunits are thought to be the basic structural components of several important ion channel types.

Taking advantage of the wealth of information on the genetics of the fruit fly, *Drosophila*, investigators have successfully cloned a K channel gene from this organism based on the observation that flies with mutations in the gene shake during exposure to ether. The cloning of this so-called Shaker K channel led to several K channel genes being cloned from a variety of tissues. The nucleotide sequences predict single polypeptides with molecular weights of only 65,000 to 70,000, each with the six hydrophobic α-helices characteristic of the basic channel subunit. Studies in which the proteins are synthesized (i.e., expressed) after introduction of the cloned nucleotide sequences in host cells suggest that four such subunits are independently produced and combine before insertion in the membrane to make a functional K channel (see Fig. 4-10). The channels can be formed from four homologous subunits or from mixtures of four heterologous subunits (Fig. 4-11). The numbers of structurally different but functionally similar K channels may be very large because of intergenetic combinations.

Protein purification and molecular cloning studies have revealed that voltage-gated Na channels are encoded by much larger nucleotide sequences than K channels. The major functional unit of voltage-gated Na channels is a single polypeptide with a molecular weight of about 250,000. Analysis of the amino acid sequence of this protein suggests that it represents four similar repeats of the prototypical six-transmembrane domain unit (Fig. 4-12). It is thought that, on insertion into the membrane, these four repeats or domains fold together to make a complete channel and that the pore is con-

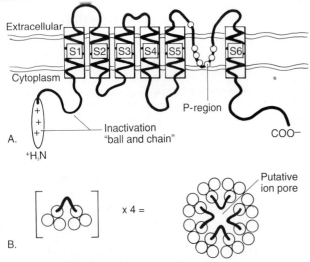

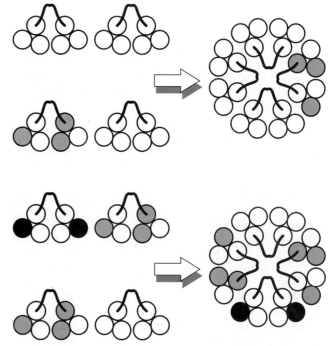

Figure 4-10

The molecular properties of some voltage-gated K channels reveal a basic protein subunit with features common to several types of voltage-gated channels. (**A**) The predicted membrane topology for the peptide sequence of the Shaker K channel cloned from the fruit fly suggests the existence of six α-helical membrane-spanning regions (S1–S6) and NH_3^+ and COO^- termini exposed on the cytoplasmic surface. Research on experimentally mutated channels expressed in foreign cells suggests several regions of functional significance. A region near the amino terminus is important for channel inactivation and apparently acts as a "ball and chain" structure, which moves into a position to occlude the open ion pathway. The fourth transmembrane helix (S4) contains a number of positively charged amino acids that sense changes in the membrane potential and induce the conformational changes associated with activation. The S5-S6 linking sequence is a crucial component of the ion pathway. Four of these basic subunits are believed to associate during synthesis and membrane insertion to form an homologous tetrameric channel. Top view of a possible model channel structure. The amphipathic S5-S6 linkers or P-regions of each monomer fold into the plane of the membrane, with hydrophilic residues forming the lining of the putative ion pore. The amino acids in this region are thought to provide the coordination sites necessary for ion movement through the channel.

Figure 4-11

It is possible for different K channel subunits to combine and form heterologous ion channels. Subunits with different amino acid sequences (*shaded circles*) can result from alternative routes of messenger RNA processing. The large number of possible combinations of these subtly different subunits provides a basis for significant diversity in the types of K channels that could be expressed in different regions of the nervous system.

structed from homologous regions to that for K channels. The same four-domain model has been suggested for the major polypeptide a subunit of voltage-gated calcium channels, which is similarly large as compared with that of K channels.

In addition to the similarity of membrane topology suggested for various voltage-gated channels, there are two particular similarities of substructure that suggest common molecular mechanisms for permeation and gating among members of this class of channels. In Na^+, Ca^+, and many K^+ channels, there is a highly conserved stretch of approximately 20 amino acids between the fifth and sixth transmembrane domains of each "subunit" that is thought to form a major part of the ion conducting pore. This region (i.e., S5-S6 linker or P region) has a number of conserved hydrophilic and charged residues that alternate with three or four more hydrophobic resi-

dues, suggesting an amphipathic peptide region, whereby one side of the peptide chain (every 3.5 residues in sequence) would be hydrophobic and the other side hydrophilic. This amphipathic linker is thought to fold into the plane of the membrane (see Fig. 4-10), with the polar hydrophilic face of each linker turned away from the other transmembrane domains. The polar faces of the linkers of the four subunits comprising a complete channel are thought to come together to form the lining of a pore, mutually stabilizing each other and being shielded from the energetically unfavorably lipid region by the other transmembrane helical domains.

Experimentally induced mutations have been engineered at various sites in the linker segments of different cloned channels. These mutant sequences can then be experimentally expressed in host cells and functionally analyzed. Such mutations have produced changes in the ion selectivity properties of the pore and in interactions of the channel with drugs (e.g., TEA^+) thought to block the entrances to the ion conductive pathway, supporting this subunit model of the channel pore.

Another structurally conserved region among many voltage-gated channels is represented by the fourth transmembrane domain (S4), which has a stretch of four or five basic (i.e., positively charged) amino acids at every third position in the sequence. It is thought that these

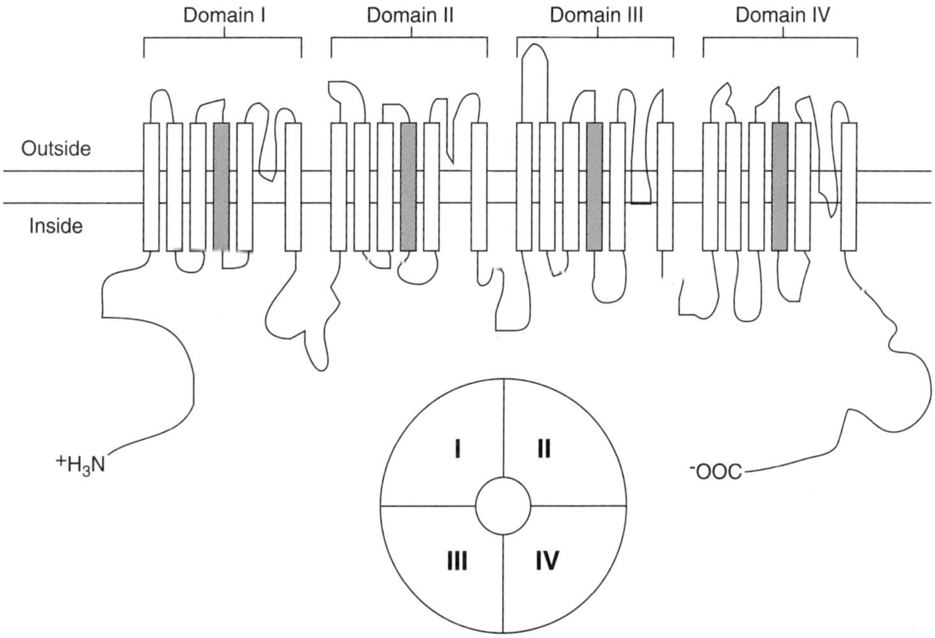

Figure 4-12
Voltage-gated Na channels are single polypeptides encoded by larger DNA sequences, but they express properties corresponding to the basic channel subunit model. The upper diagram illustrates the predicted membrane topology for the peptide sequence of a voltage-gated Na channel. The 24 transmembrane segments can be grouped into four domains of six transmembrane segments. The amino acid sequences of the S4 regions (*shaded bars*) in each domain are highly homologous and share the structural motif of positively charged residues of the S4 regions of voltage-gated K channels. It appears that the Na channels represent a precoded heterologous grouping of the basic subunit structure seen for many K channels. The bottom diagram is a hypothetical scheme for how the domains may associate in the membrane if viewed from the outside of the channel. The S5-S6 linkers from each domain are thought to form the ion pathway, as is the case for K channels.

charges form part of the critical region that senses changes in the membrane electric field triggering activation of the channels. This concept is supported by experimental mutations of several of these residues to neutral or negatively charged amino acids, which dramatically alters the voltage sensing abilities of the channels. Such mutational studies have now been performed in cloned Na, K, and Ca channels, and the results are qualitatively comparable among the channel classes. Thus, it appears that voltage-gated channels express some common structural motifs that may be important to their roles in gating and selectivity among ions.

The molecular properties underlying inactivation behavior appear more complex and varied among different channel types. In the case of several inactivating K channels, a stretch of positively charged amino acids near the NH$_3$ cytoplasmic terminus of the protein have been shown to function as a "ball and chain" to occlude the open ion pathway after activation. Experimental removal of this region destroys much of the inactivation process, and addition of a synthetic peptide corresponding to the sequence restores it. Less rapid forms of inactivation behavior appear to reside in regions associated with the carboxyl terminus of the channel protein subunit. In the case of voltage-gated Na channels, regions responsible

for inactivation may reside in only a single domain of the molecule but function in qualitatively similar ways to those of K channels.

☐ Families of genes code different ion channel types

The primary amino acid sequences of various ion channels revealed through molecular cloning efforts have indicated common properties among groups of ion channels. Furthermore, molecular genetics studies have indicated that these groups represent gene superfamilies. One superfamily of genes encodes several ligand-gated channels activated by γ-aminobutyric acid (i.e., GABA-A receptor-channel complex) or acetylcholine (i.e., nicotinic ACh receptor-channel complex). Several genes coding a family of voltage-gated K channels have been cloned and are very similar. The structural similarities between channels within a gene family from organisms as different as humans, fruit flies, and flowering plants suggests that ion channels are phylogenetically old and important molecules. Several subfamilies of K channels exhibit functional groupings across species. The combined proper-tics of subunits arising from messenger RNA variants and different genes provide the genome with the potential for tremendous diversity in K channel expression. The exact roles of such different channel subtypes and their distribution within the nervous system are largely unknown.

☐ Ion pumps harness metabolic energy to transport ions across cell membranes

The standing concentration gradients for ions between the cytoplasm and extracellular fluids are characteristic of all mammalian cells, including nerve cells. They are critical to nerve cell function as a source of energy to drive signaling processes. Dissipation of the ion gradients would stop all nervous function. Ion concentration gradients cannot arise from simple random diffusional movement of ions between compartments. The electrochemical potential of an ion in solution is proportional to the concentration of the ion. In the absence of external energy sources, net ion movement proceeds in a direction to reduce this potential energy from areas of higher to lower concentration. Without the input of energy from other sources, ions cannot remain concentrated in a given cellular compartment. In the case of neurons, metabolic energy in the phosphate bonds of ATP can be harnessed to serve this function. This energy source is tapped by yet another class of specialized membrane proteins, functionally referred to as *ion pumps*. These proteins serve to convert metabolic energy into the work of moving ions across membranes against their electrochemical gradient.

The best characterized ion pump in neurons is an integral membrane protein that exchanges Na^+ and K^+ ions between the extracellular fluid and the cytoplasm. Na^+ ions interact with sites on the cytoplasmic surface of the pump, and K^+ ions bind to extracellular exposed sites. Conformational changes in the pump protein result in the transfer of the Na^+ ions to the external fluid in exchange for K^+ ions that are transported to the cytoplasm. The exact nature of the conformational changes underlying this transport is not well understood. In conjunction with this transport function, the Na^+-K^+ pump also has enzymatic ATPase activity. The hydrolysis of the high-energy phosphate bond of ATP by the pump, yielding ADP and inorganic phosphate products, provides the energy required for ion transport. In some fashion, a conformational change in the protein occurs against normal energetic barriers, and bound Na^+ and K^+ ions are released into extracellular and cytoplasmic compartments, respectively (Fig. 4-13).

Repeated cycles of the conformational change results in a steady increase in ion concentrations. Under normal circumstances, transport of a particular ion is unidirectional (i.e., exclusively against its concentration gradient). This phenomenon arises from differences in the affinity of the pump for ions on the two sides of the membrane. For example, because the affinity of the pump for Na^+ ions is greatest on the intracellular surface, Na^+ ions do not readily coordinate with the extracellular surface of the pump protein, even at the normally high extracellular Na concentration. Under normal conditions, each transport cycle of a single pump results in the movement of three Na^+ ions in exchange for only two K^+ ions. This difference in charge transport could, in principle, contribute to the membrane potential, and the pump is said to be electrogenic. This means that under the right conditions, pump activity resulting in the outward transport of a net positive charge (the extra Na^+) contributes directly to the small charge separation across neuronal membranes responsible for the resting potential. As discussed in Chapter 5, this source of charge separation is only a minor component of cell membrane potentials.

The genes encoding several ion transporter molecules, including the Na^+/K^+ pump have been cloned. In contrast to ion channels, there is a significant diversity of the theoretical structures of transporter molecules, especially in the number of transmembrane α-helical regions. In the case of Na^+/K^+ pumps, the most common structural motif suggests that eight transmembrane segments comprise the major α-subunit of the protein. The ATPase catalytic region and intracellular Na^+ binding sites are contained in a rather long intracellular loop between the fourth and fifth transmembrane segments. The mecha-

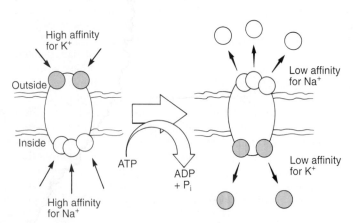

Figure 4-13
A schematic representation of the Na-K transport pump which translocates Na^+ and K^+ ions across the membrane against their normal electrochemical gradients. One conformational state of each transport protein provides high-affinity binding sites to 3 Na^+ ions on the cytoplasmic surface and to 2 K^+ ions on the extracellular surface. The Na-K pump has enzymatic ATPase activity, and the energy gained by the hydrolysis of ATP drives a series of conformational changes that result in the dissociation of K^+ ions from the protein on the cytoplasmic surface and movement of Na^+ ions to the extracellular fluid. The low affinity of the pump for external Na^+ and internal K^+ in this conformation ensures that reverse transport (i.e., in the direction of the electrochemical gradient) is not probable. A Na^+ and K^+ concentration gradient can be established and maintained by the activity of this pump protein.

nisms by which these structures transduce the exchange of Na^+ and K^+ ions is still a mystery. Evidence suggests that the protein undergoes conformational changes similar to those of ion channels. Such changes offer pathways to bound ions to exit to the opposite membrane surface. These conformational changes apparently occur at much lower rates than those of ion channels, functionally distinguishing the two types of transport proteins.

☐ Ion transport by channels and pumps differ in rate and energetics

Ion channels and ion pumps share several functional features. They both translocate ions across the neuronal cell membrane. In this function, both types of molecules provide "coordination sites" with which ions interact during the transport process. These sites exhibit selectivity among ion species in both cases. The corresponding sites in both types of protein can be saturated as the concentrations of transported ion species are increased. What functionally distinguishes the channels and transport pumps? Two characteristics appear to represent fundamental differences in the molecules. First, the rate of ion translocation is much more rapid in ion channels than pumps. As many as 10 million Na^+ ions can pass through an open Na channel in 1 second. In contrast, estimates of Na^+ transport by the Na^+/K^+ pump indicate that only about 1000 Na^+ ions are exported by a pump in 1 second.

The transport of ions in an open ion channel is thought to involve interactions that are energetically similar to the exchange of hydrating water molecules as an ion freely diffuses in solution. If this is the case, an increase in temperature should only marginally increase the rate of transport, as it does for the movement of ions in solution. This expectation is confirmed by the observation that ion flux through open channels increases only about 10% to 20% for a 10°C increase in temperature. In contrast, ion transport by many pumps increases more than threefold for a 10°C temperature increase. This could be expected, because transport probably involves complicated and energetically unfavorable changes in the protein conformation. Temperature dependence of the transport rate is a second property that mechanistically differentiates ion channels from ion pumps.

SELECTED READINGS

Aidley DJ. The physiology of excitable cells, 3rd ed. Cambridge: Cambridge University Press, 1989.

Dowling JE. Neurons and networks. An introduction to neuroscience. Cambridge, MA: Harvard University Press, 1992.

Hille B. Ionic channels of excitable membranes, 2nd ed. Sunderland, MA: Sinauer Associates, 1992.

Kaplan JH, DeWeer P (eds). The sodium pump: structure, mechanism, and regulation. New York: The Rockefeller University Press, 1991.

Clinical Correlation
DEMYELINATING DISORDERS

ROBERT L. RODNITZKY

Disorders of myelin can be divided into conditions in which there is destruction of myelin, the demyelinating disorders, and those in which there is an abnormality in the makeup of myelin, the dysmylinating disorders. Demyelination can result from a great variety of causes, including autoimmune, toxic, metabolic, infectious, or traumatic conditions. The dysmyelinating disorders are heritable conditions resulting in inborn errors of metabolism affecting myelin. Central or peripheral nervous system myelin can be affected in these conditions, but most clinical syndromes related to disorders of myelin primarily affect the brain and spinal cord.

Impaired or Blocked Impulse Conduction

Impaired or blocked impulse conduction is the cause of most clinical symptoms related to myelin disorders. Demyelination prevents saltatory conduction in normally rapidly conducting axons. Instead, impulses are propagated along the axon by continuous conduction, or in the extreme, there is total conduction block. The latter may result from the markedly prolonged refractory period in demyelinated segments and explains why rapidly arriving, repetitive impulses are especially likely to be blocked.

Demyelination may also affect conduction by rendering axon function more susceptible to changes of the internal milieu, such as alterations of pH or temperature. The enhanced effect of temperature change on the function of demyelinated central nervous system (CNS) pathways is well known and can be the cause of prominent and precipitous neurologic symptoms. The development of new neurologic symptoms after temperature elevation in multiple sclerosis is known as *Uhthoff's phenomenon*. Persons experiencing this phenomenon notice the onset of new symptoms, such as monocular blindness, profound extremity weakness, or severe incoordination, whenever they develop a mildly elevated body temperature. With more profound elevation of body temperature, such as that occurring during a warm bath or as a result of a fever, neurologic dysfunction can be still more severe. Usually, when body temperature reverts to normal, neurologic function returns to baseline. Uhthoff's phenomenon is based on the principle that conduction block supervenes in demyelinated fibers when they reach a critical threshold temperature. This threshold temperature may be just above normal body temperature, explaining why such symptoms may develop after only a slight temperature elevation such as that seen as part of normal diurnal variation.

Diseases Affecting Myelin

Many different processes can be involved in the etiopathogenesis of myelin disorders. Adrenoleukodystrophy is an example of a metabolic disorder of myelin. It is an X-linked peroxisomal disorder in which there is impaired oxidation of very-long-chain fatty acids with resultant abnormalities of myelin.

Progressive multifocal leukoencephalopathy is an infectious condition resulting in demyelination. It is caused by an opportunistic viral infection of oligodendroglial cells in immunosuppressed patients, such as those with acquired immunodeficiency syndrome. Oligodendroglial death leads to widespread demyelination, especially in the posterior portions of the cerebral hemispheres. The resultant severe neurologic function is ultimately fatal.

Radiation is an example of a form of trauma that can lead to demyelination. For this reason, patients undergoing radiation therapy in which the brain or spinal cord is in the field of treatment may later develop neurologic symptoms. Toxic agents such as cancer chemotherapy drugs can result in demyelination, especially when instilled directly into the cerebrospinal fluid.

Autoimmune mechanisms can result in demyelination. This process is thought to be important in the pathogenesis of one of the most common demyelinating disorders, multiple sclerosis (MS).

☐ **Multiple sclerosis has a distinct age, gender, race, and geographic profile**

MS is a relatively common neurologic disorder. The prevalence of MS in the United States and Northern Europe is approximately 100 per 100,000, and almost 300,000 Americans are afflicted with this condition. Women are affected 1.5 times more often than men. The age distribution of MS is distinct in that onset before the age of 15 or after the age of 45 is unusual. The white population, especially persons of northern European ancestry, has a much higher incidence of MS than Asians or African blacks.

Within the United States and to a certain degree throughout the world, there is a distinct geographic distribution of MS, with temperate zones having the highest incidence. In the United States, this distribution results in a strikingly higher prevalence of MS in the upper Midwest than in the deep South. The precise meaning of these geographic variations remains unclear, but some migration studies have suggested that environmental factors are largely operative in childhood, because emigration out of a high prevalence zone before adolescence appears to reduce the risk of developing MS.

☐ Multiple sclerosis is characterized by the presence of numerous discrete areas of demyelination throughout the brain and spinal cord

In MS, multiple plaques of demyelination are found in the CNS. Within these areas, there is evidence of inflammation, proliferation of glial tissue, and severe destruction of myelin. In these same areas, axons remain largely intact. Minimal remyelination of fibers may occur in some plaques, resulting in areas with scant myelin known as "shadow plaques." The pathogenesis of demyelination within these plaques in not fully understood, but it is generally agreed that it involves an immune response mediated by T lymphocytes that recognize myelin components of the CNS. Plaques may occur anywhere in the CNS, but they have a predilection for certain parts of the brain and spinal cord, forming the anatomic basis for the most common clinical symptoms of MS. The most common regions of involvement are the optic nerves, the cerebellum, the periventricular white matter of the cerebral hemispheres, the white matter of the spinal cord, the root entry zones of spinal or cranial nerves, and the brain stem, especially the pons.

☐ The symptoms of multiple sclerosis remit and reappear in characteristic fashion

The two clinical hallmarks of MS are the wide range of disparate neurologic symptoms that can occur in a single affected person and the propensity for these symptoms to spontaneously improve and then reappear over time. The variety of symptoms in a given patient reflects the widespread dissemination of MS plaques throughout the brain and spinal cord. In a typical MS patient, it is not usually possible to attribute all neurologic symptoms to a single anatomic locus of abnormality; instead, multiple areas of demyelination must be invoked. The second and

highly distinctive clinical feature of MS is spontaneous remittance of neurologic symptoms. Typically, serious neurologic symptoms appear and disappear over a period ranging from weeks to several months, independent of medical therapy.

Certain neurologic symptoms are much more likely to occur in MS, reflecting the predilection for certain anatomic sites of involvement alluded to above. These include monocular visual loss (the optic nerve is involved), limb incoordination (cerebellum), double vision (brain stem, medial longitudinal fasciculus), and weakness of the lower extremities (spinal cord). Other common symptoms include vertigo, extremity numbness, slurred speech, and bladder dysfunction. L'hermitte's sign is a frequent finding in MS. It consists of a shock-like sensation that travels down the spinal axis into the extremities whenever the neck is flexed forward. This phenomenon is caused by the heightened sensitivity of demyelinated dorsal column axons to the mechanical stimulation of stretch when the spinal cord is flexed by the offending neck motion.

☐ The diagnosis of multiple sclerosis can be aided by imaging studies and laboratory tests

Although the unique clinical pattern of MS often allows the diagnosis to be made with confidence, several diagnostic tests are extremely useful in equivocal cases. Magnetic resonance imaging of the brain or spinal cord is a sensitive means of demonstrating multiple areas of demyelination, many of which may prove to be clinically silent. In addition, magnetic resonance imaging can determine whether a given area of demyelination is quiescent or in an active state of evolution.

Examination of the cerebrospinal fluid (CSF) is used to screen for CNS immunoglobulin abnormalities, which are common in MS. The rate of IgG synthesis within the CNS, which is typically elevated in MS, can be determined, and the makeup of immunoglobulin in the CSF can be evaluated. When normal CSF is subjected to immunoelectrophoresis, the immunoglobulins are diffusely represented at the cathodal region. In MS, the electrophoretic pattern demonstrates several discrete bands of immunoglobulin known as oligoclonal bands. These are presumed to be specific antibodies, but the antigen(s) against which they are directed have not yet been identified. Testing for the presence of oligoclonal IgG is extremely useful in diagnosing MS because as many as 95% of individuals with proven MS have been found to exhibit this abnormality. Another useful CSF study involves myelin basic protein, a breakdown product of myelin. Because it can be detected in the CSF in the presence of active CNS demyelination, it is useful as an indicator of disease activity.

Electrophysiologic studies such as visual and somatosensory evoked potentials are used to detect subtle physiologic abnormalities in MS that are not yet sufficiently advanced to produce clinical symptoms. These tests measure the speed and completeness of conduction in sensory pathways. The technique involves delivery of a sensory stimulus, such as a flash of light and measuring the time required for an evoked response to be reflected in the cortex and recorded by a scalp electrode. In patients with overt sensory symptoms, slowed conduction can often still be demonstrated in the appropriate sensory pathway even after the symptoms remit using this technique.

☐ Immunosuppression is the most common form of therapy used in multiple sclerosis

The propensity for MS symptoms to remit spontaneously makes the scientific evaluation of potential therapies difficult. In the past, poorly designed or uncontrolled studies have led to a multitude of false claims of therapeutic efficacy for a variety of therapies, some of which were quite unorthodox. Because of its presumed immune basis, the most widely accepted, although still somewhat controversial, therapies are those involving immunosuppression. The most commonly used of this class of treatments are the corticosteroid drugs, prednisone, or methylprednisolone. These agents are usually administered during an acute flare-up of the illness. More potent immunosuppressant agents are occasionally administered over the long term to prevent future exacerbations, but treatment is sometimes limited by the toxicity of this class of drugs. One form of immunomodulation, beta-interferon, has been shown to be a useful therapy, but other forms, such as total lymphoid irradiation and antigen feeding, remain experimental and unproven.

SELECTED READINGS

Poser CM, Paty DW, Scheinberg L, et al. New diagnostic criteria for multiple sclerosis: guidelines for research protocols. Ann Neurol 1983;13:22.
Rudick RA. The value of magnetic resonance imaging in multiple sclerosis. Arch Neurol 1992;49:685.
Weinshenker BG, Sibley WA. Natural history and treatment of multiple sclerosis. Curr Opin Neurol Neurosurg 1992;5:203.

Neuroscience in Medicine, edited by P. Michael Conn.
J. B. Lippincott Company, Philadelphia © 1995

Chapter 5

*T*he Basis of Electrical Signaling in Neurons

GERRY S. OXFORD

The neuronal cell membrane and its constituent ion channels serve as the foundation for all forms of intercellular and intracellular communication associated with signaling in the nervous system. The concerted and sequenced activity of a wealth of ion channel types contributes to the rapid signaling underlying visual perception or reflexive avoidance and the slower processes involved in the less well understood deposition and retrieval of memories and instinct. An oversimplified but useful perspective on these signaling events is that they all reduce to some change in the transmembrane electrical potential in the signaling cells. The role of ion channels in the construction and dissipation of membrane potentials is the focus of this chapter.

Generation of Membrane Potentials

☐ A cell membrane exhibits properties analogous to electrical circuit components

Before considering the specific involvement of ion channels in the generation of cellular membrane potentials and the electrical signaling functions of the nervous system, we need to introduce an important historic concept in membrane physiology, the analogy between the passive electrical properties of membranes and simple electronic circuits. Our understanding of the way in which ion channel activity is harnessed to yield electrical signals encoding thoughts, sensations, and actions is based largely on a model of the neuronal membrane as an electrical circuit.

The *lipid bilayer* provides an electrically insulating barrier between cellular compartments. This exceptionally thin, nonconducting region separating two electrically conductive ionic solutions is an excellent and highly efficient electrical capacitor. This capacitor serves as a reservoir of stored charge available to do the electrical work (e.g., ion currents) of signaling. The ion channels that penetrate this bilayer provide a pathway for the ion movements between compartments and therefore provide a means of dissipating the stored charge. The resulting ion fluxes through open channels (i.e., charges/second) constitute an electrical current (I). The more channels in the membrane, the less resistive the membrane is to ion movement. The electrical resistance (R) of neuronal cell membranes is inversely proportional to the number of ion channels. The inverse of electrical resistance is conductance ($g = 1/R$). Membranes with more ion channels are more conductive to ions. Although not strictly identical to permeability, the conductance of the membrane (in units of Siemens) to an ion species is directly proportional to the permeability of the membrane to that ion. These aspects of electrical capacitance and resistance inherent in the neuronal cell membrane are often represented as an "equivalent circuit" of the cell

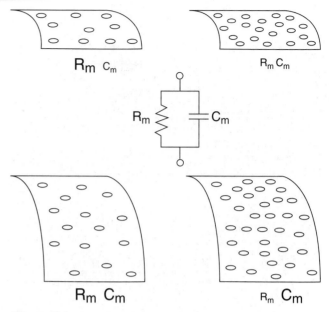

Figure 5-1
The passive electrical properties of neuronal membranes approximate the behavior of a simple equivalent electrical circuit. A simple parallel combination of a resistor (R_m) and capacitor (C_m) mimics the passive (nonexcitable) behavior of any particular patch of cell membrane. The capacitor and resistor loosely correspond to the insulating lipid bilayer and the complement of open ion channels, respectively. As the density of ion channels increases, the equivalent resistance decreases. As the surface area of the membrane increases, the equivalent capacitance increases.

membrane, a resistor and capacitor in parallel configuration (Fig. 5-1). This representation is used to explain some of the more common electrical responses of neuronal membranes to external stimuli.

☐ Ions move passively across membranes because of electrochemical forces

Concentration gradients and electrical voltage differences drive the transmembrane movement of ions. Both factors contribute to the electrochemical energy of an ion in solution. If a suitable pathway exists for passage of an ion species across the membrane, a net movement down the electrochemical gradient results. Ion channels afford such pathways, and their ability to discriminate among ion species is directly responsible for generating normal membrane voltages.

The selective movement of a single ion species across a cell membrane generates a membrane potential that can be predicted from knowledge of the concentration difference for that ion species between the intracellular and extracellular compartments. This potential is referred to as the *Nernst potential*, named for nineteenth-century physical chemist Walther Nernst, who originally described the phenomenon for experimental interfacial systems and derived an equation that formalizes the relation between ion concentration and electrical potential:

$$E_x = \frac{RT}{zF} \ln \frac{[X]_o}{[X]_i}$$

in which R is the gas constant, T is the temperature in degrees Kelvin, z is the valence of the ion, and F is the Faraday constant (i.e., the amount of charge represented by 1 mole of monovalent ions). $[X]_o$ and $[X]_i$ are the extracellular and intracellular concentrations, respectively, of ion X. For monovalent ions (z = 1) near room temperature, this equation reduces to

$$E_x = 60 \log \frac{[X]_o}{[X]_i}$$

and has the units of volts. The Nernst equation for any particular ion species separated by a membrane predicts the value of the membrane potential under the limiting condition that the membrane is exclusively permeable to only that single ion species. To gain a sense of the behavior described by the Nernst equation, consider the hypothetical case of a neuronal cell membrane expressing only K^+ selective ion channels and allowing only K^+ ions to cross the membrane (Fig. 5-2).

Because the concentration of K^+ ions in the cytoplasmic compartment is greater than that in the extracellular fluid, the passive movement of ions from the inside of the neuron to the extracellular fluid exceeds the movement in the other direction. The net outward movement of K^+ ions results in a small charge separation across the membrane as counterbalancing anions (e.g., Cl^-, glutamate) remain trapped in the cytoplasm because they cannot pass through the K^+ selective membrane. As the charge separation develops, it creates an electrical potential across the membrane that is negative on the cytoplasmic side. The developing membrane potential provides an attractive force to drive the inward movement of

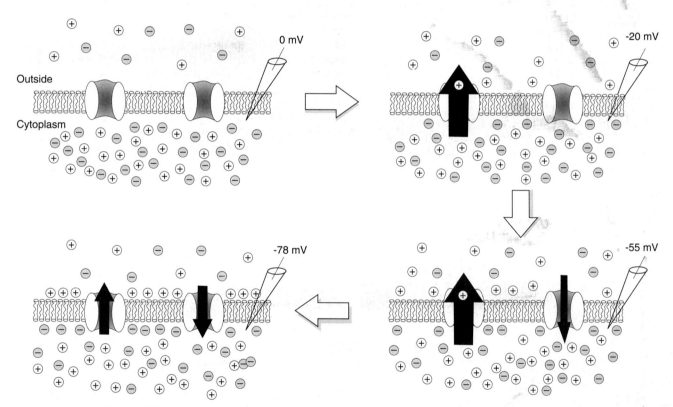

Figure 5-2

The movement of K^+ ions through selectively permeable ion channels establishes a predictable membrane potential. A hypothetical cell membrane is depicted as separating two compartments of different KCl concentrations (*upper left*). Because the cytoplasmic compartment has a higher ion concentration than the external compartment, a concentration driving force exists for K^+ and Cl^- to move out of the cell. The transmembrane potential is initially zero, because equal numbers of positive and negative charges exist in each compartment. An instant later, K^+ selective ion channels are open, providing a pathway for the efflux of K^+ ions down the KCl concentration gradient. This unbalanced movement of positive charge (*upper right*) leaves an excess of Cl^- ions in the cytoplasm, and a negative internal membrane potential develops. The developing membrane potential provides an attractive counterforce to return K^+ ions to the cytoplasm. As the channels are available, a K^+ influx develops (*lower right*). When the efflux and influx of K^+ ions are equal, a steady potential is reached (*lower left*). This condition is called the equilibrium potential for K^+ and is predictable from the Nernst equation.

positively charged cations. Because the membrane is K^+ selective, K^+ ions begin to move into the cytoplasm driven by the developing electrical force. When the efflux and influx of K^+ ions is exactly balanced, the system is in *equilibrium*. At this point, the generated voltage is defined by the Nernst equation for potassium and no net movement of K^+ ions occurs. For normal concentrations of K^+, this potential is

$$E_K = \frac{RT}{zF} \, ln \, \frac{[K]_o}{[K]_i}$$

Two important consequences of this equation should be noticed. First, the predicted equilibrium potential in this case is independent of the number of K channels (and the absolute K^+ permeability) of the membrane. Increasing the permeability of the membrane increases the total bidirectional movement of K^+, but the *net* movement is still zero at equilibrium. Second, the magnitude and sign of the membrane potential changes in a predictable manner with changes in the concentration of K^+. For example, if the extracellular concentration of K^+ were experimentally elevated to equal that of the cytoplasm, the gradient would not exist, influx would equal efflux, and the resulting membrane potential would have a value of zero.

An "equilibrium potential" defined by the Nernst equation can be calculated for any ion regardless of whether the membrane is actually permeable to that ion. Consider the case of Na^+ ions, which are at a higher concentration outside of neurons than in the cytoplasm. The calculated equilibrium potential for Na^+ ions in a typical neuron is about $+60$ mV. The positive sign predicted for this potential should be intuitively obvious as the movement of Na^+ ions into the neuron driven by the Na^+ concentration gradient would result in an excess of cytoplasmic positive charge. Although it is simple to calculate this equilibrium potential for Na^+, it is important to remember that this theoretical potential is physically realized only in a neuron in which the membrane is exclusively permeable to Na^+ and no other ions. This condition is briefly approximated during a typical nerve impulse and is the basis for the generation of the impulse.

It perhaps should not be surprising that membranes of real neurons are permeable to more than one ion at a time, given the wealth of ion channels populating neuronal membranes. This suggests that the membrane potential must represent a balance between the influences of charge separations created by the movement of several ions. An equation that predicts the potential in such a multi-ion situation was independently derived by David Goldman in 1943 and by Alan Hodgkin and Bernard Katz in 1949 and is often referred to as the Goldman-Hodgkin-Katz equation (GHK equation):

$$E_m = \frac{RT}{zF} \, ln \, \frac{P_K[K]o + P_{Na}[Na]o + P_{Cl}[Cl]i}{P_K[K]i + P_{Na}[Na]i + P_{Cl}[Cl]o}$$

in which R, T, F, z, and the concentrations of the ions have the same meanings as for the Nernst equation. P_{Na}, P_K, and P_{Cl} are the permeabilities of the membrane to Na^+, K^+, and Cl^- ions, respectively. Although other ion species may contribute, the GHK equation is usually restricted to these three highly mobile ions. This equation bears some resemblance to the Nernst equation, but there is a fundamental and important difference: whereas the membrane permeability was not a factor in the Nernst equation, the relative permeabilities of the membrane to each ion in the GHK equation determines the relative influence that the concentration gradient for each ion has on the observed membrane potential. The GHK equation, describing the relations among ion concentrations, membrane permeability, and membrane potential, predicts two general rules for the behavior of actual membrane potentials. These rules are of fundamental importance to understanding the influence of ion concentrations and permeabilities underlying electrical and chemical signaling in the nervous system.

The first rule is that *the membrane potential most closely approximates the Nernst equilibrium potential of the most permeable ion*. Consider the case in which a membrane has only K^+ selective ion channels. In this case, $P_{Na} = P_{Cl} = 0$, and the GHK equation reduces to the Nernst equation for K^+ ions. Now consider the case of a membrane equally permeable to Na^+ and K^+ ions, but impermeable to Cl^- ions. If $P_{Na} = P_K$ and $P_{Cl} = 0$, the predicted membrane potential will be midway between those predicted for a membrane selectively permeable to K^+ and one selectively permeable to Na^+. Assuming an E_K value of -78 mV and an E_{Na} value of $+60$ mV, the GHK equation predicts $(-78 + 60)/2 = -9$ mV. Considering that the typical values of actual neuronal resting potentials (-50 to -75 mV) closely approximate the Nernst potential for K^+ rather than that for Na^+, it can be concluded that the membrane is likely to be more permeable to K^+ than to Na^+ at rest. Extensive experiments on a variety of cell types have proven this conclusion to be true.

The second rule is that *the membrane potential is most influenced by changes in the concentration of the most permeable ion*. Changes in extracellular Na^+ ions have little influence on normal resting potentials of neurons, indicative of the relatively low P_{Na} at rest. In contrast, changes in extracellular K^+ produce dramatic changes in the membrane potential because the membrane is more permeable to K^+ ions. The suitability of the GHK equation to predict cellular membrane potentials is revealed by the accurate prediction of changes in the resting potential with changes in external K^+ (Fig. 5-3). The membrane permeability to Na^+ ions elevates briefly during a nerve impulse, making the membrane action potential transiently sensitive to the external Na^+ concentration.

These two concepts have successfully guided our understanding of the generation of and changes in membrane potential under a variety of conditions. We next consider basic neuronal signaling within the context of specific ion permeability changes and these two rules.

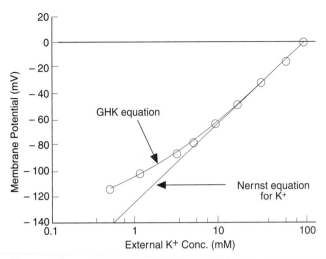

Figure 5-3
The Goldman-Hodgkin-Katz (GHK) equation provides a good description of the dependence of membrane resting potentials on the concentration of permeant ions. The graph depicts measurements of the membrane potential in an excitable cell (*open circles*) at various concentrations of extracellular K^+. The curved line that best fits the data points is calculated from the GHK equation, assuming that the P_{Na} is 100 times less than the P_K and that $P_{Cl} = 0$. The straight line is the prediction of the Nernst equation; both P_{Na} and P_{Cl} are zero. The GHK equation is a good predictor of the data. Even small permeabilities of the membrane to ions other than K^+ can influence the membrane potential as the concentration of that ion becomes large relative to that of K^+.

Action Potentials

☐ Characteristics of the basic neuronal action potential

Perhaps the most important functional property of the neuronal cell membrane is the property of electrical excitability. It is this property that underlies the basic nerve impulse, or *action potential*, which is the fundamental unit of signaling in the nervous system. The action potential is a brief transient event usually lasting from 1 to 5 milliseconds. During the action potential, the polarity of the membrane potential reverses, and the intracellular surface becomes positive (approximately $+20$ to $+40$ mV; Fig. 5-4). This event is usually initiated in the membrane of the neuronal cell body and is transmitted along the axonal process to the nerve terminal endings, where it initiates intercellular communication by the release of chemical transmitters.

The action potential waveform is characterized by several components. The early depolarizing phase is usually quite rapid as the potential reverses sign to become positive. The repolarization phase can vary somewhat among different neurons, exhibiting delays and inflections that alter the action potential duration. After the transient reversal or "spike" potential, the membrane potential in some neurons can become more negative than the normal resting potential for a few milliseconds. This event is

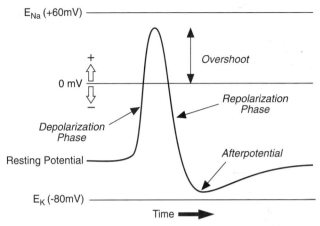

Figure 5-4
The nerve impulse or action potential is a transient reversal of the neuronal membrane potential that exhibits characteristic phases. During the phase of depolarization, the membrane potential rapidly (<1 millisecond) becomes positive, reaching a peak value just short of the sodium equilibrium potential (E_{Na}). The peak value above zero mV is referred to as the overshoot. During the phase of repolarization, the membrane potential rapidly returns to negative values and, in many neurons, can briefly become more negative than the resting potential. Only rarely does the repolarization value exceed the E_K.

referred to as a *hyperpolarizing afterpotential* and can vary in magnitude and duration among different populations of neurons.

☐ Ionic basis for the action potential

The electrochemical energy stored in the form of ion concentration gradients and membrane potentials is harnessed by the membrane through a characteristic sequence of permeability changes to monovalent cations to produce the action potential. The critical molecular components for this sequence are voltage-dependent Na and K channels in the case of the axon, but voltage-gated Ca channels often contribute to the action potential in neuronal cell bodies. Typically, the action of excitatory neurotransmitter molecules at synaptic connections between neurons or the transduction of sensory stimuli by various receptors yields membrane potential changes that can passively *depolarize* a neuron and spread from the site of initiation to only a limited region of the membrane. (Depolarization is generally meant to indicate that the membrane potential becomes less negative (e.g., going from -70 to -10 mV).

If these small depolarizing events reach a section of membrane expressing voltage-gated Na and K channels, they can trigger increases the probabilities that both types of voltage-dependent channels will open (Fig. 5-5). These channels open in response to increasingly positive membrane potentials. The responses of the two channel types differ in an important manner that is the key to the

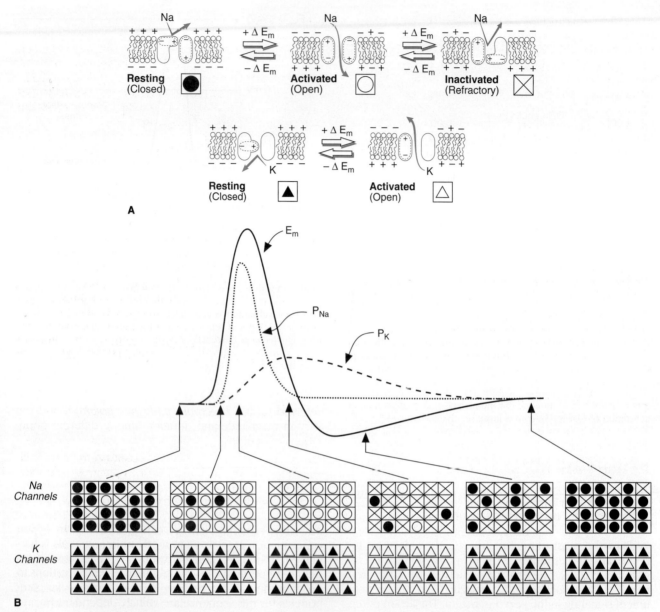

Figure 5-5

Characteristic changes in sodium (Na) and potassium (K) permeability underly the nerve action potential. The depolarization and repolarization phases of the action potential are associated with increases in P_{Na} and P_K, respectively. (**A**) The changes in voltage-gated Na and K channel conformations driven by membrane potential changes are diagrammed. With depolarization, the Na channels activate (i.e., open) and then inactivate (i.e., close), but K channels usually only activate. On repolarization, both channel types return to a resting (i.e., closed) conformation. (**B**) These transitions in each population of channels during an action potential is reflected as the time course of total membrane permeability to each ion, P_{Na} (*dotted line*) and P_K (*dashed line*). A hypothetical representation of the gating changes in a population of 24 Na channels and 24 K channels during a typical action potential is shown by the symbols that represent resting (●), activated (○), and inactivated (×) Na channels and resting (▲) and activated (△) K channels. Although the values of P_{Na} before and immediately after the action potential are almost identically small, the distribution of Na channel states is vastly different. After the action potential, most closed Na channels are inactivated rather than resting.

generation of an action potential. The Na channels respond more rapidly to the depolarizing stimulus than do the K channels. During the initial response, the membrane permeability to Na^+ rises relative to that to K^+. A similar temporal sequence of channel activation was seen under voltage clamp conditions (see Chap. 4).

During this initial response, Na^+ ions move into the neuron through the open channels driven by the concentration gradient and the negative intracellular potential. This inward flux of positive Na^+ ions further depolarizes the membrane, opening even more voltage-gated channels. At a critical potential referred to as the *threshold*, the Na^+ influx exceeds the K^+ efflux resulting from the normally high K^+ permeability at rest. At this point, an explosive cycle of Na channel opening and resultant Na^+ influx and depolarization occurs, activating more Na^+ permeability. When the membrane potential peaks at a positive level, the membrane has become about 20 to 50 times more permeable to Na^+ than to K^+. This situation is the reverse of the relative permeabilities of the resting membrane to the two ions.

At this point, we should recall Rule 1. At the peak of the action potential, when the membrane is more permeable to Na^+ than to K^+, the membrane potential should predictably be closer to E_{Na} than to E_K, which is the case (see Fig. 5-4). The peak potential should also be sensitive to changes in the extracellular Na^+ concentration (Rule 2) compared with the resting potential, which is generally insensitive to Na^+ concentration. Lowering the extracellular Na^+ should reduce the peak value of the action potential correlated with the expected decline in the value of E_{Na}. This prediction has been experimentally demonstrated in a number of neuronal cells. However, under normal physiologic conditions, the extracellular Na^+ concentration surrounding neurons does not vary over a range that may alter the action potential magnitude. Loss of almost half of the external Na^+ would be required to compromise the action potential signal. The changes actually observed in the peak of the action potential usually result from alterations in Na^+ permeability rather than changes in Na^+ concentration. For example, clinically important local anesthetic agents reduce or completely inhibit the action potential of peripheral nerve by blocking Na channels, thereby reducing P_{Na}. The exquisite and rapid sensitivity of the Na channel activation mechanism to voltage, the resultant explosive opening of Na channels, and the stability of the concentration driving force for influx of Na^+ ions combine to yield an almost invariant action potential when the threshold is reached. The high degree of reproducibility and reliability of this response for sufficiently large stimuli is often referred to as the "all or none" property of the excitable neuronal membrane. A stimulus is either of sufficient magnitude to elicit a full action potential, or no active response is produced.

The repolarizing phase of the action potential that returns the membrane potential to the resting level occurs as a result of two changes in ion permeability. As the Na^+ permeability rapidly increases during the rising phase of the action potential, the voltage-gated K channels also open, although at a somewhat slower rate. At the peak of the spike, P_K has increased above its resting level, and its value approaches that of the permeability to Na^+. As a result, the membrane potential begins to move back toward a voltage between E_{Na} and E_K (i.e., Rule 1). In addition to this K^+ permeability increase, P_{Na} begins to decline. This decrease occurs despite the fact that the membrane potential is still positive, which normally favors the open state of the Na channels. This decline in P_{Na} results from the inactivation process inherent in the gating behavior of the voltage-gated Na channel. After activation and opening, Na channels enter an inactivated or refractory state that is nonconducting. The subsequent reduction in P_{Na} during the falling phase of the action potential leaves the activated K channels to dominate the membrane permeability. Under these conditions, the enhanced outward K^+ movement returns the potential to negative values. As the changes in K^+ permeability occur slower than those to Na^+, the elevated P_K will exceed P_{Na} even more than at rest, and the potential often moves closer to E_K than at rest (i.e., Rule 1). Eventually, the voltage-gated K channels close (i.e., deactivate) when returned to more negative membrane potentials, and the original resting potential value is finally restored.

☐ A refractory period usually follows an action potential

Immediately after a single action potential, the neuronal membrane usually is refractory or insensitive to further stimuli for a brief period, usually several milliseconds. This refractoriness is manifested as an inability to generate another action potential when stimulated. There are two phases to this refractory period. For a very brief period after the first action potential, no depolarizing stimulus can elicit another action potential, regardless of its strength. The corresponding interval of this total insensitivity is referred to as the absolute refractory period. For a few milliseconds after this initial absolute insensitivity, a second stimulus can elicit another spike, but the stimulus strength must be greater than that of the initial stimulus. This phase is called the relative refractory period. These two phases can be experimentally demonstrated for a given nerve using well-controlled stimuli (Fig. 5-6). Despite this experimentally demonstrable distinction, the functional significance of these two phases to the excitability properties of the nerve is essentially identical.

The virtually constant amplitude of the all-or-none action potential in any neuron means that the refractory periods are manifested as changes in the firing probability and not in amplitude. By analogy with radio broadcasting, the basic signaling element (i.e., action potential) of the nervous system is used to encode information by frequency modulation (FM) rather than amplitude modulation (AM). The refractoriness after action potentials limits the frequency with which a given neuron can encode and transmit information as trains of action potentials. The longer the refractory period of a given nerve

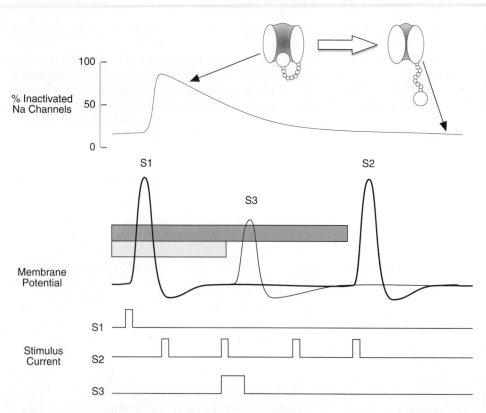

Figure 5-6

Excitable membranes are refractory or insensitive to subsequent stimuli after an action potential. Using a controlled stimulus (i.e., current injection from a micropipette electrode), a single action potential is generated in a neuron. After the initial stimulus (S1), identical stimuli (S2) are applied at various intervals. The membrane is insensitive or refractory to the first three of these, but the fourth stimulus initiates another action potential. The period of insensitivity to any stimulus with properties identical to S1 is the relative refractory period (*dark shaded bar*). In contrast, a larger stimulus (S3) can trigger a second action potential (*thin line*) in less time after the initial stimulus (S1). The period during which the membrane is insensitive to any stimulus, regardless of size, is the absolute refractory period (*light shaded bar*). The refractory periods are correlated with the time that the Na channel population remains in the inactivated conformational state after an action potential (*upper graph*).

axon, the lower is the maximal frequency of impulses the axon can transmit.

The phenomenon of refractory period results largely from the inactivation process of voltage-gated Na channels. The rapid transitions from the closed (i.e., resting) states to the open (i.e., activated) state of the Na channels account for the fast upstroke of the action potential. For an action potential to occur, a critical number of channels must be in the resting state, poised for activation by a depolarizing stimulus. Near the peak of an action potential, many open Na channels enter the inactivated state and become temporarily trapped in this conformation. Transitions between the inactivated state and other gating states are slow. In response to repolarization, the channels only slowly recover from inactivation (see Fig. 5-6), and several milliseconds must elapse before a sufficient number of channels regain the resting conformation and can be coordinately activated to generate an action potential. Changes in the inactivation properties of Na channels can result in significant changes in the ability of a neuron to encode nerve impulses.

☐ Action potentials can have different shapes and functions

Neurons from various regions of the nervous system exhibit a variety of action potential shapes and frequencies. Regardless of these differences among neurons, the basic sequence of Na$^+$ and K$^+$ permeability changes qualitatively accounts for all of the action potential behavior.

In the case of axonal processes of neurons, most evidence suggests that voltage-gated Na channels and delayed rectifier K channels are the key players in initiating and shaping the action potential as has just been described. In neuronal cell bodies, voltage-gated calcium channels often supplement the Na channels during depolarization. The different gating properties of Ca channel subtypes play major roles in defining the shape of these action potentials. In some neurons, K channels other than delayed rectifiers contribute to the action potential. For example, Ca^{2+} entry during a series of spikes in a neuron cell body can trigger the activation of Ca-activated K channels. The opening of these channels can dramatically hyperpolarize the membrane by virtue of the efflux of K$^+$ ions. These hyperpolarizations are often prolonged and can suppress spike activity for up to several seconds. For the most part, the exact role of specific action potential shapes in the function of a given neuronal populations is presently unknown.

Beyond providing interesting knowledge contributing to our understanding of neuronal function, characterization of the gating and permeability properties of ion channels underlying the action potential has provided useful tools for simulating the function of nerve membranes. For years, mathematical descriptions of the conductance changes by Na$^+$ and K$^+$ have been used in computer models of neuronal networks to accurately simulate membrane action potentials, providing better understanding of signaling mechanisms in populations of neurons in the nervous system.

Action potentials can behave as signaling elements in the nervous system

Action potentials in neuronal membranes represent the outcome of integrated input signals involved in information coding in the nervous system. For this coded information to be distributed within a neuronal population spread over great distances, the action potentials must be conducted along axonal processes from their sites of initiation to sites of neurotransmitter release. For a given axon, an action potential generally travels at constant velocity by a process of regenerative ion permeability changes in adjacent membrane segments. The basis for action potential conduction involves specific ion channel activity and passive electrical properties of the membrane.

Passive electrical properties determine the spread of excitation

Although the time-dependent and voltage-dependent properties of ion channels are important elements of nerve cell signaling, the speed and fidelity with which nerve impulses spread through the nervous system is largely determined by the passive structural features of nerve cells and their membranes. The response properties of nerve cells to subthreshold changes in membrane potential are key determinants of impulse conduction velocity and of the integration of multiple synaptic inputs.

Consider a simple cylindrical axon process subjected to an instantaneous injection of a small amount of positive charge into the cytoplasm. Such excess positive charge could arise experimentally from injection of Na^+ or K^+ ions from a glass micropipette (Fig. 5-7) or physiologically from activation of neurotransmitter receptors that open ligand-gated channels allowing Na^+ ions to move into the cell. As the positive charges are continu-

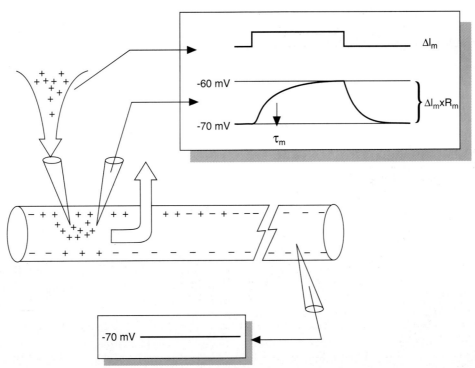

Figure 5-7
Membrane potentials respond to subthreshold current injections with a delay. When positive charges are injected into an axon from a glass microelectrode, the steady-state charge balance that exists at rest is disturbed, and the membrane depolarizes. The depolarization lags after the injection of charge, because some of the charge leaks back out through ion channels (e.g., K channels) before it can establish a new density of charge separation across the lipid bilayer. The time required to reach 37% of the final depolarization is the time constant (τ_m) and is equal to the product R_mC_m. When a new charge distribution becomes stable, the membrane potential change is steady and defined by Ohm's law. In this case, the steady current (ΔI_m) flowing through the membrane resistance (R_m) defines the voltage change (ΔV_m). At significant distances from the injection site, no change in membrane potential is observed, indicating that the subthreshold membrane potential change is localized rather than conducted along the axon. This response cannot support neuron-to-neuron signaling.

ously injected, they begin to distribute themselves along the membrane near the site of injection. Because the electrochemical balance of the resting membrane has been disturbed, the excess charge will seek pathways to return to the extracellular space and reestablish the balance. (In electrical terms, the charge flows in an opposite direction to establish a complete circuit of current). At membrane sites only a few micrometers away from the site of injection, the membrane potential is depolarized by the positive charge injection. (We assume that the amount of charge is insufficient to depolarize the membrane to threshold.) The depolarization develops with a lag, because the density of the charges accumulating on the membrane capacitance is slowed by escape of the charge through nonspecific or leakage channels in the membrane. The change in membrane potential occurs over an approximately exponential time course, eventually reaching a steady level as long as the positive current stimulus is maintained. The steady value of depolarization reflects a constant rate of escape of the charge (Q) through the membrane leakage conductance pathways (g). Recalling that conductance is the reciprocal of resistance ($I/g = R_m$) and the time rate of change of charge through the membrane is membrane current ($dQ/dt = I_m$), the steady voltage is simply defined by Ohm's law relating current and resistance to voltage.

$$\Delta V_m = \Delta I_m \times R_m$$

in which V_m and I_m are the depolarizing change in voltage and the excess current stimulus, respectively. The time delay to achieve this steady voltage change is characterized by the time required for 63% of the final value to develop. This time is called the membrane time constant (τ_m) and has a value that depends on the product of the resistance and capacitance of the patch of membrane being considered.

$$\tau_m = R_m \times C_m$$

Membranes with higher resistances tend to take more time to respond to changes in potential induced by synaptic inputs or sensory stimuli. However, a given amount of injected current yields a larger change in potential in such membranes according to Ohm's law. Both features of the membrane response help determine the magnitude and timing of responses to such stimuli.

In considering regions of the axon membrane at distances more remote from the site of the current injection, it is perhaps obvious that not all of the injected charge remains localized to the site of the initial stimulus. Some charge flows down the axon interior through the cytoplasm to more remote membrane areas as the stimulating current is continued. This current also seeks leakage pathways to the extracellular fluid and causes changes in membrane potential at these remote sites. At any arbitrary time after the membrane potential near the site of current injection has reached a steady value, the changes in membrane potential induced along the length of the axon are seen to diminish with distance from the site of the stimulus. This decrement of potential is also an ap-

proximately exponential function of distance and is characterized by a parameter, the space constant (λ_m). The space constant is that distance from the stimulus where the voltage change is only 37% of that seen at the injection site (Fig. 5-8).

The space constant for a cell or axon depends on membrane properties and structural geometry in an interesting way. To complete its electrical journey, a current representing the charge imbalance induced by the stimulus usually follows the path of least electrical resistance. In the case of charge injected at a point along an axon, the current can return to the extracellular space through the membrane or travel down a length of the cytoplasm before exiting. The fraction of charge that takes either path depends on the relative electrical resistance offered by each route. The space constant is determined by these two resistances according to an approximation of

$$\lambda_m \approx \sqrt{R_m/R_i}$$

in which R_m and R_i are the membrane resistance and axial resistance of the cytoplasmic cylinder, respectively. The distance over which a given current stimulus can produce a voltage change increases with increasing membrane resistance or with decreasing cytoplasmic resistance. This relation defines an important feature of the membrane responses to stimuli for cells of different size. As the diameter of an axon increases, R_m and R_i decrease. The increase in membrane surface area and associated increase of leakage channels for a given length of axon accounts for the decrease in R_m. The increased cross-sectional area of the cytoplasmic cylinder accounts for decreased resistance to axial current flow, much like adding lanes to a highway can speed the flow of traffic. However, for geometrical reasons R_i decreases more than R_m for a given increase in axon diameter. The space constant increases proportionally to axon diameter. This factor means that a given amount of subthreshold charge injected into a large axon usually changes the membrane potential at greater distances than the equivalent charge in a small axon (see Fig. 5-8).

This relation between the spatial influences of stimuli and structural features of nerve cells has several important functional consequences. One of the most important is the influence of axonal diameter on conduction velocity of the nerve impulse. We have just considered the spatial distribution of a subthreshold stimulus as a function of axon diameter, but the same principles influence the spread of excitation when the stimulus is sufficient to generate an action potential. Consider an action potential generated by the opening of voltage-gated Na channels at a particular point along an axon. As Na^+ ions rush in at the initiation site and drive the membrane potential to its peak value (e.g., $+40$ mV), the excess charge distributes along the axon in a manner analogous to that of a smaller subthreshold stimulus. At the instant of the action potential peak, membrane regions increasingly distant from the locus of the peak potential experience progressively smaller depolarization according to the value of the space constant (Fig. 5-9).

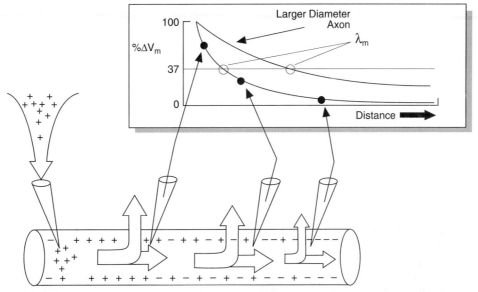

Figure 5-8

Subthreshold membrane potential changes are distributed at a distance from the initial site of stimulation. Although some of the positive charge injected into an axon leaks out near the injection site, some charge spreads along the axon interior, attracted toward regions with normal negative resting potentials. As less and less charge is available to leave the membrane at greater distances from the injection site, the associated membrane potential change becomes weaker. For larger axons with less resistance to axial current flow, the charge spread and depolarization occur at greater distances. The distance from the injection site where the voltage change is 37% of its maximum is the space constant, λ_m.

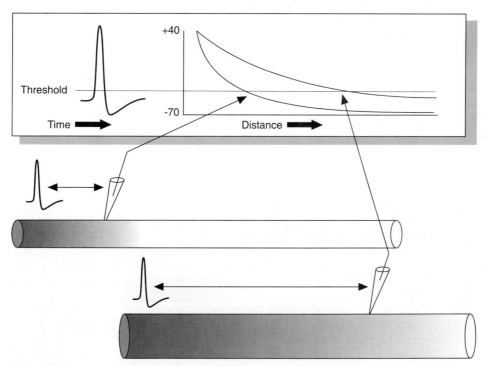

Figure 5-9

The passive spread of membrane potential in neurons influences the speed of nerve impulse conduction. As an action potential is generated in an axon, the entry of Na$^+$ ions during the depolarizing phase initiates the movement of positive charge down the axon interior and subsequent graded depolarization (*shading*). At any instant during this process, regions of the membrane ahead of the propagating action potential reach the threshold. These regions can support the regeneration of the impulse through activation of more Na channels. As the passive spread of depolarization occurs at greater distances in larger axons, the distance at which regenerative Na channel activation occurs is consequently greater. The action potential is conducted at greater velocities in larger-diameter axons.

At this instant, the excitation threshold is achieved (or exceeded) for some defined membrane region beyond the point of the activity, and voltage-gated Na channels in this region will be activated. It is the opening of these downstream Na channels and subsequent influx of Na$^+$ ions that prevents the action potential from simply dissipating because of charge leakage. The newly activated Na$^+$ influx results in an action potential and the new positive charge injection boosts or regenerates the wave of excitation. This cycle is repeated along the length of the axon as passive depolarization gives way to activation of Na channels and influx of more positive charge, which passively excites the next adjacent membrane regions (Fig. 5-10). In this manner, the action potential spreads or propagates at a constant velocity along the axon until it reaches the synaptic terminals.

Comparing this phenomenon in small-diameter and large-diameter axons reveals the fundamental basis for higher conduction velocities in large axons. By virtue of its greater space constant, at any instant, the membrane potential in a large axon can achieve (or exceed) the excitation threshold level at greater distances from an active region than is the case for a smaller axon (see Fig. 5-9).

In an equivalent amount of time, more remote Na channels are activated in the larger axon than in the smaller axon, and the action potential can be regenerated at a greater distance. As a result, the action potential moves at greater velocity (i.e., distance per unit time) along the larger axon.

Despite these general principles of propagation, the fate of action potentials that travel along neuronal processes of various diameters and with complex branching patterns is not completely understood or predictable. For example, action potentials can fail to invade synaptic terminals as a result of not propagating faithfully past branch points under certain conditions. Such propagation failures are not stable, because action potentials periodically conduct through such regions. This phenomenon can lead to variability in the total amount of excitation reaching the collective synaptic terminations of a given axon. Differences in the amount and spatial distribution of transmitter release can occur with different action potentials initiated in a single axon. These phenomena contribute to the dynamic and elusive mechanisms by which information is encoded by electrical events in the nervous system.

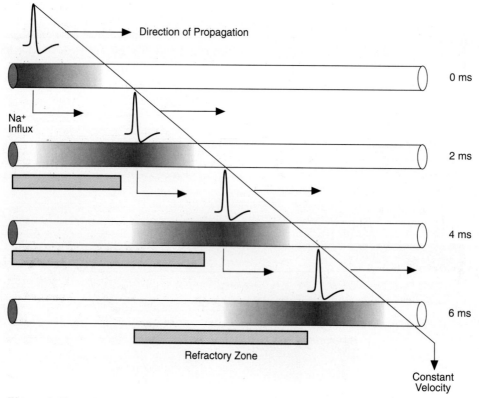

Figure 5-10

Action potentials propagate or conduct along axons at a constant velocity. An axon is depicted at four times during the conduction of an action potential. The shading represents the distribution of membrane potential along the length of the axon, with the darkest area corresponding to the peak of the action potential. As the action potential reaches a patch of membrane, the Na$^+$ influx depolarizes a downstream region of the membrane to the threshold level, stimulating a regenerative Na$^+$ influx in that region. As the action potential passes, a refractory zone occurs as the Na channels remain in an inactive configuration (*shaded bar*).

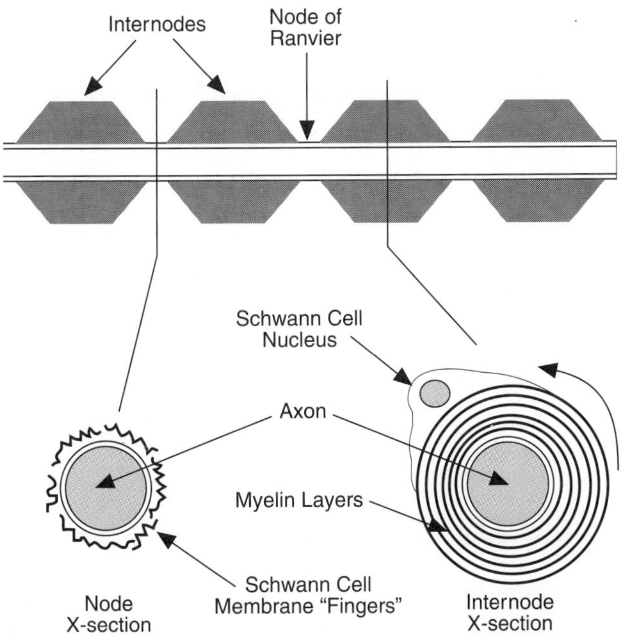

Figure 5-11

Figure 5-11

Many nerve axons are enveloped by concentric layers of glial cell membrane referred to as myelin. The multiple membrane layers are interrupted at regular intervals, referred to as nodes of Ranvier, along the course of an axon. At these nodes, bare, excitable axon membrane is exposed to the extracellular fluid, and only fragments of the myelin sheath are present. The myelinated regions between nodes are referred to as internodes.

□ **Myelin enhances the speed and efficiency of action potential signaling**

Action potential velocities along axons can vary widely. In the squid giant axon (diameters of 500 to 1000 μm), conduction velocity approaches 100 m/second. In contrast, a population of very small axons referred to as C-fibers (diameters <2 μm) that transmit information important for pain sensation in mammals propagate action potentials at only 1 to 2 m/second. Vertebrates have a clear need for high-speed action potential propagation to support reflex reaction times and to execute voluntary movements with skill and speed. For example, in humans, the axons transmitting sensory information and those delivering reflex motor commands for the clinically familiar knee-jerk or patellar tendon reflex propagate action potentials almost as rapidly as can the squid axon. Because there are many hundreds of such high-speed sensory and motor axons supporting various voluntary and reflex actions in a human limb, a paradox emerges. If very large axon diameters are required to support propagation velocities over 50 m/second, how is it possible to account for such speed in hundreds of axons packed into a single nerve bundle that is only two or three times the equivalent diameter of a single squid axon? The answer is that axons of only 10- to 20-μm diameters can achieve high propagation velocities by virtue of a property called *myelination*.

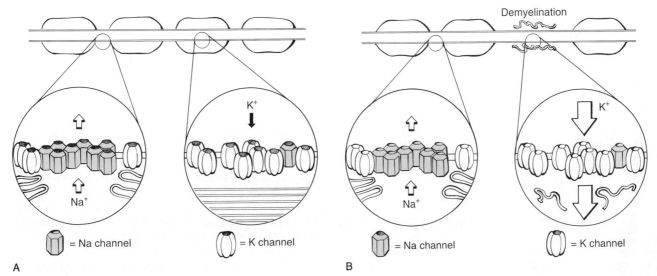

Figure 5-12

In a mammalian myelinated nerve, Na and K channels are unequally distributed between nodal and internodal membranes. Experiments indicate that Na channels are clustered in the axon membrane of the nodes of Ranvier but only rarely occur in the axon membrane of internodal regions. The K channel distribution is the mirror image, with most channels populating internodal membrane and almost absent from nodal membranes. Under pathologic or experimental conditions in which the myelin sheath degenerates (i.e., demyelination), the internodal K channels provide a pathway for the leakage of positive charge from the axon.

Myelination occurs during neuronal development by a process in which a series of cell membranes is layered around an axon and envelopes the axon at regular intervals along its length. These layers of membrane are referred to as myelin and functionally insulate the axon membrane from the extracellular fluid. This multimembrane sheath is produced during development of the axonal process by glial cells that repeatedly wrap around the axon as overlapping cell membranes build on top of each other (Fig. 5-11). In this process, the glial cell cytoplasm is displaced so that the membranes are tightly packed together. In the peripheral nervous system, the glial cells responsible for this process are the Schwann cells, and the oligodendrocytes serve this function in the central nervous system.

The myelin sheath surrounding an individual axon is interrupted at 1- to 2-mm intervals to reveal bare patches of axon membrane (see Fig. 5-10). These regions are called the nodes of Ranvier, and the myelinated regions are called internodes. The nodes are quite small, but they contain an extremely high density of voltage-gated Na channels in the axonal membrane. In contrast, much evidence suggests that very few Na channels populate the axonal membrane of the internodal regions of mammalian myelinated nerve (Fig. 5-12). The opposite distribution appears to be the case for voltage-gated K channels. K channels are almost absent from nodal membranes, but they are prominent in the paranodal and internodal regions of the axon. What are the consequences of this

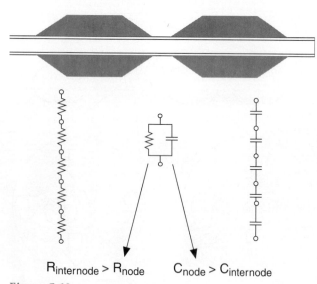

$$R_{internode} > R_{node} \qquad C_{node} > C_{internode}$$

Figure 5-13
The myelin sheath alters the effective membrane resistance (R) and capacitance (C) of the axon. The compound layers of glial cell membrane represent a linear series of membrane resistances and capacitances between the cytoplasm and extracellular fluid. The summed values of resistors connected in series yield a total effective membrane resistance greater than any single membrane. In contrast, the layers of membrane capacitance in series yield a smaller total capacitance as the transmembrane distance separating charges is increased. Less charge leaks through internodal membranes, and less charge is required to alter the internodal membrane potential during a propagating action potential.

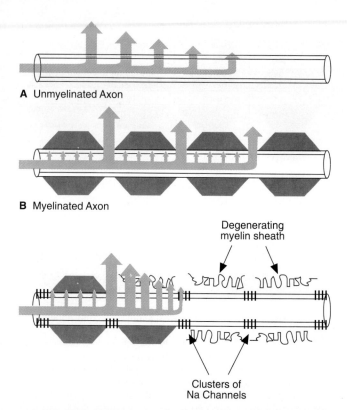

A Unmyelinated Axon

B Myelinated Axon

Degenerating myelin sheath

Clusters of Na Channels

C Demyelinated Axon

Figure 5-14
Excitatory current spreads over a greater distance in a myelinated axon than an equivalent-diameter unmyelinated axon. (**A**) The wave of excitatory positive current that spreads down an unmyelinated axon ahead of an advancing action potential leaks out of the axon and dissipates with distance. (**B**) The insulating properties of myelin retard this leakage and allow more current to be available at greater distances to depolarize and excite the membrane at the nodes of Ranvier. (**C**) Degeneration of the myelin sheath short circuits this insulation and shunts excitatory current to the extracellular space by means of K channels unmasked by the demyelination. As a consequence, insufficient charge is available to excite enough nodal Na channels to threshold levels, and action potential conduction can fail.

structural arrangement and channel distribution to action potential conduction?

The myelin sheath can be viewed in electrical terms as a group of membrane resistances and capacitances "in series" between the intracellular and extracellular compartments (Fig. 5-13). In the context of electrical circuits, the total resistance across this series of membranes is additive and is much higher than that of the bare axonal membrane. In contrast, the total capacitance of this layer of membranes is much less than any single membrane, because the insulating space between separated charges has dramatically increased.

These altered electrical properties effect changes in a propagating action potential by two means. First, as Na^+ ions enter and depolarize the axon membrane at an excited node of Ranvier, the excess positive charge distributes down the cytoplasm over a greater distance than in an equivalent diameter nonmyelinated axon. The effective membrane resistance is greatly increased while the axial resistance remains the same, and the space constant (λ_m) is greatly increased. Secondly, because the effective membrane capacitance has decreased, the amount of charge required to depolarize the internodal regions to achieve an action potential diminishes (i.e., $Q = V \times C$). The action potential seems to quickly skip from node to node (Fig. 5-14) with sufficient charge available at subsequent nodes to insure a threshold depolarization and activation of enough Na channels to sustain propagation to the next node and beyond.

There are several clinically important diseases of the nervous system that involve loss of the myelin sheath. Multiple sclerosis and Guillain-Barré syndrome are two such demyelinating diseases. The consequences of loss of the myelin sheath to the fidelity of signal transmission in the nervous system are important. If the myelin sheath is lost from sufficient lengths of an axon, propagating action potentials are slowed or completely blocked. The loss of myelin exposes stretches of internodal axon that contain K channels. These channels act as leakage paths for the positive charges that enter the node during an action potential (as Na^+ ions) to escape the cytoplasm (as K^+ ions). This leakage of charge compromises the depolarizing stimulus at the next several nodal regions where the voltage-gated Na channels reside. If a sufficient number of these channels are not activated to produce a regenerative action potential, the propagating impulse is blocked.

SELECTED READINGS

Aidley DJ. The physiology of excitable cells, 3rd ed. Cambridge: Cambridge University Press, 1989.

Dowling JE. Neurons and networks. An introduction to neuroscience. Cambridge, MA: Harvard University Press, 1992.

Nicholls JG, Martin AR, Wallace BG. From neuron to brain, 3rd ed. Sunderland, MA: Sinauer Associates, 1992.

Neuroscience in Medicine, edited by P. Michael Conn.
J. B. Lippincott Company, Philadelphia © 1995

Chapter 6

Synaptic Transmission

DANIEL GARDNER

Synapses

☐ Synapses link individual neurons

The purpose of the nervous system is the collection of information about the body and the world beyond, processing of that information, and coordination of responses to that information. This function depends on each neuron receiving input information from other nerve cells and communicating output with still others. However, the cell membrane of each neuron ensures electrical and chemical insulation from the surrounding structures, including other nerve cells. If this insulation were complete, the activity in each nerve cell would be independent of the activity of its neighbors, and each nerve cell would exist in useless isolation. This isolation must be breached selectively for the proper functioning of nerve cells, the nervous system, and the organism.

Specialized structures and mechanisms are needed for the communication of information from one neuron to another. These specialized structures are the *synapses*, and the mechanisms by which changes in the membrane potential of a nerve cell affect the membrane potential of another nerve cell are the mechanisms of *synaptic transmission*. In one sense, every interaction between neurons is synaptic (Greek *synaptein*, to join together). These interactions can have many consequences, from trophic to toxic, and many are mediated or carried by a neurochemical messenger. This chapter restricts synaptic transmission to its common meaning in neuroscience: the transfer of information from one nerve cell to another, usually by means of a chemical transmitter.

Synaptic transmission is a universal feature of the nervous system. A synapse is located wherever and whenever it is necessary for one neuron to inform or be informed by another neuron. However, synaptic transmission is not indiscriminate. In uncounted billions of places in the nervous system, the processes of different neurons pass one another without forming a synapse, secure in their mutual isolation. Synapses are found only where the design of the nervous system specifies a need for a synapse. Although some neurons may be connected to thousands of others, a typical neuron in the cerebral cortex probably makes connections with only a few percent of the neurons in its immediate neighborhood.

☐ Synapses underlie neuronal plasticity

Synapses serve an additional function, one that is related to the transmission of information but goes beyond it. Common experience provides many examples of human-designed structures for the transmission of information. This textbook is the closest at this moment, but others such as pens, telephones, tapes, disks, televisions, or computers are probably nearby. A property of most of these devices is that they transmit or process information

as appropriate but are not changed by the information that passes through them. Their functions are not altered by their experiences. The telephone does not sound especially nice after a previous conversation using harsh words. The television picture on the most-watched channel does not grow larger or more colorful with time. Change the channel, hang up the phone, eject the disks, or shut down the computer. Although the device processed a large amount of information, it will not process the next item in a different way because of its experience. Like an ideal rubber band that is unchanged by being stretched and relaxed, these information-processing structures are *elastic*.

Synapses are different. As some synapses transmit information, the information they process changes them, altering their function and sometimes altering their appearances. Like a stretched piece of chewing gum, they do not return to their prior state; they are *plastic*. This property of synapses, called *plasticity*, may be the neural mechanism responsible for some aspects of problem solving, learning, and memory. A later section of this chapter explores this property.

☐ Synapses share common principles

To emphasize the universality of synaptic transmission and to enhance the student's understanding, this chapter stresses principles at the expense of particular details. Much more electrophysiologic, ultrastructural, and neurochemical detail is known, and the interested reader is invited to consult the more advanced references listed at the chapter's end. Examples are drawn from many parts of mammalian, nonmammalian, and invertebrate nervous systems. The particular neurons or circuits from which these examples are drawn are identified only if the synapse serves as archetype or the function of the synapse is so closely related to its role that "where" and "why" become as important as "how." This chapter begins with an analysis of the first of these exceptional synapses, the neuromuscular junction.

Neuromuscular Junctions

The *neuromuscular junction* (NMJ) is the site at which information is transmitted from nerve to muscle and muscle excitation is initiated. The NMJ serves as a prototype or model for the synapse between nerve cells, in part because it is a well-understood synapse and the first to be studied in detail. Many of the physiologic features of neuromuscular transmission are common to synaptic transmission in the central and peripheral nervous systems. Much of our knowledge of neuromuscular transmission comes from a series of insightful and revealing experiments performed at University College, London, by Bernard Katz and his associates during the last 40 years. Many of these studies used the NMJ of the frog, informally called its *end plate*.

☐ The neuromuscular junction is specialized to transmit a fixed message

Although all synapses are built on a common plan, each is specialized in some way for its individual role. Several features of synaptic transmission are specialized at the NMJ to fulfill particular requirements of neuromuscular transmission. The role of synaptic transmission at the NMJ is to transmit reliably and without fail a message from the nerve to the muscle. The message is conveyed as an individual action potential in the motor neuron, from its cell body in the spinal cord, along the axon to its terminal at the NMJ. The message is then transmitted across the NMJ, producing a single action potential in the muscle and a resulting twitch.

The functioning of the system depends on each and every presynaptic action potential in the motor neuron giving rise to an action potential in the muscle fibers it innervates. The neuromuscular synapse is specialized to produce large, *suprathreshold* postsynaptic potentials. Because each conventional skeletal muscle fiber receives innervation from only a single motor neuron, the neuromuscular synapse is specialized to excite fibers in an all-or-none fashion. This faithful replication of the signal to contract transmits information across a synapse, but does not process or modify the information. Synapses in the central nervous system (CNS) are different; they process information by summing and integrating graded inputs from multiple presynaptic cells.

Because the nerve impulse is an action potential in the nerve fiber and because muscle contraction is triggered by an action potential in the muscle, why is a special mechanism needed for neuromuscular (or synaptic) transmission? Why does the nerve action potential not propagate to the muscle and excite it directly? One reason is the insulation that isolates the distinct nerve and muscle cells. Prejunctional and postjunctional membranes are separated by only 20 nm at the NMJ, but this small gap or *cleft* between nerve and muscle is adequate to insulate the two cells and prevent current from flowing between them. However, even if no gap existed and protoplasmic continuity between nerve and muscle could be maintained by gap junctions or other specializations, direct propagation of the nerve action potential into the muscle would be impossible because the muscle is too big.

It takes current to depolarize an excitable cell to a threshold that enables it to produce an action potential, and a large cell, with extensive membrane to depolarize, requires much current, about 1 μA per muscle fiber. The total amount of current available from a nerve terminal is only about 0.02 μA. These theoretical considerations are supported by experimental evidence that excludes direct electrical transmission from nerve to muscle by

showing that appropriate amounts of current injected into nerve or muscle do not flow across the NMJ to the other cell.

☐ Synaptic transmission is mediated by chemical transmitter molecules

The basic mechanism of synaptic transmission at the NMJ is shown in Figure 6-1. As at any chemical synapse, neuromuscular transmission is mediated by a *chemical transmitter substance*, which at this synapse is *acetylcholine* (ACh). The structure, synthesis, and degradation of ACh are shown in Figure 6-2.

Another feature common to all chemical synaptic transmission is that the events at the synapse can be separated into two phases: *presynaptic release* and *postsynaptic receptor-mediated events*. In the first phase at the NMJ, an action potential in the prejunctional motor nerve terminal releases the neurotransmitter ACh, which diffuses across the synaptic cleft. In the second phase, ACh acts on the postjunctional muscle cell by binding to a specialized membrane receptor. The binding of ACh to its receptor opens channels in the membrane that depolarize the muscle cell, producing an end-plate potential. The end-plate potential triggers an action potential in the muscle, leading to contraction.

Because only the presynaptic terminals are specialized to release transmitter and only the postsynaptic or postjunctional membrane is equipped with specialized receptor molecules to respond to the transmitter, this is a one-way process from nerve to muscle. Unidirectional transmission is characteristic of synaptic and junctional transmission.

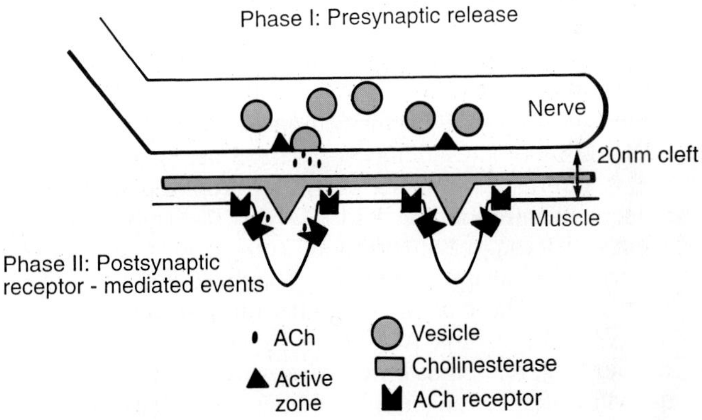

Figure 6-1

Overview of cholinergic synaptic transmission at the neuromuscular junction, showing the two major phases of synaptic transmission. Phase I shows presynaptic release of acetylcholine (ACh)-containing vesicles by active zones of the motor nerve terminal. In Phase II, ACh binds to receptors on the postsynaptic or postjunctional muscle membrane. Nerve and muscle are separated by a synaptic 20-nm cleft containing cholinesterase, which hydrolyses ACh.

Acetylcholine, a Neurotransmitter

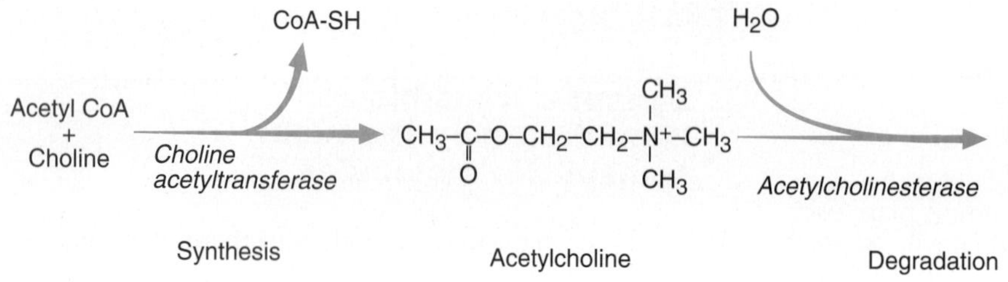

Figure 6-2
The neurotransmitter acetylcholine is synthesized by choline acetyltransferase from acetyl-CoA and choline, and it is hydrolyzed by acetylcholinesterases to acetate and choline.

The existence of neuromuscular synaptic transmission, a process interposed between nerve action potential and muscle action potential, can be demonstrated by stopping it from working. Neuromuscular transmission can be blocked by bathing a nerve-muscle preparation with saline containing a low concentration of Ca^{2+} or containing another agent such as botulinum toxin or curare. Curare, a substance brought to medical attention because of its application in envenoming arrowheads, selectively blocks neuromuscular transmission without affecting other mechanisms. It lowers the response of the postjunctional muscle membrane to the transmitter released by the nerve terminal, but it does not interfere with the generation or propagation of action potentials in nerve or muscle. Under these conditions, a nerve action potential is generated and spreads normally to the terminals, but no response is seen in the muscle. Muscle action potentials and contraction can be elicited by direct electrical stimulation of the muscle, demonstrating that there is nothing wrong with the nerve action potential, muscle action potential, or contractile machinery, despite blocked neuromuscular transmission.

☐ Acetylcholine is the neuromuscular transmitter molecule

Although ACh is the exclusive transmitter at mammalian NMJ and other sites, different synapses are specialized to use different transmitters, such as norepinephrine, glycine, or γ-aminobutyric acid (GABA). The only requirement is that the presynaptic and postsynaptic elements of a particular synapse must be specialized for the same transmitter. There is an overwhelming amount of evidence to suggest that ACh is the transmitter at the mammalian NMJ. First, motor nerves contain ACh and choline acetyltransferase, an enzyme that is responsible for its synthesis. Second, motor nerves release ACh on stimulation and in sufficient quantity to account for transmitter effects on the postjunctional membrane. Third, application of ACh by micropipette to the area of muscle just under the nerve, even if the nerve has been completely removed, produces results identical to those of nerve stimulation. Fourth, *anticholinesterases*, which prevent the hydrolysis of ACh, prolong the actions of applied ACh and the natural transmitter to the same degree.

☐ The postsynaptic membrane responds to neurotransmitter

The two phases of neuromuscular transmission, transmitter release and postjunctional interaction, are best understood when analyzed separately. The postjunctional or postsynaptic side is presented first, because much of our understanding of both prejunctional and postjunctional events derives from examination of the postjunctional cell through recordings of membrane potential, membrane and channel currents, and mechanical tension of the muscle.

The end-plate potential is a local, graded depolarization

Figure 6-3 shows a single muscle fiber innervated by a motor nerve. Two intracellular microelectrodes are inserted in the muscle fiber, one near the end-plate region where the motor nerve terminates and the other somewhat farther away. As each of the intracellular electrodes is inserted into the muscle fiber, it records the *resting membrane potential* of -90 mV. This resting potential is such a common feature of intracellular recording that it forms a baseline for the recordings of membrane potential shown as functions of time in Figure 6-3 and elsewhere. If the nerve is stimulated to produce an action potential, the processes of neuromuscular transmission give rise to an action potential in the muscle. This action potential is recorded by the first and second electrodes, but with a difference. The action potential recorded from the muscle by the first electrode is preceded by a step-like event, but the action potential recorded by the second electrode is not. This early step is not seen at either electrode if the muscle is stimulated directly, so it is not part of the muscle action potential. Because the step-like potential is recorded in the muscle only at the NMJ, or end plate, it is called the *end-plate potential* (EPP).

The effect of nerve stimulation can be mimicked by externally applying ACh in the region of the end plate. If enough ACh is applied, a muscle action potential is generated. An EPP precedes the muscle action potential recorded at the region near the end plate, but the muscle action potential recorded lacks this early potential. Application of ACh elsewhere on the muscle does not work, showing that ACh *receptors* are located only in this region. The receptor at the NMJ can be further character-

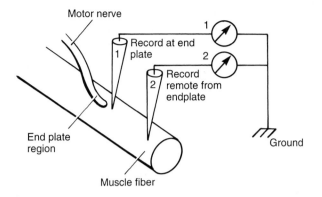

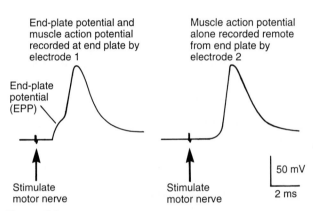

Figure 6-3
Experiment showing that the end-plate potential is an intermediate step between the arrival of the neurotransmitter acetylcholine (ACh) at the muscle and the generation of a muscle action potential. The experimental method is diagrammed above the results, which are two records of muscle membrane potential. Stimulation of the motor nerve leads to a muscle action potential, a change in membrane potential with time recorded by each of two electrodes inserted in the same muscle fiber. The end-plate potential preceding the action potential is recorded only by electrode 1, at the neuromuscular junction, and not by electrode 2, away from the end plate. Both recordings start from the baseline resting potential of −90 mV; both show a small artifact, indicating the nerve action potential. The ground in the extracellular bathing solution completes the circuits for intracellular recording. Calibration bars in the lower right indicate the scale for voltage and time.

ized as a nicotinic cholinergic receptor on the basis of its specificity to various agonists and antagonists of cholinergic transmission.

The results of many experiments imply that the EPP, a local depolarizing potential at the end plate, is an essential intermediate step between the arrival of ACh at the muscle membrane and the subsequent generation of a muscle action potential. They further suggest that the EPP is in some way generated by ACh. The EPP depolarizes the end-plate region past the threshold for generating a muscle action potential, which then propagates to all parts of the muscle fiber.

Understanding the role of the EPP requires the ability to study it without the distraction of the muscle action potential. This can be accomplished by lowering the EPP amplitude below the threshold, a requirement that can

be met by moderate treatments with the same agents that are used to block neuromuscular transmission. Neuromuscular transmission is functionally blocked when the muscle does not generate action potentials in response to a nerve action potential, but this functional block does not depend on complete elimination of the EPP. Moderate doses of curare lower the amplitude of the EPP below threshold by blocking some of the ACh receptors. Similarly, moderate reductions in the $Ca^{2+}:Mg^{2+}$ concentration ratio of the bathing solution decreases the amount of ACh released per nerve impulse without eliminating release completely. When the EPP is produced experimentally by application of ACh from a microelectrode, the amplitude of the EPP can be adjusted by reducing the amount of transmitter applied.

Any of these methods produce an EPP that is beneath the threshold for generating an action potential and allows examination of the EPP alone (Fig. 6-4). The EPP is a fast-rising depolarizing potential beginning about 0.2 to 0.5 milliseconds after the nerve action potential, reaching a peak in another 0.5 milliseconds, and then decaying back to the resting level exponentially, with a time constant of a few milliseconds.

Because the EPP is a local response, it can be seen at or near its site of generation at the end plate (electrode

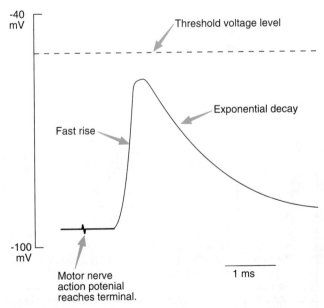

The End-Plate Potential

The amplitude of the end-plate potential has been experimentally reduced below threshold

Figure 6-4
The end-plate potential (EPP) is the postsynaptic potential at the neuromuscular junction. The amplitude of the EPP has been experimentally reduced below the threshold voltage level for the generation of a muscle action potential, revealing the time course of the potential change. The scale at the left shows muscle membrane potential; the calibration bar at the lower right gives the time scale.

1 in Fig. 6-3), but it does not propagate to all parts of the muscle membrane, and cannot be recorded at remote sites (electrode 2 in Figure 6-3). Normally, it is *suprathreshold* and gives rise to a propagated muscle action potential by depolarizing the postjunctional membrane at the end-plate region. Unlike the action potential, it is *graded*. If one half of the amount of transmitter is released, a smaller edition of the EPP is produced, and if ACh is applied for a longer time, the duration of the EPP is longer.

At directly–gated synapses, neurotransmitter binding opens postsynaptic channels

The depolarization that constitutes the EPP is a change in membrane potential (V_m). One of two causes is responsible for a change in the V_m. The concentration gradient of a membrane-permeant ion has changed, or there has been a change in the membrane conductance to some ion. In most cases, especially when the membrane potential changes rapidly, the ionic conductance has changed. If an ionic conductance changes, the answers to two questions provide an almost complete picture of the underlying events (Fig. 6-5). What turns the conductance on and off (i.e., gating)? What what ionic permeabilities underlie the conductance (i.e., ion selectivity)? These questions are the same as those used to determine the properties of conductances underlying the action potential. The differences between action potential and EPP can be accounted for by the different answers to these questions. The techniques are similar as well. To analyze the conductance change underlying the EPP, the voltage clamp is used. Chapter 4 provides a description of the voltage clamp and of its use for analysis of the conductances underlying the action potential.

With the muscle initially voltage clamped at the resting potential, no change in the current or underlying conductance is seen until the presynaptic motor fiber generates an action potential. When the motor nerve fires, inward currents are recorded from the muscle, currents that would give rise to an EPP if the muscle were not voltage clamped. The V_m remains constant. Unlike the case of the action potential, the conductance change at the EPP is not initiated by a change in V_m; instead, the event that leads to the increase in the EPP conductance (G_{EPP}) is the binding of neurotransmitter to postjunctional receptors on the muscle membrane.

Not only is G_{EPP} is not triggered by a membrane potential change, it is largely independent of membrane potential. This can be shown by repeating the previous experiment with the muscle membrane clamped at each of several values of V_m. The sizes of the resulting EPP currents (I_{EPP}) change with V_m (Fig. 6-6). Figure 6-6 also plots peak current against membrane clamp potential. The current underlying the EPP, unlike action potential currents, changes linearly with changing membrane potentials. The change arises because the EPP current is proportional to the driving forces ($V_m - E_{EPP}$) on the ions responsible for the EPP:

$$I_{EPP} = G_{EPP} (V_m - E_{EPP}).$$

The current change is linear because the conductance does not change when V_m is altered. The G_{EPP} is the slope of the current plotted against the voltage curve in Figure 6-6 and is constant over a wide voltage range, showing only a slight change at hyperpolarized V_m values.

During the EPP, the conductance of the end-plate region increases tenfold over its resting value. The magnitude of G_{EPP} is a function of the ACh concentration at the postjunctional membrane, not membrane potential. Prolonging the time during which the ACh concentration is elevated at the NMJ prolongs the EPP. There is a small secondary dependence of G_{EPP} on time and on V_m, which is most easily observed when recording individual channels. For any ionic species responsible for the conductance of the ACh channels, the channels are gated in a completely different way from the Hodgkin-Huxley action potential channels, which are gated by membrane potential and time, not by the neurotransmitter. Additionally, G_{EPP} does not show the rapid inactivation that characterizes the action potential conductances.

Conductances are Characterized by Gating and Ionic Selectivity

Gating:

Membrane conductances in neuronal and other excitable cells can be gated by:
- changes in membrane potential,
- binding of neurotransmitters or other agents, intracellular or intramembranous second messengers, or by
- combinations of any of these.
- Some conductances may be ungated; these are controlled by insertion into or removal from the membrane.

Ionic selectivity:

Channels and other conductances permit some ions to pass more readily than others:
- This preference is selective but rarely exclusive
- Some channels pass cations more readily than anions, some pass anions more readily than cations, and others admit some cations, such as Na^+, K^+, or Ca^{++}, more readily than others.
- The selectivity depends most strongly on the size, shape, and charge density of different parts of the molecule or molecules forming the ionic channel.

Figure 6-5
Membrane conductances can be characterized by gating and by ionic selectivity. Conductances are most often identified with membrane channels. These examples of gating are ways to change conductance by increasing or decreasing the number of open membrane channels. Channels also selectively pass certain ions more readily than others.

**The End-Plate Current is the Product of
an ACh-Dependent Conductance and a
Potential-Dependent Driving Force**

Currents underlying the EPP
recorded with the voltage clamp at
several values of membrane potenial

The slope of peak EPP current vs.
membrane potenial gives G_{EPP}

$$I_{EPP} = G_{EPP}(V_m - E_{EPP})$$

ΔV_m produced by the EPP at each of
several values of membrane potential

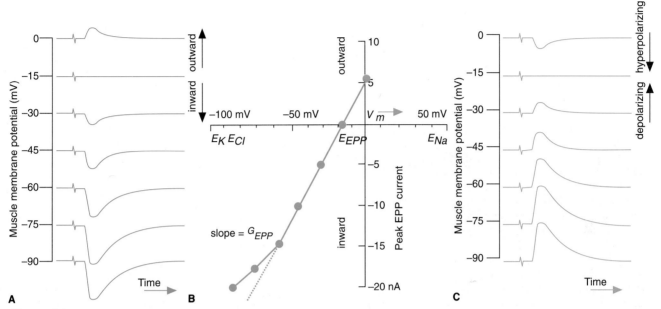

A   **B**   **C**

Figure 6-6
Recording the end-plate potential (EPP) starting from each of several different membrane potentials
shows that the EPP depends on an acetylcholine (ACh)-dependent and voltage-independent conductance and a potential-dependent driving force. (**A**) Using the voltage clamp, currents underlying the
EPP can be recorded by stimulating a motor nerve and recording ACh-mediated membrane currents
in muscle. The bottom trace shows the end-plate current recorded with the muscle clamped at its
-90-mV resting potential. Stimulating the nerve with muscle clamped at progressively more depolarized values of membrane potential (V_m) produces progressively smaller inward end-plate currents.
At the reversal potential for the EPP (E_{EPP}), no end-plate current is produced; at more depolarized
V_m, the end-plate current is reversed to an outward current. (**B**) The end-plate conductance (G_{EPP}) is
the slope of the line that results from plotting the peak end-plate current (I_{EPP}) on the vertical axis
against the membrane potential on the horizontal axis. The line is straight except at some hyperpolarized values of V_m, showing G_{EPP} to be essentially independent of V_m. The G_{EPP} is the same for each
end-plate current because it depends on the amount of ACh released by the presynaptic nerve. The
E_{EPP} is about halfway between the sodium and potassium Nernst potentials (E_{Na}, E_K) and far from E_{Cl}.
(**C**) The end-plate potentials produced in non–voltage-clamped muscle (i.e., experimentally prevented from generating action potentials) show similar dependence on V_m. The amount of depolarization produced by each EPP is smaller as V_m approaches E_{EPP}. No change in V_m is produced by an
EPP evoked at the reversal value of -10 to -15 mV, and the EPP is inverted to a hyperpolarizing
potential at still less negative values of V_m.

**The end-plate conductance is an excitatory
cation channel**

The second set of information needed to characterize the
EPP is the ionic basis of the conductance. The experiments illustrated in Figure 6-6 can provide much of this
information. Applying neurotransmitter with the muscle
clamped at more depolarized levels produced smaller
end-plate currents because of a reduction in driving
force. If the muscle membrane is clamped at -15 mV
when G_{EPP} is increased by the application of ACh, there is

no net current. Because the conductance increase produces no EPP current at -15 mV, the ions underlying the
conductance must have no net driving force on them at
this value. The reversal potential of the EPP (E_{EPP}) equals
-15 mV. In skeletal muscle, the Nernst potentials of the
major individual ions are $E_{Na} = +50$ mV, $E_K = -95$ mV,
and $E_{Cl} = -90$ mV. No single one of these ions can un-
derlic a conductance with a reversal potential of -15 mV.
This value is sensitive to experimental alteration of Na^+
or K^+, suggesting that Na^+ and K^+ are together responsible for the EPP. If the EPP channel could permit Na^+

and K^+ through in equal numbers, E_{EPP} would fall halfway between E_K and E_{Na}. The channel is actually a little more permeable to Na^+. A permeability ratio of about 1.3:1 for $G_{Na}:G_K$ gives a typical E_{EPP} of -15 mV. Other small cations get through as well, but only Na^+ and K^+ are present in significant numbers.

What ions contribute to the EPP at different potentials? At the reversal value of -15 mV, each Na^+ entering is balanced by a K^+ leaving, and there is no net charge movement, no current flow, and no EPP. If ACh is provided with the cell more negative than -15 mV (e.g., at the resting potential of -90 mV), more Na^+ enters than K^+ leaves, and a depolarizing EPP is seen. If ACh increases G_{EPP} with V_m experimentally set more positive than -15 mV, more K^+ leaves than Na^+ enters, and the EPP is hyperpolarizing (see top record, Fig. 6-6C).

On the molecular level, the binding of ACh to its receptor and subsequent channel opening is a multistate

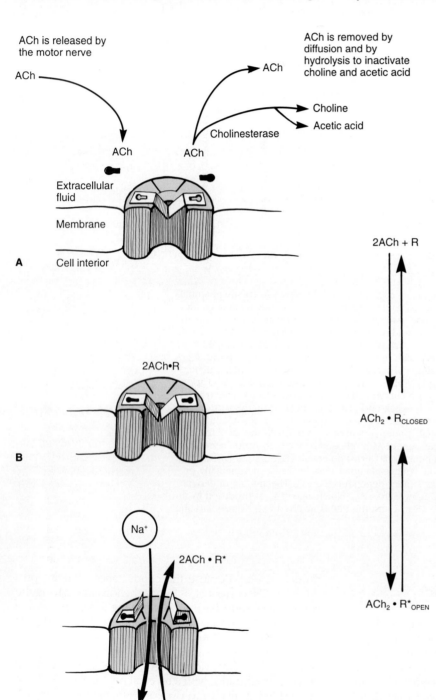

Figure 6-7
A section of muscle cell membrane at the neuromuscular junction shows an acetylcholine (ACh)-receptor–channel protein molecule (R) in each of three states. (**A**) ACh is released by the motor nerve and removed by diffusion and hydrolysis. Without ACh bound to each of the sites on the extracellular surface, the channel is closed. (**B**) ACh binding to both sites results in an intermediate state (ACh$_2$·R) in which the channel is closed. (**C**) With the ACh-bound channel in an open state (ACh$_2$·R*), small cations flow across the membrane. During the end-plate potential, the state of a typical channel progresses from top to bottom and then back to top, perhaps with several transitions between middle and bottom states. These interactions among each of the three states of the receptor-channel molecule are described algebraically on the right. Additional desensitized states are not shown.

process (Fig. 6-7). The nicotinic ACh receptor at the NMJ is a protein composed of five subunits. In postembryonic muscle, these subunits are denoted α, α, β, δ, and ε. The five subunits comprising the ACh receptor each contain four hydrophobic domains that span the membrane. This structure, which is characteristic for channels directly gated by neurotransmitters and for other structures of additional families of receptors, is further described in Chapter 7.

Each α subunit contains a binding site for ACh. Two molecules of ACh combine with each receptor to initiate channel opening. The ACh binding sites are located on regions of the α subunits exposed to the extracellular bathing solution, on the outer surface of the muscle membrane. ACh binding, which is a relatively fast process, permits the ACh-receptor complex to undergo a subsequent conformational change, opening a gate across the channel. The open channel allows passive diffusion of ions down their electrochemical gradients through the channel's pore. Contemporary studies, using the *patch clamp* described in Chapter 4, have recorded this ion movement through single channels. The patch clamp recording of Figure 6-8 traces the end-plate current, which flows through a single postjunctional ACh-activated channel molecule when the molecule undergoes the conformational change into the open state. The ACh-activated channel-receptor complex can open and close more than once, shuttling between open and closed states until one or both of the bound ACh molecules dissociates during one of the times the channel is closed, usually within 3 milliseconds after binding. This average open lifetime of the channel is shortened by depolarization. When the suprathreshold EPP in a non–voltage-clamped fiber gives rise to an action potential, the depolarization of the action potential helps to terminate the EPP.

The conductance change from this one channel opening, caused only by two molecules of ACh binding to a single receptor molecule is about 4×10^{-11} S (1 S = 1 Ω^{-1}), permitting 50,000 net positive charges to enter the muscle in a few milliseconds and depolarize it by 0.3 μV. Figure 6-8 shows such an opening and the resulting current. The design of this system can be appreciated by noticing that just two molecules of ACh, each bearing a single charge, are used to control a gate that admitted 50,000 excess charges. In addition to transmitting a message, the NMJ is specialized for *biologic amplification*. However, this response to a single pair of transmitter molecules is still far short of the depolarization needed to produce the normal EPP, which requires the summed effect of perhaps 5×10^5 molecules of ACh released from the prejunctional nerve terminal by an action potential.

The postjunctional membrane, located under the nerve terminal, contains about 10 million receptors. This is the only area of the muscle with significant ACh sensitivity; extrajunctional regions produce almost no response to iontophoretic ACh under normal conditions. However, when the motor nerve is cut or is anesthetized for a period of weeks, the ACh sensitivity gradually increases over the whole muscle surface. This spread of

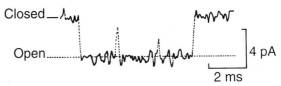

Figure 6-8

Abrupt changes in current reveal opening and closing of a single acetylcholine (ACh)-activated end-plate channel of frog muscle. The patch clamp is used to record the time course of the single-channel current. Initially closed, the channel opens about 1 millisecond after the start of the record. Two milliseconds after opening, a fast transition to the bound-but-closed ACh₂·R state is incompletely captured. Less than 4 milliseconds later, the channel closes, presumably followed by the unbinding of ACh. (From Colquhoun D, Sakmann B. Transmitter-activated bursts of openings. In Neher E, Sakmann B [eds]. Single-channel recording. New York: Plenum Press, 1983.)

denervation supersensitivity is paralleled by an increased distribution of receptors over the muscle membrane and can be alleviated by keeping the muscle active by periodic electrical stimulation.

Synaptic action is terminated by removal or degradation of the neurotransmitter

Persistent transmitter can produce continuing synaptic current. According to the scheme previously described, as long as the ACh concentration at the postjunctional membrane is high, ACh is available to bind to receptors and open EPP channels, to dissociate, and then to bind to the same or a nearby receptor, again opening EPP channels. However, the duration of the EPP and its underlying G_{EPP} is only a few milliseconds, because the transmitter concentration in the cleft is reduced within milliseconds of release.

ACh is removed from the cleft by *diffusion* and *hydrolysis* (see Fig. 6-7). The hydrolysis of ACh to inactive choline and acetic acid is accomplished by cholinesterases, including a specific *acetylcholinesterase* at the NMJ (see Figs. 6-1 and 6-2). The ACh receptor for the EPP and the ACh-binding esterase are separate and independent sites; each binds ACh, but there are molecules that bind to one but not to the other. There are about 10^7 esterase sites per end plate, located on a web in the synaptic cleft (see Fig. 6-1). The esterase helps terminate ACh action, and it provides a supply of choline, which the prejunctional terminal takes up to form more ACh.

The synapse presents distinct sites for pharmacologic manipulation

Two classes of substances can bind to the ACh receptor. *Antagonists* (e.g., curare) compete with ACh for available receptor sites. These substances can bind to the receptor but cannot undergo the conformational changes that permit the channel to open. Because they block ACh from reaching receptors and do not do anything constructive when they bind to the receptor, curare and other *com-*

petitive blocking agents lower EPP amplitude. Endogenous substances can also block receptors. For example, antibodies associated with the disease myasthenia gravis cause muscle weakness by reducing the number of available receptors.

The second class of substances, like ACh itself, are *agonists*. These substances can bind to the ACh receptor and permit the channel-opening conformational change that produces the EPP. Many of these agonists, including succinylcholine and decamethonium, also block neuromuscular transmission. The blocking mechanisms probably differ at different NMJs, but they may involve combinations of two actions: *depolarization* and *desensitization*. Because some agonists are not broken down by the specific cholinesterase concentrated at the NMJ, they can bind and unbind to ACh receptors repeatedly. This repetitive blocking may produce prolonged depolarization, which has been conjectured to prolong inactivation of muscle action potential Na^+ channels and block contraction. These depolarizing blockers also desensitize the ACh receptors, making them temporarily unavailable for agonist binding, and may also block open channels.

Cholinesterases terminate the action of ACh by hydrolyzing it into inactive products. Anticholinesterase agents such as eserine block the esterase by competing for the binding site of acetylcholinesterase. This slows the rate of ACh hydrolysis and prolongs the time of ACh action, until the ACh is removed from the cleft by diffusion. Prolonging the time after each prejunctional action potential during which ACh is available to bind to postjunctional receptors has dual consequences for neuromuscular transmission. Transmission can be enhanced by producing larger and more elongated EPP, especially if disease states or other agents have reduced some EPP amplitudes below threshold. However, large doses or prolonged administration of anticholinesterases can lead to a depolarizing or desensitizing block of neuromuscular transmission. There is evidence that anticholinesterases may exert additional effects by interacting with ACh receptors and with prejunctional terminals.

Somewhat more striking are the effects of the irreversible anticholinesterases, including the nerve gas diisopropyl fluorophosphate and the insecticide parathion. These agents bind almost irreversibly to acetylcholinesterase, and in sufficient quantities, they prevent ACh from being hydrolyzed. In addition, they deplete the cleft concentration of choline, preventing the re-uptake needed to produce more ACh. This rapidly leads to a prolonged depolarizing block of NMJs, including those of the diaphragm and intercostal muscles that are essential for respiration.

☐ Presynaptic terminals release neurotransmitters

The previous section covered the postsynaptic events produced after the arrival of neurotransmitter at the postjunctional muscle membrane. A complete analysis of neuromuscular transmission requires examining the processes of transmitter release from the presynaptic nerve. These events occupy the fraction of a millisecond between the arrival of an action potential in the motor nerve terminal and the beginning of the conductance change underlying the EPP.

At rest, transmitter release is sparse and irregular

Our understanding of how the action potential causes neurotransmitter release comes from studying the NMJ at rest. When the motor nerve is inactive, there are no EPPs, no muscle action potentials, and no contractions. However, electrical recording reveals that the muscle is not entirely quiescent (Fig. 6-9). Although there are no EPPs, recording at the end plate reveals small (<1 millisecond), randomly occurring blips, each of similar amplitude. Recording away from the junction (at electrode 2) reveals none of these blips, instead showing a flat −90-mV baseline.

These miniature potentials have the same shape and reversal potential as the EPP. Curare also causes a decrease in their amplitude, as it affects the EPP, but no agent influences their frequency of occurrence, which is random and unpredictable. Because these events resemble the EPP in everything but size and because they are recordable only at the end plate, they have been given the name *miniature end-plate potentials* (MEPP). They

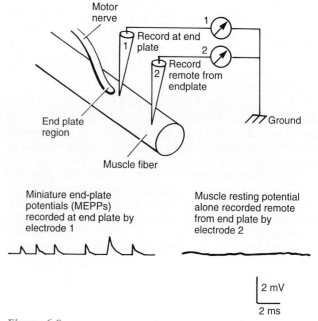

Figure 6-9
Miniature end-plate potentials (MEPPs) are recorded from the end-plate region of muscle at rest. MEPPs resemble standard EPPs, but they have an amplitude of only 1 or 2 mV and appear randomly. The experimental method for recording muscle membrane potential as a function of time is the same as Figure 6-3, except that the recording sensitivity is set to more easily record the smaller MEPP.

are produced by small quantities of ACh acting on the same receptor that produces the EPP. An individual MEPP is much larger than the conductance change produced by a single ACh receptor activated by two molecules of ACh. Each MEPP is the result of the simultaneous arrival of about 6000 molecules at the postjunctional muscle membrane, a finding indicating the existence of a structure or mechanism that releases, all at once, individual packets, each containing about 6000 ACh molecules.

Depolarization enhances transmitter release

Each MEPP is a random event, and a given MEPP neither increases nor decreases the probability of another one following it within a given time. Manipulating the membrane potential or other physiologic parameters of the muscle has no effect on the frequency of these random potentials. The packets of ACh do not arise from the postjunctional muscle; they are released by the prejunctional nerve. Depolarizing the nerve terminal increases the frequency with which MEPPs appear. As the nerve is depolarized, without producing an action potential in the nerve, MEPP frequency can be raised to the point at which two, three, or four MEPPs appear at one time. Several MEPPs appearing simultaneously summate to produce larger potentials, potentials that are two to four times larger than an individual MEPP.

Because the full-size EPP is produced by ACh released by a nerve action potential and because an action potential is a large depolarization, these findings together suggest a hypothesis: the EPP is the result of summed MEPPs, produced by the simultaneous release of many multimolecular packets or quanta of ACh from the nerve when it is depolarized by the action potential. Figure 6-10 summarizes the evidence and the resulting hypothesis.

Transmitter release is quantal and calcium-dependent

Evidence for the quantal release hypothesis is found in an elegant series of experiments in which the amplitude of the EPP evoked by nerve stimulation was lowered to

End-Plate Amplitude is Quantized

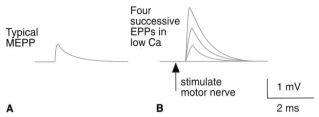

Figure 6-11
The amplitude of the EPP is quantized, restricted to particular discrete values. (**A**) A typical MEPP. (**B**) Four EPPs produced with the end plate bathed in low Ca^{2+} to reduce transmitter release are superimposed to permit comparison. Under these experimental conditions, EPP amplitude is restricted to approximate integral multiples (e.g., 0, 1, 2) of the typical amplitude of a MEPP.

the point where the EPP was the approximate size of a MEPP. This is accomplished by lowering the Ca^{2+} concentration in the medium bathing the nerve-muscle preparation. As Ca^{2+} is lowered, the frequency of randomly appearing MEPPs is reduced, and the amplitude of EPPs evoked by nerve stimulation is lowered. Because low-Ca^{2+} solutions do not affect the response to ACh applied by a microelectrode, the low levels of Ca^{2+} must be affecting release rather than the postsynaptic receptors. At very low levels of Ca^{2+} and high levels of Mg^{2+}, the EPP amplitude can be reduced to the size of a MEPP. At this point, the size of the EPP in low-Ca^{2+} medium fluctuates in a step-like fashion called *quantal* release. Although the size of the response to externally applied ACh can be continuously graded by applying slightly larger or smaller doses of ACh, the EPP is restricted to only a few discrete sizes.

In low-Ca^{2+} medium, each EPP can be about the same size as a MEPP, twice or three times the amplitude of a MEPP, or can fail to produce any response at all (Fig. 6-11). It is as if stimulating the nerve can produce one, two, three, or no packets of ACh, but not one half. The *quantal hypothesis* specifies that ACh is released in multimolecular quanta. At rest, quanta are released randomly,

---

FINDINGS

1. MEPPs resemble EPPs but are much smaller.
2. Depolarization of the nerve terminal leads to an increased frequency of MEPP.
3. MEPPs can sum to produce larger potentials.
4. The EPP in the muscle is produced by ACh released by a nerve action potential.
5. A nerve action potential depolarizes the presynaptic terminal.

HYPOTHESIS

- The EPP is the result of summed MEPPs, produced by the simultaneous release of many quanta of ACh from the nerve when it is polarized by the action potential.

Figure 6-10
Evidence for the hypothesis that the end-plate potential (EPP) is composed of summed miniature end-plate potentials (MEPPs).

producing MEPPs. Because the probability of release of any one quantum is low, MEPPs appear occasionally and rarely more than one at a time. The prejunctional action potential causes, for a short time, a many-fold increase in the probability (p) that any one packet of ACh will be released (Fig. 6-12). The transiently high probability leads to the simultaneous release of the ACh from about 100 to 200 quanta, producing the EPP.

These data permit appreciation of the EPP as an event resulting from the summed activity of a large number of similar underlying processes (Fig. 6-13). Each one of these underlying events depends on a similarly large number of an even more basic underlying process. EPPs are composed of summed MEPPs, each of which is composed of summed openings of individual ACh receptor-channel molecules. The apparently smooth and graded EPP is made of currents through varying numbers of open postsynaptic membrane channels, each of which has about the same fixed conductance, just as the varying waveform of sound from a compact disc player is composed of many small digital events. Each channel opening produces synaptic current by the summed net entry of many cations.

The action potential causes release by providing a transient depolarization required by the release process. There is a second requirement for release: a sufficient concentration of Ca^{2+} in the solution bathing the prejunctional terminal. The nerve poisons tetrodotoxin and tetraethylammonium can be applied to block the Na^+ and K^+ conductance changes of the action potential. Even in the absence of action potentials, the nerve can still be persuaded to release the ACh for an EPP, provided that two conditions are met. The nerve must be depolarized in some way, usually with injected current from an electrode, and there must be at least 10^{-4} mol/L of Ca^{2+}

Events Underlying the End-Plate Potential

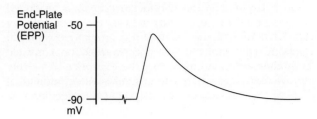

Each EPP is composed of 100 to 200 summed MEPP:

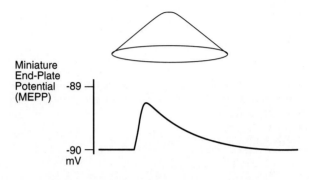

Each EPP is composed of several thousand summed channel openings:

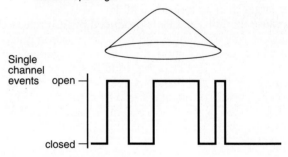

Each open channel admits several thousand ions per millisecond.

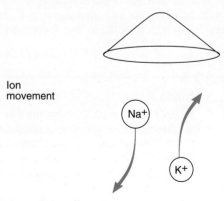

Figure 6-13
Events underlying the end-plate potential (EPP) are shown schematically. Events viewed at each of four different scales are shown with the most macroscopic at the top and the most microscopic at the bottom. EPP are the result of summed MEPPs, which are the result of summed single-channel events. Each single-channel current results from the combined net movement of many ions.

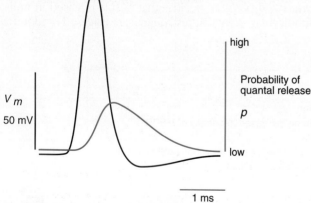

Figure 6-12
The mechanism for transmitter release by an action potential is a many-fold increase in the probability of quantal release p. The action potential and the resulting transient increase in p are displayed on the same time scale. The left vertical scale (V_m) applies to the action potential; the right vertical scale applies to p.

in the extracellular bathing solution. The Ca^{2+} must be present during the early part of the depolarization of the nerve terminal. If Ca^{2+} is provided only after the depolarization but before the time for release, no release occurs. This finding is consistent with the theory that an increased Ca^{2+} concentration in the terminal is needed for nerve-evoked release and that the needed Ca^{2+} enters the terminal from the external bathing solution through voltage-dependent Ca^{2+} channels. When the specialized Ca^{2+} channel in mammalian motor nerve terminals is blocked by toxins obtained from spiders or snails, neuromuscular transmission is abolished, even though the nerve action potential is unaltered. The required Ca^{2+} concentration is very small, and clinically observed changes in serum Ca^{2+} affect this process less than they affect action potential threshold and muscular contraction.

Quanta are vesicles

Although the quantum is a physiologic concept, it has a morphologic correlate in the *vesicle* seen in electron micrographs of presynaptic and prejunctional terminals. Combined anatomic and physiologic findings have verified that these vesicles contain quanta of neurotransmitter and that exocytosis of the contents of single vesicle is the process responsible for observed quantal release. Agents that cause sustained release of transmitter, such as black widow spider venom, reduce the number of vesicles and increase the area of the presynaptic terminal, consistent with fusion of vesicles with presynaptic cell membranes. Views of vesicles fusing to prejunctional membrane at *active zones* have been captured in quick-

freezing experiments. There is evidence for recycling of membrane and vesicle re-formation after transmitter release (Fig. 6-14).

Calculations of vesicle number and estimates of vesicular contents correlate well with physiologic data for transmitter release. The cell membranes have electrical capacitance proportional to their area, and the increase in membrane area with the fusion associated with individual exocytotic events has been observed in other cell types. Some experiments revealed additional nonvesicular release in resting nerve, a process of unknown mechanism and physiologic significance.

Exocytosis of transmitter in nerve cells is a regulated event distinct from but related to the constitutive release characterizing nonneural cell types. The action-potential–evoked transmitter release at the NMJ serves as a controlled mechanism using Ca^{2+} for the immediate and transient magnification of an exocytotic process by four orders of magnitude or more. A subsequent section of this chapter discusses the mechanism of quantal release and its dependence on Ca^{2+} and on membrane potential, because these determinants of release can be altered to affect synaptic plasticity.

Central Synapses

☐ Central synaptic actions are varied and significant

The role of a synapse in the CNS is different from that at the NMJ. These different roles require functional and structural differences between neuromuscular and cen-

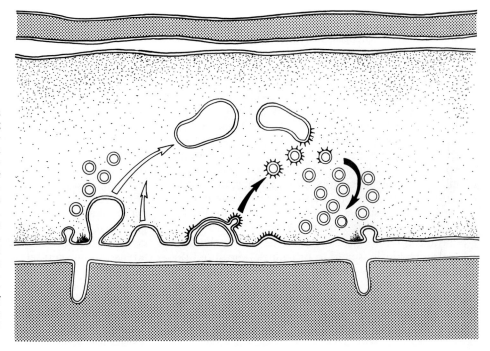

Figure 6-14
The vesicle membrane fuses with presynaptic membrane during transmitter release and then is recycled two ways. In this section through a neuromuscular junction, the nerve is shown (*lightly dotted*) above the muscle (*heavily-shaded*). Exocytosis of vesicles (◎) occurs at active zones (*right*). Recycling of membrane occurs by uptake at active zones, forming vacuoles (*left*). Clathrin-coated pits form near glial infoldings (*center*). (Drawing by Sylvia Colard-Keene redrawn from Miller TM, Heuser JE. Endocytosis of synaptic vesicle membrane at the frog neuromuscular junction. J Cell Biol 1984;98:685.)

tral *neuroneuronal synapses*. However, the similarities in function and mechanisms of the two types of synapses are more significant than their differences. The description of neuromuscular transmission earlier in this chapter applies equally well to any chemical synaptic transmission with three major variations:

1. Because individual central *postsynaptic potentials* (PSPs) are very much smaller than EPPs (usually subthreshold), summation of excitatory input is necessary to generate an action potential in the postsynaptic neuron.
2. Different central transmitter-activated postsynaptic channels have different ionic selectivities. Cation-selective channels at central synapses produce excitation, as at the NMJ. Postsynaptic ionic channels selective for Cl^- produce inhibition. Channels selective for Ca^{2+} or K^+ can influence a wide range of postsynaptic events.
3. Different synapses use different neurotransmitter molecules.

Although ACh is used by many central synapses, others are specialized to use one or more of any of dozens of other substances. Consequently, the mechanisms of termination of transmitter action and of its removal or degradation may differ.

Why do central synapses use the same chemical transmitter mechanism as the NMJ? The NMJ required such a mechanism, because the size mismatch between prejunctional nerve and postjunctional muscle mandated a chemical messenger to amplify biologically the message from the nerve. Because there is no size mismatch between CNS neurons, it would be theoretically possible to use direct current spread, called *electrotonic transmission*. The nervous system has selected chemical transmission at most central synapses, rather than the electrotonic mechanism, because chemical transmission has inherent advantages for the CNS and the organism.

There are five principal advantages (Fig. 6-15). First, chemical synapses are unidirectional; they limit information transmission to one preferred direction. Within each neuron, potential changes are limited only by neuronal geometry. Action potentials, for example, can propagate in either direction along an axon. One way in which information flow within the nervous system can be directed from input to output is to interpose specialized chemical synapses at which only the presynaptic terminal releases neurotransmitter and only the postsynaptic membrane contains receptors. Although there are exceptions to this rule, synaptic transmission usually is from presynaptic to postsynaptic cell. Second, chemical synapses can be specialized for inhibition so that activity in a neuron can reduce activity in another. Third, chemical synapses are ideally suited to provide subthreshold potentials, excitatory or inhibitory, that can be summed by the postsynaptic cell, permitting processing and integrating of synaptic information. Fourth, chemical synapses provide a delay, which often allows the arrival of information at a part of the nervous system to be timed ap-

Advantages of Chemical Synapses

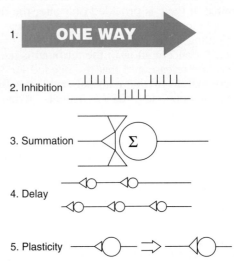

Figure 6-15
Five advantages of chemical synapses are shown schematically. Chemical synapses are unidirectional, from presynaptic to postsynaptic cell (*1*). Chemical synapses can inhibit, shown by the suppression of activity in the upper, postsynaptic trace by firing in the lower, presynaptic record (*2*). Synaptic potentials produced by chemical synapses are readily summed, as indicated by the summation sign (Σ) in the postsynaptic soma innervated by several presynaptic endings (*3*). Each chemical synapse introduces a short synaptic delay, and information can be staggered by routing it through the two-synapse or three-synapse pathways shown (*4*). Chemical synapses can undergo plastic changes to their efficacy, indicated by the change in size of the presynaptic ending (*5*).

propriately. For example, information routed through a two-synapse pathway can reach an area of the nervous system faster than other information arriving by means of a three-synapse path. The fifth major advantage of chemical synapses is that they can alter their efficacy, changing their effectiveness over time as a result of their own activity or that of other neurons. This mutability of chemical synapses is known as plasticity.

At most central synapses, whatever the transmitter, the mechanisms of transmitter release and of diffusion across the cleft are essentially the same as at the NMJ. Some chemical substance is released by the depolarization of the action potential when Ca^{2+} is present. The transmitter then diffuses the short distance from the presynaptic to postsynaptic membrane, where it combines with a receptor that is specific for the neurotransmitter at that particular synapse. Many of the differences between central synapses and the NMJ, as well as most of the variations among central synapses, arise from differences in the events after the neurotransmitter binds with its postsynaptic receptor. At some synapses, the mechanisms of transmitter release differ, specialized for continuous release or other release not triggered by action potentials.

☐ Postsynaptic potentials can excite or inhibit

Many postsynaptic effects of transmitters involve changes in the postsynaptic membrane potential. The mechanisms of these postsynaptic effects can be understood by examining the gating and ionic selectivity of the channels that are opened by the neurotransmitter or the subcellular processes controlled by neurotransmitter binding. The nature of these postsynaptic events determines whether the effects are excitatory or inhibitory. An *excitatory* effect is defined as one that tends to make it easier for the membrane potential of the postsynaptic cell to reach its threshold value and generate an action potential, and an *inhibitory* effect is one that tends to make it harder for the membrane potential of the postsynaptic cell to reach the threshold.

A prototypic CNS network, consisting of three synaptically interconnected neurons, is diagrammed in Figure 6-16. In this circuit, a postsynaptic neuron is innervated by each of two presynaptic cells. In the CNS, each cell can be postsynaptic to thousands of others and presynaptic to many more; the circuit shown is therefore greatly simplified. Each cell is shown schematically; the cell body with all its dendritic specializations and postsynaptic receptors are reduced to a circle, and the axon with all its branches is simplified to a line, so all of these structures together resemble a lollipop. Presynaptic terminals are shown as triangles synapsing on the lollipop, ignoring the anatomic specializations that underlie synaptic transmission, including the characteristic vesicles and release zones in the terminal of the presynaptic cell, the synaptic cleft, and the postsynaptic membrane clusters of receptors.

Microelectrodes in each cell permit recording of the membrane potentials of all three cells simultaneously. When a microelectrode penetrates a CNS neuron, it records a membrane potential whose resting value is approximately −60 mV. Intracellular records show changes from this −60 mV resting potential. In each presynaptic cell, changes from the resting potential consist of occasional action potentials, accompanied by the release of a

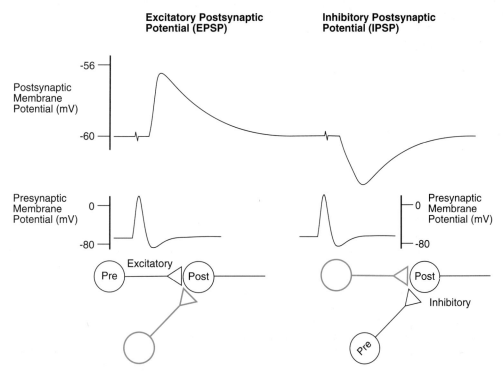

Excitatory and Inhibitory Postsynaptic Potentials

Figure 6-16
Excitatory and inhibitory postsynaptic potentials (EPSP and IPSP) produced in response to action potentials in presynaptic cells. The upper trace shows the EPSP and IPSP; the lower trace shows the presynaptic action potentials on the same time scale but with a compressed voltage scale. The EPSP is depolarizing, and the IPSP is hyperpolarizing. Each PSP changes potential rapidly after a short synaptic delay, returning to the resting potential with an exponential time course that outlasts the action potential that triggered it. The neural circuit is diagrammed below the recordings. Excitatory and inhibitory presynaptic neurons (Pre) and the common postsynaptic neuron (Post) are illustrated. The inactive presynaptic neuron in each circuit is shown in gray.

neurotransmitter. Intracellular recording from the post-synaptic cell shows that it responds differently to the two presynaptic cells.

Each action potential in one presynaptic neuron produces a small depolarizing potential in the postsynaptic cell. After a synaptic delay of about 0.5 millisecond, the potential rises quickly to a peak, usually in about 1 to 2 milliseconds, and then declines exponentially back to the resting potential, with a time constant that can vary from 2 to 100 milliseconds or more. The actual duration varies enormously from cell to cell or from synapse to synapse on the same cell, but it is important to neuronal function that these postsynaptic potentials last much longer than action potentials. The amplitude varies from 100 μV to a few millivolts for different PSPs. It is similarly important for neuronal integration that single postsynaptic potentials are too small to reach the threshold for generation of a postsynaptic action potential. Because this postsynaptic potential is excitatory, as defined previously, it is called an *excitatory postsynaptic potential* (EPSP). The EPSP is the central analog to the EPP of muscle. The EPP is a special case, at the NMJ, of an EPSP. Like the EPP, the EPSP is a local, graded potential, not a propagated potential. It is made of quantal miniature EPSPs.

Each action potential in the other presynaptic neuron again produces a small potential change in the postsynaptic cell. This event is hyperpolarizing and is an *inhibitory postsynaptic potential* (IPSP). It is like an upside-down EPSP, and although the direction in which it drives the postsynaptic membrane potential is different, it resembles the EPSP in many ways. Like the EPSP, it is produced by a transmitter from a presynaptic cell that diffuses across a synaptic cleft. It is a local, graded potential that can sum. It is quantal, composed of miniature IPSPs. There is a synaptic delay of 0.5 millisecond between presynaptic action potential and IPSP. These properties are identical to those characterizing the EPSP. The major difference is that the IPSP is hyperpolarizing, instead of depolarizing, and inhibits the cell instead of exciting it. This ultimate difference results from differences in the postsynaptic response to the various transmitters released by the excitatory and inhibitory presynaptic neurons.

☐ Neurotransmitters produce diverse postsynaptic responses

The transmitter and its actions are distinct

Certain mechanisms govern the arrival of neurotransmitter at the postsynaptic membrane, which produces a change in the membrane potential of the postsynaptic cell. At a single synapse, the presynaptic and postsynaptic elements are each specialized to use the same molecule—one to send and one to receive—and the postsynaptic elements are further specialized to react in a particular way to the arrival of the message. However, the nervous system is conservative in its selection of transmitter molecules and the specialized mechanisms needed for their use. Each transmitter is used at many

different synapses, and for any neurotransmitter, there exists more than one receptor type and more than one postsynaptic event that is produced in response to the transmitter. Similar conservation is observed for the postsynaptic processes, with similar postsynaptic channels or enzyme activated by different transmitters.

Excitation and inhibition are not properties of the neurotransmitter molecule itself; these responses are properties of individual receptors that may be activated by particular transmitters. There is nothing excitatory or inhibitory about any neurotransmitter molecule itself. Some transmitters, such as glutamate, seem to mediate only excitation, and others, such as glycine, are found at inhibitory synapses, but this is probably an arbitrary, evolutionary decision of the nervous system.

A single neurotransmitter can be used to transmit excitatory and inhibitory messages at different synapses. Acetylcholine has multiple actions, and a list of some of them provides an introduction to the diversity of postsynaptic action. At the NMJ and elsewhere, the transmitter activates an integral, *nicotinic* receptor-channel molecule, selective for cations, to produce fast excitation. Another class of ACh actions, mediated by *muscarinic* receptors, is more diverse. In the heart, the inhibitory action of ACh is effected through muscarinic, G-protein–mediated inhibition of adenyl cyclase and modulation of a potassium channel. In the brain, it works directly by inhibiting adenyl cyclase and through phospholipase C, with evidence for excitatory and inhibitory effects.

This mix-and-match design of postsynaptic action results in convergent and divergent coupling of neurotransmitter receptors to functional consequences. Individual neurotransmitters, each capable of activating multiple receptors, can produce different, even opposite, actions by selective control of channels and other postsynaptic mechanisms, and different transmitters can be used to control the same channel type through a specific receptor.

There are three styles of receptor-channel coupling

The nervous system uses combinations of many transmitters, many receptors, and many channel types and other effectors. The nervous system also varies the mechanism by which the receptor, which recognizes and binds a particular transmitter, is coupled to the channel, whose selectivity for some ion or ions determines the flow of synaptic current and produces the postsynaptic response. In some cases, the postsynaptic effect involves the activation of metabolic or other postsynaptic processes in addition to or instead of opening or closing specific ionic channels.

There are three styles for receptor-channel coupling, and each is specialized for a particular purpose (Fig. 6-17). The simplest and most direct coupling is the fastest and most specific. This is the case in which the receptor and channel are parts of the same molecule. Binding of the appropriate transmitter produces a realignment of the molecule that opens the channel, exposing the pathway for selective ion flow. Usually, subsequent channel

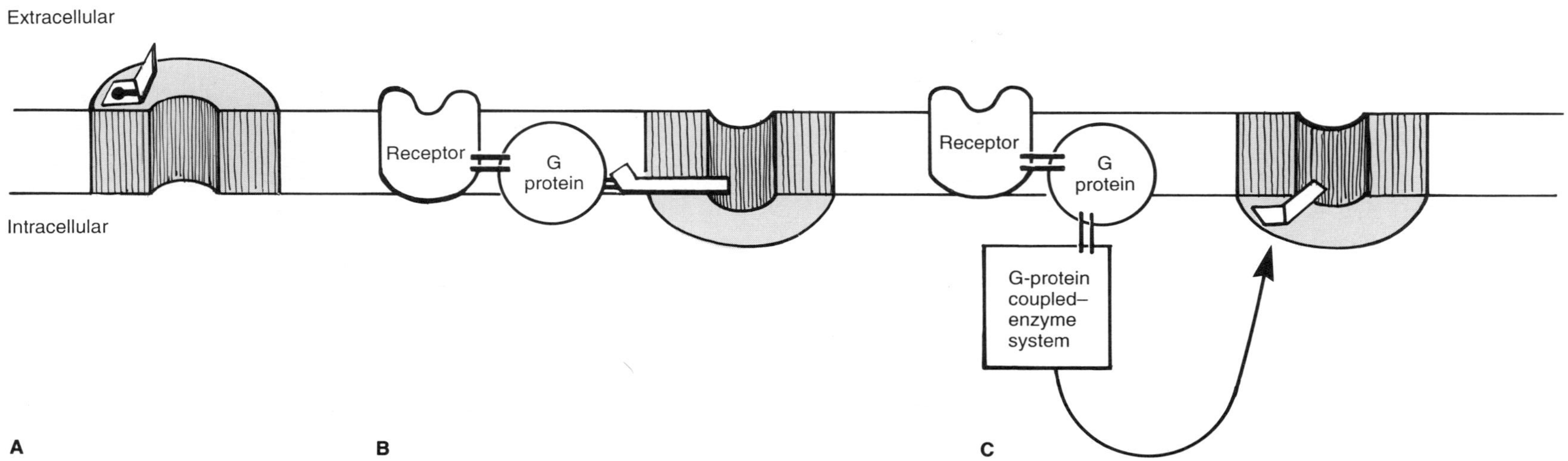

Figure 6-17

Postsynaptic receptors to neurotransmitter can be coupled to channels or other synaptic actions in any of three ways, as shown in a section through the postsynaptic membrane. (**A**) Integral channels, like the end-plate channel, consist of single receptor-channel molecules that bind transmitter and form a channel for ion passage. (**B**) Some receptors are directly coupled to a neighboring channel through an interposed G protein or a membrane-bound intermediate. (**C**) Other receptors activate a G-protein–coupled enzyme system that produces intracellular second messengers capable of modulating spatially distinct membrane channels or other cellular events.

closing and unbinding of the transmitter are tightly related processes. Many of these receptor-channel molecules have a similar structure (and significant homology) to the nicotinic ACh receptor at the NMJ, with subunits characterized by four membrane-spanning regions. This is the style found at synapses in fast pathways of the nervous system, where the information transmitted requires immediate attention or action, and the style producing the prototype EPSP and IPSP shown in Figure 6-16. Because binding and opening are directly linked on the molecular level, release and removal of the appropriate neurotransmitter tightly controls duration of the postsynaptic action. The synaptic action can be targeted to specific areas of the postsynaptic cell, allowing different actions to be spatially segregated by controlling the distribution of particular receptor-channel molecules.

This spatial specificity is shared by the second style for receptor design, in which the receptor and channel molecules are distinct but adjacent. The two molecules are coupled by an interposed G protein, possibly also using some hydrophobic molecule as a second messenger for the short distance within the membrane between the two. What is gained in this case is a much larger number of possible actions produced by the neurotransmitter, because many channel types can be coupled to almost any type of receptor by G-protein intermediates.

It is sometimes difficult to differentiate this case from the third scheme for receptor design, in which neurotransmitter binding also activates a G protein. In this case, the G protein is coupled to one of three enzyme systems that generate intracellular or transcellular second messengers. The advantage is that almost any intracellular process can be activated or modulated by any neurotransmitter. As detailed in Chapter 7, G-protein–coupled re-

ceptors differ in structure from the directly gated channels. For example, they are characterized by seven, rather than four, hydrophobic membrane-spanning domains.

Many intracellular processes are controlled through one or more of three *G-protein–coupled enzyme systems*, each of which activates one or more second messenger systems (Fig. 6-18). The first is *adenylate cyclase*, which can be activated or inhibited. Resulting changes in cyclic AMP affect cyclic AMP–dependent protein kinase, which can modify many cell proteins, including ion channels, by phosphorylation. The second activatable system is *phospholipase C*, generating diacylglycerol, which activates protein kinase C and inositol triphosphate (IP_3), which increases local intracellular calcium concentrations by releasing Ca^{2+} from intracellular stores. Protein kinase C, like cyclic AMP–dependent protein kinase, modulates many channels. The calcium concentration increased by IP_3 can control many Ca^{2+}-regulated events, including Ca^{2+}-activated potassium channels. The third system uses *phospholipase A_2*, which produces arachidonic acid and its metabolites, including prostaglandins and lipoxygenase products. In this third case, some of the metabolites may leave the cell, providing an additional chemical signal to other neurons in the region.

Many of the actions produced by G proteins modulate, rather than activate, channels; some go beyond channels to alter other features of cell biology. Unlike the directly gated channels, G-protein–coupled channels can be closed or opened as a result of neurotransmitter binding. A postsynaptic action that transiently closes a normally open inhibitory channel produces transient excitation. Neurotransmitters that transiently close an open excitatory channel produce inhibition. Phosphorylation of a channel molecule is an important result of many of these

Postsynaptic G-Proteins and Second Messengers

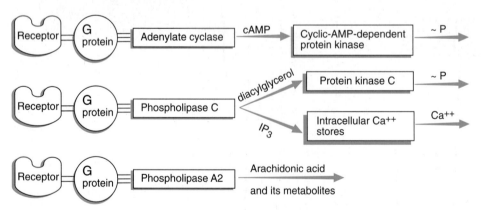

Figure 6-18

Postsynaptic G proteins can activate any of three enzyme systems, giving rise to a variety of second messengers. Adenylate cyclase activation and consequent production of cyclic AMP (cAMP) can activate cAMP-dependent protein kinase, capable of phosphorylating (~P) ion channels and other structures. Phospholipase C splits phosphatidylinositol diphosphate to produce diacylglycerol and inositol triphosphate (IP_3). Diacylglycerol, by activating protein kinase C, again leads to phosphorylation. IP_3 releases CA^{2+} from intracellular stores, regulating a variety of Ca^{2+}-dependent processes. A different G-protein–coupled phospholipase, phospholipase A_2, produces arachidonic acid and several metabolites.

G-protein–coupled steps. Phosphorylation can change the number of channels by activating dormant channels or suppressing active ones, or it can change the behavior of active channels by altering their gating. With this versatility and complexity, these systems lose the ability to respond in milliseconds that the direct channels had, and G-protein–coupled channels and other processes are not used for fast synaptic transmission in direct pathways but are restricted to modulatory systems and others where a sustained response is more important than an immediate one. Many G-protein–coupled receptors are known, and the number identified is increasing. Phosphorylation within the postsynaptic cell can also be controlled by another distinct family of receptors that activate tyrosine kinases.

Although it is convenient to think of channels as being voltage dependent or neurotransmitter dependent, some are both. Some channels, especially potassium channels, which are modulated by synaptically triggered G-protein–linked intracellular processes, are themselves voltage dependent. These channels are activated by voltage and modulated by transmitter. An important interplay between voltage and transmitter is synaptic regulation of the effective number of voltage-dependent channels, with the synapse regulating phosphorylation of a channel protein from an active to an inactive state. A subsequent section of this chapter presents another example of a channel, the N-methyl-D-aspartate (NMDA) receptor channel, which depends in a different way on transmitter and membrane potential.

Amino acid neurotransmitters produce fast PSPs

In the CNS, fast synaptic transmission is mediated primarily by one of three amino acids, each of which activates direct receptor-channel molecules. Fast excitatory synaptic transmission within the nervous system is mediated by *glutamate*, which opens cation-selective channels for only a few milliseconds, producing transient and focused depolarization and excitation. Each action potential in a glutaminergic presynaptic neuron releases a small amount of glutamate. Binding of the glutamate to an integral receptor-channel molecule produces the brief depolarizing EPSP.

Just as important to the nervous system as excitation is inhibition, and many neurons throughout the brain and spinal cord inhibit the cells they synapse on. Fast inhibitory synaptic transmission involves the amino acid neurotransmitters *glycine* and *GABA*. Each opens directly gated chloride channels, producing IPSPs that inhibit the postsynaptic cell, usually by hyperpolarization. Other channels also are activated by these transmitters.

What differentiates excitatory from inhibitory channels? The selectivity filter within the EPSP channel is fairly wide and can admit Na^+ and K^+, but it has negative fixed charges to exclude anions such as Cl^-. Because the channel is roughly equally selective for Na^+ and K^+, the reversal potential is roughly halfway between E_{Na} and E_K, or around -10 mV, resembling the EPP channel at the NMJ.

Because -10 mV is more depolarized than the threshold value of about -50 mV, the EPSP is excitatory. The channel responsible for the IPSP is more narrow and may have fixed charges, so that it is most permeable to Cl^- ions. Because Cl^- has a Nernst potential below the threshold value, opening these channels tends to keep the membrane potential away from the threshold. Consequently, the IPSP is inhibitory.

Ultrastructural distinctions between excitatory and inhibitory central synapses, such as those described in Chapter 3 characterizing Gray type I and II junctions, probably reflect several differences. These may include different structures for release and processing of glutamate or GABA and different postsynaptic receptors and other structures specialized to respond to the postsynaptic currents or voltage change.

Figure 6-19 lists sample direct receptor-channel molecules; most excite by opening cation channels or inhibit by opening chloride channels. There are receptor-channel molecules for amino acids, ACh, and a few other neurotransmitters.

Amines and peptides are slower neuromodulators

Just as some neurotransmitters, including the amino acids, are used at fast excitatory or inhibitory synapses, others are used primarily at modulatory or slower synapses, usually involving G-protein coupling. One class of these includes the catecholamines and indolamines: dopamine, epinephrine and norepinephrine, histamine, and serotonin. Each has multiple receptors and multiple actions. As a result of particular G-protein–coupled actions, specific ion conductances can be selectively increased or decreased. Perhaps the largest class of neurotransmitters is the peptides. The actions of this group are also slow or neuromodulatory, mediated through second messengers. Peptides are difficult to summarize. There are many of them and more are being discovered. They are packaged and released differently, and it is often difficult to tell what they do. Many nonneural cells also have peptide receptors.

Integral Receptor-Channel Molecules		
Gated by:	*Selective for:*	*Role or receptor:*
Glutamate	cations	CNS EPSP
Glycine	Cl^-	CNS IPSP
GABA	Cl^-	$GABA_A$ IPSP
ACh	cations	Nicotinic
Glutamate	Ca^{++}, Na^+, K^+	NMDA
Serotonin	cations	$5\text{-}HT_{M \text{ (or 3)}}$

Figure 6-19
A selection of integral receptor-channel molecules directly gated by binding of a neurotransmitter or agonist. The channels are selective for specific ions. The channel molecules determine postsynaptic action or the receptor subtype. *ACh,* acetylcholine; *GABA,* γ-aminobutyric acid; *NMDA,* N-methyl-D-aspartate; *5-HT,* 5-hydroxytryptamine or serotonin.

Unlike conventional neurotransmitters such as ACh and the amino acids, which are packaged into small, clear-cored vesicles at synaptic terminals, the dense-cored vesicles containing peptide neurotransmitters are made and filled in the soma and conveyed by axonal transport to terminals. Peptides may be released more diffusely than conventional transmitters from presynaptic endings but not at localized active zones. Peptide vesicles may be released by Ca^{2+} entering presynaptic terminals through different Ca^{2+} channels, or they may require a more widespread increase in intraterminal Ca^{2+}. It appears as if many individual presynaptic nerve cells are capable of assembling and co-releasing peptides along with a more conventional neurotransmitter. Differences in the release mechanisms of conventional or peptide transmitters suggest that different patterns of activity in a single neuron may preferentially release different amounts of the distinct transmitters. The release of monoamine neurotransmitters shares features of amino acid and peptide release.

Synaptic Integration

☐ Synaptic potentials summate

The binding of a neural transmitter substance can directly or indirectly influence many cellular properties and subcellular events in the postsynaptic neuron. The most significant effect produced by many transmitter-receptor complexes is a change in membrane potential, producing an EPSP or IPSP. At the NMJ, individual EPPs uniformly and reliably depolarize the postjunctional muscle fiber to the threshold value. The role of synaptic potentials in the CNS is more complex and requires examination of the changes they produce in the postsynaptic membrane potential and of the effect produced by this voltage change on the activity of the postsynaptic cell.

Ohm's law underlies synaptic integration

To understand what excitatory or inhibitory changes in the postsynaptic membrane potential are produced by opening or closing synaptic channels in the postsynaptic membrane, researchers make use of Ohm's law, the familiar equation that relates current, voltage, and resistance or its inverse, conductance. To calculate the size and direction of any membrane potential change (ΔV_m produced by a synapse, Ohm's law is applied in two steps, as summarized in Figure 6-20. This two-step process can also calculate the ΔV_m resulting from multiple synaptic inputs to a cell, balancing excitation and inhibition and the effect each synapse has on the cell's response to the other synapses.

The first step in applying Ohm's law is performed at each active synapse and is identical to the calculation of the amount of current that flowed through end-plate channels. For each synapse, Ohm's law is applied to reveal the magnitude and direction of the current that flows

---

Ohm's Law Calculates PSP Currents and Voltage Changes

First:

$$I_{EPSP} = G_{EPSP}(V_m - E_{EPSP}); I_{IPSP} = G_{IPSP}(V_m - E_{IPSP})$$
$$E_{EPSP} \cong -10\,mV; \quad E_{IPSP} \cong -70\,mV$$

. . . gives synaptic current magnitude and duration for *each* synapse.

Second:

$$\Delta V_m = -\frac{I_{PSP1} + I_{PSP2} + I_{PSP3} + I_{PSP4} + \cdots}{G_{non-syn} + G_{PSP1} + G_{PSP2} + G_{PSP3} + G_{PSP4} + \cdots}$$

. . . gives the change in membrane potenial due to *all* synaptic currents.

Figure 6-20
Two forms of Ohm's law are used to calculate synaptic currents and the membrane potential change produced by them. In the first form, for each active synapse, the synaptic current (I) is the product of synaptic conductance (G) and electrochemical driving force, which is the difference between membrane potential (V_m) and the synaptic reversal potential (E_{EPSP} or E_{IPSP}). In the second form, the membrane potential change (ΔV_m) produced in the postsynaptic cell by all synaptic currents depends directly on the sum of the synaptic currents and inversely on the sum of all conductances, synaptic and nonsynaptic. The minus sign is needed because excitatory synaptic currents by convention are negative, but produce depolarization.

---

across the postsynaptic cell membrane through open synaptic channels at that synapse.

For excitatory synapses: $I_{EPSP} = G_{EPSP}(V_m - E_{EPSP})$.
For inhibitory synapses: $I_{IPSP} = G_{IPSP}(V_m - E_{IPSP})$.

For many excitatory synapses, E_{EPSP} (the reversal potential) is about -10 or 0 mV, but for IPSPs, the reversal potential E_{IPSP} is perhaps -70 or -75 mV. The magnitude of the synaptic current is determined by two factors, just as the magnitude of the current at the end plate was. The first factor is the electrochemical driving force on the ion or ions, expressed as the difference between the reversal potential and V_m. The second factor is the size of the synaptic conductance change; a larger G_{PSP} means a larger synaptic current. The conductance of individual synaptic channels is largely fixed for each channel type, determined by the structure of the channel molecule. The size of the synaptic conductance change at a given synapse is proportional to the number of postsynaptic channels opened by neurotransmitter during a PSP and depends on the number of subsynaptic channels and the amount of neurotransmitter released. For conventional synapses, in which neurotransmitter binding produces an increase in conductance, G_{PSP} is zero until the transmitter is released and binds to postsynaptic receptors, at which time G_{PSP} rises to a peak value, decaying with an exponential time course to zero within several milliseconds. In contrast, some G-protein–coupled synaptic conductances are transiently reduced, rather than increased, by neurotransmitter binding.

The *direction of current flow* through individual synaptic channels is determined by whether V_m (often approximately −60 mV at rest) is above or below the reversal potential for the ions flowing through that channel. Inward currents are defined as negative and outward currents as positive to be compatible with action potential conventions. For an EPSP with $E_{EPSP} = 0$ mV, $(V_m - E_{EPSP}) = -60$ mV. When G_{EPSP} is increased by neurotransmitter binding, the resulting synaptic I_{EPSP} is an inward current that depolarizes or excites the cell. For the IPSP, $E_{IPSP} = -70$ mV, $(V_m - E_{IPSP})$ is +10 mV, and I_{IPSP} is an outward synaptic current that is capable of hyperpolarizing or inhibiting the cell. In either case, G_{PSP} is always positive or zero and is not needed to determine the direction of current flow but is required to determine its magnitude. The synaptic current, produced by and calculated for each active synapse, is largely independent of other inputs to the cell. However, the change in membrane potential produced in the cell by synaptic input depends on each of the synaptic currents and on the the total membrane conductance of the postsynaptic cell, and this conductance depends on synaptic input.

The second use of Ohm's law applies to the entire postsynaptic neuron, rather than to each individual synapse:

$$\Delta V_m = -\frac{I_{PSP1} + I_{PSP2} + I_{PSP3} + I_{PSP4} +}{G_{non\text{-}syn} + G_{SPS1} + G_{PSP2} + G_{PSP3} + G_{PSP4} + \ldots}$$

This equation reveals the direction and magnitude of the voltage change, ΔV_m, produced in the cell by the synaptic current. It can be used if there is one active synapse or one thousand. The *direction of voltage change* produced by a single PSP depends on the sign of the synaptic current, but the net voltage change produced by many synapses depends on the balance between excitation and inhibition. Because excitatory and inhibitory currents have opposite signs, excitation or inhibition must predominate, or the sum of the synaptic currents will be zero, and no voltage change will result. If excitatory, inward currents exceed the inhibitory, outward currents, the ΔV_m is depolarizing. If inhibitory synaptic currents dominate, the ΔV_m hyperpolarizes the postsynaptic neuron. Because the values of G_{PSP} are always positive, the denominator of the equation does not affect the sign of ΔV_m.

The *magnitude of* ΔV_m depends on two factors: the total synaptic current and the overall membrane conductance. The size of the synaptic voltage change is proportional to the sum of the synaptic currents given by the numerator of the second form of Ohm's law. The greater the amount by which excitatory synaptic current exceeds inhibitory or inhibitory exceeds excitatory, the greater is the resulting depolarization or hyperpolarization. The magnitude of ΔV_m is inversely proportional to the overall membrane conductance to ions. The denominator of the equation is the sum of all cell conductances, synaptic and nonsynaptic. Each active synapse transiently contributes an additional G_{PSP} to the denominator during the time that channels in the postsynaptic membrane are opened by a transmitter. Although the denominator of the equa-

tion does not affect the sign of ΔV_m, it does affect its amplitude. The larger the sum of the membrane conductances, the smaller is the voltage change, ΔV_m, produced by a given sum of synaptic currents.

To see whether these applications of Ohm's law predict the behavior of individual EPSPs and IPSPs, each of these synaptic potentials can be examined under conditions of changing membrane potential and changing driving force. Figure 6-21 shows the results of this experiment, which uses the same circuit shown in Figure 6-16. PSPs produced by presynaptic action potentials in excitatory and inhibitory neurons are recorded in a postsynaptic neuron whose membrane potential can be changed at will. For the EPSP, the first application of Ohm's law predicts that the size of the synaptic current should be reduced by changing V_m in such a way as to reduce the electrochemical driving force $(V_m - E_{EPSP})$. It further predicts that depolarization past the reversal potential, E_{EPSP}, should reverse the direction of synaptic current flow. The second application of Ohm's law predicts that the voltage change recorded (i.e., the EPSP) should be proportional to the synaptic current. The behavior of the EPSP, which resembles that of the EPP, conforms to the predictions of Ohm's law. The EPSP decreases in amplitude as the cell is depolarized, until E_{EPSP} is reached about halfway between E_{Na} and E_K, at which point no potential change is seen. Further depolarization produces an inverted EPSP. The linearity of the response confirms that the synaptic conductance change producing the EPSP depends only on transmitter, not on membrane potential.

The behavior of the IPSP conforms to Ohm's law. The value of E_{IPSP} is about −70 mV for the Cl⁻-dependent inhibitory synapse. The conductance change, G_{IPSP}, is independent of V_m, but varying V_m and thereby the driving force $(V_m - E_{IPSP})$ can change the amplitude and even the direction of the synaptic current, as predicted by the first application of Ohm's law. The second application of Ohm's law predicts the changes to V_m produced by these synaptic currents. Figure 6-21 shows that, as the postsynaptic cell is depolarized and the driving force is increased as V_m moves away from the IPSP reversal potential, the amplitudes of the synaptic current and of the IPSP are increased. Hyperpolarization from rest to about −70 mV reduces the driving force to zero, eliminating any synaptic current or voltage change from the IPSP. Further hyperpolarization inverts the IPSP to a depolarizing potential.

The IPSP is inhibitory because E_{IPSP} is below the threshold value. The second form of Ohm's law also shows that the IPSP is inhibitory for an additional reason. The IPSP, like all directly gated synapses, opens synaptic current channels across the postsynaptic cell membrane. These open channels, by increasing the total membrane conductance of the cell, increase the value of the denominator in the second form of Ohm's law. This larger denominator means that the voltage change produced by a given synaptic current, such as an excitatory current, is smaller than it otherwise would be without the IPSP. The IPSP shunts the excitation the cell receives, in addition to its hyperpolarization of the membrane, and this may be a significant contribution to synaptic integration at some

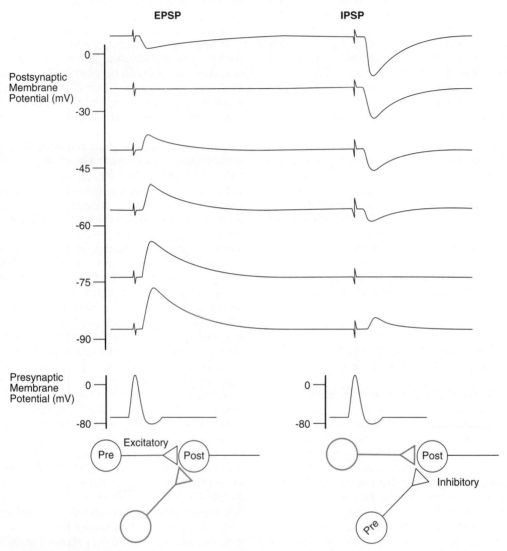

EPSP and IPSP Amplitude Depend On Electrochemical Driving Force

Figure 6-21
The amplitude and direction of the excitatory postsynaptic potentials (EPSP) and inhibitory postsynaptic potentials (IPSP) depend on the electrochemical driving force for ion movement. Although synaptic conductance changes are largely independent of postsynaptic membrane potential, synaptic currents and the ΔV_m they produce depend on the electrochemical driving force, which depends on the postsynaptic membrane potential. For this experiement, $E_{EPSP} = -15$ mV and $E_{IPSP} = -72$ mV. A PSP recorded at membrane potentials more depolarized than its reversal potential is hyperpolarizing, but a PSP recorded at membrane potentials more hyperpolarized than its reversal potential is depolarizing. The experimental method for recording postsynaptic and presynaptic membrane potentials as a function of time is the same as in Figure 6-16, except that PSPs are recorded arising from each of six different levels of postsynaptic membrane potential. Not shown are the modifications needed to suppress action potential firing and other conductances in the postsynaptic cell.

synapses. Each EPSP also shunts the effects of other inputs (i.e., reduces their effectiveness), making the EPSP somewhat less excitatory than it would otherwise be without shunting.

Multiple inputs produce spatial summation

Perhaps the most basic function of a neuron is as a machine for deciding *yes* or *no*. Membrane potentials more hyperpolarized than the threshold do not generate action potentials, but depolarizations past the threshold value do produce this electrochemical event. By their effect on the membrane potential of typical postsynaptic cells, summed synaptic potentials influence the activity of neurons and the message they convey. In some neurons, synaptic input is superseded by or combined with conductances producing spontaneous firing, a process called autoactivity.

In an idealized cell, the change in the postsynaptic membrane potential (ΔV_m) is proportional to the algebraic sum of the currents produced by synaptic inputs. Action potential generation in the postsynaptic cell depends on ΔV_m being depolarizing and, if so, large enough to depolarize the membrane to the threshold value. Because EPSPs in CNS neurons are subthreshold and because most are very small, the cell can reach the threshold only by summing synaptic currents. Summation can take one of two forms: spatial summation and temporal summation (Fig. 6-22).

Summing inputs from several different and spatially distinct presynaptic neurons is called *spatial summation*. Figure 6-22*A* illustrates an idealized cell that receives several synaptic inputs from different presynaptic cells. The EPSP produced by any one of these inputs is small, as-

sumed able to depolarize the cell only a fraction of the difference between the resting potential and threshold. A single action potential in only one presynaptic cell produces only a small, subthreshold EPSP, and the postsynaptic cell does not generate an action potential. However, if several presynaptic neurons fire action potentials at the same time, synaptic currents flow simultaneously at several postsynaptic sites. These currents sum, and as predicted by the second form of Ohm's law, the membrane potential will reach threshold, and an action potential will be generated by the postsynaptic cell. The cell can also sum inhibitory inputs, and it sums these algebraically with any excitatory inputs it is receiving. If several EPSPs sum so that ΔV_m is just large enough to reach the threshold level, a simultaneous IPSP can provide enough hyperpolarizing current and enough shunting to keep

Spatial and Temporal Summation

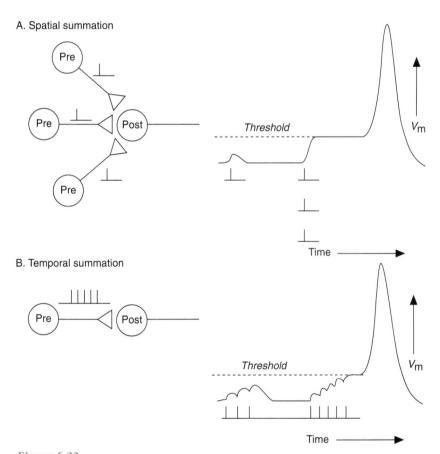

Figure 6-22
Spatial and temporal summation of individual subthreshold postsynaptic potentials (PSPs.) (**A**) Spatial summation of excitatory PSPs from three presynaptic cells (**Pre**) innervating a common postsynaptic neuron (**Post**). The change in postsynaptic membrane potential V_m in response to presynaptic action potentials is shown on the right. An action potential in only one of the presynaptic cells produces a subthreshold EPSP, but simultaneous action potentials in each of the three presynaptic cells produce simultaneous EPSPs that sum to reach the threshold for action potential generation. (**B**) Temporal summation of excitatory PSPs produced by a single repetitively firing presynaptic neuron. Slow presynaptic firing produces PSPs that fail to reach a threshold, but faster firing produces overlapping PSPs that sum to the threshold. Duration of action potential and threshold transition shown expanded in **A** and **B**.

the membrane potential below the threshold, and no action potential is generated.

Rapidly activated inputs produce temporal summation

The other basic way in which a postsynaptic cell can sum synaptic input relies on the fact that PSPs last much longer than action potentials. Although action potential duration may last 1 or 2 milliseconds, PSPs in the CNS can last from 5 to 200 milliseconds or longer in some cases. The PSPs produced by a single, rapidly firing presynaptic neuron can themselves sum, a process known as *temporal summation* (see Fig. 6-22*B*). A single action potential in an excitatory presynaptic cell produces an EPSP in a postsynaptic neuron. Before the ΔV_m produced by the initial EPSP returns to zero, while the membrane potential of the postsynaptic cell is still slightly depolarized, the presynaptic neuron can fire another action potential. The EPSP this second action potential produces can sum with the EPSP produced by the previous action potential. If action potentials from this one presynaptic neuron are produced at a rapid rate, the EPSPs they produce can sum to the threshold level, producing a postsynaptic action potential (see Fig. 6-22*B*).

☐ **Information is transmitted across synapses**

Neuronal structure influences synaptic integration

Each neuron is continually summing information spatially from many inputs and temporally from repetitively firing inputs. Each neuron is further weighing these excitatory and inhibitory influences in its decision about whether to fire an action potential. The place the decision is made is not diffusely spread throughout the cell; it is concentrated in one specialized region of the neuron: the *trigger zone*, located in the *initial segment* or *axon hillock*, just where the axon emerges from the cell body. The specialization arises from the fact that the threshold is lower at the trigger zone than at any other place of the neuron, perhaps -55 mV in the initial segment but -40 mV elsewhere in the neuron. Because it is much easier to depolarize the trigger zone than anywhere else in the neuron, action potentials are always generated first at the initial segment and spread from there to other parts of the cell.

This localization has two advantages. First, there is only one specific location in each cell that decides if the cell will generate an action potential or not; the choice is made by a decision maker instead of a committee. Second, locating this trigger zone at the initial segment allows the action potential to propagate from the trigger zone down the axon without being shunted by synaptic or nonsynaptic conductances in the cell body, which could slow down the action potential or change its amplitude. In most cells, the action potential can also propagate back into the cell body and often into the dendrites,

but the message the action potential is carrying is probably of greatest significance for the terminals at the end of the axon.

The summation of synaptic inputs and the decision about whether to produce an action potential are far more complicated in real CNS neurons than in idealized cells. The principles are the same, but their application can be quite complex. Many CNS cells receive thousands of synaptic endings from other neurons, and the cell is continually summing a large number of excitatory and inhibitory synaptic currents, some of which are quite small. Another complexity arises because the synaptic currents produced by different inputs can be of different amplitudes, and some inputs on a given cell can exert a greater influence than others. From the neuronal point of view, some inputs are more important than others. Plastic processes can also change the relative importance of synapses.

The relative importance of different inputs to a neuron also depends on the precise location of the synapse on the postsynaptic cell. The second application of Ohm's law assumed all synapses were located on one area of the postsynaptic cell, but this is an oversimplification. If synapses are far out on a dendrite, synaptic currents may be attenuated electrotonically by distance before reaching the trigger zone. As a general rule, inputs closer to the soma and to the trigger zone are given greater weight by the cell than those farther away. However, the complexity of the average nerve cell complicates this rule. One complication is that the size of the voltage change produced by a given synaptic conductance may depend on how far away it is and on the geometry of the cell process on which it is located. Longer or shorter, thinner or thicker dendrites or spines of dendrites can attenuate or enhance contributions of particular inputs. Another complication is that some cells with extensive dendritic trees may have the ability to generate action potentials in the dendrites, and these action potentials may propagate into the cell body or depolarize it by passive current spread.

Inhibition is equally important as excitation for neuronal integration. The most striking demonstration of the role of inhibition in the nervous system is the response to agents that interfere with transmission at inhibitory synapses. Poisoning with strychnine blocks inhibitory synapses that use glycine as their neurotransmitter. This action is often fatal, with death resulting from bone-breaking convulsions produced by an excess of excitation. Similar destruction of the balance between excitation and inhibition can be produced by bicuculline or picrotoxin, which block inhibitory synapses that use GABA.

Every decoding is another encoding

The mechanisms of synaptic transmission allow a cell to sum synaptic input and to decide, on the basis of the relative strength of input excitation and inhibition, whether to generate an action potential. However, messages in the nervous system rarely consist of single action potentials. Instead, information is continual and complex, usually consisting of trains or groups of action potentials. Our views of the mechanisms of information transmission, including synaptic transmission, must be expanded to appreciate how information is transmitted using these action potential trains.

Because individual action potentials are identical or similar, the information carried in a train must be conveyed by features of these trains such as the frequency or rate or pattern of action potentials. Although action potentials themselves are not propagated across synapses, the information contained in the action potential train must be transmitted from one cell to another. To carry this information, the message conveyed must be *decoded* by the presynaptic terminal and *encoded* by the postsynaptic soma or dendrite. Some mechanisms of coding were explained previously in the discussion of temporal summation. Increasing the frequency of action potentials in a neuronal axon produces more frequent release of neurotransmitter, and the resulting, closely spaced PSPs sum to produce a greater postsynaptic voltage change (see Fig. 6-22B). Assuming the synapse is excitatory, this translates as a greater depolarization of the postsynaptic soma. The message of faster or slower action potential firing is decoded as a greater or lesser postsynaptic depolarization.

To transmit the message onward, the postsynaptic neuron must encode it again as an action potential train. Ideally, the greater or lesser postsynaptic depolarization should be encoded by the generation in the postsynaptic neuron of trains of faster or slower action potentials. The processes described here and in Chapter 5 show how suprathreshold depolarization leads to the generation of single action potentials. They can also be used to reveal how prolonged suprathreshold depolarizations of various amplitudes above the threshold can be encoded as trains of action potentials of various frequencies.

The amount by which the postsynaptic cell body is depolarized above the threshold value determines the frequency at which it generates action potentials (Fig. 6-23). When summed PSPs initially depolarize the postsynaptic cell past the threshold level, an action potential is produced. If the summed PSPs result from a train of impulses in a presynaptic cell, the summed PSPs are likely to be prolonged well beyond the refractory period for action potential generation of the postsynaptic cell. During the absolute refractory period immediately after the initial action potential, the postsynaptic cell cannot fire again, no matter how strong the depolarization. During the relative refractory period, the threshold for firing of the postsynaptic cell varies. Initially very high, it declines to

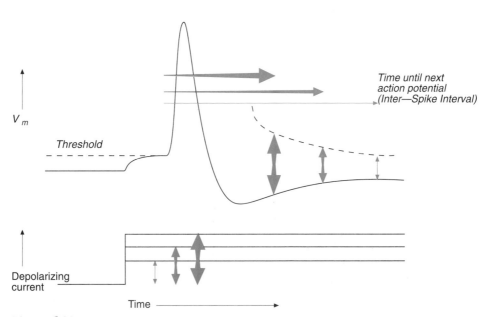

**Postsynaptic Depolarization Determines
Postsynaptic Firing Rate**

Figure 6-23
The magnitude of prolonged postsynaptic depolarization determines the rate of generation of postsynaptic action potentials; larger depolarizations produce faster firing and shorter interspike intervals. Three suprathreshold depolarizing currents are shown in the lower traces; their magnitudes are indicated by the three arrows. The upper trace shows the initial action potential generated by the suprathreshold currents and the varying threshold after the action potential (*dashed line*). The vertical arrows between the action potential and threshold indicate when during the relative refractory period each of the three suprathreshold currents is capable of generating the next action potential. The intervals between first and subsequent action potentials are shown by the horizontal arrows for each of the depolarizing currents.

Its normal value with time (see Fig. 6-23). Because the threshold varies, a greater synaptic current is able to depolarize the neuron to the threshold value earlier in the relative refractory period, but a smaller synaptic current is unable to do so until later in the refractory period. The greater the summed synaptic current, the sooner the next action potential is produced, and the shorter is the *interspike interval*. If the depolarization is maintained, a train of successive action potentials is generated by the postsynaptic neuron. The greater the amplitude of the prolonged synaptic depolarization giving rise to this train, the shorter are the interspike intervals and the greater is the frequency of firing of the postsynaptic neuron.

The threshold value depends partially on time- and voltage-dependent properties of the Hodgkin-Huxley Na^+ and K^+ channels and partially on other conductances. The major additional conductance is a voltage-dependent K^+ channel called the *A-current* (I_A) channel. Another, more colorful name for the same channel is *shaker*, so called because some mutations to the genes encoding this protein in *Drosophila* produce fruit flies that shake from repetitive muscle contractions. The ionic selectivity of this channel is similar to that of the Hodgkin-Huxley delayed-rectifier K^+ channel, but it differs in its gating and sensitivity to drugs and other agents. The voltage-dependent gating is such that moderate depolarization from below rest activates and then quickly inactivates the I_A channel, slowing the appearance of the next action potential to a prolonged depolarizing stimulus. Interspike intervals are influenced by other K^+ channels, including some activated by Ca^{2+} instead of voltage.

The information processing capability of a synapse depends on whether summed postsynaptic currents produce a depolarization above the threshold and how much above the threshold value it is. Combining this scheme for encoding with the decoding described previously, it is possible to see how messages are transformed as they are transmitted across synaptic relays. Figure 6-24 shows how such a message, encoded as a train of action potentials in the presynaptic neuron shown on the right, is successively transformed at the excitatory synapse and retransmitted postsynaptically. It is useful to assume that other inputs or cell properties place the postsynaptic membrane potential near the threshold. The initial part of the message is carried by rapid firing of the presynaptic neuron. Because the EPSPs produced in

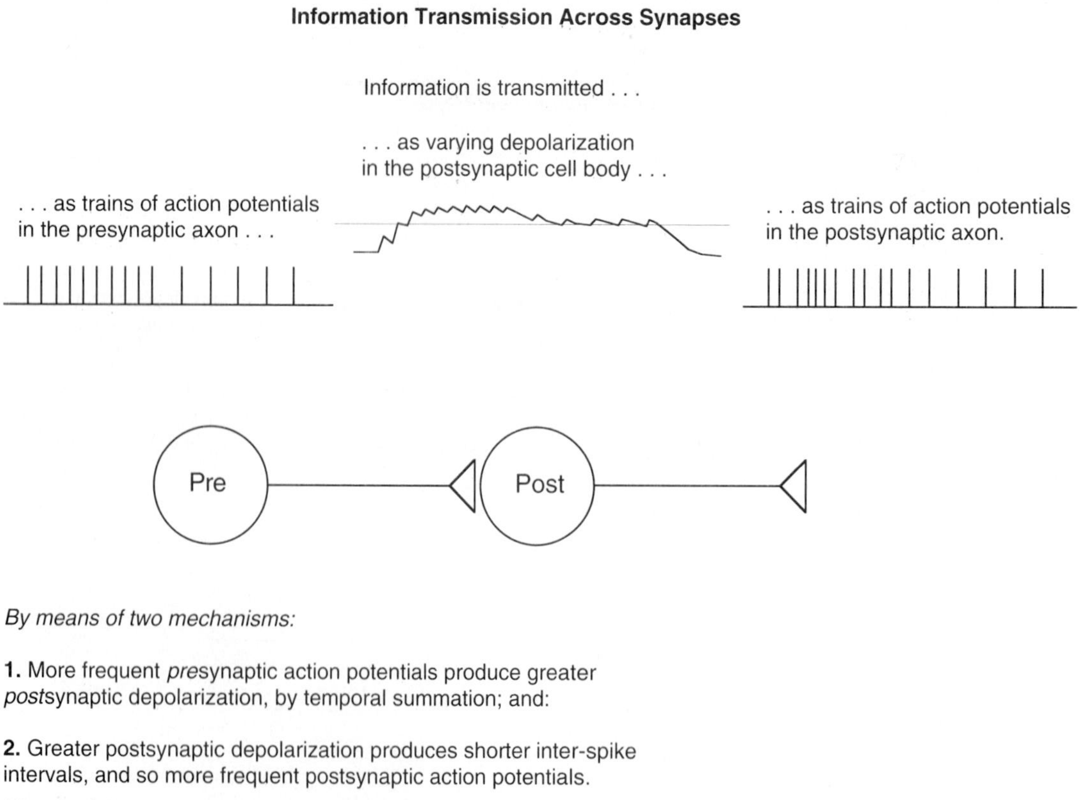

Information Transmission Across Synapses

Information is transmitted . . .

. . . as varying depolarization
in the postsynaptic cell body . . .

. . . as trains of action potentials
in the presynaptic axon . . .

. . . as trains of action potentials
in the postsynaptic axon.

By means of two mechanisms:

1. More frequent *pre*synaptic action potentials produce greater *post*synaptic depolarization, by temporal summation; and:

2. Greater postsynaptic depolarization produces shorter inter-spike intervals, and so more frequent postsynaptic action potentials.

Figure 6-24
Information is transmitted across synapses. Information, transmitted as a train or pattern of action potentials in a presynaptic axon, is decoded as varying postsynaptic depolarization and encoded as a train of action potentials in the postsynaptic axon. Each of the three traces indicates a neuronal message at the same time in one of these three stages. Presynaptic and postsynaptic trains of action potentials are shown schematically. The postsynaptic train is similar but not identical to the presynaptic train. The postsynaptic cell body trace shows the voltage change produced by summed postsynaptic potentials (PSPs) in relation to the threshold voltage level (*dotted line*); for clarity, the action potentials that these PSPs would produce have been suppressed for this trace.

the postsynaptic cell are closely spaced, they sum to produce a depolarization of the postsynaptic soma that is well above the threshold level. Resulting action potentials are generated in the postsynaptic cell with short interspike intervals, producing rapid firing in the postsynaptic cell. Later in the train, the second half of the message is encoded by slower firing in the presynaptic axon. The more widely spaced EPSPs that result sum to produce a postsynaptic depolarization that is above the threshold but smaller than that produced by the first half of the train. This lesser depolarization must wait until later in each relative refractory period to give rise to the next action potential. As a result, the action potentials generated by the postsynaptic cell have longer interspike intervals and a lower frequency. Although general features of the message have been transmitted across this hypothetical synapse, the information has been processed; the action potential train generated in the postsynaptic neuron is not an exact copy of the presynaptic train.

As information moves through the nervous system, it is successively encoded as frequency or pattern of action potentials in axons, then as amplitude of depolarization by PSPs at synapses, then as a pattern of action potentials again, and so on. The example of Figure 6-24 is highly idealized. It ignores the influence of inhibitory PSPs, which reduce the amount of depolarization produced by excitatory synapses and lower the resulting postsynaptic rate of firing. At most synapses, individual PSPs are smaller than those shown, and there is massive convergence of excitatory and inhibitory inputs, resulting in complex combinations of temporal and spatial summation. In the CNS, many neurons fire slowly under normal conditions, somewhat like an idling engine. Synaptic inputs modulate this tonic level of firing, slowing or speeding it, depending on the balance between excitation and inhibition.

The information transmitted by a synapse depends on that synapse and on the anatomy of the *neural circuit*. Circuitry can send an excitatory message through an inhibitory synapse (Fig. 6-25). In the three-cell network, cell A inhibits cell B, which inhibits cell C. Cell B is tonically active, perhaps as the result of other excitatory input. Because cell B inhibits cell C, the constant activity of B has been keeping C silent. Turning on cell A inhibits B,

preventing it from firing. When the inhibition of B is removed, cell C fires. Cell A indirectly excites cell C by *disinhibition*.

The idealized view of a CNS neuron as an integrate-and-fire device is an oversimplification. Even subthreshold membrane potential changes can influence many cellular processes and functions, including transmitter release. The action potential is a message produced by many neurons, but it is not the only product of cellular processes or even the only message. Many neurons do not rely on action potentials for signaling; these local circuit neurons encode information by graded changes in membrane potential and transmitter release.

Synaptic Plasticity

The earlier parts of this chapter described the behavior of synapses. The topic of this part might be called the synapses of behavior, because it correlates the neurobiology of synapses with the behavior of the organism. One of the central questions in the study of neuroscience is about how the nervous system changes to allow everyday learning to occur or for patterns of behavior to be modified. Although it would be possible to design a nervous system in which function would change over time in the adult by the development of new neurons and the disappearance of others, this rarely happens. Another plausible mechanism is the development of new synaptic connections between existing neurons and the concomitant removal of others. There is evidence for this process during postnatal maturation of the nervous system, as the organism learns for the first time to cope with the external world. Similar mechanisms are involved in the remodeling of the nervous system needed to compensate for injury that alters sensory input or motor expression.

Persuasive evidence from examinations of several different nervous system components suggests that one of the significant mechanisms for learning is change in the effectiveness of existing synapses. These changes may be produced purely by changes in the internal neurochemistry or neurophysiology of presynaptic or postsynaptic cell, or they may involve such neuroanatomic modifications as changes in the shape or number of postsynaptic

Disinhibition

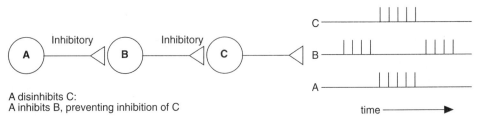

A disinhibits C:
A inhibits B, preventing inhibition of C

Figure 6-25
Disinhibition is an increase in neuronal activity produced by the suppression of an inhibitory influence. In the neural circuit shown, cell A inhibits cell B, which inhibits cell C. Activity in A, by inhibiting B, removes the inhibition of C, permitting C to fire. Cells B and C are assumed to receive excitatory input from other sources. Each of the three traces indicates activity in one of these three cells at the same time.

dendritic specializations or presynaptic terminals. Regardless of the mechanism, the consequences for neuronal integration are changes in synaptic strength. This *plasticity* of existing synapses can result in increases or decreases in the size or duration of the PSP produced by a given synapse, enhancing the effectiveness of a weak synapse or reducing the influence of a strong one. These effects could also be produced by changes in the excitability of a postsynaptic neuron, by changes in nonsynaptic conductances that affect the magnitude of the summed synaptic current, or by modulation of the ΔV_m needed to reach threshold potentials. The normal activity of the nervous system itself can produce functional plastic changes, resulting in profound and prolonged alterations of cellular and synaptic function. In some cases, these changes have been correlated with and found to be responsible for changes in behavior, including learning.

Because many of the plastic changes involve modulation of transmitter release, this section begins with the physiology of the release process. Examples follow, showing that synapses can be modified by their own actions or by the actions of other neurons. The last part of this section describes a synapse specialized for plasticity in mammalian hippocampus, a site for memory.

□ Transmitter release is variable

The quantal model underlies plasticity

Plastic modification of synaptic effectiveness—in the size of the EPSP or IPSP—can involve presynaptic or postsynaptic changes. Postsynaptic sensitivity to a transmitter can be altered by plastic modulation of the number of postsynaptic receptors or of their properties. For example, binding of the neurotransmitter can cause subsequent desensitization of postsynaptic receptors and consequent reduction in PSP amplitude. Second messenger pathways, controlled by synaptic, hormonal, or other events, can phosphorylate postsynaptic receptors, similarly affecting PSP size. However, a fundamental mechanism for plastic change involves alteration of the amount of transmitter released by each presynaptic action potential. Understanding plastic modulation of this process requires examination of quantal transmitter release.

At central synapses and at the NMJ, transmitter is released quantally in packets, each containing a few thousand molecules. Plastic changes in synaptic effectiveness usually involve changes in the number of quanta released by each action potential. Changes in the number of transmitter molecules packed within each quantum are seen only in extreme circumstances such as fatigue induced by prolonged stimulation. The number of quanta per action potential is called the *quantal content* (m) and is equal to the product of two other values: the number of releasable quanta (n) and the probability that any one quantum will be released (p) is $m = np$.

Plastic changes can be mediated through changes in either factor and sometimes in both. Because the amplitudes of PSPs or EPPs depend most directly on m, it is usually difficult or impossible to ascribe a plastic change to one particular factor or the other.

At different synapses, the factors n and p can be related to different anatomic specializations. At the NMJ, n may be thought of as the number of specialized active zones within the motor nerve terminal and p as the probability that a quantum will be released from that active zone in the short interval after an action potential. At a typical central synapse, n is most often associated with the number of synaptic buttons forming the functional synapse between two neurons and p as the average probability that one of these buttons will release a quantum of transmitter. There is some evidence that an individual button can release only one quantum at a time. According to this view, the probability of release at each button is therefore 0 or 1, and averaging all of these probabilities across all n buttons yields the value of p that describes the probability of release for the synapse. However, some buttons may contain multiple active zones, and each of these may act as an independent button. This analysis assumes a plentiful supply of vesicles filled with transmitter and available for release, but there is evidence for semiactive storage of vesicles, requiring mobilization of each vesicle before it can combine with an active zone and be released. In this view, n depends on the number of active zones and on the supply of mobilized vesicles, called the *releasable store*.

Voltage and calcium can produce plastic changes

Transmitter release at the NMJ and central synapses does not directly involve the Hodgkin-Huxley action potential conductances. Release requires depolarization and extracellular Ca^{2+}, and the depolarization is needed to open voltage-dependent Ca^{2+} channels to permit Ca^{2+} to enter the presynaptic terminal. The dependence of the processes of release on voltage and Ca^{2+} makes them possible sites for plastic modulation. The sensitivity of release, its dependence on small changes in voltage and Ca^{2+}, makes these processes attractive sites for significant control of synaptic efficacy.

The mechanisms of transmitter release are highly voltage dependent. Under experimental conditions, the synapse can be bathed in tetrodotoxin and tetraethylammonium to block the Hodgkin-Huxley conductances. The presynaptic terminal can be artificially depolarized, producing voltage changes of controllable amplitude and duration. Although small depolarizations are ineffective, depolarizations larger than a few millivolts release transmitter, producing postsynaptic potentials that can be measured (Fig. 6-26). The larger the amount of depolarization, the larger is the PSP, indicating that more transmitter is being released. Over some ranges, very small changes in the amount of depolarization produce very large changes in PSP size. Similarly, changing the duration of the artificial depolarization can affect the amount

Control of Transmitter Release

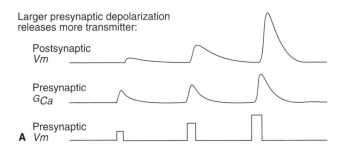

Larger presynaptic depolarization
releases more transmitter:

Postsynaptic
Vm

Presynaptic
GCa

Presynaptic
A *Vm*

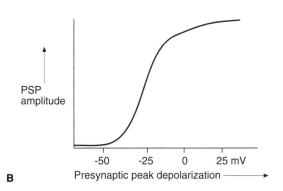

PSP
amplitude

-50   -25   0   25 mV
B   Presynaptic peak depolarization ⎯⎯⎯→

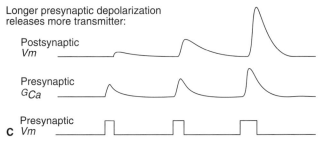

Longer presynaptic depolarization
releases more transmitter:

Postsynaptic
Vm

Presynaptic
GCa

Presynaptic
C *Vm*

Figure 6-26
Transmitter release depends on the magnitude and duration of
the depolarization in the presynaptic terminal. (**A**) Three different
amplitudes of presynaptic depolarization produce different
increases in presynaptic calcium conductance (G_{Ca}) and different
amounts of transmitter release, monitored by the size of the postsynaptic potential (PSP) change. (**B**) The curve plots the resulting
PSP amplitude for different presynaptic depolarizations. (**C**)
Three different durations of presynaptic depolarization produce
different presynaptic G_{Ca} and resulting postsynaptic potential
changes. Small changes in presynaptic V_m and G_{Ca} produce large
changes in PSP size.

of transmitter released. These findings suggest that functional plastic modulation may be produced by small
changes in voltage-dependent processes.

The voltage dependence reflects properties of voltage-
sensitive Ca^{2+} channels. Ca^{2+} influx into the terminal is
the required intermediate step between the depolarization of the terminal membrane and the release of
neurotransmitter. Ca^{2+} moves into the terminal down

its concentration gradient after depolarization opens
voltage-dependent gates across Ca^{2+} channels. Because
of the activation properties of these gates, the size of the
increase in the Ca^{2+} conductance (G_{Ca}) is determined by
several factors, including the magnitude and duration of
the voltage change during the action potential, the membrane potential preceding arrival of the action potential,
and the maximal depolarized level reached. In general,
larger or more prolonged depolarizations produce a
greater activation of G_{Ca}, a larger Ca^{2+} influx, a greater
increase in *p*, more transmitter release, and a larger PSP.
Smaller or shorter-duration voltage changes cause a
smaller activation of G_{Ca}, a smaller Ca^{2+} influx, a smaller
increase in *p*, less transmitter release, and a smaller PSP.

The mechanisms of transmitter release are highly Ca^{2+}
dependent. After each action potential, free Ca^{2+} near release sites transiently increases from the resting value of
10^{-7} mol/L to more than 10^{-4} mol/L. Ca^{2+} increases *p*,
the probability of quantal release, by aiding the fusion of
transmitter-containing vesicles with the membrane at release sites. Release depends on the fourth or higher
power of the Ca^{2+} concentration in the terminal. The extreme Ca^{2+} dependence may result from a need for four,
five, or more Ca^{2+} ions to trigger the exocytotic event.
This event may involve the Ca^{2+}-binding protein called
synaptophysin, which is contained in vesicular membrane and is likely to facilitate fusion with the neurolemmal membrane.

Most processes of neurotransmitter release have three
things in common. They use proteins, they depend on
Ca^{2+}, and they are incompletely understood. Ca^{2+} entry
is always the immediate and rapid event triggering release, but plastic changes often result from modulatory
processes that contribute to or influence release. These
findings suggest that other mechanisms for plastic modulation may depend on small changes in terminal Ca^{2+}
concentration or in its regulation.

Plastic alterations in synaptic efficacy can be produced
by changes in presynaptic Ca^{2+} concentration or in
voltage-dependent mechanisms responsible for its entry.
Small changes in Ca^{2+} concentration can have powerful
effects on transmission, making an effective synapse inoperative or a weak synapse highly effective. The activity
of a synapse itself—the nature of the synaptic message it
has conveyed—can modulate these voltage- or Ca^{2+}-
dependent mechanisms to produce plastic changes in the
efficacy of that synapse.

□ Synapses can change
   their own efficacy

Even at the NMJ, the amount of neurotransmitter that is
released by an action potential varies with the history,
especially the immediate history, of the synapse. The amplitude of the EPP is not constant; it varies with repetitive
stimulation. Increases in EPP amplitude show *potentiation* or *facilitation* of the response, and decrement of the
EPP size reflects depression. Each of these effects is

caused by different plastic changes to n or p, each of which contributes to the quantal content. Potentiation and depression depend on the prior activity of the synapse itself, rather than on external factors or experimental manipulation.

Posttetanic potentiation facilitates synaptic efficiency

A common plastic change occurs at many synapses, including the NMJ: facilitation after activity. Because this is a potentiation of the response and because it is most common after a burst of activity in the nerve that would produce a tetanic contraction in an innervated muscle, it is called *posttetanic potentiation* (PTP). Figure 6-27A shows that a PSP elicited several seconds after the end of

a tetanus is potentiated; it is larger than similar PSPs produced before the burst of activity.

Potentiation or facilitation after activity is produced by an increase in transmitter release. It is likely that activity produces a persistent increase in the baseline Ca^{2+} concentration in the presynaptic terminal, even between stimuli. This effect could result from introducing more Ca^{2+} into the presynaptic terminal than intracellular buffers and membrane transporters can remove. The transient increase in intracellular Ca^{2+} concentration produced by each action potential combines with the residual Ca^{2+} to release more transmitter than would be released by the transient G_{Ca} alone and produces the potentiated PSP.

Potentiation fits the quantal model. The probability of release, p, increases greatly during and just after the ac-

Potentiation and Depression

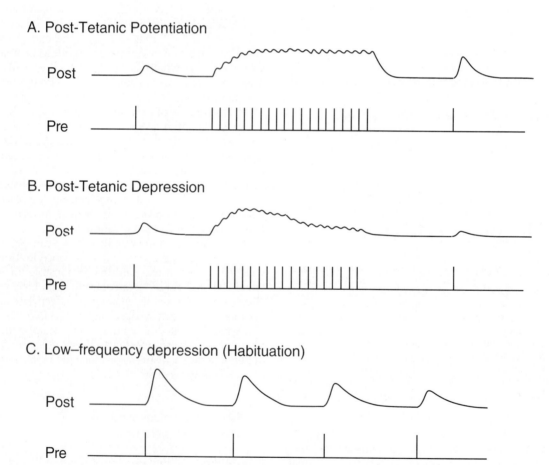

A. Post-Tetanic Potentiation

Post

Pre

B. Post-Tetanic Depression

Post

Pre

C. Low–frequency depression (Habituation)

Post

Pre

Figure 6-27
Potentiation and depression are simple forms of synaptic plasticity. (**A**) Posttetanic potentiation is seen as an increase in the amplitude of a postsynaptic potential (PSP) elicited a short time after tetanic (i.e., rapid) firing of the presynaptic cell. (**B**) Posttetanic depression is seen as a decrease in the amplitude of a PSP produced a short time after tetanic presynaptic firing. (**C**) The low-frequency depression of PSP amplitude seen with slow, repetitive, presynaptic activity in the neural analog of behavioral habituation to repeated presentation of a stimulus.

tion potential. The presence of residual Ca^{2+} after prior stimulation causes the increase in p produced by subsequent action potentials to be larger than that produced by the transient increase in G_{Ca} alone. Any action potential during this period of residual Ca^{2+} produces a facilitated PSP because of the additional rise in p.

Although the residual Ca^{2+} hypothesis is the simplest and most attractive explanation for potentiation, it is probably insufficient. Many other Ca^{2+}-dependent processes influence transmitter release and form possible sites for plastic changes. One of these may involve alterations to the mitochondrial Ca^{2+} uptake mechanism or to Ca^{2+}-binding proteins that reduce free Ca^{2+} after each action potential, again altering p. Na^+ accumulation in presynaptic terminals after repetitive activity may itself contribute to potentiation, perhaps by driving Na^+ and Ca^{2+} exchanger mechanisms in neurolemmal or organelle membranes to favor increased cytoplasmic concentrations of Ca^{2+}.

Potentiation can also result from plastic changes to n. One plausible mechanism involves enhanced mobilization of vesicles from reserve holding areas in the terminal to active zones, adding to the releasable store. Reserve vesicles are cross-linked to one another and to cytoskeletal elements by the structural protein *synapsin*. Phosphorylating synapsin by a Ca^{2+}/calmodulin-dependent protein kinase frees vesicles and permits their migration to release sites. Other mechanisms may account for PTPs at different synapses.

Posttetanic depression decreases synaptic efficacy

Repetitive activity can produce depression as well as facilitation (Fig. 6-27B). With repetitive stimulation, even though the probability of release is transiently raised by each action potential, the number of quanta that are available to be released falls with each action potential. The result of this depletion is a depression of transmitter release, seen as a decrease in the amplitude of successive PSPs. Data from experiments verify that n and p are separate processes that can be independently altered, even at the same time. During depression at the NMJ, when EPPs are reduced by a reduction in the releasable store, the frequency of MEPPs is increased, showing that p is still above normal although n is reduced.

Different frequencies of rapid stimulation can preferentially produce facilitation or depression at some synapses. In some cases, phases of facilitation and depression can be seen. Some synapses respond only with depression, and others respond with potentiation. Each of these cases represents a change in the effectiveness of the synapse caused by strong activity in its immediate past. One form of depression results from presynaptic microanatomy of central synapses. CNS presynaptic endings often branch to form multiple synaptic buttons. At these branch points, the *safety factor* or reliability of action potential transmission is reduced, so that some frequencies of repetitive or patterned stimuli may fail to invade some buttons. Lowered transmitter release, a consequence of this reduction in n or selective reduction in p, underlies the resulting depression.

Low-frequency depression resembles habituation

Depression can be produced without tetanic stimulation (Fig. 6-27C). Repetitive activity of some presynaptic neurons at slow rates, even one action potential every 10 seconds, produces a steady decline in the size of the PSP. This decline as a result of slow, repetitive activity in a pathway is called *low-frequency depression*. One cause of this depression is a reduction in n, probably mediated through a reduction in the releasable store of vesicles or of the reserve supply if stimulation is prolonged. Without sufficient high-frequency stimulation to produce an increase in residual Ca^{2+}, low-frequency depression does not show a facilitating component. Low-frequency depression of synaptic transmission is the neural analog of a basic form of behavioral plasticity called *habituation*, which is a decrease in a behavioral response to repeated presentations of a stimulus.

Each of these cases of potentiation or depression is an example of the innate plasticity of individual synapses. As a result of the prior activity of that synapse, the strength of subsequent messages conveyed by that same synapse is enhanced or diminished. Intrinsic regulating mechanisms of the presynaptic terminal alone undergo change that alters the amount of transmitter that same terminal releases later. In other cases, plastic mechanisms within the presynaptic terminal can be modified by events external to the presynaptic neuron by the influence of other neurons.

□ Synaptic efficacy can be changed by other synapses

The size of the synaptic response at some pathways can be altered by activity in other pathways. Plastic changes at these synapses depend on intrinsic mechanisms in the presynaptic and postsynaptic neurons and on additional neurons that are specialized to regulate the effectiveness of a synapse. These synapses are under the control of an outside influence, much as the gain of an an audio amplifier and the volume of the sound produced can be altered by turning the volume control.

Presynaptic inhibition selectively reduces input

Presynaptic inhibition selectively reduces input. A common example in the nervous system is a reduction in the effectiveness of a synapse between two neurons that is produced by activity in a third cell mediating *presynaptic inhibition*. Figure 6-28A shows a pathway between presynaptic and postsynaptic neurons modulated by an additional element. The size of the PSP and therefore of the efficacy of transmission between presynaptic and post-

Presynaptic Control of Transmitter Release

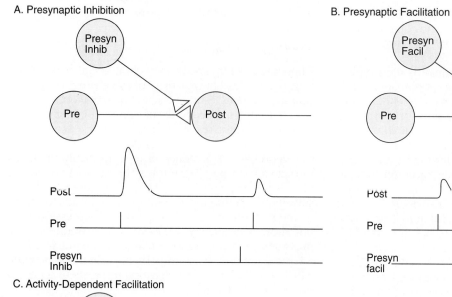

A. Presynaptic Inhibition

B. Presynaptic Facilitation

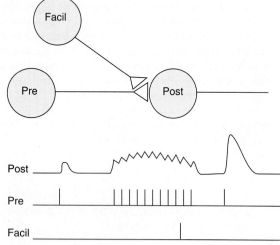

C. Activity-Dependent Facilitation

Figure 6-28

Presynaptic control of transmitter release. Each example presents a three-cell neuronal circuit and recordings of postsynaptic potentials (PSPs) in a postsynaptic neuron and action potentials in the other two cells. In each circuit, a presynaptic cell synapses on a postsynaptic neuron, and a third neuron makes a synapse on the presynaptic terminal. (**A**) Presynaptic inhibition reduces transmitter release. The size of the PSP produced in a postsynaptic cell by a presynaptic action potential is reduced by activity in a third, presynaptic inhibitory neuron just before the presynaptic neuron fires. (**B**) Presynaptic facilitation enhances transmitter release, monitored by an increase in the size of the PSP evoked by presynaptic activity that follows activity in the presynaptic facilitatory neuron. (**C**) Activity-dependent facilitation requires conjoint activity in the presynaptic cell and in the facilitatory neuron to enhance transmitter release.

synaptic neurons is controlled by the presynaptic inhibitory neuron. This third cell regulates the synapse by making a synapse of its own, innervating the *presynaptic terminal*. Activity in the presynaptic inhibitory neuron releases a transmitter, which combines with receptors on the presynaptic terminal that are specialized to decrease transmitter release.

Presynaptic inhibition is distinct in mechanism and function from the conventional postsynaptic inhibition that produces IPSPs in postsynaptic cells. Presynaptic inhibition results in a decline in excitatory transmitter release from particular presynaptic terminals. A functional advantage of presynaptic inhibition is that it is highly specific. Reducing transmitter release from one input onto a cell reduces the effectiveness of that input without altering or interfering with the effect of any other input on that same postsynaptic cell. This allows selective elimination of some inputs without affecting others, sharpening the responses to some inputs by presynaptically suppressing others. In postsynaptic inhibition, one neuron directly hyperpolarizes the cell body or dendrites of the postsynaptic cell, producing IPSPs and making it less likely that the postsynaptic cell will fire. IPSPs are nonspecific, reducing the effect of all excitatory inputs on the cell.

At some synapses, the mechanism of presynaptic inhibition is likely to involve one or more of a large number of transmitter-activated receptors that are able to reduce voltage-dependent Ca^{2+} currents in neurons. The modulation may affect voltage-dependent Ca^{2+} channels directly responsible for release or other voltage-dependent channels that admit Ca^{2+} involved in Ca^{2+}/calmodulin-dependent phosphorylation of synapsin. At other synapses, presynaptic inhibitory transmitters open transmitter-activated channels or indirectly close others, changing the membrane potential at the terminal. Subsequent action potentials produce smaller increases in G_{Ca}, reflecting the effect of the altered membrane potential on voltage-dependent Ca²⁺ channels responsible for release.

Presynaptic facilitation selectively enhances input

Presynaptic facilitation enhances transmission selectively, much as presynaptic inhibition depresses it. As in presynaptic inhibition, a third neuron makes a modulatory connection with a synaptically interconnected pair of neurons (Fig. 6-28B). The circuit for presynaptic facilitation appears similar to that for presynaptic inhibition, but the function is opposite. The size of the PSP is enhanced by activity in the presynaptic facilitatory neuron, which releases its neurotransmitter on the presynaptic terminal. Plausible mechanisms for presynaptic facilitation include direct actions on voltage-dependent Ca²⁺ channels and indirect changes produced by alterations of membrane potential or of other conductances. As a result of presynaptic facilitation of a synapse, the synaptic potential produced by the targeted presynaptic neuron is selectively enhanced, increasing its effectiveness for synaptic integration.

Activity-dependent facilitation responds to conjoint activity

Posttetanic potentiation enhances transmission at a synapse as a result of activity in that synapse's presynaptic neuron. Presynaptic facilitation of a synapse results from activity in an extrinsic innervating neuron, whose presynaptic facilitatory transmitter enhances release in a selected presynaptic terminal. Presynaptic facilitatory (or inhibitory) inputs may be made on more than one synaptic terminal. The nervous system can produce various degrees of facilitation of different terminals by a combination of these two facilitatory signals. The combination, called *activity-dependent facilitation*, provides a mechanism for the association or integration of information in the nervous system by responding to paired activity in two pathways (Fig. 6-28C). Activity-dependent facilitation responds to conjoint activity. In activity-dependent facilitation, the enhancement of vesicle mobilization or other release-related events produced by the facilitatory transmitter combines with the increase in residual Ca²⁺ in the same presynaptic terminals to further enhance transmitter release and produce a larger PSP.

☐ The NMDA receptor mediates long-term potentiation

Long-term potentiation is an analog of memory

A form of synaptic plasticity has been implicated as a mechanism for long-term learning or memory in the mammalian brain (see Chap. 30). Enhancement of synaptic transmission is produced by an increase in transmitter release, but this mechanism for plasticity relies on postsynaptic changes as well. The best-understood example of this phenomenon is found in neurons of hippocampus, a cortical region concerned with memory. Although the anatomy of the hippocampus appears relatively simple, it presents several sites for synaptic plasticity, and mechanisms differ at different sites

One of the most basic forms of memory is posttetanic potentiation, in which a synaptic response is enhanced as a result of prior activity. This physiologic analog to memory is relatively short lived, lasting perhaps a minute. There are several components of potentiation with various durations, but each is relatively short. However, in some cases, the response to facilitating or potentiating activity can last for weeks and is called *long-term potentiation*. Figure 6-29 shows the prolonged enhancement

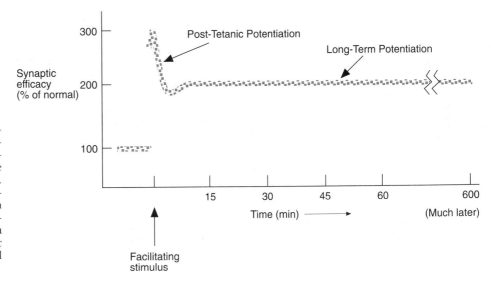

Long-Term Potentiation

Figure 6-29
Long-term potentiation of synaptic efficacy. The amplitude of postsynaptic potentials (PSPs) is plotted as a function of time before and after a facilitating stimulus. The graph is scaled so that the amplitude of PSPs before facilitation is normalized to 100%. After short-term posttetanic potentiation, a long-term potentiation of synaptic efficacy is maintained for several hours.

of synaptic transmission compared with PTP. At some sites in the hippocampus, simultaneous activity in more than one input path or strong activation of one path can produce long-term potentiation, seen as a long-term enhancement in the effectiveness of subsequent inputs into synaptic pathways. In this neuronal analog of memory, afferent input can be potentiated for hours, days, or even weeks.

NMDA receptor activation requires glutamate and depolarization

The site of the plastic synapse whose prolonged increase in effectiveness defines long-term potentiation is a dendritic spine of a hippocampal neuron (Fig. 6-30). At these excitatory synapses, the transmitter is glutamate. A single neurotransmitter can produce different postsynaptic ef-

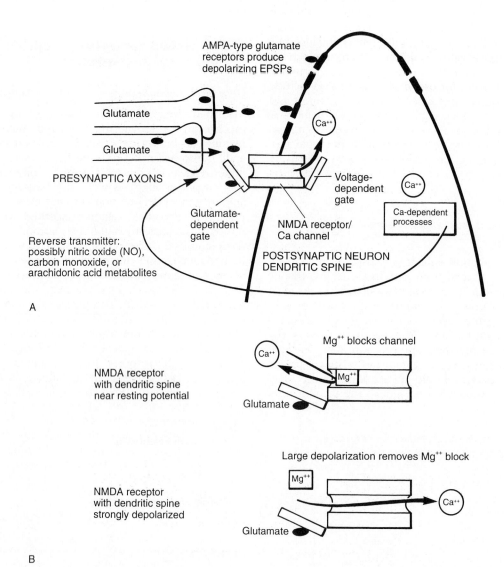

Figure 6-30

One mechanism for long-term potentiation uses postsynaptic NMDA receptors. (**A**) Glutamate-containing presynaptic axons innervate a postsynaptic dendritic spine with two classes of glutamate receptors. AMPA receptors produce conventional excitatory postsynaptic potentials (EPSPs), depolarizing the dendritic spine. NMDA receptors require this depolarization and glutamate to open their Ca^{2+}-selective channel. Postsynaptic Ca^{2+} entry activates Ca^{2+}-dependent processes that potentiate synaptic transmission. One mechanism appears to involve a reverse transmitter, released postsynaptically, that enhances presynaptic transmitter release. (**B**) The voltage dependence of the NMDA receptor arises from a voltage-dependent block by Mg^{2+} ions. With the dendritic spine near the resting potential, Mg^{2+} blocks the NMDA channel, even when the transmitter-activated gate is opened by glutamate. Only when the dendritic spine is strongly depolarized is the Mg^{2+} block removed and Ca^{2+} free to enter.

fects through its binding to different receptors. The mechanism of long-term potentiation in the CA1 region of hippocampus relies on this dual role of a single transmitter.

Individual dendritic spines have two types of postsynaptic glutamate receptors. Although each of these two receptor types is normally activated in the nervous system by glutamate (and possibly by aspartate), they are named and differentiated from one another and from other glutamate receptor types by their preferential sensitivity to other agonists, artificial analogs to glutamate. The first subtype of receptor, called AMPA (aminomethylisoxazole propionate), is the same, conventional, cation-selective glutamate receptor found throughout the nervous system and responsible for glutamate-mediated excitation by the production of EPSPs. As elsewhere in the nervous system, the AMPA receptor in the hippocampus responds to a neurotransmitter with a depolarization of the membrane. The other receptor, called an NMDA receptor channel because it is selectively activated by N-methyl-D-aspartate, is linked to a channel with two different sorts of gates across it. One gate, linked to the receptor, is opened by glutamate, the neurotransmitter. Opening the glutamate-sensitive gate is necessary but not sufficient for ion movement through the channel. The other gate is voltage-dependent, requiring the dendritic spine to be depolarized to open. The flow of current through the channel requires both gates to be opened simultaneously.

Simultaneous activation of both gates requires glutamate to bind to the NMDA receptor at the same time as the spine is depolarized. The needed depolarization of the postsynaptic cell can result from any of several processes, but the most likely is strong activity in the same or a different presynaptic pathway, releasing sufficient glutamate to depolarize by activating conventional AMPA receptors. Mild, low-frequency input is insufficient to depolarize the neuron. Similarly, depolarization alone or NMDA receptor binding alone is insufficient. To fulfill the need for two forms of activation, the neuron can *associate* two forms of input, and both together produce a long-lasting memory trace of their simultaneous activity.

The voltage-dependent gate across the NMDA receptor channel is an indirect mechanism. Voltage-dependent action potential Na^+ and K^+ channels, as well as the voltage-dependent Ca^{2+} channel involved in transmitter release, have specialized potential-sensing transmembrane segments that change the channel conformation to open the pore in response to depolarization. The NMDA receptor channel lacks such a mechanism for direct voltage sensitivity. The voltage dependence derives indirectly from the affinity of the NMDA channel for Mg^{2+} ions. Even with the glutamate-activated gate open, the extracellular opening of the channel is normally blocked by partial entry of extracellular Mg^{2+} ions, which cannot traverse the channel completely and which prevent other ions from entering. When the membrane containing the channel is depolarized, the driving force for Mg^{2+} to block the channel mouth is reduced. Only if glutamate opens the neurotransmitter-sensitive gate at the same time that the Mg^{2+} block is relieved by depolarization can ions enter the postsynaptic dendritic spine through the NMDA receptor channel.

NMDA channel opening increases intracellular calcium

The NMDA channel is Ca^{2+} selective. The result of the dual gating mechanisms of this synaptic receptor is the regulation of Ca^{2+} entry into the dendritic spine. Depolarization alone or glutamate binding to the NMDA receptor alone is insufficient to permit the influx of Ca^{2+}. Because this is a postsynaptic structure, Ca^{2+} is not being used to alter directly the Ca^{2+}-dependent mechanisms of transmitter release. The function of calcium is not well understood, but it serves as the signal to start the biologic machinery to enhance the effectiveness of inputs to this dendritic spine for a long time afterward. Evidence exists for the involvement of such Ca^{2+}-dependent processes as Ca^{2+}/calmodulin-dependent protein kinase II and protein kinase C and the subsequent induction of other enzymes.

The NMDA receptor and its ability to use a neurotransmitter to control Ca^{2+} are found at other sites in the nervous system mediating other Ca^{2+}-dependent phenomena. NMDA receptors are found in the cortex, cerebellum, spinal cord, and other sites. In addition to mediating long-term potentiation, they are involved in responses to psychoactive drugs such as opioids and antidepressants. They also contribute to neuronal death in the form of programmed cell death during development and the fatal toxic responses of neurons to excessive depolarization.

Reverse transmitters may alter presynaptic release

Reverse transmitters may alter presynaptic release. Entry of Ca^{2+} into the postsynaptic neuron triggers the induction of long-term potentiation. Increases in Ca^{2+} within a postsynaptic specialization suggests that functional changes to the postsynaptic element, such as increases in the number of EPSP-producing glutamate receptors of the AMPA subtype, may be responsible for the long-term increase in synaptic strength. Although there is some evidence for postsynaptic changes, potentiated synapses are found to release increased amounts of transmitter from presynaptic terminals. Because long-term potentiation is initiated by a postsynaptic event and carried out by increasing the amount of transmitter released by incident presynaptic neurons, a retrograde signal must be sent from postsynaptic cells to presynaptic terminals.

The most likely explanation is a *retrograde chemical messenger*, serving the same function in reverse across the synapse as glutamate does in the conventional presynaptic to postsynaptic direction. There is evidence that a short-lived compound, *nitric oxide*, which can diffuse across cell membranes without a specialized carrier or channel, serves as this messenger. Nitric oxide may be synthesized and released postsynaptically by a Ca^{2+}-dependent process, diffuse backward across the synapse, and enhance transmitter release by modulation of presynaptic mechanisms. Parallel evidence suggests a similar role for carbon monoxide. There is also evidence that arachidonic acid or its lipoxygenase metabolites may act as a retrograde messenger in later phases of long-term potentiation.

☐ How does plasticity become memory?

Most of the events of synaptic transmission appear to take place on millisecond or faster time scales. How do these plastic changes persist for minutes or hours? How do long-term plastic changes last for days or weeks? If these changes are truly analogous to memories, how do they last as long as a lifetime?

Some mechanisms of plasticity, like the mechanisms of synaptic transmission itself, may be rapid, but they possess the ability to produce more long-lasting changes in allied processes. Events such as the change in V_m due to a presynaptic action potential or PSP are truly rapid, but there is a hierarchy of longer-lasting processes that can provide some of the persistence of memory. Lasting for seconds is the elevated residual Ca^{2+} in presynaptic terminals that can underlie potentiation. Enhanced mobilization of a transmitter by freeing stores could last for minutes. Modulation, perhaps by phosphorylation, of presynaptic channels or postsynaptic receptors or by intracellular processes in either neuron can change the effective number or properties of relevant proteins for the same or longer periods. More prolonged plastic changes could be made to persist by autophosphorylation of kinases triggered by a transient message or by directed mobilization or transport of enzymes to target areas of the cell. Truly long-term modification of synaptic function probably requires structural changes in synapses, increasing the number of presynaptic release sites or the degree of postsynaptic neurotransmitter-sensitive membrane. These changes could be produced by related changes in protein synthesis and maintained by feedback interactions between nucleus and cytoplasm.

Other Aspects of Transmission

☐ Electrotonic transmission is achieved by synapses without transmitters

Throughout the nervous system, chemical synaptic transmission transmits messages from cell to cell. This is different from the arrangements for intercellular communication in nonneural tissue, in which specialized structures such as gap junctions permit direct movement of current, small ions and other molecules, and some larger molecules from one cell to another. The advantages of neural specialization for chemical transmission are as varied as its uses. At the NMJ, chemical transmission is necessary because the great size mismatch between nerve and muscle precludes the passage of sufficient current from nerve to muscle for efficient electrical transmission. In most of the CNS, the presynaptic and postsynaptic elements are about the same size, and the advantages of chemical synaptic transmission are derived from the versatility of the mechanism.

Chemical synaptic transmission, although widespread, is not universal in the CNS. In a few specialized cases, low-resistance gap junctions or similar channels provide direct current passage, called *electrotonic transmission*, between neurons. Electron-microscopic examples of gap junctions are shown in Chapter 3. Two cells with an electrotonic junction between them are illustrated in Figure 6-31. An action potential in one neuron produces current that spreads directly to its coupled partner, depolarizing it enough to generate an action potential there as well. In some electrotonic synapses, the coupling is weaker, and an action potential in one produces only a subthreshold depolarization in the other. In some electrotonic synapses, called *rectifying*, current can flow more easily in one direction than in the other, and this limits the direction of information transfer, just as for a chemical synapse. In other synapses, conduction is bidirectional or nonrectifying, and in these, current can flow from either cell to the other.

Electrotonic synapses are relatively rare in CNS and are found only in a few locations. Because there is little or no delay inherent in the mechanisms of electrotonic transmission, a group of cells all interconnected electrotonically with nonrectifying synapses tend to fire synchronously. Electrical synapses are found only in particular neural circuits, such as in the inferior olive and oculomotor nuclei, where it is advantageous to the nervous system to have fast, synchronous activity in a group of cells.

Electrotonic Synapse

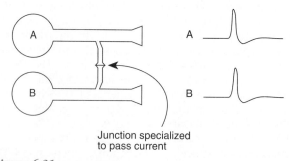

Junction specialized
to pass current

Figure 6-31
Electrotonic synapses pass current directly between cells. Two cells are interconnected with a nonrectifying electrotonic synapse, which is a junction specialized to pass current bidirectionally. Activity in either cell excites the other, as shown by the synchronous action potentials in cells A and B.

☐ Neural networks link synapses to develop intelligence

The proper and efficient functioning of the nervous system depends on appropriate synaptic connections between neurons. Studies in which synaptic physiology is correlated with the performance of the nervous system and with the response of the organism show that the strengths of synapses matter. The existence of mechanisms for changing synaptic strength (i.e., synaptic plasticity) reinforce this importance. Evidence has underscored the significance of synaptic strength by showing that changes in the efficacy of synapses can alone differentiate a functional network of neurons from an ineffective one. This evidence comes from explorations of *artificial neural network models*, composed of neuron-like elements, interconnected by synapse-like links, and simulated on computers. Although these models are not designed to reproduce the structure or function of any biologic nervous system, the information they provide about the importance of synaptic strength to nervous system function helps us to understand this aspect of neuroscience.

There are many ways to model the nervous system, and each provides some insight into its function. Neural network models, like any models, are simplified, scaled, or modified versions of a complex object or process. Because artificial network models are designed to probe the function of synaptic interconnections, many or most aspects of nervous system design or function can be disregarded. This book discusses the structure and function of subneuronal processes, neurons, synapses, circuits, nuclei, tracts, and regions; network models ignore much of this.

Artificial neural network models are characterized by three features. The first are the elements of the models, which are simplified versions of neurons and synapses. The second feature is the wiring diagram for interconnection of the "neurons" by "synapses." The third feature is functional, rather than structural. It is the set of rules for changing the strengths of the synapses to change the way the network processes information. Figures 6-32 through 6-34 detail one of these features; Figure 6-35 presents the performance of a network model assigned to perform a real-world task in neural information processing.

A *model neuron* is shown with some of its synapses in Figure 6-32. Like real neurons, these model elements receive synaptic input from many cells, sum these inputs, and if the result is above the threshold level, transmit the resulting neuronal activity to other neurons. A major simplification is that all neurons in a model are assumed to work exactly the same way; other simplifications are designed to make computing the results easier.

Although it would be possible to interconnect every neuron in a model with every other or to select pairs of neurons at random to be interconnected, neither of these schemes is representative of real neuronal circuitry in the brain or elsewhere, and neither makes a particularly efficient model. Instead, model neurons are interconnected in layers in an attempt to partially represent real nervous systems, in which neurons of different layers, laminae, or types exhibit characteristic patterns of innervation. Typical networks have three layers (see Fig. 6-33). Just as biologic nervous systems may be thought of partially as systems for processing sensory information and producing a cognitive experience or motor output, the multiple layers of a network model may be thought of as a *sensory input layer*, one or more *interneuronal layers*, and an *output layer* that specifies a motor action or selects from a number of perceptual alternatives. To permit neuronal information to flow from input to output and to maximize the chances for each neuron to receive the in-

**The Elements of Artificial Neural Network
Models are Simplified Neurons**

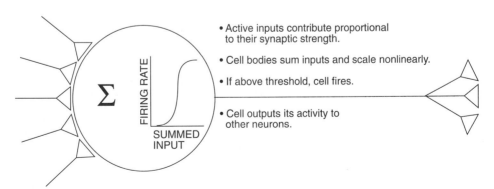

Figure 6-32

Elements of artificial neural network models are simplified versions of neurons and synapses. Synaptic inputs contribute according to their activity, proportional to their synaptic strength. Cell bodies sum this input and scale it, as shown schematically by the Σ and the graph, to produce the output of the neuron. This output is transmitted to other neurons. (From Gardner D. The neurobiology of neural networks. Cambridge, MA: MIT Press/Bradford Books, 1993.)

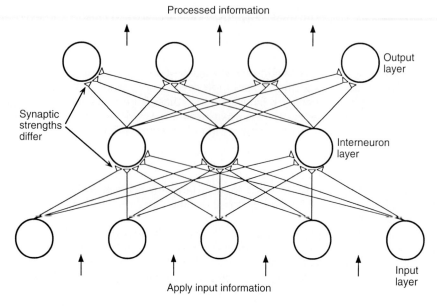

Processed information

Output layer

Synaptic strengths differ

Interneuron layer

Input layer

Apply input information

Figure 6-33
Artificial neural networks are composed of layers of model neurons, interconnected by synapses that vary in strength. Information proceeds from the input or sensory layer through one or more interneuronal layers to the output layer. Patterns of input information, coded as activity in the sensory layer, produce corresponding patterns of activity in the output layer, representing a motor command or a sensory perception. (From Gardner D. The neurobiology of neural networks. Cambridge, MA: MIT Press/Bradford Books, 1993.)

formation it needs to contribute its part to the information processing task that the network is designed to carry out, each neuron in a layer makes a synaptic connection on every neuron in the next layer (see Fig. 6-33). Input patterns representing sensory information are presented to the input or sensory layer. Activity characterizing these inputs is successively transmitted to and processed by higher layers, producing an pattern of activity in the final output layer that represents a sensory discrimination or motor pattern.

The pattern of activity produced to an input depends most of all on the strengths of the interconnecting synapses. Any input produces an output, but the output is likely to be inappropriate or meaningless. For the network to process information as real nervous systems do, the synaptic strengths must be precisely adjusted. In a network, the synaptic strength is how much of a postsynaptic change should result from a particular pattern of activity in a presynaptic element. The necessity of adjusting synaptic strengths appropriately can be understood from examining the performance of a just-constructed

network. Because it is not possible to know how strong each of the synapses should be made when the network is initially designed, these strengths are set randomly. With these arbitrary strengths, outputs are random, and the network is not capable of performing any meaningful information processing.

Although it would be possible for a designer to study each model network and adjust synaptic strengths to try to increase the capability of the network, it turns out that the network itself is capable of altering its own plastic synapses, using a few simple rules. The network is trained to process information by presenting it with a series of sensory input patterns and, with each, the desired output pattern. Because the network can calculate the difference or error between the ideal and the real and can assign the credit or blame for this error to improper synaptic strengths, it can adjust these synaptic strengths to minimize the error and optimize the output. Figure 6-34 shows how reducing the erroneous output of one neuron depends on adjusting each of the synapses through which it receives input information.

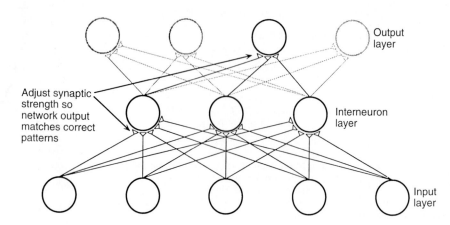

Output layer

Adjust synaptic strength so network output matches correct patterns

Interneuron layer

Input layer

Figure 6-34
Information processing by artificial neural networks requires appropriate combinations of synaptic strengths. The processing power of a network is derived from combinations of these synaptic strengths, which can be tuned by the network itself in a trial-and-error process in response to repeated presentation of inputs and the desired output patterns. The cells and synapses shown in black represent all those that need to be adjusted to produce the correct response from the indicated unit in the output layer. (Modified from Gardner D. The neurobiology of neural networks. Cambridge, MA: MIT Press/Bradford Books, 1993.)

Adjusted Synaptic Strengths Permit Neural Networks to Process Information

- ## As training begins:

```
Text input:         i_couldnt_get_any_food._i_mean_there_was_this_
Correct phonemes: A-k-U-dn--gEt-EnI_fu-d._A-mi-n-Q-^r---^--d-Is-
Network output:   n-n*n--nn-bpn-bn---*----t-b--n-d-----b-n-d----
```

- ## After training with 25 passes:

```
Text input:         there_this_one_night_i_couldnt_get_any_food.
Correct phonemes: D-Er--D-Is-^n--nA--t-A-k-U-dn--gEt-EnI_fu-d.
Network output:   D-Er--D-Is-^n--nA--t-A-k-U-dn--gEd-EnI_fu-d.
```

- ## The network is tested with new material:

```
Text input:         _when_my_brother_was_little_he_went_to_
Correct phonemes: _w-Em-mA-br^D-^r-w^z-lId-l-_hi-wEn--tu_
Network output:   _w-Em-mA-br^D-^r-w^z-lid-l---i-wEnt-t^_
```

Figure 6-35

Adjusted synaptic strengths are necessary for artifical neural networks (ANN) to process information appropriately. The performance of an ANN converting text to phonemes is shown at each of three stages of training. In each example, three lines show the English text input, the correct phonemes to represent this text as speech, and the phonemes output by the network. As training begins (*top*), the network produces random output. After 25 repetitions of the training material (*center*), the network is almost perfect. Testing the network after training with new material (*bottom*) also produces appropriate output. (Modified from Sejnowski TJ, Rosenberg CR. NETtalk: a parallel network that learns to read aloud. Baltimore: Johns Hopkins University/Electrical Engineering and Computer Science Technical Report No. JHU/EECS-86/01, 1986.)

Neural network models, like real nervous systems, are capable of performing many information-processing tasks. For example, network models have been used in medicine for differential diagnosis of myocardial infarction in emergency room patients presenting with anterior chest pain. Another example of the ability of these networks to learn by adjusting synaptic strength is shown by a network designed to carry out a task that the human nervous system also learns by doing: reading aloud. A major aspect of this task, technically text-to-speech conversion, can be carried out by a neural network presented with words of English text and asked to output the sets of phonemes (i.e., sound elements) characterizing the words. Because the rules for text-to-speech conversion are many and complex, with many exceptions, this task requires a network with 300 neurons and more than 18,000 synapses.

The importance of synaptic strength and of its careful adjustment is illustrated in Figure 6-35. The figure provides three samples of the performance of the network, each comparing text input, correct phonemes, and the phonemes output by the network. With randomly set synaptic strengths, the output is random, and the network babbles. After training with each of 1000 words and their correct phonemes, each presented many times, the network has adjusted its synaptic strengths to appropriate values, and the output phonemes are correct. This adjustment also permits the network to pronounce new text that was not used in its original training. The only difference between the initial babbling of the network and its later ability to speak is that the strengths of synapses have been adjusted.

In addition to their usefulness in demonstrating the importance of synaptic strength, network models have other relevance for neurobiology. When presented with the typical input patterns for groups of cells in the mammalian nervous system, network models trained to produce appropriate outputs do so by adjusting synaptic strength to produce interneurons with receptive fields and field patterns resembling those of real CNS interneurons. They may also serve as models for understanding how the nervous system distributes a complex task among many relatively simple elements, a mechanism called *distributed processing*. The study of artificial neural networks is likely to be of increasing importance in neuroscience.

☐ Synapses are the links in an unimaginably complex net

The operation of the nervous system, the most powerful and sophisticated computer known, relies principally on the synaptic mechanisms described in this chapter. Transmitter release, PSPs, Ohm's law, and other synaptic events provide the links between individual neurons that make possible higher processes such as sensation, motor control, behavior, learning, and thought. Simple changes in ionic conductances can cause changes in the size of synaptic potentials and make an ineffective pathway important or render a functioning pathway less effective. Changes in the effectiveness of established synaptic pathways can produce some simple forms of learning and behavior modification.

We do not understand everything about synapses and the nervous system. Even if all properties of every synapse in the body were known, there are 10^{15} synapses in each of our nervous systems. If this chapter summarized information about each of the 10^{15} synapses in a single line of text, the chapter would have to be about a million kilometers thicker than it is. Moreover, this giant catalog would be almost useless. As important as the properties of each synapse is the ability to process information that emerges from the synaptically interconnected network of neurons that is the nervous system.

SELECTED READINGS

Aidley DJ. The physiology of excitable cells, 3rd ed. Cambridge, UK: Cambridge University Press, 1989.

Gardner D. The neurobiology of neural networks. Cambridge, MA: MIT Press/Bradford Books, 1993.

Hille B. Ionic channels of excitable membranes, 2nd ed. Sunderland, MA: Sinauer Associates, 1992.

Jessel TM, Kandel ER. Synaptic transmission: a bidirectional and self–modifiable form of cell–cell communication. Cell 72/Neuron 10 (suppl) 1993;1.

Junge D. Nerve and muscle excitation, 3rd ed. Sunderland, MA Sinauer Associates, 1992.

Katz B. The release of neural transmitter substances. Springfield, IL: Charles C Thomas, 1969.

Stevens CF. Quantal release of neurotransmitter and long-term potentiation. Cell 72/Neuron 10 (suppl) 1993;55.

Neuroscience in Medicine, edited by P. Michael Conn.
J. B. Lippincott Company, Philadelphia © 1995

Chapter 7

*P*ostsynaptic Receptors

DAVID K. SUNDBERG

Receptor Structure and Function

A receptor is a cell component, usually a protein, that combines with a drug, a transmitter, or a hormone to alter the cellular response.

☐ Receptors may be classified anatomically, pharmacologically, or mechanistically

There have been several classifications of receptor systems. These include anatomic, pharmacologic, and since the advent of molecular biologic techniques, mechanistic classifications. The earliest anatomic classification systems were based on the location of specific types of receptors. Examples include somatic or autonomic, parasympathetic or sympathetic, and postsynaptic or presynaptic (i.e., having the receptor on the nerve terminal or dendrite and serving neuromodulatory function) classification systems.

The pharmacologic classification of receptors has been developing since the late 1800s, when the transmitter substances were first being isolated and chemically characterized. Receptors are classified according to transmitter groups and their response to drugs. Receptors that respond to the catecholamines (e.g., dopamine, norepinephrine, epinephrine) are called catecholaminergic or sometimes "adrenergic" receptors. Because some respond better to norepinephrine than to epinephrine, they have been further subdivided into α- and β-adrenergic receptors. Both of these receptor subtypes have been further characterized into types 1 and 2 (e.g., β_1, β_2). These types can be demonstrated when relatively small chemical differences in the transmitter structures selectively affect only one population of receptors. This extensive system has made it necessary for the student to learn a plethora of receptor subtypes, but the subtypes have been convenient for the purposes of research, therapy, and the pharmaceutical industry. For example, it is possible to develop a drug (e.g., β_2-adrenergic agonist) that dilates bronchial smooth muscle in an asthmatic that does not have significant effects on the β_1 receptor of the heart, which would greatly increase the heart rate and cause cardiac palpitations.

A mechanistic method for the classification of receptors is based on the molecular mechanism that activates the cell to respond after the transmitter has combined with its receptor. Much of this information was developed when the genes for the cellular machinery (e.g., membrane and cytoplasmic proteins) were "cloned" and sequenced. Cloning of genes refers to the process of inserting copies of the gene for a specific protein (e.g., a receptor) into bacteria and allowing the natural reproduction and growth of the bacteria to produce the gene, mRNA, or protein of interest for harvest in bulk form. Using the mechanistic classification of receptors, several groups or "families" of receptors have been uncovered. One of the major families includes the ion channel-gated receptors. These receptors are part of a channel or pore in the membrane through which various ions can travel. Transmitter coupling can open or close the membrane pores.

Another mechanistic family of receptors is the G-protein–coupled receptors. The receptor protein, which spans the membrane seven times, activates a G protein, so called because it hydrolyzes GTP. This protein initiates a series of events (i.e., second messenger system) that produces the appropriate effect within the cell. This may be the opening or closing of an ion channel, the production of another second messenger, such as cyclic AMP or cyclic GMP, or the stimulation of genetic translation and transcription.

Although it seems logical that a given member of a pharmacologic class of receptors (e.g., cholinergic) would fall within the same gene family of the mechanistic class of receptors, this is not the case. The nicotinic cholinergic receptor belongs to the ion-gated family of receptors, and the muscarinic cholinergic receptor, which also responds to acetylcholine, belongs to a second messenger–coupled G-protein receptor. To this extent, the muscarinic cholinergic receptor and the adrenergic receptor that responds to norepinephrine belong to the same family of membrane proteins.

Other receptor families include the membrane receptors that respond to growth factors and the intracellular cytoplasmic receptors that respond to the steroid hormones (e.g., estrogen, testosterone, cortisone) and thyroxine. The membrane-bound receptors that respond to growth factors are themselves enzymes (i.e., kinases) that elicit the cellular actions. The cytoplasmic domain of the receptor for atrial natriuretic factor is itself a guanylate cyclase.

In this chapter, specific receptors are discussed within the framework of the pharmacologic method of classification. This method is important in terms of medical practice and is the basis for therapeutics. However, we further define the receptors in terms of their molecular mechanisms and the major gene families to which they belong.

☐ Receptors can be mathematically characterized using biochemical kinetic parameters that were developed to understand enzymes

The receptors are like enzymes that bind to a substrate and catalyze a cellular activity. In the case of enzymes, the activity is usually the addition or deletion of an atom or group of atoms to or from the substrate. In the case of the receptor, the transmitter is not chemically changed. These membrane-bound proteins, when activated by their transmitters, catalyze an activity within the membrane, which may be the opening of an ion channel or regulation of other intracellular steps that inhibit or stimulate the specific tissue.

The concept of the cellular receptor was developed in the early 1900s by Langley, who ascribed the effects of drugs and hormones to an interaction with a receptive substance (now called receptor). Later, A.J. Clark and other scientists, including Hill and Gaddum, mathematically described these interactions and developed methods for characterizing them pharmacologically.

The attractive forces between the transmitter or drug and its receptor can be chemically mediated by covalent, ionic, hydrogen binding, or Van der Waals forces. High-energy covalent bonding is rare. Covalent bonding occurs between the α-adrenergic receptor blocker, phenoxybenzamine, and the α receptor. This type of binding is irreversible and noncompetitive. Ionic attractions occur between cationic (i.e., positive) and anionic (i.e., negative) charges of different molecules and provide an attractive force that enables the transmitter to find the receptor. Hydrogen binding is the attraction between hydrogen atoms and unpaired electrons in nitrogen or oxygen atoms and is important for attracting the ligand to the receptor. Van der Waals forces are weak dipolar interactions, but because of the number of atomic interactions and the complementarity or "fit" between the ligand and receptor, these forces are important for defining the tightness of binding (i.e., affinity) and specificity of the receptor for its transmitter.

The interactions of most transmitters with their receptors follow Michaelis-Menten kinetics, as do biochemical enzymatic reactions. Figure 7-1 shows two response curves generated by measuring the effect of a transmitter on an effector system. This could be the effect of increasing concentrations of acetylcholine on the firing rate of a neuronal pathway in the forebrain. In the graph on the left (often called a Michaelis-Menten curve), increasing concentrations of the transmitter lead to an increased response that eventually plateaus. This plateau has been

shown by kinetics to represent the maximal activity of the system when all of the receptors are occupied or saturated. The transmitter concentrations when one half of the maximal effect is achieved yields a number called the K_d or dissociation constant. This number describes the affinity of the receptor and is roughly defined as the amount of substrate needed to activate or occupy one half of all of the receptors. If this number is low (e.g., in the nM range), the receptor-transmitter interaction is said to be a high-affinity interaction. If the concentration of the transmitter needed to cause one half of the maximal effect is high (e.g., μM or mM range), the receptor is said to have a low affinity for the ligand or agonist.

In the graph of Figure 7-1, the x-axis is modified so that the logarithm of the transmitter concentration is plotted against the effect. This method, commonly called the dose-response curve, consistently gives a sigmoidal plot and is most useful for investigating pharmacologic interaction. The semilog plot minimizes the amount of graph paper required to represent the relation and gives the log of the K_d. The Michaelis and dose-response curve provide information on the affinity of transmitter-receptor interactions and the efficacy of that effect. Efficacy is a pharmacologic term referring to the maximal response of a system to a drug or transmitter. In these diagrams, it is represented by the height of the response reached to a maximal dose of transmitter.

The Scatchard plot and the Hill plot are other pharmacologic tools for determining receptor-transmitter interactions. Figure 7-2 shows representative examples of these plots. In the Scatchard plot, the degree of binding of a radiolabeled transmitter to its receptor is measured and plotted as shown on the left side of Figure 7-2. This plot is similar to the Lineweaver-Burk plot of enzyme kinetics, and can tell the number or receptors present in

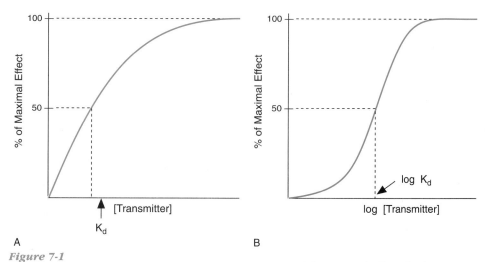

Figure 7-1

(**A**) Graph of a Michaelis-Menten plot, in which the percent of the maximal effect of a drug or transmitter is plotted against the dose. The concentration of transmitter that causes 50% of the maximal response (ED_{50}) represents the dissociation constant (K_d). (**B**) Graph of a typical dose-response curve, in which the percent of the maximal effect is plotted against the log of the dose. In this case the ED_{50} gives the log of the K_d.

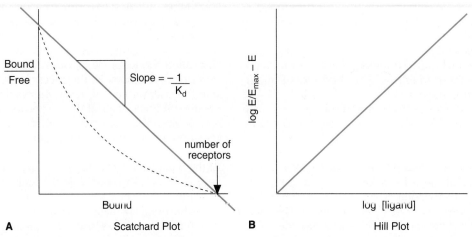

Figure 7-2
(**A**) The graph represents a typical Scatchard plot. The ratio of the amount of a labeled ligand or transmitter bound to a membrane preparation over the total free amount of radiolabeled ligand added is plotted against the amount of ligand bound. This plot gives the total number of receptors and their dissociation constant (K_d), which is determined from the slope. If the response is curvilinear (*dotted line*), it indicates that there are two or more populations of receptors with different K_d values. Using this method, spare receptors can be measured, because the transduction system is unnecessary for ligand binding. (**B**) The Hill plot compares the log of the effect (E) of a dose of drug corrected for the maximal effect (E_{max}) to the log of the ligand or transmitter concentration. This relation reveals if there is positive or negative cooperativity between the binding sites on a given receptor. If the slope is equal to 1, there is no cooperativity.

a particular tissue and the K_d of those receptors ($-1/$slope). Briefly, the amount of a radioactive ligand that is bound to a membrane preparation (x-axis) is plotted against the ratio of the bound to free ligand present in the reaction (y-axis). If this relation is not linear, as shown by the dotted line, it indicates that more than one population of receptors is present that have high and low dissociation constants.

Using these kinetic techniques, it has often been found that the K_d of the physiologic effect measured by the dose-response curve is much less than the K_d measured by ligand binding to membranes using the Scatchard plot. This discrepancy led to the concept of spare receptors. It appears that there are many more receptors than there are transducer systems. Not all receptors are coupled to a viable G-protein system, and if using ligand studies in which all of the receptors are measured while using dose-response curves, only those receptors that are coupled to G-protein systems are seen.

The Hill plot can be used to determine whether a receptor-agonist interaction displays positive or negative cooperativity. Cooperativity means that the binding of one transmitter molecule to a receptor influences the binding and action of another molecule of the same transmitter. The effect (E) of a given amount of a drug is corrected by the maximal effect of that drug (E_{max}) and plotted against the log of its dose (see Fig. 7-2). If the slope of this interaction is 1, there is no cooperativity. If the slope is greater than 1, there is more than one binding site for the drug causing positive cooperative interaction.

☐ Receptors are a part of an ion channel in the membrane or modulate an effector system through phosphorylation and second messengers

Much of the current knowledge about what cellular receptors look like and how they work has come from studies in which the receptors were solubilized from the lipid membrane, chromatographically purified, and sequenced. This work was tedious but crucial to our understanding of the receptor complex. The second major scientific approach that has helped us understand the receptor has been the molecular biologic methods needed to "clone" receptor genes and sequence them to determine the amino acids of the expressed proteins. The use of "homology" cloning and sequencing (i.e., isolating unknown genes by using molecular probes for known genes or proteins) has uncovered a wide variety of receptors, sometimes for totally unrelated transmitter systems. Such was the case for the muscarinic receptor, which was found to have a high degree of homology (i.e., degree of structural similarity) to the adrenergic-type, single-subunit receptors. Two receptors that received much of the early interest fall into each of the major classes of receptor families. The nicotinic acetylcholine receptor is a ligand-gated (i.e., directly triggered) ion channel receptor type, and the β-adrenergic receptor is a G-protein–mediated (i.e., second messenger) receptor complex.

Ligand-gated ion channel receptors

Ligand-gated ion channel receptors are composed of five protein subunits, each of which has four membrane-spanning domains. The nicotinic cholinergic receptor was studied largely because of its accessibility. The electric organ of the eel and ray is a tissue that is rich in cholinergic nicotinic receptors. Much effort has gone into isolating and characterizing the receptor from the membrane of this tissue. The isolated receptor appeared to be "oligomeric" in that it had several protein subunits. The receptor subunits were isolated and purified, and part of the amino acid sequence was determined. The possible DNA sequences (i.e., DNA probe) coding part of the receptor protein were estimated and synthesized. Using these small DNA probes, called oligoprobes, a library of genes was screened for homologous sequences. A messenger RNA was isolated and sequenced that was similar to the isolated protein. Surprisingly, many of the cloned subunits demonstrated a great deal of homology. This was particularly true in the very hydrophobic segments that were thought to span the lipid membrane. These methods have confirmed that the nicotinic receptor contained several protein subunits that together surrounded an ionic channel through the membrane. These subunits do have many differences. An α subunit contained an extracellular sequence on the amino-terminal end that bound the transmitter acetylcholine. Two of these α subunits were contained in each of the muscle-type nicotinic receptors.

In all of the ligand-gated receptor proteins (e.g., nicotinic, γ-aminobutyric acid, glycine, and glutamate receptors), one of the membrane-spanning domains of each subunit contains a repeating sequence of the polar amino acids, threonine and serine. These are thought to line the channel surface itself and serve as the lipid-water interface within the channel. Each receptor has five of these channel-lining domains, α-helical sections called the M2 domains, that are contributed by each of the five subunits. Our concept of the receptor subunits that combine

to make up the pentameric nicotinic receptor is shown in Figure 7-3. The M2 spanning domain that lines the channel is shown in color.

Since the elucidation of the nicotinic receptor, several other ligand-gated ionic channel systems have been described. These include the γ-aminobutyric acid (GABA) and the glycine receptors, which form channels for the chloride ion and hyperpolarizes or inhibits nerve cells; the several subtypes of the glutamate receptor that make up an intrinsic channel for calcium, sodium, and potassium ions, similar to the nicotinic receptor; and the serotonin (5-HT₃) receptor that is an intrinsic channel for cations. Table 7-1 lists the known ligand-gated ion channels and some characteristics of these receptors. All of these receptors are related in that they are pentameric, have four membrane-spanning domains per subunit, and have an M2 spanning domain that lines the ion channel.

When the receptive subunit of these receptors is occupied by a neurotransmitter, a conformational change of the five subunits allows the channel to open and the appropriately sized ion to enter the cell (Fig. 7-4). The type of ion that can traverse the channel seems to depend on the molecular radius of the ion, how tightly it combines with water molecules (i.e., free energy of hydration), and the strength of polar sites within or just outside the channel. For example, the sodium ion is smaller than the potassium ion but is excluded from the potassium channel. This is probably because it has a higher free energy of hydration and affinity for water molecules and cannot be dehydrated by the relatively weak ionic charges of the amino acids making up the potassium channel. All of the cationic (i.e., Na^+, K^+, Ca^{2+}) or positively charged ion channels have groupings of negatively charged amino acids (i.e., glutamate and aspartate) close to the mouth of the ion channel. These amino acids are probably important in removing the water from the cations (i.e., dehydrating) and producing Na^+ or K^+ channel selectivity.

In the case of the chloride channel for GABA or glycine, positively charged amino acids (i.e., arginine and

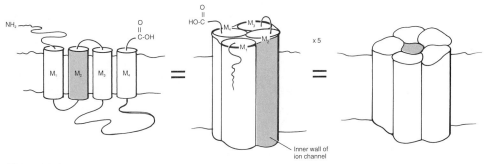

Figure 7-3
The structural components of a ligand-gated ionic channel receptor. The graph shows the subunit protein with the four membrane-spanning domains (*left*). This protein clusters to form a single subunit of the receptor (*center*). Five of these subunits fit together to form a ligand-gated ion channel (*right*). The M2 subunit (*shaded*) has repeating sequences of charged amino acids and makes up the inner wall of the ionic channel.

Table 7-1
ATTRIBUTES OF LIGAND-GATED ION CHANNELS AND THEIR RECEPTORS

Receptor	Subtype	Number of Subunits/ Transmembrane Domains	Comment	Effector[§]
GABA	A	5/4		Cl^-
Glutamate	NMDA	5/4	Glycine*	$Na^+/K^+/Ca^{2+}$
	AMP	5/4		$Na^+/K^+/Ca^{2+}$
	Kainate	5/4		$Na^+/K^+/Ca^{2+}$
Glycine		5/4	Strychnine†	Cl^-
Serotonin	5-HT$_3$	5/4		Cation
Nicotinic	Muscle	5/4	50 ps‡	$Na^+/K^+/Ca^{2+}$
	Neural	5/4	15–40 ps‡	$Na^+/K^+/Ca^{2+}$

*The NMDA glutamate receptor has a glycine binding site.
†The glycine receptor is strychnine sensitive.
‡The muscle nicotinic receptor is open longer than the neural receptor; time is given in picoseconds.
§The effector pathways represent the opening of ionic channels.

lysine) are localized near the mouth of the channel. This is important in allowing the anions to enter the channel. The polar serine and threonine residues that line the M2 domain of the channels do not appear to determine ionic specificity of the channel but are essential for maintaining the aqueous interface.

G-protein–coupled receptors

The G-protein–coupled receptors are composed of only one protein unit that has seven membrane-spanning domains. The prototypical G-protein receptor is represented by the β_1-adrenergic receptor of the heart. This receptor has received much attention because of the development of specific drugs that interact with them and their therapeutic importance in heart disease, asthma, and hypertension. Studies on the cardiac β receptor in-

dicated that it was associated with a membrane-bound protein that required guanylate nucleotidase for activity and hydrolyzed GTP to GDP. Unlike the ligand-gated ion channel receptor, the β receptor has only a single protein subunit. It was painstakingly solubilized from the membrane and parts of it sequenced. Eventually, this led to the cloning of a gene for a protein that had seven membrane-spanning domains. Figure 7-5 shows the current conception of the G-protein–associated receptor. The amino-terminal portion of the protein is found extracellularly, and the carboxyl-terminal end is inside the

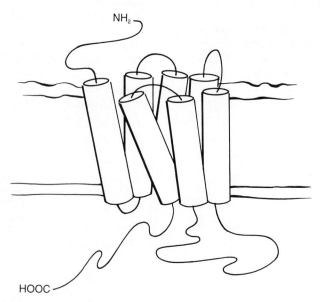

Figure 7-5
The guanine nucleotide-binding regulatory protein (G protein)–coupled receptor is composed of seven hydrophobic membrane-spanning domains with three intracellular and extracellular loops of amino acids. The amino-terminal end of the protein is extracellular, and the carboxyl-terminal end is in the cytoplasm of the postsynaptic neuron. The third intracellular loop is thought to be responsible for activating the appropriate G protein.

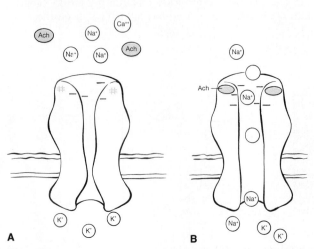

Figure 7-4
(**A**) Resting and (**B**) stimulated nicotinic receptors. Combination of the α subunits of the receptor with acetylcholine causes the channel to open and allows sodium to enter the neuron and depolarize it. This receptor is drawn roughly to scale to show that about two thirds of the receptor protein is extracellular.

cell. There is a large intracellular loop of protein between spanning domains five and six. This loop is thought to interact and regulate the G-protein "transducer" that is responsible for initiating the postsynaptic neural event.

Homology cloning of these G-protein–coupled receptors uncovered a wide variety of different receptors. Many of these are shown in Table 7-2. When genetic probes were made against portions of the known G-protein receptors and compared with a gene library, several previously unrecognized genes for receptor subtypes were uncovered. For example, five separate genes were identified for the muscarinic cholinergic receptor. Table 7-2 demonstrates important characteristics of many G-protein–coupled receptors.

G proteins. A major receptor transducing element is called the guanine nucleotide (G) protein and is composed of three protein subunits. An integral part of this family of receptors is the G protein, which represents the transduction element between the receptor and the second messenger system that ultimately induces the cellular response. Much of the functional specificity of the receptors resides in the G-protein type. For example, the α_1- and α_2-adrenergic receptors respond to endogenous norepinephrine but initiate entirely different cellular events. The α_1-receptor activation activates a G protein, which initiates the activity of membrane phospholipase (G_p), eventually resulting in increased intracellular calcium. The α_2-receptor, although similar in terms of the ligand-binding unit, activates what is called a G_i protein (i.e., i for inhibitory), which inhibits adenylate cyclase activity and generally causes inhibitory neuromodulatory activity. The β receptors are coupled to G_s proteins (i.e.,

s for stimulatory), which activate adenylate cyclase and activate many cellular processes.

Other G proteins are involved in sensory modulation. In the nasal mucosa, odorant molecules bind to G-protein–coupled receptors that have similar protein structures with seven membrane-spanning domains. These receptors activate a G_o protein, sometimes called G_{olf} for olfactory G protein. One of the earliest characterized G proteins was called transducin (G_t). This molecule is responsible for initiating the signal from stimulation of the rhodopsin molecule by light. Rhodopsin is a receptive protein with seven membrane-spanning domains that is covalently bound to 11-*cis*-retinal.

The G-protein complex is actually composed of three protein subunits called the α, β, and γ subunits. There are an excess of α subunits in the cytoplasm that probably compete for the appropriate β and γ membrane-bound subunits. The β and γ subunits are highly conserved membrane proteins, and they anchor the α subunit near the receptor. The β-γ complex also stabilizes the binding of GDP to the α subunit and inhibits its activation. When the receptor is occupied by a transmitter, the α subunit is released and becomes activated. The α subunit is capable of binding GTP, hydrolyzing it to GDP, and activating or inhibiting the effector or second messenger system, such as adenylate cyclase. The G-protein system allows for a high degree of amplification of the signal. Each receptor typically activates 10 to 20 G proteins.

Many bacterial toxins interfere with G-protein systems. The agent responsible for cholera, cholera enterotoxin, irreversibly activates the G_s protein, increasing the intracellular cyclic AMP. Pertussis toxin irreversibly inhibits the G_i and G_o proteins.

Table 7-2
CHARACTERISTICS OF G-PROTEIN–COUPLED RECEPTORS

Receptor Group	Subtype	AA/TMD*	Effector Pathways[†]
Adenosine	A_1, A_2	326/7	↓ cAMP, K^+, Ca^{2+}
Adrenergic	Alpha$_1$ (3)	466/7	IP_3 DAG
	Alpha$_2$ (3)	450/7	↓ cAMP, K^+, Ca^{2+}
	Beta$_1$	477/7	↑ cAMP
	Beta$_2$	413/7	↑ cAMP
Cannabinoid	THC	472/7	↓ cAMP
Dopamine	D_1	466/7	↑ cAMP
	D_2	443/7	↓ cAMP, K^+, Ca^{2+}
Glutamate/aspartate	Metabotropic	872/7	↓ cAMP
Histamine	H_1	491/7	IP_3/DAG
	H_2	359/7	↓ cAMP
Serotonin	5-HT$_{1A}$	421/7	↓ cAMP, K^+
	5-HT$_{1C}$	460/7	IP_3/DAG
	5-HT$_{1D}$	377/7	↓ cAMP
	5-HT$_2$	471/7	IP_3/DAG
Muscarinic	M_1	460/7	IP_3/DAG
	M_2	466/7	↓ cAMP, K^+

*Number of amino acids (AA) in the receptor/number of transmembrane-spanning domains (TMD).
[†]The effector pathways demonstrate changes in adenylate cyclase and resulting changes in cyclic 3'5'-AMP, modulation by opening (↑) or closing (↓) of potassium (K^+) or calcium (Ca^{2+}) channels, or modulation of phospholipase activity with changes in inositol trisphosphate (IP3) and diacylglycerol (DAG) as second messengers.

Phosphorylation. Many effector mechanisms ultimately result in the phosphorylation of a protein with inhibition or excitation of the postsynaptic cellular event. The final steps in the postsynaptic response include the activation of a second messenger system by the α subunit of the G protein and triggering of a phosphorylation system. Several second messenger systems are involved, including the cyclic nucleotide system, cyclic AMP and cyclic GMP, calcium-calmodulin, the production of nitric oxide from arginine, and the phospholipase C products of inositol triphosphate (IP3) and diacylglycerol (DAG). Figure 7-6 shows several G-protein and second messenger or effector pathways for sensory stimuli such as smell and vision and for neurotransmission by norepinephrine.

A novel effector system that may also represent a transducer or a transmitter is the simple molecule nitric oxide. Although muscarinic vasodilation had been studied for decades, the exact mechanism of this action of acetylcholine was unknown. The muscarinic receptor in the vasculature is not innervated, but it can be activated by circulating choline esters. In the early 1980s, Furchgott found that the muscarinic receptor of the vasculature occurred on the endothelium rather than vascular smooth muscle. The well-known vasodilation induced by stimulation of this receptor turned out to be a secondary effect of the production of nitric oxide by the endothelium. This small, very-short-lived molecule easily traverses lipid membranes, enters the smooth muscle, and activates a guanylate cyclase that results in activation of myosin light chain kinase and relaxation of the muscle. Nitrate vasodilators, such as nitroglycerin, have been used for years to alleviate angina pectoris. This is another example of the body having an endogenous chemical that resembles drugs that have been used for centuries.

Originally called endothelial-derived relaxing factor, nitric oxide was generated by the action of an enzyme that used the amino acid arginine as a substrate. The enzyme has been called nitric oxide synthetase and has been found widely distributed in the brain. In the future, it will be interesting to see whether this compound is given transmitter, transducer, or effector status in the overall scheme of how the brain works.

Receptors as part of the effector pathway

Some receptors are part of a second messenger generating system and directly stimulate the effector pathway. These receptors are themselves the transducers and the generators of the second messengers. Some of these are membrane-bound proteins that include receptors for growth factors, atrial natriuretic peptide, and activin. Others, such as the steroids, bind to soluble cytoplasmic receptors that interact with specific promoter regions of genes to initiate or inhibit genetic translation; this group of receptors is not discussed in this chapter. In the case of growth factors and the polypeptide activin, the receptors themselves are enzymes. Activation by these molecules causes receptor phosphorylation that allows them to function as a protein tyrosine kinase or protein serine-threonine kinase, respectively. The atrial natriuretic peptide receptor is a guanylate cyclase that results in increased intracellular cyclic GMP. Each of these membrane-bound receptors has only one membrane-spanning domain that anchors it to the membrane, unlike the seven spanning domains of the G-protein–coupled receptors.

Neurotransmitters and Their Receptors

Mushrooms, herbs, and bacterial and snake toxins have contributed much to our knowledge of neurotransmitters and their receptors. Early in the development of human civilization, pharmacology was mostly the jurisdiction of herbalists and alchemists. Plant and animal products have provided useful sources of drugs for various ailments and research tools to advance our knowledge of the function of the body and brain. Some of these remedies were quite sophisticated; the plant belladonna was burned in an open fire and inhaled by the ancient Hindus for allergies and asthma; the plant contains the alkaloid atropine. Pilocarpine, the acetylcholinesterase inhibitor, was used in Africa; the nicotinic blocker, curare, from the Amazonian jungles was used to paralyze game; and willow bark (i.e., salicylate) was used to cure fever and pain. Bacterial toxins from *Vibrio cholerae* and *Bordetella pertussis*, which causes whooping cough, interact with G proteins. The snake toxin α-bungarotoxin was essential for binding and purifying the nicotinic receptor. These natural products proved to be important for our present understanding of the brain transmitters and receptors that mediate bodily function and cognition.

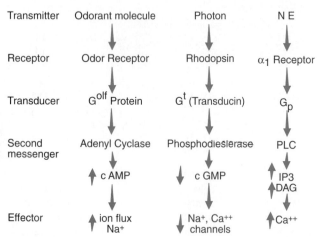

Figure 7-6
There are several G-protein–stimulated pathways, from the receptor responses of stimuli in the nose (ie, odorant), eye (ie, photon), and a representative transmitter to activation of the effector system. PLC, phospholipase C; cAMP, cyclic 3′5′-adenosine monophosphate; cGMP, cyclic 3′5′-guanine monophosphate; IP3, inositol trisphosphate; DAG, daicylglycerol. The arrows indicate activation (↑) or inhibition (↓).

☐ Cholinergic receptors fall into both main genetic families of transmitter receptors

Like the neurotransmitters that were discussed in the previous chapter, cholinergic receptors were the first receptor types to be identified and pharmacologically classified in the early 1900s. The subdivision of these receptors grew out of an interest in the herbs that have been used for centuries as medicines or poisons. The major subdivisions of the cholinergic receptor systems are shown in Figure 7-7 with a representative drug that selectively acts on that system.

Nicotinic receptors

Nicotinic receptors are ligand-gated ionic channels for sodium and selectively respond to the active ingredient in tobacco. The first major subdivision of the cholinergic receptor was named after a plant alkaloid, nicotine, that was introduced into western European cultures by Native Americans. Nicotine selectively stimulates the cholinergic receptor at the neuromuscular junction and the autonomic ganglia. This receptor is also found in the brain, although it is much less prevalent than the muscarinic cholinergic receptor.

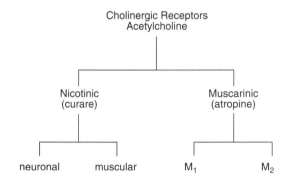

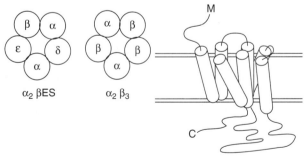

Figure 7-7
The family of cholinergic receptors. This flow chart shows the breakdown of cholinergic receptors and the names of their subtypes. The drugs in parentheses are antagonists that block those receptors. The diagram below the chart shows the proposed subunit configuration of the two nicotinic receptors and the structure of the muscarinic receptor.

Early studies by Dale and Langley showed that, although semipurified acetylcholine stimulated skeletal muscle and the visceral smooth muscle, nicotine was selective only for the neuromuscular junction and the ganglia. Langley was able to map out the sympathetic and parasympathetic autonomic ganglia by applying a tincture of nicotine to peripheral nerve ganglia and observing the visceral effects in the frog.

Further subdivisions of the nicotinic receptor were made during the 1930s and 1940s. Patton and Zamus noticed that the nicotinic receptor of the neuromuscular junction was blocked by a molecule having two quaternary nitrogens separated by 10 carbon atoms (i.e., decamethonium) but that the autonomic ganglia was selectively blocked by a similar molecule having a separation of only a six carbon atoms (i.e., hexamethonium). These molecules looked like two acetylcholine molecules linked back to back but separated by different distances. A second subdivision of the cholinergic receptor was then postulated: muscle and neuronal or ganglionic nicotinic receptors. Figure 7-8 shows the structures of hexamethonium, decamethonium, and acetylcholine and the subunit makeup of the two nicotinic receptors.

Much of the molecular biology of the nicotinic receptor was discussed in the previous section on ligand-gated ion channels. The nicotinic receptor was originally isolated from the electric organ for the eel. To facilitate its purification, the snake venom toxin, which tightly binds to the α subunit of the nicotine receptor, α-bungarotoxin, was used. By radiolabeling this toxin, the purification of the nicotinic receptor could be followed through various chromatographic steps.

With the help of molecular biologic techniques, we know that the human neuromuscular nicotinic receptor is very similar to the electric eel receptor. It contains five subunits that surround a membrane channel that, when open, allows positively charged ions to flow into the cell. Two of the five protein subunits of the nicotinic receptor have binding sites for acetylcholine (i.e., α subunit). The makeup of the other three subunits may affect the distances between the α subunits of muscle-type and ganglia-type nicotinic receptors, which could explain the pharmacologic specificity of the receptors (see Fig. 7-8). The neuronal or ganglionic receptor appears to have two α and three β subunits rather than the α (2), β (1), γ (1), and δ (1) makeup of the muscle receptor.

Muscarinic cholinergic receptors

Muscarinic cholinergic receptors are G-protein–coupled receptors that are activated by muscarine, a substance found in the mushroom, *Amanita muscaria*. The receptor is found on the visceral organs. Euripides, an ancient Greek, was the first to describe the poisonous effects of *Amanita muscaria*, to which he lost several family members. The general syndrome of mushroom poisoning has become known as the SLUDE syndrome, which is an acronym for salivation, lacrimation, urination, defecation, and emesis. This acronym describes quite accurately the toxicologic effects of parasympathetic overstimulation by

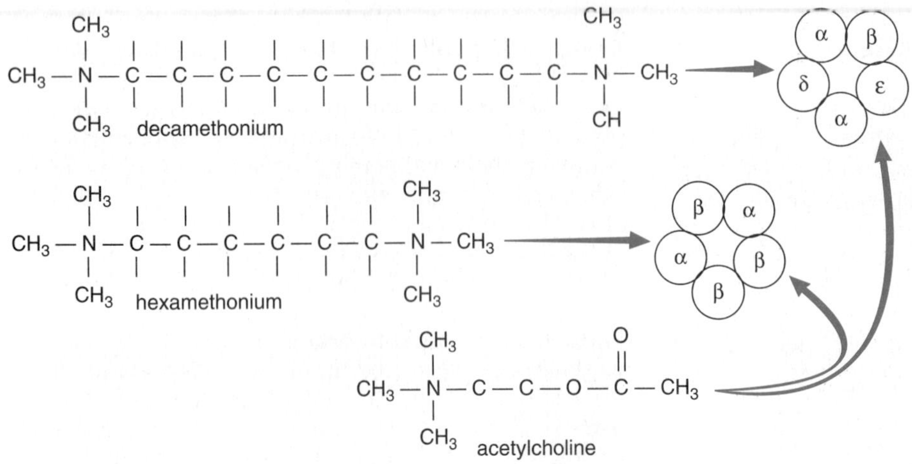

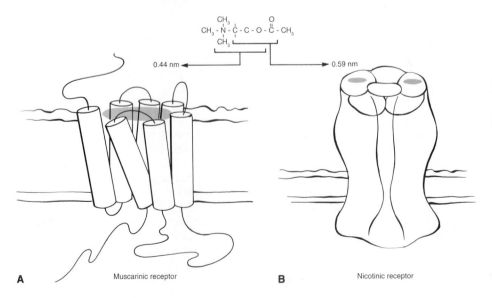

Structures of decamethonium, hexamethonium, and acetylcholine with receptor subunit diagrams.

Figure 7-8
Structures of the selective neuromuscular and ganglionic (i.e., neural) nicotinic receptor antagonists. Decamethonium selectively blocks the muscle receptor, and hexamethonium blocks the neural receptor. The makeup of the pentameric subunits and the distance between the two subunits may partially explain this specificity. Acetylcholine stimulates both of the receptors.

acetylcholine-like agonists, muscarine, and acetylcholinesterase inhibitors. This is the receptor found on the postganglionic parasympathetic effector organs, in eccrine sweat glands, and is the most common cholinergic receptor found in the brain; it is 10 to 100 times more prevalent than nicotinic receptors.

How can the nicotinic and muscarinic receptors respond to the same neurotransmitter but also possess the capacity to respond selectively to different drugs? One explanation has been that acetylcholine is not a rigid molecule, because the two methyl groups separating the quaternary nitrogen of choline and the acetyl groups can freely rotate. This conformation allows the quaternary nitrogen and an atom capable of donating a pair of electrons to exist at various distances. When they are in their closest configuration (0.44 nm), they can "fit" into the muscarinic receptor, and when they are in their more distant configuration (0.59 nm), they can fit into the nicotinic receptor (Fig. 7-9). Examination of the structures of nicotine and muscarine have shown that these rigid molecules have chemical characteristics similar to acetylcholine and are separated by similar intermolecular distances.

Molecular biologic techniques have shown other important differences between the nicotinic and muscarinic receptor that explain their pharmacologic and physiologic uniqueness. Unlike the ligand-gated ion channel nicotinic receptor, the muscarinic receptor is a G-protein–coupled receptor that uses a second messenger system to accomplish its actions. Although acetylcholine is a fast transmitter (1 to 2 milliseconds) when stimulating the nicotinic receptor, it acts like a slow transmitter (100 to 250 milliseconds) when activating the muscarinic receptor. Like the β-adrenergic receptor, the muscarinic receptor is composed of a single membrane-bound protein that has seven membrane-spanning domains within its structure. Five separate muscarinic receptor genes have been cloned, two of which have pharmacologic and therapeutic importance.

The M_2 receptor is the most commonly found peripheral muscarinic receptor. It is blocked by the drug atropine and is found in the heart, bronchi, gastrointestinal

Figure 7-9
Diagram of how acetylcholine can interact with the muscarinic (left) and nicotinic (right) receptor. This theory contends that the distance between the quaternary nitrogen of acetylcholine and an atom capable of donating a pair of electrons is the structural requirement for an agonist. Because acetylcholine has two methyl groups separating the nitrogen and the oxygen, it can exist in both configurations. The muscarinic receptor requires a distance of about 0.44 nm, and the nicotinic receptor requires 0.59 nm.

A Muscarinic receptor

B Nicotinic receptor

tract, and bladder smooth muscle. Stimulation of this receptor affect the appropriate G protein, depending on the tissue type. This can lead to a decrease in cyclic AMP or activation of K^+ channels. In the smooth muscle of the bronchi and gastrointestinal tract, the G_i-mediated decrease in cyclic AMP will lead to contraction. In the heart, activation of non–ligand-gated potassium channels decreases the heart rate and automaticity.

The M_1 muscarinic receptor is widely found in the brain, in the autonomic ganglia, and on hydrochloric acid-secreting cells of the stomach. This receptor uses the IP3-DAG second messenger system and is selectively blocked by the drug pirenzepine. Atropine blocks this receptor as well. The pharmacologic differences in these muscarinic receptors may be important therapeutically for the treatment of peptic ulcers.

☐ The catecholaminergic receptors are G-protein–coupled receptors whose specificities result from differences in ligand binding and effector systems

Soon after the discovery of the sympathetic transmitters in the early 1900s, it was noticed that their postsynaptic actions were extremely variable. Although they stimulated some systems, they were found to inhibit others. They stimulated the heart and most vascular smooth muscle, but they inhibited bronchial, some vascular, and gastrointestinal smooth muscle. To explain these differences, Cannon proposed in the early 1920s that there were two sympathetic adrenergic transmitters: sympathin E excited smooth muscle, and sympathin I inhibited smooth muscle. With the discovery of epinephrine in the adrenal and norepinephrine in the sympathetic nerves, this hypothesis held for decades. In 1948, Raymond Alquist proposed

that the diversity of physiologic actions of catecholamines could be explained by the existence of distinctly different receptor subtypes. He called these adrenergic receptors the α and β receptors.

Alquist's hypothesis was based on the actions of several adrenergic agents and derivatives on visceral activity; among them were norepinephrine, epinephrine, and isoproterenol. The excitatory actions of the catecholamines were ascribed to α-receptor activation. The agents that were most potent in stimulating these actions were norepinephrine and epinephrine. α-Adrenergic activation by these agents included vasoconstriction and pupillary dilation. β-Receptor stimulation tended to be predominantly inhibitory and included bronchial dilation, decreased gastrointestinal motility and bladder smooth muscle tone, vasodilation of some vascular beds, and cardioacceleration. The agents that stimulated these β-receptor functions tended to have larger substitutions on the amine group of the catecholamine. Isoproterenol and epinephrine were much more potent stimulators of these receptors than norepinephrine, which has no methyl groups on the amine terminal. Many of β-receptor agonists and antagonists have been synthesized, strengthening the hypothesis that β-receptor selectivity is largely based on a bulky amino-terminal substitution of the biogenic amine. Figure 7-10 gives a breakdown of the various therapeutically important subtypes of adrenergic receptors.

α-Adrenergic receptors

α-Adrenergic receptors can stimulate postsynaptic events ($α_1$) or through presynaptic receptors can inhibit neuronal activity ($α_2$). The α-adrenergic receptors were pharmacologically subdivided in the 1970s when the actions of some drugs were found to interact with presynaptic alpha-like receptors that inhibited neuronal activity. Molecular biologic techniques have shown that there are

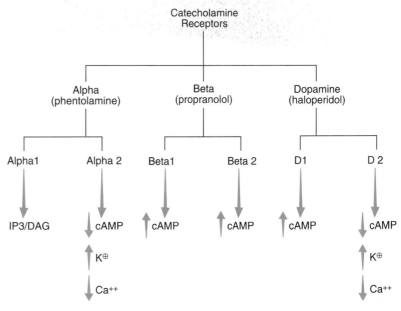

Figure 7-10
The family of catecholamine receptors are shown with their subtypes and major effector pathways. The drugs in parentheses are selective blockers for the different receptors.

five separate genes for various α receptors. All of these receptors and all of the known β and dopamine receptors belong to the G-protein–coupled family, and each is a single membrane-bound protein that has seven membrane-spanning domains. Figure 7-11 describes the shared molecular mechanisms of many of the adrenergic receptors.

The α_1 receptor, as described by Alquist, is an excitatory receptor. It is coupled to a G protein that activates the enzyme phospholipase C. This enzyme is able to cleave the membrane lipid, phosphatidylinositol bisphosphate (PIP_2), into two separate second messengers. DAG remains within the membrane and activates protein kinase C. This kinase is capable of phosphorylating serine and threonine residues of a number of substrates within the membrane or in the cytoplasm. The other PIP_2 second messenger is IP_3. This water-soluble molecule is released into the cytoplasm, where it binds to several intracellular organelles to cause the release of Ca^{2+}. Most of this is contributed by the endoplasmic reticulum. Ca^{2+} then activates tissue-specific secondary effector systems. The ability of a single transducer system with protein kinase C to generate two second messengers from the same substrate provides additional amplification of the original transmitter and α-receptor interaction.

The elucidation of the α_2-adrenergic receptor helped to explain several confusing observations related to the autonomic nervous system and the central nervous system. Many of the side effects of α-receptor blockade seemed to resemble activation of the sympathetic and the parasympathetic nervous system. For example, the side effects of the nonselective α-blocking drug phentolamine included increased gastrointestinal motility (e.g., diarrhea) and elevated circulating levels of the sympathetic transmitter norepinephrine. This plasma norepinephrine is still able to activate the unblocked β receptors. With the development of α_2-type agonists, such as clonidine, the opposite effects were found. This type of drug seemed to turn off sympathetic and parasympathetic activity. Clonidine is useful for high blood pressure because it decreases sympathetic norepinephrine release. However, one of its side effects is decreased gastrointestinal motility.

The α_2-type receptors are found predominantly on neurons within the brain or presynaptically on sympathetic and parasympathetic nerves. The α_2 receptor is coupled to a G_i protein, which results in a decrease in cyclic AMP, inhibition of Ca^{2+} channel opening, and increased potassium channel activity (see Fig. 7-11). The resulting effect is inhibition of neuronal activity. α_2-Adrenergic agonists have been found useful for opiate and nicotine withdrawal symptoms.

β-Adrenergic receptors

Both of the major β-adrenergic receptors are coupled to a G_s protein, which results in an increase in cyclic AMP. β-Adrenergic receptor responses in the peripheral autonomic nervous system include dilation of a variety of smooth muscles, stimulation of metabolic function such as lipolysis and glycogenolysis, and an increase in heart rate. From a therapeutic standpoint, drugs selective for the β receptors of different tissues are important. For example, a drug that would selectively stimulate the β receptor of bronchial smooth muscle (causing bronchodilation) could be useful in asthma, and a drug that could selectively block the cardiac β receptor might be beneficial for angina or high blood pressure. The search for selective β-adrenergic drugs did lead to the discovery of specific β_1 blockers (e.g., atenolol) and β_2 agonists (e.g., terbutaline) and the cloning of three different β receptors that range in size from 402 to 477 amino acids. All of these β receptors use the same transducer and effector systems. As shown in Figure 7-12, they activate a G_s protein that stimulates cyclic AMP formation, leading to phosphorylation of various cellular serine and threonine residues by a cyclic AMP-dependent protein kinase.

The β receptor is the prototypical G-protein–coupled receptor that was the first to be isolated and cloned. The β_1-adrenergic receptor is the subtype responsible for increased heart rate and lipolysis, and most of the other receptors, including those involved in bronchodilation, are defined as β_2. (This can be easily remembered, because there are one heart [β_1] and two lungs [β_2]). There are also many β receptors in the central nervous system, although the subclassifications have not been worked out.

Dopamine receptors

Central dopamine receptors are involved in the mechanisms of movement disorders such as Parkinson's disease and mental illness. Two specific dopamine receptors have been pharmacologically defined, and five have been cloned. Most of the actions of dopamine receptors occur in the central nervous system, although some do mediate peripheral functions. The central actions include vasodilation in the kidney vasculature (D_1 receptor), inhibition of sympathetic ganglionic activity (D_2), and inhibition of

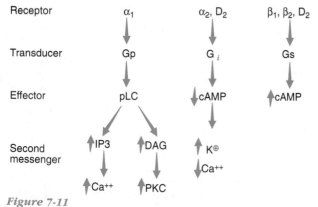

Receptor	α_1	α_2, D_2	β_1, β_2, D_2
Transducer	Gp	G_i	Gs
Effector	pLC	↓cAMP	↑cAMP
Second messenger	↑IP3 ↑DAG ↓ ↓ ↑Ca++ ↑PKC	↑ K⊕ ↓Ca++	

Figure 7-11
Shared effector pathways. Some G protein and effector pathway mechanisms are shared by different types of catecholamine receptors. PLC, phospholipase *C*; *IP3*, inositol trisphosphate; *DAG*, diacylglycerol. The arrows indicate activation (↑) or inactivation (↓) of a pathway.

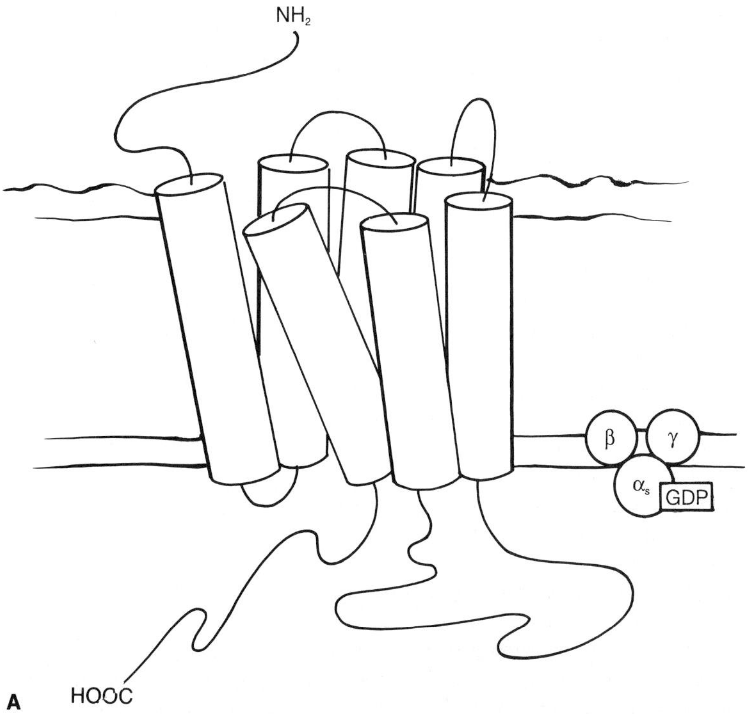

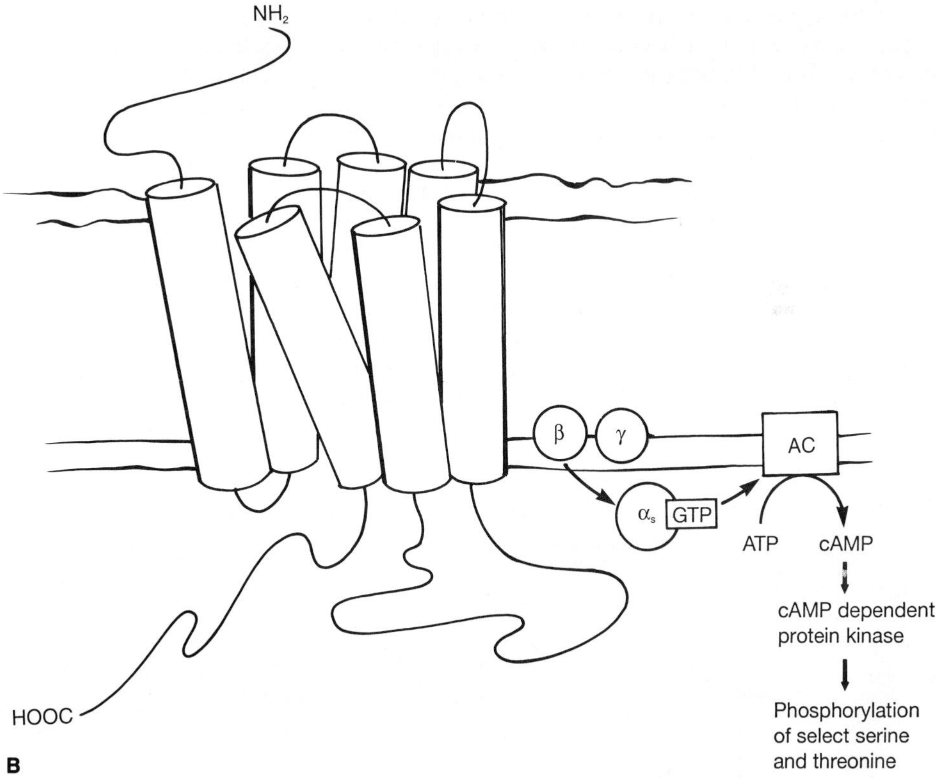

Figure 7-12
Conceptualized G-protein–coupled receptor. The β-adrenergic receptor is coupled to the G protein complex. (**A**) The G protein is composed of β and γ subunits, which are anchored in the membrane, and an α subunit, which binds GTP and can be released into the cytoplasm to activate the second messenger system. (**B**) In this case, the second messenger is cyclic AMP, which is generated when adenylate cyclase (AC) is activated by the G protein.

prolactin secretion by the anterior lobe of the pituitary (D_2). The D_2 receptor also inhibits acetylcholine release from the striatum. This is the transmitter function that is damaged in Parkinson's disease and demonstrates why L-DOPA, the precursor of dopamine, and atropine, the cholinergic muscarinic blocker, are effective in alleviating the tremors associated with this disorder.

The D_1 receptor stimulates cyclic AMP formation, and the D_2 receptor inhibits the formation of this second messenger. This is not the only mechanism; both receptors can affect phosphoinositide turnover in some brain regions. The D_2 receptor has an extremely high affinity for its transmitter, dopamine, which occurs in concentrations in the nM range. The D_1 receptor and the postsynaptic D_2

D₁ receptor

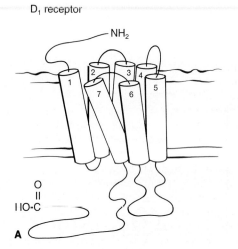

D₂ receptor

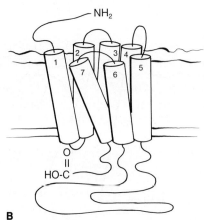

Figure 7-13
The dopamine receptor family subtypes are drawn roughly to scale, showing the structure of the two dopamine (D_1 and D_2) receptors. The D_1 receptor stimulates adenylate cyclase and has a short third cytoplasmic loop between transmembrane domains 5 and 6. It also has a longer carboxyl-terminal end. The D_2 receptor, which inhibits adenylate cyclase, has a long third cytoplasmic loop and short carboxy-terminal end. The seven membrane-spanning domains exhibit a high degree of homology.

receptor appear to have a lower affinity for dopamine (μM range).

The dopamine D_2 receptor gene was the first to be cloned and was discovered using a probe against the β receptor. The receptor is found in high concentrations in the caudate, nucleus accumbens, and olfactory tubercle. The D_1 receptor gene has recently been cloned, and the receptor is found in the caudate, nucleus accumbens, olfactory tubercle, cerebral cortex, limbic system, and hypothalamus. The structure of the two receptors is shown in Figure 7-13. The major differences are found in the size of the cytoplasmic loop between the membrane-spanning domains 5 and 6 and the size of the cytoplasmic carboxyl terminal. The D_2 receptor that inhibits adenylate cyclase has a long loop between the spanning domains 5 and 6 and a short carboxyl terminal. The opposite is true for the D_1 receptor. These cytoplasmic, intracellular loops of protein are thought to interact with their specific G protein to mediate the appropriate effector mechanisms.

☐ Serotonin receptors are diverse

Serotonin receptors are diverse, similar to the cholinergic receptors, and include G-protein–coupled and ligand-gated ion channels. Although serotonin concentrations are very high in the periphery, particularly the gastrointestinal tract and platelets, only a small percentage of the total body serotonin is found in the brain. Nonetheless, brain serotonin is widely distributed and serves a multiplicity of functions. Figure 7-14 shows the cloned serotonin receptors and their second messenger system. Four of the serotonin (5-HT) receptors belong to the G-protein–coupled family, and one is a ligand-gated ion channel. The first 5-HT receptor to be cloned was found using a DNA probe for the G-protein–activating domain of the β receptor.

Early binding studies showed that the 5-HT_1 receptors tended to be inhibitory, and the 5-HT_2 receptor was excitatory. The 5-HT_{1A} and 5-HT_{1D} receptors are coupled to a G_i transducer and inhibit cyclic AMP. The 5-HT_{1A} recep-

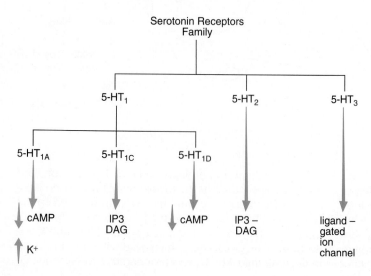

Figure 7-14
The serotonin (5-HT) family of receptors, with the breakdown of serotonin receptor subtypes and the effector pathways that they use. Notice the 5-HT_3 receptor is a ligand-gated ion channel, but the remainder are G protein gated. *DAG*, diacylglycerol; *IP3*, inositol triphosphate.

tor appears to be an autoreceptor and is found in highest concentrations in the raphe nucleus, where it inhibits serotonergic activity. The 5-HT$_{1D}$ receptor also inhibits cyclic AMP formation and is most prevalent in the basal ganglia, where it may be involved in control of voluntary muscle. The 5-HT$_2$ receptor activates phosphoinositol metabolism and causes depolarization of neurons in the cortex. All of these 5-HT receptors have extensive homology with the other G-protein–coupled receptors. They show the greatest amount of variability in the third cytoplasmic loop, between the fifth and sixth membrane-spanning domains, which is an area thought to initiate the G-protein response.

The 5-HT$_3$ receptor is a ligand-gated ion channel and causes depolarization and excitation of neurons in the peripheral nervous system, entorhinal cortex, and the area postrema. Like other ligand-gated channels, the 5-HT$_3$ receptor is composed of five protein subunits that have four transmembrane domains that surround an ionic channel. These receptors are important in sleep-wake cycles, cognitive function, and mental health problems, such as schizophrenia and depression.

☐ Specific histamine receptor-blocking drugs are therapeutically useful

Specific histamine receptor–blocking drugs are therapeutically useful in motion sickness, as sedatives, and for allergic disorders. Receptor binding and molecular biologic techniques have demonstrated the existence of two histamine receptors that respond to two completely different effector pathways. As the case with most transmitters, histamine has peripheral and central actions.

Histamine mediated through H$_1$ receptors is a potent vasodilator, causing a fall in blood pressure with flushing of the face and engorgement of mucosal vasculature. This receptor is coupled to phosphoinositol metabolism, causing an increase in the second messengers IP$_3$ and DAG. It is on this receptor that the classic antihistamine drugs, such as diphenhydramine, work. Figure 7-15 shows the general scheme in which a neurotransmitter can activate a phospholipase C, which causes a branched reaction by liberating DAG and IP$_3$ from phosphoinositide bis-phosphate. The DAG can then activate protein ki-

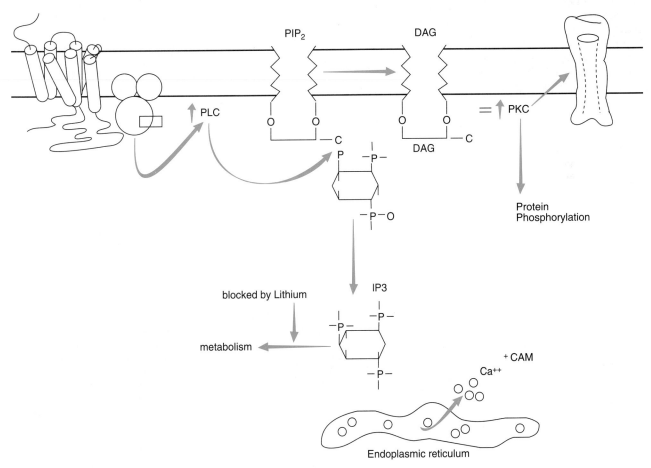

Figure 7-15
G-protein–activated phospholipase-C (PLC) second messenger systems. PLC cleaves phosphoinositol bisphosphate (PIP2) to diacylglycerol (DAG) and inositol trisphosphate (IP3). DAG activates protein kinase C (PKC) to phosphorylate a voltage-gated receptor or cytoplasmic protein. IP3 combines with receptors on the endoplasmic reticulum to release calcium that can act by means of the calmodulin effector pathway. Lithium, a drug used in treating manic depression, inhibits the breakdown of IP3.

nase C, and IP$_3$ releases calcium from the endoplasmic reticulum to interact with calmodulin. Blockade of histamine H$_1$ receptors in the brain causes sedation and inhibits motion sickness, which is the basis of a large market for over-the-counter antihistamines. There is also evidence that the histaminergic pathway activates phospholipase A, which cleaves arachidonic acid from membrane phospholipids, activating lipoxygenase and cyclooxygenase pathways.

The histamine H$_2$ receptor increases cyclic AMP and is a single membrane-bound protein with seven membrane-spanning domains. In the heart, there is an H$_2$ receptor that causes an increased heart rate that is not blocked by the β-blocking agent propranolol. In the stomach, an H$_2$ receptor is responsible for stimulating the parietal cell to secrete hydrochloric acid. The selective H$_2$ blockers, such as ranitidine, block this action of histamine and have revolutionizes the treatment of peptic ulcers. An H$_3$ receptor appears to be an autoreceptor.

☐ Receptors for amino acids provide the major excitatory and inhibitory pathways within the central nervous system

The amino acids glutamate and GABA occur in concentrations 1000 times higher than the better-characterized biogenic amines. In terms of the number and complexity of receptors, the system appears to be proportionally important. Most of the amino acid receptors fall into the category of ligand-gated ion channels, but there are exceptions. The GABA-B receptor and the metabotropic glutamate receptor are G-protein–coupled receptors that have the single-unit, seven trans–membrane structure.

GABA receptors

Both GABA receptors are inhibitory, acting as ligand-gated chloride channels (GABA-A) or by modifying cyclic AMP and changing potassium and calcium channels (GABA-B). GABA serves as the major inhibitory pathway in the brain and is usually contained within a specific structure, such as the cortex or cerebellum. These neurons have relatively short axons. The GABA-A receptor is a pentameric complex containing four different types of proteins. All of the subunits have four transmembrane domains and, along with the glycine receptor, have extensive homology with the nicotinic cholinergic receptor. Many subunit types have been cloned, including six α, four β, three γ, and one δ subunit. The number of possible combinations of these receptors offers a mind-boggling array of putative GABA receptors.

The GABA-A receptor is pharmacologically interesting, because it has specific binding sites for a number of seemingly unrelated drugs and agents. The α subunit has a specific benzodiazepine binding site that facilitates the inhibitory action of GABA on the receptor that augments opening of the chloride channel. This explains much of the sedative and antianxiety action of benzodiazepines such as Valium or Librium. The ability of the GABA receptor to bind to the benzodiazepines, as determined by affinity chromatography, was used to isolate and characterize these receptors. Much effort has gone into isolating and characterizing an endogenous benzodiazepine that may serve this function. This story is similar to the isolation of opiate peptides that bind to the "morphine receptor." Other specific binding sites on the GABA-A receptor exist for the barbiturates, alcohol, and certain anesthetic steroids that may be endogenous in the brain. All of these agents augment the action of GABA and induce inhibition of central nervous system activity, sedation, and sleep.

Figure 7-16 shows a diagram of the two GABA receptors and their effector pathways. In the case of the GABA-

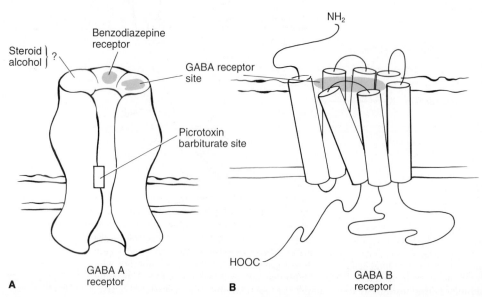

Figure 7-16
Diagram of the GABA receptors. (**A**) The GABA A receptor is a pentameric receptor that forms a chloride ion channel. (**B**) The GABA B receptor is a G-protein–coupled receptor that has seven membrane-spanning domains. GABA can stimulate both receptors, but the exact transmitter binding subunit on the GABA A receptor is not defined. The GABA A receptor also has binding sites for benzodiazepines (eg, librium, valium), steroid anesthetics, alcohol, barbiturates, and picrotoxin. Benzodiazepines, steroids, alcohol, and barbiturates augment or amplify the action of GABA on the GABA A receptor.

A receptor, activation by GABA opens the chloride channel, and this effect can be augmented by benzodiazepines, alcohol, or barbiturates. There is also a specific picrotoxin-sensitive subunit that is the binding site for this channel-blocking toxin. The GABA-A receptor can be selectively activated by the drug muscimole and inhibited by bicuculline. The GABA-B receptor is coupled to a G_i protein that inhibits the formation of cyclic AMP. Like the α_2 and D_2 receptors, this leads to the opening of potassium and closing of calcium channels. The GABA-B receptor can be selectively activated by the drug baclofen, which is an effective muscle relaxant.

Glycine receptor

The glycine receptor is found in the spinal cord and brain stem and is inhibited by the poison strychnine. Although the GABA receptor is the principle inhibitory receptor in the brain and brain stem, glycine serves that role in the spinal cord. The inhibitory glycine receptor is a pentameric ligand-gated receptor that forms a chloride channel. Activation of this channel leads to the entry of chloride ion into the cell, hyperpolarization of the membrane, and inhibition of neural activity. One of the best understood glycine pathways in the spinal cord is that through the Renshaw cell, which inhibits the α motor neuron. When this important feedback inhibition is blocked by the glycine receptor antagonist strychnine, seizures can lead to death. This is the basis of the use of strychnine in rat poison. Sensory information in the spinal cord is also probably filtered through glycine receptors in the cord, because low doses of strychnine increase tactile sensations and have been complexed with a variety of drugs throughout history. There is a glycine binding site on one of the glutamate receptors. Glycine binding to this N-methyl-D-aspartate subtype of the glutamate receptor increases the frequency of cation channel opening (i.e., excitation). This effect is not antagonized by strychnine.

Glutamate receptors

The glutamate receptors respond to a variety of agents found in hot peppers that are often used in spicy foods. The acidic amino acids glutamate and aspartate stimulate almost all brain neurons. This action was thought to be nonspecific because it was so prevalent, but at least five different excitatory amino acid receptors are now known to exist, and specific neural pathways that use these acidic amino acids as their transmitters are beginning to be defined. The glutamate receptor system does appear to be the major stimulatory transmitter system in the brain. These pathways seem to have long axonal projections and communicate information between major brain structures, such as the cortex to midbrain areas.

Most of the glutamate receptor subtypes belong to the ligand-gated ionic channel family, with the exception of the "metabotropic" glutamate receptor that is coupled to a G protein. These ligand-gated receptors seem to belong to a different gene family than the nicotinic, GABA, and glycine receptors. In the glutamate receptors that have been cloned, the extracellular amine terminal of the subunit peptides are much longer, containing almost 500 amino acids rather than the 200 found in the nicotinic receptor. Like the other ligand-gated channels, the glutamate receptors are pentameric and have four transmembrane domains within each subunit (Fig. 7-17).

Activation of the glutamate receptors by an excitatory amino acid opens a cationic channel that depolarizes and stimulates the postsynaptic neuron. In the case of the AMPA and kainate receptor subtypes, this channel is permeable to sodium and potassium. They are stimulated by the drugs quisqualate and kainate. Only the NMDA receptor has a channel that is also permeable to calcium. The NMDA receptor is relatively slow and has an unusual form of voltage regulation. The channel has a Mg^{2+} binding site that inhibits ion flow when occupied. When the neuron is depolarized by other neural mechanisms, the Mg^{2+} can be displaced and the receptor channel opened by glutamate or NMDA. A unique aspect of the NMDA receptor is that it only functions adequately in the presence of glycine.

The NMDA receptor is distributed throughout the brain, particularly in the cortex and hippocampus. Al-

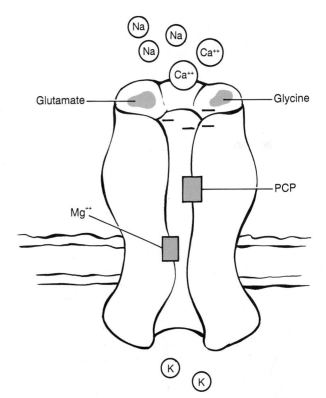

Figure 7-17
NMDA subtype of the glutamate receptor. The diagram of the glutamate N-methy-D-aspartate (NMDA) receptor shows ligand binding sites for glutamate, glycine, magnesium, and the drug phencyclidine (PCP). When the cell becomes depolarized, magnesium is released from the interior of the channel, which allows glutamate to open the receptor channel. A glycine binding site is also needed for channel activation. The NMDA receptor is unique in that calcium can enter the channel, as do sodium and potassium.

Table 7-3
NEURAL PEPTIDES

Peptide	Subtype	Effector Pathways	AA/TMD[†]
Angiotensin II		IP_3/DAG[*]	359/7
Atrial natriuretic peptide	A, B	↑ cGMP	1029/1
Bradykinin	B_1, B_2	IP_3/DAG	360/7
Neuropeptide Y	Y_1, Y_2	↓ cAMP/Ca^{2+}	349/7
Neurotensin		IP_3/DAG ↓ cAMP	427/7
Opiate receptors	μ	↓ cAMP, K^+	
	δ	↓ cAMP, K^+	372/7
	κ	Ca^{2+}	
Vasopressin	V_1	IP_3/DAG	395/7
	V_2	↑ cAMP	371/7
Oxytocin	OT_1	IP_3/DAG	388/7
Vasoactive intestinal peptide	V_1P_1,V_1P_2	↑ cAMP	457/7

*IP_3/DAG means that the effector system uses the inositol trisphosphate and diacylglycerol effector pathway.
†Number of amino acids (AA) in the receptor/number of hydrophobic transmembrane domains (TMD) in the receptor protein. The atrial natriuretic peptide receptor has only one TMD and is itself a guanylate cyclase.

most all cells have NMDA receptors, and they appear to be important in synaptic plasticity. The distribution of the AMPA receptor parallels that of the NMDA receptor, but the kainate receptor is localized in specific regions of the brain, including the hippocampus. Overactivation by glutamate and the application of large concentrations of the selective agonists (e.g., kainate, ibotenic acid) is neurotoxic. It appears that overstimulation of the ligand-gated glutamate channels causes postsynaptic neuronal death.

The metabotropic glutamate receptor is linked to a second messenger system that uses IP_3. This G-protein–coupled receptor may be important in developmental plasticity.

☐ Peptide receptors demonstrate a diversity of biologic activity

Just as there was an explosion in the discovery of neuroactive peptides around 1970, the advent of molecular biologic techniques stimulated the discovery of a vast array of neuropeptide receptors. Almost all of these belong to the G-protein–coupled receptor family, but the exceptions are important. Atrial natriuretic peptide is also present in the brain and acts on a protein receptor that has only one membrane-spanning domain. The cytoplasmic extension of this receptor is itself a guanylate cyclase, which is activated when the receptor is occupied by the agonist. Table 7-3 lists representative neural peptides, their subtypes, the effector pathways, and the number of amino acids and membrane-spanning domains within their structures.

Conclusion

There are numerous neurotransmitter receptors that belong to only a few major gene families. However, a given transmitter often has several subtypes of receptors that use different G-protein coupling and effector systems. In many cases, transmitter receptors belong to entirely different families, representing ligand-gated channel and G-protein–coupled receptor subtypes. The receptor system offers extensive diversity for biologic actions and remarkable specificity despite using a limited number of chemical messages. The eventual cellular activation that is mediated by the receptor represents changes in ionic current by ligand-gated channels or simple protein phosphorylation that activates diverse effector mechanisms.

SELECTED READINGS

Bjorklund A, Hökfelt T, Kuhar MJ (eds). Classical transmitters and receptors in the CNS, part I and II. In Handbook of chemical neuroanatomy, vols 2 and 3. Amsterdam: Elsevier, 1984.

Cooper JR, Bloom FE, Roth RH. The biochemical basis of neuropharmacology, 6th ed. New York: Oxford University Press, 1991.

Evans CJ, Keith DE, Morrison H, Magendro K, Edwards RH. Cloning of a delta opioid receptor by function expression. Science 1992;258:1952.

Glennon RA. Serotonin receptors: clinical implications. Neurosci Biobehav Rev 19990;14:35.

Gillman AG, Rall TW, Nies AS, Taylor P (eds). Goodman and Gilman's the pharmacological basis of therapeutics, 8th ed. New York: Pergamon Press, 1990.

Kenakin TP, Bond RA, Bonner TI. Definition of pharmacological receptors. Pharmacol Rev 1992;44:351.

Sibley DR, Monsma FJ. Molecular biology of the dopamine receptors. Trends Pharmacol Sci 1992;13:61.

Sivilotti L, Nistri A. GABA receptor mechanisms in the central nervous system. Prog Neurobiol 1991;36:35.

Spedding M, Paoletti R. Classification of calcium channels and the sites of action of drugs modifying channel function. Pharmacol Rev 1992;44:363.

Watson S, Girdlestone D (eds). TIPS receptor and ion channel no-menclature supplement 1994. Trends in Pharmacological Sciences 1994; (suppl) vol 15.

Zorumski CF, Thio LL. Properties of the vertebrate glutamate receptors: calcium mobilization and desensitization. Prog Neurobiol 192;39:295.

SURVEY OF STRUCTURE AND FUNCTION

Neuroscience in Medicine, edited by P. Michael Conn.
J. B. Lippincott Company, Philadelphia © 1995

Chapter 8

Neuroembryology and Differentiation

JEAN M. LAUDER

Understanding how the nervous system develops is an important first step in studying the complex neuroanatomic and functional relations present in the adult. This is especially true for the central nervous system (CNS), in which early embryonic structures undergo regional growth and differentiation to form the major subdivisions of the brain and spinal cord. It is also important to understand the principles of neuroembryology, because many neurologic disorders and birth defects originate from abnormal embryonic development.

Neuroembryology

☐ The neural plate is the precursor of the nervous system

At the end of the second gestational week, the embryo consists of a bilaminar disc, called the *embryonic disc* or *epiblast*, composed of ectoderm and endoderm. During the third week, the process of *gastrulation* takes place,

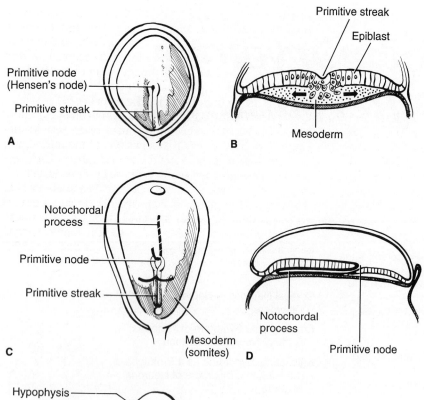

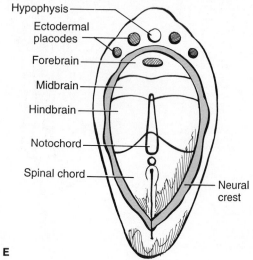

Figure 8-1

Drawings of the presomite human embryo illustrate the appearance of the primitive streak from (**A,C,E**) dorsal, (**B**) transverse, and (**D**) midline aspects. (**B,C**) Invagination (i.e., inward migration) of epiblast cells to form the intraembryonic mesoderm and notochordal process. (**E**) The developmental fates of the ectoderm after invagination and induction of the neural plate (i.e., neural induction).

beginning with formation of the *primitive streak* on the dorsal surface of the epiblast, which establishes the longitudinal axis. Migrating cells from the epiblast invaginate through the primitive streak to form the intraembryonic mesoderm and endoderm (Fig. 8-1*A,B*). Cells remaining in the epiblast become the ectoderm. Cells migrating through the primitive streak laterally become intraembryonic mesoderm, and cells migrating through the *primitive node*, also called Hensen's node, rostrally become the *notochordal process* (Fig. 8-1*C,D*), which develops into the notochord. Simultaneously, the ectoderm begins to thicken and form the neural plate. This process, known as *neural induction*, is thought to involve signaling between invaginating cells or their derivatives and overlying ectoderm. Neural induction causes irreversible specification of the neural plate, committing certain regions to form particular parts of the nervous system (Fig. 8-1*E*).

On day 18 of gestation, the thickened neural plate begins to develop into the *neural tube*. This process, called *neurulation*, begins by bending of the neural plate to form the neural groove at the point where it overlies the notochord (possibly under its influence) and elevation of the lateral edges of the neural plate to form the *neural folds* (Fig. 8-2*A,B*). Once elevated, the neural folds begin to approach each other and fuse in the midline to form the neural tube. Fusion begins first in the cervical region and proceeds in rostral and caudal directions (Fig. 8-2*C*). Until fusion is completed, the rostral (i.e., cranial) and caudal ends of the neural tube remain open (Fig. 8-2*D*) in the form of cranial and caudal neuropores, providing communication between the lumen of the neural tube and the amniotic cavity. The cranial and caudal neuropores normally close at 25 and 27 days of gestation, respectively. Failure of cranial neuropore closure leads to *anencephaly*, a condition of open cranial neural folds. This birth defect is commonly associated with failure of caudal neural tube closure, called *rachischisis*, which is a form of spina bifida, a term meaning a cleft of the spinal column.

☐ The neural tube forms the brain and spinal cord

After closure, the cranial end of the neural tube expands because of differential cell proliferation to form the three primary brain vesicles (Fig. 8-3), the *prosencephalon*, *mesencephalon*, and *rhombencephalon*, and bends at three locations, the *cephalic, mesencephalic,* and *cervical flexures*. These flexures divide the mesencephalon from the rhombencephalon and the rhombencephalon from the spinal cord. That part of the neural tube located adjacent to the somites becomes the spinal cord.

Through further cell proliferation, neuronal migration, and differentiation, the three primary brain vesicles develop into the five major subdivisions of the adult brain (see Fig. 8-3). The lumen of the neural tube of each vesicle becomes one of the brain *ventricles*. Brain structures developing around each ventricle remain associated with that ventricle in the adult. This organizational scheme is useful in thinking about the three-dimensional associations of adult brain regions. The cortex, limbic system, and basal ganglia develop around the lateral ventricles

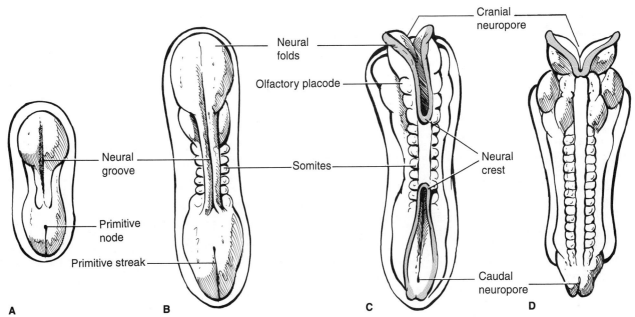

Figure 8-2
Dorsal view drawings of human embryos show (**A**) the neural plate at approximately 18·days, (**B**) elevation of the neural folds at approximately 20 days, and (**C,D**) and neural tube closure at approximately 22 and 23 days.

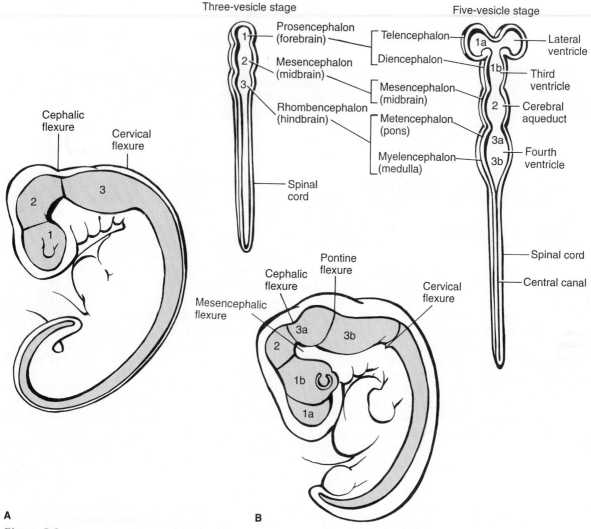

Figure 8-3
Formation of the brain vesicles, ventricles, and flexures. (**A**) Dorsal and lateral views of a 4-week-old embryo (three-vesicle stage). (**B**) Dorsal and lateral views of an embryo at the beginning of the sixth week (five-vesicle stage). (Modified from Martin JH. Neuroanatomy, text and atlas. New York: Elsevier, 1989)

(i.e., from the telencephalic vesicle), and the thalamus and hypothalamus develop from the diencephalic vesicle and surround the third ventricle. The cerebral aqueduct passes through the midbrain, and the pons lies ventral to the fourth ventricle.

The *spinal cord* grows longitudinally and does not form vesicles. The vertebral column also lengthens and eventually outdistances the spinal cord, such that by birth, the caudal end of the cord reaches only as far as the third lumbar vertebra (Fig. 8-4). This differential growth continues postnatally, and by adulthood, the spinal cord extends only to the first lumbar vertebra. This arrangement requires that the dorsal and ventral rootlets of the lumbar and sacral segments of the spinal cord must travel down to the first sacral vertebra to exit the spinal column. These rootlets are called the *cauda equina,* which is Latin for horse's tail. The space surrounding the

cauda equina, called the *lumbar cistern,* is filled with cerebrospinal fluid (CSF) and covered by meninges. In clinical practice, this region represents a relatively safe area from which to withdraw CSF by lumbar puncture (i.e., spinal tap) or inject anesthetics (i.e., spinal block).

☐ The neural crests and ectodermal placodes form the peripheral nervous system

As the neural tube closes, neural crest cells at the lateral edges of the neural folds (Fig. 8-5*A,B*; also Fig. 8-1*E* and Fig. 8-2*C,E*) separate from the rest of the *neuroectoderm* and locate between the neural tube and overlying ectoderm (Fig. 8-5*C*). These cells proliferate and migrate to

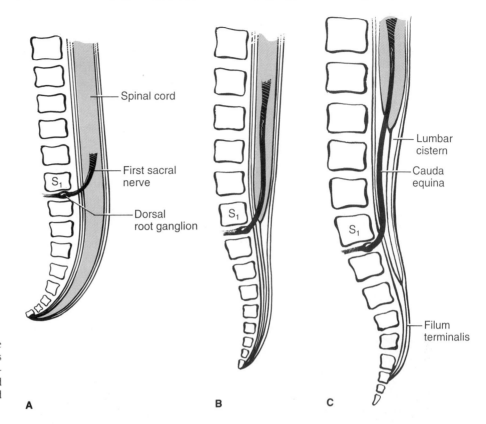

Figure 8-4
The spinal cord in relation to the vertebral column at different stages of development, showing formation of the cauda equina. (**A**) Third month, (**B**) end of fifth month, and (**C**) newborn.

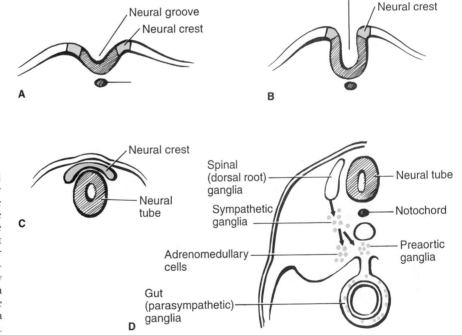

Figure 8-5
Migration of neural tube and neural crest cells in the trunk of successively older embryos. (**A,B**) The cells of the neural crest are initially located at the edges of the neural folds. (**C**) These cells detach from the neural tube as it closes and form an intermediate layer between the neural tube and ectoderm. (**D**) They then migrate laterally to form the spinal (dorsal root) ganglia and cranial nerve ganglia, or migrate ventrally to form sympathetic ganglia or adrenal (suprarenal) chromaffin cells.

various destinations. Depending on the level of the neur-axis where they originate and their route of migration, neural crest cells differentiate into different types of neural cells (Fig. 8-5), including peripheral neurons of the spinal (dorsal root) ganglia, sensory ganglia of the cranial nerves, autonomic ganglia, and glial cells of the peripheral nerves, such as Schwann cells and satellite cells. Other neural crest cells differentiate into nonneural phenotypes such as melanocytes, adrenal (suprarenal) chromaffin cells, cardiac mesenchyme, meninges, and craniofacial skeletal elements. Ectodermal placodes, which arise from nonneural ectoderm (see Fig. 8-1), also contribute neurons to some cranial nerve ganglia and produce the sensory epithelia of the olfactory and au-ditory systems. Abnormal neural crest migration can lead to Hirschsprung's disease, a condition of absent parasympathetic ganglia of the lower bowel with

associated megacolon, and familial dysautonomia, pro-ducing abnormal sympathetic and parasympathetic func-tion.

Spinal nerves develop by the coordinated growth of axons into the dorsal horn from neural crest–derived dorsal root ganglia and outgrowth of axons from spinal motor neurons in the ventral horn (Fig. 8-6*A,B*). These axons grow toward the periphery, where they synapse with the motor end plates of striated muscles. Pregan-glionic sympathetic neurons from the intermediolateral horn grow axons toward the periphery, where they form synapses with neural crest–derived postganglionic neu-rons in the sympathetic ganglia of the trunk (Fig. 8-6*C*). Preganglionic parasympathetic neurons from the cervical and sacral regions of the spinal cord grow axons toward neural crest–derived postganglionic neurons in the wall of the gut.

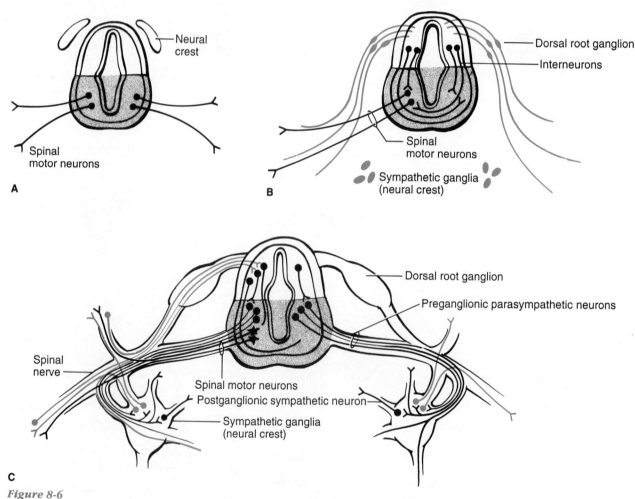

Figure 8-6
Formation of spinal nerves. (**A,B**) Spinal nerves develop by the coordinated growth of axons into the dorsal horn and outgrowth of axons from the spinal motor neurons in the ventral horn. (**C**) Preganglionic sympathetic neurons from the intermediolateral horn grow axons toward the periph-ery, where they form synapses with neural crest–derived postganglionic neurons located in the sym-pathetic ganglia of the trunk.

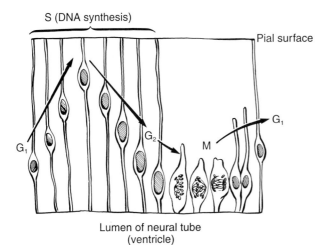

Figure 8-7
Cell proliferation in the neuroepithelium of the recently closed neural tube. The wall of the neural tube is composed entirely of proliferating neuroepithelial cells at this stage and appears as a pseudostratified epithelium in histologic sections. This effect is created by interkinetic nuclear migrations occurring during the G_1 to S (DNA synthesis), and G_2 phases of the cell cycle. During mitosis (M), the cells retract their distal processes, become rounded, and divide next to the lumen of the ventricle.

Differential Proliferation, Migration, and Differentiation

☐ Cell proliferation occurs in germinal zones

After neural tube closure, cell proliferation occurs in *germinal zones* located adjacent to the ventricles of the entire neuraxis (i.e., ventricular zones). At this time, the neural tube consists entirely of proliferating *neuroepithelial cells* and has the appearance of pseudostratified epithelium (Fig. 8-7) because of the pattern of cell proliferation, which involves *interkinetic nuclear migration*. During the cell cycle, neuroepithelial cells extend processes from the ventricular lumen to the external limiting membrane. As these cells enter DNA synthesis (i.e., S phase of the cell cycle), their nuclei migrate away from the ventricle. After S phase, they descend toward the lumen, where they retract their distal process, become rounded, and undergo mitosis. During the early phases of development, both sister cells repeat this cycle, generating increasing numbers of proliferating neuroepithelial cells.

After neuronal differentiation begins, the wall of the neural tube undergoes several transitions (Fig. 8-8), which occur at particular times at different levels of the neuraxis. Increasing numbers of cells leave the proliferative cycle after mitosis and begin to differentiate into

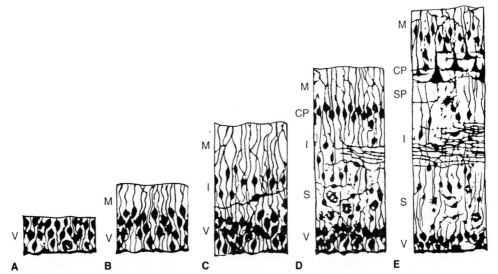

Figure 8-8
Developing wall of the telencephalic vesicle. (**A,B**) Early stages of neural tube development, showing the ventricular zone (V) and marginal zone (M). (**C**) In most brain vesicles and the spinal cord, differentiating neurons accumulate in the intermediate zone (I), which is also called the mantle layer. (**D,E**) In the telencephalic vesicle, neurons destined for the cerebral cortex migrate through the intermediate zone and accumulate in the cortical plate (CP) and the subplate (SP), a transient layer of neurons that serve as temporary synaptic targets for cortical afferents. These plates form layers II through VI of the cortex. The subventricular zone (S) is a secondary germinal zone that gives rise to small interneurons and glia. (Modified from Sidman RL, Rakic P. Neuronal migration, with special reference to developing human brain. Brain Res 1973;62:1.)

neurons or glia. Neurons are permanently postmitotic after they leave the cell cycle. Glial cells are only temporarily postmitotic; they resume proliferation at a later stage of development, when they give rise to other types of CNS glia, including astrocytes. For some time, glia remain attached to the ventricular surface and external limiting membrane, giving them the name *radial glia*. Postmitotic neurons move away from the lumen as if they were going to begin another cell cycle and accumulate distal to the ventricular zone, where they form the *intermediate zone*, also called the mantle zone (Fig. 8-8*C*). In the spinal cord and brain stem, the intermediate zone develops into the alar and basal plates. Distal to the intermediate zone is the *marginal zone* (Fig. 8-8*B–E*), an acellular layer consisting of the processes of neuroepithelial cells, neurons, and radial glia. In the developing cerebral cortex, most neurons migrate through the intermediate zone and accumulate in the cortical plate (Fig. 8-8*D,E*), which forms layers II through IV of the cortex. The *subplate* is a transient layer of neurons that serve as temporary synaptic targets for cortical afferents. Further details of cortical development are given in Chapter 22.

☐ Neurons use radial glia as guides for migration

At early stages of neural tube development (see Fig. 8-8*B*), it is controversial whether neurons actually undergo true migrations or move to the intermediate zone by interkinetic nuclear migration, followed by detachment of their basal process from the ventricular surface. The work of Rakic indicates that after the wall of the neural tube begins to become more complex (see Fig. 8-8*C–E*), radial glia serve as migrational guides for neurons, which often migrate over relatively long distances (Fig. 8-9). Rakic (1988) has proposed that groups of neurons generated in particular sectors of the telencephalic vesicle migrate as cohorts along radial glia, directly transposing a map of the ventricular zone onto the developing cortical plate. This process could provide one basis for the development of cytoarchitectonic areas in the cerebral cortex.

Cortical malformations such as lissencephaly (i.e., small, less convoluted cortex), pachygyria (i.e., abnormally large and small gyri), polymicrogyria (i.e., highly convoluted, thin cortex with many small gyri), and neuronal ectopias (i.e., clusters of neurons that do not reach their final destinations) may be caused by abnormalities in germinal cell proliferation or neuronal migration.

☐ The cerebellum develops from the rhombic lips of the pons

The rhombic lips are transient structures arising from the roof of the metencephalon (i.e., pons) along the rostrodorsal edge of the fourth ventricle. Through cell proliferation in the underlying ventricular zone and bending of the brain stem at the pontine flexure, the rhombic lips

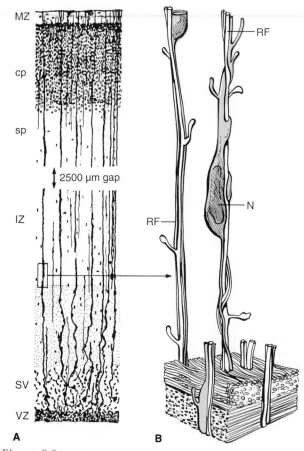

Figure 8-9
Radial glia as guides for neuronal migration. (**A**) Camera lucida drawing of the occipital lobe of the cerebral cortex of the midgestation monkey fetus stained by the Golgi method showing neurons migrating along radial glial fibers (RF). (**B**) Drawing of migrating neuron (N) made from three-dimensional reconstructions of semiserial electron microscopic sections (*right*) through the boxed area in **A**. *M,* marginal zone; *C,* cortical plate; *SP,* subplate; *I,* intermediate zone; *SV,* subventricular zone; *V,* ventricular zone. (Modified from Rakic, P, Defects of neuronal migration and the pathogenesis of cortical malformations. Prog Brain Res 1988;73:18.)

join to form the *cerebellar plate* (Fig. 8-10), which develops into the cerebellar vermis and hemisphere (Fig. 8-10*B*). Large neurons of the cerebellar cortex and deep cerebellar nuclei are formed by outward migration from the ventricular zone beneath the cerebellar plate. Small interneurons (e.g., granule, basket, and stellate cells) are formed by inward migration from a secondary germinal zone located on the surface of the cerebellum, called the external granular layer (EGL; Fig. 8-10*C*). The EGL forms from cells that migrate from the lateral edges of the rhombic lip over the surface of the cerebellar plate and proliferate. Granule cells begin their differentiation in the deep layer of the EGL, where they grow T-shaped axons (i.e., parallel fibers) and begin to migrate inward along radially oriented processes of Bergmann glia. After they reach their final destination in the internal granular layer, granule cells grow claw-like dendrites that form

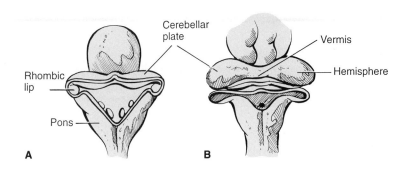

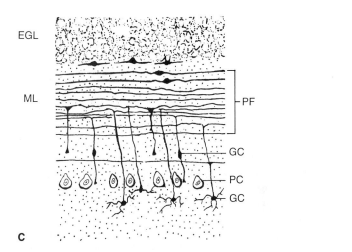

Figure 8-10

Development of the cerebellum. (**A,B**) The cerebellar plate develops from the rhombic lip of the metencephalon (pons) along the edges of the fourth ventricle (IV). (**C**) A three-dimensional view of the developing cerebellar cortex shows immature granule cells (GC) leaving the external granular layer (EGL), forming T-shaped axons called parallel fibers (PF), and then descending through the molecular layer (ML) past the Purkinje cells (PC). Differentiating GC in the internal granular layer (IGL) develop claw-shaped dendrites that synapse with incoming afferents and axons of Golgi neurons (not shown). Basket cells (BC) and stellate cells (not shown) remain in the molecular layer, enmeshed in a grid of parallel fibers. They also develop from the EGL and are oriented perpendicular to the axons of the GC.

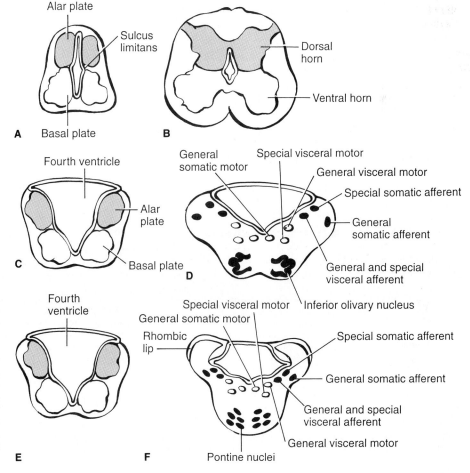

Figure 8-11

The alar and basal plates give rise to the sensory and motor nuclei of the (**A,B**) spinal cord, (**C,D**) medulla, and (**E,F**) pons. The spatial relations of the alar and basal plates are similar in the (**A**) early spinal cord and (**C,E**) brain stem. In the more mature brain stem, expansion of the fourth ventricle causes the nuclei derived from the alar plate to end up lateral to those from the basal plate. (**D,F**) Two sets of nuclei become displaced ventrally in the mature brain stem, one from the alar plate, called the general somatic afferent (*open circles*), and one from the basal plate, called the special visceral motor (*closed circles*). If this is taken into account, the relative positions of the functional columns of cranial nerve nuclei derived from the alar and basal plates can still be appreciated in the adult brain. (Modified from Martin JH. Neuroanatomy, text and atlas. New York: Elsevier, 1989.)

Table 8-1

FUNCTIONAL CLASSES OF SPINAL NERVES

Classification	Function	Structure Innervated
Afferent Fibers		
General somatic	Tactile, pain/temperature proprioception	Skin, mucous membranes, skeletal muscles
General visceral	Mechanical, pain/temperature	Viscera, cardiovascular, respiratory, genitourinary tract
Motor Fibers		
General somatic	Skeletal muscle control (somites)	Limb and axial musculature
General visceral	Autonomic control	Sweat glands, gut

Martin JH. Neuroanatomy: text and atlas. New York: Elsevier, 1989:483.

synapses with Golgi cell axons and incoming afferents (mossy fibers).

Other small interneurons, the basket and stellate cells, originate from the EGL but do not truly migrate. After leaving the EGL, they grow processes oriented at right angles to the parallel fibers, which traps them in the molecular layer. The timing of their exit from the EGL determines where they will be located in the molecular layer. Basket cells, which form first, are found at the bottom of the molecular layer, where parallel fibers pile on top of them. Stellate cells are deposited nearer the top of the developing molecular layer (towards the EGL). As more parallel fibers develop and the EGL disappears, stellate cells are found throughout the outer half of the molecular layer (see Fig. 8-10*C*).

☐ The alar and basal plates give rise to sensory and motor nuclei

Sensory relay and motor nuclei of the cranial nerves and spinal cord are formed from accumulations of neurons in the intermediate zones of the dorsal and ventral sectors of the neural tube. These clusters of neurons are called the *alar and basal plates* and are separated by an acellular region known as the *sulcus limitans* (Fig. 8-11). Spatial relations of alar and basal plates are similar in spinal cord and brain stem during the early stages of development (Fig. 8-11*A,C,E*). As development proceeds, the sensory nuclei of the brain stem alar plate become located lateral to motor nuclei of the basal plate because of expansion of the fourth ventricle. The middle nuclei in each set become displaced ventrally (Fig. 8-11*B,D*). These rearrangements explain why sensory nuclei are always located lateral to motor nuclei in the adult brain stem.

☐ Cranial nerve nuclei are classified according to the embryologic origin of their target tissues

Cranial nerve nuclei form discontinuous columns that extend through the brain stem. These columns are described according to their function and the embryologic origin of the tissues they innervate. This organizational scheme comes from terminology devised for the spinal

Table 8-2

FUNCTIONAL CLASSES OF CRANIAL NERVES

Classification	Function	Structure Innervated	Cranial Nerve*
Afferent Fibers			
General somatic	Tactile, pain/temperature proprioception	Skin, mucous membranes, skeletal muscles of head and neck	V, VII, IX, X
General visceral	Mechanical, pain/temperature proprioception	Oral cavity, pharynx, larynx	V, VII, IX, X
Special somatic	Audition, balance	Cochlea, labyrinth	VIII
Special visceral	Taste, olfaction	Taste buds, olfactory epithelium	I, VII, IX, X
Motor Fibers			
General somatic	Skeletal muscle control (somites)	Extraocular and tongue muscles	III, IV, VI, XII
General visceral	Autonomic control	Tear glands, sweat glands, gut	III, VII, IX, X
Special visceral	Skeletal muscle (branchiomeric)	Facial expression, jaw	V, VII

*The optic nerve (II) is not considered in this table because it does not contain the axons of primary sensory neurons but rather those of third-order neurons in the visual pathway. The visual system is considered to be part of the special somatic afferent class.
Martin JH. Neuroanatomy: text and atlas. New York: Elsevier, 1989:483.

cord (Table 8-1). Afferent fibers from dorsal root ganglia to the dorsal horn are classified as general somatic or general visceral afferents, depending on whether they bring sensory information from mesodermal (i.e., somatic) or endodermal (i.e., visceral) derivatives, respectively. Motor fibers from the ventral horn are classified as general somatic or general visceral efferents. Cranial nerve nuclei are classified in a similar manner, but they have an additional designation, that of "special," such as special somatic afferent (Table 8-2). When applied to sensory nuclei, special indicates that information is coming from organs of special senses (e.g., eye, ear, nasal epithelium, tongue). For motor nuclei, special indicates that cells innervate muscles of the branchial arches. Because functional columns are located in the same relative positions throughout the brain stem (Fig. 8-11*D,F*), knowl-edge of this classification scheme can be useful in learning functional and anatomic associations of the cranial nerve nuclei in the adult brain.

SELECTED READINGS

Martin JH. Neuroanatomy: text and atlas. New York: Elsevier, 1989:483.

Rakic P. Defects of neuronal migration and the pathogenesis of cortical malformations. Prog Brain Res 1988;73:15.

Shatz CJ, Chun JJM, Luskin MB. The role of the subplate in the development of the mammalian telencephalon. In Peters A, Jones EG (eds). Cerebral Cortex, vol 7. New York: Plenum Publishing Corp, 1988:35.

Sidman RL, Rakic P. Neuronal migration, with special reference to developing human brain. Brain Res 1973;62:1.

Clinical Correlation
DISORDERS OF NEURONAL MIGRATION

ROBERT L. RODNITZKY

Abnormalities in the process of neuronal migration during embryogenesis result in CNS structures that are dysfunctional, architecturally abnormal, or totally absent. Several distinct mechanisms can be operative in the etiopathogenesis of clinical disorders related to disrupted neuronal migration. Some neuronal migrational disorders are heritable, and others are presumed to be the result of ischemic, toxic, or metabolic damage during the perinatal period.

Cortical Neurons in Neuronal Migration Disorders

The most common neuronal migration abnormalities involve the neocortex. Normal neuronal migration to the cerebral cortex takes place during the 6th to 24th weeks of gestation. Neurons originating deep in the brain along the surface of the ventricles migrate to the cortex along a network of extensions of glial cells known as radial glia. The first neurons to arrive populate the deepest layer of the cortex, and successive groups of arriving cells occupy progressively more superficial positions. It has been postulated that some structural abnormalities of the cortex are the result of perinatal insults that have caused the death of radial glia or disruption of the network they normally form. The end result is a group of conditions characterized by abnormal laminar or columnar neuronal cortical architecture. In these conditions, a variety of abnormalities are often apparent on gross inspection of the brain, including decreased overall brain size and weight (i.e., microcephaly), abnormally small cortical gyri (i.e., polymicrogyria), or more commonly, enlarged gyri (i.e., macrogyria). Microscopically, abnormal clusters of misplaced neurons (i.e., heterotopias) can also be seen. When the overall size and weight of the brain is subnormal, it suggests that there may have been an insufficient number of cells migrating rather than merely misguided migration. Abnormal formation of the corpus callosum is commonly associated with disordered neuronal migration and can be readily appreciated on gross inspection of the cut brain.

Many of these gyral and callosal abnormalities can be identified in life through the use of magnetic resonance imaging. This enhanced ability to identify such abnormalities has proven that they are much more common than once thought. Before modern neuroimaging, only patients with severe neurologic dysfunction (e.g., epilepsy, mental retardation, weakness and incoordination) were identified as being afflicted with a neuronal migration disorder, usually at autopsy. It is now understood that similar, but much milder clinical syndromes can occur and affect a far greater number of persons.

Epilepsy as a Symptom of Abnormal Cortical Neuronal Migration

Neuronal migration abnormalities can be so severe that they are incompatible with life or so mild that they are asymptomatic. Among those who survive with these disorders, epilepsy is the most common neurologic symptom. Even small foci of abnormally placed cortical neurons can sufficiently disrupt normal interneuronal physiology to result in epilepsy. The convulsive seizures associated with neuronal migration disorders often are relatively refractory to medical therapy. The modern assessment of a newborn infant demonstrating failure to thrive and intractable epilepsy includes a neuroimaging evaluation to investigate the possible presence of one of the cortical dysplasias related to abnormal neuronal migration.

Kallmann's Syndrome as a Prototype of the Heritable Disorders of Neuronal Migration

It is estimated that between 5% and 20% of neuronal migration abnormalities are genetic in origin. Kallmann's syndrome is an inherited condition limited to males, characterized by an inability to smell and underdevelopment of gonadal function. The impaired olfaction is related to lack of development of the olfactory bulbs and tract, and the hypogonadism is caused by deficiency of a hypothalamic hormone, gonadotropin-releasing hormone, that is critical to the development of the male gonads. The abnormalities are caused by defective neuronal migration of olfactory neurons and of neurons producing gonadotropin-releasing hormone. Both classes of neurons probably share a common origin in the migration pathway, which explains the linkage of these seemingly disparate anomalies.

The isolation of a gene from the critical chromosomal region to which the syndromal locus had been assigned may shed considerable light on the pathogenesis of inherited disorders of neuronal migration. The predicted protein of this gene shows similarity to neuronal cell adhesion and axonal path-finding molecules, which suggests that the gene may influence neuronal migration. This notion has been supported by the discovery that this gene was deleted in a patient with Kallmann's syndrome.

SELECTED READINGS

Barth P. Migrational disorders of the brain. Curr Opin Neurol Neurosurg 1992;5:339.

Franco B, Guioli S, Pragliola A, et al. A gene deleted in Kallmann's syndrome shares homology with neural crest cell adhesion and axonal path-finding molecules. Nature 1991;353:529.

Kazee AM, Lapham LW, Torres CF, Wang DD. Generalized cortical dysphasia: clinical and pathological aspects. Arch Neurol 1991;48:850.

Neuroscience in Medicine, edited by P. Michael Conn.
J. B. Lippincott Company, Philadelphia © 1995

Chapter 9

*T*he Vasculature of the Human Brain

JULIO H. GARCIA
SURESH C. PATEL
SIMONE WAGNER

The vasculature of the brain and spinal cord consists of an arterial input and a venous drainage system only; there are no lymphatic vessels in the central nervous system (CNS).

Intracranial blood vessels are involved in two types of diseases of the brain: *infarcts*, which are sites of tissue destruction caused by insufficient blood supply, and *hemorrhages*, which are sites of spontaneous rupture of arterial branches resulting in subarachnoid hemorrhage or parenchymal bleeding.

The intracranial vasculature differs from that found in the rest of the body in several ways. For example, arteries and veins supplying the brain are bathed by cerebrospinal fluid after they pierce the arachnoid membrane. All intracranial veins drain into dural sinuses, and these do not exist outside the skull. There are also differences in structure, innervation, and functional responses to injury. For example, subarachnoid hemorrhage may have profound and long-lasting effects on the cerebral arteries. The presence of blood in the spinal fluid may induce sustained vasoconstriction in neighboring arteries, but blood applied to the tunica adventitia of extracranial arteries does not induce sustained vasospasm. Many fine details about these differences are incompletely characterized, partially because most observations on the nature of neurogenic responses in intracranial vessels have been carried out in animals, and vascular responses to pharmacologic and electrical stimuli vary significantly among species.

Arterial System

☐ Multiple anastomoses connect with each other branches of the four major arteries that supply the intracranial contents

Four large *extracranial arteries* supply the brain: two common carotid arteries and two vertebral arteries. The most common pattern of origin of these vessels from the aortic arch is illustrated in Figure 9-1. A common variation in the origin of these vessels is shown in Figure 9-2.

Intracranial arteries are of two types: extradural or epidural branches of the external carotid artery and intradural ramifications derived from the internal carotid artery and the vertebral arteries.

The histologic structure of the arteries supplying the brain changes as these vessels penetrate the dura mater. The elastic fibers, which in extracranial arteries are distributed throughout the entire width of the arterial wall (Fig. 9-3A,B), condense into a subendothelial elastic lamina after the arteries penetrate the dura mater (Fig. 9-

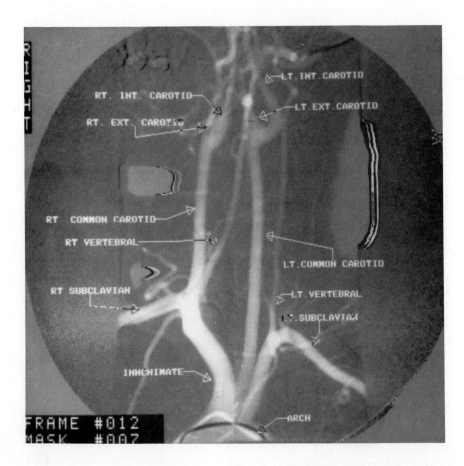

Figure 9-1
Aortic arch, left anterior oblique projection of a normal contrast arch aortogram.

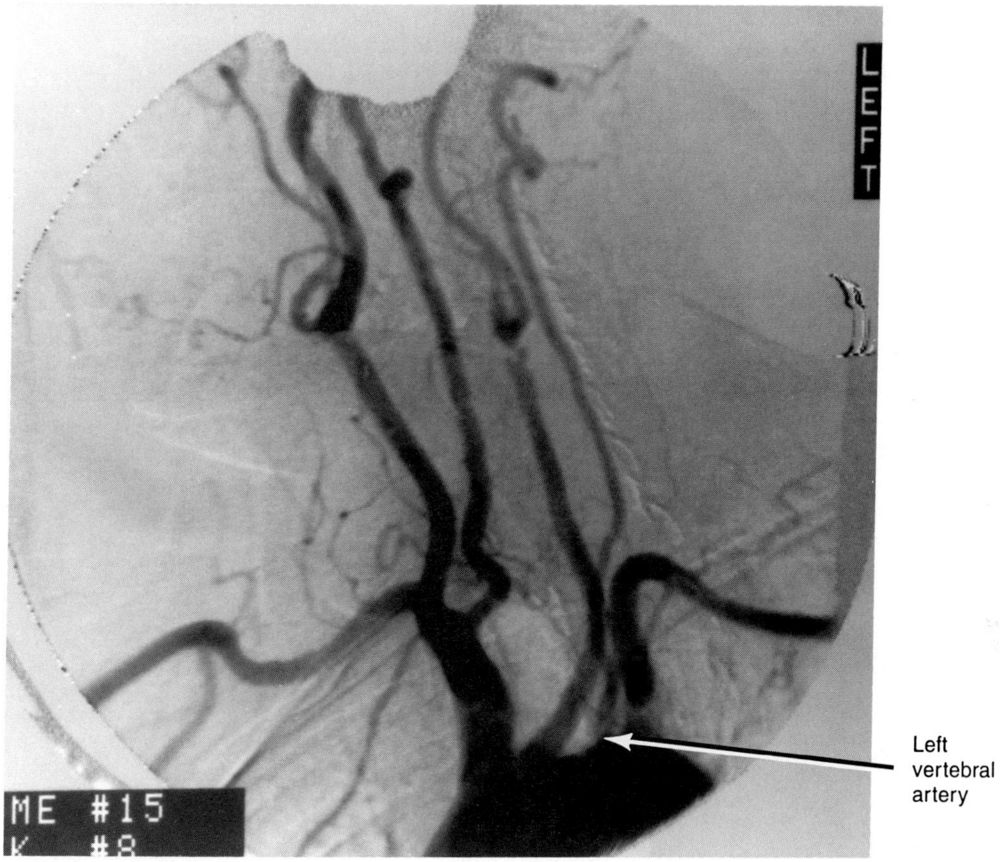

Left
vertebral
artery

Figure 9-2
Left anterior oblique projection, showing a common variation in the origin of the left vertebral artery
(*arrow*) arising directly from the aortic arch between the left common carotid artery and the left sub-
clavian artery.

3C,D). The thickness of the tunica media is decreased in intracranial vessels compared with arteries of the same caliber located outside the skull.

Collateral circulation

Arteries supplying the brain possess abundant end-to-end anastomoses, forming multiple networks that facilitate collateral circulation. This system of collateral or alternate circulation allows distal branches of an occluded artery to fill in a retrograde fashion through the end-to-end anastomoses that connect neighboring vessels and to compensate for the changes in blood flow.

The main anastomoses between the anterior (i.e., carotid) circulation and the posterior (i.e., vertebrobasilar) circulation are provided by the posterior communicating arteries that connect the internal carotid artery with the ipsilateral posterior cerebral artery (Fig. 9-4). Connections across the midline are provided by the anterior communicating artery and its branches. Credit for the correct description of the arterial anatomic network located at the base of the brain is given to Thomas Willis, and the *circle of Willis* designates this arterial network.

These anastomotic connections are significant because of the collateral circulation they provide. In many instances, local circulatory abnormalities created by occluding a single artery at points proximal to the circle of Willis can be adequately compensated through the collateral circulation, and symptoms of cerebral ischemia do not develop.

The circle of Willis and its feeding branches constitute a symmetric structure in only about 40% of adults examined postmortem. A significant number of anatomic variations, which in most cases reflect the persistence of embryonal or fetal vascular patterns, are found in about 60%. Among the most common variations in the anatomy of the circle of Willis are marked atrophy of one vertebral artery and of the contralateral anterior cerebral artery or a posterior cerebral artery originating from the internal carotid artery instead of the basilar artery.

Anastomoses among branches of the internal and external carotid arteries exist through the ophthalmic artery (i.e., large branch of the internal carotid artery; see Fig. 9-4) and its end-to-end connections with facial branches of the ipsilateral external carotid artery. Many additional, direct anastomoses connecting large branches of the internal and external carotid arteries have been demonstrated by contrast angiographic methods.

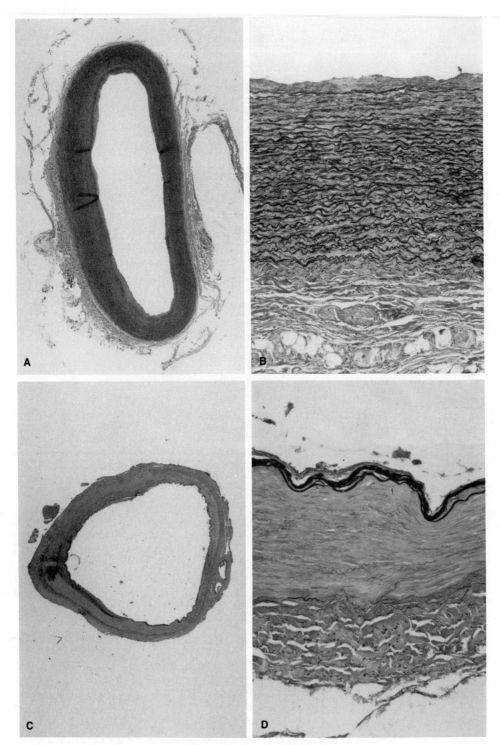

Figure 9-3
(**A, B**) Normal common carotid artery, and (**C, D**) normal intradural portion of the internal carotid artery. The black lines correspond to elastic fibers. The elastic fibers condense into a subendothelial elastic lamina after the arteries penetrate the dura matter.

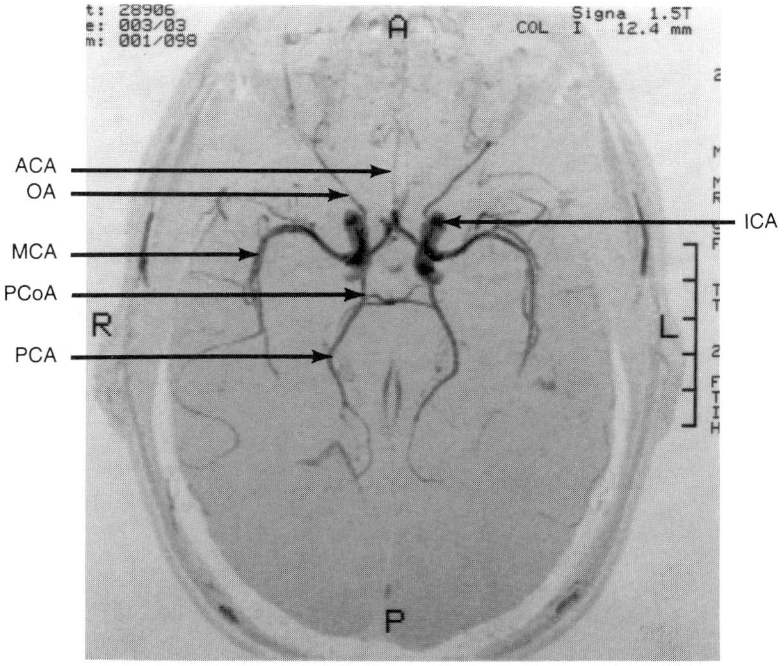

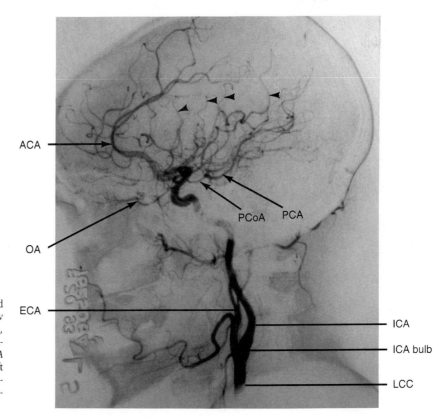

Figure 9-4
Axial projection of a two-dimensional time-of-flight magnetic resonance angiogram showing the circle of Willis. *ACA,* anterior cerebral artery; *ICA,* internal carotid artery; *MCA,* middle cerebral artery; *OA,* ophthalmic artery; *PCA,* posterior cerebral artery; *PCoA,* posterior communicating artery.

The anterior, middle, and posterior cerebral arteries, which are the three main arteries supplying each cerebral hemisphere, are illustrated in Figures 9-5 through 9-7. The vertebrobasilar arteries and their branches to the brain stem and cerebellum are shown in Figures 9-8 and 9-9. The main branches of the external carotid artery are identified in Figure 9-10.

Leptomeningeal or pial vessels

Leptomeningeal or pial vessels describe the branches of the circle of Willis arteries that course through the surface of the brain for various distances before penetrating into the brain parenchyma. Numerous arterial anas-
Text continues on p. 158

Figure 9-5
Lateral projection of a left common carotid angiogram. The arrowheads indicate a few branches of the middle cerebral artery. *ACA,* anterior cerebral artery; *ECA,* external carotid artery; *ICA,* internal carotid artery; *ICA bulb,* internal carotid artery bulb; *LCC,* left common carotid artery; *PCA,* posterior cerebral artery; *PCoA,* posterior communicating artery; *OA,* ophthalmic artery.

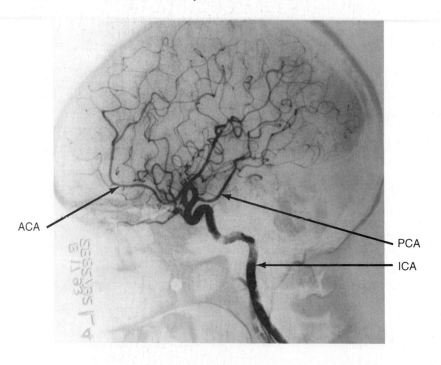

ACA

PCA

ICA

Figure 9-6
Lateral projection of a left internal carotid angiogram. Notice the origin of the left posterior cerebral artery from the internal carotid artery. *ACA,* anterior cerebral artery; *ICA,* internal carotid artery; *PCA,* posterior cerebral artery.

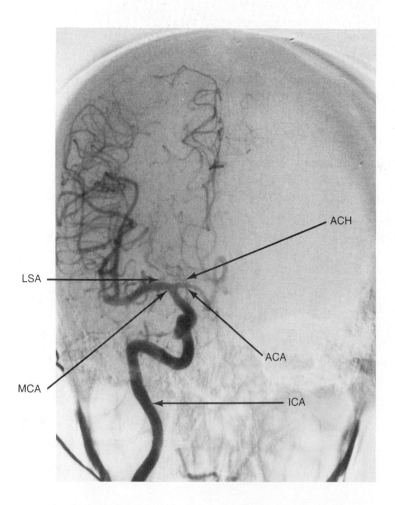

ACH

LSA

ACA

MCA

ICA

Figure 9-7
Frontal projection of a right internal carotid angiogram. *ACA,* anterior cerebral artery; *ACH,* anterior choroidal artery; *ICA,* internal carotid artery; *LSA,* lenticulostriate arteries; *MCA,* middle cerebral artery.

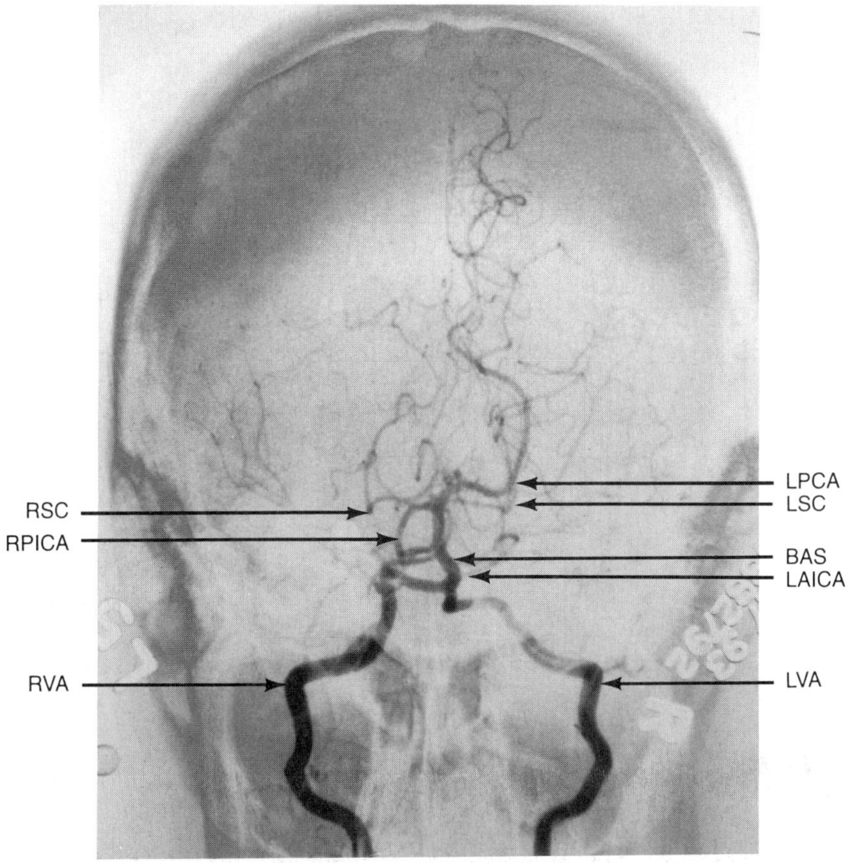

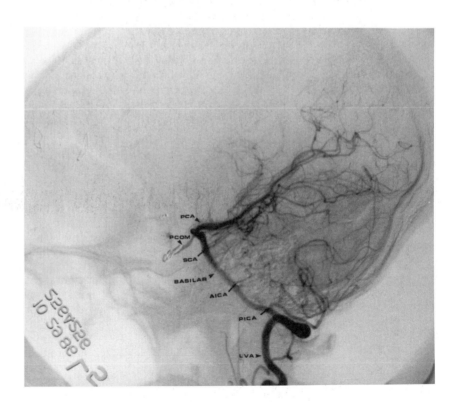

Figure 9-8
Frontal view of a right vertebral angiogram.
BAS, basilar artery; *LAICA*, left anterior inferior cerebellar artery; *LPCA*, left posterior cerebral artery; *LSC*, left superior cerebellar artery; *LVA*, left vertebral artery; *RPICA*, right posterior inferior cerebellar artery; *RSC*, right superior cerebellar artery; *RVA*, right vertebral artery.

Figure 9-9
Lateral projection of a left vertebral angiogram.

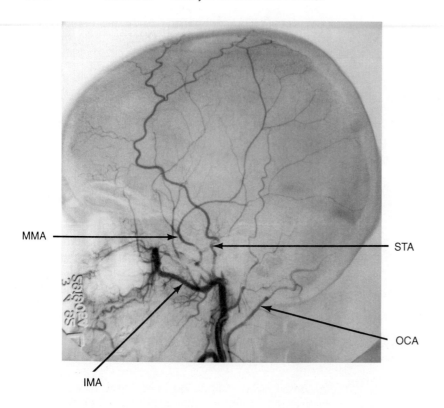

MMA

IMA

STA

OCA

Figure 9-10
Lateral projection of a left external carotid angiogram. *IMA,* internal maxillary artery; *MMA,* middle meningeal artery; *OCA,* occipital artery; *STA,* superficial temporal artery.

tomoses exist among these leptomeningeal or pial branches. Two types are found: large-diameter end-to-end anastomoses that connect branches from two different arterial stems (e.g., branches of the middle and anterior cerebral arteries) and extremely-small-diameter anastomoses connecting branches from the same or a different parent artery. The diameter of the largest anastomoses joining arterioles end to end varies from 25 to 90 μm. The average diameter of the small, straight anastomoses is 10 μm.

Intraparenchymal or penetrating arteries

Cortical or leptomeningeal branches divide into small pial vessels before penetrating the cortex. Each penetrating cortical artery forms a vascular palisade that supplies the respective capillary bed. Major vessels supplying the cortex enter the gyral surface at a perpendicular angle; those with the largest caliber and longest course supply the deepest cortical layers. Anastomoses forming a continuous horizontal layer appear in layer 3 among the large pyramidal cells, but the greatest number of these connecting vessels is visible in layers 4 and 5, and few are visible in layer 6.

Central arteries at each gyrus always have a large diameter (260 to 280 μm) at their point of origin. Peripheral arteries have an average diameter of 150 to 180 μm. On the cortical surface, all arterioles measuring 50 μ or less penetrate the cortex or anastomose with neighboring ones. Penetrating vessels into the cerebral cortex are

short arterioles (<100 μm in diameter) devoid of continuous internal elastic lamina and having a tunica media composed of one or two layers of smooth muscle cells. Most penetrating arterioles have a diameter of approximately 40 μm.

Some of the large, *penetrating arteries* are branches originating directly from the trunk of a large vessel, such as the middle cerebral artery and the basilar artery. These branches supply the basal ganglia and the thalamic nuclei, respectively. Penetrating arteries at these sites are long, muscular vessels (100–400 μm in internal diameter) endowed with internal elastic lamina and three or four layers of smooth muscle fibers in the tunica media.

One of the few groups of vessels supplying gray matter structures that lack anatomic anastomoses among themselves are the *lenticulostriate branches* originating from the main trunk of the middle cerebral artery (see Fig. 9-7). The lenticulostriate vessels have been implicated as being the source of the most common type of nontraumatic intracerebral hemorrhage.

Forty percent of the total vascular resistance in the CNS can be traced to the penetrating parenchymal arteries that are less than 200 μm in diameter. Chronic arterial hypertension increases the resistance in these small arteries, protecting the microvasculature (i.e., arterioles, capillaries, venules) from the effects of sustained high blood pressure. The vascular resistance in extracranial organs is primarily a function of arterioles.

Blood flow to subcortical white matter fibers, also called U-fibers, is supplied by arteries and arterioles, but the deeper white matter structures (i.e., centrum semiovale) are supplied only by nonanastomosing, long, radial arteries.

Long, transcortical vessels traverse the cortex without branching and, on entering the subjacent white matter, form a cascade of vessels that terminates in a periventricular plexus. The long, penetrating radial arteries that supply the cerebral hemispheric white matter do not interconnect with one another. The name *terminal arteries* has been applied to these and other nonanastomosing vessels.

☐ The structure of intradural arteries is different from that of other arteries

Intradural arteries and veins are bathed in spinal fluid, even after the arteries penetrate the cerebral parenchyma; this is made possible by perivascular sheath-like extensions of the subarachnoid space into the brain parenchyma. The anatomic, perivascular structures that contain the brain vessels in the subarachnoid space are known as the Virchow-Robin spaces. The arteries and arterioles in the Virchow-Robin spaces lack tunica adventitia; in these vessels, the muscular tunica media is surrounded by a single cell layer of leptomeningeal origin. The Virchow-Robin space disappears when the *glia limitans*, a subpial structure formed mainly by the end-feet of astrocytes and attached to the pial membrane, fuses with the basal lamina of the smallest arterioles and capillaries.

Intracranial arteries frequently exhibit at bifurcation sites (i.e., distal carina) *medial defects* or interruptions in the continuity of the smooth muscle layers. These are the same sites where saccular aneurysms develop in persons with inherited disorders of connective tissue metabolism.

At the branching sites of intracranial arteries (i.e., proximal carina), there are *intimal cushions* or sites where well-demarcated intraluminal protrusions are formed by subendothelial aggregates of smooth muscle fibers. These intimal cushions become more apparent with increasing age, but their functional significance has not been elucidated. Normal intracranial arteries in persons younger than 20 years of age probably lack vasa vasorum; aging and diseases associated with this process, such as atherosclerosis, may induce the development of a few vessels in the tunica adventitia of intradural arteries.

☐ Three types of nerve fibers have endings on the walls of large intracranial arteries

Pain-sensitive fibers

Pain-sensitive fibers were first demonstrated in humans by electrical stimulation of arteries attached to the dura (i.e., branches of the external carotid artery) and arterial branches of the internal carotid artery located within the subarachnoid space. Substance P is the main tachykinin involved in the transmission of nociceptive information; using immunohistochemical methods, substance P has been identified on the tunica adventitia of large intracranial vessels. Many intracranial and extracranial blood vessels supplying the brain are surrounded by axonal terminals originating from the trigeminal and the upper dorsal root ganglia. The densest network of these fibers exists on the tunica adventitia of the major arterial branches as they emerge from the circle of Willis.

The caudal portion of the basilar artery and both vertebral arteries and their tributaries are innervated by fibers originating in the upper cervical dorsal root ganglia. The central projections of the fibers originating from the trigeminovascular system are incompletely known.

Bleeding in the subarachnoid space is one of the best known causes of severe headache; this is a unique response limited to the intracranial vessels. The trigeminovascular system of nerve fibers around arteries, which includes substance P–dependent fibers, is thought to be intimately involved in the mechanism of pain and migrainous headaches.

Sympathetic fibers

Sympathetic fibers supplying large intracranial vessels originate primarily from the superior cervical ganglion, the middle cervical sympathetic ganglion, and stellate ganglion. In addition to norepinephrine, sympathetic nerve terminals also secrete the vasoconstrictor neuropeptide Y. Sympathetic fibers are not involved in the regulation of the cerebral blood flow, except that sympathetic stimulation blunts anticipated rises in cerebral blood flow during severe arterial hypertension. The most important role of the sympathetic innervation in intracranial arteries is to protect small arteries from the effects of blood pressure surges within the physiologic range.

Parasympathetic fibers

Parasympathetic fibers to the cerebral blood vessels originate mainly from sphenopalatine, otic, and associated miniganglia. The most dense network of fibers exists around the proximal segments of arteries branching out of the circle of Willis. In addition to cholinergic fibers, parasympathetic nerve endings contain vasoactive intestinal polypeptide (VIP). Nitric oxide synthase (NOS) has been co-localized in the same fibers that contain VIP. NOS may be the same as the endothelium-dependent relaxing factor. Cholinergic mechanisms have a minimal influence in the control of the normal cerebral blood flow.

None of the cerebral perivascular networks of nerve fibers plays a significant role in the normal autoregulation of blood flow to the brain. Autoregulation is primarily the result of myogenic responses to changes in blood pressure or in neuronal metabolism. Sensory, parasympathetic, and sympathetic nerves contribute to the cerebral blood flow regulation and preserve the integrity of the blood vessel wall only in pathologic conditions such as sustained hypertension and chronic hypotension.

☐ Arteries can be visualized in three ways

In addition to conventional angiography, which requires intravascular injection of contrast medium, in vivo studies of human intracranial vessels can be obtained with Doppler ultrasonography. Magnetic resonance has been adapted for imaging arteries. Time-of-flight techniques have become the mainstay in noninvasive magnetic resonance angiographic evaluation of arteries supplying the brain (Fig. 9-11).

Capillaries

The perivascular sheath of cerebral spinal fluid surrounding penetrating arteries and arterioles disappears where the glia limitans merges with the basal lamina of the brain capillaries. These vessels, which in humans are 4 to 7 μm

in diameter, are composed of one endothelial cell layer, resting on a basal lamina that completely encircles a pericyte. Pericytes do not form a continuous layer around the endothelial cell layer; individual pericytes are found at infrequent intervals on the abluminal side of the capillary wall. Encircling the basal lamina of the endothelial cell or the pericyte are numerous processes of astrocytes joined to one another by *gap junctions*.

☐ The blood-brain barrier refers to a complex array of physical, metabolic, and transport properties of the capillary endothelium

Circulating macromolecules, such as globulins and albumin, do not cross the endothelial lining of brain capillaries. This contrasts with the ready escape of circulating

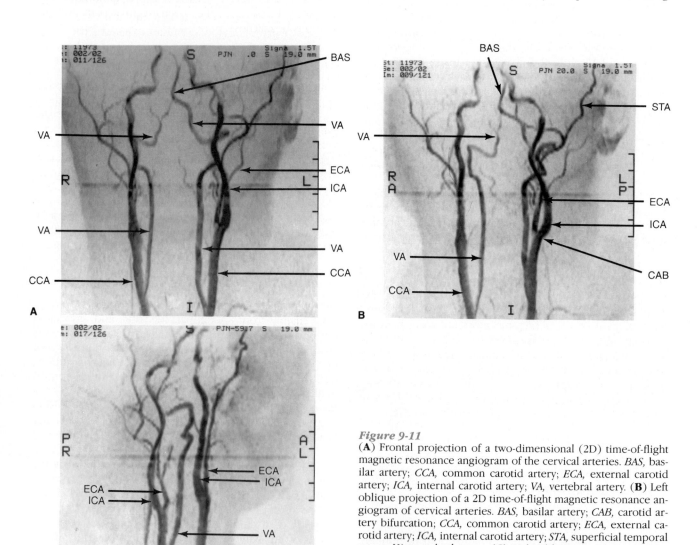

Figure 9-11
(**A**) Frontal projection of a two-dimensional (2D) time-of-flight magnetic resonance angiogram of the cervical arteries. *BAS*, basilar artery; *CCA*, common carotid artery; *ECA*, external carotid artery; *ICA*, internal carotid artery; *VA*, vertebral artery. (**B**) Left oblique projection of a 2D time-of-flight magnetic resonance angiogram of cervical arteries. *BAS*, basilar artery; *CAB*, carotid artery bifurcation; *CCA*, common carotid artery; *ECA*, external carotid artery; *ICA*, internal carotid artery; *STA*, superficial temporal artery; *VA*, vertebral artery. (**C**) Right oblique projection of a 2D time-of-flight magnetic resonance angiogram of the cervical vessels. *CCA*, common carotid artery; *ECA*, external carotid artery; *ICA*, internal carotid artery; *VA*, vertebral artery.

macromolecules that normally occurs in most extracranial tissues. The original description of the blood-brain barrier is attributed to Ehrlich who, in 1885, observed that intravenous injections of Evans blue, a dye that circulates bound to albumin, result in the diffuse distribution of the dye to almost every organ and tissue, except the brain and spinal cord.

The concept of a blood-brain barrier describes the inability of circulating macromolecules to enter the extracellular space or interstitial fluid of the brain and spinal cord. The mechanical component of the barrier has been traced primarily to structural characteristics of the endothelial capillary lining of the brain and spinal cord that are lacking in the endothelial lining of capillaries in other organs. A first important feature is that endothelial cells lining capillaries and venules in the CNS are joined at the luminal portion by *zonulae occludentes* or pentalaminar structures that represent the fusion of the outer most layers of two apposing endothelial cell membranes (Fig. 9-12). The second factor preventing escape of circulating macromolecules in the brain is the paucity of endocytotic pits in the endothelium of most vessels in the CNS. In contrast, the endothelial lining of capillaries and venules in extraneural tissues has abundant endocytotic pits and sizable gaps or fenestrae through which circulating particles such as 40-kDa horseradish peroxidase or 445-kDa

apoferritin readily escape into the surrounding interstitial fluids.

Cerebral endothelium may become abnormally permeable to circulating macromolecules by several mechanisms: enhanced transcytosis or transport of molecules across the endothelial cytoplasm by means of endothelial vesicles; separation of the endothelial junctions; formation of tubular channels by fusion of endothelial vesicles; and loss of the negative charge on the endothelial surface, particularly loss of the terminal sialic group on the luminal side of the endothelial plasma membrane.

☐ Capillaries at the circumventricular organs are permeable to circulating macromolecules

The circumventricular organs are seven small, well-circumscribed areas located at the ependymal border of the third and fourth ventricles where capillaries are permeable to hydrophilic solutes. These sites are the pineal body, median eminence, neurohypophysis, subcommisural organ, area postrema, subfornical organ, and organum vasculosum laminae terminalis (Fig. 9-13).

The circumventricular organs are endowed with permeable capillaries that have fenestrated endothelium, with the exception of the subcommisural organ. The functions of the circumventricular organs are not understood, although some investigators suggest that macromolecular permeability at these sites may be related to the involvement of the respective neuronal groups in the regulation of neuroendocrine functions.

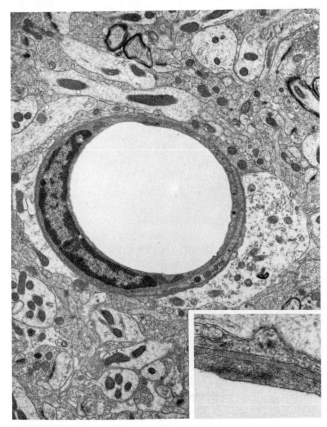

Figure 9-12
Normal rat brain capillary (original magnification × 7000). The inset shows a close-up view of the capillary wall to demonstrate a tight junction (original magnification × 32,200).

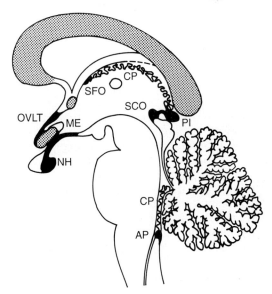

Figure 9-13
Circumventricular organs. *AP,* area postrema; *CP,* choroid plexus; *ME,* median eminence; *NH,* neurohypophysis; *OVLT,* organum vasculosum laminae terminalis; *PI,* pineal body; *SCO,* subcommissural organ; *SFO,* subfornical organ.

Intracranial Venous System

For didactic purposes, the description of the venous system that drains the contents of the intracranial cavity is artificially divided into three parts. This division is justified by the fact that the nature of the clinical syndromes associated with thrombosis of intracranial sinuses or veins depends on the location and caliber of the occluded sinus or vein.

☐ The Galenic venous system drains primarily veins originating in midline structures

The system of the deep cerebral veins (i.e., Galenic system) includes several veins that drain anatomic structures of the cerebral hemispheres located near the midline; most of these veins converge into the great vein of Galen. Veins from the deep cerebral structures, such as the thalamus, join the subependymal veins draining into the internal cerebral vein, the basal vein of Rosenthal, the internal occipital vein, and other vessels contributing to the great vein of Galen (Fig. 9-14). The great vein of Galen, which is 2.0 cm long, is close to the pineal body, quadrigeminal plate, and dorsum of the superior cerebellar vermis (see Fig. 9-14). The main tributaries of the vein of Galen include the septal, internal cerebral, and thalamostriatal veins. The deep middle cerebral veins drain the insular cortex and portions of the adjacent opercular surface; these veins usually receive lenticulostriate veins that drain the inferior portion of the basal ganglia. The anterior caudate, septal, and midatrial veins are illustrated in Figure 9-14*B*. Identification of these deep veins with contrast angiography is extremely useful for the localization of lesions, especially tumors, involving midline structures such as the thalamus and the pineal body.

The vein of Galen also drains anatomic structures related to the optic chiasm, uncinate gyrus, parahippocampal gyrus, portions of the ventricular temporal horn, and upper brain stem. Venous flow originating from these sites enters the vein of Galen through the basal veins of Rosenthal. The basal vein of Rosenthal originates near the optic chiasm and courses around the cerebral peduncles to terminate into the great vein of Galen (see Fig. 9-14). Shortly after its origin, the basal vein is covered by the uncus and the parahippocampal gyrus before it circles around the cerebral peduncle.

The vein of Galen and the inferior sagittal sinus, a structure running parallel to the dorsal surface of the corpus callosum, join one another to form the straight sinus, which occupies a midline position in the cerebellar tentorium. The inferior sagittal sinus courses above the corpus callosum along the free edge of the falx cerebri. This sinus receives numerous veins that drain the roof of the corpus callosum, the cingulate gyrus, and adjacent structures of the cerebral hemisphere; most superior cerebellar veins also drain into the straight sinus. The straight sinus ends in the torcula, a dura mater structure located at the site of the internal occipital protuberance (see Fig.

9-14). The torcula is the site of convergence for the straight sinus and the superior sagittal sinus, which sometimes remain separate from one another. The straight sinus usually drains into the left transverse sinus, and the superior sagittal sinus continues into the right transverse sinus (Fig. 9-15).

The transverse sinuses drain into the sigmoid sinuses, which originate at the posterior petrous portion of the temporal bone. The sigmoid sinus traverses the jugular foramen, where it continues into the internal jugular vein. The superior petrosal sinus commonly drains into the proximal portion of the sigmoid sinus. This sinus also receives veins from the cerebellum, the lateral pons, and the medulla. The two internal jugular veins converge to form the superior vena cava (see Fig. 9-15). Selected veins draining the brain stem and cerebellar structures are identified in Figure 9-16.

☐ Superficial cerebral veins drain into either the superior sagittal sinus or transverse sinuses

Superficial cerebral veins designate the network of venous channels visible on the surface of each cerebral hemisphere. Superficial cerebral veins coalesce on the pial surface and convey blood from the outer 1 or 2 cm of the cortex and underlying superficial white matter. The superficial veins are divided into *superior* veins that drain into the superior sagittal sinus, and *inferior* veins whose flow is directed into the transverse sinus; a *middle* group of vessels may drain superiorly or inferiorly. All the superficial veins normally empty into one or more of the major dural sinuses. The inferior superficial veins drain territories located primarily on the ventral surface of the temporal and occipital lobes.

The main draining avenue for most superficial cerebral veins, the superior sagittal sinus, originates in the foramen cecum of the frontal bone, courses anteroposteriorly along the midline of the dura mater, and ends in the torcula (see Fig. 9-14*A–C*).

Most superficial veins follow their course through the subarachnoid space over the surface of the various cerebral gyri, and the superior superficial veins usually converge on each side to form four to six large veins that, after piercing the arachnoid membrane, terminate in the walls of the superior sagittal sinus. The short segments of these veins located in the subdural space are designated as superficial cortical or *bridging veins* (see Fig. 9-14*B,C*). In most instances of subdural bleeding, the hemorrhage is thought to originate from tears in the bridging segment of the superficial cerebral veins.

One of the largest superior superficial veins, usually running a course approximately parallel to the central sulcus, is called Rolandic vein. Another large superficial vein, running an anteroposterior course and establishing connections between the superior and inferior groups of veins is sometimes designated the anastomotic vein of Labbé (see Fig. 9-14*A*).

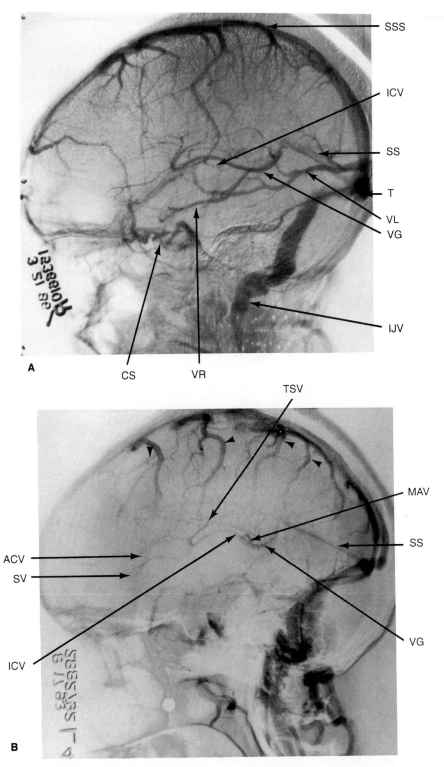

Figure 9-14
(**A**) Lateral projection (venous phase) of a right carotid angiogram. *CS,* cavernous sinus; *ICV,* internal cerebral vein; *IJV,* internal jugular vein; *SS,* straight sinus; *SSS,* superior sagittal sinus; *T,* torcula; *VG,* vein of Galen; *VL,* vein of Labbe; *VR,* vein of Rosenthal. (**B**) Lateral projection (late venous phase) of a left carotid angiogram. The arrowheads indicate superficial cortical or bridging veins. *ACV,* anterior caudate vein; *ICV,* internal cerebral vein; *MAV,* midatrial vein; *SS,* straight sinus; *SV,* septal vein; *TSV,* thalamostriate vein; *VG,* vein of Galen. (**C**) Lateral projection (venous phase) of a left carotid angiogram. The arrowheads indicate superficial cortical or bridging veins. *CS,* cavernous sinus; *IPS,* inferior petrosal sinus; *ISS,* inferior sagittal sinus; *SIS,* sigmoid sinuses; *SSS,* superior sagittal sinus; *T,* torcula; *TS,* transverse sinus.

continued

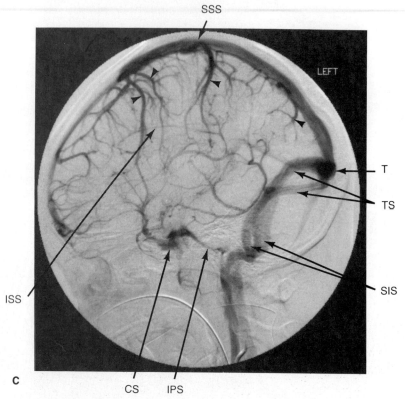

Figure 9-14 Continued

☐ Most veins from the orbit, pituitary area, anterior hypothalamus, and paranasal sinuses drain into the cavernous sinuses

Before reaching the right side of the heart through the superior vena cava, the intracranial venous system drains on each side into three or four major sinuses whose flow eventually goes into the internal jugular veins (see Fig. 9-14A–C).

All intracranial sinuses are endothelium-lined structures bounded by thick layers of collagenous tissues derived from the dura mater. On the convexity and midline of the brain are the superior and inferior sagittal sinuses and the straight sinus, all of which converge into the torcula. At the base of the skull there is, on each side of the pituitary fossa, a large cavernous sinus that primarily drains veins from the orbit, the pituitary gland, anterior hypothalamic structures, and some of the paranasal sinuses. The cavernous sinuses are located on the lateral surface of the body of the sphenoid bone. The sinus receives blood from the ophthalmic veins, the sphenoparietal sinus, the inferior petrosal sinus, and the basilar plexus of veins. The basilar veins receive blood from the inferior petrosal sinus, which also drains posteriorly into the marginal sinus and the anterior internal vertebral venous plexus.

The cavernous sinus drains into the sigmoid sinus by means of the superior and inferior petrosal sinuses, two structures that run parallel to the petrous portion of the temporal bone. Both the superior and inferior petrosal sinuses communicate anteriorly and medially with the cavernous sinus. The superior petrosal sinus courses along the bone with the attached margin of the cerebellar tentorium, and the inferior petrosal sinus runs inferior and parallel to the superior sinus in the petro-occipital suture. The superior petrosal sinus drains the ventral surface of the temporal lobe and the dorsal surface of the cerebellum; a few veins from the brain stem may also reach this sinus.

The major source of disease associated with intracranial veins and sinuses is the occlusion of these draining channels by thrombi or by adjacent structures, such as tumors. Intracranial venous thrombosis is causally related to infectious processes involving structures located near the cavernous sinus, such as the paranasal sinuses, eye globe, teeth, skull, and scalp. Infectious processes involving any of the anatomic structures located near the cavernous sinus, such as the intraorbital contents, can lead to the thrombotic occlusion of this sinus.

There are several venous channels connecting extracranial and intracranial veins, known as *emissary veins*. The flow in these and many other cranial venous structures can be easily reversed, because there are no intraluminal valves or structures to ensure unidirectional flow.

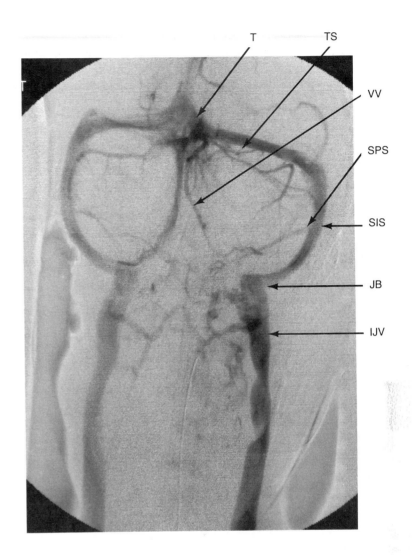

Figure 9-15
Frontal projection (venous phase) of a vertebral angiogram. *IJV,* internal jugular vein; *JB,* jugular bulb; *SIS,* sigmoid sinus; *SPS,* superior petrosal sinus; *T,* torcula; *TS,* transverse sinus; *VV,* vermian veins.

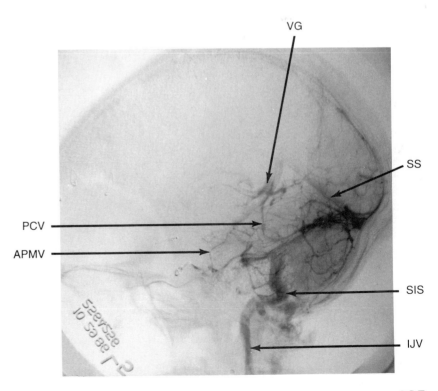

Figure 9-16
Lateral projection (venous phase) of a vertebral angiogram. *APMV,* anterior pontomesencephalic vein; *IJV,* internal jugular vein; *PCV,* precentral cerebellar vein; *SIS,* sigmoid sinus; *SS,* straight sinus; *VG,* vein of Galen.

Microscopically, the walls of the brain venules are almost indistinguishable from capillary walls. Venous walls consist of a continuous lining of endothelial cells. The endothelium is nonfenestrated, and the junction of two apposed plasma membranes is separated only by a narrow cleft, except at the luminal surface, where the membranes form zonulae occludentes. The walls of cerebral veins consist of an endothelium-lined tunica intima surrounded by an adventitial layer. Smooth muscle cells are not a common component of the venous wall. The wall of dural sinuses consist of an inner lining of endothelium and an outer layer having essentially the same architecture as dura mater. This outer layer consists chiefly of fibroblasts and large interlaced bundles of collagenous fibers. The arachnoid villi appear shortly after birth, and with advancing age, they form cauliflower-like clusters referred to as pacchionian granulations. They are primarily located along the walls of the superior sagittal sinus and the lateral lacunae; with aging, lateral lacunae may become large enough to form impressions on the overlying bone and become visible on skull x-ray films. The arachnoid villi are intimately involved in the process of transferring spinal fluid from the subarachnoid space to the superior sagittal sinus.

Intracranial veins do not collapse, even at transmural pressure of 1.0 mm Hg, which contrasts with what happens in extracranial veins.

SELECTED READINGS

Capra NF, Anderson KV. Anatomy of the cerebral venous system. In Kapp JP, Schmidek HH (eds). The cerebral venous system and its disorders. Orlando: Grune & Stratton, 1984:1.

Damasio H. A CT guide to the identification of cerebral vascular territories. Arch Neurol 1983;40:138.

De Reuck J. The human periventricular blood supply and the anatomy of cerebral infarctions. Eur Neurol 1968;5:321.

Edvinsson L. Innervation of the cerebral circulation. Ann N Y Acad Sci 1987;529:334.

Heistad DD, Marcus M. Cerebral blood flow: effects of nerves and neurotransmitters. New York: Elsevier Scientific Publishers, 1982.

Macfarlane R, Moskowitz MA. The innervation of pial blood vessels and their role in cerebral vascular regulation. In Caplan LR (ed). The scientific basis of stroke. New York: Springer-Verlag, 1995.

Van der Eecken HM. The anastomoses between the leptomeningeal arteries in the brain. Springfield: CC Thomas, 1959.

Weibel J, Fields WS. Atlas of arteriography in occlusive cerebral vascular disease. Philadelphia: WB Saunders, 1969.

Wilkins RH. Cerebral vasospasm. Crit Rev Neurobiol 1990;6:51.

Young RF. The trigeminal nerve and its central pathways. Physiology of facial sensation and pain. In Rovit RL, Murali R, Jannetta PJ (eds). Trigeminal neuralgia. Baltimore: Williams & Wilkins, 1990:27.

Clinical Correlation
STROKE

ROBERT L. RODNITZKY

Stroke is the third leading cause of death in the United States, with approximately 500,000 new cases and 60,000 fatalities each year. Stroke refers to the neurologic dysfunction that results from a reduction of blood supply to the brain. The neurologic symptoms of a stroke may be temporary or permanent. Cerebral ischemia is a potentially reversible alteration of brain function that results from inadequate delivery of critical blood-borne substrates such as oxygen and glucose. Cerebral infarction occurs if ischemia is severe enough to kill cells. In this circumstance, there is a high likelihood of permanent dysfunction. The brain is particularly sensitive to severe ischemia, with irreversible cell death resulting in 2 to 3 minutes in some species and in approximately 6 to 8 minutes in humans.

Pathologic Mechanisms of Stroke

☐ Cellular energy failure is at the root of most neurologic symptoms resulting from a stroke

Cell death in stroke results from the failure to synthesize ATP and other nucleoside triphosphates. Without an adequate energy source, cell survival is threatened in several ways. As a result of intracellular acidosis related to enhanced anaerobic glycolysis, mitochondrial respiration is further depressed, free radical formation is enhanced, and lipid peroxidation occurs. Ion homeostasis is disrupted, with a resultant influx of calcium, sodium, and chloride ions along with water. The enhanced cellular water content further damages individual cells and leads to regional brain swelling, which compresses neighboring blood vessels, further reducing blood supply to the injured area. Cell structure degenerates, because ATP is required to support resynthesis of macromolecules in the normal course of cell maintenance.

☐ Normal homeostatic mechanisms in the cerebral vascular system may be lost because of cerebral ischemia

The process of vascular autoregulation maintains a relatively constant cerebral blood flow despite variations in mean arterial pressure. This system is generally effective as long as the mean arterial pressure does not fall below 60 mm Hg or rise above 150 mm Hg. As a result of autoregulation, cerebral blood flow increases when there is an elevation of arterial $Paco_2$. In an area of infarcted brain, autoregulation is typically lost, and the cerebral blood flow passively mirrors changes in systemic blood pressure so that there is virtually no compensation for extremely low or extremely high pressures. The lost ability to respond to increased levels of $Paco_2$ prevents a normal compensatory increase in blood flow in response to potentially damaging acidosis.

☐ The normal blood supply to the brain can be disrupted by several mechanisms

Blood flow to the brain is diminished by any process that significantly narrows or occludes a nutrient blood vessel. Narrowing of a blood vessel without total occlusion is referred to as stenosis. In a large cerebral blood vessel such as the carotid artery, a substantial stenosis of 50% to 75% is required before blood flow is seriously diminished. Even under these circumstances, cerebral blood flow can remain relatively normal if collateral circulatory pathways compensate by shunting more blood to the affected region of the brain. Narrowing or occlusion of a cerebral artery is most commonly the result of lipid-laden atherosclerotic deposits attached to the inner surface of the vessel.

If the atherosclerotic deposit does not itself occlude an artery, it may serve as a nidus for the formation of a superimposed occluding blood clot, called a *thrombus*. Occlusion of a critical blood vessel can occur when a blood clot from a distant site, such as the heart, travels through the arterial system and lodges in a blood vessel of smaller caliber within the brain. A clot from a distant origin is referred to as an *embolus*. Cholesterol particles that have broken loose from an area of atherosclerosis and traveled downstream in the vascular system constitute another form of embolus.

A drop in cerebral perfusion pressure can result in diminished blood flow to the brain despite fully patent vessels. Such a drop, usually related to low blood pressure, can be the cause of watershed or border zone infarctions. The terms watershed and border zone refer to areas of the brain between the terminal distributions of two adjacent arteries, such as the anterior and middle cerebral arteries. Because they are at the end of the pipeline, such regions are subject to low, marginally adequate arterial pressure under normal circumstances. They are therefore the first to fail when blood pressure in the system drops further. If a drop in blood pressure is sufficiently severe and sustained, the entire brain is affected, resulting in *global cerebral ischemia*.

Cerebral hemorrhage, one of the most severe forms of stroke, results from the spontaneous rupture of the wall of a blood vessel that has been weakened by longstanding high blood pressure or from the rupture of a cerebral *aneurysm*, a balloon-like outpouching of the layers of an arterial wall. In the former case, bleeding usually occurs directly into the brain, resulting in an intracerebral hemorrhage. In the latter circumstance, hemorrhage may also occur within the brain substance, but because aneurysms are more typically found on the surface of the brain, hemorrhage into the cerebrospinal fluid contained within the subarachnoid space is much more common. Intracerebral and subarachnoid hemorrhages are extremely serious, and either can be fatal, one because of mass effect and compression of adjacent structures within the brain and the other because of associated severe arterial spasm provoked by blood in the cerebrospinal fluid.

☐ Transient ischemic symptoms often precede cerebral infarction

Under certain circumstances, the symptoms of cerebral ischemia may last only several minutes and then resolve. These episodes are known as *transient ischemic attacks*. By definition, a transient ischemic attack is an episode in which symptoms persist for less than 24 hours before resolving, but most last less than 15 minutes. A common cause of a brief ischemic episode is the passage of an arterial embolus downstream, resulting in the temporary occlusion of a smaller blood vessel until the embolus breaks up and is flushed downstream. These emboli can arise from the heart or from an atherosclerotic lesion in a large blood vessel. The point at which the common carotid artery bifurcates into its internal and external branches is a common site of severe atherosclerosis and often serves as a staging ground for fibrin-platelet or cholesterol emboli. Emboli from this site can travel to the brain, producing sensory, motor, or language dysfunction, or to the arterial supply of one eye, producing the classic syndrome of amaurosis fugax (i.e., fleeting blindness) before they break up. In this condition, the affected person describes a brief episode during which it appears that a fog or a cloud has descended like a window shade over the entire visual field of one eye.

Transient ischemic attacks may also result when a region of the brain being perfused by a severely stenotic artery is temporarily subjected to abnormally low perfusion pressure during an episode of systemic hypotension. This further reduces already marginal blood supply, resulting in transient neurologic symptoms until the blood pressure returns to normal.

Transient ischemic attacks can be repetitive and stereotyped. Because they are often the forerunner of a subsequent episode which produces a permanent neurologic deficit, a transient ischemic attack mandates immediate investigation to try to identify and correct its underlying cause and prevent a more devastating, irreversible ischemic event.

☐ The neurologic deficit in ischemic stroke depends on which blood vessel is involved

The brain is perfused by the paired carotid arteries and the vertebral basilar system of blood vessels. The neurologic signs and symptoms associated with a stroke depend on which of these blood vessels or their branches are involved.

One of the most common vascular territories to be involved in stroke is that of the *middle cerebral artery*. The middle cerebral artery has deep (i.e., lenticulostriate) and superficial (i.e., pial) branches. The deep branches perfuse the corona radiata, portions of the internal capsule, and parts of the globus pallidus and caudate nucleus. The pial branches provide blood supply for most of the lateral surface of the frontal, temporal, and parietal lobes. The clinical syndrome resulting from middle cerebral artery stenosis or occlusion depends on which of its branches are most involved. Among the most common clinical findings associated with middle cerebral artery involvement are contralateral limb paralysis and contralateral sensory loss, both involving the arm much more than the leg. The relative sparing of the lower extremity reflects the fact that its representation in the primary motor and sensory cortex is on the medial surface of the frontal and parietal lobes, respectively, areas which are outside the middle cerebral artery perfusion zone. Because the speech area is within the middle cerebral artery territory, aphasia is common in dominant hemisphere infarctions. In nondominant hemisphere infarction, especially those involving the parietal lobe, there may be severe disturbances of spatial function, such as *hemispatial neglect*, which is a tendency not to attend to objects or stimuli located in space on the side opposite the infarction. Separate and distinct from hemispatial neglect, there can be contralateral *hemianopia*, which is a loss of half the visual field because of involvement of the optic radiations coursing through the temporal and parietal lobes.

The *anterior cerebral artery* perfuses the anterior frontal lobe and the parts of the frontal and parietal lobe on the medial surface of the hemisphere. Because the lower-extremity portions of the motor and sensory homunculus are located on the medial hemispheric surface, anterior cerebral artery infarction preferentially produces paralysis and sensory loss in the lower extremity. Aphasia or visual field loss are not typically part of the syndrome. The eyes may become deviated toward the side of the lesion because of destruction of the frontal lobe gaze center responsible for directing rapid eye movements in the horizontal plane to the contralateral side. When this center is damaged, only the intact gaze center in the opposite hemisphere continues to drive gaze, and the eyes are involuntarily directed toward the side of the frontal lobe infarction.

Occlusion of the *internal carotid artery* can result in infarction of the entire anterior two thirds of the hemisphere, which constitutes the perfusion area of its two major branches, the anterior cerebral and middle cerebral arteries. Because the anterior cerebral artery often receives significant collateral flow from the opposite anterior cerebral artery through a vessel connecting the two, an internal carotid artery occlusion commonly results in damage confined largely to areas perfused by the middle cerebral artery.

Occlusion of one *vertebral artery* may go unnoticed if the opposite vertebral artery is patent and allows adequate blood flow into the basilar artery. However, vertebral artery occlusion can result in an infarction of the structures perfused by one of its branches, the *posterior inferior cerebellar artery*. Occlusion of this artery results in infarction of the lateral medulla, and the resultant constellation of neurologic signs, known as *Wallenberg's syndrome*, is one of the most striking examples of predictable clinicoanatomic correlation in clinical neurology. Structures that are affected in a lateral medullary infarction include the spinal tract of the trigeminal nerve, the spinothalamic tract, the nucleus ambiguous, the inferior cerebellar peduncle, and the sympathetic fibers descending through the brain stem from the hypothalamus.

The neurologic deficit occurring in Wallenberg's syndrome coincides exactly with the function of these structures. Sensory symptoms consist of loss of pain and temperature perception but not touch perception on the ipsilateral side of the face (i.e., uncrossed spinal tract of trigeminal nerve) and loss of pain and temperature sensation on the contralateral side of the body (i.e., crossed spinothalamic tract). There is ipsilateral limb incoordination (i.e., cerebellar peduncle) and a raspy, breathy voice due to paralysis of the ipsilateral vocal cord (i.e., nucleus ambiguous). In the ipsilateral eye, a small pupil and a drooping eyelid (i.e., Horner's syndrome) are caused by interruption of the descending sympathetic fibers.

Wallenberg's syndrome also illustrates an important principle of clinical neuroanatomic localization. Crossed motor or sensory symptoms, such as involvement of one side of the face and the other side of the body, when caused by brain infarction usually imply brain stem pathology.

Occlusion of the *basilar artery* can result in infarction of the entire upper brain stem and both occipital lobes. The massive brain stem dysfunction is often fatal. The paired posterior cerebral arteries originate from the bifurcation of the terminal portion of the basilar artery. Each posterior cerebral artery has a hemispheric branch that supplies the occipital cortex and penetrating branches that perfuse the midbrain in concert with similar branches from the basilar artery. Occlusion of the hemispheric branches on one side causes loss of the opposite half of the visual field (i.e., hemianopia), which is identical (i.e., homonymous) in both eyes. An occlusion of the right posterior cerebral artery resulting in a right occipital infarction causes loss of the left half of the visual field from the right and left eyes.

Occlusion of the hemispheric branches to both occipital cortices results in a form of total visual loss known as *cortical blindness*. Cortically blind persons often deny their visual disability. Each posterior cerebral artery also perfuses the splenium of the corpus callosum. When this structure is infarcted along with the primary visual cortex of the dominant hemisphere, the syndrome of *alexia without agraphia* results. The syndrome of *hemiachromatopsia*, which is an inability to perceive color, occurs when the inferior, medial occipital lobe is infarcted. The cells in this area of the visual cortex are wavelength selective in response to light and form the basis of color recognition. As is the case with hemianopia, the visual field opposite the side of the lesion is affected.

☐ Lacunar infarctions do not conform to the distribution of major cerebral arteries

Lacunar infarctions are small lesions, usually less than 15 mm in diameter. They are thought to result from occlusion of small, penetrating arteries that have been damaged by chronically elevated arterial blood pressure. Although very small, lacunar infarctions may occur in strategic areas, such as the internal capsule or the pyramidal tract in the pons, where they can cause severe hemiparesis. A lacunar infarction involving the posterior ventral nucleus of the thalamus can cause isolated, severe contralateral sensory loss.

Treatment of Stroke

The treatment of stroke involves prevention of blood clot formation and improvement of blood vessel patency. For persons who have experienced an episode of cerebral ischemia, agents such as aspirin, which inhibit platelet aggregation, are of major benefit in preventing recurrent ischemic events. More potent anticoagulant drugs such as warfarin can be used to prevent larger clots in the heart or blood vessels from forming or, once formed, from breaking loose and entering the cerebral circulation. A patient who has had a transient ischemic event or complete stroke as a result of severe carotid artery stenosis often benefits from surgical removal of the atherosclerotic deposit responsible for the arterial narrowing. Experience suggests that stroke patients having arteries with the highest degree of blockage, in excess of 70%, benefit the most from surgical treatment, with a definite reduction in the risk for future ischemic events. Accessibility considerations limit such surgical procedures largely to the common carotid artery and the extracranial portion of the internal carotid artery, the most common site for surgery.

The treatment of intracerebral hemorrhage centers on surgical removal of blood clots that are of sufficient mass to dangerously compress adjacent vital brain structures. A ruptured aneurysm is treated by occluding or tying off the weakened arterial bleb so that it cannot bleed again. Medical therapy is directed at preventing blood vessels from going into spasm in response to the presence of blood in the subarachnoid space. Certain calcium channel-blocking agents are useful for this purpose.

SELECTED READINGS

Biller J. Vascular syndromes of the cerebrum. In Brazis P, Masdeu J, Biller J (eds). Localization in clinical neurology. Boston: Little, Brown and Company, 1990:429.

Scheinberg P. The biological basis for the treatment of acute stroke. Neurology 1991;41:1867.

Seisjo BK. Pathophysiology and treatment of focal cerebral ischemia. J Neurosurg 1992;77:169.

Neuroscience in Medicine, edited by P. Michael Conn.
J. B. Lippincott Company, Philadelphia © 1995

Chapter *10*

Ventricles and Cerebrospinal Fluid

CONRAD E. JOHANSON

Characterization of Cerebrospinal Fluid

Cerebrospinal fluid (CSF), actively secreted into the ventricles by the choroid plexus epithelium, helps to establish a stable and specialized extracellular fluid environment for neurons. The anatomic relation of the choroid plexus–CSF system to the brain and spinal cord is depicted in Figure 10-1. The choroidal tissues, which are suspended in the ventricular cavities, generate a nonvascular, nonlymphatic percolation of fluid that has been referred to as the third circulation. The continual formation and drainage of CSF allows this unique circulatory system

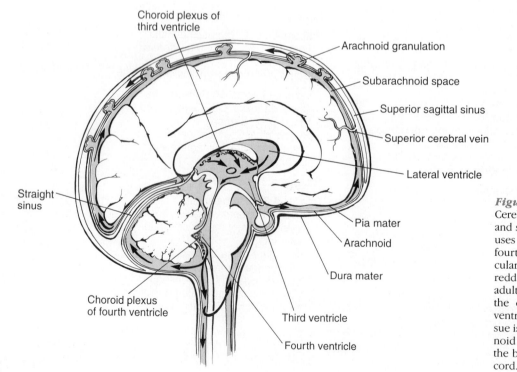

Choroid plexus of
third ventricle

Arachnoid granulation

Subarachnoid space

Superior sagittal sinus

Superior cerebral vein

Lateral ventricle

Straight
sinus

Pia mater

Arachnoid

Dura mater

Choroid plexus
of fourth ventricle

Third ventricle

Fourth ventricle

Figure 10-1
Cerebrospinal fluid (CSF) is formed and secreted by the choroid plexuses in the lateral, third, and fourth ventricles. The great vascularity of the plexuses imparts a reddish cast to these tissues. In adult humans, the total weight of the choroid plexus in the four ventricles is 2 to 3 g. Choroidal tissue is not present in the subarachnoid CSF space that surrounds the brain hemispheres and spinal cord.

Table 10-1
ROLES OF CEREBROSPINAL FLUID IN SERVING THE BRAIN

Cerebrospinal Fluid Functions	Examples
Buoyancy effect	Because brain weight is effectively reduced by more than 95%, shearing and tearing forces on neural tissue are greatly minimized.
Intracranial volume adjustment	CSF volume can be adjusted, increasing or decreasing acutely in response to blood volume changes or chronically in response to tissue atrophy or tumor growth.
Micronutrient transport	Nucleosides, pyrimidines, vitamin C, and other nutrients are transported by the choroid plexus to CSF and eventually to brain cells.
Protein and peptide supply	Macromolecules like transthyretin, insulin-like growth factor, and thyroxine are transported by the choroid plexus into CSF for carriage to target cells in brain.
Source of osmolytes for brain volume regulation	In acute hypernatremia there is bulk flow of CSF with osmolytes, from ventricles to surrounding tissue. This promotes water retention by shrunken brain, i.e., to restore volume.
Buffer reservoir	When brain interstitial fluid concentrations of H^+, K^+, and glucose are altered, the ventricular fluid can help to buffer the extracellular fluid changes.
Sink or drainage action	Anion metabolites of neurotransmitters, protein products of catabolism or tissue breakdown, and xenobiotic substances are cleared from the CNS by active transporters in the choroid plexus or by bulk CSF drainage pathways to venous blood and the lymphatics.
Immune system mediation	Cells adjacent to ventricles have antigen-presenting capabilities. Some CSF protein drains into cervical lymphatics, with the potential for inducing antibody reactions.
Information transfer	Neurotransmitter agents like amino acids and peptides may be transported by CSF over distances to bind to receptors in the parasynaptic mode.
Drug delivery	Some drugs do not readily cross blood-brain barrier but can be transported into the CSF by endogenous proteins in choroid plexus epithelial membranes.

to perform many functions for the adult brain. The physical and chemical attributes of CSF are summarized in Table 10-1.

☐ A buoyancy effect is imparted by CSF

The buoyant properties of the watery CSF help to protect the brain against forces associated with acceleration or deceleration. The comparative specific gravities of the CSF (1.007) and nervous tissue (1.040) are such that a 1400-g brain weighs only about 45 g as it is suspended in CSF in situ. Injury is avoided because the CSF sufficiently reduces the momentum of the brain in response to stresses and strains inflicted on the head during the course of everyday living. However, the angular acceleration that often occurs in severe trauma may override the normally protective effect of CSF buoyancy and cause tearing or herniation of cerebral tissues.

☐ Intracranial volume adjustment of CSF to alterations in brain bulk

CSF volume can be increased or decreased to stabilize the intracranial pressure (ICP). In response to elevated ICP, the CSF volume decreases relatively rapidly by means of enhanced absorption into venous blood. The CSF volume can be increased if its venous absorption diminishes when the ICP is reduced. The ability of the CSF volume to freely adjust to alterations in ICP is a reflection of the Monro Kellie doctrine. This doctrine recognizes that the brain, CSF, and intracranial blood are encased in a rigid chamber. Because these tissue and fluid contents are practically incompressible, a change in volume in any single constituent must be balanced by an almost exactly equal and opposite effect in one or another of the remaining components. Except in pathologic conditions in which nervous tissue volume is changed, the most common displacements of CSF volume occur in response to blood volume alterations, secondary to vasodilation or constriction of the cerebrovascular bed.

☐ Micronutrients are transported from blood to CSF

The choroid plexus, by means of CSF secretion, transports many water-soluble substances such as micronutrients into ventricular CSF for eventual carriage to target cells in brain. A micronutrient is a substance essential to the brain that is needed in relatively small amounts over extended periods. The following micronutrients are transported into CSF: vitamin C, folates, deoxyribonucleosides, vitamin B_6, and some trace elements. Active transport pump-like carriers in the choroid plexus pull these micronutrients from blood and transport them into CSF at relatively high concentrations. By bulk flow and diffusion, these substances are distributed across the ventricular wall (i.e., ependymal lining) into the brain parenchyma.

☐ Distribution of peptides and growth factors between CSF and the brain

CSF has a dynamic function as a neuroendocrine pathway for communication and integration within the brain. The choroid plexus has an important part in the CSF's role in neuroendocrine signaling. Receptors in the choroid plexus for arginine vasopressin, atriopeptin, and angiotensin II suggest that centrally released peptides transported into CSF can act on the plexus to modulate its secretion. The choroid plexus also constitutes a pathway for endocrine communication between the periphery and brain. Within the central nervous system (CNS), the choroid plexus is the main site of synthesis of insulin-like growth factor-II (IGF-II) and transthyretin. On secretion into the CSF, IGF-II can reach neurons and glia on which metabolic and tropic effects may be exerted. Transthyretin, known also as the thyroid hormone transport protein, is involved in T_4 transport from blood to choroid plexus to CSF. Transthyretin occurs early in development, and its deficiency seems to be incompatible with life in higher vertebrates. The choroid plexus–CSF distribution system for transthyretin and other tropic factors is critical for brain cell development and metabolism.

☐ CSF ions provide an osmolyte source for the brain

CSF has a relatively high concentration of sodium chloride (NaCl). Under certain conditions, the Na^+ and Cl^- in CSF can serve as inorganic osmolytes to help restore brain volume that has been decreased as the result of the net loss of water to blood. In acute hypernatremia, for example, the brain shrinks because there is a net loss of water from the CNS into hypertonic plasma. This initial compensatory phase, involving the movement of CSF with its inorganic ions from ventricles into brain parenchyma, precedes the osmotic adjustment phase occurring several days later, when there is net accumulation by brain cells of organic osmolytes like inositol.

☐ CSF buffers sudden changes in brain interstitial fluid solute concentrations

Increments in ion concentrations in brain interstitial fluid are minimized by the ability of ventricular CSF to receive and buffer excessive amounts of ions that can build up in brain interstices. Some pathophysiologic conditions can promote the buildup in interstitial fluid of acids (e.g., ischemia) or K^+ (e.g., seizures). Excesses of these ions in the brain interstitial fluid leads to diffusion, down their respective concentration gradients, into the CSF. The large-cavity CSF reservoir can be viewed as a buffer to accommodate such spillovers from brain fluids. By volume dilution and by transport mechanisms in the choroid plexus that actively remove K^+ from CSF or effectively neutralize H^+ in the CSF by choroidal secretion of HCO_3^-, the oscillations of certain ions in CNS extracellular fluid are dampened.

☐ Metabolites and toxic substances are excreted from CSF to blood

In addition to supplying substances for brain anabolism and maintenance, the CSF excretes metabolic products of neuronal and glial reactions. For example, there is a drain from brain interstitial fluid into CSF of the organic anions 5-OH-indole acetic acid and homovanillic acid, which are the metabolites of serotonin and dopamine, respectively. In CSF, these organic anions are actively reabsorbed by the choroid plexus into blood, or they are cleared convectively by bulk flow of CSF through the arachnoid villi into venous blood. This type of "sink action" occurs for numerous organic anions and cations and for proteins and other macromolecules. The iodide ion, which is especially toxic to brain tissue, is rapidly removed from CSF by the choroid plexus. Unfortunately, some antibiotics also are actively cleared from CSF by the plexus, reducing CSF concentrations to subtherapeutic levels.

☐ CSF mediates immune responses in the central nervous system

The choroid plexus–CSF system plays a role in the immunologic communication between the brain and the periphery. The plexus epithelium is able to present antigen to and stimulate proliferation of peripheral helper T lymphocytes. Moreover, proteins in CSF drain by bulk flow along the subarachnoid space that envelops optic and olfactory nerves. Because such CSF drainage routes eventually pass through lymphatic tissue (e.g., cervical nodes), there is the potential that proteinaceous antigenic material in CSF may elicit antibody reactions in the nodes. Such immunologic responses to proteins draining from CSF have implications for interactions between the CNS and immune system in pathologic states, such as multiple sclerosis or allergic encephalitis, in which certain proteins in CSF display antigenicity.

☐ Drug delivery via CSF can circumvent the blood-brain barrier

In treating brain cancer and other neural disorders, it is difficult to get water-soluble drugs to target cells inside the CNS. Many strategies, with limited success, have been employed to facilitate drug delivery to brain parenchyma. A particularly promising approach is to promote drug passage across the blood-brain barrier (i.e., cerebral capillary endothelium) or blood-CSF barrier (i.e., mainly choroid plexus) by using therapeutic agents transportable by endogenous protein carriers in barrier membranes. An important example is azidothymidine (AZT), currently used in treating acquired immunodeficiency syndrome (AIDS). AZT is a nucleoside with affinity for the nucleoside transport system in choroid plexus. After gaining access to the CSF, AZT or a similar agent is able to penetrate the ependymal lining that interfaces the brain interstitial (i.e., extracellular) fluid with ventricular CSF. Amino acid derivatives may also be therapeutically useful, because such drugs are carried by amino acid transporters from the plasma across the cerebral capillary wall and into the brain interstitial fluid. In the interstitial fluid, a drug, if it is not taken up by cells, may eventually reach the contiguous CSF of the ventricles or subarachnoid space.

Early Development of the Human Cerebrospinal Fluid System

Embryologically, the development of the ventricular system begins when the neural groove closes to form a tube. The earliest fluid within the neural tube precedes the appearance of the choroid plexus, and it is not a true CSF. Ciliary action within the fetal ventricles mixes the fluid and promotes the diffusional exchange of materials across the wall of the neural tube. Early fetal brain fluid is contained within the ventricles because it cannot escape into the meningeal fluid spaces, such as the subarachnoid space.

All of the major components of the ventricular systems—the first and second (i.e., lateral) and the third and fourth cavities—are present during the early stages of brain growth (Fig. 10-2A). The lateral ventricles are spherical and close to the midline at 2 months of gestation. During the second trimester, parts of the first and second ventricles expand laterally as the cerebral hemispheres enlarge. Posterior and inferior expansion of the

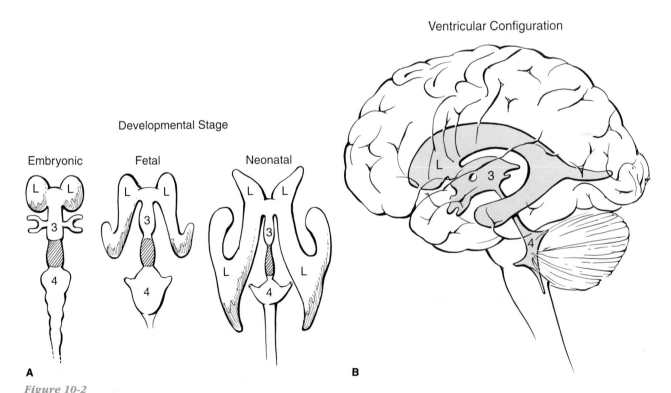

Figure 10-2
(**A**) Shape of the ventricular system during early stages of development. Even by the second month of the first trimester, all of the major components of the ventricles are present. In the 5-month fetus, the first and second ventricles grow laterally as the cerebral hemispheres enlarge. By birth, the general configuration of the ventricular system is similar to that of an adult. (Adapted from diagram by Martin J and Jessel T. In Kandel E, Schwartz J, Jessel T (eds), Principles of Neural Science, 3rd ed. Norwalk, CT: Appleton-Lange, 1991.) (**B**) Physical configuration of the cerebroventricular system of mammalian brain. The right and left lateral ventricles are located in the medial portion of their respective hemispheres. The third ventricle, which is smaller and situated in the midline, is physically contiguous with the anterior horns of the lateral ventricles above and the fourth ventricle below.

Table 10-2
CHANNELS OR NARROW DUCTS IN THE CEREBROSPINAL FLUID SYSTEM

Name of Channel	Location and Significance
Foramina of Monro	Connect each lateral ventricle to the third ventricle; tissue adhesions may block channels
Cerebral (sylvian) aqueduct	Connects the third ventricle with the fourth ventricle; narrowest passageway in ventricular CSF flow route and therefore the most likely site of obstruction leading to hydrocephalus
Foramina of Luschka	Two exits located in the lateral recesses of the fourth ventricle permit access to basal cisterns
Median foramen of Magendie	Midline at the caudal end of the fourth ventricle; direct access to the cisterna magna

brain forces the cortex into a C shape, and several underlying structures, such as the lateral ventricles, caudate nucleus, and hippocampal formation of the limbic system, are also molded into a C shape. At birth, the general shape of the entire ventricular system is similar to that in adulthood (Fig. 10-2*B*).

Choroid plexus tissue first appears in the ventricles during the second month of intrauterine life. Several stages of differentiation of choroidal tissue have been described. By the third gestational month, the choroid plexuses almost fill both of the lateral ventricles. The fetal choroid plexus is proportionately larger, relative to brain size, than that of an adult human being and fills more of the ventricular space. This implies that the ventricular fluid of the choroid plexus–CSF system has a particularly prominent role in providing nutrients to neural tissue during early development, when brain capillary density and blood flow are sparse.

Adult Cerebrospinal Fluid System

Knowledge about the size and shape of the CSF system is essential for understanding phenomena such as the kinetics of drug distribution among CNS regions, neuroendocrine integration of fluid balance within brain regions, and extracellular aspects of the inactivation or promotion of neurotransmitter and peptide signaling. In the interior or proximal portion of the CSF system, the cerebroventricles, CSF is generated by the choroid plexuses, and the fluid percolates down through the ventricles and out into the subarachnoid space. The more exterior or distal part of the CSF system, subarachnoid space, lacks the fluid-generating choroid plexuses and is involved largely with convection of fluid into drainage sites near the venous sinuses and lymph glands.

The ventricles are linked to each other by channels or foramina. Ventricular CSF flows from the telencephalon to rhombencephalon, and it finally mixes with fluid in the subarachnoid space at the base of the brain, where CSF flows out of foramina in the fourth ventricle into the cisterna magna and basal cisterns. The cisterna magna is formed by an arachnoid membrane bridge over the space between the cerebellar hemispheres and the medulla. Table 10-2 summarizes the various channels or pathways that allow communication among the large-cavity compartments of CSF.

☐ Configuration of the ventricular and subarachnoid cavities

Ventricles

In higher vertebrates, the cerebroventricular system is composed of four interconnecting cavities, each of which contains a choroid plexus (see Fig. 10-2*B*). The two lateral ventricles, which are more or less symmetric with each other, are the most prominent in size. A choroid plexus lies as a narrow band of tissue on the floor of each lateral ventricle. A thin layer, the septum pellucidum, separates the lateral ventricles, which are in the lower medial portion of the cerebral hemispheres. The lateral ventricles are not physically contiguous, but both communicate with the third ventricle by way of the interventricular foramina of Monro.

Each lateral ventricle consists of a main body and three horn-shaped recesses. The most rostral part of the lateral ventricle is commonly called the *anterior horn*. It is angled downward into the frontal lobe, and its apex curves around the anterior portion of the caudate nucleus. The *inferior horn* bends around the posterior end of the thalamus, extends backward and then laterally downward within the temporal lobe. The *posterior horn* runs laterally and projects backward into the occipital lobe. The region where the body divides into the inferior and posterior horns is known as the *trigone*.

The third ventricle lies below the body of the lateral ventricles, and it houses the smallest choroid plexus tissue, a thin cleft in the midline between the two thalami. The third ventricle receives CSF from the lateral ventricle and then passes the fluid downward into the sylvian aqueduct. The irregularly shaped third ventricle has four prolongations or recesses (see Fig. 10-2*B*). The front, lower part of this ventricle has adjacent recesses designated as *optic* and *infundibular*. The back, upper part of the third ventricle has recesses, named *pineal* and *suprapineal* because of their proximity to the pineal gland.

The fourth ventricle occupies the most caudal part of the cerebroventricular system. Lying well below the lateral and third ventricles, it is bounded by the pons, medulla oblongata, and cerebellum. The fourth ventricle has the shape of a rhombus. It is convenient to describe this ventricle as having a roof and a floor. The V-shaped roof is composed of thin laminae of white matter between the cerebellar peduncles. A median opening at the caudal end of the roof is called the foramen of Magendie, and it is hydrodynamically significant because CSF flows through this aperture into the subarachnoid space. Part of the roof of the fourth ventricle is occupied by choroid plexus. The fourth ventricular choroid plexus is T shaped, with the vertical portion lying in the midline.

The floor of the fourth ventricle is divided into symmetric halves by the *median sulcus.* Running perpendicular to this sulcus are delicate strands of transverse fibers, the *striae medullares* of the fourth ventricle. Other neuroanatomic features of the floor are the median eminence and the sulcus limitans. The *median eminence* is a longitudinal elevation that flanks both sides of the median sulcus. The *sulcus limitans* lies lateral to this eminence.

The analog to the ventricular system in the spinal cord is the spinal or central canal. This canal ends within the *filum terminale.* Imaging studies of normal human adults reveal that CSF in the fourth ventricle does not readily exchange with CSF in the central canal.

Subarachnoid space

The subarachnoid space lies between the arachnoid membrane externally and the *pia mater* internally. The subarachnoid space in adults provides a route by which the fluid can flow to absorptive sites of exit from the CNS. The subarachnoid space extensively covers the convexities of the cerebral hemispheres and forms a circumferential sleeve around the spinal cord (see Fig. 10-1). Figure 10-3 depicts the architecture of a small section of the subarachnoid space.

Because the pia intimately hugs the external contour of nervous tissue and the arachnoid membrane bridges the sulci of the brain and cord, relatively large pockets of subarachnoid space are formed between these structures. The large bridged spaces are called cisterns, and they are found mainly at the base of the brain. One of the largest is the *cisterna magna*, situated between the inferior surface of the cerebellum and the medulla. Because of its readily accessible location at the foramen magnum, the cisterna magna has been widely used in experimental animals as a convenient site for sampling CSF. The *cisterna ambiens* is the pocket of CSF that lies dorsal to the midbrain. Lying between the base of the brain and the

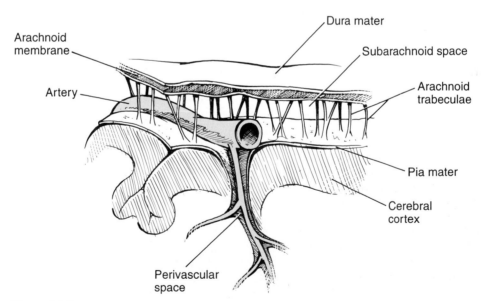

Figure 10-3
Meningeal aspects of the subarachnoid space (SAS). The roof of the SAS is the arachnoid membrane, and the floor of the SAS is the intima pia or the pia-glia, the external limiting membrane of the CNS. The arachnoid and pia-glia, which are ectodermal in origin, are bridged to each other by the arachnoid trabeculae. Cerebrospinal fluid percolates through the SAS. Blood vessels entering and leaving the nervous tissue carry with them arachnoid and pia-glia, which form a cuff around each major vessel. This cuff is called the Virchow-Robin space, which serves as a conduit that allows fluid movement between the brain extracellular space and the SAS. [Adapted from Carpenter M and Sutin J (eds), Human neuroanatomy, 8th ed. Baltimore: Williams and Wilkins, 1983.]

Table 10-3

EXTRACELLULAR FLUIDS IN THE CENTRAL NERVOUS SYSTEM

Fluid	Location and Characteristics
Extracellular fluid	Two main types: CSF in the ventricles and subarachnoid space and the ISF that intimately bathes the parenchymal cells of brain (e.g., neurons, glial cells)
Nascent CSF	Secreted across the apical membrane of the choroid plexuses (i.e., lateral, third, and fourth ventricles) into the ventricular space, referred to as nascent or newly formed CSF; active secretion
Ventricular CSF	Contained in the four cerebral ventricles and aqueduct; consists mainly of nascent CSF with some exchange of contents with the ependymal lining and underlying brain tissue as CSF flows down the ventricular axis
Subarachnoid CSF	Cranial or spinal subarachnoid space; mixture of ventricular fluid that has flowed into the subarachnoid space and brain fluid that has gained access to the subarachnoid spaces; subarachnoid and ventricular CSF regarded as large-cavity CSF
Brain interstitial fluid	Actively secreted by endothelial cells in walls of capillaries in brain and spinal cord (i.e., blood-brain barrier), modified by the water and solutes that are exchanged with brain neurons and glia; undergoes transependymal exchange with ventricular fluid; chemical composition similar to that of CSF
Cerebral endothelial secretion	Endothelium in brain capillaries, in conjunction with astrocyte foot processes on the vascular wall, actively secrete ions and solutes from plasma into the interstices of the brain; fluid-secretory capacity of the blood-brain barrier is less than that of the choroid plexuses

CSF, cerebrospinal fluid; *ECF*, extracellular fluid; *ISF*, intracellular fluid.

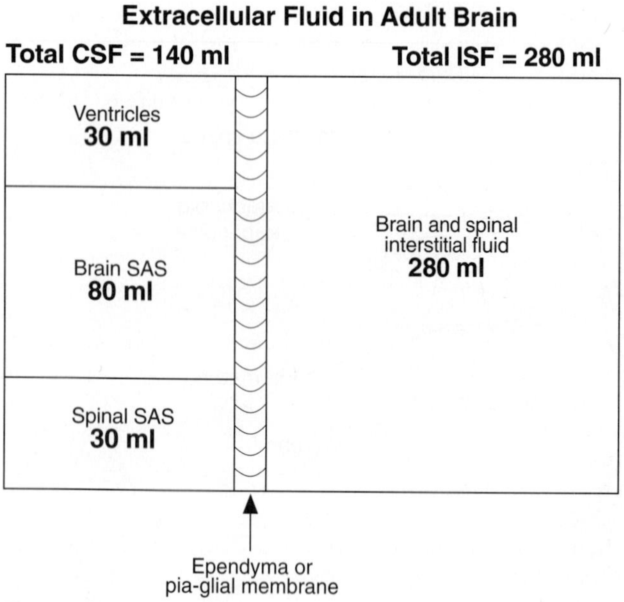

Extracellular Fluid in Adult Brain

Total CSF = 140 ml   **Total ISF = 280 ml**

Ependyma or pia-glial membrane

Figure 10-4
An analysis of the respective volumes of the fluids in the cerebrospinal compartment and in the interstitial space of the adult human brain. Almost one third of the fluid in the CNS lies outside of cells. A typical adult brain weighing 1400 g contains about 140 mL of CSF (10% of its weight) and approximately 280 mL of interstitial fluid (ISF) that intimately bathes the neurons and glial cells. The cerebrospinal fluid and ISF are in more or less free communication with each other across the permeable interfaces that separate them (i.e., the ependymal lining in the ventricular system, the pia-glial membrane in the subarachnoid system). *SAS*, subarachnoid space.

floor of the cranial cavity are the pontine, chiasmatic, and interpeduncular cisternae.

☐ Volume of CSF in the ventricular and subarachnoid spaces

The total volume of CSF in normal adult humans is about 140 mL. The volume of the ventricular system has been estimated by casting techniques, computed tomography (CT) scans, and radioisotope distributions. By averaging the data from several techniques, the mean volume of the ventricular system is estimated to be approximately 30 mL. The composite volume of the four ventricles is about 2% of the brain volume. Studies have found no correlation between ventricular volume (range, 10 to 60 mL) and brain volume.

Most of the total CSF volume of 140 mL is composed of the 110 mL of CSF in the subarachnoid spaces of the brain and spinal cord. The volume of CSF surrounding the spinal cord is at least 30 mL. The largest compartment of CSF is the 80 mL in the subarachnoid spaces and cisterns enveloping the cerebral and cerebellar hemispheres.

CSF is only one of the extracellular fluids in the CNS (Table 10-3). The other major type of extracellular fluid is the interstitial fluid that bathes the neurons and glia in the brain parenchyma. The CSF and interstitial fluid have similar concentrations of many substances. Figure 10-4 illustrates the quantitative relation between the respective volumes of the CSF and interstitial fluid. A typical

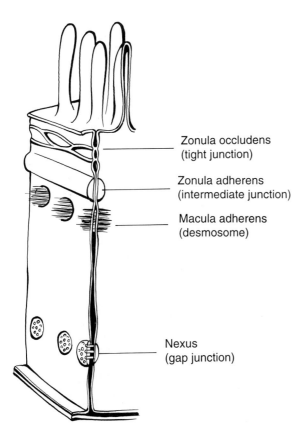

Zonula occludens
(tight junction)

Zonula adherens
(intermediate junction)

Macula adherens
(desmosome)

Nexus
(gap junction)

Figure 10-5
Ultrastructure of the intercellular junctions in choroid plexus and in ependymal membranes. An integral part of the blood-cerebrospinal fluid (CSF) barrier is the tight junctions or zonulae occludens. These tight junctions are located near the apical (CSF-facing) borders of the choroidal epithelium, where the cells abut each other. The tight junctions are multilayered membranes that completely envelop the cells, offering a physical restriction to the diffusion of most solutes between the plasma and CSF. Gap junctions are between the cells of the ependymal and pia-glial linings. The gap junctions form incomplete belts around the cells, and these intercellular junctions are more "leaky" than their tight junction counterparts in the choroid plexus. Overall, the composition of CSF resembles brain interstitial fluid more than plasma because the gap junctions in the ependyma permit unrestricted diffusion, but the tight junctions in the choroid plexus do not.

1400-g adult brain has about 280 mL of interstitial fluid and 140 mL of CSF, for a total of 420 mL of extracellular fluid. This can be compared with the approximately 800 mL of fluid in the whole brain intracellular compartment.

Cellular Linings Demarcating the Cerebrospinal Fluid System

CSF is contained within and surrounds the brain and spinal cord. The CSF that bathes the inside and outside surfaces of the brain is separated by a membrane composed of a single layer of cells. In the interior of the brain, the thin ependymal lining separates the ventricular CSF from the underlying nervous tissue. On the exterior of the

brain, the pia-glial membrane is the interface between CSF in the subarachnoid space and the adjacent cortical tissue. A third membranous interface is the choroidal epithelium, a single, frond-shaped layer of epithelial cells that separates ventricular CSF from the blood coursing through the vascular plexus. The epithelial parenchyma of the choroid plexus has ultrastructural characteristics distinct from those of the ependyma and pia-glia (Fig. 10-5).

The choroid plexus epithelial membrane is the source of CSF

The epithelium of the choroid plexus in all four ventricles consists of tightly packed cuboidal cells with a finely granular cytoplasm. A distinctive feature of the choroidal epithelium is the *zonula occludens,* or tight junction, located in apical regions between adjacent cells. This tight junction slows down or blocks the blood-to-CSF passage of many hydrophilic molecules and ions. The electrical resistance associated with the choroid plexus is not as great as in the blood-brain barrier or extra-CNS tissues such as the distal tubule or urinary bladder. Consistent with its transporting function, the typical choroid cell has a high density of mitochondria, a rich Golgi apparatus, and an extensive microvilli system on the apical, CSF-facing membrane. The basal region of the cell is characterized by elaborate infoldings and interdigitations, similar to the ultrastructure of the proximal renal tubule and other fluid-transporting epithelia. In general, the ultrastructure of the choroid cellular organelles reflects the brisk metabolic and transport activity associated with CSF secretion and ion homeostasis.

The ependymal cell lining is a permeable membrane separating CSF from the brain

Embryologically, the ependymal lining begins as a layer of spongioblasts lining the neural tube. In late fetal stages, the ependymal lining is multilayered, sometimes attaining a thickness of six to seven layers. In the neonate, the lining is attenuated to two or three layers. In early postnatal development, some ependymal cells known as tanycytes send long processes, extending from their bases, out into the neuropil. By adulthood, the tanycytes have largely disappeared, and the ependymal layer becomes a single layer of cuboidal or columnar cells.

Even within a given species, the structure of the ependymal lining in adults is not uniform. Great variations in cellular morphology occur, especially in the third ventricle where the ependyma is intimately associated with regions such as the hypothalamus and subcommissural organ. Emanation of cilia from the apical surface is common. In a few specialized regions in the membrane, there are tight junctions between ependymal cells. Most ependymal cells have gap junctions in the intercellular regions. The gap junctions do not completely envelop the cells, and the intercellular clefts are readily permeable to macromolecules. The functional hallmark of the ependyma lining is its great permeability to virtually all ions and molecules. After a drug or endogenous substrate is in the CSF, it can easily permeate the ependymal wall to reach neurons and glia.

☐ The pia-glial membrane and the arachnoid membrane form the leptomeninges

The pia-glial membrane is more like the ependymal lining than the choroid plexus. Discontinuous gap junctions between the pia-glial elements allow relatively free bidirectional exchange of solutes between the cortical interstitial fluid and CSF in the subarachnoid space. Large protein markers like ferritin and horseradish peroxidase move rapidly across the pia-glial membrane and then penetrate into subpial tissue. The Virchow-Robin spaces, which are perivascular cuffs of pia-glia and arachnoid that envelop major vessels as they penetrate deeply into the

brain, promote the uptake of materials from the subarachnoid fluid extending down into the brain substance. The facility of passage of large molecules from the subarachnoid CSF into underlying brain gives the impression that the pia-glial membrane is more permeable than even the relatively leaky ependymal walls encompassing the ventricular CSF.

The pia mater and the arachnoid membrane are known as the leptomeninges. The pia mater, a thin connective tissue membrane, hugs the contours of the brain surface and carries blood vessels supplying nervous tissue. The arachnoid is a thin, avascular membrane between the pia mater and dura mater. The arachnoid is separated from the overlying dura by the subdural space and from the underlying pia mater by the subarachnoid space, which contains CSF. In the dura mater are venous sinuses into which CSF is drained.

Circumventricular Organs

The ventricular wall is the site of several small, individual organs with somewhat similar structures but distinct albeit interrelated functions. These small organs are circumventricular (i.e., surrounding the ventricles), and

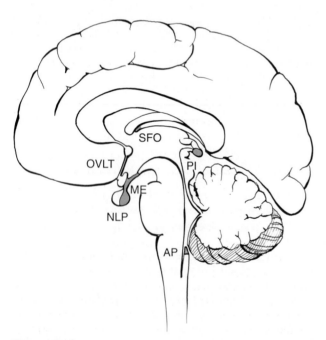

Figure 10-6
Sagittal view of the anatomic associations of the circumventricular organs (CVOs) located on the midline of the brain. The CVOs are situated at apparently strategic positions on the surface of the cerebroventricular system to perform neuroendocrine functions. The diminutive CVOs are highly vascular and have various numbers of neurons. Generally not protected by a blood-barrier, the CVOs are sites containing central receptors for peripherally circulating factors (e.g., peptides). Neuronal processes extend into the large perivascular spaces of the CVOs. Each CVO is encompassed by a ring of glial cells (i.e., tanycytes) with their tight junctions that help to isolate the CVO from surrounding brain tissue. The area postrema (AP) and subfornical organ (SFO) are attached to choroid plexuses, with which they have vascular shunts. *ME,* median eminence; *PI,* pineal gland; *OVLT,* organum vasculosum of the lamina terminals.

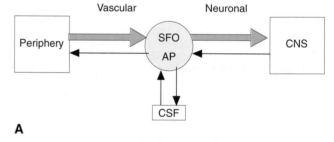

A

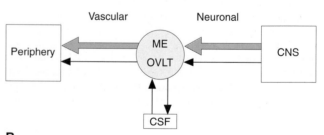

B

Figure 10-7
Schematic diagrams of the vascular, neuronal, and ependymal components of circumventricular organs (CVOs). The subfornical organ (SFO) and area postrema (AP) receive a prominent vascular input, (**A**), but the median eminence (ME) and organum vasculosum of the lamina terminalis (OVLT) have a substantial vascular output (**B**). Neuronal output is strongest in the SFO and AP, and neuronal input is substantial in the ME and OVLT. (Adapted from diagram by M. Paklovits, in Gross P, Circumventricular organs and body fluids. Boca Raton: CRC Press, 1987.)

Table 10-4
FUNCTIONS AND ANATOMIC ASSOCIATIONS OF SOME CIRCUMVENTRICULAR ORGANS

Organ	Location	Projections	Functions
Subfornical organ (SFO)	Attached to anterior dorsal wall of third ventricle, between the interventricular foramina of lateral ventricles	*Afferent:* Central input poorly characterized but is probably significant *Efferent:* Projects into the preoptic area and hypothalamus (i.e., paraventricular and supraoptic nuclei)	Induction of drinking behavior, mediated by angiotensin signals; SFO can modulate fluid homeostasis by many mechanisms through multiple projections to endocrine, autonomic, and behavioral areas of CNS
Area postrema (AP)	Lies at caudal extent of fourth ventricle on the dorsal medulla in contact with the nucleus of the solitary tract	*Afferent:* Input from underlying nucleus of the solitary tract and the dorsal motor nuclei of vagus; hypothalamus also innervates AP *Efferent:* Projections to major relay nuclei for ascending visceral sensory information; major projection to the parabrachial nucleus of the pons	Modulates interoceptive information that reaches it through visceral sensory neurons or humorally by way of its permeable capillaries; directly affects motor outflow of the dorsal motor nucleus; stimulation of a chemotaxic center causes vomiting
Organum vasculosum of the lamina terminalis (OVLT)	Lies in the anterior ventral extent of the third ventricle along the lamina terminalis	Connectivity of the OVLT is poorly understood but seems to have a greater afferent input than efferent outflow	Implicated in water balance because damage to it and surrounding structures affects drinking behavior and vasopressin release
Median eminence (ME)	Forms the ependymal floor of the third ventricle in the central portion of the tuber cinereum in the hypothalamus	*Afferent:* ME receives neuronal input from the arcuate nucleus and medial areas of preoptic nucleus and anterior hypothalamus *Efferent:* No efferent projections to the brain; portal circulation carries hormones to anterior pituitary	Represents the final common pathway for the neural control of hormone production in and secretion from cells of the adenohypophysis (i.e., anterior pituitary)

they include the area postrema, subfornical organ, pineal gland, median eminence, organum vasculosum of the lamina terminalis, and neural lobe of the pituitary (Fig. 10-6). The term *neurohypophysis* refers to the median eminence and the neural lobe of the pituitary. Unlike most regions of the CNS, the circumventricular organs have highly permeable capillaries that permit the diffusion of polypeptides and proteins into circumscribed, highly specialized regions of the brain. These circumventricular organs are capable of receiving macromolecular chemical signals from blood. The humoral signals are involved with the integration of pathways that mediate fluid and electrolyte homeostasis inside and outside of the CNS.

Most of the circumventricular organs are closely associated with the diencephalon. Each circumventricular organ has an ependymal interface and a highly permeable capillary interface (Fig. 10-7). Collectively, the ependymal and capillary surface areas of the circumventricular organs are relatively small, about 1% of the size of the ventricles or brain capillary bed. Despite these diminutive transport interfaces, the circumventricular organs and certain neuropeptide systems (e.g., angiotensin, vasopressin) carry out important functions to maintain fluid balance in the brain and whole organism (Table 10-4).

The circumventricular organs have structural and functional connectivity. The subfornical organ, which inte-grates water balance through angiotensin signaling, communicates with the other circumventricular organs. The area postrema, for example, is physically contiguous with fourth ventricle choroid plexus, which also has characteristics of a circumventricular organ. Lateral ventricle choroid plexus blood flow is markedly altered when the area postrema is stimulated, suggesting a mechanism by which circumventricular organ function can alter CSF formation. In general, the complex anatomic connections of the circumventricular organs with each other and with the pituitary and autonomic nervous system enable these tiny CSF-adjacent organs to modulate endocrine and homeostatic processes that maintain stability of the *internal milieu.*

Elaboration of Cerebrospinal Fluid

Fluids are generated at multiple sites in the adult CNS. The sources of most of the fluid are the choroid plexuses, which elaborate the CSF proper. Extrachoroidal sites of fluid production include a CSF-like secretion by the cerebral capillary wall and the metabolic generation of water by the complete oxidation of glucose by brain parenchyma. Because choroid plexus tissue is the preponderant fluid-production site, representing 75% or more

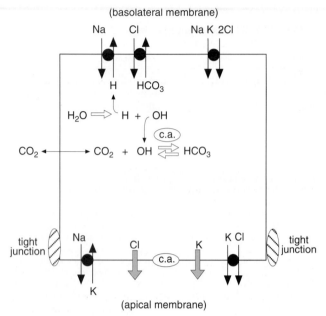

Figure 10-8
Schema for ion transport processes in choroid plexus epithelium (box diagram) that underlie CSF secretion. There is coordinated activity of ion transporters and channels in the basolateral (plasma-facing) and apical (CSF-facing) membranes that are sites of vectorial transport of Na, K, Cl, HCO_3 and water across choroid plexus. Active transporters in the membranes are depicted as 'filled' circles. Channels are the wide, 'unfilled' arrows. Direction of transport is indicated by arrow points. Primary driving force for CSF secretion is Na-K pumping in the apical membrane, which keeps choroid cell [Na] much lower than extracellular [Na]. As a result, there is an inwardly-directed Na gradient promoting basolateral Na uptake by the cell in exchange (*antiport*) for cytoplasmic H ion, or along with extracellular Cl (*cotransport*). Cl is also actively transported from plasma into the cell across the basolateral face, by an antiporter and a cotransporter. At the CSF side of the cell, Na is actively pumped into the ventricles. In addition, K and Cl, as well as HCO_3 [generated in the cell from carbonic anhydrase (c.a.) catalyzed hydration of CO_2], leave the cell via channels in the apical membrane. Water movement across the apical membrane into the ventricles is likely associated with the active release of K and Cl (cotransport) into CSF. This unified model has been constructed with transport data from amphibians (Wright EM, The choroid plexus as a route from blood to brain, Ann N Y Acad Sci 1986;481:214) and mammals (Johanson and Murphy, Acetazolamide and insulin alter choroid plexus epithelial cell [Na], pH and volume, Am J Physiol 1990;258:F1544).

of the total fluid formed, it is customary to regard the true CSF to be choroidal in origin.

As an active secretion by the choroidal epithelium, the CSF is not just a filtration of fluid across membranes at the blood-CSF barrier. The high rate of secretion of CSF (i.e., about 0.5 mL/minute/g of choroid plexus) depends on a brisk vascular perfusion of the plexus. The blood flow to the plexuses of approximately 5 mL/minute/g is about tenfold faster than the average cerebral blood flow. The great vascularity of the plexuses is evident from the reddish cast imparted to the tissue by its relatively large content of blood.

CSF is continually replenished, and the normal volume of 140 mL turns over about three times every 24 hours.

Nuclear magnetic resonance studies of humans have provided evidence that CSF production may actually increase at night. On average, the net production of approximately 0.35 mL of CSF each minute indicates a total 24-hour formation of at least 500 mL in adults.

The initial step in the CSF secretory process is the filtration of plasma across the choroidal capillaries. The fenestrated endothelium is not a barrier to the movement of macromolecules across the capillary wall into the interstitial fluid of choroid plexus. The interstitium offers minimal restriction to the delivery of ions and other substrates to transporters on the basolateral membrane of the epithelium. Several ion transporters have been identified, and their vectorial properties are summarized in Figure 10-8.

A distinctive feature of mammalian CSF is its relatively low concentration of protein and certain organic substrates. The total protein concentration in CSF is two to three orders of magnitude lower than in plasma. Glucose and urea concentrations in CSF are held at concentrations of 60% to 70% of those in arterial plasma. Many amino acids are in CSF at levels of only 10% to 20% of corresponding levels in plasma. The pH of CSF (7.35) is typically slightly less than arterial blood (7.40), in part because CSF PCO_2, like cerebral venous PCO_2, is higher than arterial PCO_2. CSF osmolality is slightly higher than that of plasma (by a few mOsm/L), mostly because of the relatively high concentrations of Na^+ and Cl^- in CSF.

CSF is normally a crystal-clear, colorless fluid; any degree of coloration is pathologic. Compositional data obtained from several mammalian species indicate that human CSF is remarkably similar to that of laboratory animals.

☐ Formation of CSF by the choroid plexuses

Quantitatively, the main constituents of CSF are Na^+, Cl^-, and HCO_3^-, and knowledge of the transport of these ions across the choroid plexus is essential to understanding the key elements of CSF production. The apically located Na^+-K^+ pump (i.e., ATPase) has a pivotal role (see Fig. 10-8). Primary active pumping of Na^+ from choroid cells to CSF keeps the intracellular concentration of Na^+ low (20 to 30 mmol/L) compared with the extracellular (i.e., interstitial) concentration of Na^+ of about 140 mmol/L. This sets up a substantial, inwardly directed transmembrane gradient for Na^+, which acts as a strong driving force for the secondary active transport of Na^+ into the cell by means of Na^+-H^+ exchange and Na^+-Cl^- cotransport in the basolateral (blood-facing) membrane. The coordinated activity of the basolateral Na^+ uptake and apical Na^+ extrusion systems ensures a continual net flux of Na^+ across the entire choroidal membrane from the blood to CSF. Through flow driven according to an osmotic gradient, water can follow the actively transported Na^+ and accompanying anions.

Cl^- uptake by choroid plexus occurs by two types of secondary active transport. In one mechanism, the inward transport of Cl$^-$ across the blood-facing membrane is driven by exchange with intracellular HCO$_3^-$. The HCO$_3^-$ is amply generated in the cell from the hydration of CO$_2$, which is catalyzed by carbonic anhydrase. In another mechanism, the basolateral co-transport takes advantage of the favorable, inwardly directed concentration gradient for Cl$^-$ and Na$^+$. The efflux of Cl$^-$ and HCO$_3^-$ from the choroid cell to ventricular CSF involves anion movement through channels and perhaps transport proteins that facilitate extrusion of anions (see Fig. 10-8).

Overall, the net transcellular movement of ions in CSF secretion is fueled by ATP which, when hydrolyzed by its ATPase, provides the energy for creating the essential Na$^+$ gradient that directly or indirectly drives several transport processes. The obligatory movement of water, which comprises 99% of CSF, occurs secondarily to the net transport of anions and cations into the ventricular cavities.

☐ Composition and homeostasis of CSF cations and anions

Nascent fluid

Newly formed fluid collected directly from the apical surface of choroid plexus epithelium has a high concentration of Na$^+$ (158 mEq/kg H$_2$O), Cl$^-$ (138 mEq/kg H$_2$O), and HCO$_3^-$ (25 mEq/kg H$_2$O), and lower concentrations of K$^+$ (3.3 mEq/kg H$_2$O), Ca^{2+} (1.7 mEq/kg H$_2$O), and Mg^{2+} (1.5 mEq/kg H$_2$O). Due to active transport processes in choroid plexus, the Cl$^-$ concentration in CSF is consistently greater than in plasma. The CSF concentration of K$^+$ is normally about 1 to 1.5 mEq/kg H$_2$O less than in plasma. The nascent CSF levels of Ca^{2+} and Mg^{2+} are steadfastly held slightly lower and higher, respectively, than the corresponding normal concentrations in plasma. Stability of CSF concentrations of K$^+$ and Ca^{2+} is essential, because relatively small deviations in these ion concentrations can alter CNS excitability. Table 10-5 relates the ionic concentrations in newly formed CSF to those in other fluids.

Mixing of choroidal secretion with brain interstitial fluid

As CSF flows from its site of origin in the interior of the CNS to more distal regions in the subarachnoid space on the exterior surface of the brain and cord, there are 5% to 10% modifications in the ionic composition of CSF (compare nascent with cisternal CSF in Table 10-5). This occurs because the relatively permeable ependymal and pia-glial linings permit free exchange of true CSF in the ventricles and subarachnoid space with brain interstitial fluid. The source of the interstitial fluid is presumably a slow but steady secretion by the cerebral capillary wall (i.e., blood-brain barrier) of a fluid similar in composition to CSF. In addition, the exchange of ions and nonelectrolytes across the external limiting membranes of neurons and glia contributes to the composition of interstitial fluid.

Table 10-5
CONCENTRATIONS (mEq/kg H$_2$O) OF IONS IN FLUIDS DERIVED FROM PLASMA

Fluid*	Cl$^-$	Na$^+$	K$^+$	Ca^{2+}	Mg^{2+}
Plasma	132	163	4.4	2.62	1.35
Plasma ultrafiltrate†	136	151	3.3	1.83	0.95
Choroid plexus fluid (nascent CSF)	138	158	3.28	1.67	1.47
Cisterna magna fluid	144	158	2.69	1.50	1.33

*Fluids were collected from cats (Ames A, Sakanoue M, Endo S, J Neurophysiol 1964;27:674).
†Plasma ultrafiltrate data, obtained from dialysis experiments, are values expected if the CSF were formed by passive distribution phenomena rather than by an active secretory process in the choroid plexus.

There is an approximately twofold difference in protein concentration gradients within regions of the CSF system. Lumbar CSF has about twice as much immunoglobulin G and albumin as ventricular fluid (Table 10-6). Such differences reflect regional variations in secretory or reabsorptive phenomena at the blood-CSF barrier. However, the roughly similar compositions of CSF and interstitial fluid ensure that, even after their mixing, the extracellular fluid of the CNS has a characteristic, if not completely uniform, composition. As the CSF flows down the neuraxis and exchanges with brain tissue, the content of protein and ions in cisternal CSF, spinal fluid, and interstitial fluid, although altered, still resembles closely the nascent ventricular CSF rather than plasma.

Regional sampling of cerebrospinal fluid and interstitial fluid

It is often desirable to sample CSF or brain interstitial fluid to evaluate the extracellular environment of the neurons. The various sites of CNS extracellular fluid sampling for experimental and clinical analyses are nascent CSF, large-cavity ventricular CSF, cisternal and lumbar

Table 10-6
REGIONAL DIFFERENCES IN THE CONCENTRATIONS OF PROTEINS AT VARIOUS CEREBROSPINAL FLUID SAMPLING SITES

Protein	Ventricular (n=27)	Cisternal (n=33)	Lumbar (n=127)
Total protein	25.6 ±1.1	31.6 ±1.0	42.0 ±0.5
Albumin	8.3 ±0.5	12.7 ±0.7	18.6 ±0.6
IgG	0.9 ±0.1	1.4 ±0.1	2.3 ±0.1

Values are means ± standard errors, given in units of mg/dL. The data, which are summarized by Fishman (p. 206), demonstrate a gradient of protein concentration from ventricular to spinal fluid. Protein concentrations in CSF are two to three orders of magnitude less than in plasma. Newly secreted CSF has a protein concentration of about 10 mg/dL. A CSF protein content of >500 mg/dL can indicate a lesion that is blocking the subarachnoid space.

fluids, and brain interstitial fluid. Nascent or fresh CSF can be collected by pipette as it exudes from the choroid plexus, but complex surgical procedures are necessary to isolate the secreting tissue. CSF sampled from the lateral, third, and fourth ventricles is a mixture of nascent CSF and brain interstitial fluid that has percolated across the ependymal lining; it can be sampled by invasive stereotactic procedures that may limit application. The most straightforward and common CSF sampling procedures involve removal of subarachnoid CSF from the cisterna magna (i.e., mainly in laboratory animals) or the lumbar region (i.e., spinal taps in humans).

Experimental neuroscience has benefited from recent technical advances in microprobe dialysis. With this technique, a tiny probe is inserted into a discrete brain region for the continuous collection of brain interstitial fluid as dialysate samples. Microdialysis has been especially fruitful in assessing microregional differences in neurotransmitter concentrations. Microprobes have also been placed in the cisterna magna to analyze CSF by an approach that does not disturb brain tissue or cause alterations in the normal volume of endogenous CSF.

Although relatively small regional differences in CSF or interstitial fluid concentrations have been demonstrated for inorganic ions and organic substrates such as urea and glucose, fairly large differences in regional concentrations have been observed for neuropeptides secreted by specific cell groups. For example, relatively high concentrations of angiotensin are found in hypothalamic interstitial fluid and in nearby third ventricle CSF; concentrations of this peptide dissipate with increasing distances from the hypothalamic source.

Homeostasis of cerebrospinal fluid composition

The hallmark of CSF is its ability to keep a stable composition of solutes despite severe excesses or deficiencies of plasma ions and molecules. CSF ion homeostasis is critical because even small alterations in CSF concentrations of K^+, Mg^{2+}, Ca^{2+}, and H^+ can affect respiration, blood pressure, heart rate, muscle tone, and emotional state. CSF composition has been extensively analyzed after acute and chronic perturbations of systemic acid-base parameters and ion concentrations. Analyses of nascent and cisternal CSF samples have revealed an impressive ability of the CSF system to maintain fairly even levels of K^+, Mg^{2+}, Ca^{2+}, and H^+ ions when challenged with wide swings in the concentrations of these ions in plasma. Direct analyses of the choroid plexus indicate that this tissue plays a major role in ensuring that only minor changes occur in the CSF ionic concentrations. Even certain organic substrates in CSF, such as the water-soluble vitamins B and C, are adequately maintained when the organism is suffering from vitamin-deficient states.

Two factors undergird the ability of the blood-CSF barrier to buffer changes in CSF composition effectively. First, the choroid plexus with its tight junctions between epithelial cells acts as a permeability barrier, thwarting bidirectional diffusion of substances between blood and ventricular fluid. Secondly, numerous transporters for ions and organic molecules enable the plexus to regulate the passage of substances across the barrier. CSF homeostasis is a function of the finely controlled movement of solutes by active transport or by facilitated diffusion (i.e., secondary active) systems that do not directly expend energy.

In hypovitaminosis C, for example, the low levels of ascorbate in plasma are avidly scavenged by an active transporter in the blood-facing membrane of choroid plexus for uptake into the epithelium. Concentrated in the cytoplasm of the plexus, the ascorbate can then move out of the cell by facilitated diffusion across the apical, CSF-facing membrane. Coordinated transport by these two mechanisms at opposite poles of the cell, working in series, enables the CSF to concentrate vitamin C to a level fourfold above that of plasma. Because the basolateral active transporter for ascorbate is a one-way path into the cell, the vitamin C is not leached from CSF when the plasma level is severely reduced. Comparable mechanisms for other micronutrients, the transport of which is also a function of plasma substrate concentration and choroid plexus carrier affinity, ensure stable CSF concentrations.

Neurohumoral regulation of cerebrospinal fluid secretion

Neurotransmitters and neuropeptides can modulate the choroidal secretion of ions, water, and proteins. Choroid plexus tissue has a high density of receptors for norepinephrine (α and β subtypes), serotonin ($5\text{-}HT_{1c}$ subtype), angiotensin II (AT_1 subtype), vasopressin (V_1 subtype), and other hormones. These receptors have been localized to the vasculature and to the choroidal epithelium. An investigation of cultured choroid plexus cells showed that serotonin was able to stimulate the secretion of transthyretin, a quantitatively important protein involved in the transport of hormones such as T_4 across the blood-CSF barrier. However, there is no evidence for neurohumoral modulation of choroidal secretion of proteins that can adjust the viscosity of CSF, as in the manner of cholinergic stimulation of salivary glands to secrete protein over wide ranges of concentration in the saliva.

In vitro transport studies of the choroid plexus (no blood flow) found that several neurohumoral agents, such as serotonin, vasopressin, and angiotensin, inhibit the release of Cl^- from the epithelial cells into an artificial CSF bathing medium. Because Cl^- transport from the in situ plexus to the ventricular fluid is an integral part of CSF formation, the in vitro studies are consistent with the known effects of serotonin, vasopressin, and angiotensin in reducing the CSF formation rate in intact animals. Neuropeptides administered in vivo can also curtail CSF formation by markedly reducing the rate of blood flow to the choroid plexuses. Substantial reductions in the vascular perfusion of the plexuses limit the rate of delivery of water and ions to the secreting epithelium.

An important and well-understood aspect of the neurohumoral modulation of CSF secretion is the involve-

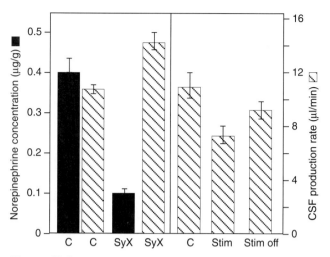

Figure 10-9

Regulation of cerebrospinal (CSF) formation by the sympathetic nervous system. The involvement of the sympathetic nerves in rabbit choroid plexus norepinephrine concentration (*solid bars*) and CSF production (*bars with diagonal lines*) was demonstrated in experiments involving denervation and electrical stimulation of nerves. One week after sympathetic denervation (SyX) of the choroid plexus, there was a substantial decrease in norepinephrine concentration, concomitant with a significant increase in the rate of CSF formation, compared with nondenervated controls (C). The electrical stimulation (Stim) of both superior cervical ganglia (which send sympathetic fibers to the choroid plexus) resulted in a statistically significant reduction in CSF production. After the stimulation was stopped (Stim off), there was a tendency toward normalization of the rate of CSF formation. Bars indicate the means ± standard errors. (From Nilsson C, Lindvall-Axelsson M, Owman C. Neuroendocrine regulatory mechanisms in the choroid plexus-cerebrospinal fluid system. Brain Res Rev 1992; 17:109.)

ment of the sympathetic nervous system (Fig. 10-9). The superior cervical ganglia send adrenergic fibers to the choroid plexuses. Resection of these sympathetic fibers reveals a significant increase in the rate of CSF formation, strongly implying that the sympathetic effect on the choroid plexuses is normally inhibitory and that, when this braking action is released by blocking the sympathetic signals, there is a resultant enhancement of fluid output by the choroidal epithelium. The findings from the denervation experiments have been bolstered by pharmacologic analyses establishing that α- and β-adrenergic agonists have the ability to inhibit CSF production.

Pharmacologic manipulation of cerebrospinal fluid formation rate

The clinical need for selective agents to lower ICP has spurred research to find drugs that can slow CSF production. Agents from many different pharmacologic classes have been used to assess dose-response phenomena in the choroid plexus–CSF system. Acetazolamide, an inhibitor of carbonic anhydrase that is abundant in choroid plexus tissue, consistently reduces the CSF formation rate by 50% to 60%. Short-term treatment with acetazolamide is therapeutically effective in unloading augmented CSF

pressure, but chronic use leads to attenuated efficacy and undesirable systemic acidosis as a side effect. Cardiac glycosides such as ouabain markedly reduce CSF production by inhibiting Na^+,K^+-ATPase, but they have limited therapeutic value because of their poor access to the CSF and their ability to raise the CSF concentration of K^+. Amiloride, by inhibiting the Na^+-H^+ exchange in the basolateral membrane of choroid plexus, can decrease CSF formation appreciably if hefty doses (75 to 100 mg/kg) are employed. Other diuretic agents can slow down CSF formation, but they introduce the expected complications caused by the urinary loss of water and electrolytes. Other agents need to be identified to complement the moderately effective use of acetazolamide.

Functional Interaction Between the Choroid Plexus—Cerebrospinal Fluid System and the Blood-Brain Barrier

The blood-CSF barrier of the choroid plexus epithelium and blood-brain barrier of the cerebral capillary endothelium work in concert to impart a relatively stable composition and volume to brain interstitial fluid (Fig. 10-10). The microenvironment of neurons depends on the transfer of materials at the two barrier interfaces. Each barrier has distinctive transport and permeability characteristics. The integrity of both barriers is essential for the homeostasis of CNS extracellular fluid. Disruption of barrier function, individually or in tandem, can have deleterious effects on brain parenchyma, brought about by alterations in fluid composition or pressure. Normally, the integrated flow of solutes and water across the respective barrier interfaces sustains an optimal environment for neurons and glia.

☐ Biochemical composition of brain interstitial fluid: vitamins and proteins

CSF formed by the choroid plexus membrane and brain fluid manufactured at the level of the capillary wall in the cerebrum, cerebellum, and spinal cord are not simply ultrafiltrates of plasma. The finely controlled fluids generated across these barriers are the result of active secretory processes. Once secreted into the CNS, the CSF and the endothelial-derived fluid mix with each other (Fig. 10-11). Mixing is promoted by bulk flow or diffusion, depending on the direction and magnitude of hydrostatic pressure or solute concentration gradients, respectively. Fluid mixing occurs mainly at two interfaces: the ependymal lining and the pia-glial membrane. Solute and water movements between the CSF and brain can occur bidirectionally. The vectorial distribution of a substance in a particular direction is a function of favorable driving force.

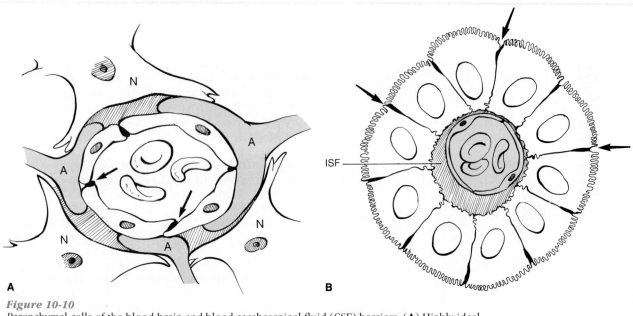

Figure 10-10
Parenchymal cells of the blood-brain and blood-cerebrospinal fluid (CSF) barriers. (**A**) Highly idealized model for the components of the blood-brain barrier. The endothelial cells of the cerebral capillaries lack fenestrations and are tightly joined by zonulae occludentes (*arrows*). Astrocyte foot processes extensively abut the outside surface of the endothelium. The darkened area is the interstitial space surrounding the capillary wall. *N,* neuron, *A,* astrocyte foot process. (**B**) Cross section of a choroidal villus. A ring of choroid epithelial cells surround the interstitial fluid (ISF) and adjacent vascular core. The basolateral surface of the cells has interdigitations, and the outer CSF-facing apical membrane has an extensive microvilli system. Arrows point to the tight junctions between cells at their apical ends.

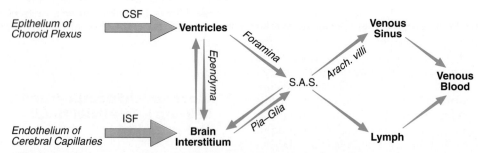

Figure 10-11
Schema for fluid formation, exchange, and drainage routes in the central nervous system. Cerebrospinal fluid (CSF) is derived from constituents of plasma ultrafiltrate in the plexuses by an active secretion occurring in the choroid epithelia. The plexus-generated fluid percolates down the ventricular system and then out into the subarachnoid space in the cisterna magna. From this great cistern, CSF continues to flow in the subarachnoid space (S.A.S.) overlying the hemispheres and cord. The S.A.S. fluid is reabsorbed into venous blood by a hydrostatic pressure–dependent mechanism in the arachnoid villi contained within the dura mater and into the lymphatics through the cranial and spinal nerves. CSF formation occurs at the blood-CSF barrier (choroid plexus) as the cerebral interstitial fluid (ISF) is slowly produced by endothelia of the blood-brain barrier. Once formed, ISF undergoes bulk flow exteriorly across the pia-glial membrane into the S.A.S. and interiorly into the ventricles, across the ependymal lining. Fluid flow is usually unidirectional through the ventricular foramina and arachnoid villi, but it is potentially bidirectional across the ependyma and pia-glia. For example, in pathophysiologic states such as hydrocephalus, in which CSF pressure is elevated, fluid can move from the ventricles into the brain tissue. The fluid in the S.A.S. is a mixture of CSF and ISF. Subarachnoid fluid drains into blood across arachnoid villi and through lymphatic tissue in the eyes and nose, which receive fluid draining along nerve roots to these organs. (Adapted from Andus KL, Raub TJ, eds, Pharmaceutical biotechnology. New York: Plenum Publishing, 1993;5:467.)

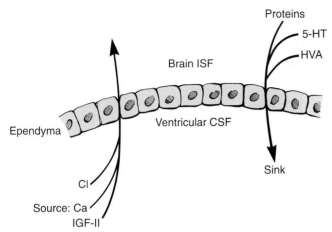

Figure 10-12
Cerebrospinal fluid (CSF) as a "source" or a "sink" for the brain. Depending on the prevailing concentration gradient for diffusion across the ependymal wall, the CSF can supply or remove solutes. The choroid plexus secretes ions, proteins, and various micronutrients into the ventricles. These transported solutes are derived from plasma or the choroidal epithelium, which has the capability of synthesizing certain proteins. After gaining access to the ventricular system, these solutes are distributed by bulk flow in the CSF, which acts as a supplier of materials for target cells in the brain. Conversely, the CSF functions like a drain for solutes such as the metabolites of neurotransmitters, protein catabolites, or iodide that are breakdown products of cerebral metabolism or that leak across barriers from blood. Once in the CSF, these potentially harmful materials are actively reabsorbed by choroid plexus or cleared from the CNS by bulk flow drainage mechanisms. *IGF,* insulin-like growth factor; *5-HT,* 5-hydroxytryptamine; *HVA,* homovannillic acid.

With large-cavity CSF as the reference fluid, the CSF can be regarded as a "source" or a "sink" for the brain (Fig. 10-12). The brisk secretory activity by the choroid-plexuses furnishes generous amounts of Na^+, Cl^-, Ca^{2+}, vitamins B and C (and other micronutrients), transthyretin, IGF-II, and additional tropic substances to ventricular fluid, allowing the latter to be a source of these materials for brain cells. Because the nascent CSF is normally low in protein and many metabolites, the ventricular fluid, as it sweeps down the ventriculocisternal axis, acts as a drain or sink to remove potentially harmful proteinaceous material and catabolites that have a higher concentration in brain interstitial fluid because of brain metabolism and blood-brain barrier leaks than in the CSF. Overall, the brain interstitial fluid composition is kept within narrow limits by virtue of the transependymal fluxes of solutes, the concentration gradients of which are set by transport phenomena at the blood-CSF and blood-brain barriers.

Stability of Brain Volume

Regulation of brain water content, and therefore of the volume, is critical for maintaining the ICP within tolerable limits. Brain volume is affected by many physiologic parameters. The influxes of water across the blood-brain and blood-CSF barriers are important determinants of water balance among the CNS compartments. Brain volume must be stabilized at two levels: the interstitial (i.e., extracellular) fluid and the neuronal and glial intracellular fluid.

Interstitial volume

The interstitial fluid volume of the brain is mainly the extracellular water content, and it comprises about 15% to 20% of the total tissue weight. Water is transported across the relatively impermeable cerebral capillary wall more slowly than across the more permeable peripheral capillaries. The Starling hypothesis, describing the role of passive hydrostatic and osmotic pressure gradients in driving net filtration in many vascular beds, does not apply to fluid exchange in the CNS between cerebral capillaries, which have low hydraulic conductivity, and the interstitial space that surrounds them. Normally, the water that gains access to brain by slow permeation across cerebral capillaries eventually flows out into the CSF system, and the pressure and volume of interstitial fluid are maintained at levels compatible with CNS functions.

Water is also transported into the CNS by means of the choroid plexuses as an integral part of CSF formation. Most of the water generated in the CNS originates from the four choroid plexuses. Once formed, the CSF moves by bulk flow predominantly along pathways of least resistance, such as through the ventricular and subarachnoid spaces rather than through the less compliant brain tissue (Fig. 10-13). The orderly flow of CSF and interstitial fluid along defined pathways normally keeps the extracellular fluid volume and the ICP relatively constant.

Intracellular volume

Brain volume is intimately related to water content. The water content of cells depends directly on the total amount of osmotically active solutes. Neurons and glial cells are continually exchanging ions and organic molecules across their external, limiting membranes. Cotransporters move solutes like Na^+, K^+, Cl^-, inositol, and taurine into and out of cells, depending on the concentration gradients and capacities of the respective transporters. Water osmotically follows the net movement of transported solutes. In this manner, cell volume can be stabilized even when extracellular tonicity is altered.

As brain tissue swells or shrinks, as in acute hyponatremia or hypernatremia, the activity of cellular cotransporters is appropriately modified by up- or downregulation. The result is that cell volume is rapidly reestablished. Ischemic or pharmacologic disruption of cellular transporters can cause swelling of parenchyma and the whole brain. Some of the states that alter the size of the intracellular and extracellular compartments of the CNS are discussed in subsequent sections of this chapter.

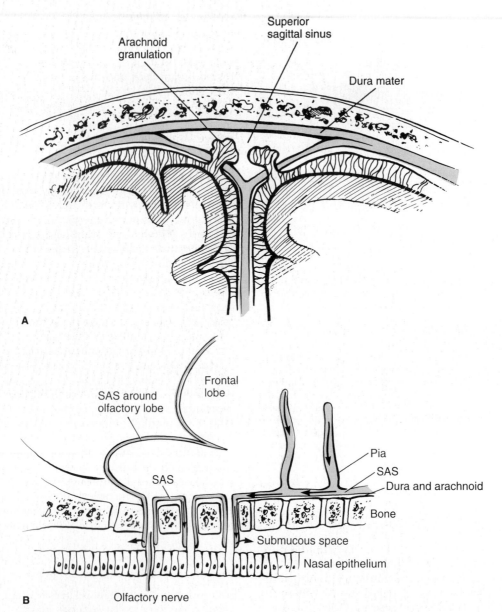

Figure 10-13
Drainage of cerebrospinal fluid (CSF) by bulk flow across arachnoid villi in the dura mater and across lymphatic capillaries in the eye and nose. (**A**) A hydrostatic pressure gradient drives the CSF in the subarachnoid space across valve-like structures in the arachnoid villi out into the venous sinus. These valves are big enough to allow protein and even cells in the CSF to pass, normally only unidirectionally from CSF to blood. After the CSF with its macromolecules has flowed into the venous sinus, it has effectively been cleared from the brain. The villi increase in size and number with advancing age. (**B**) Another route for CSF drainage is along the outside of cranial nerves down the submucosal tissue of the nose and eye. CSF can flow through sleeve-like extensions of the subarachnoid space around the nerve through the cribriform plate to reach the submucosal tissue of nose. Lymphatic capillaries drain the submucous spaces and convey fluid to lymph nodes in neck. (Adapted from Bradbury MW, Cserr HF, Westrop RJ. Drainage of cerebral interstitial fluid into deep cervical lymph of the rabbit. Am J Physiol 1981;240:F335.)

Effects of Fluid Imbalances

Consideration of relative and absolute volumes of the brain and CSF is essential for understanding normal cerebral functions and those caused by derangements in these fluid systems. Alterations in the size of the brain interstitial space affects excitability phenomena. Severe contraction of the ventricular volume may compromise the ability of the CSF to function as a sink for brain metabolites. These are examples of how brain and CSF volume changes can affect physiologic functions.

A substantial increase in the brain content of water can bring about pathologic problems such as tissue herniation and intracranial hypertension. Edema may be generalized or local (e.g., surrounding a tumor or infarct). In localized edema, herniation of tissue can involve the cerebellar tonsils through the foramen magnum or the temporal lobe uncus across the tentorium. Edematous states can also be roughly classified as affecting mainly the interstitial or the intracellular compartments of brain water.

☐ Vasogenic edema results from a breakdown of the blood-brain barrier

The most prevalent type of brain edema is the vasogenic disturbance, which commonly occurs in regions bordering ischemic zones. Vasogenic edema is caused by increased permeability of the blood-brain barrier, which allows plasma with its content of proteins and ions to leak across the endothelial wall. The resulting increase in brain interstitial fluid volume can raise the ICP, slow the electroencephalographic pattern, and impair consciousness. White matter is particularly affected. Vasogenic edema frequently occurs in cases of head trauma and meningitis and can be visualized by magnetic resonance imaging (MRI) or CT scans. A substantial increase in brain volume from vasogenic edema occurs at the expense of the ventricles, which decrease to mere slits on clinical scans because of CSF displacement.

☐ Interstitial edema is caused by excess CSF in the hydrocephalus

Another edematous state in which the brain extracellular compartment undergoes swelling is interstitial edema. Elevated ICP in obstructive hydrocephalus promotes the spread of ventricular water and sodium across the ependymal lining into the adjacent white matter. Axial MRI reveals the periventricular edema as a white rim around the frontal and occipital horns of the ventricles. Interstitial edema caused by chronic hydrocephalus can be relieved by surgically shunting the CSF to another cavity in the body. Hydrocephalus-induced interstitial edema is associated with ventricular enlargement.

☐ Cytotoxic edema is a consequence of drug poisoning or altered brain cell metabolism in disease states

Cytotoxic edema usually involves intracellular swelling of glia, endothelia, and neurons. The result of the expanded volume of cells is an attenuation of the size of the interstitial space. There are many different causes of cell swelling, including drug poisoning, water intoxication, hypoxia from asphyxia, and acute hyponatremia. Under such conditions, there is a net shift of water from the extracellular space to the interior of brain cells. Cytotoxic edema may coexist with other forms of edema that occur in encephalitis and meningitis. The brain swelling attendant to severe cytotoxic edema leads to marked reduction in the size of the ventricular system and basal cisterns. Such distortion of the ventricular and subarachnoid spaces can interfere with CSF circulatory dynamics.

Circulation of Cerebrospinal Fluid

Cerebrospinal fluid has been referred to as a third circulation. CSF is derived from an arterial supply consisting of the anterior and posterior choroidal arteries. Choroidal venous drainage occurs largely through the vein of Galen. There are no lymphatic capillaries in the choroid plexuses. However, once secreted into the ventricles, the continuous flow of CSF acts like a quasilymphatic system. This unique anatomic arrangement allows brain extracellular fluid to circulate along circumscribed pathways without a true lymphatic system.

In addition to serving as clearance conduits, the CSF circulatory pathways provide a means for distributing tropic substances and micronutrients to target cells in the brain. As it heads for multiple venous drainage sites, the human CSF percolates relatively slowly by bulk flow through the ventricular and subarachnoid spaces at a rate of about 0.35 mL/minute.

☐ Gradients of hydrostatic pressure between CSF and blood

Several driving forces propel CSF. Formation of CSF by the choroid plexuses occurs with a hydrostatic pressure of about 15 cm H_2O, providing a force for the forward movement of the newly formed fluid. Another force promoting CSF circulation is the strong pulsation of the blood coursing through the choroidal vasculature. The beating of cilia extending from the apical surface of the choroidal epithelium and some ependymal cells impart an additional thrust to the ventricular CSF. The higher pressure of CSF relative to that in the dural venous sinuses creates a favorable pressure gradient of 7 to 8 cm H_2O that promotes clearance of CSF by bulk flow from subarachnoid space to blood.

☐ Direction of currents of CSF flow

Ventricular fluid moves from the lateral ventricles, through the paired foramina of Monro into the anterior third ventricle. After leaving the posterior third ventricle, the CSF continues its movement along the aqueduct of Sylvius, eventually emptying into the fourth ventricle. Fourth ventricle CSF seeps into the subarachnoid space by three different apertures. Two exits are the foramina of Luschka at the extreme lateral portions of the fourth ventricle. The third major exit is the foramen of Magendie in the roof of the ventricle, which communicates with the *cisterna magna* or the *cisterna cerebellomedullaris*.

CSF flows from the basal areas of the brain over the hemispheric convexities until it reaches the arachnoid villi in the walls of the superior sagittal sinus. CSF from the foramina of Luschka flows forward from the cisterna pontis to the *interpeduncularis* and *chiasmatis cisterns*, from which it sweeps upward over the outer surfaces of both lateral hemispheres. It progresses anteriorly, upward along the longitudinal fissure, over the corpus callosum, along the sylvian fissure, and over the temporal lobes. At the most distal end of the flow route, the cranial subarachnoid CSF encounters the arachnoid villi.

CSF is also directed from the *cisterna magna* down the posterior or dorsal surface of the spinal cord. CSF fills a sleeve of subarachnoid space around the spinal cord and extends below the end of the cord into the region of the second sacral vertebra. Although the spinal subarachnoid CSF is in effect an anatomic blind pocket, a slow mixing of spinal CSF with cranial CSF is induced by postural changes.

Drainage of Cerebrospinal Fluid

The nature of CSF drainage is of particular interest to neurosurgeons because of the ICP problems created when CSF is not cleared adequately from the CNS. Normally, the CSF drainage keeps pace with CSF formation, and ICP is stabilized. There are two different bulk flow mechanisms (Fig. 10-13) by which CSF leaves the CNS: drainage directly into the venous blood in the dural sinuses and drainage along the cranial and spinal nerves into the prelymphatic tissue spaces and then into the lymphatics.

☐ Arachnoid villi facilitate CSF drainage into venous blood

Most of the CSF is returned passively along hydrostatic pressure gradients across arachnoid villi that separate the subarachnoid space from the venous blood in the sinuses of the dura mater. The arachnoid villi have a high degree of hydrodynamic permeability, much more so than that in peripheral capillaries. If there is a sufficient hydrostatic pressure gradient of at least 3 cm H_2O and in the right direction (CSF to blood), even large proteins and erythrocytes are able to penetrate the one-way valves that regulate CSF flow into the venous sinuses. At about 12 cm H_2O of CSF pressure, the rate of CSF absorption is equal to the rate of CSF production; with additional elevation in CSF pressure, the absorption rate is augmented in proportion to the increment in CSF pressure.

Vacuoles in the mesothelial cells lining the arachnoid villi suggest an additional dynamic process of vacuolation that is pressure sensitive. The formation of vacuoles in the valves may constitute transendothelial channels. If venous pressure exceeds that of the CSF, the valves close and prevent reflux of blood into the subarachnoid CSF system.

☐ Lymphatics outside the brain receive CSF from the subarachnoid space

Using another drainage pathway, the CSF must first pass through lymphatic tissue before reaching venous blood. Sleeves of subarachnoid CSF surround the optic and olfactory nerves in particular and the other cranial and spinal nerves. A positive-pressure gradient promotes the flow of CSF along this perineural space that extends into the submucosal tissue of the eye and nose, in which the lymphatic capillaries resorb the CSF and convey it to the cervical lymph glands. CSF draining by this route is first exposed to the immune system before reaching the venous blood. CSF containing antigenic material (e.g., products of myelin breakdown) can induce antibody reactions, which can promote reactionary immune phenomena in the CNS.

Cerebrospinal Fluid Pressure-Volume Associations

Because the bony skull is virtually incompressible, ICP rises when there is a significant increase in any one of the three major constituents of the intracranial space: brain parenchyma, CSF, and vascular tissue. The CSF constituent can be regarded as a potential liability and an asset. Life-threatening increases in ICP can ensue if CSF drainage is blocked, but CSF can be readily shunted to unload ICP, and the CSF pressure is useful for measuring the response to therapy.

Markedly elevated intracranial or CSF pressure can irreversibly injure the CNS. Normally, the CSF confers protection against impact pressures and acute changes in arterial and venous pressure, but tumors, infections, neurosurgical procedures, trauma, and diseases can lead to serious elevations in ICP. There are several sites at which ICP can be monitored, and there are numerous treatment modalities available to lower the ICP.

☐ Normal ranges of intracranial pressure as measured in the lumbar region

It is common to assess ICP by measuring CSF pressure in the lumbar region. In patients with no pathologic lesions and with normal blood pressure, the average pressure of subarachnoid CSF in the lumbar region of reclining individuals is about 10 cm H_2O. The generally accepted range for lumbar CSF pressure is 5 to 15 cm H_2O, which is about 4 to 11 mm Hg. CSF production is relatively constant over the normal range of CSF pressure, but the production of CSF may decrease when the CSF pressure is substantially elevated, as in severe hydrocephalus. The augmented CSF pressure reduces filtration of plasma across choroidal capillaries, which is the initial step in CSF formation.

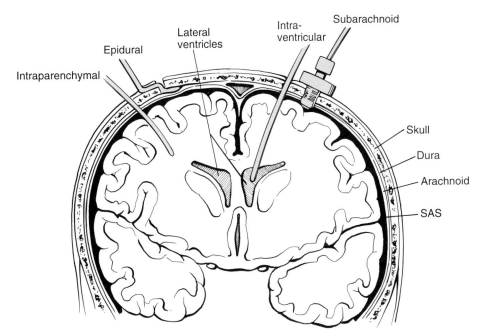

Figure 10-14
Various sites for monitoring intracranial pressure. The intraventricular catheter is inserted through a burr hole in the frontal lobe, down into a lateral ventricle near the foramen of Monro. Placement of a probe in the epidural space carries minimal risk for brain infection because the dura remains intact. The subarachnoid bolt is placed in the subarachnoid space, (*SAS*) but it often needs irrigation with saline to remain patent. Intraparenchymal microtransducers can usually be inserted 2 to 3 cm into white matter without complications. (From Lyons MK, Meyer FB. Cerebrospinal fluid physiology and the management of increased intracranial pressure. Mayo Clin Proc 1990;65:684.)

☐ Measurement of intracranial pressure by invasive and noninvasive probes

The pressure on the intracranial contents can be assessed by placing probes at various depths under the skull bone, epidurally or in the subarachnoid space, brain parenchyma, or a lateral ventricle. These four sites for monitoring ICP are depicted in Figure 10-14. The standard for measuring ICP has been the intraventricular catheter connected to a manometer by way of a fluid-filled tube. For placement of the catheter, a burr hole is drilled over the frontal lobe. After piercing the dura, the catheter is directed into the lateral ventricle with the tip close to the foramen of Monro. Accurate ICP assessments can be made with ventricular probes, but not without occasional complications (Table 10-7).

Relatively noninvasive pressure probes can be applied epidurally. The epidural pulsations, an indirect measure of ICP, can be monitored by fiberoptic, strain gauge, or pneumatic systems. However, this approach does not have sufficient accuracy to warrant regular use. Pressure recorded in the subarachnoid space over the convexities is another way to estimate ICP, but this method lacks reliability. The subarachnoid screw or bolt is a hollow tube that is secured to the calvaria. The device is connected to a fluid-filled system that has an externally located pressure transducer. Intraparenchymal pressure recording is useful when a catheter misses entering the ventricle. The Camino fiberoptic system can be used for intraparenchymal recording and for epidural and intraventricular monitoring. Infection rates for any of the pressure-measuring systems is minimal (<1%) if the monitor is left in place for 4 days or less. Table 10-7 compares strengths and limitations of four main locations for measuring ICP.

Table 10-7
DEVICES AND LOCATIONS FOR MONITORING INTRACRANIAL PRESSURE

Location	Advantages	Disadvantages
Intraventricular	Reliable measure of ICP, allows CSF drainage, good pressure waveform	Invasive, risk of infection, need to enter ventricles, can obstruct
Epidural	Less invasive, dura remains intact, less risk of infection	No CSF drainage, poor measure of ICP, no waveform
Subarachnoid	Brain theoretically not penetrated, lower risk of infection, ease of placement	No CSF drainage, brain tissue easily obstructed, fluid-filled system but waveform usually poor
Intraparenchymal	Non–fluid-filled system, fairly reliable measure of ICP	Invasive, risk of infection, no CSF drainage

CSF, cerebrospinal fluid; *ICP*, intracranial pressure. Adapted from Lyons MK, Meyer FB. Cerebrospinal fluid physiology and the management of increased intracranial pressure. Mayo Clin Proc 1990;65:684; and from J.G. McComb (personal communication).

☐ CSF pulsations and pressure waves in normal and pathologic states

Normally, CSF pulsates as the result of continuous changes in venous and arterial pressures. Intracranial venous pressure decreases and increases, respectively, during respiratory inspiration and expiration. Such alterations in venous pressure during the respiratory cycle are concomitantly transmitted to the CSF pressure. Arterial systolic pulsations, particularly in the choroid plexus, are normally reflected as synchronous elevations in CSF pressure.

Sustained intracranial hypertension can result in pathologic plateau waves. Instability of vasomotor control, such as a loss of cerebrovascular autoregulation and reduction of cerebral blood flow, can trigger a sudden onset of plateau waves. These clinically significant plateau waveforms may last 5 to 20 minutes and can be associated with an ICP of more than 100 cm H_2O. The elevated CSF pressure is a consequence of enhanced cerebral blood volume. Plateau waves occur in advanced stages of intracranial hypertension and may indicate potential damage to the CNS.

☐ Cerebral perfusion pressure is a function of intracranial pressure and cerebrovascular parameters

Marked elevations in ICP can reduce the arterial blood supply to the brain, causing irreversible damage to nervous tissue. *Cerebral perfusion pressure* is a critical parameter, and it is defined as the difference between mean systemic arterial blood pressure and the ICP. When the ICP rises to the level of systolic blood pressure, blood flow to the CNS ceases because the perfusion pressure, which is the driving force, becomes negligible. If prolonged, this can result in brain death.

Substantial elevations in ICP are capable of compromising the delivery of oxygen and nutrients to the brain. Because of the rigidity of the skull, the relatively fixed volume of the intracranial space cannot accommodate increases in brain tissue mass, such as those occurring with space-occupying lesions (e.g., tumors, blood clots) or as the result of edematous fluid accumulation secondary to trauma. In such cases, failure to adequately lower the ICP and maintain perfusion can be fatal.

☐ Elevated CSF pressure can be life-threatening

Trauma commonly causes a rise in ICP. Cerebral edema or bleeding into CSF can markedly elevate ICP. There are surgical and nonsurgical ways to alleviate intracranial hypertension. Surgical decompression is feasible in many patients with hematomas located epidurally, subdurally, or even in the parenchyma. Drainage of CSF through ven-

triculostomy is the primary method of lowering ICP if its elevation is secondary to obstruction of CSF drainage.

If surgery is not indicated, there are postural and several pharmacologic strategies available for diminishing ICP. Elevation of the head above the heart facilitates venous drainage in some patients. To minimize cerebral edema from injured cerebral vessels, fluid restriction is useful if the patient does not have diabetes insipidus. Extremely high levels of blood glucose should be avoided because hyperglycemia, which often occurs in cases of head injury, can be detrimental to cerebral function if ischemia also occurs. Ventilatory support in the form of hyperventilation can rapidly lower ICP in many patients; this beneficial effect results from constriction of the cerebrovascular bed, which effectively lowers the volume of blood in the brain.

Diuretic administration has been widely employed to reduce the water content of the CNS, which decreases ICP. Furosemide and acetazolamide curtail the choroid plexus output of CSF into the ventricles and function as renal diuretics, eliminating fluid from the body. Mannitol has been widely used as an osmotic diuretic agent because it slowly permeates the blood-brain and blood-CSF barriers. As the result of the osmotic gradient established between blood and brain, mannitol is able to "pull" water from nervous tissue. The dehydrating effect of mannitol may be prolonged by the concurrent use of loop diuretics such as furosemide. Because mannitol may be cleared from blood faster than from CNS, a rebound intracranial hypertension can occur if the agent is not carefully administered.

Corticosteroids and barbiturates have limited use in controlling ICP. High doses of glucocorticoids, such as dexamethasone and methylprednisolone, decrease ICP in some patients with large brain tumors by suppressing edema formation and stabilizing cellular membranes in the edematous brain adjacent to the tumor. However, the efficacy of steroid hormones has not been demonstrated in head-injured patients. Barbiturates such as pentobarbital have been used when conventional modalities fail to control ICP. The barbiturates confer protection by decreasing cerebral blood flow and cellular metabolism, thereby reducing ICP.

☐ Compression of cerebrospinal fluid and the optic nerve can cause blindness

The increasing mass of a CNS tumor can occlude drainage of CSF, with a resultant increase in ICP. Tumors distant from the ventricles may not significantly obstruct CSF flow until the mass attains a relatively large size. Tumor growths in the posterior fossa of the cerebellum exert pressure on the roof of the fourth ventricle, obstructing CSF flow into the subarachnoid space. Brain tumors that compress the optic nerve can cause papilledema by choking the optic disc because of the high pressure inside the sleeve of dura mater surrounding the nerve. The sustained papilledema can severely damage the optic nerve and cause blindness. Papilledema can also result from *pseudotumor cerebri*, a condition producing benign intracranial hypertension in which there is an increased rate of CSF production and a greater content of water in the brain interstitium. Benign intracranial hypertension, which markedly elevates ICP in young, obese women, can sometimes be treated with effective resolution in several weeks.

☐ Continuously elevated intracranial pressure can cause enlargement of the ventricles

Hydrocephalus, which is an increase in CSF volume within the cranial cavity, can occur with or without an elevation in ICP. *Compensatory hydrocephalus* occurs without an increase in ICP, and it represents an increase in CSF volume as a compensation for cerebral atrophy due to primary CNS disease. *Normal-pressure hydrocephalus* results from impaired CSF absorption into venous blood. CSF composition and pressure are normal, and although the ventricles are enlarged, there is no alteration in the size of the cerebral cortex or subarachnoid space. In normal-pressure hydrocephalus, the shunting of CSF can relieve the characteristic triad of symptoms: unsteady gait, dementia, and urinary incontinence.

Hydrocephalus with increased ICP can be categorized as communicating or noncommunicating (i.e., obstructive). In *communicating hydrocephalus*, there is free communication between ventricular and subarachnoid fluids. Hydrocephalus can be caused by altered CSF dynamics (i.e., production or absorption) or by obstruction to the flow of CSF through the subarachnoid space. In *obstructive hydrocephalus*, something impedes the percolation of fluid within the ventricles, aqueduct, or fourth ventricle outlets. As the result of CSF flow blockage by developmental abnormalities, inflamed tissue, or tumors, fluid is retained in the ventricles, which elevates the ICP.

Continuously elevated ICP in hydrocephalic states with accumulation of CSF causes ventriculomegaly, an enlargement of the ventricles that compresses cells and their processes. During the early stages of hydrocephalus, there is damage to the ependyma and periventricular white matter. Two to three weeks of severe hydrocephalus can bring about compression of the cortical mantle, sometimes to 25% of its original thickness. Cytologic and cytoarchitectural studies of brain cells in animal hydrocephalus models have revealed greatly shrunken somata and an abundance of vacuoles. In severe hypertensive hydrocephalus, there is a decrease in the number of axons and blood vessels. Surgical shunting of CSF to the peritoneal cavity reduces ventriculomegaly and decompresses the cerebral cortex. If the shunting is done early enough, much of the structural and functional damage to cells can be reversed.

Cellular Composition of Cerebrospinal Fluid

A distinguishing feature of CSF, especially compared with blood, is its paucity of cellular elements. CSF usually contains no more than 4 mononuclear cells or lymphocytes/mm^3. Leukocyte counts as low as 5 to 10/mm^3 occasionally signify a pathologic condition. An elevated cell count in CSF can be the result of brain injury, central inflammatory processes, or spreading tumor cells. Cytologic examination of CSF is becoming more effective with the application of polyclonal and monoclonal antibodies to identify specific pathologic processes.

☐ Cellular content of normal CSF

The relatively high water content of CSF (>99%) reflects the very low cell count. The only cells routinely observed in normal CSF are a few B and T lymphocytes and monocytes. It is rare to find sloughed-off choroid epithelial, ependymal, or arachnoidal cells in CSF samples that otherwise appear normal. The relatively impermeable barriers prevent the penetration of formed elements of the blood into the CSF compartment of healthy humans. The occasional appearance of erythrocytes in CSF samples (usually tapped from the lumbar region) indicates contamination of the collected CSF specimen with blood.

☐ Meningitis and AIDS alter the cellular profile of CSF

Cerebrospinal fluid pleocytosis is common in cases of acute infections of the CNS. In fungal infections, the predominant type of cell in CSF is the lymphocyte, and in bacterial infections, it is the neutrophil. In acute bacterial meningitis, 90% or more of the cells in CSF may be neutrophils. With severe infections, especially if an abscess ruptures, the CSF cell counts can exceed 20,000/mm^3.

The appearance of ependymal and choroidal cells in CSF, along with leukocytes, can be a manifestation of neurologic diseases and infections. The mumps virus is unique in that it causes ependymitis. Such destruction of the ependymal lining can lead to narrowing of the cerebral aqueduct, with resulting hydrocephalus.

The cellular composition of CSF in AIDS patients is highly variable. In one study of human immunodeficiency virus-1 infection, only a small percentage of patients developed lymphocytic pleocytosis in CSF, even when the full picture of AIDS was present. AIDS patients have a high frequency of opportunistic infections. Their CSF profiles of immune cells are similar to those caused by the opportunistic invader, although the magnitude of the response may be blunted by the immunodeficient state.

☐ Exfoliation of tumor cells from primary vs. metastatic neoplasms

CSF sampling can be used for the management of patients with brain tumors. Primary neoplasms around the brain stem, cerebellum, and spinal cord can abut CSF pathways and shed tumor cells, which appear in the sediment of CSF samples. Medulloblastomas arising from the external germinal layer of the cerebellum may shed many tumorous cells into CSF. Meningiomas shed few malignant cells into CSF. Meningiomas are firm, arachnoidal elements that do not readily exfoliate cells, which accounts for the low frequency of positive CSF specimens.

Metastatic neoplasms (i.e., carcinomatosis) have a greater propensity to exfoliate cells than do most of the primary tumors. There is a high yield (20% to 50% positive) from cytologic examination of malignant cells in CSF of patients with cerebral metastases. Carcinomas of the lung, stomach, and breast are among the most common to metastasize into the CSF. The less frequently occurring melanoma usually metastasizes rapidly to the brain and CSF and is cytologically characterized by pigmentation. Occasionally, CNS and CSF metastases occur before the primary peripheral tumor is discovered. Inflammatory cells are often interspersed with tumor cells in the CSF. When the exfoliated tumor cells in CSF are sufficiently characteristic, the primary site of origin outside the CNS can sometimes be identified.

Thorough treatment of leukemia depends on CSF cytologic analysis. Because many chemotherapeutic agents do not readily penetrate the blood-CSF and blood-brain barriers, malignant cells inside the CNS are often left intact after treatment. Because surviving tumor cells in the subarachnoid CSF can be a reservoir contributing to a later systemic relapse, it is essential to ascertain the presence of even a small number of leukemic cells in CSF. The technique of flow cytometry combined with monoclonal antibody staining enables the detection of small numbers of malignant cells. CSF examination is important in leukemia management for deciding whether irradiation or intrathecal chemotherapy is indicated.

Clinical Use of Cerebrospinal Fluid Profiles

Many neurologic disorders are associated with changes in the chemical composition of CSF. As metabolic processes in brain tissue are altered by disease or trauma, the numerous cellular metabolites that are released into surrounding interstitial fluid eventually gain access to the CSF. CSF biochemical profiles are often altered during illness or injury. Such changes in CSF composition can sometimes mirror modifications in cerebral chemistry. Clinicians have learned to use CSF findings to guide diagnoses and management. Limitations to this approach must be recognized, but if CSF sample contents are appropriately interpreted, the biochemical and cellular analyses can be diagnostically valuable.

The subarachnoid space in the lumbar cord is a convenient penetration point for sampling CSF after the potential value and risks of the tapping have been considered. Although CT scans have obviated the need for some lumbar punctures (e.g., suspected subarachnoid hemorrhage), there are many indications for procuring CSF samples.

☐ Diagnostic information can be gained from lumbar CSF samples

Lumbar punctures are done mainly for diagnostic reasons. Analyses of CSF can strengthen the diagnoses of neurosyphilis, multiple sclerosis, and many inflammatory diseases of the brain and its meningeal coverings. Inexplicable seizures should also prompt the analysis of CSF biochemistry. In myelographic procedures, the CSF removed at the initial stage of the procedure should be characterized to serve as a baseline for future comparisons. Access to the CSF system by way of a lumbar tap also permits assessment of intracranial hypertension.

☐ Intrathecal or intraventricular drug administration bypasses the blood-brain barrier

Some drugs that cross the blood-brain and blood-CSF barriers too slowly to be therapeutically useful can be continuously administered intraventricularly through lateral ventriculostomy or intrathecally by infusion into spinal subarachnoid fluid (Fig. 10-15). Injection of drugs directly into human nervous tissue is usually not feasible, but agents infused into the CSF of patients move by bulk flow eventually to spread into the brain and cord by diffusion across permeable interfaces like the ependyma of the ventricles and pia-glial cells of the subarachnoid space.

CSF infusion circumvents the usual barriers and exploits pharmacokinetic factors. Drug metabolism and binding to proteins is usually less of a problem in CSF than in plasma. Central administration of drugs largely avoids renal-hepatic metabolism. Because CSF has a relatively low content of protein, there is much less chance that a drug effect will be diminished by extensive binding to a large reservoir of extracellular albumin or other protein.

Implantable pump systems can deliver drugs into the CSF. Some implanted pumps hold as much as 50 mL and deliver at rates of 75 to 100 µL/minute. The variation in delivery rate is about ±5%. Drug dose is regulated by adjusting concentration in the reservoir. Stability of the therapeutic agent is required for many weeks at 37°C in a CSF-like buffer. Diluents (i.e., vehicles) are confined to a certain pH an osmolality and cannot contain a solubilizing agent potentially harmful to the exquisitely sensitive brain tissue.

An advantage of the CSF drug delivery approach is the ability to localize the desired pharmacologic effect. This is significant because many neurologic diseases are cir-

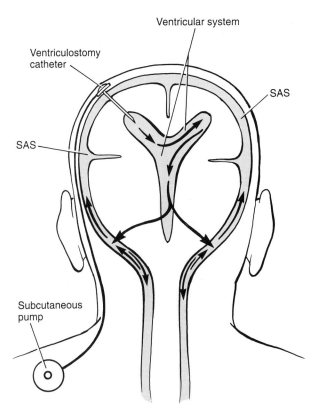

Ventricular system

Ventriculostomy
catheter

SAS

SAS

Subcutaneous
pump

Figure 10-15
Cerebrospinal fluid infusions of drugs through a totally implanted system for constant intraventricular drug infusion. The
components of the system include a subcutaneously implanted
Infusaid pump, Silastic catheter, and ventriculostomy reservoir
and catheter. The amount of drug for the refill is calculated as
the product of dose rate times pump capacity, divided by flow
rate of pump. *SAS,* subarachnoid space. (From Dakhil S, Ensmin-
ger W, Kindt G, et al. Implanted system for intraventricular drug
infusion in central nervous system tumors. Can Treat Rep
1981;65:405.)

cumscribed to a specific region. Drug delivery inside the
CNS is most effective when there is an optimal blending
of pharmacodynamics and pharmacokinetics. Intrathecal
morphine, used successfully in pain control, avoids systemic narcotic effects like anorexia and oversedation. Intraventricularly administered bethanechol, an acetylcholine agonist, has led to some improvement in Alzheimer's
patients. Additional clinical investigations of CNS diseases
will reveal how higher brain center functions can be effectively modified with intraventricular infusions of therapeutic agents.

☐ Future clinical strategies may involve manipulation of CSF levels of growth factors and cytokines

There is a surge of interest in the CSF concentration of
neuropeptides, transport proteins, and growth factors.

The source of these materials and the kinetics of their
distribution throughout the ventricular system and surrounding brain tissue have numerous implications for
developmental neurobiologists and clinicians. The choroid plexus and certain hypothalamic nuclei have their
respective abilities to synthesize and secrete certain proteins and peptides into the extracellular milieu of the
brain. CSF bulk flow routes through the ventricular and
subarachnoid systems then promote the paracrine distribution of peptides and proteins to various regions in the
CNS.

Vasopressin, angiotensin, and atriopeptin are manufactured by cells within the CNS. Release of these centrally
synthesized peptides into brain interstitial fluid and CSF
is stimulated by alterations in extracellular osmolality or
Na^+ concentration or by perturbations associated with
ischemia and trauma to the brain. Target cells with receptors for these peptides are located in the choroid plexus,
in circumventricular organs, and at the blood-brain barrier. Vasopressin and angiotensin, which can alter CSF
formation rate and blood-brain barrier permeability,
have a modulatory role in fluid and ·electrolyte balance
in the brain. Neuroendocrine regulation of brain volume
is probably mediated by physiologic actions induced by
these peptides.

Some transport proteins facilitate the passage of hormones from blood to CSF. One such protein is transthyretin (i.e., prealbumin), elaborated by the choroid plexus
for secretion into CSF. The choroid plexus is the only site
within the CNS that manufactures and secretes transthyretin. Transthyretin helps to carry thyroxine (i.e., T_4)
across the choroid plexuses into ventricular CSF. Molecular studies by Schreiber and colleagues have demonstrated mRNA in choroid plexus for other transport proteins, such as ceruloplasmin and transferrin, that have
been implicated in the carriage of trace elements such as
copper and iron from plasma to the CSF. This transport
path may represent an important supply route for the
brain by means of the CSF.

The choroid plexus–CSF nexus is of great importance
to the brain as it is developing and again later in life as it
is undergoing repair when injured. The plexus produces
IGF-II and binding proteins for growth factors. Research
is needed to elucidate target cells in brain for this and
other growth factors made in the epithelium of the
blood-CSF barrier. During fetal development, cells that
form the cerebral cortex arise from a layer of neuroepithelial cells surrounding ventricular CSF. The roles of
growth factors and other tropic proteins in modulating
cells in the subventricular zone and in the cellular migration processes await future ontogenetic investigations.
The ability of adult brain to repair itself may also depend
partially on growth factor availability from the choroid
plexus–CSF system. Ischemia models have revealed that
nerve growth factor or transforming growth factor administered intraventricularly minimize damage of the
brain.

A substantial information base indicates that certain
proteins and other substances in CSF have a tropic effect
on the brain. The secretions of the choroid plexus, in

concert with the secreted products of glial cells, create an extracellular milieu capable of bestowing many beneficial effects on neurons.

SELECTED READINGS

Bito LZ, Davson H, Fenstermacher JD (eds). The ocular and cerebrospinal fluids. London: Academic Press, 1977:561.

Boulton AA, Baker GB, Walz W (eds). Neuromethods, 9: neuronal microenvironment. Clifton, NJ: Humana Press, 1988:732.

Bradbury M. The concept of a blood-brain barrier. Bath, UK: Pitman Press, 1979.

Butler AB. Alteration in CSF outflow in experimental acute subarachnoid hemorrhage. In Gjerris F, Borgesen SE, Sorensen PS (eds). Outflow of cerebrospinal fluid. Copenhagen: Munksgaard, 1989:6975.

Cserr HF, Fenstermacher JD, Fencl V (eds). Fluid environment of the brain. New York: Academic Press, 1975:289.

Davson H, Welch K, Segal MB (eds). The physiology and pathophysiology of the cerebrospinal fluid. London: Churchill Livingstone, 1987.

Fishman RA. Cerebrospinal fluid in diseases of the nervous system. Philadelphia: WB Saunders, 1980:384.

Gross PM (ed). Circumventricular organs and body fluids, vol 1. Boca Raton: CRC Press, 1987:203.

Herndon RM, Brumback RA (eds). The cerebrospinal fluid. Boston: Kluwer Academic Publishers, 1989:306.

Johanson CE. Biological barriers to protein delivery. Tissue barriers: diffusion, bulk flow and volume transmission of protein within the brain. In Audus KL, Raub TJ (eds). Pharmaceutical biotechnology. New York: Plenum Publishing, 1993;5:467.

Johanson CE. Choroid plexus. In Adelman G (ed). Encyclopedia of neuroscience, vol I. Boston: Birkhäuser, 1987;236.

Johansson BB, Owman C, Widner H (eds). Pathophysiology of the blood-brain barrier. Amsterdam: Elsevier, 1990:610.

Jones HC, Keep RF. The control of potassium concentration in the cerebrospinal fluid and brain interstitial fluid of developing rats. J Physiol 1987;383:441.

Katzman K, Pappius HM (eds). Brain electrolytes and fluid metabolism. Baltimore; Williams & Wilkins, 1973:419.

Lyons MK, Meyer FB. Cerebrospinal fluid physiology and the management of increased intracranial pressure. Mayo Clin Proc 1990;65:684.

McAllister J, Cohen M, O'Mara K, Johnson M. Progression of kaolin-induced hydrocephalus in neonatal kittens and decompression with ventriculoperitoneal shunts: correlation of radiologic findings and gross morphology. Neurosurgery 1991;29:329.

McComb JG, Zlokovic BV. Choroid plexus, cerebrospinal fluid, and the blood-brain interface. In Chick WR (ed). Pediatric neurosurgery, imaging of the developing nervous system, 3rd ed. Philadelphia: WB Saunders, 1993.

Milhorat TH. Cerebrospinal fluid and the brain edemas. New York: Neuroscience Society of New York, 1987:168.

Netsky MG, Shuangshoti S (eds). The choroid plexus in health and disease. Charlottesville: University Press of Virginia, 1975:351.

Neuwelt EA (ed). Implications of the blood-brain barrier and its manipulation. New York: Plenum Medical Books, 1989:403, 633.

Nilsson C, Lindvall-Axelsson M, Owman C. Neuroendocrine regulatory mechanisms in the choroid plexus-cerebrospinal fluid system. Brain Res Rev 1992;17:109.

Pollay M. Formation of cerebrospinal fluid. J Neurosurg 1975;42:665.

Rapoport SI. Blood-brain barrier in physiology and medicine. New York: Raven Press, 1976:316.

Rosenberg GA. Brain fluids and metabolism. New York: Oxford University Press, 1990:207.

Spector R, Johanson CE. The mammalian choroid plexus. Sci Am 1989;260:68.

Spector R. Micronutrient homeostasis in mammalian brain and cerebrospinal fluid. J Neurochem 1989;53:1667.

Wood JH (ed). Neurobiology of cerebrospinal fluid, vol 1. New York: Plenum Press, 1980:768.

Wright EM. The choroid plexus as a route from blood to brain. Ann N Y Acad Sci 1986;481:214.

Neuroscience in Medicine, edited by P. Michael Conn.
J. B. Lippincott Company, Philadelphia © 1995

Chapter *11*

Spinal Cord

MARION MURRAY

Sensory information from the body is transmitted into the central nervous system (CNS) through the dorsal roots projections into the spinal cord. Motor neurons in the spinal cord project their axons into the periphery to innervate muscles and autonomic ganglia. Somatic perceptions, coordinated movements, and autonomic functions depend on the integrity of the spinal cord and its projections. The neurons and fiber bundles within the spinal cord are organized in a simpler and more uniform way that other parts of the CNS, but the features of organization and function are similar to those that are encountered in more complexly organized regions of the brain.

Segmental Organization of the Spinal Cord

During early embryonic life, the spinal cord extends almost the whole length of the vertebral canal. As development proceeds, the body and the vertebral column grow at a much greater rate than the spinal cord. As a result, in newborns the spinal cord extends only as far caudally as the midlumbar vertebral levels, and in adults, it extends only to the level of the first or second lumbar vertebra (Fig. 11-1). Dorsal and ventral roots enter and leave the vertebral column through intervertebral foramina at vertebral segments corresponding to the spinal segment; because the vertebral column is longer than the spinal cord in the adult, the caudal roots are longer than the more rostral roots. At its caudal end, the cord tapers markedly to form the *conus medullaris*, and the lumbar, sacral, and coccygeal roots extending to their appropriate vertebral levels form bundles, the *cauda equina* (i.e., "horse's tail"), surrounding the conus.

The spinal cord is organized into 31 continuous spinal segments. They are divided into cervical (C1 through C8) segments, including those that supply the arms, thoracic (T1 through T12) segments innervating the trunk and sympathetic ganglia, lumbar (L1 through L5) segments supplying the legs, and sacral (S1 through S5) and coccygeal (1 segment) segments supplying the saddle region, the buttocks, and pelvic organs.

A *segment* is defined by dorsal roots that enter and ventral roots that exit the cord (Fig. 11-2). The axons in the dorsal roots arise from dorsal root ganglion cells located in ganglia lateral to the vertebral column.

A pair of *dorsal root ganglia* are associated with each segment, except C1, which may have no ganglia. Each dorsal root ganglion cell gives rise to two axonal processes. One *axonal process* enters a spinal nerve, and these peripheral sensory axons innervate sensory receptors in the body; the strip of skin supplied by the peripheral process from cells in one dorsal root ganglion is called a *dermatome* (Fig. 11-3). Adjacent dermatomes overlap considerably so that any portion of the skin is likely to be supplied by sensory axons in several peripheral nerves; damage to a single dorsal root therefore results in little sensory loss. The other axonal process aris-

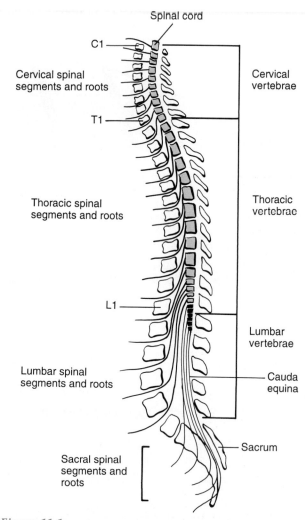

Figure 11-1
The organization of the spinal cord into cervical, thoracic, lumbar, and sacral segments. Notice the exit of the lumbar and sacral roots through intervertebral foramina located caudal to the spinal segment with which the roots are associated.

ing from the dorsal root ganglion cell projects through the dorsal root to terminate on neurons within the CNS.

Because sensory information from the body is relayed to the CNS through the dorsal roots, axons originating from dorsal root ganglion cells are sometimes called *primary afferents*. A small population of dorsal root axons enter the spinal cord through the ventral root (i.e., ventral root afferents), but their functional significance is unclear. Most axons in the ventral roots arise from motor neurons in the spinal cord and innervate skeletal muscle or autonomic ganglia. These ventral root axons join with the peripheral processes of the dorsal root ganglion cells to form the mixed spinal nerves, which merge to form peripheral nerves. The axons in the peripheral nerves are often classified according to diameter, and this provides a useful correlation with the functions they serve (Table 11-1).

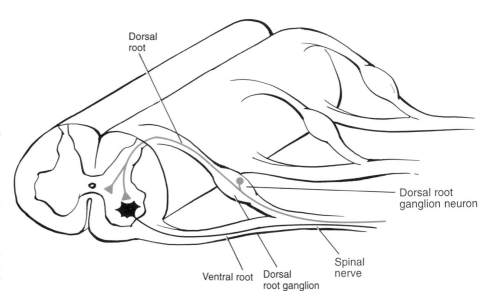

Figure 11-2
In the diagram of two segments of spinal cord, three dorsal roots enter the dorsal lateral surface of the cord, and three ventral roots exit. The dorsal root ganglion contains dorsal root ganglion cells, whose axons bifurcate; one process enters the spinal cord in the dorsal root, and the other extends peripherally to supply the skin and muscle of the body. The ventral root is formed by axons from motor neurons in the spinal cord.

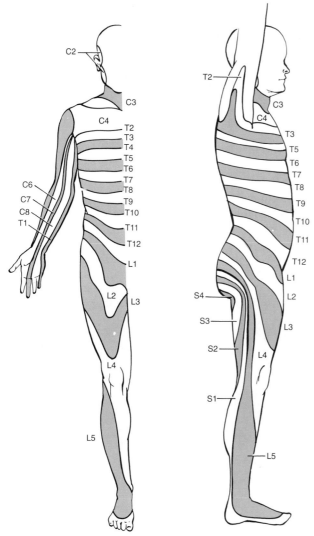

Figure 11-3
A dermatome is the area of skin supplied by axons from a single dorsal root ganglion.

Spinal Neurons Organized Into Nuclei and Laminae

In cross sections, the spinal cord is composed of a butterfly-shaped core of *gray matter* (i.e., cell bodies and their processes) surrounded by *white matter* (i.e., axons, most of which are myelinated). The gray matter is subdivided into a sensory portion, the *dorsal or posterior horn*, and a motor portion, the *ventral or anterior horn,* separated by an *intermediate zone* (Fig. 11-4). The neurons of the spinal gray matter are also classified according to their projections. They include *sensory relay neurons,* which receive dorsal root input and whose axons contribute to the ascending pathways, *motor neurons,* whose axons exit in the ventral roots, and *propriospinal* cells, spinal interneurons whose axons do not leave the spinal cord. Propriospinal cells account for about 90% of spinal neurons.

As in other areas of the nervous system, many of the neurons of the gray matter are organized in functionally related clusters called *nuclei.* These nuclei may extend the length of the spinal cord, forming columns of functionally related cells. In cross sections, the neurons in the gray matter demonstrate a laminated distribution, particularly in the dorsal horn. Because the histologic differences between laminae reflect functional differences, the spinal gray matter is sometimes also classified into laminae. These different schemes for classifying spinal neurons are compared in Table 11-2.

The dorsal horn and intermediate zones (laminae I through VII) contain sensory relay nuclei, including the *marginal* and *proprius* nuclei and the *substantia gelatinosa.* Motor neurons are subdivided functionally into *somatic* and *visceral* motor neurons; all somatic motor neurons are located in the ventral horn (laminae VIII and IX) and innervate striated muscle. All visceral motor neurons are located in the intermediate zone (lamina VII) and innervate neurons in autonomic ganglia.

Table 11-1
AXONS IN PERIPHERAL NERVES

Fibers	Diameters (μm)	Conduction Velocity* (m/s)	Role or Receptors Innervated
Sensory			
Ia (A-α)	12–20, myelinated	70–120	Muscle spindle afferents
Ib (A-α)	12–20, myelinated	70–120	Golgi tendon organ; touch and pressure receptors
II (A-β)	5–14, myelinated	25–70	Secondary afferents of muscle spindle; touch, pressure, and vibratory sense receptors
III (A-δ)	2–7, myelinated	10–30	Crude touch and pressure receptors; pain and temperautre receptors; viscera
IV (C)	5–1, unmyelinated	<2.5	Pain and temperature receptors; viscera
Motor			
Alpha (A-α)	12–20	15–20	Alpha motor neurons innervating extrafusal muscle fibers
Gamma (A-γ)	2–10	10–15	Gamma motor neurons innervating intrafusal muscle fibers
Preganglionic autonomic fibers (B)	>3	3–15	Lightly myelinated preganglionic autonomic fibers
Postganglionic autonomic fibers (C)	1	2	Unmyelinated postganglionic autonomic fibers

Ascending and Descending Tracts in White Matter

The white matter is subdivided into dorsal, lateral, and ventral *funiculi*, demarcated by the dorsal medial sulcus, the dorsal root entry zone, the ventral roots, and the ventral medial sulcus (Fig. 11-5). The axons in the white matter form pathways that are classified as ascending (i.e., sensory), descending (i.e., motor), and propriospinal; individual pathways or tracts run in specific funiculi.

Cells in the dorsal root ganglia and spinal gray matter give rise to the axons that form the pathways that ascend in the dorsal, lateral, and ventral funiculi. Cell bodies giving rise to the axons in the descending tracts are located in many parts of the brain. Their axons descend in the lateral and ventral funiculi and terminate on motor neurons or on interneurons that project to motor neurons within the spinal gray matter. Descending systems are important in the control of movement and posture. Some descending axons terminate on sensory relay neurons

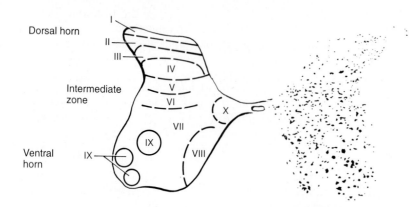

Figure 11-4
Composite figure compares the classification schemes for the neurons in the gray matter of the spinal cord. On the right, a section is stained to show cell distribution and location of the nuclei in the lumbar spinal cord, and on the left are the related laminar boundaries.

Table 11-2
CLASSIFICATION OF SPINAL NEURONS

Gray Matter Subdivision	Lamina	Nuclei Included in Laminae
Dorsal horn	Lamina I	Marginal nucleus
	Lamina II	Substantia gelatinosa
	Lamina III, IV	Nucleus proprius
	Lamina V	Reticular nucleus
	Lamina VI	Commisural nuclei
Intermediate zone	Lamina VII	Clarke's, intermediolateral nuclei
	Lamina VIII	Medial motor nuclei
Ventral horn	Lamina IX	Lateral motor nuclei
	Lamina X	Central Gray

in the dorsal horn and can modify sensory input to the CNS.

Propriospinal pathways originate and terminate in the spinal cord itself. These axons may be very short, connecting one cell with its neighbor; they may cross the midline (i.e., commissural fibers) dorsal or ventral to the central canal; or they may ascend or descend in the white matter and connect distant segments of the cord.

☐ Sensory tracts ascend

The input to the central nervous system from the body must be organized in such a way that information about modality (i.e., type of sensation) and the location of peripheral stimulation can be used to produce appropriate spinal reflexes and be transmitted to appropriate parts of the brain for sensory processing. Information about a painful stimulus to the skin is distributed in pathways that are different from those transmitting information about nonpainful stimuli, such as light pressure.

Two classes of dorsal root ganglion cells can be recognized: larger cells whose axons are myelinated, called groups I and II or A-α and A-β fibers, and smaller cells whose axons are unmyelinated or thinly myelinated, called groups III and IV or A-δ and C fibers (see Table 11-1). The small dorsal root ganglion cells are further differentiated by their synthesis of a variety of peptides (e.g., substance P, somatostatin) that are used as neuromodulators or neurotransmitters.

At the dorsal root entry zone, the dorsal root fibers separate into lateral and medial divisions. The lateral division contains finer myelinated and unmyelinated fibers that transmit responses to nociceptive (i.e., painful) and thermal stimulation of the skin and viscera. The medial division contains large-caliber fibers whose peripheral receptors lie in muscle, joints, and skin. These fibers relay information about muscle length and tension to interneurons and motor neurons at segmental levels, which provides the basis for spinal reflexes, and information about somesthesis and joint position to the brain, which provides the basis for stereognosis.

Lateral division

Axons in the lateral division form a bundle, called the tract of Lissauer (Fig. 11-6), at the dorsal root entry zone. These axons may ascend or descend in Lissauer's tract for several segments before entering into the lateral portion of the dorsal horn to synapse on cells in the dorsal horn, such as the marginal nucleus, substantia gelatinosa, nucleus proprius, and laminae I through IV. This distribution of collaterals rostrally and caudally means that activation of axons in one segment of the lateral division may stimulate dorsal horn cells over several adjacent segments.

Neurons in the dorsal horn relay sensory information received from dorsal root axons to nuclei in the brain. These second-order neurons transmit this information through axons that ascend in the lateral and ventral funiculi. These sensory relay neurons receive input from the dorsal roots and from various descending (i.e., motor) tracts directly or by means of interneurons. In this way, the cell's ability to respond to sensory input is modified by the brain.

The location of an important ascending tract is shown in Fig. 11-7. Most axons in the *spinothalamic tract* arise

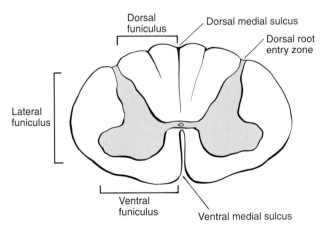

Figure 11-5
Cross section of cervical spinal cord shows major landmarks and divisions of the white matter.

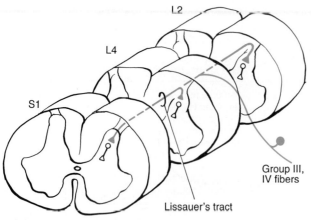

Figure 11-6
The diagram shows the entry and central course of axons of the lateral division from one dorsal root ganglion. These unmyelinated and thinly myelinated axons may branch and ascend or descend several segments in the tract of Lissauer before entering the gray matter and synapsing on second-order neurons in the dorsal horn.

from the marginal cells in lamina I and from the nucleus proprius in lamina III and IV and transmit nociceptive and thermal information. The axons of these second-order neurons cross to the contralateral spinal cord through the ventral commissure and ascend in the white matter as the spinothalamic tract to terminate in the thalamus. The spinothalamic tract, along with other pathways ascending to the reticular formation (i.e., spinoreticular tract) and mesencephalon (i.e., spinomesencephalic and spinotectal tracts), are located in the ventral or anterior lateral portion of the white matter and are sometimes referred to collectively as the *anterolateral system*. The function of these tracts of the anterolateral system appears to be similar to that of the spinothalamic tract, relaying pain, temperature, and crude touch sensations to the brain.

Axons contributing to the anterolateral system join the ascending pathways at each spinal segment. As the contribution from one segment enters the white matter, it displaces the fibers from lower segments in a dorsolateral direction. These tracts also become laminated; the fibers carrying sensation from the lower limbs are dorsal and lateral to those representing the upper limbs.

Medial division

Dorsal root axons in the medial division have local targets in their segments of entry and collaterals that ascend to terminate in more distant targets in somatosensory relay and cerebellar relay nuclei.

The axons with local segmental targets enter the gray matter at the level of entry of the dorsal root, synapse on interneurons or motor neurons at that segmental level, and subserve reflex organization (Fig. 11-8). The Ia fibers of the medial division whose peripheral processes innervate receptors in muscle send central processes into the ventral horn that synapse on the dendrites of motor neurons and on interneurons. This monosynaptic connection

between sensory axons and motor neurons is the anatomic basis for the stretch reflex.

Other medial division axons, including collaterals of axons with segmental targets, enter the white matter in the dorsal funiculus, where they ascend to terminate in relay nuclei for somatosensory or cerebellar pathways. These relay nuclei receive sensory information from the skin, fascia, muscles, joints, and periosteum and transmit it to nuclei that project to the thalamus and then to the cortex, where it is consciously appreciated, or to the cerebellum, where it contributes to the control of posture and movement without being consciously perceived.

Dorsal root fibers destined for the somatosensory relay nuclei enter the dorsal funiculus at each segment, displacing medially the fibers originating from more caudal ganglia. As a result, fibers become laminated. In the cervical region, fibers from sacral dorsal roots are found nearest the midline, and those from cervical roots are nearest the dorsal root entry zone. Fibers representing the lower half of the body (sacral to T5) ascend in the *gracile fasciculus*; those from the upper half (T5 to C2) comprise the *cuneate fasciculus*. Axons in both bundles terminate ipsilaterally in nuclei of the medulla, after which they are named (i.e., nucleus gracilis and cuneatus); these nuclei relay impulses to the thalamus (Fig. 11-9).

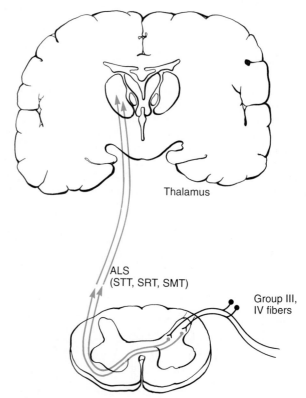

Figure 11-7
Formation of the anterolateral system (ALS). Lateral division axons entering the dorsal root synapse on second-order neurons, giving rise to axons that cross the spinal cord. Some axons forming the spinothalamic tract (SST) ascend to terminate in the thalamus. Other axons terminate in the reticular formation (i.e., spinoreticular tract [SRT]) or mesencephalon (i.e., spinomesencephalic tract [SMT]).

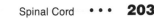
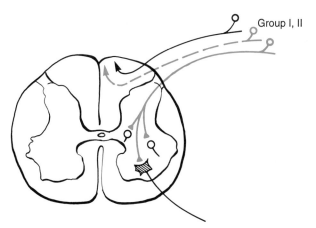

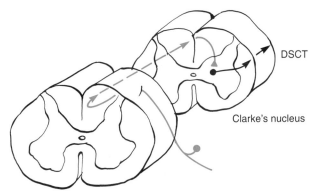

Figure 11-8
Axons forming the medial division of the dorsal root enter the spinal cord and then continue into the gray matter at their level of entry, making reflex connections with motor neurons and interneurons at the level of entry, or ascend in the dorsal columns to terminate at more rostral spinal or brain stem levels.

Figure 11-10
Axons conveying proprioceptive and muscle information form part of the medial division. Axons from lumbar and caudal thoracic dorsal root ganglia enter the spinal cord and ascend in the dorsal columns to the thoracic level, where they terminate on second-order neurons in Clarke's nucleus. The axons of Clarke's neurons ascend in the lateral funiculus as the dorsal spinocerebellar tract (DSCT), which terminates on third-order neurons in the cerebellum.

Other dorsal root axons enter the dorsal funiculus to ascend for several segments before terminating in relay nuclei for cerebellar pathways. The axons arising from dorsal root ganglion cells in caudal thoracic, lumbar, and sacral regions ascend in the dorsal columns to terminate in *Clarke's nucleus* (Fig. 11-10), located in segments T1

to L2, and those arising from more rostral ganglia ascend in the dorsal columns to terminate in the *lateral or external cuneate nucleus* in the medulla.

The *dorsal spinocerebellar tract* arises from neurons located in Clarke's nucleus. It ascends in the white matter ipsilaterally and is therefore uncrossed. The dorsal spi-

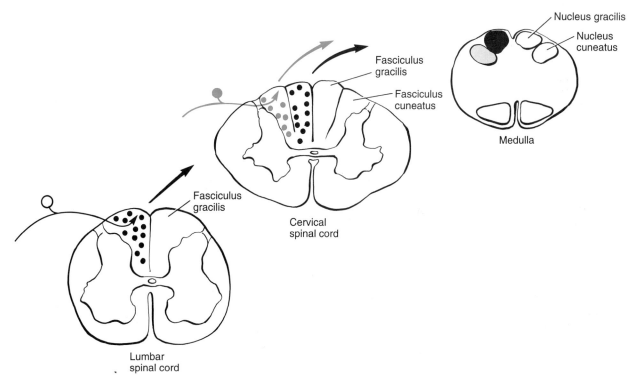

Figure 11-9
Axons mediating fine tactile sensibility form part of the medial division. They ascend in the dorsal columns to the brain stem, where they terminate on second-order neurons in the dorsal column nuclei. The axons arising from lumbar and low thoracic dorsal root ganglia ascend in the fasciculus gracilis and terminate in the nucleus gracilis. Axons arising from upper thoracic and cervical ganglia ascend in the more laterally located fasciculus cuneatus and terminate in the nucleus cuneatus located lateral to the nucleus gracilis in the medulla.

nocerebellar tract terminates in the cerebellum (see Fig. 11-10). Similarly, the *cuneocerebellar tract* carries information from the lateral cuneate nucleus to the cerebellum. Both nuclei relay sensory information from the periphery to the cerebellum. There is a third pathway to the cerebellum, the ventral spinocerebellar tract. Cell bodies whose axons form the ventral spinocerebellar tract are distributed throughout the dorsal horn and intermediate zone; their axons cross in the ventral commissure to ascend in the contralateral ventral spinocerebellar tract to the cerebellum, where they cross again before terminating. Although the course of the ventral spinocerebellar tract differs from that of the dorsal spinocerebellar tract and cuneocerebellar pathways, their functions appear to be similar.

☐ Central motor pathways descend

The axons that form the central motor pathways arise from cell bodies at all levels of the brain stem and the cerebral cortex. Many of these tracts undergo partial or complete decussation (i.e., crossing) before entering the cord, but some descend ipsilaterally, enter the gray matter, and then cross in the spinal commissures to terminate on neurons contralateral to their origin. Still other tracts are primarily ipsilateral.

The descending tracts are located in the lateral and ventral funiculi. The locations of four important descending tracts, the corticospinal, rubrospinal, vestibulospinal, and reticulospinal tracts, are shown in Figure 11-11. Most of the axons in these tracts terminate on interneurons that project to motor neurons; very few terminate directly on motor neurons. Some descending axons terminate on sensory relay neurons, providing central control over sensory processing.

The *corticospinal tract* arises from motor, premotor, and somatic cortex, descends to the spinomedullary junc-

tion, where 90% of the axons cross, and then continues to descend as the lateral corticospinal tract in the lateral funiculus contralateral to the cell bodies of origin. The axons that do not cross at the spinal medullary junction descend in the spinal cord as the ventral corticospinal tract but then cross in the spinal cord before their termination. The *rubrospinal tract* arises from neurons in the red nucleus; it is virtually completely crossed. The *reticulospinal tracts* contain axons that arise from reticular nuclei ipsilaterally and contralaterally. Most of the axons in the *vestibulospinal tract* are uncrossed.

Topographic organization of somatic motor neurons

Somatic motor neurons express the activity of the CNS. Sherrington called them the final common pathway. Axons from somatic motor neurons exit in the ventral roots and innervate striated muscle. These neurons are located exclusively in the ventral horn (laminae VIII and IX). A medial motor nucleus is present in the ventral horn throughout the length of the cord. These motor neurons innervate axial musculature. There are also prominent groups of nuclei located laterally in the ventral horn that are particularly well developed in the segments supplying the limbs. The lateral group of nuclei is subdivided functionally into the ventral nuclei, which innervate extensor muscles, and the dorsal nuclei, which innervate flexors. Within these two subdivisions, the nuclei innervating proximal muscles are located medially; those that supply the distal muscles are located more laterally (Fig. 11-12).

Motor neurons are among the largest in the spinal cord. Although the motor neuron cell bodies are localized into discrete nuclei, their dendrites extend into the intermediate zone, the dorsal horn, and even into the white matter. This allows considerable convergence of input onto a motor neuron, and a single large motor neu-

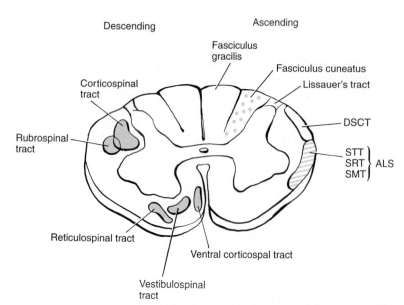

Figure 11-11
The approximate location of ascending tracts is shown on the right and of descending tracts on the left.

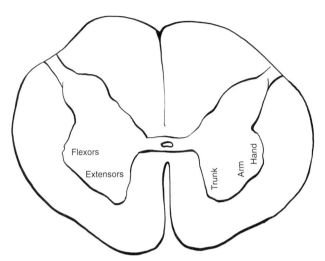

Figure 11-12
Somatic motor neurons are organized topographically in the ventral horn. Axial musculature is supplied by motor neurons located medially and limb musculature by motor neurons located laterally in the ventral horn.

ron may receive as many as 10,000 axon terminals from many different sources.

All motor neurons use the excitatory transmitter acetylcholine. The axons of many motor neurons give off a collateral before exiting in the ventral root. These short *recurrent collaterals* terminate on interneurons. Some of these interneurons are part of a disynaptic pathway that inhibits motor neurons (i.e., motor axon collateral → interneuron → motor neuron). This interneuron is called the *Renshaw cell*, and it uses the inhibitory transmitter glycine to inhibit the postsynaptic motor neuron. This feedback circuit permits the regulation of activity in motor neurons by the motor neurons themselves (Fig. 11-13).

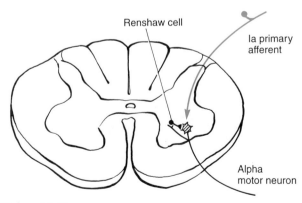

Figure 11-13
Renshaw cells. A Ia primary afferent axon makes a monosynaptic contact with an alpha motor neuron, whose axon is innervating a somatic muscle. The motor neuron also emits a collateral axon that innervates an interneuron, the Renshaw cell, situated near the alpha motor neuron that inhibits (i.e., recurrent inhibition) the motor neuron.

Visceral motor neurons in the intermediate zone

Visceral motor neurons innervate neurons in autonomic ganglia, and the the autonomic ganglion cells innervate viscera. The preganglionic neurons in the *intermediolateral nucleus* in lamina VII at C8 through L3 send axons to the sympathetic chain and provide central regulation of the sympathetic nervous system. Neurons in the intermediate zone at levels at S2 through S4 form the more poorly defined *sacral parasympathetic nucleus*. Their axons innervate the sacral parasympathetic ganglia and provide central control of the sacral portion of the parasympathetic system.

Regional specializations in the gray matter

At all segmental levels of the spinal cord, the dorsal horn (laminae I through IV), the intermediate zone (laminae V through VII), and the ventral horn (laminae VIII and IX) can be recognized. Regional specializations modify the butterfly shape of the gray matter in characteristic ways. In the spinal segments that innervate the limbs (i.e., cervical and lumbar enlargements), the number of neurons is greatly increased, and the dorsal and ventral horns are concomitantly expanded, but two nuclei in the intermediate zone, Clarke's nucleus and the intermediolateral nucleus, are prominent only in segments that do not innervate the limbs (T1 though L2). The gray matter has different silhouettes that are characteristic of the particular cervical, thoracic, lumbar, or sacral spinal levels (Fig. 11-14).

The amount of white matter also differs according to segment. Ascending pathways become larger more rostrally because axons from dorsal root ganglia or from sensory relay nuclei are added at each segment. Most descending tracts send fibers to terminate in gray matter at all segments. A descending tract becomes smaller at more caudal levels, because it is continuously depleted of fibers.

Spinal Reflexes

The activity of dorsal root axons is expressed, monosynaptically or polysynaptically, on motor neurons whose axons form the ventral root. This arrangement comprises the *segmental organization* of the spinal cord, and it is this organization that determines the reflex activity of the spinal cord, i.e., the reflex activity that persists even after a spinal transection that separates the brain from the spinal cord. *Spinal reflexes* are stereotyped responses (e.g., contraction, relaxation) made by somatic muscles in response to stimuli that excite receptors in muscle, tendon, or skin. These reflexes may be more readily demonstrable in a transected preparation than in an organism with an intact spinal cord, but they nevertheless contribute to providing muscle tone, posture, and voluntary movement for the organism.

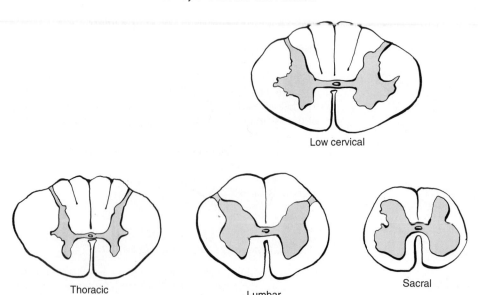

Low cervical

Thoracic

Lumbar

Sacral

Figure 11-14
The configuration of gray and white matter at different levels of the spinal cord differs in characteristic ways.

The stretch reflex determines muscle tone

When a muscle is passively stretched it responds by contracting. This response is the result of stimulation of receptors in muscle spindles. This *myotactic reflex* provides the basis for muscle tone, the slight resistance to stretch found in all healthy, innervated muscles.

A striated muscle is composed of two types of fibers: a large number of *extrafusal muscle fibers* and a smaller number of highly specialized sensory structures, the *muscle spindles* (Fig. 11-15). The muscle spindle is a fusiform structure containing several modified muscle fibers, the intrafusal fibers, and the axons that innervate them; each spindle is encased in a connective tissue capsule.

There are also two types of somatic motor neurons: *alpha and gamma motor neurons.* Alpha motor neurons innervate the extrafusal muscle fibers. They are grouped into nuclei that innervate individual muscle groups.

These nuclei also contain small gamma motor neurons that innervate the intrafusal muscle fibers and control the sensitivity of the muscle to stretch. Gamma motor neurons are located near the alpha motor neurons, which innervate extrafusal fibers of the same muscle.

The central portion of the intrafusal fiber is noncontractile; this zone is innervated by Ia afferent fibers, which make annulospiral endings around the nonconductive central zone, and by group II afferents, which make another type of ending, called flower spray endings, on them. The polar regions of the intrafusal fiber are contractile. Each pole receives innervation from gamma motor neurons. The polar regions of intrafusal fibers respond to gamma efferent stimulation with a slow, maintained tonic contraction, stretching the central regions of the intrafusal fiber, and stimulating the Ia and II afferent axons.

The central processes of the Ia and II fibers terminate on cells in the spinal gray matter, including the alpha motor neurons supplying the muscle from which the Ia fiber

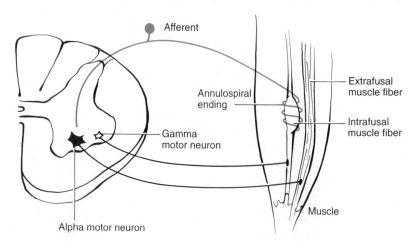

Afferent

Annulospiral ending

Gamma motor neuron

Alpha motor neuron

Extrafusal muscle fiber

Intrafusal muscle fiber

Muscle

Figure 11-15
The gamma loop. A small gamma motor neuron, located near an alpha motor neuron, innervates the poles of intrafusal fibers. The alpha motor neuron innervates extrafusal muscle fibers in the same muscle. Activity in the gamma motor neuron causes contraction of the poles of the intrafusal fiber, which stretches the central zone and activates the peripheral process of the Ia afferent. The Ia afferent is shown innervating the intrafusal fiber peripherally and the alpha motor neuron centrally (i.e., monosynaptic reflex).

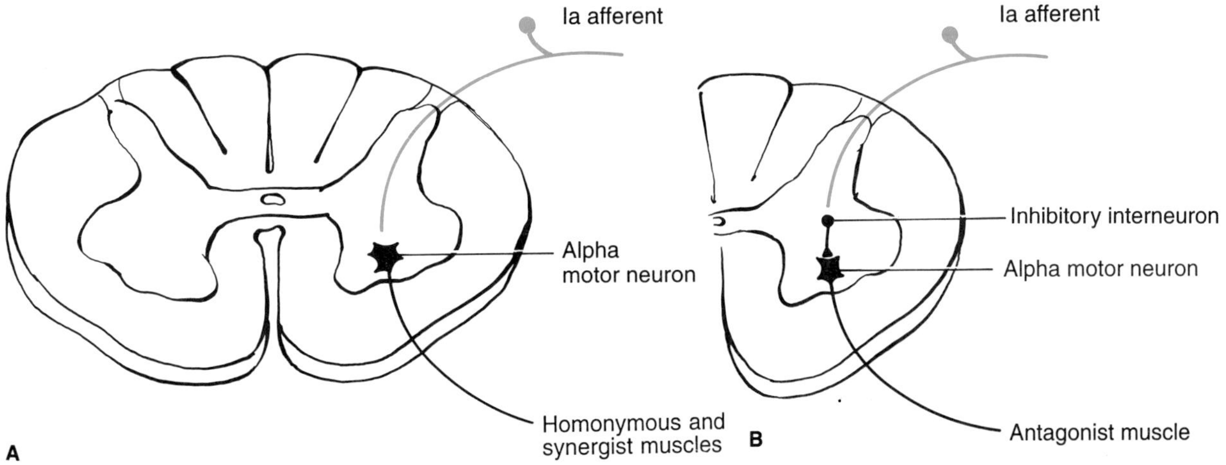

Ia afferent   Ia afferent

Inhibitory interneuron

Alpha
motor neuron   Alpha motor neuron

A

Homonymous and
synergist muscles   **B**   Antagonist muscle

Figure 11-16

(**A**) Monosynaptic reflex pathway. Ia afferents supplying muscle spindles in somatic muscles make monosynaptic excitatory contacts on alpha motor neurons supplying the same muscle and synergistic muscles. Stretch of the muscle produces contraction of that muscle and its agonists. (**B**) Disynaptic inhibitory reflex pathway. Ia afferents also make excitatory contacts on inhibitory interneurons that make synaptic contact with alpha motor neurons supplying muscles that are antagonists to the muscle supplied by the Ia afferent. Stretch of the muscle also produces relaxation of opposing muscles.

arises (Fig. 11-16*A*). There are also Ia terminals on motor neurons whose muscles are synergists of that muscle. Other important terminations of the Ia fiber are on interneurons, some of which inhibit motor neurons that innervate the muscles that act antagonistically to that of the Ia fiber. There is one synapse in the excitatory path (i.e., monosynaptic pathway) and two synapses (i.e., disynaptic pathway) or more involved in the inhibitory pathway (Fig. 11-16*B*). Passive stretch of a muscle facilitates contraction of that muscle and its synergists and, with a slight delay, inhibits contraction of the antagonist muscles.

☐ The inverse myotatic reflex limits the stretch reflex

When a muscle contracts, another type of muscle receptor, the *Golgi tendon organ*, is stimulated (Fig. 11-17). These receptors are located in tendons close to their junctions with muscle and are stretched during muscle contraction. Golgi tendon organs principally measure

muscle tension. They receive sensory innervation from Ib dorsal root axons but do not receive motor innervation. The central process of the Ib fiber terminates on interneurons that inhibit motor neurons of the muscle of origin (i.e., homonymous muscles) and facilitate the antagonists (i.e., heteronymous muscles). This reflex pathway limits or "brakes" the contractions of muscle.

☐ The flexor and crossed extensor reflex constitute another protective mechanism

Noxious or thermal stimulation of the skin or deep tissues excites group III and IV axons in the peripheral nerves. This information is transmitted to motor neurons through interneurons. These polysynaptic pathways permit a divergence of the sensory stimulation so that neurons in many ipsilateral segments and the contralateral side of the cord may be recruited. This type of stimulation produces ipsilateral excitation of flexors and inhibi-

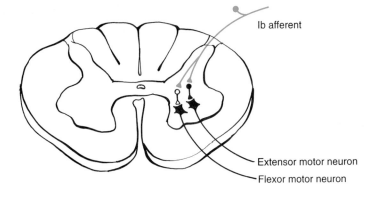

Figure 11-17

The inverse myotatic reflex. The Ib afferent innervates a Golgi tendon organ from an extensor muscle peripherally; centrally, this Ib afferent makes excitatory contacts with two interneurons. The excitatory interneuron synapses on motor neurons supplying flexor muscles, and the inhibitory interneuron synapses on motor neurons supplying flexor motor neurons. Continued stretch of the muscle produces relaxation.

Ib afferent

Extensor motor neuron

Flexor motor neuron

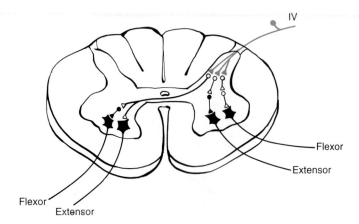

IV

Flexor

Extensor

Flexor

Extensor

Figure 11-18
The flexor and crossed-extensor reflex. Painful stimulation activates small-diameter axons that make contacts with interneurons within the gray matter. Through polysynaptic pathways, using interneurons that inhibit and excite flexor and extensor motor neurons on both sides of the spinal cord, the ipsilateral limb flexes, a protective response, and the contralateral limb extends, providing postural support

tion of extensors and contralateral inhibition of flexors and excitation of extensors. This reflex pattern is known as the *flexor and crossed-extensor reflex* (Fig. 11-18).

☐ Pathologic reflexes have several causes

Spasticity is an abnormal increase in muscle tone due to loss of inhibition of gamma motor neurons. *Rigidity* is an abnormal increase in muscle tone due to loss of inhibition of alpha motor neurons. *Flaccidity* is an absence of muscle tone that occurs after loss of peripheral nerve innervation.

Spasticity

Passive stretch of a muscle evokes a reflex contraction of that muscle. The strength of this monosynaptic stretch reflex is controlled by the gamma motor neurons that innervate that muscle, and these neurons are under excitatory or inhibitory drive from the CNS. Normally, there is a balance of excitation and inhibition that is reflected in some tension in the polar regions of the intrafusal fibers. This tension stimulates Ia fibers, producing a basal level of excitation of the alpha motor neurons; this balance produces muscle tension or tone. Increased excitation or decreased inhibition of gamma motor neurons increases tension on the intrafusal fibers, which become hypersensitive to stretch of the whole muscle, resulting in an abnormal increase contraction in response to stretch, the phenomenon known as spasticity.

Under such circumstances, changes in reflexes are clearly observable clinically. If the stretch is applied continuously to a spastic muscle, the resistance increases but then suddenly gives way because of inhibition by the Golgi tendon organ system. This sudden collapse of resistance is called the lengthening reaction or clasp-knife effect. It is only revealed when muscle tone is abnormally increased, usually by some pathologic state, but its basis, activation of the Golgi tendon organ, normally contributes to muscle tone.

Spasticity is seen in muscle groups innervated by motor neurons below the level of a spinal transection. An attempt to elicit a tendon reflex produces a much-increased response. The mild stimulus of stroking the lateral border of the foot also evokes the classic extensor-plantar (i.e., Babinski) response in which the toes spread and extend instead of flexing as they would normally do.

Rigidity

Rigidity is an increased activation of the alpha motor neurons, seen after damage to some descending pathways. Rigidity is seen as an increased resistance to movements in all directions and does not depend on the dorsal root innervation of muscle spindles. This resistance to passive motion that is felt when examining the patients has been likened to the feeling of bending a lead pipe. Occasionally, the resistance has a phasic quality that is called cogwheel rigidity. Parkinson's disease, which results from loss of input to neurons in the basal ganglia that indirectly project to spinal cord, produces the purest form of clinical rigidity.

Flaccidity

A lesion to motor neurons or their axons produces a flaccid paralysis of the muscle innervated by those neurons. No voluntary movement is possible, there is no resistance to passive movement, and no reflexes can be elicited. If axonal regeneration does not occur, the muscle may atrophy.

Spinal Lesions

☐ Spinal transection interrupts neuronal transmission

Transection of the spinal cord interrupts descending input from the brain to spinal levels below the level of transection and input to the brain ascending from sensory structures located below the level of transection. A patient whose spinal cord is transected undergoes a period of spinal shock, characterized by flaccid paralysis of muscles innervated by neurons below the transection. Voluntary control of muscles innervated by motor neurons and perception of sensory events arising below the level of the lesion are permanently lost. Over time, the flaccid paralysis is followed by a bilateral, symmetric spasticity below the level of the lesion. This hyperactivity after recovery from spinal shock is caused by the loss of descending inhibitory influences on gamma motor neurons.

Spinal hemisection generates both ipsilateral and contralateral impairment

Hemisection, also called the Brown-Sequard syndrome, commonly results from a slow-growing mass in the spinal cord that deforms the cord, compressing descending and ascending axons on one side of the cord and producing functional disorders. Because many ascending systems cross, the sensory impairments arising from hemisection may be referred to the side of the body opposite the lesion. Because important descending pathways cross at supraspinal levels, the motor impairments are ipsilateral and below the level of the hemisection. Motor function and discriminatory tactile and kinesthetic sense are lost ipsilaterally at levels below the lesion, and pain and thermal sensitivity are lost contralaterally below the lesion. A neurologic examination reveals spastic paralysis and diminished touch sensation on one side of the body plus a loss of pain and temperature sensation on the other side of the body.

Syringomyelia produces bilateral impairment

Syringomyelia refers to a pathologic enlargement of the central canal. As the cavity expands dorsally and ventrally, it interrupts the fibers that cross through the anterior and ventral commissures in the spinal cord, including some ascending sensory pathways, and may involve muscles of the medial motor neuron groups. In the cervical spinal cord, where syringomyelia is prone to occur, interruption of decussating sensory pathways results in bilateral loss of pain and temperature sensation but with preservation of tactile and kinesthetic sensation. With enlargement of the cavity, wasting and motor dysfunction, particularly of muscles innervated by medial motor neurons, represents encroachment on the ventral horn.

SELECTED READINGS

Houk JC, Rymer WZ. Neural control of muscle length and tension. In Brooks VB (ed). Handbook of Physiology. Bethesda, MD: American Physiological Society, 1981:257.

Matthews PBC. The human stretch reflex and the motor cortex. Trends Neurosci, 1991:14;87.

Nathan DW, Smith MC. Effects of two unilateral cordotomies on the motility of the lower limbs. Brain, 1973:96;471.

Renshaw B. Activity in the simplest spinal reflex pathways. J Neurophysiol, 1940:3;373.

Sherrington C. The integrative action of the nervous system, 2nd ed. New Haven: Yale University Press, 1947.

Clinical Correlation
DISORDERS OF THE SPINAL CORD

ROBERT L. RODNITZKY

Clinical abnormalities of the spinal cord lead to great disability resulting from severe disruption of motor, sensory, and autonomic functions. The distinct longitudinal organization of ascending and descending fiber tracts in the spinal cord coupled with the segmental groupings of neurons with motor, sensory, and autonomic functions often allow relatively precise localization of the lesion responsible for a spinal cord syndrome.

Causes of Spinal Cord Disorders

Clinical spinal cord syndromes are more commonly the result of structural damage than metabolic dysfunction. Although relatively minor aberrations of the metabolic milieu of the brain can result in severe neurologic dysfunction, clinical symptoms related to spinal cord dysfunction are much less likely to be of metabolic origin. Neurologic symptoms can result from spinal cord ischemia but not as readily as is the case in cerebral ischemia. Spinal cord ischemia severe enough to produce symptoms is commonly the result of atherosclerotic occlusion of the aorta or its branches, embolization to these blood vessels, or vascular malformations within the substance of the spinal cord.

Most clinical spinal cord syndromes result from mechanical disruption or physical degeneration of the constituent cells and fiber tracts of the cord. Typical examples of conditions resulting in mechanical disruption of spinal cord elements include spinal trauma (e.g., spinal column fracture, bullet wound), spinal cord tumors, and intervertebral disc herniation causing spinal cord compression. Degenerative processes causing spinal cord symptoms include demyelination of long tracts (e.g., multiple sclerosis) and neuronal degeneration (e.g., motor neuron disease).

Localization of the Causative Lesion

Precise localization of the causative lesion is extremely important in treating spinal cord disorders. Even in the era of modern neuroimaging, a thorough neurologic examination is essential in localizing the abnormality within the spinal cord. Although magnetic resonance imaging or computerized tomography of the spinal cord can be performed in most communities, it is critical to know which part of the spinal cord should be imaged, and it is essential to understand whether the pathology demonstrated by the procedure can explain the patient's symptoms.

The most important clinical finding in attempting to localize spinal cord pathology is the sensory level. A sensory level is a horizontal demarcation of sensory loss on the trunk or a diagonal linear demarcation on an extremity that corresponds to the segmental innervation of skin by the paired spinal nerves leaving the spinal cord. Classically, sensation is lost below the level of a spinal cord lesion and is intact above it because of the interruption of ascending impulses in sensory spinal pathways such as the spinothalamic tracts or dorsal column pathways by the offending lesion. If the lesion involves the entire spinal cord at a given level, sensation is lost equally on both sides of the body below that level. A sensory level to pain or touch sensation is demonstrated by applying the appropriate stimulus to the skin below the level of the suspected lesion and gradually ascending up the trunk until the stimulus is felt. A sensory level of vibratory perception can be similarly established by applying a vibrating tuning fork at bony landmarks beginning with the toes, proceeding to the ankles, knees, and hips, and ultimately moving up the spinous processes of the vertebrae until a normal sense of vibration is perceived.

☐ **Sparing of sacral sensation may help differentiate a lesion in the center of the spinal cord from one on its circumference**

After the fibers that transmit pain and temperature sensations enter the spinal cord, they cross the midline and ascend on the opposite side, where they are progressively displaced laterally by incoming fibers entering and crossing at higher levels. The fibers entering at the lowest (i.e., sacral) level, which are those that are the first to enter, cross, and ascend, come to occupy the most lateral position in the lumbar, thoracic and cervical spinal cord as ascending fibers enter the spinothalamic tracts at higher levels. The extreme lateral position of fibers transmitting sacral sensation protects them from lesions deep within the center of the spinal cord, such as a midline tumor. In such cases, there may be profound loss of sensation below the level of the lesion, with the exception of the saddle or perianal area, subserved by sacral fibers, where sensation is spared. This can be extremely useful to the clinician in determining whether spinal cord pathology is intrinsic to the spinal cord or is causing symptoms by compression from outside.

In cases of external compression, the most laterally placed fibers (i.e., those subserving sacral sensation) may be involved first, and the initial sensory level may be considerably below the level of the lesion early in the illness and ascend to the true level with the passage of time as more fibers are compressed. Another clue to the possibility that a lesion may be compressing the spinal cord from the outside the spine is the presence of radicular pain. This is sharp, stabbing pain that follows the cutaneous distribution of a single dorsal spinal root. It is caused by irritation of the nerve as it enters the spinal cord. This type of pain is even more likely to occur when the offending lesion is attached to the spinal nerve, as is the case with nerve sheath tumors. Such tumors are rare, and herniated lumbar or cervical intervertebral discs are by far the most common cause of irritation of a spinal root with resultant radicular pain.

☐ Lesions involving half the spinal cord produce a distinct clinical syndrome

The Brown-Sequard syndrome results from interruption of motor and sensory tracts on only one side of the spinal cord. In this condition, the functions of uncrossed fibers are lost on the side of the body ipsilateral to the lesion, and the functions of crossed fibers are lost on the contralateral side. Pain and temperature sensation are lost on the side opposite of the lesion because of the interruption of the crossed lateral spinothalamic tract, and weakness and loss of joint position and vibratory sensation appear on the same side because of interruption of the corticospinal tract and dorsal column fibers, respectively, which are uncrossed in the spinal cord. Because fibers transmitting pain and temperature sensations may ascend one or two spinal cord segments on the side of entry before crossing the midline to join the contralateral spinothalamic tract, there may be loss of pain and temperature sensation ipsilateral to the side of the lesion over an expanse of skin corresponding to one or two dermatomes at the level of the offending lesion.

☐ A cape-like sensory loss may also be a clue to the localization of a spinal cord lesion

Small lesions involving the region immediately adjacent to the central canal of the spinal cord may involve the anterior white commissure. Because this structure contains fibers conveying pain and temperature sensation from both sides as they cross to join the contralateral spinothalamic tract, these modalities may be preferentially lost in the corresponding dermatomal segments on both sides, often symmetrically. Simple touch, position sense, and vibratory sensation are preserved in the same segments, because fibers conveying these sensations do not pass near the center of the spinal cord. More importantly, all sensory modalities above and below the involved segments remain intact. The most common condition affecting this region of the spinal cord is syringomyelia. The centrally located spinal cord cyst of syringomyelia is commonly found in the lower cervical and upper thoracic spinal cord. In this condition, the area of symmetric sensory loss often includes the arms and upper thorax, giving rise to the term cape-like sensory loss.

☐ The anterior portion of the spinal cord can be preferentially involved in ischemic lesions

The anterior spinal artery perfuses the anterior two thirds of the spinal cord, excluding the dorsal columns. Occlusion of this artery causes infarction of the anterior spinal cord and results in loss of all sensory and motor functions below the level of the infarction, with the exception of vibratory and proprioceptive sensation, both of which are subserved by the intact dorsal columns. These fiber tracts receive their blood supply instead from the posterior spinal arteries.

☐ Pathology below the L1—L2 vertebral level affects the cauda equina but not the spinal cord

In normal adults, the spinal cord does not extend the entire length of the vertebral column. It does not extend below the second lumbar (L2) vertebral level. Because of this difference in length between the spinal cord and the vertebral column, spinal nerves in the cervical and upper thoracic regions enter or exit at almost right angles, but the lower thoracic, lumbar, and sacral spinal nerves originate at increasingly downward oblique angles. The lumbar and sacral spinal roots originate at the level of the lower thoracic and upper lumbar vertebrae and then descend in the spinal canal to the corresponding lower lumbar and sacral vertebrae, where they exit. This mass of descending spinal nerves within the spinal canal but below the spinal cord forms the cauda equina. Lesions within the spinal canal below the L2 vertebrae cannot involve the spinal cord; they involve only the cauda equina.

Lesions in this region of the spinal canal neither produce upper motor neuron symptoms nor patterns of sensory loss related to interruption of ascending spinal cord fiber tracts. The lesions in this region produce symptoms that are related to involvement of lumbosacral spinal nerves, such as radicular pain, sensory loss involving the lower extremities, or significant bladder and sexual dysfunction, the last two abnormalities reflecting involvement of the autonomic fibers contained in the sacral spinal nerve roots.

At the lower lumbar vertebral level, there is no concern about a spinal needle impaling the spinal cord. The spinal cord does not descend below the L2 level, allowing the removal of cerebrospinal fluid to be accomplished safely with a lumbar puncture of the lower lumbar spinal canal. In this procedure, a sterile needle is passed through the space between adjacent lower lumbar vertebrae into the spinal subarachnoid space. Although the component roots of the cauda equina are coursing through the subarachnoid space within the spinal canal at this vertebral level, an appropriately placed spinal needle nudges them aside without causing damage.

Diagnostic Neuroimaging Procedures and Electrophysiologic Tests

Myelography had been the premier radiologic diagnostic modality used to investigate spinal cord disorders for several decades until the introduction of computerized tomography (CT) and magnetic resonance imaging (MRI) scanning. Myelography involves the introduction of a radiopaque dye into the spinal fluid through a lumbar puncture so that it outlines the spinal cord, spinal nerves, and cauda equina. CT scanning is sometimes used, but MRI has emerged as the most definitive technique for imaging the spinal cord because of its excellent contrast and its ability to display the cord in axial and sagittal arrays.

Somatosensory evoked potentials (SEPs) are performed by stimulating a peripheral nerve in the leg or arm, typically the tibial or median nerve, and recording the evoked potential over the upper cervical region or scalp. The elapsed time between the application of the stimulus and the arrival of the evoked potential is recorded, as is the amplitude of the response. These potentials reflect passage of the evoked impulse through spinal cord pathways, largely the posterior columns. An abnormality within these pathways may affect the latency of the response or its amplitude. SEPs are useful in identifying subtle pathology of the spinal cord even in the absence of clinical symptoms. They are also used to monitor spinal cord function intraoperatively during procedures involving significant manipulation of the spinal cord. Scoliosis surgery, during which there may be stretching of the spinal cord, is one example of a procedure during which SEPs are routinely performed.

SELECTED READINGS

Brazis PW, Masdeu JC, Biller J. Localization in clinical neurology, 2nd ed. Boston: Little, Brown, 1990:69.

Neuroscience in Medicine, edited by P. Michael Conn.
J. B. Lippincott Company, Philadelphia © 1995

Chapter *12*

The Cerebellum

JAMES R. WEST

The cerebellum functions as a comparator and a coordinator. It compares movement intention with performance and coordinates equilibrium, posture, and muscle tone necessary for smooth, coordinated motor activity. The cerebellum is present in all vertebrates. It receives considerable input from sensory systems, but it functions as a part of the motor system. The cerebellum contributes few direct connections to motoneurons, but it projects profusely to all major motor control regions in the cerebrum. It is a key component for sensorimotor coordination, but cerebellar damage typically produces neither sensory impairment nor decreased muscle strength. To influence motor performance, the cerebellum must receive and process an enormous amount of information about the position and state of muscles and tendons and the equilibrium of the body, and it must continuously integrate that data with information sent to the muscles from the motor cortex. The cerebellum performs these complex functions automatically, without conscious effort.

General Organization of the Cerebellum

☐ Gross morphology of the cerebellum: convolutions increase surface area

The word cerebellum means little brain, and it is an appropriate name. The cerebellum weighs about 150 g in men, comprising about 10% of the total weight of the brain. The cerebellum occupies the posterior cranial fossa, separated from the occipital lobes of the cerebral hemispheres by a conspicuous thickening of the dura, the *tentorium cerebelli*. The cerebellum is the largest part of the hind brain. It overlies a substantial portion of the posterior surface of the pons and medulla oblongata. Developmentally, the cerebellum is derived first from the germinal cells of the alar lamina and later, toward the end of the embryonic period, from the rhombic lip and therefore from sensory precursors.

The cerebellum is a highly convoluted, ovoid structure that is constricted in the middle. The superior surface is flattened, and the inferior surface is convex. The cere-

bellum consists of two *cerebellar hemispheres* joined by a narrow, median, longitudinal strip, the vermis (Fig. 12-1). The *vermis* is separated from each hemisphere by longitudinally running grooves, but these are not prominent on the superior surface. On the inferior surface, the hemispheres are separated by a deep, wide furrow, the *vallecula*. The medial portion of the hemispheres on each side of the vermis is called the *paravermis or intermediate zone*. The primary mass of the cerebellum (all except the flocculonodular lobe) is called the corpus cerebelli (Fig. 12-2). The cerebellum has a distinct three-layered cortex, a medullary center, four pairs of nuclei embedded deep within the structure, and three pairs of peduncles. A thin, white lamina, the *superior medullary velum*, projects between the superior cerebellar peduncles, forming the cranial portion of the roof of the fourth ventricle (Fig. 12-3).

The entire outer covering of the cerebellum is a highly convoluted layer of gray matter covering a thick medullary core of white matter. The surface of the cerebellum is conspicuously different from the cerebrum, most notably because of the many thin, transverse-running folia formed by many parallel fissures, which allow for extensive folding so that approximately 85% of its surface is hidden. Despite the comparatively small size of the cerebellum, its cortex is surprisingly extensive; if it were laid out as a flat sheet, the cerebellar cortex would be almost 1 meter long and have an area of approximately 50,000 mm².

☐ The surface of the cerebellum is characterized by lobes and fissures

The cerebellum is divided by many transversely oriented fissures, but only two, the primary fissure and the posterolateral fissure, are of significance. They divide the cerebellum into three main lobes, the anterior lobe, the posterior lobe, and the flocculonodular lobe (see Figs. 12-2 and 12-3).

The anterior lobe is the superior part of the cerebellum rostral to the V-shaped *primary fissure*. The posterior lobe, sometimes called the middle lobe, is the largest of the three lobes, and it is positioned between the primary

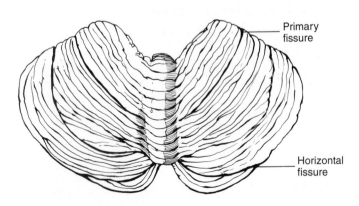

Primary fissure

Horizontal fissure

Figure 12-1
The superior surface of the cerebellum.

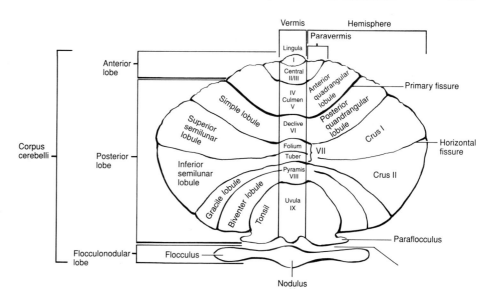

Figure 12-2
Schematic of the flattened cerebellum, showing the fissures, lobes, and lobules of the cerebellum. The terms on the left side of the diagram refer to terminology used for the human cerebellum. The terms on the right side refer to terminology used for animals. The Roman numerals designate the ten lobules of the vermis.

fissure and *posterolateral* (i.e., prenodular) *fissure*. The deep *horizontal fissure* is easily identified, but it has no apparent functional significance. Rostral to the posterolateral fissure, at the rostral edge of the inferior surface of the cerebellum, is the smallest lobe, the flocculonodular lobe. The flocculonodular lobe is composed of the two flocculi, the small, semidetached portions of each cerebellar hemisphere, which continue medially into the *nodulus*, the rostral pole of the inferior vermis. The lobes are divided into lobules, which are further subdi-

vided into many folia. The lobes and most of the folia run continuously in a predominantly transverse, but somewhat curved direction from hemisphere to hemisphere so that each has a vermal and two hemispheric components. Traditionally, the lobules are identified by names, and in the vermis, the lobules also are identified with roman numerals I through X. The lobular organization can best be distinguished in the vermis (see Fig. 12-3), because only there are all ten lobules present. Because the separate lobules have no known functional signifi-

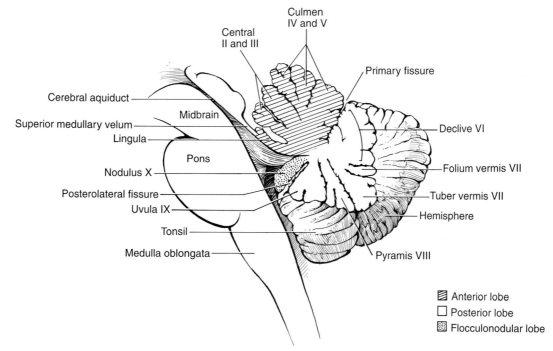

Figure 12-3
A midsagittal section through the cerebellar vermis and brain stem.

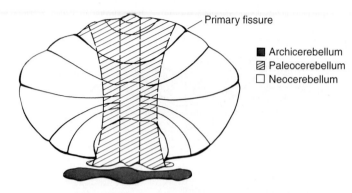

Primary fissure

■ Archicerebellum
▨ Paleocerebellum
☐ Neocerebellum

Figure 12-4
The functional divisions of the cerebellum are based on phylogenetic development and the termination of afferents to the cerebellar cortex.

cance, they are of limited use clinically. However, they are commonly used for descriptive purposes in experimental studies.

☐ The cerebellum is functionally divided into three parts

Based on a combination of criteria, including its phylogenetic development, and on experimental studies of fiber connections, the cerebellum can be divided into three basic functional divisions (Fig. 12-4). The *archicerebellum* or *vestibulocerebellum* is phylogenetically the oldest, and as one of its names implies, it is related functionally with the vestibular system. It receives direct connections from the vestibular nerve and it has reciprocal connections with the vestibular nuclei in the medulla oblongata. It controls balance and coordinates eye movements with movements of the head. The *paleocerebellum* or *spinocerebellum,* consisting of most of the vermis, paravermis, and most of the anterior lobe, is phylogenetically the next region of the cerebellum to appear. The paleocerebellum receives a variety of sensory inputs from the spinal cord, which it uses for the control of posture, muscle tone, and synergy during stereotyped movements such as walking. The *neocerebellum* or *pontocerebellum* consists of most of the large, lateral parts of the cerebellar hemispheres. It is the largest and phylogenetically the youngest part of the cerebellum in humans. The neocerebellum receives projections from the pontine nu-

clei that have been relayed from the cerebral cortex. The neocerebellum coordinates the planning of movements, functioning in the coordination of muscle tone required for accurate nonstereotyped (i.e., learned) movements.

☐ The cerebellar peduncles convey fibers in and out of the cerebellum

The cerebellum is connected to the posterior surface of the lower three segments of the brain stem by three thick pairs of fiber bundles: the superior, middle, and inferior cerebellar peduncles (Fig. 12-5). All of the inputs and outputs of the cerebellum are routed through the peduncles.

The *superior cerebellar peduncle*, the major portion of which is called the *brachium conjunctivum*, connects the cerebellum with the lower portion of the midbrain. It contains the principal efferent pathways leaving the cerebellum from the globose, emboliform, and dentate nuclei. The superior cerebellar peduncle also conveys afferent fibers into the cerebellum from the ventral spinocerebellar tract, the tectocerebellar tract, the rubrocerebellar tract, and a small noradrenergic projection from locus coeruleus.

The *middle cerebellar peduncle*, also called the *brachium pontis*, is the largest of the peduncles, and it connects the pons with the cerebellum. It is formed exclusively by the (cortico)pontocerebellar tract, consisting of afferent fibers projecting to the cerebellum from the con-

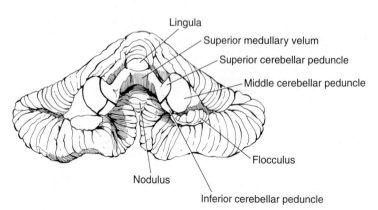

Lingula
Superior medullary velum
Superior cerebellar peduncle
Middle cerebellar peduncle
Flocculus
Nodulus
Inferior cerebellar peduncle

Figure 12-5
The anteroinferior surface of the cerebellum, illustrating the cut ends of the three cerebellar peduncles.

Table 12-1
PRINCIPAL INPUTS TO THE CEREBELLUM

Tracts	Origin*	Termination Zone*	Peduncle
Ventral spinocerebellar	Spinal cord	V, P	Superior
Tectocerebellar	Sup and Inf colliculi	V, P	Superior
Aminergic	LC, RP, VMT	V, P, L	Superior
Corticopontocerebellar	Pontine nuclei	V, P, L	Middle
Dorsal spinocerebellar	Spinal cord	V, P	Inferior
Olivocerebellar	Inf and acces olive	V, P, L	Inferior
Cuneocerebellar	Lat cuneate nucleus	V, P	Inferior
Vestibulocerebellar	Vestibular organ,	F, N, U	Inferior
	vestibular nuclei	F, V	Inferior
Reticulocerebellar	Reticular formation	V, P	Inferior
Arcuatocerebellar	Arcuate nucleus	F	Inferior
Trigeminocerebellar	Trigeminal nerve (CNV)	V, P	Inferior

*V, vermal zone; P, paravermal zone; L, lateral zone; F, flocculus; N, nodulus; U, uvula; LC, locus coeruleus; RP, raphe nuclei; VMT, ventral mesencephalic tegmentum, Inf, inferior; Sup, superior; Lat, lateral; Acces, accessory; CNV, cranial nerve V.

tralateral pontine nuclei. These fibers relay information from all parts of the cerebral cortex. No efferent fibers leave the cerebellum through the middle cerebellar peduncle.

The *inferior cerebellar peduncle* is a large bundle of input and output fibers that connects the cerebellum with the medulla oblongata. Seven distinct afferent pathways enter the cerebellum through the inferior cerebellar peduncle (Table 12-1). It is composed of two parts, the larger *restiform body*, which conveys afferent fibers to the cerebellum from the spinal cord and brain stem, and the smaller, medially positioned *juxtarestiform body*, which carries a small bundle of vestibular fibers to and from the cerebellum. Important efferent fibers that pass out through the inferior cerebellar peduncle are the cerebellovestibular, cerebello-olivary, and the cerebelloreticular fibers.

☐ Deep cerebellar nuclei relay cerebellar output to other parts of the CNS

Purkinje cells are the projection neurons of the cerebellar cortex, but few of them leave the cerebellum directly. Most Purkinje cell axons synapse in the deep cerebellar nuclei. In humans and the highest primates, there are four pairs of nuclei embedded within the white matter of the cerebellum. From medial to lateral, they are the *fastigial, globose, emboliform,* and *dentate nuclei* (Fig. 12-6). In monkeys and lower mammals, there are only three pairs of cerebellar nuclei; the emboliform and globose nuclei are combined into one cell mass on each side, and the unit is called the *interposed* or *interpositus nucleus*. This can be a source of confusion, because the term interposed nucleus is often used in reference to humans to refer to the two nuclei. Inputs to the deep cerebellar nuclei are derived from two sources, the Purkinje

cells of the cerebellar cortex and the extracerebellar sources.

Together with the vestibular nuclei, the deep cerebellar nuclei relay the entire output from the cerebellum to other parts of the central nervous system. The fastigial nucleus is sometimes called the "roof nucleus" because it is located in the roof of the fourth ventricle. It receives inputs from Purkinje cells in the vermis. The fastigial nucleus gives rise to three efferent bundles that project through the inferior cerebellar peduncle primarily to the vestibular nuclei and to the reticular formation. The globose nucleus consists of two or three very small clumps of cells. The emboliform nucleus is a small oval nucleus lateral to the globose nucleus and medial to the hilus of the dentate nucleus. The globose and emboliform nuclei receive inputs from Purkinje cells in the paravermis. The efferent fibers of the globose and the emboliform nuclei exit the cerebellum through the superior cerebellar peduncle and project to numerous motor control areas. The

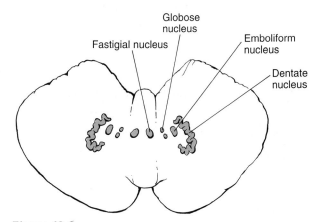

Figure 12-6
The diagram demonstrates the four pairs of deep cerebellar nuclei.

dentate nucleus is the largest, most laterally placed, and most conspicuous of the deep cerebellar nuclei. In cross section, it has the appearance of a crumpled band of cells, similar to that of the inferior olive. Careful dissection has shown that it is composed of numerous nodules or fingers of gray matter. The dentate nucleus receives inputs from Purkinje cell axons from the lateral portions of the cerebellar hemispheres. The efferent fibers from the dentate nucleus represent a major component of the superior cerebellar peduncle. The main outputs project from the hilus of the dentate nucleus to the red nucleus and thalamus, the reticular formation, dorsal medial nucleus, and oculomotor nucleus, all on the contralateral side. All of the deep cerebellar nuclei send fibers to the inferior olive and the reticular formation.

In addition to distributing cerebellar information to other parts of the brain, collateral axons from deep cerebellar nuclei project back to the same areas of the cerebellar cortex from which they received Purkinje cell projections. The deep nuclei also receive collateral inputs from climbing fibers and mossy fibers projecting to the cerebellar cortex. The outputs from the deep cerebellar nuclei therefore represent the main feedback loops from the cerebellum to areas associated with the control of the cerebellar function, including the cerebellum itself.

The blood supply to the cerebellum is derived from three arteries

The blood supply to the cerebellum is from three paired arteries. A *posterior inferior cerebellar artery* arises from each vertebral artery just before joining the basilar artery. The posterior inferior cerebellar arteries supply most of the inferior surface of the cerebellum, including the inferior vermis. The *anterior inferior cerebellar artery*

arises bilaterally from the caudal end of the basilar artery and supplies the more anterior portion of the inferior cerebellum (e.g., flocculus). The *superior cerebellar arteries* arise bilaterally from the rostral end of the basilar artery. They supply the superior (i.e., dorsal) portion of the cerebellum. All three pairs of arteries contribute branches to the deep cerebellar nuclei. The venous drainage of the cerebellum is achieved through various venous sinuses and the great cerebral vein.

Cytoarchitecture of the Cerebellar Cortex

☐ There are three well-defined layers of the cerebellar cortex

The cerebellar cortex is a highly convoluted sheet of gray matter that is about 1 mm thick. Unlike the cerebral cortex, the cerebellar cortex exhibits uniformity in thickness and in organization. It exhibits a relatively simple but highly organized cytoarchitecture that is remarkably similar in all mammalian species. The cerebellum has three distinct layers, and it contains five major types of neurons. The neurons exhibit a remarkable degree of uniformity in their organization throughout the cortex in respect to the transverse and longitudinal axes and to the pial surface (Fig. 12-7).

The outermost layer, just beneath the pial surface, is called the *molecular layer*. It is packed with dendrites and axons but has relatively few neurons. Only two neuronal types, *basket cells* and *stellate cells*, have their cell bodies in the molecular layer. During development, there is a transient germinal layer several cells thick between the pial surface and the molecular layer. Because this is where granule cells are generated, it is called the *external granular layer*. Soon after they are formed, the granule cells migrate through the molecular and Purkinje

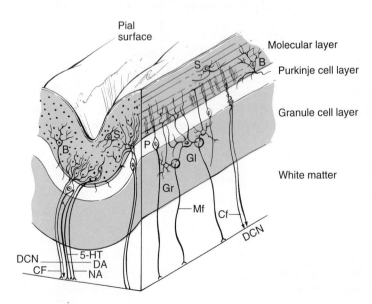

Figure 12-7
Basic cytoarchitecture of the cerebellar cortex, represented by a single cerebellar folium. Three kinds of inputs, climbing fibers (Cf), mossy fibers (Mf), and aminergic fibers are illustrated: *DA,* dopaminergic; *5-HT,* serotonergic; *NA,* noradrenergic. Five main types of neurons are identified: *P,* Purkinje cell; *Go,* Golgi cell; *Gr,* granule cell; *B,* basket cell; *S,* stellate cell; *Pf,* parallel fiber; *Gl,* glomerulus; *DCN,* deep cerebellar nuclei. The orientation of the diagram is such that the surface on the left side represents the transverse plane, and the right side of the diagram represents the longitudinal plane of a folium.

cell layers to reach the granular layer. When granule cell neurogenesis is complete, the external granular layer disappears.

The middle layer, also called the *Purkinje cell layer*, is composed of a single layer of Purkinje cell bodies. The Purkinje cells send their dendrites into the molecular layer and their axons through the *granular layer* into the medullary white matter.

The deepest of the three cerebellar cortical layers is the granular layer. It is the thickest layer, densely packed with small *granule cells*. The granular layer also contains a few *Golgi cells* that are located just below the Purkinje cell layer.

☐ There are five basic types of neurons in the cerebellar cortex

Purkinje cells

Purkinje cells are highly differentiated neurons, whose cell bodies form a monolayer sandwiched between the molecular and granular layers. The large, flask-shaped cell bodies give rise to large, fan-shaped dendritic arbors that fill the molecular layer in a distinctive pattern that is broad in the transverse plane of the folia and flattened in the horizontal plane (see Fig. 12-7). Purkinje cells are often called the principal neurons of the cerebellum because they provide the sole output from the cerebellar cortex; all other neurons in the cerebellar cortex are intrinsic neurons. As a Purkinje cell axon passes through the granule cell layer, it gives off one or more collaterals directed back near the Purkinje cell layer. Purkinje cell axons are myelinated, and they synapse on neurons in the deep cerebellar nuclei to be relayed to other regions of the brain. A few Purkinje cell axons from the vermis bypass the deep cerebellar nuclei and synapse directly onto neurons in the lateral vestibular nucleus. γ-Amino butyric acid (GABA) is thought to be the inhibitory neurotransmitter of the Purkinje cells.

Granule cells

There is an immense population of cerebellar granule cells. Cerebellar granule cells are the smallest, most densely packed, and by far the most numerous neuron type in the entire brain. They form the mass of the thick granular layer deep beneath the Purkinje cell layer. Granule cells, provide the largest of the two principal inputs to the Purkinje cells, and theirs is the only excitatory input among the intrinsic neurons of the cerebellum.

The claw-like terminal branches of granule cell dendrites act as a functional relay to enable mossy fiber information to reach the cerebellar cortex. After passing through the Purkinje cell layer into the molecular layer, each of the small-diameter, nonmyelinated granule cell axons bifurcate into its characteristic T-shaped junction. The two branches travel 1 to 1.5 mm in each direction horizontally through the molecular layer along the long axis of the folium, side by side with many thousands of other similar fibers, accounting for their name, *parallel fibers* (see Fig. 12-7).

There is a geometric relation between the position of the granule cell body and the position of the parallel fiber in the molecular layer. The more superficial the granule cell body is in the granular layer, the more superficial are its parallel fibers in the molecular layer. A parallel fiber makes multiple synaptic contacts with the dendrites of numerous Purkinje cells. This organization allows each granule cell to synapse with dendritic spines of several hundred Purkinje cells. Parallel fibers also make synaptic contact with basket, stellate, and Golgi cells. The functional output of granule cells is excitatory, and their neurotransmitter is thought to be glutamate.

Basket cells

Basket cells are small neurons found in the deeper parts of the molecular layer near the Purkinje cell bodies. Their dendritic trees are unremarkable except that they are oriented as thin wafers; they are broad only in the transverse plane, similar to the orientation of the dendritic trees of Purkinje cells. Basket cells get their name from the characteristic pericellular basket that their axons form around Purkinje cell bodies (see Fig. 12-7). The projection field of basket cell axons is in the transverse plane and is the source of several collateral branches that descend to the cell bodies of Purkinje cells. Each basket cell primary axon synapses with about one dozen Purkinje cells, but it does not synapse with the one closest to it. Collateral branches travel in the longitudinal direction, reaching an additional three to six rows of Purkinje cells on each side of the primary axon. It is estimated that this allows each basket cell to contact a patch of 100 to 200 Purkinje cells. The functional output of basket cells is inhibitory, and the evidence indicates that GABA is the inhibitory neurotransmitter.

Stellate cells

Stellate cells are so named because of the star-shaped appearance of their dendrites. They are usually located in the outer two thirds of the molecular layer. Their axons synapse only on Purkinje cell dendritic shafts (see Fig. 12-7). Stellate cells are functionally similar to the basket cells in that they receive the same inputs, and, like the basket cells, they inhibit Purkinje cell firing. However, because they synapse farther from the Purkinje cell body than the basket cells, the influence of stellate cells is less than that of basket cells. It is thought that stellate cells use taurine as a neurotransmitter.

Golgi cells

Golgi cells are large neurons scattered in the superficial part of the granular layer immediately below the Purkinje cell bodies. Estimates of the size of the Golgi cell population range from a 1 : 10 ratio of Golgi cells to Purkinje cells (Eccles et al., 1967) to almost a 1 : 1 ratio (Ito, 1984). Golgi cell bodies are about the same size as those of Purkinje cells. Golgi cell dendrites are not as elaborate as those of Purkinje cells, but their dendritic fields are about three times more extensive (see Fig. 12-7). Unlike Purkinje cells, Golgi cell dendrites are not compressed into the transverse plane of the folia. Golgi cell dendrites project to the molecular and granular layers. Because the dendrites that enter the molecular layer overlap the dendritic fields of three Purkinje cells in each plane, they act as a central point in a functional hexagon, influencing about 10 Purkinje cells. Golgi cells receive inputs from mossy fibers. The dendrites that stay in the granular layer contribute to a complex synaptic structure called a *glomerulus.*

Golgi cell axons are conspicuous. They are short and branch profusely, and they contribute to the formation of the glomerulus. The output from Golgi cells is inhibitory, and the cells probably use GABA as a neurotransmitter.

☐ The intracortical circuitry of the cerebellum produces a modulated inhibitory output

Based on data from a variety of experimental sources, much of the functional circuitry of the cerebellar cortex has been determined. The connections among mossy fibers and granule cells, granule cells and Purkinje cells, and climbing fibers and Purkinje cells all are excitatory (Fig. 12-8). However, the excitation of Purkinje cells is modulated by several feedback circuits that inhibit Purkinje cell activity and therefore suppress transmission from the cortex to the deep nuclei. The inhibitory circuits tend to limit the area of the cerebellar cortex and the degree of excitation produced by an incoming signal. The output of the Purkinje cells that reaches the deep cerebellar nuclei is a finely calibrated inhibitory signal.

Input and Output Systems in the Cerebellum

☐ There are three basic categories of input fibers to the cerebellum

Inputs to the cerebellum are directed mainly to the cerebellar cortex. There are numerous inputs to the cerebellum, but classically, they have been combined into two major pathways: *climbing fibers* (i.e., olivocerebellar tract) and *mossy fibers* (i.e., all other inputs). A third small group of *aminergic fibers* that reach the cerebellar cortex has been reported. All of the afferents to the cerebellum arrive through one of the three cerebellar peduncles, but primarily through the inferior and middle peduncles (see Table 12-1).

Climbing fibers

All climbing fibers have been described as projecting from neurons in the inferior olive to the cerebellar cor-

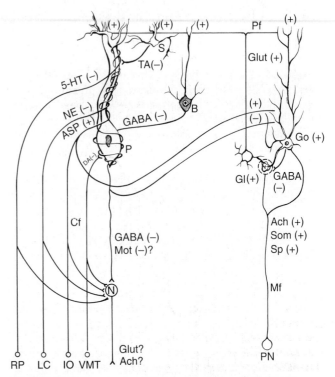

Figure 12-8

Essential circuitry of the cerebellar cortex. *5-HT*, serotonin; *Ach*, acetylcholine; *Asp*, aspartate; *B*, basket cell; *Cf*, climbing fiber; *DA*, dopamine; *DCN*, deep cerebellar nuclei; *GABA*, gamma-amino butyric acid; *Gl*, glomerulus; *Glut*, glutamate; *Go*, Golgi cell; *Gr*, granule cell; *IO*, inferior olive; *LC*, locus coeruleus; *Mf*, mossy fiber; *Mot*, motilin; *NE*, norepinephrine; *P*, Purkinje cell; *Pf*, parallel fiber; *PN*, precerebellar neurons (from various sources); *RP*, raphe nuclei; *S*, stellate cell; *Som*, somatostatin; *SP*, substance P; *TA*, taurine; *VMT*, ventral mesencephalic tegmentum; *VN*, vestibular nuclei.

tex (see Figs. 12-7 and 12-8). Although studies of animals have challenged that notion, most arise from the inferior olive and accessory olivary nuclei. Climbing fibers are named for the manner in which they attach to Purkinje cells and ascend through the molecular layer, entwining themselves in ivy fashion around Purkinje cell dendritic trees. Climbing fibers ramify in the granular layer, synapsing with granule and Golgi cells, and then ascend into the molecular layer.

Climbing fibers are myelinated until they reach the level of the Purkinje cell layer. In the molecular layer, each climbing fiber makes several hundred synaptic contacts with the smooth dendritic branches of a single Purkinje cell, and there is no convergence or divergence of the input. Climbing fibers exert a powerful excitatory influence on individual Purkinje cells, but through collateral branches, they also provide an excitatory, albeit weaker, input to Golgi, basket, and stellate cells and to the deep cerebellar nuclei (see Fig. 12-8).

Climbing fibers have been postulated to play roles other than influencing motor activity, including a trophic function during development and in plastic reorganization. The neurotransmitter of climbing fibers is thought to be aspartate.

Mossy fibers

The mossy fiber projections to the cerebellum are extensive (see Figs. 12-7 and 12-8). They include all of the cerebellum afferents except the climbing fibers and the aminergic projections. They enter the cerebellum through all three peduncles. Myelinated mossy fiber axons bifurcate many times in the white matter and then lose their myelination as each branch gives off numerous collaterals in the granular layer. Large clump-like swellings called rosettes are specialized synaptic endings that occur repeatedly along the course of the axon branches.

Each mossy fiber rosette forms the nucleus of a complicated structure called a *glomerulus* (see Fig. 12-8). A glomerulus is a large synaptic complex composed of one mossy fiber rosette, dendritic contacts from many granule cells, the proximal portions of Golgi cell dendrites, and terminal branches of Golgi cell axons. The entire structure is encapsulated in a glial sheath. In Nissl-stained tissue, glomeruli appear as irregularly-shaped "islands" where granule cells are absent. Each mossy fiber synapses with several hundred granule cells, and each granule cell receives mossy fiber input from several different mossy fibers, resulting in considerable divergence and convergence of input. Mossy fibers also send collaterals to the deep cerebellar nuclei. Unlike climbing fibers, mossy fibers activate Purkinje cells indirectly by activating granule cells.

Given the wide variety of their sources, it is not surprising that their neurotransmitters are unknown. However, all mossy fibers are thought to be excitatory, and the postulated neurotransmitters include acetylcholine and the neuropeptides, substance P and somatostatin.

Aminergic fibers

Studies of animals have identified amine-containing projections to the cerebellum (see Fig. 12-8). A projection of noradrenergic fibers from the locus coeruleus (i.e., A6 cell group) that travels through the superior cerebellar peduncle to synapse with Purkinje cell dendrites in all parts of the cerebellar cortex has been identified. Dopamine-containing fibers from the A10 cell group of the ventral mesencephalic tegmentum project to Purkinje and granular cell layers and to the interposed and dentate nuclei. Another input from cell groups B5 and B6 in the raphe nuclei in the brain stem enters the cerebellum through the middle cerebellar peduncle to provide serotonergic fibers to the granular and molecular layers. Little is known about the functional significance of the aminergic inputs.

☐ There are three primary sources of cerebellar inputs

Although information reaches the cerebellum from a variety of sources (see Table 12-1), the three major sources of afferents are the spinal cord, vestibular system, and the cerebral cortex.

Inputs from the spinal cord

Information from the spinal cord reaches the cerebellum through the ventral spinocerebellar tract, dorsal spinocerebellar tract, and the cuneocerebellar tract. They provide information on the position and condition of muscles, tendons, and joints. The ventral spinocerebellar tract transmits proprioceptive information from all parts of the trunk and limbs, ascending on both sides in the white matter of the spinal cord. The contralateral fibers then recross to the ipsilateral side, and all of them enter the cerebellum through the superior cerebellar peduncle, terminating as mossy fibers. The dorsal spinocerebellar tract conveys proprioceptive information from the lower trunk and lower limbs. This projection of mossy fiber inputs ascends in the spinal cord on the ipsilateral side and enters the cerebellum through the inferior cerebellar peduncle. The cuneocerebellar tract transmits proprioceptive information from the upper trunk and upper limbs. Axons in the cuneate nucleus in the medulla oblongata enter the cerebellum through the inferior cerebellar peduncle to terminate as mossy fibers.

Data from experimental studies have demonstrated that the inputs from proprioceptive and tactile stimuli to the cerebellar cortex are organized in two somatotopic maps (i.e., homunculi), one centered in the vermis of the anterior lobe and the other (a double one consisting of bilateral mirror images) in the posterior lobe. These cerebellar homunculi are considerably less precise than those in the cerebral cortex.

Inputs from the vestibular system

Primary fibers from the vestibular labyrinth by means of cranial nerve VIII reach the ipsilateral cerebellum through the juxtarestiform body. A much larger bundle of fibers is directed to the vestibular nuclei in the brain stem and is relayed from the vestibular nuclei through the juxtarestiform body to terminate bilaterally in the flocculus and most of the vermis as mossy fibers. The vestibular fibers provide important information about equilibrium conditions of the body.

Inputs from the cerebral cortex

Information from multiple regions of the cerebral cortex reaches the cerebellum indirectly from three different pathways: the (cortico)pontocerebellar tract, the (cortico)olivocerebellar tract, and the (cortico)reticulocerebellar pathway. This information that is destined for the cerebellum originates from cells in various regions of the cerebral cortex and synapses in the pontine nuclei, the inferior olivary nuclei, and the reticular formation. The pontocerebellar tract continues from neurons in the pontine nuclei that give rise to fibers that cross the midline and enter the cerebellum on the contralateral side through the middle cerebellar peduncle. This huge fiber tract projects to the cerebellum as mossy fibers. The olivocerebellar tract relays information from the cerebral cortex that projects to the inferior olive. Inferior olivary fibers cross the midline, enter the cerebellum through the inferior cerebellar peduncle, and terminate as climbing fibers. The reticulocerebellar tract also relays information from the cerebral cortex. Neurons in the reticular formation give rise to fibers that enter the cerebellum on the ipsilateral side through the inferior cerebellar peduncle to end as mossy fibers.

☐ The outputs from the cerebellum are less complex than the inputs

There are fewer outputs from the cerebellum than inputs. The sole output from the entire cerebellar cortex is from the axons of Purkinje cells. However, information from most of the Purkinje cells is relayed through neurons in the deep cerebellar nuclei before exiting the cerebellum (Fig. 12-9). All outputs from the cerebellum exit through the superior or inferior cerebellar peduncles.

Most outputs from the cerebellum exit through the superior cerebellar peduncle. The small projection of Purkinje cell axons that do leave the cerebellum directly travel through the juxtarestiform body of the inferior cerebellar peduncle and synapse in the vestibular nuclei. The output from the other Purkinje cells project to the deep cerebellar nuclei in a specific pattern (see Fig. 12-9), and the principal efferents from the cerebellum originate from cells in these deep nuclei. Purkinje cells in the vermis project to the fastigial nucleus; Purkinje cells in the intermediate zone project to the globose and emboliform nuclei; and those in the lateral cerebellar hemispheres project to the dentate nucleus. Some Purkinje cells from the flocculus and nodulus project to the fastigial and dentate nuclei and to the vestibular nuclei.

The outputs from the fastigial nucleus project mainly to the brain stem, where they synapse in the vestibular nuclei bilaterally, and in the reticular formation, where the projection is primarily contralateral. The ipsilateral projections from the fastigial nucleus travel through the juxtarestiform body. The fibers destined for contralateral

regions cross the midline while still within the cerebellum, hook around the superior cerebellar peduncle (hence the name *hook bundle* or *uncinate fasciculus*), and then exit through the juxtarestiform body on the contralateral side. Small contralateral projections also reach the cervical spinal cord and the thalamus.

The largest output from the cerebellum consists of axons passing through the superior cerebellar peduncle (i.e., brachium conjunctivum). The cells of origin are in the globose, emboliform, and dentate nuclei. Some of these cells project to the inferior olive and reticular formation (i.e., descending limb of the superior cerebellar peduncle). However, most fibers in the peduncle project rostrally and then cross to the contralateral side in the midbrain at the level of the inferior colliculus. Most of these axons originate in the emboliform and globose nuclei. With a few fibers from the dentate nucleus, they synapse in the red nucleus. The remaining axons from the dentate nucleus (i.e., *dentatorubrothalamic tract*) continue on to the thalamus, where they terminate in the red nucleus and thalamus.

With the exception of a small bundle of Purkinje cell axons that project directly to the vestibular nuclei in the brain stem, all outputs of the cerebellum consist of axons whose cell bodies are located in the deep cerebellar nuclei. The fastigial nucleus projects mainly to the vestibular nuclei; the globose and emboliform nuclei project mainly to the red nucleus; and the dentate nucleus projects primarily to the thalamus.

Cerebellar Dysfunction

In contrast to the detailed knowledge of the specific connections of the relatively simple and uniform cortical organization of the cerebellum, surprisingly little is known about its functions in relation to other regions of the brain. Cerebellar function is best appreciated by considering the consequences of lesions to specific parts of the cerebellum. With that goal in mind, two plans of functional organization have been proposed. Although they are similar, they do differ somewhat in the resultant functional zones that they represent. Various investigators use one or the other, or both, schemes to localize cerebellar function.

Based in part on data from clinical and experimental studies, three "cerebellar syndromes" can be differentiated: flocculonodular syndrome, anterior lobe syndrome, and neocerebellar syndrome. This system has been used by several researchers to help correlate structural and functional associations in the cerebellum and to explain deficits resulting from cerebellar lesions. The flocculonodular syndrome is characterized by problems with maintaining equilibrium, but there is no ataxia of the limbs, hypotonia, or tremor. The anterior lobe syndrome is characterized by increased postural reflexes. The neocerebellar syndrome is distinguished by ataxia and hypotonia, producing clumsy movements. Although it is possible to demonstrate these syndromes experimentally in animals, pure cerebellar syndromes, espe-

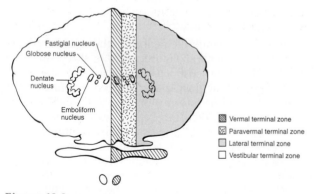

Fastigial nucleus
Globose nucleus
Dentate nucleus
Emboliform nucleus

☒ Vermal terminal zone
☒ Paravermal terminal zone
☐ Lateral terminal zone
☐ Vestibular terminal zone

Figure 12-9
The zones of organization of the cerebellum are based on the longitudinal termination of inputs to the cerebellar cortex and their projections to the deep cerebellar nuclei. Compare with Figure 12-4.

cially the anterior lobe syndrome, are seldom manifested in humans.

Because the local circuitry is essentially identical throughout the cerebellar cortex, a strict localization of function does not exist within the cortex itself. Functional localization in the cerebellum is instead the result of differences in the location of the termination of its afferent and efferent projections. From a clinical perspective, cerebellar function and dysfunction can best be understood by the organization of three longitudinal zones based on the terminal fields of major inputs to the cerebellar cortex (see Fig. 12-9). These three longitudinal (i.e., sagittal) bands are the vermal, paravermal, and lateral zones. This organization represents a crude parcellation of afferent termination zones in the cerebellar cortex, but there is considerable overlap. The cerebellar cortex continues the longitudinal organization by projecting from the three longitudinal cortical zones to corresponding deep cerebellar nuclei (see Fig. 12-9).

Damage to the cerebellum or its associated afferent and efferent systems produces distinctive symptoms and signs, usually on the same side of the body as the lesion. The remarkable uniformity of the cerebellum suggests that its basic functioning is the same throughout the cerebellar cortex. It is not surprising that lesions of deep nuclei or superior cerebellar peduncle (i.e., lesions that interrupt cerebellar outflow) produce more severe signs than lesions restricted to parts of the cerebellar cortex. Signs of cerebellar dysfunction in humans are usually the result of lesions that involve more than one specific region of the cerebellum. They manifest themselves as deficits in somatic motor control; the most common symptoms are related to disturbances of gait. For organizational purposes, it is worthwhile to consider separately the results of damage to the vermal, paravermal, and lateral zones. There have been several excellent descriptions and reviews of the disorders of cerebellar function (Holmes, 1939; Dow and Moruzzi, 1958; Gilman et al., 1981) to which interested readers are referred.

☐ Vermal zone cerebellar damage is associated with disorders of posture and gait, head position, and eye movement

The vermal or midline longitudinal zone consists of the vermis, flocculi, the fastigial nuclei, and related input and output fibers. The principal afferents to this zone are from the vestibular organ and nuclei, from the trunk and neck through the spinal cord, and from the reticular formation. Some primary inputs from the vestibular part of the vestibulocochlear nerve (i.e., cranial nerve VIII) enter the cerebellum as part of the juxtarestiform body and terminate in the flocculus and nodulus, with a small projection to the uvula and lingula of the vermis and collaterals to the fastigial nucleus. A larger projection of secondary fibers from the vestibular nuclei also projects to most of the vermal zone. Proprioceptive information from the head, particularly from the temporomandibular joint and

muscles of mastication, are transmitted through the trigeminocerebellar tract, which originates from the mesencephalic trigeminal nucleus and projects to parts of the vermis, paravermis, and flocculus by means of the restiform body.

Functionally, the vermal zone is associated with posture, head position, locomotion, and extraocular movements. Clinical signs resulting from midline cerebellar damage produce disorders of posture and gait, head positions, and eye movements (e.g., nystagmus). Because of its connections with the vestibular system, damage to the flocculus, nodule, and uvula result in pronounced loss of equilibrium, including ataxia (i.e., swaying while standing or staggering while walking with a tendency to fall, usually backwards) and a compensatory wide-based gait. There is an inability to incorporate vestibular information with body and eye movements. When damage is restricted to the vermal zone, these deficits and trunk asynergia are usually present without the disruption in other limbs when tested separately.

A clinical picture corresponding to the vermal or flocculonodular syndrome seen in children between 5 and 10 years of age is caused by a specific type of tumor, medulloblastoma. It usually occurs in the nodulus of the vermis, is characterized by unsteady gait and frequent unexplained falls, and is sometimes accompanied by nystagmus. The medulloblastoma is a neoplasm that is derived from persistent clumps of germinal cells from the external granular layer. This germinal field normally disappears after about 2 years of age, which helps to explain why this tumor is the most common type of central nervous system tumor in children but never occurs in adults.

☐ Paravermal zone cerebellar damage is associated with altered patterns of muscle movement

The functional paravermal or intermediate zone consists of the cerebellar cortex just lateral to the vermis, the emboliform and globose nuclei, and interconnected inputs and outputs. The principal source of information to the paravermal zone comes from the limbs, but afferents are from a variety of sources in the spinal cord, brain stem, and cerebral cortex. Outputs project to rostral and caudal portions of the nervous system. Damage restricted solely to the paravermal zone is found only in experimental animals, but this region seems to be involved in the modulation of velocity, force, and the pattern of muscle movement (e.g., hypertonia, hypotonia).

☐ Lateral zone cerebellar damage is associated with movement disorders

The large lateral zone consists of most of the cerebellar hemispheres (including much of the anterior lobe), the dentate nuclei, and related inputs and outputs. The inputs are from numerous sources, but most important is the projection from the cerebral cortex that is relayed through the pons. The outputs course to the brain stem and thalamus. This region is involved in the planning of voluntary movements in conjunction with the cerebral cortex. Damage to the large lateral hemispheric regions affects the initiation of coordinated movements. The most common type of cerebellar disease involves the hemispheres or some part of their efferent projections. Damage to the lateral zone produces movement disorders of the limbs and difficulty in posture and gait. If the lesion is unilateral, the signs are ipsilateral; the patient tends to stagger and deviate to the affected side.

☐ Cerebellar symptoms of damage include several common disturbances

There are several common disturbances associated with damage to the lateral cerebellar zone. *Ataxia* is a general term for disturbances or clumsiness of motor activity related to voluntary movement. *Asynergia* or limb ataxia is the reduced ability to execute smooth, coordinated sequential movements. *Hypotonia* is a reduced resistance to passive movement caused by a loss of cerebellar influence. *Asthenia* is an increased propensity for muscle fatigue, which is often associated with hypotonia. *Intention tremor* is an oscillating movement of a limb that is particularly pronounced at the end of the movement (in contrast to the Parkinson's "resting" tremor). *Adiadochocinesia* (i.e., dysdiadochokinesia) refers to the irregular movements after attempts at rapid alternating movements, such as pronating and supinating the forearm. In *dysmetria*, the judgment of distance is impaired. This is evident by the inaccurate control of the range and direction of movement, with undershot or overshot of the desired point. There can be ocular signs, especially *nystagmus*, which is of greatest amplitude when looking to the same side as the lesion. *Decomposition of movement* is a term given to the breaking down of a complex movement into its various components, which appear jerky and uncoordinated, rather than appearing as a smooth, flowing movement. *Dysarthria* refers to disorders in the mechanical component of the articulation of speech due to ataxia of the muscles of the larynx.

Relatively small lesions to the cerebellar cortex produce little discernible deficits. Even the diagnosis of substantial cerebellar damage is often difficult because the effects are often transient due to the remarkable functional plasticity that can occur. Generally, the more rapid the onset of the pathologic process, the older is the patient, the more widespread is the damage, and the better defined and longer lasting are the symptoms. The compensatory plasticity can be so great in response to cerebellar damage in fetal and neonatal life that major signs of cerebellar damage can be lacking, even in cases where an entire cerebellar hemisphere is completely absent.

The cerebellum is vulnerable to damage from a variety of sources, including developmental defects, degenerative diseases (i.e., hereditary and nonhereditary), infectious processes, chronic alcoholism, toxic and metabolic effects (including hypoxia), thrombosis of the cerebellar arteries, trauma, and tumors. Damage can occur from direct or indirect sources; tonsillar herniation, for instance, results in the cerebellar tonsils being squeezed out of the base of the skull through the foramen magnum in response to a tumor or hemorrhaging. Considerable experimental data indicate that severe malnutrition during development and fetal alcohol exposure produce permanent structural damage to the cerebellum.

Lesions of the cerebellum produce ipsilateral disturbances. Lesions restricted to the vermis produce disturbances of the trunk (e.g., stance, gait), and lesions of the lateral parts of the cerebellar hemispheres produce disturbances chiefly related to voluntary movements of the limbs. Typically, lesions of the deep cerebellar nuclei and superior cerebellar peduncles produce significantly greater deficits than lesions of the cerebellar cortex. The most prevalent clinical symptoms of cerebellar damage are associated with disruptions of standing and walking. The identification of the loci of damage based on the analysis of cerebellar symptoms is often difficult. More than one basic cerebellar region is usually affected. Symptoms tend to dissipate with time, particularly if the damage occurs in childhood or develops slowly over a long period. Cerebellar damage can have many causes. Secondary effects such as pressure, tissue dislocation, or vascular changes due to associated circulatory disturbances can produce serious cerebellar damage.

Other Functions of the Cerebellum

In addition to the roles played by the cerebellum in assisting the rest of the brain in controlling and coordinating muscle activity, studies have identified other cerebellar functions. The cerebellum helps to regulate some autonomic functions, including respiration, cardiovascular functions, and pupillary size.

Another area thought to be influenced by the cerebellum is motor learning. Many studies have generated convincing data demonstrating that cerebellar circuitry can be altered functionally by experience and that certain types of motor learning can be prevented by cerebellar lesions. The cerebellum also plays a role in some types of classical conditioning.

Taken together, these and related studies indicate that, despite the voluminous experimental literature focused on the cerebellum, cerebellar function is not understood completely.

SELECTED READINGS

Adams JH, Corsellis JAN, Duchen LW (eds). Greenfield's neuropathology, 4th ed. New York: John Wiley & Sons, 1984.

Bloom FE, Hoffer BJ, Siggins GR. Studies on norepinephrine-containing afferents to Purkinje cells of rat cerebellum. I. Localization of the fibers and their synapses. Brain Res 1971;25:501.

Brodal A. Neurological anatomy in relation to clinical medicine, 3rd ed. New York: Oxford University Press, 1981.

Dobbing J. The later development of the brain. In Davis JA, Dobbing J (eds). Scientific foundations of paediatrics, 2nd ed. London: William Heinemann Medical Books, 1981:744.

Dow RS, Moruzzi G. The physiology and pathology of the cerebellum. Minneapolis: University of Minnesota Press, 1958.

Eccles JC, Ito M, Szentçgothai J. The cerebellum as a neuronal machine. New York: Springer-Verlag, 1967.

Gilman S. Cerebellum and motor dysfunction. In Asbury AK, McKhann GM, McDonald WI (eds). Diseases of the nervous system, vol I. London: William Heinemann Medical Books, 1986:401.

Gilman S, Bloedel JR, Lechtenberg R. Disorders of the cerebellum. Philadelphia: Davis, 1981.

Gluhbegovic N, Williams TH. The human brain: a photographic guide. Hagerstown, MD: Harper & Row, 1980.

Holmes G. The cerebellum of man. Brain 1939;62:1.

Ito M. The cerebellum and neural control. New York: Raven Press, 1984.

Larsell O. Anatomy of the nervous system, 2nd ed. New York: Appleton-Century-Crofts, 1951.

McCormick DA, Thompson RF. Cerebellum: essential involvement in the classically conditioned eyelid response. Science 1984;223:296.

O'Leary JL, Dunsker SB, Smith JM, Inukai J, O'Leary M. Termination of the olivocerebellar system in the cat. Arch Neurol 1970;22:193.

Simon H, LeMoal ML, Casal A. Efferents and afferents to the ventral tegmental-A10 region studied after local injection of (^{3}H) leucine and horseradish peroxidase. Brain Res 1979;178:17.

Takeuchi Y, Kimura H, Sano Y. Immunohistochemical demonstration of serotonin-containing nerve fibers in the cerebellum. Cell Tissue Res 1982;226:1.

West JR (ed). Alcohol and brain development. New York: Oxford University Press, 1986.

Neuroscience in Medicine, edited by P. Michael Conn.
J. B. Lippincott Company, Philadelphia © 1995

Chapter *13*

The Brain Stem Reticular Core and Monoamine System

HAROLD H. TRAURIG
BRUCE E. MALEY

The reticular formation consists of an extensive aggregation of several subtypes of interconnected neurons extending throughout the brain stem tegmentum. The reticular formation integrates sensory, visceral, limbic, and motor functions, and its efferent connections project throughout the central nervous system. It exerts important influences on autonomic regulation of vital organ systems, behavior, somatic motor activities, sleep cycles, alertness, and pain modulation.

The Reticular Formation and Neural Functions

The term *reticular formation* was adopted by early anatomists to differentiate what appeared to be diffusely interconnected neurons in the brain stem tegmentum from those forming more anatomically distinct nuclei, such as hypoglossal or spinal trigeminal nuclei. This characterization of the reticular formation suggested that it was poorly organized and served only primitive functions. However, some early investigators, notably Cajal, recognized organization and specificity in the reticular formation based on neuron morphology, location in the brain stem, and connectivity. Despite Cajal's work, the reticular formation evoked little research interest until the electrophysiologic studies by Moruzzi and Magoun in the 1940s and 1950s demonstrated that the maintenance of consciousness and alertness depended on input from the sensory pathways to the brain stem reticular formation.

Later studies demonstrated that inputs to specific components of the reticular formation also originate from the cerebral cortex, striatum, limbic system, hypothalamus, cerebellum, and other central neural structures. Certain components of the reticular formation project to these same structures, often through the dorsal thalamus, and to the somatic and visceral motor nuclei of the brain stem and spinal cord. Subgroups of neurons or terminals in the reticular formation contain specific combinations of transmitters and peptides, such as serotonin, norepinephrine, and enkephalin, suggesting that reticular formation functions depend on complex neurochemical interactions.

The reticular formation provides an important matrix for neural integration. Ascending reticular formation projections play key roles in alertness, behavior, and affect. Brain stem reticular formation circuits, in concert with limbic and hypothalamic input, regulate cardiovascular and respiratory rhythms and other visceral responses through influences on cranial nerve nuclei and descending connections with autonomic centers in the spinal cord. Reticulospinal tracts also influence transmission of sensory modalities by dorsal horn neurons and activities of alpha and gamma motor neurons, modulating pain transmission, skeletal muscle tone, and somatic reflexes.

Anatomic Characteristics of Reticular Formation Neurons

The reticular formation has been subdivided into several nuclei based on their anatomic locations. The more prominent of the reticular nuclei are described here, but it is convenient to characterize the reticular formation as three columns of neurons that extend throughout most of the brain stem tegmentum (Fig. 13-1). Although these columns of neurons are not separated by unambiguous anatomic boundaries, they are distinct from one another based on neuron morphology, circuitry, and location in the mediolateral plane. The unpaired *raphe* nuclei (i.e., *median column*) lie in the midline of the *brain stem tegmentum*. (Raphe means seam and refers to the midline of the brain stem.) The paired *medial and lateral columns* of nuclei are found in the central (immediately lateral to the midline) and lateral portions of the tegmentum, respectively.

Reticular formation neurons possess an aggregate of characteristics that differentiate them from most other brain stem neurons. Many reticular formation neurons have extensive dendritic arborizations oriented in a plane perpendicular to the long axis of the brain stem and the ascending and descending sensory and motor tracts (Fig. 13-2*B*). Other brain stem neurons typically have restricted dendritic fields.

Reticular formation neuron dendrites typically intermingle with axons in motor and sensory tracts, suggesting functional interactions. Axons of many reticular formation neurons are long, form numerous collaterals, and may have ascending and descending branches. A single reticular neuron could influence brain stem and spinal cord neurons associated with several different somatic and visceral functions.

The inputs to reticular neurons typically originate from many and varied sources. Taken together, these anatomic arrangements reflect the modulatory and integrative roles of the reticular formation.

☐ The median column of the reticular formation lies in the midline of the brain stem

The *raphe of the brain stem* is characterized by numerous decussating axons, such as those forming the medial lemnisci, olivocerebellar, and secondary auditory system tracts. Raphe neurons are scattered among these decussating axons, most prominently in the medulla and pons, but they also extend into the midbrain (see squares in Fig. 13-1 and 13-2*A*, and asterisks in Fig. 13-3). The following raphe nuclei have been described based on neuron morphology: *raphe obscurus* and *raphe pallidus* in the medulla (see Fig. 13-1*A,B*); *raphe magnus*, *raphe pontis*, and *central superior nucleus* in the pons (see Figs. 13-1*C,D* and 13-3); *dorsal tegmental*

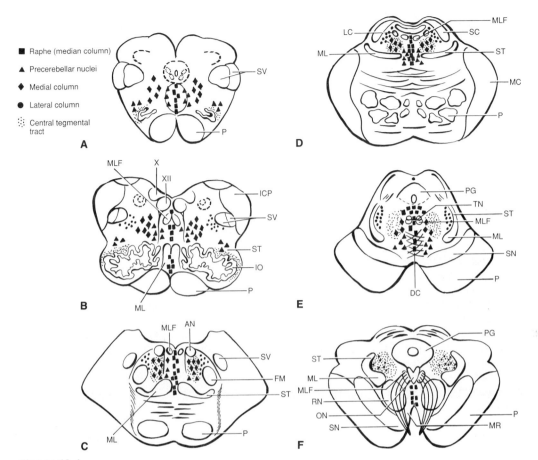

Figure 13-1
Representations of brain stem levels, localizing the major nuclei of the reticular formation. The anatomic boundaries of these nuclei often overlap, and there are numerous named subdivisions based on neuron morphology and neurochemical characteristics: (**A**) caudal medulla, level of the sensory decussation; (**B**) midmedulla, level of the vagal and hypoglossal nuclei; (**C**) caudal pons, level of the facial and abducens nuclei; (**D**) rostral pons, level of the locus ceruleus; (**E**) midbrain, level of the inferior colliculus; and (**F**) midbrain, level of the superior colliculus. The median column or raphe nuclei (*squares*) consist of the (**A, B**) raphe obscurus and raphe palladus in the medulla; (**C, D**) raphe magnus and central superior in the pons; and (**E, F**) nucleus linearis and dorsal tegmental nucleus in the midbrain. The medial column (*diamonds*) includes (**A, B**) the ventral reticular nucleus in the medulla and (**C, D, E**) the gigantocellular and pontine reticular nuclei in the medulla and pons. (**E**) The rostral (i.e., oral) portion of the pontine reticular nucleus extends into the midbrain. The lateral column nuclei (*large dots*) are (**B, C**) the parvicellular nucleus in the medulla and pons; (**D, E**) parabrachial nucleus in the pons and caudal midbrain; and (**E, F**) the pedunculopontine and cuneform nuclei in the midbrain. The percerebellar nuclei (*triangles*) include (**A, B, C**) the paramedian reticular and (**D, E**) pontine reticulotegmental nuclei. (**A, B**) The lateral reticular nucleus is located in the lateral aspect of the medulla. *AN*, abducens nucleus; *DC*, decussation of the superior cerebellar peduncles; *FM*, facial motor nucleus; *IC*, inferior cerebellar peduncle; *IO*, inferior olivary nucleus; *LC*, locus ceruleus; *MC*, middle cerebellar peduncle; *ML*, medial lemniscus; *MLF*, medial longitudinal fasciculus; *MR*, mesolimbic region; *ON*, oculomotor nucleus and root; *P*, pyramidal tract; *PG*, periaqueductal gray; *RN*, red nucleus; *SC*, superior cerebellar peduncle; *SN*, substantia nigra; *ST*, spinothalamic tracts; *SV*, spinal nucleus and tract of the trigeminal; *TN*, trochlear nucleus; *X*, dorsal motor nucleus of the vagus; *XII*, hypoglossal nucleus.

nucleus and *nucleus linearis* (see Fig. 13-1*E,F*) in the midbrain.

Many of the raphe neurons contain *serotonin* (i.e., 5-hydroxytryptamine) and probably use this indoleamine as a transmitter (see Fig. 13-2*A*). Some neurons of the medial column of the reticular formation also contain serotonin.

Raphe serotonergic neurons of the caudal brain stem project prominently to the spinal cord (Fig. 13-4); those ending in the dorsal horn are thought to play a key role in modulation of pain transmission to conscious centers. Many pontine and midbrain serotonergic raphe neurons have rostral projections directly and through the diencephalon to the cerebral cortex and the limbic system

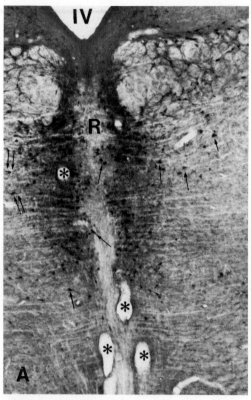
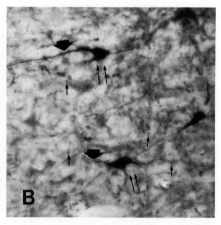

Figure 13-2

(**A**) Immunocytochemical localization of a molecular marker for serotonin in the pons of the monkey. The dorsal surface and the fourth ventricle (IV) are toward the top of the picture. R, raphe; *, blood vessels. The dark-staining reaction localizes serotonin-like immunoreactivities in neuron cell bodies (*arrows*) and terminals (*dark background* surrounding the neurons) in the nucleus raphe magnus. (Original magnification × 21) The neurons (*double arrows*) are enlarged in **B**. (**B**) The molecular marker for serotonin demonstrates some aspects of the dendritic arborizations of the nucleus raphe magnus neurons (*large arrows*) and the numerous serotonergic terminals (*small arrows*) in the nucleus raphe magnus. (Original magnification × 100)

(Fig. 13-5). The mesolimbic region, which is located in the ventromedial aspect of the midbrain (see Fig. 13-1*F*), receives serotonergic terminals from raphe and other reticular neurons. Mesolimbic neurons are dopaminergic and project to the mediobasal frontal cortex, hippocampus, amygdala, nucleus accumbens, and other limbic and cortical regions. Evidence supports the concept that some schizophrenic symptoms may result from inap-

propriate activation of the mesolimbic dopamine projections. Because activities of some dopaminergic neurons are facilitated by serotonin, it is likely that serotonergic raphe neurons are an important component of the circuitry that regulates limbic system functions. One action of certain antipsychotic drugs is to down-regulate serotonin receptors on dopamine neurons.

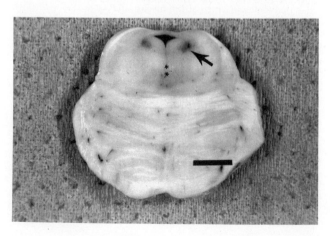

Figure 13-3

Fresh section of human rostral pons, demonstrating the pigmented neurons of the locus ceruleus (*arrow*). The arrow overlies the central tegmental tract. Asterisks indicate the raphe region. Bar = 5 mm.

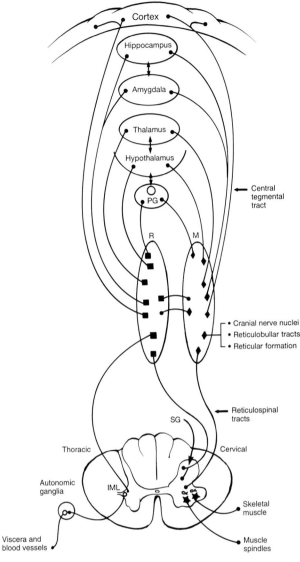

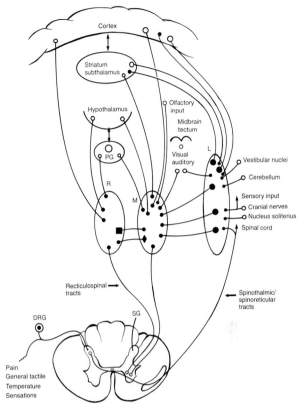

Figure 13-5

Representation of the principal afferent projections to the reticular formation. The raphe (i.e., median column) nuclei of the brain stem are represented by a square, the medial column nuclei are represented by diamonds, and the lateral column nuclei are represented by large dots. The open circles indicate the cell bodies of other neurons that project to the reticular formation. Small dots represent neuron terminals. *DRG*, dorsal root ganglion; *PG*, periaqueductal gray; *SG*, substantia gelatinosa of the dorsal horn.

Figure 13-4

Representation of the principal efferent projections of the reticular formation. The raphe (i.e., median column) nuclei of the brain stem are represented by squares, and the medial column nuclei are represented by diamonds. Small dots represent neuron terminals. *IML*, intermediolateral nucleus (i.e., preganglionic sympathetic neurons) of the spinal cord; *PG*, periaqueductal gray; *SG*, substantia gelatinosa of the dorsal horn.

with those projecting rostrally. These projections terminate on other regions of the reticular formation, the thalamus (i.e., interlaminar nuclei), cranial nerve nuclei, and the spinal cord (see Fig. 13-4). The effects of the medial column or effector neurons of the reticular formation are conveyed by reticulobulbar and reticulospinal tracts to numerous and varied regions of the central nervous system.

☐ The medial column is the effector portion of the reticular formation

The *medial column of the reticular formation* occupies the central portion of the medullary and pontine tegmentum (see diamonds in Fig. 13-1) and consists of the *ventral reticular* (see Fig. 13-1A,B), *gigantocellular*, and *pontine reticular nuclei* (see Fig. 13-1C–E). Most of these neurons are large and have extensive dendritic arborizations oriented perpendicular to the long axis of the brain stem. Long ascending and descending axons with numerous collateral branches originate from the neurons of the medial column. The arrangement of these collaterals is such that neurons projecting caudally interact

☐ The lateral column is the afferent component of the reticular formation

The principal *lateral column nuclei* (see circles in Fig. 13-1) are composed mostly of small neurons and include the *parvicellular nucleus* in the medulla and pons (see Fig. 13-1B,C), *parabrachial nucleus* in the pons and midbrain (see Fig. 13-1D,E), and *pedunculopontine* and *cuneiform nuclei* in the midbrain (see Fig. 13-1E,F). Lateral column components in the caudal brain stem are located between the medial reticular column medially and the spinal trigeminal system laterally; rostrally, the lateral column is just medial to the superior cerebellar peduncle and ventral to the midbrain colliculi.

The neurons of the lateral column are characterized by prominent inputs from collaterals of the sensory tracts and from direct spinoreticular tracts. The sensory input to lateral column neurons is relayed medially to the medial column of the reticular formation (see Fig. 13-5). For example, the parvicellular nucleus in the medulla and pons relays inputs from the ascending spinothalamic, trigeminothalamic, and auditory pathways to the medial column and raphe nuclei. Functionally, this relay provides activating influences through the reticulospinal and reticulobulbar pathways that modulate sensory transmissions and through ascending reticular pathways that maintain alertness.

Connections of the parabrachial nucleus in the pons and midbrain suggest roles in visceral and limbic functions. The midbrain components of the lateral column—the pedunculopontine and cuneiform nuclei—project to the motor cortex, subthalamus, striatum, substantia nigra, and precerebellar and raphe nuclei, implying roles in motor functions.

The *locus ceruleus* is a compact group of neurons in the rostral pons ventrolateral to the fourth ventricle and dorsolateral to the pontine reticular nucleus (see Figs. 13-1*D* and 13-3). Locus ceruleus neurons contain *norepinephrine*, a catecholamine transmitter. The locus ceruleus has two prominent ascending projections; one projects to the thalamus and all areas of the cerebral cortex, including the hippocampus and amygdala, and the other projects to the periaqueductal gray and hypothalamus. Descending axons from the locus ceruleus terminate in the cerebellum, reticular formation, and spinal cord.

The substantia nigra of the midbrain consists of neurons that also use a catecholamine, dopamine, as a transmitter. The locus ceruleus and substantia nigra neurons contain *melanin-like pigment* (see Fig. 13-3), which permits the identification of these nuclei with the unaided eye in fresh brain stem sections.

Other Brain Stem Nuclei

There are numerous other brain stem nuclei that have some of the characteristics of reticular formation nuclei or are functionally closely related. A few are briefly described here; they are considered in detail elsewhere.

The *red nucleus* (see Fig. 13-1*F*) occupies the central midbrain tegmentum and is frequently included as a reticular formation component. However, its functional role is exclusively related to contralateral motor regulation; it receives inputs predominantly from the ipsilateral motor cortex and contralateral cerebellum. It forms contralateral rubrobulbar and rubrospinal tracts, which terminate on brain stem reticular neurons, cranial nerve motor nuclei, and spinal cord. Another major projection through the central tegmental tract (see stippling in Figs. 13-1 and 13-3) courses to the ipsilateral inferior olivary nucleus (see Fig. 13-1*A,B*), which projects to the contra-

lateral cerebellum. These circuits provide a basis for interactions between the reticular formation and motor activities.

The *inferior olivary nucleus* (see Fig. 13-1*A,B*) is a prominent structure in the ventral tegmentum of the medulla; its relation to the motor system was described earlier. Olivocerebellar axons cross through the medullary reticular formation, but virtually all terminate in the contralateral cerebellum.

Three reticular formation nuclei in the caudal brain stem that are collectively called the *precerebellar reticular nuclei* (see triangles in Fig. 13-1) also have restricted functional roles related to the motor system; they project almost exclusively to the cerebellum. Two of these—the *paramedian reticular* (see Fig. 13-1*A,B,C*) and the *pontine reticulotegmental nuclei* (see Fig. 13-1*D,E*)—form a parallel column of neurons between the medial column of the reticular formation and the brain stem raphe. The third precerebellar reticular nucleus, the *lateral reticular nucleus* (see Fig. 13-1*A,B*), is a conspicuous cluster of large neurons in the lateral tegmentum of the medulla between the spinal trigeminal system and the inferior olivary nucleus.

The *periaqueductal gray* (see Fig. 13-1*E,F*) surrounds the cerebral aqueduct and consists of numerous axonal terminals and subgroups of neurons containing many different transmitters and peptides. The larger neurons have extensive dendrites that intermingle with the midbrain reticular formation. The periaqueductal gray receives its most prominent inputs from the hypothalamus, frontal cortex, and limbic structures (see Fig. 13-4); afferents also originate from the parabrachial and other reticular nuclei, nucleus tractus solitarius, and the ascending sensory pathways. Periaqueductal gray provides reciprocal projections to all of these structures and to the serotonergic raphe nuclei of the medulla (see Fig. 13-5). The periaqueductal gray, in concert with the reticular formation, plays a key role in integrating limbic, autonomic, and nociceptive actions.

Clinical Implications of Functional Interactions of the Reticular Formation

Reticular formation morphology, circuitry, and chemical transmission characteristics provide an important matrix for the integration of nervous system activities. The control of vital functions such as regulation of blood pressure depend on integrated circuitry among the cortex, diencephalon, reticular formation, spinal cord, and peripheral tissues. The inputs converge on the reticular formation from almost all somatic and visceral sensory pathways and from the cortex, hypothalamus, striatum, limbic structures, and spinal cord. Moreover, the reticular formation provides prominent projections to these same structures. Because of this integrated circuitry, particular components of the reticular formation influence many neural functions (see Figs. 13-4 and 13-5), and certain reticular formation regions or nuclei are closely associated with important neural responses.

☐ The reticular formation participates in arousal

Early studies demonstrated that stimulation of the reticular formation evoked changes in cortical activity characteristic of the arousal induced by sensory stimulation. Later studies revealed that the ascending spinothalamic and trigeminothalamic tracts provided collateral input to the lateral column of the reticular formation (see the parvicellular nucleus in Fig. 13-1*B,C*). The lateral column neurons project to the medial column in the medulla and pons (i.e., ventral reticular, gigantocellular, pontine reticular nuclei; see Fig. 13-1*A–E*). The medial column forms a prominent ascending projection that follows the central tegmental tract (see stippling in Fig. 13-1 and Fig. 13-3), terminating in the hypothalamus and the intralaminar (i.e., centromedian) nuclei of the dorsal thalamus.

Thalamocortical fibers relay activation to all areas of the cerebral cortex (Fig. 13-4). This projection is the *ascending reticular activating system* (ARAS) that provides several important functions. The flow of sensory stimuli through ARAS activates the hypothalamic and limbic structures, which regulate emotional and behavioral responses (e.g., response to pain). More important, the flow of sensory stimuli exerts facilitating effects on cortical neurons. Other examples of ARAS activity include the alerting responses to a sudden loud sound, a flash of light, smelling salts, or a splash of cold water in the face. Without cortical activation by ARAS, a person is less able to detect new specific stimuli, and the level of consciousness is diminished.

This discussion should not suggest that the reticular formation is the center of consciousness. Experimental and clinical observations imply that *consciousness*, which is a person's ability to be aware of self and the environment and to orient toward new stimuli, results from the integrated activities of a number of neural structures, including the reticular formation.

There are important clinical implications related to ARAS function. The reticular formation projections in the ARAS traverse the midbrain tegmentum (see Fig. 13-1*E,F*), and some of these projections follow the central tegmental tract (see Figs. 13-1, 13-3, and 13-4). Lesions of the midbrain can interrupt the ARAS, leading to *altered levels of consciousness or coma* due to the diminished facilitation of cortical neurons. Lesions frequently affecting the midbrain include cerebrovascular accidents (i.e., stroke) and head trauma.

Head trauma can induce increased intracranial pressure due to collection of blood (i.e., hematoma) between the skull and the brain or accumulation of edema fluid in the injured brain. Because the brain is encased in the skull, the *increased intracranial pressure* causes the medial aspect of the temporal lobe (i.e., uncus of the hippocampal gyrus) to herniate through the incisura of the tentorium. The midbrain passes from the posterior cranial fossa through the incisura to become continuous with the diencephalon. This situation is a medical emergency, because the herniating temporal lobe can exert pressure directly on the lateral aspect of the midbrain or interfere with its blood supply. These lesions often de-

stroy ARAS pathways in the midbrain, resulting in permanent coma or a *persistent vegetative state*.

Certain drugs and metabolic imbalances are thought to induce altered states of consciousness through their actions on the reticular formation. Cerebrovascular accidents related to the diminished blood supply to the brain stem may alter consciousness because of diminished oxygen supply to the reticular formation. Some anesthetics alter consciousness through their actions on the reticular formation while conduction of sensations by the somatic pain pathways to the cortex remains unaffected.

The molecular structure of *lysergic acid diethylamide* (LSD) is similar to serotonin. Available evidence suggests that LSD exerts its hallucinogenic effects by inhibition of the raphe serotonergic projections to the cortex and limbic system.

☐ Sleep is an active neural process requiring participation of the reticular formation

The reticular formation plays a prominent role in the elaboration of components of the *normal sleep cycle* through its circuitry with the cortex and diencephalon. Lesions or stimulations of certain regions of the hypothalamus and frontal lobe also affect sleep cycles.

The *slow wave component* of the sleep cycle is characterized by synchronized cortical activity. Data demonstrate that pontine raphe serotonergic neuron activities diminish as slow wave sleep progresses. Serotonergic neurons (see Fig. 13-2) also activate the *paradoxical sleep component*, which is characterized by *rapid eye movements (REM)*, diminished muscle tone, and a desynchronized electroencephalogram similar to that of the waking state. This activation involves the locus ceruleus (see Fig. 13-3), whose norepinephrine-containing terminals are distributed to the reticular formation and all parts of the cerebral cortex. Cholinergic mechanisms also appear to be involved in the activation of paradoxical sleep.

Destruction of the serotonergic neurons (see Fig. 13-2) or inhibition of serotonin synthesis produces insomnia. Lesions of the locus ceruleus and surrounding reticular formation alter the induction of the paradoxical or REM stage of the sleep cycle. The slow wave and paradoxical stages recur several times during a sleep period. The sleep cycle depends on a complex neurochemical circuitry linking particular components of the reticular formation and other neural structures.

☐ The raphe nuclei of the medulla modulate pain transmission

Axons of the ascending spinothalamic and trigeminothalamic tracts, which convey pain sensations from the visceral and body surface, provide collaterals to the lateral column of the reticular formation (see Fig. 13-5). This input becomes part of the ARAS. Also activated by nociceptive input are the serotonergic neurons of the raphe and medial column nuclei in the caudal brain stem; these include the *raphe magnus* and *gigantocellular nuclei* (see Fig. 13-1C,D). The axons of some of these serotonergic neurons distribute to the spinal trigeminal nucleus by means of reticulobulbar tracts, and others descend in the dorsolateral aspect of the spinal cord white matter by means of reticulospinal tracts to terminate in the dorsal horn (see Figs. 13-4 and 13-5). The *serotonin* released in the dorsal horn and spinal trigeminal nuclei modulates pain transmission by inhibiting sensory transmission neuron activities. This inhibition is accomplished partially through activation of *inhibitory interneurons that use enkephalin* as their transmitter. Evidence suggests that a descending noradrenergic pathway may also play a role in pain modulation.

Stimulation of the *periaqueductal gray* or the caudal brain stem raphe nuclei induces *analgesia*. Periaqueductal gray neurons project to the serotonergic neurons of the caudal brain stem and probably involve opiate and nonopiate mechanisms (see Fig. 13-5). In certain patients with intractable pain, brief activation of electrodes implanted in the periaqueductal gray can provide an analgesic effect that lasts for hours or longer.

☐ The reticular formation participates in the regulation of skeletal muscle tone, reflexes, and body posture

The reticular formation influences motor activities through its reciprocal connections with the red nucleus, substantia nigra, subthalamus, striatum, motor cortex, and cerebellum (see Figs. 13-4 and 13-5). The influences of the motor system over *alpha and gamma lower motor neuron activities* are largely conveyed by reticulospinal and reticulobulbar tracts that originate primarily in the medial column of the reticular formation. The pontine reticular nucleus (see Fig. 13-1C–E) projects to the ipsilateral spinal cord by way of the pontine reticulospinal tract and facilitates axial and limb extensors. Bilateral projections arise from the gigantocellular nucleus (see Fig. 13-1C–E), and by means of reticulospinal tracts, they primarily inhibit lower motor neurons innervating axial extensors but facilitate motor neurons that innervate limb flexors (see Fig. 13-4).

The inputs from reticular formation, vestibulospinal projections, and other motor system components are integrated to provide a continuously adapting background of muscle tone and body posture to facilitate voluntary motor actions.

☐ The reticular formation participates in the integration of conjugate eye movements

A portion of the medial column of the reticular formation in the pons called the *paramedian pontine reticular formation* (PPRF) overlaps the pontine reticular nucleus and integrates horizontal eye movements (see Fig. 13-1C). The PPRF receives inputs from the superior colliculus, vestibular nuclei, reticular formation, and the frontal eye fields. It projects primarily to the ipsilateral abducens nucleus and, by the medial longitudinal fasciculus, to the portion of the contralateral oculomotor nucleus that innervates the medial rectus muscle. The PPRF integrates horizontal *conjugate eye movements* in response to head and body position, reflex responses to light, and cortical activity. A group of neurons have been located in the rostral midbrain that regulate conjugate eye movements in the vertical plane.

☐ The reticular formation regulates vital visceral responses

Reticular formation neurons that participate in the regulation of cardiovascular, respiratory, and other visceral functions are intermingled with those serving the other functions described earlier. Terms such as "inspiratory center" reflect the observation of a physiologic response after a particular form of reticular formation stimulation rather than an anatomically defined cluster of neurons that serve only inspiration. Nonetheless, certain areas of the caudal brain stem reticular formation have been shown to influence visceral functions.

The evidence suggests that visceral sensory input to the reticular formation is relayed in a polysynaptic manner through collaterals from ascending spinal cord sensory pathways and the *nucleus tractus solitarius*. Other afferents related to visceral functions reach the reticular formation through descending fibers from the hypothalamus and limbic system, specifically the *dorsal longitudinal fasciculus, medial forebrain bundle*, and *mammillotegmental tract* (see Fig. 13-5).

Reticular formation neurons influencing *cardiovascular responses* are located primarily in the medulla. The cardiac rhythm and blood pressure are diminished after stimulation of the raphe and medial column nuclei, primarily the gigantocellular nucleus (see Fig. 13-1C). Stimulation of the parvicellular nucleus (see Fig. 13-1B,C) in the lateral column of the medullary tegmentum yields opposite effects. Axons from these "cardiovascular" neurons are thought to provide facilitory or inhibitory influences on parasympathetic preganglionic neurons in the medulla associated with the vagus nerve (i.e., nucleus ambiguus) and on the sympathetic preganglionic neurons of the thoracic spinal cord (i.e., intermediolateral nuclei).

Reticular formation neurons influencing *respiratory rhythms* are widely distributed in the brain stem. Inspiratory responses are obtained after stimulation of *gigantocellular nucleus* (see Fig. 13-1C), but expiratory responses are activated from the *parvicellular nucleus* (see Fig. 13-1B,C). These "respiratory centers" overlap the "cardiovascular centers" described previously. Another cluster of neurons influencing respiratory rhythms is located in the region of the *parabrachial nucleus* (see Fig. 13-1D) in the pons. Efferent fibers from these respiratory neurons directly or indirectly influence the activities of *preganglionic sympathetic* and *parasympathetic nuclei* in the brain stem and spinal cord and motor neurons associated with the phrenic and intercostal nerves. In some newborn infants, the immaturity of these respiratory circuits may alter neural control of respiratory rhythms, placing the child at risk for *sudden infant death syndrome.*

Caudal brain stem lesions that destroy neurons involved with the integration of cardiovascular and respiratory functions present life-threatening situations. For example, cerebrovascular accidents involving the intrinsic vasculature of the medulla or pons can destroy these neurons, resulting in distortions of cardiac and respiratory rhythms and death.

Increased intracranial pressure in the posterior cranial fossa can result in herniation of the cerebellar tonsils through the foramen magnum. In this situation, the pressure exerted on the caudal medulla diminishes its blood supply and interrupts the reticulospinal fibers that link the reticular formation with autonomic nuclei in the spinal cord. For patients with high cervical spinal cord lesions to survive the acute stages of their injury, prompt external support to maintain cardiac and respiratory functions is required.

Certain drugs, including anesthetics, can depress activity in these neurons, altering or interrupting cardiovascular and respiratory rhythms.

The hypothalamus plays a predominant role in maintaining homeostasis through the integration of somatic, visceral, and endocrine functions. It forms prominent functional connections with cortical areas, the olfactory system, and limbic system structures that generate emotions and behavioral patterns. The cortex, limbic system, and hypothalamus have prominent reciprocal connections with the reticular formation (see Figs. 13-4 and 13-5). These anatomic and functional interactions imply that some aspects of emotional display and elaboration of behavior are conveyed through reticular formation influences on autonomic nuclei in the brain stem and spinal cord. It is likely that reticular formation interactions with the striatum and motor neurons influence neurons that elaborate the body postures and muscle tone appropriate for emotional expression and behavior (see Fig. 13-4).

SELECTED READINGS

Brodal A. The reticular formation and some related nuclei. In Neurological anatomy in relation to clinical medicine, 3rd ed. New York: Oxford University Press, 1981:394.

Carpenter MB, Sutin J. Human neuroanatomy, 8th ed. Baltimore: Williams & Wilkins, 1983:315.

Hobson JA, Brazier MA. The reticular formation revisited: specifying function for a non-specific system. In International brain research organization monograph series, vol 6. New York: Raven Press, 1980.

Kinney HC, Burger PC, Harrell FE, Hudson RP. Reactive gliosis in the medulla oblongata of victims of sudden infant death syndrome. Pediatrics 1983;72:181.

Martin GF, Holstege G, Mehler WR. Reticular formation of the pons and medulla. In Paxinos G (ed). The human nervous system. New York: Academic Press, 1990:203.

Paxinos G, Törk I, Halliday G, Mehler WR. Human homologs to brainstem nuclei identified in other animals as revealed by acetylcholinesterase activity. In Paxinos G (ed). The human nervous system. New York: Academic Press, 1990:149.

Quattrochi JJ, Baba N, Liss L, Adrion W. Sudden infant death syndrome (SIDS): a preliminary study of reticular dendritic spines in infants with SIDS. Brain Res 1980;181:245.

Scheibel AB. The brain stem reticular core and sensory function. In Smith D (ed). Handbook of physiology: the nervous system III, vol 1. Bethesda: American Physiological Society, 1984:213.

Scheibel ME, Scheibel AB. On circuit patterns of the brain stem reticular core. Ann N Y Acad Sci 1961;89:857.

Clinical Correlation
DISORDERS OF THE AUTONOMIC NERVOUS SYSTEM

ROBERT L. RODNITZKY

Clinical symptoms of autonomic dysfunction can arise from abnormalities of autonomic pathways within the central or peripheral nervous systems. Mild autonomic dysfunction can occur in otherwise normal persons with advancing age. More substantial autonomic symptoms arise when there is structural disruption of autonomic pathways or when they are involved by degenerative or metabolic conditions.

Peripheral Nerve Dysfunction

Nerve fibers subserving autonomic function are easily damaged in diseases of the peripheral nerves. The peripheral neuropathies that are most likely to result in autonomic symptoms are those causing acute demyelination or involving small myelinated or unmyelinated fibers. Among the most common neuropathies in this category are those caused by diabetes, amyloidosis, or porphyria and the Guillian-Barré syndrome. In a neuropathy such as that associated with Friedreich's ataxia, in which the large fibers are reduced in number but small fibers are preserved, there are virtually no autonomic symptoms. Between these two extremes are a variety of neuropathies in which autonomic symptoms are present but which do not constitute the primary disability. Examples include those caused by alcohol abuse, nutritional deficiency, leprosy, chronic kidney failure, and acquired immunodeficiency syndrome.

Central Nervous System Disorders

Abnormalities of the brain or spinal cord can result in autonomic dysfunction. The autonomic dysfunction can be the primary clinical abnormality or part of a wider syndrome affecting many parts of the nervous system. *Progressive autonomic failure* is a primary autonomic disorder in which there is gradual development of symptoms such as bladder dysfunction, decreased tear production, erectile dysfunction, reduced sweating, and orthostatic hypotension (i.e., decreased blood pressure on standing). The primary pathologic finding is a loss of sympathetic preganglionic cell bodies in the intermediolateral column of the spinal cord. This same syndrome may occur with more diffuse involvement of the central nervous system and other nonautonomic symptoms. If autonomic failure is associated with signs of Parkinson's disease, it is assumed that nigrostriatal degeneration has also occurred, and the condition is called the *Shy-Drager syndrome*. If, in addition, other nervous system structures are involved, such as the cerebellum, the entire conglomerate of autonomic, parkinsonian, and cerebellar symptoms is referred to as the *syndrome of multiple system atrophy*.

Among spinal cord disorders, severe transverse lesions are most likely to be associated with prominent autonomic dysfunction. Such lesions at or above the midthoracic spinal cord level typically result in severe orthostatic hypotension. In such cases, there is inadequate control of the sympathetic outflow from the spinal cord below the lesion. This results in inadequate reflex constriction of blood vessels in response to the normal drop in blood pressure associated with assuming the standing position. Persons with transverse spinal cord lesions are typically paraplegic and unable to stand on their own, but caution must be exercised in moving them passively to the upright position because of the risk of severe hypotension.

Bladder function can also be affected by a variety of spinal cord lesions. A lesion involving the conus medullaris or the cauda equina results in an autonomous neurogenic bladder characterized by an inability to initiate micturition and marked urinary retention. In this circumstance, the afferent or efferent arc of the micturition reflex, both of which travel through cauda equina parasympathetic nerves, or the center for micturition located in the second, third, or fourth segments of the sacral spinal cord have been involved by the offending lesion. With

236

spinal cord lesions above the sacral parasympathetic center for bladder control, the bladder reflex remains intact but cannot be controlled by descending inhibitory influences from supraspinal centers. In this situation, the voiding reflex goes unchecked, and there is frequent, spontaneous, and precipitous micturition. Patients with spinal cord lesions of this type can be taught to precipitate this reflex voluntarily by stroking the skin in sacral-innervated areas or by gently compressing the bladder through the abdomen. Using this technique, they regain some control over the timing of voiding.

Abnormalities of Neurotransmitter Metabolism

A deficiency or an excess of neurotransmitters subserving autonomic function can occur. A deficiency of dopamine β-hydroxylase, the enzyme responsible for the final reaction in the synthesis of norepinephrine, has been documented in some persons and appears to be inherited as an autosomal recessive trait. Marked reduction in plasma and cerebrospinal fluid norepinephrine levels can be documented in these patients, who often suffer from symptoms of sympathetic autonomic dysfunction, especially orthostatic hypotension. The opposite circumstance prevails in patients with a catecholamine-producing tumor of the adrenal gland known as a pheochromocytoma. In this condition, excessive production of norepinephrine results in hypertension.

Plasma norepinephrine levels can be measured. In the case of pheochromocytoma, the levels of norepinephrine and several other catecholamines measured over a 24-hour period are typically elevated. Plasma norepinephrine levels can be useful for other purposes. The response of plasma norepinephrine levels to standing has been used as a means of differentiating preganglionic from postganglionic sympathetic failure. Because most plasma norepinephrine is thought to be derived from sympathetic postganglionic nerve terminals, resting levels of plasma norepinephrine are expected to be normal in a preganglionic sympathetic lesion. When these persons stand, the otherwise normal postganglionic neuron cannot be activated to produce the normal rise in norepinephrine with the expected spillover into the plasma. In postganglionic sympathetic lesions, even the resting level is diminished because of inadequate numbers of norepinephrine-releasing nerve terminals. In this circumstance, the elevation in plasma norepinephrine levels after standing is also less than normal, depending on how widespread the postganglionic dysfunction is.

Clinical Tests Useful in Documenting Autonomic Dysfunction

Recording the blood pressure in the supine and standing positions while determining the concurrent heart rate is a simple and extremely useful test for sympathetic autonomic dysfunction. If after two minutes of standing the systolic blood pressure has fallen by 30 mm Hg or more, a diagnosis of orthostatic hypotension is justified. Absence of acceleration in the heart rate in the face of this hypotension further confirms the presence of autonomic dysfunction.

A variety of provocative tests can be performed to assess the sympathetic mechanisms that control heart rate and blood pressure. These include measuring the extent to which a sustained hand grip, immersing a hand in ice water, or performing difficult mental arithmetic stimulate the sympathetic outflow and elevate the heart rate and blood pressure. One mechanism of evaluating parasympathetic control of heart rate is to have the patient briefly execute the Valsalva maneuver (i.e., forcefully attempt to expel breath against a closed glottis). This results in a transient increase in intrathoracic pressure with resultant decreased cardiac filling and a lowering of blood pressure. The normal baroreflex response to lower blood pressure should be an elevation in heart rate. On release, there should be an overshoot of the normal blood pressure and slowing of the heart rate. Absence of these reactions is a sign of parasympathetic dysfunction. The response to the Valsalva maneuver is actually much more complex than this and can be divided into three or four component phases, but in its simplest form, it can be a useful screening test for parasympathetic function.

Sweating or sudomotor function is a sympathetic function that can be assessed by applying Quinazarin-red powder to the skin. When the body temperature is elevated, the powder, which is initially white, becomes red wherever it is in contact with perspiration. This can be particularly striking in conditions such as loss of sweating (i.e., anhidrosis) on one side of the face.

Extremely localized autonomic functions can be measured. A classic example is the assessment of sympathetic innervation of the pupil and eyelid. When this innervation is interrupted by a lesion in the sympathetic chain, a syndrome consisting of ptosis (i.e., drooping) of the eyelid, meiosis (i.e., a smaller than normal pupil), and anhidrosis on the same side of the face may occur. Collectively, these neurologic signs are known as *Horner's syndrome*. A series of pharmacologic tests can determine whether Horner's syndrome is present, and if so, whether it is due to involvement of the first- or second-order neuron in the sympathetic chain (i.e., preganglionic lesion) or the third-order neuron (i.e., postganglionic lesion). An initial test that can determine if Horner's syndrome is present involves the instillation of a dilute solution of cocaine into the conjunctival sac of the involved eye and the opposite normal eye. Cocaine inhibits re-uptake of synaptic norepinephrine. In the normal eye, this causes enhanced sympathetic stimulation of the iris dilator muscle, resulting in enlargement of the pupil. On the side with sympathetic dysfunction, there has been little or no synaptic norepinephrine elaborated, and the pupil dilates considerably less than that on the normal side. This effect occurs regardless of the location of the lesion within the sympathetic pathway.

The next step differentiates a preganglionic lesion from a postganglionic lesion. Paredrine, an amphetamine derivative, is instilled into both eyes. Because amphetamine enhances the release of norepinephrine from sympathetic terminals, the expected normal response is pupillary dilatation, but this only occurs if there is a healthy third-order (i.e., postganglionic) neuron. A normal response to Paredrine in a patient with Horner's syndrome suggests that the causative sympathetic lesion is located in the first- or second-order neuron. An abnormal response to Paredrine indicates that the pathology is in the postganglionic neuron.

SELECTED READINGS

Kahn R. Autonomic nervous system testing. Muscle Nerve 1992; 15:1158.
Mcleod JG. Autonomic dysfunction in peripheral nerve disease. Muscle Nerve 1992;15:3.

Neuroscience in Medicine, edited by P. Michael Conn.
J. B. Lippincott Company, Philadelphia © 1995

Chapter 14

The Trigeminal System

HAROLD H. TRAURIG
BRUCE E. MALEY

The *trigeminal nerve* (i.e., cranial nerve V) provides sensory innervation to the face and structures in the oral and nasal cavities, and its motor component innervates the muscles of mastication. Discriminatory tactile, general or light tactile, proprioceptive, thermal, and pain sensory modalities are conveyed to the trigeminal nuclei in the brain stem. The axons from the sensory trigeminal nuclei contribute to important reflex circuits and relay integrated sensory information to the thalamus. The thalamus provides further integration and relays information to the cerebral hemispheres.

The trigeminal system frequently is involved in several important clinical conditions because the peripheral and central components of the trigeminal system have extensive anatomic distributions in the face, cranial cavity, and brain stem.

The Trigeminal Nerve and Its Three Divisions

The trigeminal nerve central root axons form a large sensory component and a smaller, superiorly positioned motor portion (Figs. 14-1 and 14-2). The axons exit the lateral aspect of the midpons, interdigitating with the fibers of the middle cerebellar peduncle, and then course obliquely and superiorly through the posterior cranial fossa to the apex of the petrous portion of the temporal bone. The root then crosses the petrous ridge into the middle cranial fossa, passing inferior to a ligamentous (attached) portion of the tentorium and the superior petrosal sinus, and becoming continuous with the trigeminal ganglion. The trigeminal ganglion and the origins of the three divisions of the trigeminal nerve—ophthalmic

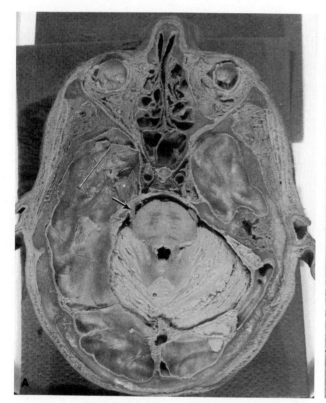

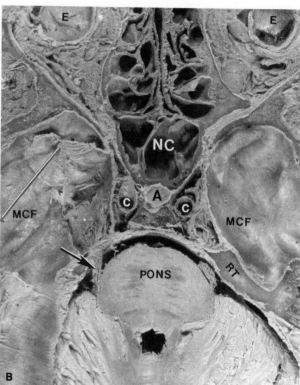

Figure 14-1

(**A**) Horizontal section of the head, showing the position of the trigeminal ganglion (*white star*) and root (*arrow*) on the anterior slope of the petrous portion of the temporal bone in the middle cranial fossa (MCF). The anterior direction is toward the top of the picture. The cut is inclined superiorly about 1 cm off the horizontal plane toward the right. (**B**) A close view of the central portion of **A**. The section cut through the apex of the tentorium reveals the pons and cerebellum in the posterior cranial fossa. The trigeminal ganglion is located in a dural sleeve; in this preparation, the dural sleeve is opened, reflected anteriorly, and held by the pin, revealing the trigeminal ganglion (*star*). The trigeminal ganglion underlies the medial inferior aspect of the temporal lobe, which has been removed. The trigeminal root (*arrows*) exits the pons, crosses the petrous ridge, and passes inferior to the superior petrosal venous sinus to join the trigeminal ganglion. *A*, adenohypophysis in the sella turcica; *C*, internal carotid arteries are evident just lateral to sella turcica as they course through the cavernous sinuses; *E*, globes of the eyes; *NC*, nasal cavity; *RT*, ridge of the petrous portion of the temporal bone.

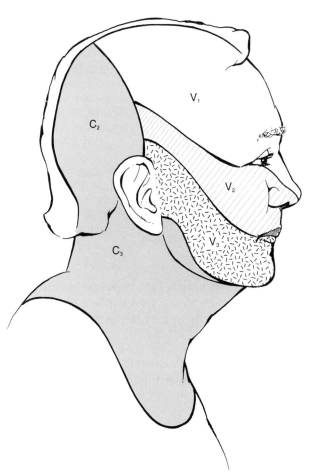

Figure 14-2
Sensory fields of the ophthalmic (V_1), maxillary (V_2), and mandibular (V_3) divisions of the trigeminal nerve. The sensory fields of the upper cervical spinal nerves (C2, C3) are also shown. The skin of the auricle is innervated by C2, C3, V_3, and the vagus nerve. The sensory innervation of the external auditory canal and lateral (i.e., external) surface of the tympanic membrane is conveyed by the auriculotemporal branch of V_3 and the auricular branch of the vagus nerve. The medial surface of the tympanic membrane is innervated by the glossopharyngeal nerve.

(V_1), maxillary (V_2), and mandibular (V_3)—are located in a dural pocket (i.e., Meckel's cave) on the anterior slope of the petrous bone (see Fig. 14-1).

The central sensory axons of the mandibular division are located posterolateral in the root, and ophthalmic division axons are in a anteromedial position. The sensory axons originate from several subgroups of unipolar neurons in the trigeminal ganglion.

Near its exit point from the brain stem, the trigeminal root is anatomically closely related to the superior cerebellar artery and the superior petrosal vein (i.e., vein of Dandy). Compression or irritation of the root by these vessels could evoke the pain syndrome known as trigeminal neuralgia (i.e., tic douloureux).

The *trigeminal ophthalmic division* passes forward from the trigeminal ganglion in the lateral wall of the cavernous sinus and then through the superior orbital

fissure to enter the orbit where it divides into its peripheral branches. V_1 conveys sensory innervation from the supratentorial dura, globe, cornea, upper eyelid, regions of the face and scalp, and the mucosa of the ethmoidal, sphenoidal and frontal sinuses (see Fig. 14-2).

The *trigeminal maxillary division* passes forward from the trigeminal ganglion on the floor of the middle cranial fossa and exits the cranial cavity through the foramen rotundum. V_2 then courses through the pterygopalatine fossa, which is bounded by the palatine bone medially, the pterygoid process posteriorly, and the maxilla anteriorly; the fossa also contains the terminal branches of the maxillary artery and the parasympathetic pterygopalatine ganglion. Branches of V_2 pass through the pterygopalatine ganglion and probably provide some sensory collateral terminals to the autonomic neurons that innervate visceral and vascular structures of the nasal and maxillary regions. Peripheral sensory fibers supply maxillary teeth, gingiva, and sinuses and part of the nasal cavity and palate. Cutaneous fibers follow the infraorbital nerve to innervate skin of the midface, including the lower eyelid and upper lip (see Fig. 14-2).

The *trigeminal mandibular division* consists of sensory and motor fibers. It courses inferiorly from the trigeminal ganglion and then exits the cranial cavity through the foramen ovale. Near the base of the skull, the parasympathetic otic ganglion is attached to the medial aspect of V_3. Sensory innervation is conveyed from the dura, mandibular teeth and gingiva, mucosa of the floor of the oral cavity, and the anterior two thirds of the tongue (except taste). Cutaneous branches innervate skin of the lower lip and cheek (see Fig. 14-2).

The trigeminal motor root innervates the muscles of mastication: lateral and medial pterygoids, temporalis, masseter, mylohyoid, and anterior belly of the digastric. It also innervates the tensor veli palatini and tensor tympani. It is important to note that the skeletal muscles of facial expression and the tongue are innervated by the facial and hypoglossal nerves, respectively.

The trigeminal (i.e., gasserian or semilunar) ganglion contains most of the primary sensory cell bodies associated with the sensory trigeminal nerves (Figs. 14-1 and 14-3). Other primary sensory trigeminal neurons are located in the trigeminal mesencephalic nucleus. The unipolar sensory neurons vary in size, and a significant portion give rise to myelinated axons. Several subsets of neurons have been described in a number of species based on their content of peptides (e.g., substance P, calcitonin gene-related peptide, galanin), some of which may serve as transmitters.

□ The spinal trigeminal tract in the brain stem consists of a subset of central axons of trigeminal nerve primary sensory neurons

Trigeminal central axons, which form the spinal trigeminal tract, enter the pons with the trigeminal root and

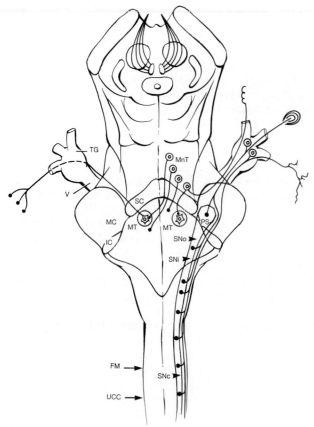

Figure 14-3
Posterior view of the brain stem with the cerebellum removed and the locations of the major components of the trigeminal system illustrated. The motor component is represented on the left, and sensory components are on the right. The trigeminal ganglion (TG) has three major divisions. The foramen magnum (FM), mesencephalic nucleus (MNT), trigeminal motor nucleus (MT), principal sensory nucleus (PS), upper cervical cord (UCC), and spinal tract of the trigeminal (ST) are depicted. The spinal nucleus (SN) extends through the medulla into the upper cervical cord and consists of the subnuclei oralis (o), interpolaris (i), and caudalis (c). Compare with Figure 14-4.

form a well-defined bundle in the lateral tegmentum. They extend caudally through the medulla into the upper cervical cord, where they interdigitate with the dorsolateral fasciculus (Figs. 14-3 and 14-4*A–D*). The spinal trigeminal nucleus lies just medial to the spinal trigeminal tract throughout its course. Many spinal trigeminal tract axons are well-myelinated, large-diameter fibers that bifurcate, sending synaptic terminals to neurons of the principal sensory trigeminal nucleus and, through descending branches, to all subnuclei of the spinal trigeminal nucleus. These fibers are probably associated with mechanoreceptors and tactile sensations. Most of the spinal trigeminal tract axons are small-diameter, unmyelinated axons derived from primary sensory neurons in the trigeminal ganglion that have been associated with thermal and nociceptive sensations. These axons descend in the spinal trigeminal tract to terminate in the spinal trigeminal nucleus (predominantly in subnucleus

caudalis). Some of these axons descend as far caudally as the third cervical level before terminating (Figs. 14-3 and 14-4*A*).

Primary sensory axons associated with the facial, glossopharyngeal, and vagus nerves also join the spinal trigeminal tract and terminate in the spinal trigeminal nucleus. The sensory fibers following these cranial nerves innervate receptors associated with skin of the external auditory canal, tympanic membrane, and mucosa of the middle ear cavity (see Fig. 14-2).

Axons in the spinal trigeminal tract maintain a precise arrangement in their descending course. Those originating from V_1 are most ventral; V_2 axons are intermediate; and V_3 axons, with those from the facial, glossopharyngeal, and vagus nerves, are positioned dorsally in the tract.

☐ Trigeminal nuclei extend from the midbrain through the upper levels of the cervical spinal cord

The *spinal trigeminal nucleus* is the largest of the trigeminal nuclear components, extending as a continuous column of neurons from midpontine to the upper levels of the cervical spinal cord (see Figs. 14-3 and 14-4*A–D*). The spinal trigeminal nucleus is composed mostly of small neurons. Three subnuclei are recognized: *subnucleus oralis* (see Fig. 14-4*D*), which is anatomically continuous with the principal sensory trigeminal nucleus (see Fig. 14-4*E,F*); *subnucleus interpolaris* (see Fig. 14-4*B*) in the medulla; and *subnucleus caudalis* (see Fig. 14-4*A*), which overlaps the substantia gelatinosa in the dorsal horn of the upper cervical spinal cord (see Fig. 14-3). The detailed cytomorphology, circuitry, and functions of the trigeminal spinal nucleus, especially the caudal portion, are similar to those of the substantia gelatinosa in the dorsal horn of the cervical spinal cord.

The ipsilateral face and head are represented in the spinal trigeminal nucleus in an inverted orientation, with the jaw posterior and the forehead anterior. The rostral portion of the nucleus, the subnucleus oralis (see Fig. 14-4*D,E*), receives stimuli originating predominantly in the nasal and oral cavities; the caudal component of the spinal trigeminal nuclear complex, the subnucleus caudalis (see Fig. 14-4*A*), which overlaps the dorsal horn of upper cervical spinal cord, is functionally related primarily to nociception from the more posterior aspects of the face. This anatomic arrangement could explain the onionskin pattern of sensory loss occasionally seen as a result of lesions affecting the caudal medulla and upper cervical spinal cord. For example, a tumor exerting pressure on dorsolateral aspect of the upper cervical cord may interrupt pain and thermal sensations from the posterior portions of the ipsilateral face but spare these sensations in the perioral region (see Fig. 14-3).

The principal sensory trigeminal nucleus receives primary sensory input activated by fine and general tactile and pressure sensors in the sensory fields innervated by all three trigeminal nerve divisions (see Figs. 14-3 and

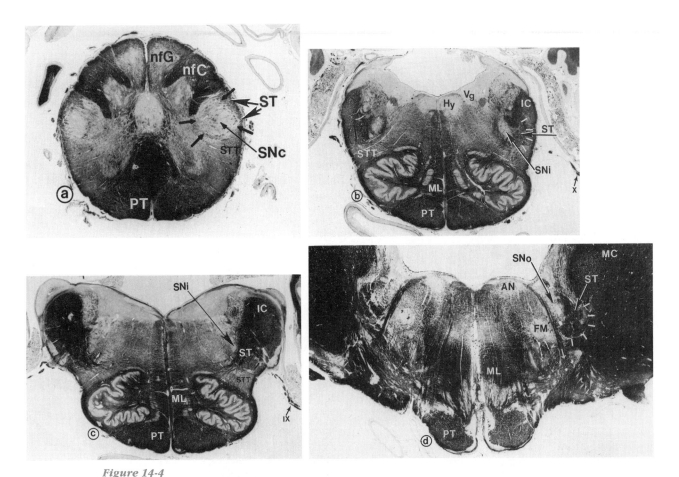

Figure 14-4

Brain stem sections demonstrate the components of the human trigeminal system. Sections are from the same brain and were photographed at the same original magnification (×10). (**A**) Medulla-cervical spinal cord junction. The spinal trigeminal tract (ST), with the dark-staining fibers indicated by the large arrows, at this level consists mostly of descending, small-diameter fibers that convey pain from the ipsilateral face. ST fibers are the central processes of primary sensory neurons whose cell bodies are in the trigeminal ganglion. These ST fibers interdigitate with the dorsolateral fasciculus in the upper cervical spinal cord and terminate on the neurons of trigeminal spinal nucleus pars caudalis (SNc), outlined by the small arrows. The SNc is prominent at this level and merges with the substantia gelatinosa of the upper cervical spinal cord. These neurons give rise to axons that cross the midline of the upper cervical cord and medulla to collect as the ventral trigeminothalamic tract and ascend to the thalamus. The pyramidal tract (PT), nucleus, and faciculus gracilis (nfG), and nucleus and faciculus cuneatus (nfC) are also shown. The spinothalamic tracts convey pain, temperature, and general tactile sensations from the contralateral body to the thalamus (STT). Compare with Figures 14-3 and 14-5. (**B**) Medulla at the level of the vagal (Vg) and hypoglossal (Hy) nuclei. The ST and SN are located in the lateral tegmentum. The lateral extent of the descending fibers of the ST is marked by the white bars. The spinal trigeminal nucleus pars interpolaris (SNi) is the light-staining area just medial to the ST. The medial lemniscus (ML) and inferior cerebellar peduncle (IC) are also shown. The spinothalamic tracts convey pain, temperature, and general tactile sensations from the contralateral body to the thalamus (STT). Rootlets of the vagus nerve exit the lateral aspect of the brain stem at this level. (**C**) Rostral medulla, showing the locations of the descending fibers of the ST and SNi embedded in the medial aspect of the IC. The ST and SNi are obscured by the crossing olivocerebellar fibers, which join the IC peduncle. The spinothalamic tracts convey pain, temperature, and general tactile sensations from the contralateral body to the thalamus (STT). Rootlets of the glossopharyngeal nerve exit the lateral aspect of the brain stem at this level (IX). (**D**) Pons-medulla junction, showing the location of the ST (*white bars*) as a distinct bundle of descending fibers embedded in the middle cerebellar peduncle (MC) and trigeminal spinal nucleus pars oralis (SNo) just medial to the ST. The emerging root fibers of the facial nerve (*white arrow*) separate the SNo from the facial motor nucleus (FM). The abducens nucleus (AN) is also depicted. (**E**) Pons at the level of the trigeminal motor nucleus (MT) and principal sensory trigeminal nucleus (PS). The PS merges with the SNo at this level. The trigeminal root fibers (V) are associated with the trigeminal nuclei. (**F**) Pons at the level of the locus ceruleus (LC), showing the rostral extent of the MT and PS. The mesencephalic tract and nucleus (MNT, *small arrows*) are just lateral to the superior cerebellar peduncle (SP). The location of the dorsal trigeminothalamic tract (DTT) is indicated in the tegmentum near the floor of the fourth ventricle. The fibers of the ventral trigeminothalamic tract (VTT) accompany the medial lemniscus (ML) and spinothalamic tracts (STT). The DDT and VTT ascend to the dorsal thalamus. The spinothalamic tracts convey pain, temperature, and general tactile sensations from the contralateral body to the thalamus. Compare with Figure 14-5.

continued

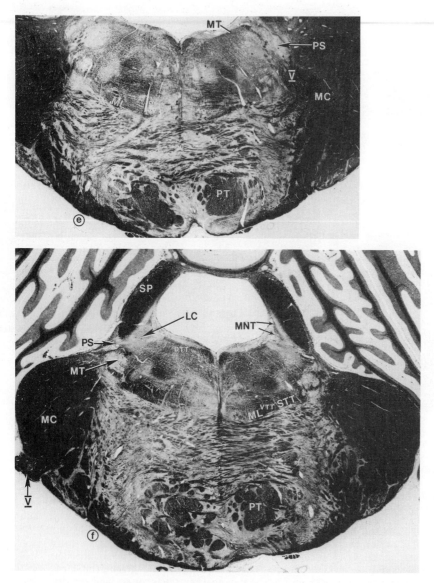

Figure 14-4 *(Continued)*

14-4*E,F*). Anatomically, the principal sensory nucleus is continuous caudally and overlaps to some extent with the subnucleus oralis of the spinal trigeminal nucleus (see Fig. 14-4*E*).

The separation of sensory modalities in the trigeminal system nuclei is not as sharp as once conceived; nevertheless, most neurons of the principal sensory trigeminal nucleus respond to mechanical stimulation in a manner suggesting perception of fine (i.e., discriminatory) tactile sensations. Moreover, at least some neurons in all sensory trigeminal nuclei respond when mechanical, thermal, or nociceptive stimuli are applied to their receptive fields or primary sensory input fibers. Some trigeminal neurons respond to more than one modality, and some nuclear subdivisions respond more prominently to a particular sensory modality. For example, tactile stimuli corresponding to light touch or deflection of hair on the skin

activate neurons in the principal sensory trigeminal nucleus and in all subnuclei of the spinal trigeminal nucleus. Most neurons in the principal sensory nucleus and the spinal trigeminal subnucleus oralis are activated by stimuli applied to well-defined sensory fields and conveyed by large-diameter, well-myelinated primary sensory fibers associated with rapidly adapting mechanoreceptors (see Figs. 14-3 and 14-4*D–F*).

Neurons of spinal trigeminal subnucleus interpolaris are thought to project trigeminal sensory activation to the cerebellum and brain stem reticular formation (see Fig. 14-4*B,C*). The spinal trigeminal subnucleus caudalis, which extends from caudal medulla and overlaps the dorsal horn through the third cervical level of the spinal cord, is most sensitive to facial pain (i.e., nociceptive) and thermal stimuli (see Figs. 14-3 and 14-4*A*). Prominent internuncial connections between the trigeminal sensory nuclei have been revealed, implying that complex integration of sensory modalities occurs in these structures.

The *trigeminal mesencephalic nucleus* extends from midpontine levels to the rostral midbrain and lies just lateral to the periaqueductal gray (see Figs. 14-3 and 14-4F). It consists of unipolar primary sensory neuron cell bodies receiving input through large-diameter, myelinated fibers from the proprioreceptors (i.e., pressure and stretch receptors) of the orofacial regions. This nucleus is the only example of primary sensory neuron cell bodies located in the central nervous system. The trigeminal mesencephalic nucleus, by means of its closely associated tract of the trigeminal mesencephalic nucleus (see Fig. 14-4F), provides bilateral input to the trigeminal motor nuclei, the surrounding reticular formation, and to the cerebellum through the superior cerebellar peduncle (see Fig. 14-3). Evidence suggests that most of the proprioceptive stimuli originating from extraocular muscles is conveyed by primary sensory neurons, whose cell bodies are in the trigeminal ganglion.

The *trigeminal motor nucleus* is a well-defined cluster of neurons located in the lateral tegmentum of the pons, ventromedial to principal sensory trigeminal nucleus; it extends from the abducens nucleus caudally to the level of the superior colliculus (see Figs. 14-3 and 14-4E,F). The trigeminal motor nucleus consists mostly of large and some small multipolar neurons; the somas of the larger neurons are almost completely covered with axosomatic synaptic contacts, and the smaller neurons are probably interneurons. The large motor neurons are the lower motor neurons for the skeletal muscles that are innervated by the trigeminal nerve and participate in important reflexes and other responses. Cortical control (i.e., upper motor neurons) of skeletal muscles innervated by the trigeminal motor neurons (i.e., lower motor neurons) originates from the face region of the precentral gyrus and other cortical motor areas. There are about equal numbers of ipsilateral and contralateral cortical projections (i.e., corticobulbar tracts) to the trigeminal motor nuclei and surrounding reticular formation. Direct corticobulbar tracts innervate trigeminal motor neurons, and indirect input originates from internuncial neurons in the surrounding reticular formation and from the hypothalamus. Diffuse connections from limbic structures through the reticular formation with other cranial nerve motor nuclei activate the trigeminal motor nucleus to produce facial expressions in response to emotion.

Extensive inputs to the trigeminal motor nucleus, usually through internuncial connections, are derived from all sensory trigeminal nuclei and from the sensory nuclei of other cranial nerves. Sensory and motor responses are integrated through the reticular formation and the medial longitudinal fasciculus to provide for reflex activities such as chewing swallowing and speaking.

The input to the motor trigeminal nucleus from the trigeminal mesencephalic nucleus is direct and may form the sensory limb of a two-neuron reflex circuit that could be the basis for the jaw jerk reflex (see Fig. 14-3).

Functions and Malfunctions of the Trigeminal System

☐ Nociception in the trigeminal system involves the spinal trigeminal nucleus and the serotonergic neurons of the reticular formation

The spinal trigeminal nucleus, specifically the subnucleus caudalis, provides for integration and relay of nociceptive stimuli to higher centers. Nociceptive stimuli, relayed through the reticular formation and the thalamus (Fig. 14-5), reach limbic system structures; these connections are thought to evoke the disagreeable characteristics of pain.

Small-diameter, unmyelinated axons in the spinal trigeminal tract end mostly in spinal trigeminal subnucleus caudalis (see Figs. 14-3 and 14-4A). These terminals contain substance P, enkephalin, and other putative transmitters that have been associated with nociception (Fig. 14-6). Enkephalinergic and serotonergic neurons, located in the midbrain periaqueductal gray and nucleus raphe magnus of the reticular formation, provide input to the trigeminal sensory nuclei. These neurons are thought to play a prominent role in modulation of pain transmission (see Chap. 13).

☐ Ascending trigeminal system pathways project to the dorsal thalamus

The principal sensory trigeminal nucleus neurons convey tactile and mechanical information through axons that join the contralateral medial lemniscus and ascend to the ventral posteromedial nucleus of the dorsal thalamus; some axons form an ipsilateral ascending pathway, located in the dorsal tegmentum (see Fig. 14-5). These bundles of axons are the ventral and dorsal trigeminothalamic tracts, respectively (see Fig. 14-4F). The related thalamocortical axons project to the postcentral gyrus and other cortical areas through the posterior limb of the internal capsule. The cortical projections provide conscious appreciation of the character and precise location of tactile and proprioceptive sensations originating in the trigeminal sensory field.

Axons from the spinal trigeminal nucleus, especially the subnucleus caudalis and surrounding reticular formation, form an ascending pathway that transmits pain, thermal, and some general tactile (i.e., light touch) sensations to the contralateral thalamus, which relays information to the cerebral cortex. These axons cross the midline through the upper cervical cord and tegmentum of the medulla, joining the medial lemniscus and spinothalamic tracts in the pons and terminating in the ventral posteromedial thalamic nucleus (see Fig. 14-5). At the midpontine and more rostral levels, the ascending tri-

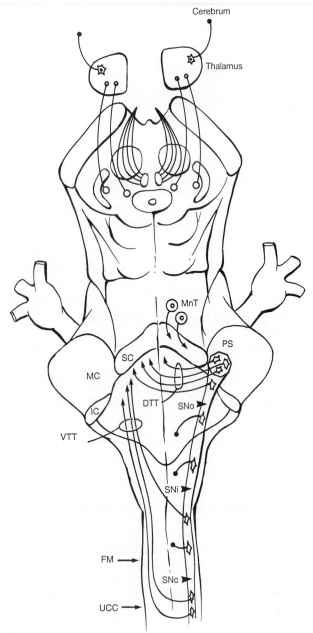

Figure 14-5
Representation of the ascending pathways of the trigeminal system. The locations of the dorsal (DTT) and ventral trigeminothalamic (VTT) tracts are also indicated in Figure 14-4F. *FM*, foramen magnum; *IC*, inferior cerebellar peduncle; *MC*, middle cerebellar peduncle; *Mnt*, mesencephalic tract and nucleus; *PS*, principal sensory trigeminal nucleus; *SC*, superior cerebellar peduncle; *SNc*, spinal nucleus pars caudalis; *SNi*, spinal nucleus pars interpolaris, *SNo*, trigeminal spinal nucleus pars voralis; *UCC*, upper cervical cord.

geminal pathways originating from the contralateral principal sensory and spinal trigeminal nuclei are in anatomic proximity and are collectively referred to as the ventral *trigeminothalamic tract* or *trigeminal lemniscus* (see Fig. 14-4F). All sensory modalities, including taste, are

integrated first in the ventral posteromedial nucleus of the thalamus before projection to the postcentral gyrus.

☐ Trigeminal peripheral sensory and motor distributions have important functional and diagnostic implications

The peripheral trigeminal sensory distribution has been described, but several points deserve emphasis. As indicated in Figure 14-2, the forehead, anterior scalp, upper eye lid, bridge of the nose, and cornea are innervated by branches of V_1. The upper lip, alar region of the nose, most of the cheek, and the lower eye lid are innervated by V_2. V_3 innervates the lower lip, chin, and mandibular region. However, a significant portion of the mandibular region, the posterior scalp, and neck is innervated by branches of C2–C3 spinal nerves.

The boundaries of the sensory regions innervated by each division of the trigeminal nerve on the face and scalp are sharp and display little overlap with one another or across the midline. In contrast, there is considerable overlap in the sensory regions (dermatome patterns) innervated by the spinal nerves on the body surface.

A discussion of the causes of headache are beyond the scope of this chapter, but it should be recalled that most of the meninges, falx, and tentorium receive sensory innervation from the trigeminal system. The major intracranial blood vessels are also innervated by trigeminal sensory nerves. Intracranial lesions could evoke pain by causing tension or displacing these structures. However, the pain perceived by the patient is often not an accurate reflection of the anatomic location or size of the lesion.

Axons forming the root and main divisions of the trigeminal nerve follow a course partly in the posterior and middle cranial fossae and could be affected by lesions in either of these anatomic compartments (see Fig. 14-1).

The *corneal response* (i.e., blink) is frequently used to test the integrity of the peripheral components of the trigeminal and facial nerves and their internuncial brain stem connections. The examiner touches the cornea with a wisp cotton; the normal response is that both eyes blink. The afferent limb of the corneal response consists of the primary sensory axons in the ipsilateral V_1 division, which terminate on the sensory trigeminal nuclei serving tactile sensations (see Fig. 14-3). Secondary axons arising from the sensory trigeminal nuclei project to both the ipsilateral and contralateral facial motor nuclei in the caudal pons (see Fig. 14-4D). Axons originating from the facial motor nucleus follow the facial nerve peripheral distribution and innervate the respective orbicularis oculi, providing the efferent limb of the corneal response. Touching one cornea normally evokes both direct (in the ipsilateral eye) and consensual or indirect (in the contralateral eye) responses.

Lesions affecting V_1 or the trigeminal root result in no or diminished direct and consensual responses if the ip-

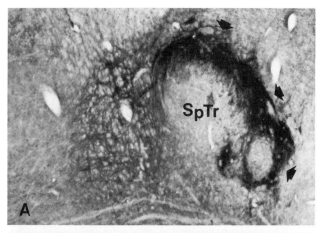

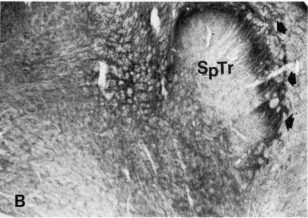

Figure 14-6

Immunocytochemical preparations of adjacent sections of the pons of the monkey. The dorsal direction is toward the top, and the lateral direction is to the right. Immunohistochemical localization of (**A**) substance P and (**B**) enkephalin in the spinal trigeminal tract and superficial laminae of the spinal trigeminal nucleus (SpTr, *arrowheads*). The dark-staining reaction product in **A** represents a substance P-like peptide in the central axons and terminals of the primary sensory trigeminal neurons, whose cell bodies are in the ipsilateral trigeminal ganglion (original magnification × 26). The reaction product in **B** localizes an enkephalin-like peptide in the terminals of intrinsic neurons of the spinal trigeminal nucleus (SpTr) and in the extrinsic neurons that provide afferents to SpTr. The terminals are restricted to the superficial laminae of the spinal trigeminal nucleus (original magnification × 20). The observation that primary sensory terminals containing substance P intermingle with enkephalin-containing terminals of intrinsic brain stem neurons in the spinal trigeminal nucleus suggests a functional interaction related to pain modulation.

silateral cornea is touched. However, if the contralateral cornea is touched, both eyes blink because of the bilateral input to the facial motor nuclei.

The corneal responses are also altered in patients with facial motor nucleus or facial nerve lesions. Touching the cornea evokes no direct blink response if the ipsilateral facial nerve is lesioned, but the contralateral eye does respond. Alternately, touching the contralateral cornea activates a direct but no consensual response.

Other trigeminal sensory reflexes are activated by sensory trigeminal nuclei through the reticular formation and other pathways. Through these circuits, the brain stem and spinal nuclei regulating emesis, lacrimation, swallowing, and other responses are influenced by trigeminal inputs (see Fig. 14-5).

The *masseter* or *jaw jerk reflex* is activated by depressing the mandible or tapping the chin, resulting in bilateral contraction of the masseter and temporalis muscles that are innervated by the trigeminal nerve and close the jaw. The response involves the trigeminal mesencephalic nucleus and direct activation of the trigeminal motor nucleus (see Fig. 14-3).

☐ Trigeminal system components can be involved in neurologic conditions

Skull fractures involving the petrous portion of the temporal bone could damage the trigeminal root as it passes between the posterior to the middle cranial fossae (see Fig. 14-1). In this case, all sensory modalities, except taste, would be lost on the ipsilateral face. Opening the mouth would result in deviation of the jaw toward the side of the lesion because of a lower motor neuron-type paralysis of the ipsilateral muscles of mastication, specifically of the lateral pterygoid muscle. The muscles of facial expression and the tongue are innervated by axons originating from neurons in the facial (see Fig. 14-4*D*) and hypoglossal (see Fig. 14-4*B*) nuclei, respectively, and these would not be affected.

A fracture traversing the foramen ovale on the floor of the middle cranial fossa (see Fig. 14-1) may damage the V$_3$ division and its motor component but spare the V$_1$ and V$_2$ functions. This would result in diminished cutaneous sensations of the ipsilateral lower lip, chin, and jaw region and in a lower motor neuron-type paralysis of the ipsilateral muscles of mastication.

Space-occupying lesions (e.g., tumors, aneurysms) in the posterior and middle cranial fossae may exert pressure on the trigeminal nerve root (see Fig. 14-1) or its divisions and alter functions on the ipsilateral face. Tumors in the posterior cranial fossa (e.g., acoustic neurinoma, cerebellopontine angle tumor) may exert pressure on the lateral aspect of the brain stem and on the spinal trigeminal tract and nucleus (see Figs. 14-3 and 14-4*A–D*). Pain and thermal sensations from the ipsilateral face could be diminished or exacerbated, but tactile sensations, most of which relay in the principal sensory trigeminal nucleus, would be spared (see Fig. 14-5). The tumor could also influence the functions of the ipsilateral spinothalamic tracts, inferior cerebellar peduncle, and facial, statoacoustic, glossopharyngeal, or vagus nerves (see Fig. 14-4B–D). Alternately, a tumor could exert pressure on the trigeminal root. An early sign of this type of tumor is decreased sensitivity of the ipsilateral cornea. Asymmetry is observed when eliciting the corneal response.

Trigeminal neuralgia or *tic douloureux* is characterized by the sudden onset of excruciating pain lasting for seconds to minutes. It does not respond to common pharmacologic methods of pain control. The pain is often evoked by tactile stimulation of a particular location, called the *trigger zone*, on the face. The trigger zone is usually in the perioral region or oral cavity, and bursts of pain may be evoked by jaw movements or food in the mouth. The pain is restricted to the territory of one division of the trigeminal nerve, usually the maxillary but rarely the ophthalmic (see Fig. 14-2). Episodes of pain tend to become more frequent and present a progressively debilitating condition for the patient. Some pharmacologic therapies are helpful, but surgical treatment is usually required for relief. Surgical treatment options include various methods for lesioning a small portion of the trigeminal ganglion containing the primary sensory neurons innervating the affected region of the face. The lesions are produced by injection of alcohol or coagulation through the foramen ovale, obviating the need for a major neurosurgical procedure. In some procedures, the neurons that convey pain are most susceptible, and some tactile sensations on the affected portion of the face may be spared. Transection of the spinal trigeminal tract in the caudal medulla is a neurosurgical procedure only rarely used to treat patients with trigeminal neuralgia.

Pseudobulbar palsy is caused by brain stem vascular lesions that interrupt the corticobulbar tracts (i.e., upper motor neurons) that innervate cranial nerve somatic motor nuclei, including the motor trigeminal nucleus. The resulting paresis of the skeletal muscles innervated by the cranial nerves interferes with ocular, facial, jaw, and tongue movements.

Lateral medullary syndrome (i.e., posterior inferior cerebellar artery syndrome, Wallenberg's syndrome) is caused by a cerebrovascular accident involving the intrinsic vasculature supplying the lateral aspect of the medulla (see Fig. 14-4A–D). It may result in the destruction of several neural structures in the lateral medulla, leading to the dysfunctions characteristic of this syndrome. Specific to the trigeminal system, the spinal trigeminal nucleus and tract are interrupted, resulting in an ipsilateral loss of pain and thermal sensations in the face, but tactile sensations and trigeminal motor function, which depend on the principal sensory trigeminal and trigeminal motor nuclei in the pons, are unaffected (see Fig. 14-4E,F).

SELECTED READINGS

Brodal A. Cranial nerves—the trigeminal nerve. In Neurological anatomy in relation to clinical medicine, 3rd ed. New York: Oxford University Press, 1981:508.

Carpenter MB, Sutin J. Human neuroanatomy, 8th ed. Baltimore: Williams & Wilkins, 1983:393.

Gundmundsson K, Rhoton AL, Rushton JG. Detailed anatomy of the intracranial portion of the trigeminal nerve. J Neurosurg 1971; 35:592.

Johnson LR, Westrum LE, Henry MA. Anatomic organization of the trigeminal system and the effects of deafferentation. In Fromm GH, Sessle BJ (eds). Trigeminal neuralgia: current concepts regarding pathogenesis and treatment. Boston: Butterworth-Heinemann, 1991:27.

Paxinos G, Törk I, Halliday G, Mehler WR. Human homologs to brain stem nuclei identified in other animals as revealed by acetylcholinesterase activity. In Paxinos G (ed). The human nervous system. New York: Academic Press, 1990:149.

Selby G. Diseases of the fifth cranial nerve. In Dyck JP, Thomas PK, Lambert EH, Bunge R (eds). Peripheral neuropathy, vol II. Philadelphia: WB Saunders, 1984:1224.

Sessle B. Physiology of the trigeminal system. In Fromm GH, Sessle BJ (eds). Trigeminal neuralgia: current concepts regarding pathogenesis and treatment. Boston: Butterworth-Heinemann, 191:71.

Young RF. The trigeminal nerve and its central pathways. In Rovit RL, Murali R, Jannetta PJ (eds). Trigeminal neuralgia. Baltimore: Williams & Wilkins, 1990:27.

Neuroscience in Medicine, edited by P. Michael Conn.
J. B. Lippincott Company, Philadelphia © 1995

Chapter 15

*T*he Oculomotor System

ROBERT F. SPENCER

The oculomotor system is the efferent limb for the visual system and vestibular system. The extraocular muscles are controlled by precise connections with several brain stem structures that are related to eye movements in the vertical and horizontal planes. The constituent muscle fiber types and their elegant organization in the extraocular muscles are unmatched by any of the skeletal muscles that are controlled by spinal cord motor systems. Because these premotor structures also have connections with motor neurons in the cervical spinal cord that innervate muscles of the neck, the motor behavior produced by these structures is called *gaze*. Gaze is a combination of coordinated eye and head movements and is the result of a complex interaction between the visual and vestibular systems.

The mechanisms related to eye movements and gaze can best be appreciated by understanding the brain stem, cerebellar, and cortical connections of the preoculomotor structures. It is sometimes clinically possible to separate the different visual and vestibular components such that lesions in different portions of the cerebral cortex, the brain stem, and the cerebellum produce specific eye movement or gaze deficits.

Types of Eye Movements

Eye movements are classified into five categories: vestibular, optokinetic, smooth pursuit, saccadic, and vergence. Of these, vestibular, optokinetic, smooth pursuit, and saccadic eye movements are *conjugate movements*, in which both eyes move in the same direction at the same time. Vergence eye movements are *disjunctive movements*, in which the eyes move in opposite directions at the same time.

Vestibular eye movements are elicited by movements of the head that activate the semicircular canals. The *vestibulo-ocular reflex* (VOR) is a compensatory eye movement that replicates a head movement, but in the opposite direction. The function of the VOR is to maintain the stability of the visual field while the head moves. For example, rotation of the head to the right produces a compensatory conjugate eye movement to the left so that images in the visual field remain stationary on the retina.

Optokinetic eye movements are tracking movements elicited by movement of the whole visual field, such as occurs while looking out of the window of a moving train. *Optokinetic nystagmus* is characterized by a slow-phase eye movement in the direction of a moving stimulus and a quick-phase return eye movement in the opposite direction when the excursion limit of the oculomotor range has been reached. The optokinetic system works synergistically with the vestibular system to stabilize images in the visual field on the retina during movements of the head.

Smooth pursuit eye movements track images that move across the visual field. These movements maintain the focus of moving targets in the visual field on the fovea of the retina. Smooth pursuit eye movements that are ac-

companied by movement of the head in the same direction require cancellation of the VOR.

Saccadic eye movements typically are rapid scanning movements that change foveal fixation from one point in the central visual field to another point in the periphery. Intentional saccades, such as those made to a remembered target, are differentiated from reflexive saccades, which are made in response to a novel object that appears in the peripheral visual field.

Vergence eye movements are associated with changing the point of foveal fixation from a distant object to a near object. Vergence movements are disjunctive, because they are produced by contraction of the medial rectus muscles in both eyes. Vergence eye movements also are associated with changes in the shape of the lens of the eye (i.e., accommodation) and constriction of the pupil (i.e., miosis) as a part of the near response.

Extraocular Motor Nuclei

Motor neurons in the extraocular motor nuclei (i.e., cranial nerves III, IV, and VI) are the final common pathway on which inputs converge from several brain stem premotor structures that are related to the control of different types of eye movements. The cranial nerves with which these motor neurons are associated provide *general somatic efferent* innervation of the extraocular muscles.

□ The oculomotor complex contains somatic and visceral motor neurons

The somatic division of the oculomotor complex (i.e., cranial nerve III) contains motor neurons that innervate the ipsilateral medial rectus, inferior rectus, and inferior oblique muscles and innervate the superior rectus and levator palpebrae superioris muscles bilaterally with contralateral predominance. Axons of the latter decussate in the vicinity of the oculomotor nucleus before coursing ventral to exit the brain stem with the remainder of the third cranial nerve from the ventral surface of the mesencephalon through the interpeduncular fossa. At this location, the nerve passes between the superior cerebellar and posterior cerebral arteries, a clinically important anatomic configuration.

The motor neurons that innervate the different muscles are arranged in a precise topographic organization within the oculomotor nucleus. The medial rectus and inferior rectus motor neurons are in proximity to each other, and the superior rectus and levator palpebrae superioris motor neurons occupy adjacent subdivisions. Inferior oblique motor neurons are located in the vicinity of superior rectus motor neurons. This arrangement is largely a reflection of the synergistic actions of the muscles and the common inputs to the different populations of motor neurons. Levator palpebrae superioris motor neurons are considered to occupy the *caudal central nucleus* subdivision of the oculomotor complex.

In addition to motor neurons, the somatic oculomotor nucleus and the overlying supraoculomotor region contain *internuclear neurons* that have descending brain stem connections with the abducens nucleus and facial nucleus. At least some of the internuclear neurons are contacted by motor neuron axon collaterals and may be involved in the coordinated activity of extraocular and facial muscles, such as can occur during blinking.

The nucleus of Perlia is an archaic term for the presumed midline location of motor neurons that are related to accommodative vergence. A distinct collection of neurons is not apparent cytoarchitectonically, although a region of the dorsomedial oculomotor nucleus contains coexistent motor neurons that innervate the medial rectus and inferior rectus muscles. Some preganglionic parasympathetic neurons that innervate the ciliary body are located in the immediate vicinity of this subgroup, but they also extend ventrally along the midline of the oculomotor nucleus.

The *anteromedian and Edinger-Westphal nuclei* are a collection of midline *preganglionic parasympathetic neurons* that overlie the rostral portion of the somatic oculomotor nucleus and curve ventrally rostral to the somatic nucleus. The *general visceral efferent* axons of these neurons course via the third cranial nerve to synapse in the ciliary ganglion, with postganglionic fibers distributed primarily to the sphincter pupillae muscle of the iris and the ciliary body. The post ganglionic fibers that innervate the sphincter pupillary muscle are related to the *pupillary light reflex,* while those that innervate the ciliary body control lens accomodation. Another population of coexistent neurons located in the same region projects to the upper thoracic spinal cord and may be involved in the coordination of parasympathetic-constriction and sympathetic-dilation control of pupillary function.

The combination of somatic and parasympathetic components in the third cranial nerve forms the basis for characteristic deficits that are associated with a third nerve palsy. The loss of parasympathetic innervation is manifested by pupillary dilatation (i.e., mydriasis), with a complete loss of the direct and consensual light reflexes for the affected eye, and anisocoria. In the resting position, the apparent motility deficits are manifested as exotropia (i.e., external strabismus due to unopposed lateral rectus) and ptosis (i.e., drooping eyelid). The loss of innervation to the vertical eye muscles becomes apparent only when the patient is asked to move the eyes upward or downward.

The trochlear nucleus innervates the contralateral superior oblique muscle

The trochlear nucleus (i.e., cranial nerve IV) contains motor neurons that innervate predominantly the contralateral superior oblique muscle, with approximately 10% of the motor neurons innervating the ipsilateral muscle. The axons exit the dorsomedial portion of the nucleus, course around the medial longitudinal fasciculus and periaqueductal gray, decussate in the anterior medullary velum, and exit the dorsal surface of the brain stem immediately caudal to the inferior colliculus. The small size of the fourth cranial nerve and its anatomic relation to the tentorium cerebelli make it especially vulnerable clinically in trauma. Lesions of the fourth nerve (e.g., fourth nerve palsy) are associated with an overacting inferior oblique muscle, which usually is manifested only if the the lesion is bilateral (i.e., V-shaped pattern).

The abducens nucleus is the center for conjugate horizontal eye movement

Approximately 70% of the neurons of the abducens nucleus (i.e., cranial nerve VI) are motor neurons that innervate the ipsilateral lateral rectus muscle. The axons exit the ventral surface of the brain stem at the pontomedullary junction, just lateral to the pyramids as they emerge from the basilar pons. The sixth cranial nerve has the longest intracranial course of any of the cranial nerves, and because of it size, it is clinically vulnerable (e.g., sixth nerve palsy), especially where it courses ventral to the basilar pons and where it traverses the petrous portion of the temporal bone.

The remaining 30% of neurons in the abducens nucleus are internuclear neurons, whose axons cross the midline at the level of the abducens nucleus, ascend in the medial longitudinal fasciculus (MLF), and establish extensive excitatory synaptic connections with medial rectus motor neurons in the contralateral oculomotor nucleus. The postulated location of these abducens internuclear neurons formerly was believed to be in the vicinity of the abducens nucleus, explaining the archaic term of parabducens nucleus. These neurons, however, coexist with the motor neurons throughout the entire extent of the abducens nucleus. The abducens nucleus is known as the *center for conjugate horizontal eye movement,* because it controls the ipsilateral lateral rectus muscle directly and the contralateral medial rectus muscle indirectly.

Lesions of the abducens nucleus and the abducens nerve produce quite different deficits in ocular motility. Because the abducens nucleus contains motor neurons and internuclear neurons, a lesion involving the nucleus produces paralysis of conjugate horizontal eye movements toward the side of lesion. Lesions of the sixth nerve are manifested by esotropia (i.e., internal strabismus due to unopposed medial rectus) at rest and a paralysis of ipsilateral attempted abduction.

Preoculomotor Nuclei

Four brain stem premotor areas individually are important for the control of eye movement and gaze in the vertical and horizontal planes. These structures have direct, monosynaptic connections with motor neurons in the extraocular motor nuclei.

☐ The vestibular nuclei control the vestibulo-ocular reflex

The superior vestibular nucleus and rostral portions of the medial and inferior vestibular nuclei have afferent connections with semicircular canal-related primary vestibular axons and with the flocculus and fastigial nuclei of the cerebellum. Canal-specific differential efferent projections of second-order vestibular neurons with motor neurons in the extraocular motor nuclei provide the basis for the VOR. Each semicircular canal is related to two pairs of muscles in each eye through reciprocal excitatory and inhibitory synaptic connections with the extraocular motor neurons. The basis of this interaction is explained by the relation of the spatial orientation of the semicircular canals with the pulling actions of the extraocular muscles.

Axons from second-order anterior and posterior canal-related superior vestibular neurons ascend the ipsilateral MLF and are inhibitory to all vertical motor neurons in the oculomotor and trochlear nuclei (Fig. 15-1). Axons from second-order anterior canal-related neurons in the superior vestibular nucleus and from posterior canal-

related neurons in the medial and inferior vestibular nuclei ascend the contralateral MLF and underlying tegmentum and are excitatory to all vertical motor neurons in the oculomotor and trochlear nuclei. Horizontal canal-related medial or inferior vestibular neurons are inhibitory to ipsilateral abducens motor neurons and internuclear neurons and are excitatory to contralateral abducens neurons. Medial rectus motor neurons in the oculomotor nucleus receive only a small direct ipsilateral excitatory input from neurons located in the ventral portion of the lateral vestibular nucleus, whose axons ascend via the ascending tract of Deiters, which is lateral to the MLF.

The basic three-neuron chain (i.e., first-order vestibular ganglion neuron ⤳ second-order vestibular nucleus neuron ⤳ extraocular motor neuron) is necessary but insufficient for the normal functional operation of the VOR. When a compensatory eye movement is made in response to rotation of the head, the head velocity signals of vestibular neurons are incapable of maintaining gaze in the new position. A fundamentally important role in gaze holding is played by the *prepositus hypoglossi nucleus*, which is located in the periventricular dorsal aspect of the medulla extending from the rostral pole of

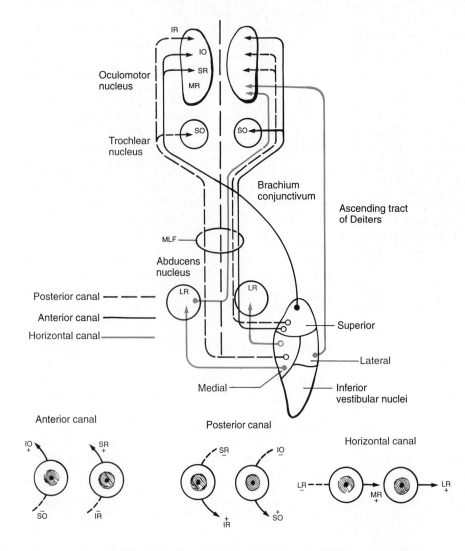

Figure 15-1
The vestibulo-ocular pathways that relate each of the semicircular canals to the activation of specific pairs of extraocular muscles. Excitatory neurons are indicated by filled circles; inhibitory neurons are indicated by open circles. Excitatory fibers ascend contralateral to their cells of origin, and inhibitory fibers ascend ipsilateral to their cells of origin. The actions of extraocular muscles influenced by activation of the individual canals are indicated at the bottom of the drawing.

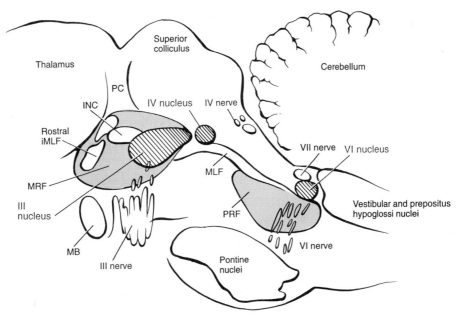

Figure 15-2
A midsaggital view of the brain stem, indicating the locations of premotor neurons that are related to the control of gaze. The mesencephalic reticular formation (MRF) rostral to the oculomotor (III) nucleus contains the interstitial nucleus of Cajal (INC) and the rostral interstitial nucleus of the medial longitudinal fasciculus (rostral iMLF), both of which are related to the control of vertical upward and downward gaze. The paramedian pontine reticular formation (PRF) rostral and ventral to the abducens (VI) nucleus contains neurons that are related to the control of horizontal gaze. *MB,* mamillary body.

the hypoglossal nucleus to the abducens nucleus (Fig. 15-2). The prepositus hypoglossi nucleus has extensive reciprocal connections with the vestibular nuclei and is regarded as the *neural integrator* that is responsible for converting the head velocity signals of vestibular neurons to eye position signals that are carried by extraocular motor neurons. Consistent with its function related primarily to horizontal eye movements, the excitatory and inhibitory connections of prepositus hypoglossi neurons are directed predominantly to the abducens nucleus. A similar neural integrator function has been postulated for neurons in the interstitial nucleus of Cajal in the rostral midbrain that control vertical gaze.

Lesions of the vestibular nerve or nuclei and of the MLF produce quite different deficits. Vestibular nerve or nucleus lesions are characterized by spontaneous nystagmus, an enhanced vertical nystagmus induced by caloric stimulation of the contralateral ear, and positional nystagmus. These deficits underscore the importance of balanced inputs from the semicircular canals and otolith organs on both sides. However, deficits associated with peripheral lesions are only transient because of compensation mediated by commissural connections between the vestibular nuclei on each side.

Lesions of the MLF in the pons and caudal midbrain are responsible for the clinical syndrome of *internuclear ophthalmoplegia,* which is characterized by paralysis of ipsilateral adduction on attempted conjugate horizontal eye movements to the opposite side, nystagmus in the abducted eye, but preservation of vergence. In this case, the innervation to the medial rectus and lateral rectus

muscles is intact and muscle function is normal, but the axons of abducens internuclear neurons have been disrupted, and signals related to conjugate horizontal eye movements are not relayed to the oculomotor nucleus. Vergence eye movements are unaffected, because the premotor neurons and motor neurons that are responsible for these movements are located in the midbrain, rostral to the lesion.

☐ The mesencephalic reticular formation controls vertical gaze

The region of the mesencephalic reticular formation in the vicinity of the oculomotor complex contains two structures that are related intimately to the control of vertical upward and downward gaze (see Fig. 15-2). The *rostral interstitial nucleus of the medial longitudinal fasciculus* is located at the junction of the mesencephalon and diencephalon, lateral to the periventricular gray in the region of the subthalamus and the field H of Forel. Extending caudally from this location into the rostral midbrain, the *interstitial nucleus of Cajal* is lateral to the MLF at the level of the rostral portion of the somatic and visceral oculomotor nuclei. Neurons in both structures have afferent synaptic connections with the superior colliculus and the vestibular nuclei. Excitatory and inhibitory efferent connections are established bilaterally with vertical motor neurons in the oculomotor nucleus and trochlear nuclei (Fig. 15-3). Contralateral projections that are re-

Visuomotor Connections of the Superior Colliculus and Vestibular Nuclei

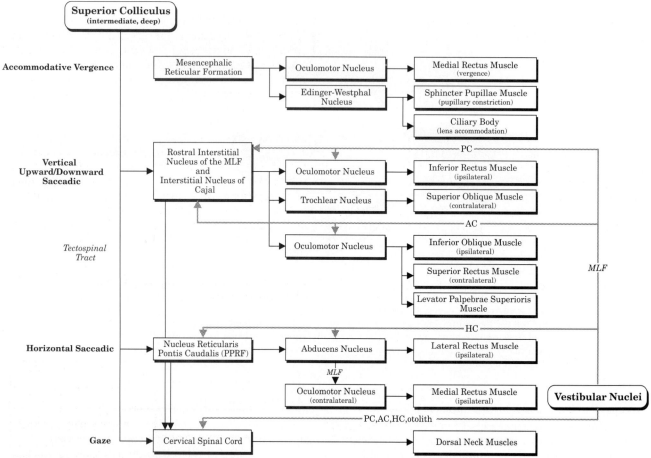

Figure 15-3
Flow diagram of the major efferent connections of the superior colliculus and the vestibular nuclei that are related to the control of gaze. Connections related to accommodative vergence also are indicated. *AC,* anterior semicircular canal; *HC,* horizontal semicircular canal; *MLF,* medial longitudinal fasciculus; *PC,* posterior semicircular canal.

lated specifically to vertical upward eye movements cross the midline through the *posterior commissure*. The rostral interstitial MLF, which contains burst neurons that discharge before the initiation of vertical saccadic eye movements, is physiologically differentiated from the tonic neurons in the interstitial nucleus of Cajal, whose activity is related to eye position. Many of the neurons in these structures also project to the spinal cord, forming the basis for their role in the control of vertical gaze.

Lesions of the rostral interstitial MLF at the mesodiencephalic junction produce paralysis of vertical upward or downward gaze (i.e., Parinaud's syndrome), the direction of which depends on the extent of the lesion. Lesions of the posterior commissure have a selective effect on vertical upward saccadic eye movements in addition to producing deficits in the pupillary light reflex as a result of the proximity to the pretectal area (i.e., pretectal syndrome).

☐ The pontine reticular formation is the center for conjugate horizontal gaze

The paramedian zone of the pontine reticular formation in the vicinity of the abducens nucleus contains neurons that discharge with a burst of activity before the initiation of conjugate horizontal gaze. The *nucleus reticularis pontis caudalis* is rostral to the abducens nucleus and contains *excitatory burst neurons* that project to ipsilateral abducens neurons, and to the cervical spinal cord. *Inhibitory burst neurons* are located caudal to the abducens nucleus and project to contralateral abducens neurons. Both populations of neurons receive inputs from the superior colliculus and vestibular nuclei. A third type of neuron, called *omnipause neurons*, also is located in the vicinity of the burst neurons. The tonic activity of omnipause neurons inhibits the burst neurons, except during saccades. The region of the paramedian pontine reticular formation is known as the *center for conjugate horizontal gaze*. Lesions of this area result in paralysis of ipsilateral conjugate horizontal gaze, in contrast to lesions of the abducens nucleus, which affect only conjugate horizontal eye movements.

Role of the Cerebral Cortex and Cerebellum in Eye Movement

Although many of the functions of the oculomotor system appear to involve mainly brain stem structures, the cerebral cortex has a considerable role in their proper operation. Despite the direct projections from the retina to the superior colliculus and pretectum, the binocularity and directional-selectivity features of the neuronal receptive fields in these structures rely on descending connec-

tions from the primary and secondary visual areas of the cortex. Association areas of the cortex (e.g., frontal eye fields, occipitotemporal and posterior parietal cortices) are largely responsible for the motor behaviors mediated by subcortical structures.

In contrast to the unambiguous role of the descending corticospinal and corticobulbar control from Brodmann's area 4 over spinal and other brain stem motor nuclei (i.e., trigeminal, facial, and hypoglossal), the cortical motor control of eye movement, which originates from area 8 of the prefrontal cortex (i.e., "frontal eye fields"), is less direct. This cortical region has afferent corticocortical connections with visual cortical areas and is a major source of cortical input to the intermediate and deep layers of the superior colliculus. Lesions of this region produce only transient deficits in eye movements. The main deficit appears to be in eye movements that require attention to a particular stimulus in the visual field. Only combined lesions of the frontal eye field and the superior

Afferent and Efferent Connections of the Superior Colliculus

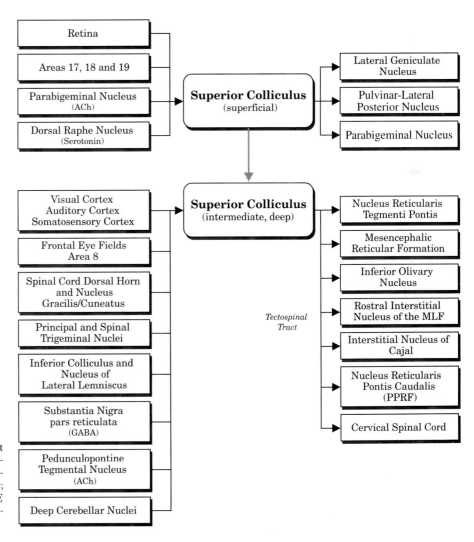

Figure 15-4
Flow diagram of the major afferent and efferent connections of the superficial and deep layers of the superior colliculus. *ACh,* acetylcholine; *GABA,* γ-aminobutyric acid; *PPRF,* paramedian pontine reticular formation.

Visual-Vestibular Convergence in the Cerebellum

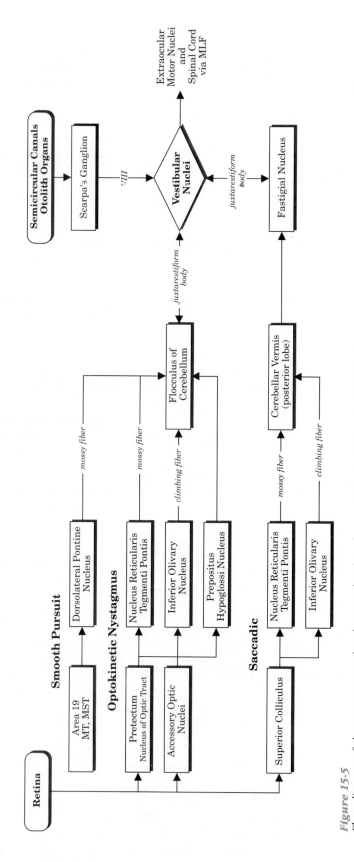

Figure 15-5

Flow diagram of the structures and pathways related to the convergence of visual and vestibular information in the cerebellum and their role in smooth pursuit, optokinetic, and saccadic eye movements. The superior colliculus, pretectum, and accessory optic nuclei project to precerebellar relay nuclei in the basilar pons (nucleus reticularis tegmenti pontis, dorsolateral pontine nucleus) and medulla (inferior olivary nucleus). These regions project visual information to the same areas of the cerebellum that receive input from neurons in the vestibular ganglion and vestibular nuclei. The output of the cerebellum is directed to neurons in the vestibular nuclei that project to the extraocular motor nuclei and the spinal cord through the MLF.

colliculus have a significant effect on voluntary eye movements, which is manifested by deficits in visually guided saccadic eye movements, particularly in the horizontal plane.

The *superior colliculus* is regarded as a site of sensorimotor transformation, and its function is related to orientation behavior. The superficial layers of the superior colliculus are associated predominantly to visual inputs from the retina and the visual cortex. The deeper layers of the superior colliculus receive somatosensory, auditory, and visual inputs from a variety of cortical and subcortical areas (Fig. 15-4). The maps of the visual field, auditory space, and the body are in register with one another and are superimposed on a motor map. Stimulation of different points in the superior colliculus produces saccades that differ in amplitude and direction as function of the retinotopic location.

The output neurons in the deeper layers of the superior colliculus project by way of the *tectospinal tract* to brain stem premotor areas in the mesencephalic and pontine reticular formation that control vertical and horizontal saccadic eye movements and to motor neurons in the cervical spinal cord that innervate muscle of the neck (see Fig. 15-3). The activity of these deep layer efferent neurons encodes information about the direction, velocity, and amplitude of saccades.

Cortical projections to the *pretectum* (i.e., nucleus of the optic tract) and *accessory optic nuclei* are involved in optokinetic nystagmus. Regions of the occipitotemporal and parieto-occipital cortex appear to be important for the generation of optokinetic nystagmus, because the bilateral, direction-specific response of optokinetic nystagmus can be altered by lesions of the posterior parietal cortex. Cortical projections from the same regions to the *dorsolateral pontine nucleus* are associated specifically with smooth pursuit eye movements.

Total cerebellectomy produces persistent deficits in smooth pursuit, optokinetic nystagmus, and holding eccentric positions of gaze. The vestibulocerebellum, which includes the flocculus, nodulus, and uvula, controls eye movements that stabilize images on the retina whether the head is still (i.e., smooth pursuit) or moving (i.e., cancellation of the VOR). The *flocculus* of the cerebellum is the site of convergence of primary vestibular fibers and axons from brain stem precerebellar nuclei (e.g., nucleus reticularis tegmenti pontis and dorsolateral pontine nucleus ⤳ mossy fiber; inferior olivary nucleus ⤳ climbing fiber) that are synaptically related to the pretectal nucleus of the optic tract, the accessory optic nuclei, and the occipitotemporal cortex (Fig. 15-5). The *fastigial nucleus* of the cerebellum is primarily related to the visual portion of the vermis of the posterior lobe, which also has afferent connections from the same brain stem nuclei, but to those regions that are synaptically related to the superior colliculus (see Fig. 15-5). The dorsal cerebellar vermis and fastigial nuclei are important for the control of saccadic amplitude and accuracy. The cerebellum can be considered as the principal site of convergence of visual and vestibular information, which is transmitted to the oculomotor system by means of cerebellovestibular projections from the flocculus and fastigial nuclei to the vestibular nuclei.

Role of the Basal Ganglia in Eye Movement

Diseases of the basal ganglia, including parkinsonism and Huntington's chorea, are characterized by deficits in saccadic eye movements. The γ-aminobutyric acid–mediated inhibitory projection from the pars reticulata of the substantia nigra to the intermediate layer of the superior colliculus provides one pathway through which the basal ganglia may influence oculomotor control. This region of the substantia nigra receives input from the caudoputamen (i.e., striatum) and the subthalamus and forms part of an indirect corticotectal circuit by which sensory information from widespread regions of the cerebral cortex gain access to the superior colliculus. Neurons in the substantia nigra and the caudate nucleus discharge before intentional saccades.

Another link between the basal ganglia and the oculomotor system may be provided by an excitatory cholinergic projection to the superior colliculus, which arises from a collection of neurons in the caudal mesencephalic and rostral pontine tegmentum that partially corresponds with the pedunculopontine tegmental nucleus or cholinergic cell group Ch5. This region receives afferent connections from the pars reticulata of the substantia nigra and has efferent connections with the pars compacta of the substantia nigra and the subthalamus. This region of the tegmentum has been the only location from which the resting tremor that is the hallmark of Parkinson's disease has been produced by experimental lesions.

SELECTED READINGS

Bender MB. Brain control of conjugate horizontal and vertical eye movements. A survey of the structural and functional correlates. Brain 1980;103:23.

Büttner-Ennever JA (ed). Neuroanatomy of the oculomotor system. Rev Oculomot Res 1988;2:3.

Bogousslavsky J, Meienberg O. Eye-movement disorders in brain-stem and cerebellar stroke. Arch Neurol 1987;44:141.

Fukushima K, Kaneko CRS, Fuchs AF. The neuronal substrate of integration in the oculomotor system. Prog Neurobiol 1992;39:609.

Kennard C, Lueck CJ. Oculomotor abnormalities in diseases of the basal ganglia. Rev Neurol (Paris) 1989;145:587.

Leigh RJ, Zee DS. The neurology of eye movements, 2nd ed. Philadelphia: FA Davis, 1991.

Pierrot-Deseilligny C. Motor and premotor structures involved in eye movements. Bull Soc Belge Ophthalmol 1989;237:259.

Ranalli PJ, Sharpe JA, Fletcher WA. Palsy of upward and downward saccadic, pursuit, and vestibular movements with a unilateral midbrain lesion: pathophysiologic correlations. Neurology 1988; 38:114.

Tusa RJ, Zee DS. Cerebral control of smooth pursuit and optokinetic nystagmus. Curr Neuro-ophthalmol 1989;2:115.

Wurtz RH, Goldberg ME (eds). The neurobiology of saccadic eye movements. Rev Oculomot Res 1989;3:3.

Clinical Correlation
DISORDERS OF OCULAR MOTILITY

ROBERT L. RODNITZKY

Normal eye movements depend on intact supranuclear control mechanisms, brain stem gaze centers, certain cranial nerves, extraocular muscles, and neuromuscular transmission. Dysfunction of any of these components impairs ocular motility and produces variety of distinct visual and neurologic signs and symptoms.

Causes of Eye Movement Disorders

☐ Several supranuclear systems guide eye movement

The *saccadic system* allows visual refixation on an object of interest seen in the periphery by moving the eyes rapidly until the image of the object is moved onto the fovea and into central vision. Any voluntary movement of the eyes required for changing gaze from one object to another is also accomplished by a rapid refixation movement, also called a saccade. The *pursuit system* moves the eyes at the appropriate speed to hold the image of a slowly moving object on the fovea. The *vergence system* allows the eyes to move apart or toward one another so that images of objects at various distances can be kept on the fovea of both eyes simultaneously. The *vestibular system* moves the eyes to compensate for head movements so that images remain on the fovea during such activity.

Abnormalities of saccadic eye movement can result from lesions of the frontal eye fields, pontine and mesencephalic reticular nuclei, the cerebellum, or the basal ganglia, which constitute some of the major structures involved in the system responsible for generation of this form of ocular movement. Acute lesions of the eye fields in the frontal lobe result in the inability to voluntary initiate a saccade in the direction opposite the lesion. In a patient with a stroke involving the right frontal lobe, the eyes are deviated to the right and cannot voluntarily be moved to the left beyond the midline. However, by evoking reflex movements of the eyes, they can be moved to the left, indicating that the cranial nerve nuclei, cranial nerves, and extraocular muscles used for movement in this direction are still intact. This form of eye movement

disturbance in the face of intact cranial nerves and nuclei is referred to as a *supranuclear gaze palsy*. Another way in which saccadic movements can be abnormal is in relationship to their speed. In certain degenerative conditions of the cerebellum or basal ganglia, saccades become slowed. In Huntington's disease, for example, the appearance of slow saccades may be one of the earliest signs of neurologic dysfunction.

Like the saccadic system, the pursuit system incorporates several CNS structures, including the striate and extrastriate cortex, pontine nuclei, vestibular nuclei, and cerebellum. Lesions in a variety of anatomic sites may interfere with pursuit movement. When pursuit fails, the speed of eye movement does not keep up with a moving target. To make up this shortfall, corrective saccades are periodically generated, lending the otherwise smooth eye movements a jerky or ratchet-like character. These jerky movements are referred to as *saccadic pursuit*. A lesion of the occipital lobe on one side can impair pursuit, resulting in saccadic pursuit when following objects moving toward the side of the lesion. From the ipsilateral occurrence of pursuit abnormalities, it is apparent that, unlike the control system for saccades, the pursuit system pathway is uncrossed.

☐ Abnormalities of conjugate gaze imply brain stem pathology

When both eyes move in the same direction, at the same speed, maintaining a constant alignment, gaze is said to be conjugate. The center controlling horizontal conjugate gaze is located in the pons in the vicinity of the abducens nucleus and is known as the paramedian zone of the pontine reticular formation. A lesion involving the pontine parareticular formation on one side results in inability to move either eye beyond the midline toward that side. Unlike supranuclear gaze paralysis due to frontal lobe lesions, this form of gaze paresis cannot be overcome by inducing involuntary reflex eye movements. The vertical conjugate gaze center is located in the midbrain. Lesions in this region, such as a tumor of the pineal gland compressing the dorsal midbrain, produce paralysis of conjugate vertical gaze, which in this case is an inability to gaze upward.

The cranial nerve nuclei on both sides of the brain stem subserving conjugate eye movements are connected by the MLF fiber tract. An isolated lesion of this tract produces a specific abnormality of ocular motility. Because the cranial nerve nuclei are left intact by a MLF lesion, eye movements in all directions are still possible, but because the nuclei are disconnected, the eyes do not move in a conjugate manner in the horizontal plane. When left lateral gaze is attempted by an individual with a right MLF lesion, the right eye, which must move inward, cannot cross the midline, although the outward moving left eye responds normally. The same eye that could not move inward during attempted voluntary lateral gaze can do so when both eyes converge on a near target, proving the intactness of the third cranial nerve and medial rectus muscles underlying this movement. This pattern of deficient inward eye movement due to a MLF lesion is referred to as an *internuclear ophthalmoplegia*. This abnormality can be caused by any lesion interrupting the MLF, but multiple sclerosis is the most common cause, especially when the MLF is affected on both sides.

☐ Distinct patterns of impaired ocular motility result from lesions of individual cranial nerves

Abnormalities of the third, fourth, or sixth cranial nerves produce distinct patterns of impaired ocular motility. In an oculomotor nerve palsy, the superior, inferior, and medial recti are paralyzed, as is the inferior oblique muscle. The affected eye deviates downward and laterally. This position results from the action of the two functioning ocular muscles, the lateral rectus and the inferior oblique, which are innervated by other cranial nerves. In a complete oculomotor nerve palsy, the pupil is dilated and the eyelid droops. The patient complains of double vision (diplopia) in several directions of gaze because of the number of different muscles paralyzed. The oculomotor nerve can be damaged by ischemia in diabetes, be impinged on by an arterial aneurysm on the surface of the brain, or be compressed by brain tissue being forced downward by an expanding mass such as a tumor within the cranium. The latter turn of events often signals imminent death from compression of other vital brain structures if the causative mass effect is not corrected.

In an abducens nerve palsy, only the lateral rectus is weakened. The affected eye deviates medially, and the globe cannot be moved laterally beyond the midline. The patient complains of double vision that is worse on lateral gaze toward the side of the lesion. The abducens nerve pursues a long intracranial course and is angled over bony structures at the base of the skull. Because it pursues this path, the nerve is susceptible to stretch after any displacement of the brain stem to which it is attached. An abducens nerve palsy may appear in patients with increased intracranial pressure of any cause if the pressure

results in slight downward displacement of the brain stem or after blunt trauma to the skull, for the same reason.

In a trochlear nerve palsy the superior oblique muscle is weak, resulting in inability to depress the globe, especially in the adducted (i.e., medial) position. In this syndrome, there may be diplopia that is improved by tilting the head in the direction of the normal eye. The presence of head tilt, especially in a child, is often the first indication of the presence of a trochlear nerve palsy.

☐ Disorders of muscle or neuromuscular transmission can affect ocular motility

Ocular motility is involved in some disorders of muscle. This is especially true of the muscle disorder associated with thyroid disease and that seen in the mitochondrial disorders. In these two conditions, the degree of weakness of ocular muscles may be much greater than that seen in appendicular or truncal musculature.

Abnormalities of ocular motility are extremely common and often the presenting sign in myasthenia gravis, a disorder of neuromuscular transmission. In this condition, any one or all of the ocular muscles can be involved. Most commonly, there is associated weakness and drooping of the eyelid. Myasthenia is characterized by abnormal fatigability, and unlike motility disturbances related to central nervous system disorders, the degree of motility impairment may vary from day to day and hour to hour, depending on the use of the eyes. For example, patients with myasthenia gravis often notice diplopia only after sustained gaze in the same direction, as occurs while watching TV or looking at an object in the sky. When impaired ocular motility and eyelid weakness appear in a myasthenic, a diagnosis of oculomotor nerve palsy may be mistakenly made, but the absence of pupillary abnormality and the response to cholinesterase-inhibiting medications clearly establish myasthenia as the cause.

☐ Nystagmus is an abnormal pattern of repetitive eye movements

Nystagmus is a rhythmic to and fro oscillation of the eyes. It can occur in the vertical or horizontal plane and sometimes can be rotatory. It is usually phasic; the oscillation in one direction is faster than in the other direction. In pathologic states, nystagmus may occur when the eyes are in the primary position or after they are moved to their limit in one direction. *Gaze-evoked nystagmus* is usually seen in patients receiving sedative drugs and also occurs in cerebellar diseases. It is only seen when gaze is directed away from the middle position in the vertical or the horizontal plane. *Downbeat nystagmus*, evident while the eyes are in the primary position, is most commonly caused by lesions in the vicinity of the craniocervical junction.

Vestibular nystagmus results from dysfunction of the vestibular apparatus or vestibular nerve. If the dysfunction is unilateral, the nystagmus has its fast phase directed away from the side of the lesion. There is often a rotatory component to the nystagmoid movement. Vertigo, a spontaneous hallucination of movement (often a sense of spinning), commonly accompanies vestibular nystagmus. Vestibular nystagmus can result from causes as diverse as a tumor of the eighth cranial nerve or viral labyrinthitis. Vestibular nystagmus can be readily provoked in normal persons by instilling ice water into the ear canal. This caloric stimulation sets up convection currents in the endolymph within the semicircular canals, resulting in nystagmus toward the side of stimulation and a profound sense of vertigo. This procedure can be used as a test of the intactness of the vestibular system.

Pendular nystagmus differs from most other types in that it is not phasic; the speed of the nystagmoid movement is equal in both directions. It is often congenital, in which case it may be associated with poor vision. If pendular nystagmus is acquired, it usually indicates brain stem pathology, most often related to multiple sclerosis or stroke. *Physiologic nystagmus* appears at the extremes of gaze in normal persons, especially when the eye muscles are fatigued or when the eyes are held at the extreme of lateral or vertical gaze for an extended period. Approximately 5% of the population is capable of producing *voluntary nystagmus,* consisting of a 10- to 25-second burst of extremely rapid back-and-forth horizontal movements.

SELECTED READINGS

Glaser JS (ed). Neuro-opthalmology, 2nd ed. Philadelphia: JB Lippincott, 1990:299, 361.

Neuroscience in Medicine, edited by P. Michael Conn.
J. B. Lippincott Company, Philadelphia © 1995

Chapter *16*

The Hypothalamus

MARC E. FREEMAN

Functions of the Hypothalamus

The hypothalamus is the part of the limbic portion of the brain in vertebrates that regulates the internal milieu of the cells within narrow limits as it compensates for external conditions, such as variations in temperature, energy, or defensive requirements. The constancy of the internal environment resulting from the fine adjustments made by the hypothalamus is referred to as *homeostasis*. The hypothalamus maintains homeostasis by exerting control over the two regulatory systems of the organism: the nervous system and the endocrine system.

In regulating the nervous system, the hypothalamus controls autonomic processes such as cardiovascular, thermoregulatory, and visceral functions. Behavioral processes, such as ingestive, sexual, maternal, and emotional behaviors, are also regulated by the hypothalamus. The role of the hypothalamus in the regulation of the nervous system is summarized in Table 16-1.

In regulating the function of the endocrine system, the hypothalamus exerts control of the two subdivisions of the pituitary gland. The *anterior pituitary* or *adenohypophysis* synthesizes hormones that regulate adrenal, thyroid, and gonadal function and that regulate growth and lactation. The synthesis and secretion of anterior pituitary hormones are regulated by peptides and amines that are synthesized by and secreted from specific hypothalamic neurons and are transported to the adenohypophysis through a microscopic vascular route, the *hypothalamo-hypophysial portal system* (Fig. 16-1), to stimulate or inhibit the synthesis and secretion of specific hormones of the adenohypophysis. These peptides and amines are known collectively as *releasing* or *release-inhibiting hormones* (Table 16-2).

The hormones of the *posterior pituitary* or *neurohypophysis* are hormones that are synthesized by specific hypothalamic neurons and transported to the neurohypophysis axonally by the hypothalamohypophysial tract (see Fig. 16-1) and released into sinusoids and ultimately into the peripheral circulation to regulate blood pressure, water balance, and milk ejection (see Table 16-2).

The fact that certain neurons can serve two functions—as typical nerve cells that receive and transmit electrical information and as endocrine cells that secrete products into a minute blood supply to regulate the adenohypophysis or into the neurohypophysis and ultimately into the peripheral circulation to regulate visceral processes—led to the concept of neurosecretion and ultimately to the birth of the science of neuroendocrinology.

The hypothalamus integrates the control of physiologic processes. For example, thermoregulatory processes are governed by the autonomic nervous system and the endocrine system. Exposure to extremes of temperature results in adjustment of blood flow through autonomic processes and metabolic adjustments through the regulation of thyroid hormone secretion. Both of these seemingly unrelated controls are under the influence of the hypothalamus.

Historic Perspective

Although there are indications that some ancient cultures may have appreciated the vital role of higher centers in regulating normal physiology, an understanding of the role of the hypothalamus did not begin to crystallize until a series of clinical observations were made during the late nineteenth and early twentieth centuries. Most of the early studies focused on hypothalamic control of pituitary function, because pituitary abnormalities were the most obvious. Because the connection between the hypothalamus and pituitary gland was not appreciated at that time, many of the early observations were incorrectly attributed directly to "pituitary tumors."

In 1901, Alfred Fröhlich, a Viennese physician, correctly reported a case of adiposogenital dystrophy in a 14-year-old boy suffering from a pituitary tumor compressing the optic tract and hypothalamus, which was subsequently relieved by surgery. Soon thereafter, Erdheim described gonadal atrophy and obesity that was caused by hypothalamic damage without damage of the pituitary gland. In 1913, Camus and Roussay demonstrated polyuria in dogs bearing surgical lesions of the hypothalamus without damage to the pituitary gland. These were the first direct observations of hypothalamic control of the pituitary.

In 1925, the development by Philip Smith of a parapharyngeal surgical procedure (i.e., hypophysectomy) to remove the pituitary gland of rats prompted a flurry of studies of the pituitary and brain. It was appreciated at that time that the pituitary gland must remain intact with the brain for coitus to induce ovulation in rabbits. In 1936, the classic, albeit crude, experiments of Marshall and Verney demonstrating ovulation induction in rabbits by the passage of an electrical current through the brain were followed in 1937 by the experiments of Geoffrey Harris showing that more localized stimulation of the hypothalamus also led to ovulation induction in rabbits.

Table 16-1
NEURAL PROCESSES REGULATED BY THE HYPOTHALAMUS

Process	System or Activity	Effect
Autonomic	Cardiovascular	Blood flow ($\downarrow$ or $\uparrow$)* Vasodilation or vasoconstriction
	Thermoregulatory	Blood flow, shivering, panting
	Visceral	Digestive acid secretion ($\uparrow$)
Behavioral	Sexual	Sexual receptivity ($\uparrow$)
	Maternal	Nest building
	Emotional	Aggression ($\uparrow$)
	Ingestive	Eating and drinking ($\uparrow$ or $\downarrow$)

* $\uparrow$, increase; $\downarrow$, decrease

Figure 16-1

Control of the neurohypophysis and adenohypophysis by the hypothalamus. Neuron 1 is a peptidergic magnocellular neuron from the supraoptic or paraventricular nuclei of the hypothalamus that secretes oxytocin or vasopressin into the sinusoids in the neurohypophysis. The axons of these two nuclei travel to the neurohypophysis in the hypothalamohypophysial tract. Neuron 2 could be a hypophysiotropic peptidergic or aminergic neuron terminating adjacent to the short portal vessels, which represent a potential route of communication between the neurohypophysis and adenohypophysis. The hypothalamic release or release-inhibiting peptidergic neurons are of this type. The tuberohypophysial dopaminergic neurons are also of this type. Neuron 3 could also be a hypophysiotropic peptidergic or aminergic neuron. In this case, the neuron terminates on the primary capillary bed of the median eminence. It also secretes into portal blood release or release-inhibiting peptides that reach the adenohypophysis by way of the long portal vessels. The tuberoinfundibular dopaminergic neurons are also of this type. Neurons whose cell bodies lie within the arcuate and periventricular nuclei and terminate on the primary capillary bed in the median eminence comprise the infundibular tract. The link between the rest of the brain and the pituitary gland is represented by neurons 4 and 5. These neurons secrete catecholamines (and in some cases peptides) that act as neurotransmitters or neuromodulators on the hypophysiotropic neurons. The termination of neuron 4 is axodendritic or axosomatic, and that of neuron 5 is axoaxonic.

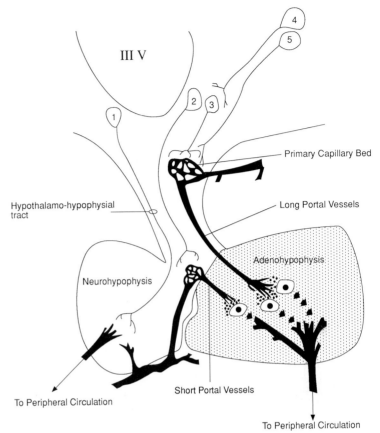

Table 16-2
NEUROENDOCRINE REGULATORS OF HYPOTHALAMIC ORIGIN

System Regulated	Site of Regulation	Action	Hypothalamic Regulator
Thyroid gland	Adenohypophysis	Stimulation of thyrotrophin secretion	Thyrotrophin-releasing hormone
Adrenal cortex	Adenohypophysis	Stimulation of adrenocorticotrophin secretion	Corticotrophin-releasing hormone
Gonads	Adenohypophysis	Stimulation of luteinizing hormone and follicle-stimulating hormone secretion	Gonadotropin-releasing hormone
Muscle, bone, liver	Adenohypophysis	Stimulation or inhibition of growth hormone secretion	Growth hormone–releasing hormone, somatostatin
Milk synthesis and secretion from the mammary gland	Adenohypophysis	Inhibition of prolactin secretion	Dopamine
Cardiovascular, renal	Vascular smooth muscle, renal tubule	Vasoconstriction, water reabsorption	Vasopressin (antidiuretic hormone)
Mammary gland, uterus	Smooth muscle of mammary ducts and uterus	Increase intramammary pressure, inducing milk ejection; increase uterine contraction in labor	Oxytocin

Later in 1937, Westman and Jacobsohn found that coitus would not result in ovulation in the rabbit if the pituitary stalk was cut and a foil barrier placed between the hypothalamus and pituitary with the (mistaken) intention of preventing regrowth of severed "nerves." It was later understood that coitus stimulated the release of the decapeptide gonadotropin-releasing hormone (GnRH) into portal blood and that the hormone's role was to stimulate the release of an ovulation-inducing amount of luteinizing hormone into the peripheral circulation.

Possibly the most significant early contribution to the science of neuroendocrinology was the development of the concept of *neurosecretion* by the husband and wife team of Ernst and Berta Scharrer. Beginning in the early 1930s, they proposed that cells of the hypothalamus must have a function distinct from other brain cells based on their multinucleated appearance; the abundance of protein-containing, colloid-like vacuoles; and the unique proximity between these cells and the surrounding capillary network. The Scharrers proposed that these nerve cells must therefore have a glandular function. In 1930, Popa and Fielding described the vascular connection between the hypothalamus and adenohypophysis in rabbits, although they mistakenly surmised that the direction of blood flow was from the gland toward the hypothalamus. The first report of flow toward the pituitary was described for toads by Houssay in 1935.

The developing concept of neurosecretion coupled with the description of the portal vasculature opened the door for an innovative series of experiments demonstrating that the hypothalamus controlled the adenohypophysis with messages transported over a vascular route. However, experiments involving transection of the stalk connecting the hypothalamus with the pituitary gland produced various results, leading some to doubt the importance of a vascular connection until 1947, when Green and Harris suggested that the cut portal vessels could regenerate. In 1950, Harris showed that reproductive function was restored to a degree proportional to portal vessel regeneration in stalk-sectioned rats. The elegant experiments of Harris and Jacobsohn in 1952 convincingly demonstrated the primacy of the hypophysial-portal vasculature in anterior pituitary function. In these experiments, rats were hypophysectomized, and adenohypophyses from their newborns were transplanted to the temporal lobe of the brain or, by a transtemporal route, immediately beneath the median eminence of the hypothalamus. Only the animals whose transplants beneath the median eminence were revascularized by the portal vasculature showed a resumption of reproductive function.

In 1958, Nikitovich-Weiner and Everett autografted anterior pituitaries to the kidney capsule and demonstrated a loss of thyrotropin-stimulating hormone (TSH), corticotropin (ACTH), follicle-stimulating hormone (FSH), and LH secretion but an enhancement of prolactin secretion from the transplants. These transplants were subsequently removed and placed under the temporal lobe of the brain or beneath the median eminence. Only the rats

bearing transplants to the median eminence showed a resumption of normal anterior pituitary function.

The dawning of the science of neuroendocrinology was completed with the focus on the chemical nature of the activities of the adenohypophysis and neurohypophysis. After the studies of Van Dyke and associates in 1941 established the existence of separate oxytocic and pressor principals, du Vigneaud identified the structure of oxytocin (1950) and vasopressin (1954). These discoveries were followed by a flurry of work from the mid-1960s through the 1970s by Andrew Schally and Roger Guillemin's laboratories on the chemical nature of the hypothalamic neuropeptides that control the secretion of TSH, LH, FSH, and ACTH and the control of growth hormone from the anterior pituitary. The maturation of the science of neuroendocrinology was recognized by Nobel Prize awards to these two investigators in 1977.

Anatomy of the Hypothalamus

☐ The boundaries of the hypothalamus are distinctly defined

The hypothalamus is in the lowermost portion of the diencephalon (Figs. 16-2 and 16-3). The human hypothalamus has well-defined boundaries. The rostral border is limited by a vertical line drawn through the anterior border of the anterior commissure, lamina terminalis, and optic chiasm. The hypothalamus is bordered caudally by a vertical line drawn through the posterior border of the mammillary body as it bounds the interpeduncular fossa. The superior border of the hypothalamus is the hypotha-

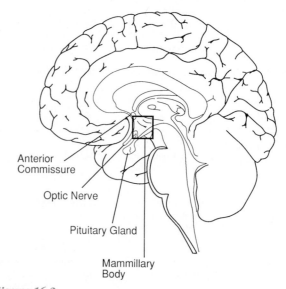

Figure 16-2
The position of the hypothalamus and pituitary gland of the human relative to the rest of the brain. The hypothalamus is bounded by the dark-bordered box.

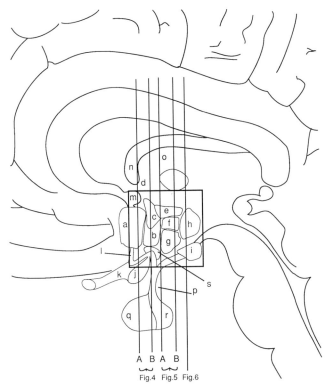

Figure 16-3
The position of hypothalamic nuclei and adjacent structures are
viewed in sagittal section. The vertical lines represent the planes
of the frontal views of Figures 16-4A, B, 16-5A, B, and 16-6. The
box outlines an area that corresponds to the area of Figure 16-2.
a, preoptic nucleus; *b,* anterior hypothalamic area; *c,* paraventricular nucleus; *d,* hypothalamic sulcus; *e,* dorsal hypothalamic area;
f, dorsomedial nucleus; *g,* ventromedial nucleus; *h,* posterior hypothalamic area; *i,* mammillary body; *j,* optic chiasm; *k,* optic
nerve; *l,* lamina terminalis; *m,* anterior commissure; *n,* fornix; *o,*
thalamus; *p,* infundibulum; *q,* adenohypophysis; *r,* neurohypophysis; *s,* suprachiasmatic nucleus.

lamic sulcus as it borders the thalamus, and the inferior
boundary is the bulging tuber cinereum, which tapers to
form the infundibulum, the funnel-shaped functional
connection between the hypothalamus and the pituitary.
Laterally, the borders are ill defined because of the blending of the hypothalamic gray matter with adjacent structures. By convention, the lateral borders are confined by
the internal capsule and its caudal limits.

□ **The divisions of the
hypothalamus are described
as functional groupings**

The hypothalamus is divided into clusters of perikarya
embedded in gray matter. These clusters are referred to
as *nuclei.* There are several problems inherent in this
designation. In most cases, the nuclei are not morphologically distinct structures whose boundaries are distinct
in histologic preparations. The dendrites and axons of

these neurons may extend for distances beyond the limits
of the nucleus. Moreover, chemically and functionally, the
nuclei may be heterogeneous to various degrees. It is
possible to infer only vague functional and anatomic
boundaries for hypothalamic nuclei.

The groupings of the nuclei can be described in a rostrocaudal direction as *anterior* or *supraoptic* (Fig. 16-4),
located between the lamina terminalis and the posterior
edge of the optic chiasm; *medial* or *tuberal* (Fig. 16-5),
located between the optic chiasm and the mammillary
bodies; and *posterior* or *mammillary* (Fig. 16-6), including the mammillary bodies and the structures just dorsal
to them. The hypothalamus can also be described longitudinally in mediolateral zones (see Figs. 16-4 through
16-6) known as the *periventricular zone*, bordering the
third ventricle; the *medial zone*, comprising the major
hypothalamic nuclei that are sites of limbic system projections; and the *lateral zone*, which is separated from
the medial zone by the *fornix*, a large, C-shaped tract that
interconnects limbic system structures.

The most anterior hypothalamic areas are ill-defined.
Rather than being grouped as diencephalic structures,
the *medial preoptic* and *septal areas* are actually part of
the telencephalon (see Fig. 16-4A). However, modern
embryology has shown that the preoptic area has the
same embryonic origins as many diencephalic structures.
The preoptic area is often considered part of the hypothalamus. The medial preoptic area (see Fig. 16-4A) possesses sexually dimorphic features. Uniquely stained
groups of neurons form in this area in organisms exposed to testosterone prenatally or neonatally. The *lateral preoptic area* (see Fig. 16-4A) is not morphologically
distinct from the medial preoptic area but has uniquely
distinct physiologic roles. More caudally, the *anterior
and lateral hypothalamic areas* appear (see Fig. 16-4B).
The cells of these areas are small with few dendritic
branches. The lateral hypothalamic area receives fibers
from the medial forebrain bundle. Chemical lesioning of
this area leads to aphagia. This area plays a stimulatory
role in feeding behavior.

The *paraventricular nucleus* (see Fig. 16-5A) is wedge
shaped and lies adjacent to the third ventricle. The deeply
staining neurons are of two types: magnocellular neurons
with large perikarya and parvicellular neurons with small
perikarya. The axons of the magnocellular neurons terminate in the neurohypophysis, and the axons of the
parvicellular neurons terminate on the primary capillary
bed of the hypophysial portal vasculature in the median
eminence. The *supraoptic nucleus* is found directly
above the beginning of the optic tracts and consists of a
large anterolateral subnucleus and a smaller posteromedial subnucleus connected by a thin strand of cells (see
Fig. 16-5A). As in the paraventricular nucleus, the neurons of the supraoptic nucleus stain darkly and consist of
magnocellular perikarya whose axons terminate in the
neurohypophysis. The axons of the supraoptic and paraventricular nuclei travel in a bundle, the hypothalamohypophysial tract, to the neurohypophysis (see Fig. 16-1).
The magnocellular and parvicellular cells of these re-

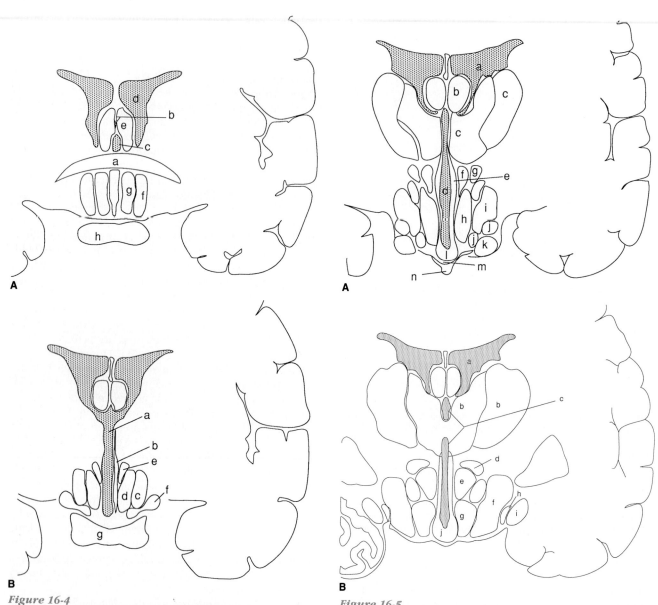

Figure 16-4

The hypothalamic and adjacent structures of the anterior or supraoptic groupings. (**A**) *a*, anterior commissure; *b*, septal area; *c*, third ventricle; *d*, lateral ventricle, *e*, column of fornix; *f*, lateral preoptic area; *g*, medial preoptic area; *h*, optic chiasm. (**B**) *a*, third ventricle; *b*, periventricular nucleus; *c*, lateral hypothalamic area; *d*, anterior hypothalamic area; *e*, paraventricular nucleus; *f*, supraoptic nucleus; *g*, optic chiasm.

Figure 16-5

The hypothalamic and adjacent structures of the medial or tuberal groupings. (**A**) *a*, lateral ventricle; *b*, body of fornix; *c*, thalamus; *d*, third ventricle; *e*, periventricular nucleus; *f*, paraventricular nucleus; *g*, dorsal hypothalamic area; *h*, anterior hypothalamic area; *i*, lateral hypothalamic area; *j*, supraoptic nucleus; *k*, optic tract; *l*, arcuate nucleus; *m*, median eminence; *n*, infundibulum. (**B**) *a*, lateral ventricle; *b*, thalamus; *c*, third ventricle; *d*, dorsal hypothalamic area; *e*, dorsomedial nucleus; *f*, lateral nucleus; *g*, ventromedial nucleus; *h*, supraoptic nucleus; *i*, optic tract; *j*, arcuate nucleus.

gions produce vasopressin and oxytocin. The *suprachiasmatic nuclei* are distinctly staining paired structures (in rodents) overlying the optic chiasm. In humans, the suprachiasmatic nuclei are not strikingly morphologically distinct. In all mammals, the cells of this area receive retinohypothalamic input and are thought to be the "circadian clock" that controls the temperature cycle, sleep-wake cycle, and the circadian changes in the timing of certain hormone systems, such as the pituitary hormones that control the adrenal cortex (i.e., ACTH) and the reproductive system (i.e., LH, prolactin). The *periventricu-*

lar nuclei (see Fig. 16-5) have small perikarya that contain some of the release and release-inhibiting factors controlling the anterior pituitary gland.

Associated with the anterior hypothalamic area are the morphologically indistinct telencephalic structures known as *circumventricular organs*. One is the organum vasculosum of the lamina terminalis (OVLT), and the other is the subfornical organ. These are areas where the

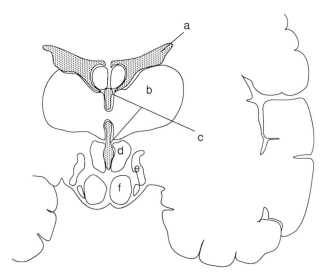

Figure 16-6
The hypothalamic and adjacent structures of the posterior or mammillary groupings. *a*, lateral ventricle; *b*, thalamus; *c*, third ventricle; *d*, posterior hypothalamic area; *e*, lateral hypothalamic area; *f*, mammillary body.

blood-brain barrier is absent and that can sense plasma osmolality. These areas play a role in regulation of blood pressure and thirst because they contain angiotensin II receptors and may even be able to synthesize their own angiotensin II. The OVLT has been implicated in the control of LH secretion by GnRH. The other "leaky" circumventricular organs are the subcommissural organ and the area postrema.

In the medial area, the optic tracts separate, the lateral hypothalamus continues caudally, and the caudal termination of the supraoptic nucleus is found. The anterior hypothalamic area ends (see Fig. 16-5*A*) in this area and is replaced by two distinct nuclei, the *dorsomedial* and *ventromedial nuclei* (see Fig. 16-5*B*). The two may be separated by the small cells of the dorsomedial nucleus and the dense grouping of the neurons in the ventromedial nucleus. Both of these nuclei play a role in food intake. Lesions of both nuclei cause hyperphagia, because this area regulates food intake in an inhibitory manner. The ventromedial nucleus in particular contains glucose-sensitive cells that are thought to be the site through which caloric intake is monitored. The cells of the ventromedial nucleus are also rich in receptors for the gonadal steroids estrogen and testosterone and are thought to play a major role in reproductive behavior and regulation of hormone secretion from the adenohypophysis.

The *arcuate nucleus* begins in the medial hypothalamic area (see Fig. 16-5). The parvicellular neurons in this area have short axons, some of which terminate on the primary capillary bed of the hypothalamohypophysial portal system. The primary capillary bed is found in the underlying median eminence (see Fig. 16-5). The neurons of the arcuate nucleus have several functions. In mammals such as guinea pigs, humans, most monkeys, bats, ferrets, cows, horses, cats, dogs, and rabbits, GnRH

neurons are in the basal portion of the medial hypothalamus. However, in others, such as rats and sheep, this area is devoid of GnRH neurons, or small numbers of such neurons are found in a so-called cell-poor zone. In species in which the arcuate nucleus contains GnRH neurons, the fibers continue to the median eminence, and some continue through the infundibular stalk and into the neurohypophysis (e.g., in humans).

A second function of the arcuate nucleus is in the control of prolactin and growth hormone secretion. This area is populated by cells that contain dopamine, the prolactin-inhibiting hormone, and growth hormone–releasing hormone, the peptidergic stimulator of growth hormone secretion. The arcuate nucleus is abundant in cells that contain β-endorphin, the endogenous opioid that is a cleavage product of the larger peptide, proopiomelanocortin. These neurons project to various hypothalamic and forebrain sites and are thought to play a role in emotional behavior and endocrine function.

The posterior hypothalamic area contains the continuation of the lateral hypothalamic area and the posterior hypothalamic nuclei and mamillary bodies (see Fig. 16-6). The posterior hypothalamic nucleus contains small and large cell bodies that give rise to efferent fibers descending through the central gray matter and the reticular formation of the brain stem. It is thought that these neurons play a role in temperature regulation, because they respond to cooling with the induction of shivering and the burning of brown adipose tissue. The mammillary nucleus is actually a complex consisting of medial and lateral nuclei. The mammillary bodies are critical circuits linking the hypothalamus with the limbic forebrain and midbrain structures lying rostral and caudal, implying a role in hypothalamic activity.

☐ The afferent and efferent connections are the information pathways of the hypothalamus

The afferent and efferent connections of the hypothalamus reveal that this part of the brain is a complex integration center for somatic, autonomic, and endocrine functions.

Intrinsic tracts

There are two main intrinsic tracts in the hypothalamus (see Fig. 16-1). The *infundibular tract* arises from neurons in the arcuate nucleus and periventricular nucleus with terminals on capillaries within the median eminence. These axonally transport substances such as dopamine to the portal vessels. The *hypothalamohypophysial tract* arises in the supraoptic and paraventricular nuclei and terminates in the neurohypophysis. These axons transport vasopressin and oxytocin, respectively. Both tracts transfer information unidirectionally from the hypothalamus to the pituitary.

Extrinsic tracts

The lateral hypothalamus is reciprocally connected with the thalamus, the paramedian mesencephalic area (i.e., limbic midbrain area), and the limbic system. The medial hypothalamus also receives connections from the limbic system (Fig. 16-7). The higher cortical centers communicate with the hypothalamus through the limbic system. In addition to the hypothalamus, the limbic system includes the hippocampus, the amygdala, the septal area, the nucleus accumbens (part of the striatum), and the orbitofrontal cortex. Anatomically, the hypothalamus is intimately related to the amygdala, which sits in the temporal lobe just rostral to the hippocampus. Efferents from the amygdala enter the hypothalamus by means of the ventral amygdalofugal pathway. The rostral amygdalofugal fibers form the diagonal band of Broca. More caudally, these fibers fan out and enter the hypothalamus, and many terminate near the ventromedial nucleus. An afferent to the hypothalamus arises from the corticomedial amygdala. This pathway, the stria terminalis, terminates near the ventromedial nucleus of the hypothalamus.

The other major limbic afferent to the hypothalamus arises from the hippocampus. The body of the hippocampus gives rise to the columns of the fornix, which courses toward the anterior commissure and then splits into two portions. The postcommissural fornix terminates in the mammillary bodies at the caudal end of the hypothalamus. Arising from the mammillary bodies is the *mammillothalamic tract*, which courses to the anterior nuclei of the thalamus and from there projects to the cingulate gyrus, the parahippocampal gyrus, and then back to the hippocampus.

A second efferent projection from the mammillary bodies, the *mammillotegmental tract*, turns caudally to the ventral tegmentum. A reciprocal pathway from the ventral tegmentum to the mammillary bodies is the *mammillary-pudendal tract*. The *dorsal-longitudinal fasciculus of Schutz* are efferents from the periventricular nuclei of the hypothalamus that terminate in the mesencephalic central gray. Stimulation of this fiber bundle produces fear and aversive reactions.

There is a subset of ganglion cells in the retina that project to the suprachiasmatic nucleus of the hypothalamus by way of the *retinohypothalamic tract*. This tract transmits lighting periodicity information to be transduced by the suprachiasmatic nucleus. The *hypothalamospinal tract* originates in the supraoptic and paraventricular nuclei (parvicellular division) that project down the spinal cord to the thoracic level and terminate in the intermediolateral column and from there to the preganglionic sympathetic nerves. Based on the anatomy, this pathway must be an important route over which the hypothalamus influences autonomic function.

The other major hypothalamic fiber tract is the *median forebrain bundle*. This is a collection of tracts with ascending and descending fibers that run in the lateral hypothalamus between the midbrain reticular formation and the basal forebrain. The descending fibers originate from structures in the basal forebrain, including the olfactory cortex, the preoptic area, the septal area, the accumbens, and the amygdala. The ascending portion comes from the spinal cord and reticular formation, visceral and taste nuclei in the brain stem, and monoaminergic centers in the brain stem.

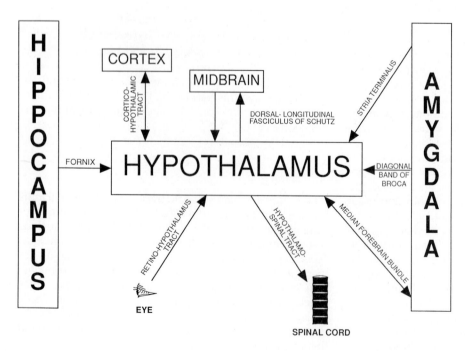

Figure 16-7
Some afferents and efferents of the hypothalamus.

☐ Blood flow as a means of communicating hypothalamic information

The key neurohumoral link between the hypothalamus and the pituitary gland is the *hypothalamohypophysial portal* vasculature (Fig. 16-8). It arises from a primary capillary plexus that extends from the median eminence to the neurohypophysis. This plexus is supplied with blood from three sources: rostrally by the superior hypophysial artery, caudally by the inferior hypophysial artery, and mediorostrally by the anterior hypophysial artery (or trabecular artery). All three of these arteries arise from the internal carotid artery. In some species (e.g., rat, rabbit, cat), they unite to form a single artery that supplies the infundibular stem. These arteries encircle the median eminence. The inferior hypophysial artery also supplies the neurohypophysis. The primary capillary plexus in the median eminence is drained by the fenestrated long portal vessels that course to the adenohypophysis, where they branch to a secondary capillary plexus. The primary capillary plexus is the site at which axon terminals converge to release their quanta of hy-

pophysiotropic peptides into portal blood. After transport through the long portal vessels, they are released from the secondary capillary plexus to the surrounding cells of the adenohypophysial cells. A set of short portal vessels arise from the anterior hypophysial artery. These connect the infundibular stem, the neurohypophysis, and the intermediate lobe of the pituitary to the adenohypophysis. The short portal vessels are the route over which neurohypophysial and intermediate lobe peptides travel to the anterior pituitary.

☐ The chemiarchitecture describes the functions of the hypothalamus

Monoamines

There are essentially three monoamines of importance to hypothalamic function: dopamine, norepinephrine, and serotonin.

Dopamine. There are two major dopaminergic systems with long axons that originate from outside the hypo-

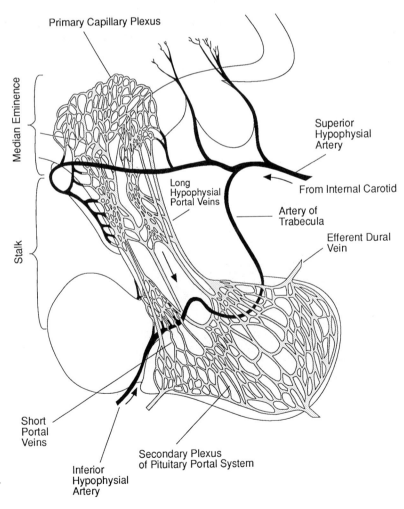

Primary Capillary Plexus

Median Eminence

Stalk

Superior Hypophysial Artery

From Internal Carotid

Long Hypophysial Portal Veins

Artery of Trabecula

Efferent Dural Vein

Short Portal Veins

Inferior Hypophysial Artery

Secondary Plexus of Pituitary Portal System

Figure 16-8
Human hypothalamohypophysial portal vasculature.

thalamus (Fig. 16-9). One of these, the *nigrostriatal system*, has cell bodies in the substantia nigra with long axons that terminate in the caudate-putamen complex and globus pallidus. A second, the *mesolimbic system*, has cell bodies in the ventral tegmentum that send projections through the hypothalamus and terminates in areas of the limbic system such as the nucleus accumbens, olfactory tubercle, cingulate cortex, and frontal cortex. Axons of both of these areas travel through the medial forebrain bundle.

The hypothalamus contains three intrinsic dopaminergic pathways with short axons. The cell bodies of the *incertohypothalamic neurons* are in the caudal hypothalamus, zona incerta, and rostral periventricular nucleus with axons terminating in the dorsal hypothalamus, preoptic area, and septum. The cell bodies of the *tuberoinfundibular neurons* are in the arcuate and periventricular nuclei with short axons that terminate in the median eminence. These converge on the primary capillary bed of the hypophysial portal system and play a direct role in the release of hormones from the adenohypophysis. The *tuberohypophysial neurons* have cell bodies in the rostral arcuate and periventricular nuclei with axons terminating in the intermediate and posterior lobes of the pituitary. In the neurohypophysis, these axons are in proximity to vascular spaces, neurosecretory axons, and pituicytes (i.e., modified astroglial cells). Within the intermediate lobe, these axons terminate on secretory

cells known as melanotropes. It is thought that a portion of the dopamine acting within the adenohypophysis originates from the axon terminals in the intermediate and posterior lobes and ultimately reaches the anterior pituitary by means of short portal vessels. Within the hypothalamus, the incertohypothalamic dopaminergic neurons appear to play a neuromodulatory role while the tuberoinfundibular, and tuberohypophysial neurons serve a neuroendocrine role.

Norepinephrine. The noradrenergic cell bodies of greatest importance to the hypothalamus are in the *locus ceruleus* (Fig. 16-10). The efferents course toward the hypothalamus as the large dorsal noradrenergic or tegmental bundle and the rostral limb of the dorsal periventricular pathway. The former pathway joins the ascending ventral noradrenergic bundle from the lateral tegmental noradrenergic cell groups. The dorsal and ventral noradrenergic pathways unite in the median forebrain bundle to enter the amygdala (i.e., dorsal path) and hypothalamus (i.e., ventral path).

Serotonin. Two groups of serotinergic cell bodies are found in the brain in the *dorsal and medial raphe nuclei* (Fig. 16-11). Axons from the dorsal raphe nucleus form the ventral ascending serotinergic pathway that sweep ventrally and then curve rostrally through the ventral tegmentum to join noradrenergic fibers of the median forebrain bundle in the lateral hypothalamic area. Two large

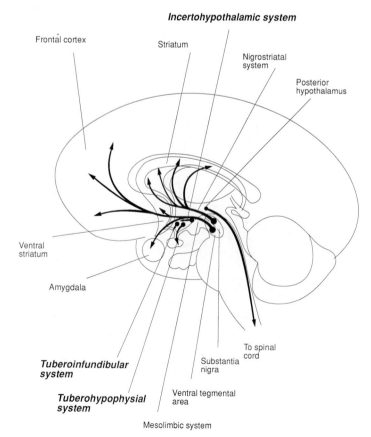

Incertohypothalamic system

Frontal cortex

Striatum

Nigrostriatal system

Posterior hypothalamus

Ventral striatum

Amygdala

Tuberoinfundibular system

Tuberohypophysial system

Substantia nigra

Ventral tegmental area

To spinal cord

Mesolimbic system

Figure 16-9
The dopaminergic system. Two dopaminergic systems originate and terminate outside of the hypothalamus: the nigrostriatal and mesolimbic. The hypothalamus contains three intrinsic dopaminergic systems: the incertohypothalamic system with cell bodies in the caudal hypothalamus, zona incerta, and rostral periventricular nucleus and with terminals in the rostral preoptic area and septum; the tuberoinfundibular system with cell bodies in the arcuate and periventricular nuclei and terminals in the external zone of the median eminence adjacent to the primary capillary bed; and the tuberohypophysial system with cell bodies in the rostral arcuate and periventricular nuclei and terminals in the intermediate and posterior lobe of the pituitary gland. The tuberoinfundibular and tuberohypophysial dopaminergic systems are responsible for delivering dopamine to the adenohypophysis through the portal vasculature.

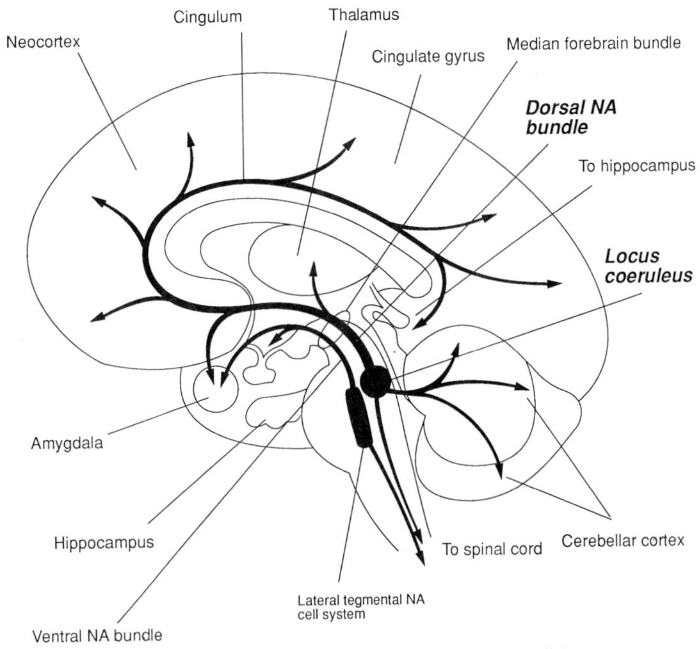

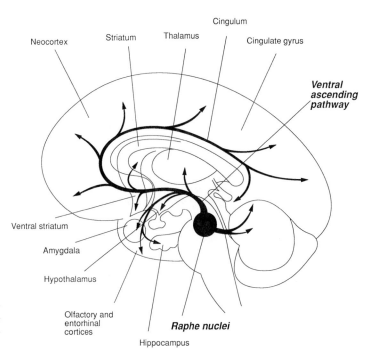

Figure 16-10
The noradrenergic system. The cell bodies of greatest importance to the hypothalamus are located in the locus ceruleus, whose efferents course toward the hypothalamus as the dorsal noradrenergic bundle. They terminate in the paraventricular and arcuate nuclei of the hypothalamus and in the preoptic area.

fiber groups leave the ventral ascending pathway as it courses through the lateral hypothalamus. One is directed laterally, and the other courses ventromedially. The ventromedial fibers innervate several hypothalamic areas, including the lateral, medial preoptic, and anterior hypothalamic areas; the dorsomedial and ventromedial nuclei; the infundibulum; and the suprachiasmatic nuclei. The OVLT also is rich in serotonin terminals.

Peptides

There are time-honored steps that must be taken to identify peptides as physiologically significant in the hypothalamus (Table 16-3). First, a quantitative bioassay must be established. A specific, dose-dependent relation must be established between the amount of peptide and the biologic response. Second, evidence must be provided

Figure 16-11
The serotinergic system. The cell bodies important to hypothalamic function are found in the dorsal and medial raphe nuclei. Axons from the dorsal raphe form the ventral ascending serotinergic pathway and enter the hypothalamus ventromedially to terminate in the anterior hypothalamus, dorsomedial and ventromedial nuclei, the suprachiasmatic nuclei, and the infundibulum. Serotinergic terminals also are found in the lateral and medial preoptic areas and the organum vasculosum of the lamina terminalis.

Table 16-3
STRATEGIES IN THE ANALYSIS OF A HYPOTHALAMIC NEUROPEPTIDE

Development of quantitative bioassay
Proof of peptidic nature
Development of extraction scheme
Chemical and physical characterization
Synthesize peptide and determine its bioactivity
Produce antibodies to the peptide
Use antibodies for immunocytochemical localization and radioimmunoassay
Isolate the cDNA encoding the precursor of the peptide

that the biologically active material is peptidic in nature. This can be established by demonstrating that proteolytic enzymes diminish or destroy the biologic activity. Third, a scheme for extraction and separation of maximal yields of the purified peptide must be devised. Fourth, chemical and physical characterization of the peptide must be performed. This consists of molecular weight characterization and amino acid composition and sequencing. Fifth, after the sequence is known, the peptide must be synthesized and the synthetic product tested for biologic activity in the bioassay. Sixth, antibodies to the peptide must be produced, and the purified antibodies must be characterized using synthetic analogs of the peptide. Seventh, immunologic approaches with the antibodies must be developed. This consists of immunocytochemistry for visualization of peptides in neural tissue and radioimmunoassay for quantitation of the concentration of the peptide in neural tissue and portal blood. The cDNA that encodes the precursor of the peptide must be isolated, and methods such as in situ hybridization histochemistry and Northern blotting developed for detecting the mRNA of the precursor are applied. Most of these approaches have been taken to identify the peptides of importance to hypothalamic function.

Six arbitrary classes of peptides are involved in hypothalamic function (Table 16-4): hypophysiotropic peptides (Fig. 16-12), which affect the function of the adenohypophysis; neurohypophysial peptides (Fig. 16-13), which control blood pressure, water retention, milk ejection, and smooth muscle contraction; brain-gut peptides (Fig. 16-14), which mostly serve neuromodulatory functions; the proopiomelanocortin (POMC)-derived peptides (Fig. 16-15), which also serve neuromodulatory roles in hypothalamic function; and the enkephalins (Fig. 16-16), which serve modulatory roles. Other peptides, such as angiotensin II, neuropeptide Y, galanin, and the endothelins (Fig. 16-17), could stand as a class by themselves because they may play a variety of neuroendocrine, neuromodulatory, neurotransmitter, and hormonal roles.

The following sections describe the peptides that are found in the hypothalamus and those found outside of the hypothalamus that affect hypothalamic function.

Hypophysiotropic peptides. Of all the neuropeptides that control the function of the adenohypophysis, the chemiarchitecture of *gonadotropin-releasing hormone* (GnRH; also luteinizing hormone–releasing hormone) is perhaps the most widely studied. GnRH-positive cell bodies fibers and terminals are not restricted to the hypothalamus. They are scattered over a continuum extending from the septal region to the premammillary region (Table 16-5). These are found in the medial septal nucleus, the nucleus of the diagonal band of Broca, the bed nucleus of the stria terminalis, the OVLT, the medial preoptic nucleus, the anterior hypothalamic area, the supraoptic nucleus, and the arcuate nucleus.

Cell bodies in the arcuate nucleus and the adjacent medial median eminence area give rise to short axons that pass to the infundibulum and form a dense plexus around the primary capillary bed of the hypothalamohypophysial portal system in the median eminence. These are the cells that control the secretion of LH from the

Table 16-4
CLASSES OF HYPOTHALAMIC PEPTIDES

Class	Function	Example
Hypophysiotropic peptides	Regulate adenohypophysis	TRH, GnRH, GHRH, CRH, somatostatin
Neurohypophysial peptides	Regulate water retention, blood pressure, milk ejection, uterine contraction	Vasopressin, oxytocin
Brain-gut peptides	Neuromodulatory, neuroendocrine	VIP, CCK, substance P
POMC-derived peptides	Neuromodulatory, neuroendocrine	Endorphins, ACTH
Dynorphin-derived peptides	Neuromodulatory, neuroendocrine	Met-enkephalin, leu-enkephalin
Other peptides	Neuromodulatory, neuroendocrine	Angiotensin II, NPY, galanin, endothelins

ACTH, corticotropin (ACTH); *CCK*, cholecystokinin; *CRH*, corticotropin-releasing hormone; *GnRH*, gonadotropin–releasing hormone; *GHRH*, growth hormone–releasing hormone; *NPY*, neuropeptide Y; *POMC*, proopiomelanocortin; *TRH*, thyrotropin–releasing hormone; *VIP*, vasoactive intestinal polypeptide.

Figure 16-12
Amino acid sequences of the hypothalamic hypophysiotropic peptides, so named because they all regulate pituitary function directly. The superscript 1 represents the amino terminus, and the carboxyl terminus is designated with larger superscript numbers. The Glu at position 1 of luteinizing hormone-releasing hormone (LHRH) and thyroid-releasing hormone (TRH) is designated pyro (p). LHRH, TRH, cortocotropin-releasing hormone (CRH), and growth hormone-releasing hormone (GHRH) all are amidated at their carboxyl termini.

LHRH: pGlu^1-His-Trp-Ser-Tyr-Gly-Leu-Arg-Pro-Gly^{10}-NH$_2$

TRH: pGlu^1-His-Pro^3-NH$_2$

CRH: Ser^1-Gln-Glu-Pro-Pro-Ile-Ser-Leu-Asp-Leu-Thr-Phe-His-Leu-Leu-Arg-Glu-Val-Leu-Glu-Met-(ovine)
Thr-Lys-Ala-Asp-Gln-Leu-Ala-Gln-Gln-Ala-His-Ser-Asn-Arg-Lys-Leu-Leu-Asp-Ile-Ala^{41}-NH$_2$

GHRH: Tyr^1-Ala-Asp-Ala-Ile-Phe-Thr-Asn-Ser-Tyr-Arg-Lys-Val-Leu-Gly-Gln-Leu-Ser-Ala-Arg-Lys-Leu-Leu-Gln-Asp-Ile-Met-Ser-Arg-Gln-Gln-Gly-Glu-Ser-Asn-Gln-Glu-Arg-Gly-Ala-Arg-Ala-Arg-Leu^{44}-NH$_2$

Somatostatin: Ala^1-Gly-Cys-Lys-Asn-Phe-Phe-Trp-Lys-Thr-Phe-Thr-Ser-Cys^{14}

adenohypophysis. In some mammals, these same cells have axons terminating directly in the neurohypophysis. These are of unknown function, but their termini in the neurohypophysis adjacent to the adenohypophysis suggests that they may play a role in adenohypophysial function.

GnRH-positive cells in the medial preoptic nucleus send projections to the suprachiasmatic nucleus and median eminence. Cells in the medial septal nucleus and the diagonal band of Broca also contribute to this projection. Other GnRH-positive cells from these areas terminate in the OVLT to form a dense plexus around the capillary network, suggesting another vascular route through which the peptide reaches the anterior pituitary. GnRH-positive cells originating in the medial preoptic nucleus, the bed nucleus of the stria terminalis, and the septum send fibers that terminate on the ependymal linings of the third and lateral ventricles. This suggests that the cerebral spinal fluid may be an additional vehicle for transport of GnRH. In the medial septal nucleus, the diagonal band of Broca, and the olfactory tubercle, a few groups of GnRH-positive cells terminate on blood vessels in this area. In addition, GnRH-positive cells in the medial preoptic nucleus, medial septal nucleus, and the diagonal band of Broca terminate in the external plexiform and glomerular layers of the olfactory bulb. The medial preoptic nucleus also provides fibers to the amygdala by way of the stria terminalis.

Although the GnRH neurons originating in the preoptic and arcuate nuclei that terminate in the median emi-

Arginine Vasopressin: Cys^1-Tyr-Phe-Glu-Asn-Cys-Pro-Arg-Gly^9-NH$_2$

Oxytocin: Cys^1-Tyr-Ile-Glu-Asn-Cys-Pro-Leu-Gly^9-NH$_2$

Figure 16-13
Amino acid sequences of the neurohypophysial peptide hormones. They share common features: both are nonapeptides, have an intramolecular disulphide bond, and are amidated at the carboxyl terminal. The differences that account for their different bioactivities are found at positions 3 and 8. In addition, a closely related pressor peptide, lysine vasopressin, differs from arginine vasopressin by substituting Lys for Arg in position 8.

nence have been shown to play a direct role in cyclic LH release, termini in other areas, such as the olfactory system, the amygdala, the habenula, and the mesencephalic gray, have not been assigned a firmly established functional role in reproductive processes. However, it is probable that these areas help to control pituitary hormone secretion and the behaviors related to reproduction that are controlled by GnRH as a neuromodulator instead of as a neurohormone.

GnRH is not confined to the central nervous system. In sympathetic ganglia of the bullfrog, a peptide that resembles GnRH and is co-localized with acetylcholine elicits prolonged excitatory postsynaptic potentials with long latencies. This would categorize GnRH as a neurotransmitter.

The *thyrotropin-releasing hormone* (TRH), the physiologic stimulator of TSH release from the adenohypophysis, is found widely throughout the nervous system of mammals (see Table 16-5). Approximately one third of the total amount of the peptide found in the brain is localized to the hypothalamus. TRH is thought to play neuromodulatory and neuroendocrine roles. The highest concentration of TRH in the hypothalamus is found in the median eminence, with significant levels in the dorsomedial, ventromedial, and arcuate nuclei. Extrahypothalamic structures, such as the preoptic and septal areas and the motor nuclei of some cranial nerves, also contain significant levels of TRH. Within the hypothalamus, TRH-positive cell bodies are found in the periventricular area, the paraventricular nucleus, the dorsomedial and ventromedial nuclei, and the arcuate nucleus and median eminence area. TRH-positive nerve terminals are found in the greatest abundance in the external layer of the median eminence with dense fiber networks in the parvicellular part of the paraventricular nucleus, the periventricular hypothalamic area, the dorsomedial nucleus, the perifornical region, the nucleus accumbens, and the bed nucleus of the stria terminalis. TRH-positive fibers are also found in the spinal cord.

Corticotropin-releasing hormone (CRH), the physiologic stimulator for the release of ACTH from the adenohypophysis, has been localized in hypothalamic and extrahypothalamic sites (see Table 16-5). CRH-positive

Substance P: Arg^1-Pro-Lys-Pro-Glu-Glu-Phe-Phe-Gly-Leu-Met^{11}-NH$_2$

VIP: His^1-Ser-Asp-Ala-Val-Phe-Thr-Asp-Asn-Tyr-Thr-Arg-Leu-Arg-Lys-Glu-
Met-Ala-Val-Lys-Lys-Tyr-Leu-Asn-Ser-Ile-Leu-Asn^{28}-NH$_2$

CCK (8): Asp^1-Tyr-Met-Gly-Trp-Met-Asp-Phe^8-NH$_2$
                                                    |
                                                  SO$_3$

NT: pGlu^1-Leu-Tyr-Glu-Asn-Lys-Pro-Arg-Arg-Pro-Tyr-Ile-Leu^{13}

Figure 16-14
Amino acid sequences of the hypothalamic brain-gut peptides, so named because they are localized and active in the brain and gastrointestinal system. *CCK*, cholecystokinin; *NT*, neurotensin; *VIP*, vasoactive intestinal polypeptide.

cells are found in greatest abundance in the paraventricular nucleus of the hypothalamus. Although most of the cells are parvicellular, some are members of the magnocellular population of this nucleus. Axons from the CRH-positive cells of the paraventricular nucleus project to a neurohemal area comprising the external zone of the median eminence. These are the cells that provide CRH to the portal vasculature bathing the adenohypophysis and stimulating the release of ACTH and β-endorphin. CRH-positive cell bodies have been identified in the supraoptic and arcuate nuclei in the hypothalamus, with additional groups of cell bodies in the dorsal raphe nucleus, the hippocampus, and the OVLT. Scattered CRH cell bodies are found throughout the lateral preoptic lateral hypothalamic continuum, and numerous cell bodies have been identified in the medial preoptic nucleus and the bed nucleus of the stria terminalis.

In the midbrain, CRH-staining cells are found in the reticular formation and the periaqueductal gray. Moreover, CRH has been co-localized with the catecholamines in the locus ceruleus. CRH-immunoreactive fibers originating from the bed nucleus of the stria terminalis enter the lateral and medial septal nuclei. Numerous fibers found within the stria terminalis and the ventral amygdalofugal pathway connect the rostral hypothalamus and basal telencephalon with the amygdala. Fibers originating

from the telencephalon or diencephalon course caudally through the median forebrain bundle and split into a dorsal pathway throughout the brain stem and a ventral pathway to the lateral part of the reticular formation. Although a well-founded physiologic role has been ascribed to the hypothalamic paraventricular–median eminence pathway, a role for the other pathways has not been similarly characterized. Because ACTH-immunoreactive cell bodies and fibers of unknown function have been found throughout the brain localized closely with CRH, it is possible that this reflects the same regulatory relation described for pituitary ACTH.

Growth hormone–releasing hormone (GHRH), the physiologic stimulator of growth hormone secretion from the adenohypophysis, has been found in hypothalamic, nonhypothalamic, and even nonneural sites (see Table 16-5). The concentration of immunoreactive GHRH in the hypothalamus is highest in the arcuate nucleus and median eminence area, which is probably a reflection of its neuroendocrine role in the pituitary. Immunocytochemical localization studies have revealed GHRH-positive cell bodies in the arcuate nucleus with short axons terminating in the arcuate nucleus and the median eminence. Some of these GHRH cell bodies also contain neurotensin, and others contain galanin. The physiologic significance of this dual packaging is unknown. Surgical

ACTH: Ser^1-Tyr-Ser-Met-Glu-His-Phe-Arg-Trp-Gly-Lys-Pro-Val-Gly-Lys-Lys-Arg-Arg-Pro-Val-
Lys-Val-Tyr-Pro-Asn-Gly-Ala-Glu-Asp-Glu-Leu-Ala-Glu-Ala-Phe-Pro-Leu-Glu-Phe^{39}

β-LPH: Glu^1-Leu-Thr-Gly-Gln-Arg-Leu-Arg-Glu-Gly-Asp-Gly-Pro-Asp-Gly-Pro-Ala-Asp-Asp-
Gly-Ala-Gly-Ala-Gln-Ala-Asp-Leu-Glu-His-Ser-Leu-Leu-Val-Ala-Ala-Glu-Lys-Lys-Asp-
Glu-Gly-Pro-Tyr-Arg-Met-Glu-His-Phe-Arg-Trp-Gly-Ser-Pro-Pro-Lys-Asp-Lys-Arg-Tyr-
Gly-Gly-Phe-Met-Thr-Ser-Glu-Lys-Ser-Gln-Thr-Pro-Leu-Val-Thr-Leu-Phe-Lys-Asn-Ala-
Ile-Ile-Lys-Asn-Ala-Tyr-Lys-Lys-Gly-Glu^{89}

α-MSH: Ac-Ser^1-Tyr-Ser-Met-Glu-His-Phe-Arg-Trp-Gly-Lys-Pro-Val^{13}-NH$_2$

γ-LPH: Glu^1-Leu-Thr-Gly-Gln-Arg-Leu-Arg-Glu-Gly-Asp-Gly-Pro-Asp-Gly-Pro-Ala-Asp-Asp-
Gly-Ala-Gly-Ala-Gln-Ala-Asp-Leu-Glu-His-Ser-Leu-Leu-Val-Ala-Ala-Glu-Lys-Lys-Asp-
Glu-Gly-Pro-Tyr-Arg-Met-Glu-His-Phe-Arg-Trp-Gly-Ser-Pro-Pro-Lys-Asp^{56}

β-END: Tyr^1-Gly-Gly-Phe-Met-Thr-Ser-Glu-Lys-Ser-Gln-Thr-Pro-Leu-Val-Thr-Leu-Phe-Lys-
Asn-Ala-Ile-Ile-Lys-Asn-Ala-Tyr-Lys-Lys-Gly^{31}

β-MSH: Asp^1-Glu-Gly-Pro-Tyr-Arg-Met-Glu-His-Phe-Arg-Trp-Gly-Ser-Pro-Pro-Lys-Asp^{18}

γ-MSH: Tyr^1-Val-Met-Gly-His-Phe-Arg-Trp-Asp-Arg-Phe-Gly^{12}

Figure 16-15
Amino acid sequences of the hypothalamic peptides derived from proopiomelanocortin. *ACTH*, adrenocorticotropic hormone; *END*, endorphin; *LPH*, lipotropin hormone; *MSH*, melanocyte-stimulating hormone.

Met-Enk: Tyr^1-Gly-Gly-Phe-Met^5

Leu-Enk: Tyr^1-Gly-Gly-Phe-Leu^5

Figure 16-16
Amino acid sequences of the enkephalins (ENK). The sole difference between them is in position 5.

isolation of the medial basal hypothalamus does not lead to a significant decline in the concentration of GHRH in the arcuate nucleus median eminence area. Virtually all of the hypothalamic GHRH originates from cells in this area.

There are also significant amounts of GHRH found in some nonneural sites. GHRH messenger RNA and newly synthesized GHRH are found in somatotropes of the adenohypophysis. This has led to the belief that some degree of growth hormone secretion is intrinsic, functioning as an autocrine agent. Other locations of GHRH cells lack a compelling physiologic explanation. GHRH has been found in the ovary and the placenta, sites at which a role for GHRH has yet to be described.

Growth hormone release-inhibiting hormone or *somatostatin* inhibits the secretion of growth hormone or somatotropin from the adenohypophysis. However, the name does not fully represent the variety of roles played by this neurohormone. Somatostatin also inhibits the release of thyrotropin and prolactin from the adenohypophysis. In addition to its location in the hypothalamus, somatostatin is widely distributed throughout the central nervous system (see Table 16-5), suggesting that it may be a neurotransmitter or neuromodulator and a neurohormone. Of interest to control of growth hormone secretion, somatostatin-positive cell bodies are quite abundant in the preoptic–anterior hypothalamic area. Parvicellular somatostatin-positive cells are also found in the paraventricular nucleus. These areas send axons as a group caudally to terminate in the suprachiasmatic nucleus and the area of the arcuate nucleus and median eminence. Fibers pass from the preoptic area and terminate on the primary capillary bed of the median eminence. This source of somatostatin directly inhibits

growth hormone secretion from the adenohypophysis, and these same preoptic somatostatin fibers synapse on GHRH cell bodies in the arcuate nucleus. This suggests two levels of inhibition of growth hormone secretion by somatostatin: directly at the somatotrope and secondarily at the GHRH neuron.

Somatostatin can be found outside of the nervous system. Within the endocrine pancreas, a specific cell type, the delta cell, synthesizes and secretes somatostatin, which is identical to that made by hypothalamic neurons. Pancreatic somatostatin plays numerous roles in the gastrointestinal tract, and somatostatin directly affects pancreatic insulin and glucagon secretion.

Neurohypophysial hormones. *Arginine-vasopressin* (AVP), which is also known as *antidiuretic hormone* (ADH), and another neurohormone, *oxytocin*, are produced in magnocellular neurons, whose cell bodies are located in the supraoptic (i.e., AVP production) and paraventricular (i.e., oxytocin production) nuclei of the hypothalamus (Table 16-6). They are synthesized as prohormones, or precursor proteins, in the cell body (Fig. 16-18). These are large molecules that consist of packaging peptides and specific axonal transport peptides, *neurophysins* (NP), and the bioactive fragment of AVP or oxytocin, which is ultimately found in the peripheral circulation. There are two forms of the NPs. NP-I or estrogen-linked neurophysin is associated with oxytocin, and NP-II or nicotine-linked neurophysin is associated with AVP. Both NPs are 9.5 to 10 kDa.

The translation product of the AVP gene is a 21-kDa glycoprotein known as *preprovasopressin* (see Fig. 16-18) that in humans consists of a 19-amino acid signal peptide at the amino terminal, vasopressin (9 amino acids), a 3-amino acid spacer sequence, NP-II (93 amino acids), another spacer sequence (1 amino acid), and a 39-amino acid peptide at the carboxyl terminal. *Provasopressin* (19 kDa) is the peptide with the 19-amino acid signal sequence cleaved and carbohydrate added posttranslationally to the carboxyl-terminal peptide. Oxytocin is synthesized in a similar fashion, except that there is no glycopeptide at the carboxyl terminus and the prohormone is 15 kDa (see Fig. 16-18). In both cases, the NP is required as a carrier protein to the axon terminal and

Angiotensin II: Asp^1-Arg-Val-Tyr-Ile-His-Pro-Phe^8

Human NPY: Tyr^1-Pro-Ser-Lys-Pro-Asp-Asn-Pro-Gly-Gln-Asp-Ala-Pro-Ala-Gln-Asp-Met-Ala-Arg-Tyr-Tyr-Ser-Ala-Leu-Arg-His-Tyr-Ile-Asn-Leu-Ile-Thr-Arg-Gln-Arg-Tyr^{36}-NH$_2$

Rat Galanin: Gly^1-Trp-Thr-Leu-Asn-Ser-Ala-Gly-Tyr-Leu-Leu-Gly-Pro-His-Ala-Ile-Asp-Asn-His-Arg-Ser-Phe-Ser-Asp-Lys-His-Gly-Leu-Thr^{29}-NH$_2$

Endothelin 1: Cys^1-Ser-Cys-Ser-Ser-Leu-Met-Asp-Lys-Glu-Cys-Val-Tyr-Phe-Cys-His-Leu-Asp-Ile-Ile-Trp^{21}

Endothelin 2: Cys^1-Ser-Cys-Asp-Ser-Trp-Leu-Asp-Lys-Glu-Cys-Val-Tyr-Phe-Cys-His-Leu-Asp-Ile-Ile-Trp^{21}

Endothelin 3: Cys^1-Thr-Cys-Phe-Thr-Tyr-Lys-Asp-Lys-Glu-Cys-Val-Tyr-Tyr-Cys-His-Leu-Asp-Ile-Ile-Trp^{21}

Figure 16-17
Amino acid sequences of some of the other peptides localized in the hypothalamus and controlling hypothalamic function. Among the unusual features of this group, the endothelins have two intramolecular disulphide bonds. *NPY,* neuropeptide Y.

Table 16-5
LOCALIZATION OF THE HYPOPHYSIOTROPIC PEPTIDES

Peptide*	Cell Bodies	Fibers	Terminals	Portal Blood†
GnRH	Medial septal nucleus, nucleus of the diagonal band of Broca, bed nucleus region of stria terminalis, OVLT, medial preoptic nucleus, anterior hypothalamic area, arcuate nucleus, median eminence, olfactory tubercle	Continuum from septal region to premamillary nucleus	Median eminence, neurohypophysis, suprachiasmatic nucleus, ependymal lining of ventricles, olfactory bulb, amygdala	+
TRH	Periventricular area, paraventricular nucleus, dorsomedial/ventromedial nuclei, arcuate nucleus	Paraventricular nucleus, periventricular hypothalamic area, dorsomedial nucleus perifornical region, nucleus accumbens, bed nucleus of stria terminalis, spinal cord	Median eminence	+
CRH	Paraventricular nucleus, supraoptic nuclei, arcuate nuclei dorsal raphe nucleus, hippocampus, OVLT, medial preoptic nucleus, bed nucleus of stria terminalis, locus ceruleus	Septal nuclei, stria terminalis, median forebrain bundle	Median eminence	+
GHRH	Arcuate nucleus		Median eminence	+
SS	Preoptic/anterior hypothalamic area, paraventricular nucleus		Median eminence, Suprachiasmatic nucleus	+

*ACTH, corticotropin (ACTH); CCK, cholecystokinin; CRH, corticotropin-releasing hormone; GnRH, gonadotropin–releasing hormone; OVLT, organum vasculosum of the lamina terminalis; SS, somatostatin; TRH, throtropin-releasing hormone; VIP, vasoactive intestinal polypeptide.
† +, occurs in concentrations greater than in peripheral blood.

presumably prolongs the half-life of the neuropeptide. The peptide-NP complex is exocytosed into fenestrated capillaries where the respective NP is cleaved from oxytocin or AVP. It was once thought that the cell bodies of the supraoptic nucleus contained exclusively AVP and that of the paraventricular nucleus contained exclusively oxytocin. We now know that both types of nuclei are found in each area.

Axons from each of these nuclei form the hypothalamohypophysial tract that terminate on sinusoids in the neurohypophysis. Oxytocin and AVP can also be transported from the neurohypophysis to the adenohypophysis through the short portal vessels connecting these two areas. The paraventricular nucleus also sends fibers to the median eminence where they terminate on the primary capillaries from which the long portal vessels project to bathe the adenohypophysis. This is the pathway by which AVP reaches the corticotrope and stimulates ACTH secretion. AVP is an accessory ACTH-releasing factor of hypothalamic origin. The caudal portion of the paraventricular nucleus contains parvicellular or small neurons that contain mostly oxytocin but some AVP. These neurons also project to the median eminence and other parts of the brain and spinal cord. Significant quantities of oxytocin are found in portal blood. The parvicellular part of the paraventricular nucleus also sends fibers to the locus ceruleus, the parabranchial nuclei, the dorsal motor vagal nucleus, the nucleus of the solitary tract, the midbrain central gray, and the dorsal horn of the spinal cord. Localization of AVP and oxytocin fibers in the lower

Table 16-6
LOCALIZATION OF NEUROHYPOPHYSIAL PEPTIDES

Peptide	Cell Bodies	Fibers	Terminals	Portal Blood*
Oxytocin	Paraventricular nucleus, supraoptic nucleus	Hypothalamohypophysial tract	Neurohypophysis, median eminence	+
Arginine vasopressin	Supraoptic nucleus, paraventricular nucleus, suprachiasmatic nucleus, bed nucleus of stria terminalis, nucleus of diagonal band of Broca, amygdala	Hypothalamohypophysial tract, septum, thalamus, hippocampus	Neurohypophysis, median eminence organum vasculasum of the lamina terminalis, dorsomedial nucleus	+

* +, occurs in concentrations greater than in peripheral blood.

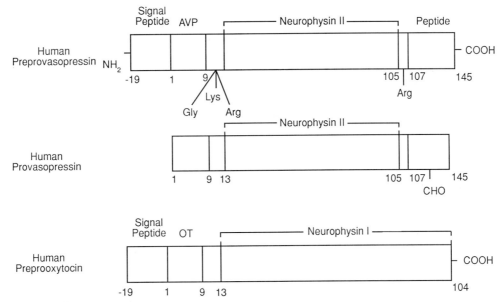

Figure 16-18
The structure of human preprovasopressin, provasopressin, and preprooxytocin. Preprovasopressin is 21 kDa and consists of a signal peptide, the AVP sequence, a tripeptide spacer, neurophysin II (also called nicotine-linked neurophysin), a spacer glycosylation signal (Arg), and a 39-amino acid carboxyl terminal. Provasopressin is a 19-kDa product of preprovasopressin, from which the signal sequence has been deleted and carbohydrate (CHO) added to the C-terminal peptide after translation. Preprooxytocin is smaller than preprovasopressin with a different neurophysin, and the post translational modifications do not include glycosylation.

brain stem and autonomic centers may reflect their roles in such peripheral processes as regulation of blood pressure and lactation. Other parvicellular neurons are found in the suprachiasmatic nucleus. These contain almost exclusively AVP and project to the OVLT, the dorsomedial hypothalamic nucleus, and the thalamus.

Outside the supraoptic and paraventricular nuclei, AVP cells have been found in the bed nucleus of the stria terminalis, nucleus of the diagonal band of Broca, lateral septum, anterior amygdala, lateral habenular nucleus, mesencephalic central gray, and locus ceruleus. Location of AVP fibers in the septum is compatible with a role of the peptide in thermoregulation, and the presence of fibers in the mediodorsal thalamic nucleus, the hippocampus, and the neocortex is compatible with a role for this peptide in learning and memory.

Brain-gut peptides. Brain-gut peptides are so named because their activities have been characterized by immunocytochemistry within the gastrointestinal tract and the hypothalamus (Table 16-7). Many of them have been localized in the hypothalamus and have had neuroendocrine roles identified.

Substance P (SP) has been found in extracts of brain and intestine. Its structure is known, and it has been implicated in pain perception, baroreception, and chemoreception. In the monkey, SP cell bodies have been found in the most lateral portions of the arcuate nucleus. Fibers pass to the external zone of the median eminence and to the neurohypophysis. In the rat, SP-positive cells are found in the medial and lateral preoptic areas, the ante-

rior hypothalamic area, and the dorsomedial and ventromedial nuclei. SP-containing afferent pathways project to the supraoptic and paraventricular nuclei and to the arcuate nucleus. Despite the location of SP fibers in the neurohemal zone of the median eminence, the level of SP in portal blood is essentially equivalent to that of peripheral blood, eliminating it as a neurohormone. SP and its receptors are also found within the adenohypophysis. SP may be found in the anterior lobe as a paracrine agent rather than as a neurohormone. SP plays a role in the secretion of all the important anterior pituitary hormones by acting directly on the specific cells of the gland or indirectly as a neuromodulator, affecting the release of the releasing hormones of hypothalamic origin.

Vasoactive intestinal polypeptide (VIP) occurs in large quantities throughout the gastrointestinal tract, where it plays multiple roles in digestive processes. It is vasodilatory, glycogenolytic, and lipolytic. It enhances insulin secretion, inhibits gastric acid production, and stimulates secretion from the exocrine pancreas and small intestine. Because it exerts some of its effects through vascular pathways, it fits the description of a true hormone. Neuronal VIP is found in highest concentration in the cerebral cortex, where it acts as an excitatory neurotransmitter or neuromodulator. Within the hypothalamus, the suprachiasmatic nuclei contain dense concentrations of VIP-positive cell bodies. Efferents from these course dorsally and then split into a dense rostrodorsal component and a less dense caudal component. The rostrodorsal fibers terminate on the paraventricular nucleus, and the caudal fibers terminate at the dorsomedial, ventromedial,

Table 16-7
LOCALIZATION OF BRAIN-GUT PEPTIDES

Peptide*	Cell Bodies	Terminals	Portal Blood†
Substance P	Arcuate nucleus, preoptic area, anterior hypothalamic area, dorsomedial/ventromedial nuclei	Median eminence, neurohypophysis, supraoptic nucleus, paraventricular nucleus, arcuate nucleus	−
VIP	Suprachiasmatic nucleus	Paraventricular nucleus dorsomedial/ventromedial nuclei	+
CCK	Cortex, striatum, amygdala, supraoptic nucleus, paraventricular nucleus, neurohypophysis, preoptic area, dorsomedial nucleus	Median eminence	−
NT	Preoptic area, anterior hypothalamic area, medial preoptic area, paraventricular nucleus, dorsomedial nucleus, arcuate nucleus	Median eminence, neurohypophysis	+

*CCK, cholecystokinin; NT, neurotensin; VIP, vasoactive intestinal polypeptide.
†+, occurs in concentrations greater than in peripheral blood; −, occurs in concentrations that are the same or lower than in peripheral blood.

and premammillary nuclei. Like SP, VIP is found in high concentrations in the adenohypophysis. Unlike SP, it is also found in high concentrations in portal blood. There is evidence that it may be synthesized in the adenohypophysis, and VIP has been shown to affect adenohypophysial hormone secretion as a transmitter or neuromodulator, as a neurohormone, and as a local autocrine or paracrine agent.

Cholecystokinin (CCK) is synthesized in the duodenum and stimulates the secretion of pancreatic enzymes and the ejection of bile from the gall bladder. Unfortunately, CCK occurs in multiple molecular forms, which complicates the description of its tissue distribution. Duodenal CCK is composed of 33 or 39 amino acids. The carboxyl-terminal octapeptide of CCK (CCK8), which has full biologic activity, shares a pentapeptide sequence with gastrin, another gastrointestinal hormone. Although CCK8 occurs throughout the central nervous system, gastrin-like peptides are found only in the pituitary and hypothalamus. The highest concentrations of immunoreactive CCK8 are found in the cortex, striatum, and amygdala, with lesser amounts found in the hypothalamus. CCK-like immunoreactivity has been found in the magnocellular systems of the supraoptic and paraventricular nuclei of the hypothalamus and in the neurohypophysis. In some neurons, oxytocin and CCK are co-localized. Physiologic perturbations that stimulate the release of AVP and oxytocin lower neurohypophysial CCK. Parvicellular CCK-positive cells project axons that terminate in the median eminence. CCK cells also are found in the medial preoptic area and the dorsomedial and supramamillary nuclei. Functionally, CCK has been implicated as a neuromodulator in the control of pituitary hormone release, and as such, it probably facilitates the release of

the hypothalamic-releasing hormones, oxytocin, and AVP. It has also been shown that CCK may be co-packaged with many of the hypothalamic peptides. There is no solid direct evidence implicating CCK as a neurohormone.

Neurotensin (NT) is a peptide consisting of 13 amino acids that was first isolated from brain tissue and later from the intestine. In general, NT is a peptide neurotransmitter that is found in highest concentrations in the hypothalamus and seems to play a role in the control of release of the adenohypophysial hormones. NT-positive cell bodies are found in the preoptic and anterior hypothalamic areas, the medial preoptic nucleus, the magnocellular and parvicellular zones of the paraventricular nucleus, the arcuate nucleus, and the dorsomedial nucleus. In the arcuate nucleus, NT is co-localized in tuberoinfundibular dopaminergic neurons. The biologic significance of this common packaging is not fully understood, but there is the suggestion that NT mediates the release of dopamine into the portal vasculature. NT-positive fibers are also localized in the external zone of the median eminence. NT occurs in portal blood and may assume the role of a classic neurohormone. Although NT placed into the brain can affect the secretion of prolactin, growth hormone, TSH, and LH, placement directly into pituitary cell cultures is only effective at supraphysiologic doses. This suggests that NT acts within the hypothalamus as a neuromodulator or neurotransmitter and perhaps not at the pituitary cell directly. In addition to NT fibers terminating at the external zone of the median eminence, some NT-positive axon terminals are found in the neurohypophysis. These probably originate in the paraventricular nucleus and play a role in the release of oxytocin and AVP. The adenohypophysis contains cells that stain NT-positive. It seems unlikely that this material arises from neuronal sources but probably is synthesized directly in the pituitary and plays a paracrine or autocrine role in the secretion of one or more of the adenohypophysial hormones.

Proopiomelanocortin-derived peptides. Proopiomelanocortin (POMC) is a large-molecular-weight precursor protein (265 amino acids) that in humans (i.e., there are subtle differences among mammals) is posttranslationally cleaved into moieties ACTH (39 amino acids), β-lipotropin (β-LPH; 89 amino acids), and a 16-kDa amino-terminal fragment of 76 amino acids (Fig 16-19). Each of these is further cleaved enzymatically to yield α-melanophore-stimulating hormone (α-MSH = ACTH$_{1-13}$), and corticotropin-like intermediate lobe peptide (CLIP = ACTH$_{18-39}$). β-LPH is further cleaved into γ-LPH (β-LPH$_{1-56}$) and the endogenous opioid β-endorphin (β-endorphin = β-LPH$_{59-89}$). β-MSH is a cleavage product of LPH (amino acids 39–56) and γ-MSH is a fragment of the 16-kDa amino-terminal peptide. These are widely distributed in the central nervous system (Table 16-8).

Outside the adenohypophysis, ACTH is found in cells of the arcuate nucleus and median eminence area. The cells are diffusely distributed throughout this area that extends rostrally to the retrochiasmatic area, caudally to the submammillary region, and dorsally to the area between the ventricular surface and the ventromedial nucleus of the hypothalamus. That this activity is not a product of the corticotropes of the adenohypophysis is emphasized by the fact that hypophysectomy does not influence the amount of ACTH in this area. ACTH is formed in the brain.

Because ACTH and other peptides are common products of POMC, it is not surprising that β-LPH, α-MSH, β-MSH, and β-endorphin are co-localized with ACTH in these neurons. The neurons give rise to numerous ACTH-positive fibers that are distributed widely throughout the brain. Within the hypothalamus, ACTH fibers terminate in the anterior, mediobasal, and periventricular areas of the hypothalamus. In the periventricular area, the fibers penetrate the ependymal lining of the third ventricle. ACTH is measurable in the cerebral spinal fluid. ACTH terminals are found in the dorsomedial nucleus, the magnocellular and parvicellular portions of the paraventricular nucleus, and the OVLT. Other ACTH terminals are found in the external zone of the median eminence close to the portal capillaries and within the neurohypophysis. Terminals are also found in the medial preoptic area. β-LPH cell bodies are found in the arcuate nucleus and median eminence area, with fibers projecting to various areas of the brain. ACTH-positive fibers and β-LPH fibers usually project to the same areas of the brain. β-LPH is also found in the corticotropes of the adenohypophysis. These common distribution patterns are not surprising given the common precursor of both.

α-MSH is secreted from the intermediate lobe of the pituitary in all vertebrates. Although it has a dramatic skin-coloring effect in poikilotherms, the role for α-MSH in homeotherms is uncertain. There is some suggestion that it may participate in hormone secretion from the adenohypophysis. Within the hypothalamus, α-MSH–positive cell bodies and fibers are found in the same areas of the arcuate nucleus and median eminence that were described for ACTH. Most of the fibers project to the same areas as the ACTH-positive groups. There is a second group of α-MSH–positive cells that are distinct from those in the medial basal hypothalamus. These cells are concentrated in the area between the dorsomedial nucleus and in the fornix and in the lateral hypothalamic area. These cells do not co-localize α-MSH with any other of the POMC-derived peptides, suggesting that the biosynthetic route of α-MSH in these cells may be different from that of the mediobasal hypothalamus or the intermediate lobe of the pituitary. Fibers from this group project to the caudate-putamen complex, the neocortex, and various parts of the hippocampus.

As with the POMC-derived peptides, β-endorphin–positive cell bodies are most numerous in the arcuate nucleus and median eminence area of the hypothalamus,

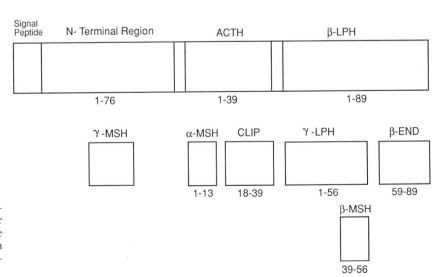

Figure 16-19
Human proopiomelanocortin and its cleavage products. The numbers below each bar represent the number of amino acids of the parent peptides (*upper bar*) or the position in the parent peptides from which the products are cleaved (*middle and lower bars*).

Table 16-8
LOCALIZATION OF PROOPIOMELANOCORTIN-DERIVED PEPTIDES

Peptide*	Cell Bodies	Terminals	Portal Blood[†]
ACTH	Arcuate nucleus, median eminence	Anterior hypothalamic area, periventricular area, dorsomedial nucleus, paraventricular nucleus, OVLT, median eminence, preoptic area	+[‡]
β-LPH	Same as ACTH		ND
γ-MSH	Same as ACTH		+[‡]
β-END	Same as ACTH		+

*ACTH, corticotropin (ACTH); β-LPH, β-lipotropin; γ-MSH, γ-melanophore-stimulating hormone; β-END, β-endorphin; OVLT, organum vasculosum of the lamina terminalis.
[†] +, occurs in concentrations greater than in peripheral blood; ND, not determined.
[‡] Probably by retrograde blood flow from the pituitary.

and the course of their efferent projections is similar to ACTH, LPH, and MSH. Similarly, β-endorphin is found in the intermediate lobe of the pituitary. β-endorphin is also found in significant concentrations in hypophysial portal blood. β-endorphin qualifies as a neurotransmitter, neuromodulator, and neurohormone. Because β-endorphin binds to opiate receptors throughout the nervous system, it has been classified an endogenous opioid.

Enkephalins. The opioid peptides are derived from three different precursors. The derivation of β-endorphin from POMC was described earlier. Smaller opioids, the enkephalins are pentapeptides derived from larger molecules known as *proenkephalins* (Fig. 16-20). One, proenkephalin A (50 kDa, 267 amino acids), contains four copies of methionine-enkephalin (met-enkephalin), one copy of leucine-enkephalin (leu-enkephalin), and one copy each of a met-enkephalin carboxyl-terminal hepta-

peptide and a met-enkephalin carboxyl-terminal octapeptide. The other, proenkephalin B (i.e., prodynorphin) contains three copies of leu-enkephalin.

The most widely distributed opioid peptides are the enkephalins, with met- and leu-enkephalin found in the same areas. Met-enkephalin usually is found in higher concentrations than leu-enkephalin. In the hypothalamus (Table 16-9), enkephalin-positive cell bodies are found in the supraoptic and paraventricular nuclei. Enkephalin-positive efferents project from these areas and terminate in the external zone of the median eminence adjacent to the portal vessels and in the neurohypophysis. Met-enkephalin has been found in portal blood. Cells in the intermediate lobe of the pituitary contain the hepta- and octapeptide of met-enkephalin but not free met-enkephalin. In the adenohypophysis, gonadotropes contain all forms of met-enkephalin. These are not the same gonadotropes that co-localize β-endorphin. Moreover, there is

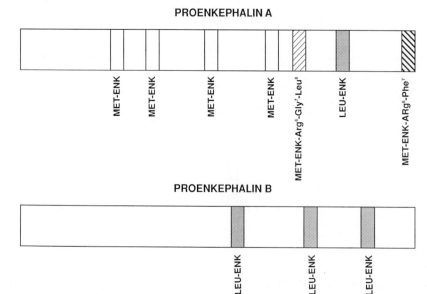

Figure 16-20
The structure of proenkephalin A and proenkephalin B. Notice the four repeating Met-ENK sequences, the single Leu-ENK, and each single Met-ENK heptapeptide and octapeptide characteristic of proenkephalin A. Proenkephalin B is characterized by three intramolecular Leu-ENK sequences.

Table 16-9
LOCALIZATION OF ENKEPHALINS

Peptide	Cell Bodies	Terminals	Portal Blood*
Met-enkephalin	Supraoptic nucleus, paraventricular nucleus preoptic area, dorsomedial/ventromedial nuclei	Median eminence, neurohypophysis	+
Leu-enkephalin	Same as met-enkephalin		+

*+, occurs in concentrations greater than in peripheral blood.

a population of gonadotropes that also contain prodynorphin. The medial preoptic nucleus and the dorsomedial and ventromedial hypothalamic nuclei also contain enkephalin-positive cell bodies.

The enkephalins appear to regulate pituitary hormone secretion by acting as neuromodulators or neurotransmitters. In the neurohypophysis, enkephalin inhibits the release of oxytocin and AVP. In the adenohypophysis, enkephalin inhibits LH and stimulates prolactin, growth hormone, and ACTH secretion by acting within the hypothalamus as a neuromodulator or neurotransmitter. The co-localization of the enkephalins with pituitary hormones suggests a paracrine or autocrine role of enkephalin, but the only direct effect has been shown in the neurohypophysis on the inhibition of oxytocin and AVP.

Other hypothalamic peptides. Several neuropeptides have been described, on the basis of their neuroanatomic location and pharmacologic studies, as possible important regulators of hypothalamic function (Table 16-10). Although the evidence for their physiologic significance is incomplete, they should be mentioned as potential physiologic regulators of hypothalamic function.

One peptide, *angiotensin II*, plays a role in vasoconstriction, sodium retention, antidiuresis, and drinking behavior through direct actions on peripheral structures such as the adrenal cortex and kidney; action on the circumventricular organs such as the OVLT, the area postrema, and the subfornical organ; and direct action on the hypothalamus. There is physiologic evidence that angiotensin II affects the secretion of LH and prolactin from the adenohypophysis through neuromodulator or neurotransmitter, neurohumoral, and paracrine or autocrine roles.

Angiotensin II is formed by the action of a renal proteolytic enzyme, renin, acting on a peptide produced in the liver, *angiotensinogen*, to form a circulating decapeptide, *angiotensin I*. Angiotensin I is cleaved by an *angiotensin-converting enzyme* (ACE), produced in the lungs to form the biologically active octapeptide, angiotensin II. Peripherally, angiotensin II acts on smooth muscle in arterial walls to promote vasoconstriction and raise blood pressure. Application of angiotensin II directly to the circumventricular organ also evokes an increase in blood pressure, secretion of AVP, and short-latency drinking behavior. The circumventricular organs are outside the blood-brain barrier and possess a significant number of angiotensin II receptors. Angiotensin II stimulates the adrenal cortex to secrete aldosterone, the hormone that promotes sodium retention by the nephron. Some hypothalamic and extrahypothalamic structures bear angio-

Table 16-10
LOCALIZATION OF OTHER HYPOTHALAMIC PEPTIDES

Peptide	Cell Bodies	Terminals	Portal Blood*
Angiotensin II	Paraventricular nucleus, Supraoptic nucleus, Adenohypophysis	Median eminence, neurophypophysis, dorsomedial nucleus	ND
Neuropeptide Y	Arcuate nucleus, median eminence, dorsomedial nucleus, locus ceruleus	Medial preoptic area, anterior hypothalamic area, periventricular area; suprachiasmatic, supraoptic, paraventricular, arcuate, and ventromedial nuclei; median eminence	+
Galanin	Supraoptic, paraventricular, and arcuate nuclei	Median eminence, neurohypophysis	+

*ND, not determined; +, occurs in concentrations greater than in peripheral blood.

tensin II receptors and are sensitive to the application of angiotensin II. The preoptic area contains a large number of angiotensin II receptors, and its cells increase their firing rate when angiotensin II is applied microiontophoretically. These cells mediate the dipsogenic effects of angiotensin II. The paraventricular nucleus is also sensitive to angiotensin II. It is not clear how peripheral angiotensin II gains access to these centers, but there is abundant evidence for the existence of angiotensin II–producing elements within the central nervous system. Angiotensin II–producing cell bodies are found within the magnocellular cells of the paraventricular nucleus and within the supraoptic nucleus. The efferent projections of these cells terminate within the median eminence and within the neurohypophysis. Angiotensin II terminals are also concentrated in the dorsomedial nucleus of the hypothalamus and scattered throughout the medial basal hypothalamus.

It has been reported that angiotensin II and AVP are co-packaged and that angiotensin II, renin, and oxytocin are co-packaged. In the adenohypophysis, angiotensin II has been reported to be packaged in gonadotropes. These various locations of angiotensin II–positive cells and terminals explain the neuromodulator or neurotransmitter roles (e.g., circumventricular organ), the neuroendocrine role (e.g., supraoptic, paraventricular nuclei; neurohypophysis), and the paracrine or autocrine role (e.g., gonadotrope of the adenohypophysis) of angiotensin II.

Neuropeptide Y (NPY) is a highly conserved 36-amino acid peptide that is widely distributed in the central nervous system in many mammals. Of particular importance to hypothalamic function are the extensive networks of NPY-positive fibers and terminals within the medial preoptic area; the periventricular and anterior hypothalamic areas; the suprachiasmatic, supraoptic, and paraventricular nuclei; the arcuate and ventromedial nuclei; and the median eminence. Significant concentrations of NPY are found in hypophysial portal blood. Within the hypothalamus, NPY-positive cells are distributed in the arcuate nucleus and median eminence area and in the dorsomedial nucleus. Much of the hypothalamic NPY originates from noradrenergic cells outside of the hypothalamus. These cells are found in the lateral reticular medulla, the nucleus tractus solitarius region, the locus ceruleus, and subceruleus. Transection of ascending noradrenergic fibers does not eliminate but substantially decreases NPY immunoreactivity in various hypothalamic areas.

NPY has multiple actions within and outside the central nervous system. Most striking is its effect on the adenohypophysis and on some behaviors. NPY plays a role in the regulation of gonadotropin secretion by acting within the hypothalamus as a neuromodulator or neurotransmitter affecting GnRH secretion and a neurohormone directly affecting LH secretion. NPY has also been implicated in the control of secretion of ACTH, growth hormone, and prolactin from the pituitary. It may act as a neurotransmitter affecting the secretion of AVP from the neurohypophysis. NPY is synthesized in a subpopulation of thyrotropes in the adenohypophysis, suggesting a par-

acrine or autocrine role. NPY also controls eating behaviors through intrahypothalamic pathways.

Galanin is a highly conserved 29-amino acid peptide that is widely distributed throughout the central and peripheral nervous system. Within the hypothalamus, galanin-positive cell bodies are found in the supraoptic and paraventricular nuclei and the arcuate nucleus. Dense efferent fibers from these areas terminate in the external and internal layer of the median eminence and in the neurohypophysis. Galanin-like immunoactivity and its message is expressed in somatotropes, lactotropes, and thyrotropes within the adenohypophysis. The expression in thyrotropes and expression in lactotropes is positively regulated by thyroid hormones and estrogen, respectively. Galanin has been implicated as a neurotransmitter or neuromodulator in the control of growth hormone, ACTH, TSH, and LH and prolactin secretion. In fact it has been found to co-localize with CRH and GnRH in the hypothalamus. Galanin is secreted from cells of the adenohypophysis and is found in portal blood. Galanin affects pituitary hormone secretion as a neurotransmitter or neuromodulator, as a neurohormone, and as a paracrine or autocrine factor.

Endothelins are a family of regulatory peptides with vasoconstrictor activity originally isolated from the incubation media of vascular endothelial cells. The one designated ET-1 is a 21-residue peptide that contains two intramolecular disulfide bonds. Two related peptides, ET-2 and ET-3, differ by 2 and 6 amino acid residues, respectively. The localization of the endothelins in the supraoptic and paraventricular nuclei of the hypothalamus and in the adenohypophysis and neurohypophysis, and the presence of endothelin receptors in the hypothalamus and the adenohypophysis and neurohypophysis have prompted inquiry into the role of the endothelins in pituitary hormone secretion. The endothelins act within the hypothalamus to enhance LH secretion through stimulation of GnRH. Within the pituitary, they are capable of stimulating LH, FSH, TSH, and ACTH secretion directly. They are potent inhibitors of prolactin secretion but exert no action on growth hormone secretion from the pituitary. It is not known if the endothelins are found in the portal circulation. These data suggest that the endothelins can act as neuromodulators or neurotransmitters or even in a paracrine or autocrine manner to affect pituitary hormone secretion.

Techniques for Studying Hypothalamic Function

Many techniques have been employed to study hypothalamic function. The following sections offer descriptions of some of the approaches available.

☐ Sampling methods

The responses to stimuli suspected of involving the hypothalamus are studied at several levels. The endpoint or dependent variable may be the response itself. It is possible to manipulate the hypothalamus and observe alterations of behaviors, temperature regulation, or pituitary hormone secretion. The alterations of the biogenic amines and neuropeptides accompanying natural or artificial stimuli can be measured directly. In making these measurements, the investigator is confronted with the problems of using a method of collecting the sample that will not affect the quantitation of material and of using reliable measuring tools.

The content of neurotransmitters and neurohormones in hypothalamic tissue can be measured (Table 16-11). Although there are several brain microdissection techniques available, a particularly innovative and useful approach is the *micropunch technique*. This involves punching from fresh or frozen sections of brain areas as small as nuclei for subsequent measurement of content of biogenic amine or neuropeptide. The area is "punched" with needles constructed from stainless steel tubing. The dimension of the punched area is determined by the size of the needle. The pellet is blown out into a dish or tube for homogenization and extraction. Using this technique, large numbers of samples can be rapidly processed.

Although the micropunch technique represents a highly useful method for estimating tissue content of neurotransmitters and neurohormones, its utility is limited by the fact that a single animal can only be sampled at a single point in time. This has been overcome by two related methodologies: push-pull perfusion and microdialysis (see Table 16-11). Each method has the distinct advantage of multiple sampling over time. In *push-pull perfusion*, concentric stainless steel cannulae are implanted so that the tip of the inner cannula is located at the desired site of study (Fig. 16-21). Artificial cerebral spinal fluid (CSF) lacking the biogenic amines and neuropeptides is pushed through the inner cannula and instantaneously pulled through the outer cannula into an appropriate receptacle. Assuming the push-pull rates are matched, the pulled CSF should be rich in biogenic amines and neuropeptides.

The related technique, *microdialysis*, has been likened to the implantation of an artificial blood vessel in tissue. A probe bearing a small piece of semipermeable dialysis membrane at the end is implanted into the hypothalamus. The end is localized in the area of interest for study. Artificial CSF or saline is pumped through the probe and recovered. Theoretically, amines or peptides can diffuse

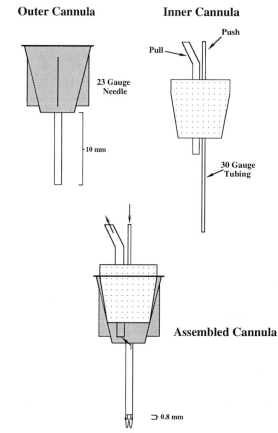

Figure 16-21
The structure of the inner, outer, and assembled push-pull perfusion cannula. The arrows indicate the direction of flow through the assembled cannula.

from the area of higher concentration in the brain to the area of lowest concentration on the probe side of the dialysis tubing. The size exclusion selectivity of the membrane determines the size of the molecule diffusing into the probe, and the length of the membrane determines the amount of tissue sampled.

The concentration of biogenic amines and neuropeptides can be measured in the microscopic vessels of the hypothalamohypophysial portal vessels connecting the median eminence with the adenohypophysis (see Table 16-11). Collection of the blood in these vessels is free from dilution by peripheral blood and involves a com-

Table 16-11
SAMPLING METHODS FOR STUDYING HYPOTHALAMIC FUNCTION

Methods	Use
Micropunch	Determine amine or peptide content of various areas of the hypothalamus
Push-pull perfusion	Measure amine or peptide dynamics in cerebrospinal fluid
Microdialysis	Measure amine or peptide dynamics in cerebrospinal fluid
Collection of hypophysial portal or peripheral plasma	Measure neurohormones in portal blood and peripheral plasma

plex ventral surgical approach to expose the median eminence. Once exposed, the stalk is cut and placed inside a polyethylene cannula. Using this procedure, blood is collected from all of the cut portal vessels simultaneously. Although this procedure involves experimental manipulation and collection of portal blood under anesthesia in rats, surgical approaches have been developed to collect portal blood from unanesthetized sheep and monkeys.

□ Methods of quantitating hypothalamic function

Assay techniques

There are two modern assay techniques used to measure catecholamines and indolamines in tissue, media, CSF, and blood (Table 16-12). One technique, the *microradioenzymatic assay*, takes advantage of the fact that the catecholamines and indolamines are methylated in vivo. Dopamine, norepinephrine, and epinephrine are *O*-methylated in vivo to form methoxytyramine, normetanephrine, and metanephrine, respectively, and serotonin is *O*-methylated and *N*-acetylated to form melatonin (Fig. 16-22).

Radioenzymatic assays exploit these metabolic fates by incorporating ^{3}H into the structure of the *O*-methylated derivatives. In the case of the catecholamines, this is accomplished by using the enzyme catechol-*O*-methyl transferase and, as methyl donor, ^{3}H-*S*-adenosyl methionine (^{3}H-SAM). Similarly, the serotonin assay is based on the ability of hydroxyindole-*O*-methyl transferase to catalyze the transfer of ^{3}H-methyl group of SAM to *N*-acetyl serotonin and form melatonin. Because the amount of radiolabeled product is proportional to unlabeled substrate, the isolation of these products by thin layer chromatography represents a quantitative estimate of substrate.

The other widespread approach to quantitating the catecholamines is the use of *high-performance liquid chromatography* coupled with *electrochemical detection*

(HPLC-EC; see Table 16-12). Using this technique, the separation of catecholamines is achieved with an analytical column packed with C_{18}-reverse-phase material. This material allows resolution of catecholamines, their precursors, and metabolites and of serotonin and its metabolites. Resolution of sample molecules takes place by their differential interactions with the mobile-phase solvent and the column packing material. Distinct bands of solute form during passage through the column. Resolution of the solutes is controlled by pH, ionic strength, the nature and concentration of aqueous phase, and the concentration of the organic components of the mobile organic phase. Quantitation is achieved by eluting the resolved solutes through the electrochemical detector. The potential applied to the detector's cell favors oxidation of the catecholamine. For a given set of operating conditions, the oxidative current is directly proportional to the concentration of electroactive species in solution.

Related to the HPLC-EC procedure is a method for determining catecholamine flux in vivo in neural tissue. This procedure, known as *in situ voltammetry*, involves stereotaxic placement of a carbon-based microelectrode. As a potential is applied and increased, the catecholamines in a thin surface adjacent to the electrode are oxidized. The magnitude of the oxidizing current generated is a function of the concentration of electroactive species in solution. The potential at which the current appears is specific for particular catecholamines. Unfortunately, this technique cannot differentiate subtle differences in side-chain groups and dopamine, norepinephrine, and epinephrine cannot be adequately differentiated.

The neuropeptide content in various areas of the hypothalamus or its concentration in portal blood is routinely measured by *radioimmunoassay* (RIA; see Table 16-12). RIAs also are used to measure the hormones of the anterior and posterior pituitary, which the hypothalamus controls. RIAs depend on the ability of relatively specific antibodies to recognize the unlabeled species of neuropeptide or hormone after it has been labeled with a radioactive tag such as ^{125}I. Because the binding sites on the antibody are the same and specific, there is a competition between labeled and unlabeled species for the binding site. The amount of binding of labeled peptide is inversely proportional to the amount of unlabeled peptide with which it competes. The unlabeled peptide would be represented by known amounts of standard or unknown amounts extracted from tissue or blood.

With the advent of modern molecular biologic techniques, it is possible to measure the amount of peptide in the hypothalamus and to measure the regulation of neuroendocrine peptide gene expression by quantitation of specific messenger RNA. Most of the methods for measuring neuroendocrine peptide are some variant of a *hybridization assay* (see Table 16-12). The basis for the assay is to manipulate the animal in vivo or the cells in vitro with a specific treatment and then to isolate cell nuclei. During the isolation, RNA polymerases remain bound to the genes being transcribed. The nuclei are then incubated in vitro with radiolabeled nucleotide triphosphates, and the polymerases continue to transcribe the

Table 16-12
METHODS OF QUANTITATING HYPOTHALAMIC FUNCTION

Methods	Use
Microradioenzymatic assay	Measure catecholamines or indoleamines
High-performance liquid chromatography/electrochemical detection	Measure catecholamines or indoleamines
In situ voltammetry	Measure catecholamines or indoleamines
Radioimmunoassay	Measure neuropeptides or pituitary hormones
Hybridization assay	Measure peptide message

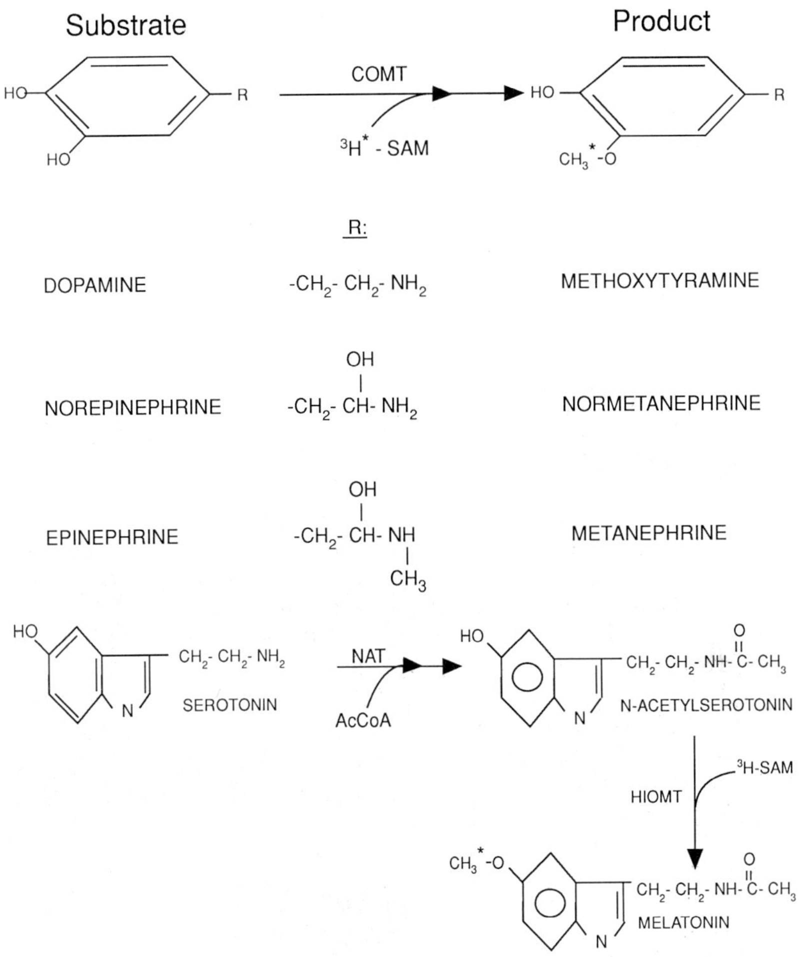

Substrate   **Product**

COMT

$^3H^* - SAM$

R:

| DOPAMINE | $-CH_2- CH_2- NH_2$ | METHOXYTYRAMINE |

NOREPINEPHRINE $-CH_2- \overset{OH}{\underset{|}{CH}}- NH_2$ NORMETANEPHRINE

EPINEPHRINE $-CH_2- \overset{OH}{\underset{|}{CH}}- \underset{|}{NH}$ METANEPHRINE
$\qquad\qquad\qquad\qquad\qquad\qquad CH_3$

Figure 16-22
The critical reactions in the microradioenzymatic assay for catecholamines (i.e., dopamine, norepinephrine, epinephrine) and indoleamines (i.e., serotonin). The amount of the tritiated methylated products (e.g., methoxytyramine, normetanephrine, metanephrine, melatonin) is proportional to the amount of the respective starting substrate (e.g., dopamine, norepinephrine, epinephrine or serotonin). *COMT*, catecholamine-o-methyl transferase; *^{3}H-SAM*, tritiated S-adenosyl methionine; *NAT*, N-acetyl transferase; *AcCoA*, acetyl CoA; *HIOMT*, hydroxyindole-o-methyl transferase; *position of ^{3}H donated by ^{3}H-SAM.

genes for several hundred nucleotides. The newly synthesized transcripts are labeled. The critical requirement of this type of assay is the availability of specific cDNA probes for the neuropeptide under study. The specific RNA transcripts are hybridized to the cDNA probes bound to an inert matrix, such as nitrocellulose filter, that are subsequently washed and counted in a scintillation counter. The amount of radioactivity counted is proportional to the amount of specific mRNA.

Electrophysiology

Electrophysiologic studies of the hypothalamus have provided much information about the firing patterns of hypothalamic neurons. However, this information is most valuable only when the firing patterns are correlated with a hypothalamic-dependent event, such as pituitary hormone secretion, or with behavioral responses. In general, hormone release from cells or neuropeptide release from axon terminals in the hypothalamus follows the influx of calcium through a membrane depolarized during action potential activity.

Extracellular recordings from hypothalamic nuclei in vivo can be used for topographic or functional identification (Table 16-13). For example, magnocellular neu-

rons in the paraventricular nucleus can be excited antidromically and identified by electrically stimulating the neurohypophysis. The topographic origin of the neurons terminating in the neurohypophysis can be identified by this means. Similarly, these neurons can be recorded from orthodromically after application of a suckling stimulus and correlated with the release of oxytocin into peripheral plasma. The relation suggests, but does not prove, a functional role for these neurons in the control of pituitary hormone secretion.

Several in vitro electrophysiologic approaches can be used to study neurotransmitter effects on hypothalamic neurons or the effects of hypophysiotropic substances of

Table 16-13
ELECTROPHYSIOLOGIC METHODS FOR STUDYING HYPOTHALAMIC FUNCTION

Methods	Use
Extracellular recordings	Topography of neurons
Hypothalamic slice recordings	Characterize neuronal excitation
Voltage clamp or current clamp	Neurosecretory mechanisms

hypothalamic origin on target cells (see Table 16-13). Slice preparations of whole hypothalami allow the introduction of drugs or other factors by superfusion close to the neuron being recorded. Excitable membrane properties of pituitary cells can be studied after application of suspected neurohormones of hypothalamic origin. *Voltage clamp* or *current clamp* approaches are used to study the effects of hypophysiotropic substances on secretory function of pituitary cells (see Table 16-13). These yield valuable information about the identity of ion channels involved in secretion.

☐ Neuroanatomic approaches

Immunocytochemistry is the favored approach for visualizing neuropeptides in neuronal cell bodies, dendrites, axons, and axon terminals (Table 16-14). The technique involves saturating histologically prepared sections of hypothalamus with antisera that are specific for the neuropeptide in question. The reaction complex is then treated with an anti-immunoglobulin that has been bonded with an enzyme that causes a visible precipitate. Alternatively, the antibody can be conjugated with a fluorescent chromogen that produces a distinctive fluorescent color.

Localization of a peptide in different parts of neurons requires different approaches and produces different information. For example, to study neuropeptides in cell bodies, axonal transport must be blocked with agents, such as colchicine, that disrupt microtubules. Cell body density can then be estimated. Three-dimensional localization of neuropeptide-containing neurons can be determined. Axons are most difficult to visualize by immunocytochemistry, because neuropeptides are transported from the cell body to the nerve terminal by fast axoplasmic flow and not enough material is available for immunostaining. Nerve terminals can be visualized by immunocytochemical methods under the light microscope.

Tract tracing techniques can be employed alone or in combination with immunocytochemical approaches to determine the path taken by axons of particular neurons (see Table 16-14). Two approaches can be used. Tracts can be visualized from the neuronal cell body to the axon terminal (i.e., anterograde) or from the nerve terminal to the cell body (i.e., retrograde). Horseradish peroxidase, an enzyme, is a glycoprotein capable of catalyzing the oxidation of some chromogens and can be used for antero-

grade and retrograde tracings. Fluorogold is a fluorochrome that is used specifically for retrograde tracings.

Neurons can also be labeled for activity. The product of the protooncogene *FOS* can be detected in the nuclei of active neurons by immunocytochemistry (see Table 16-14). Neuronal activity can also be quantitated autoradiographically with the 2-deoxyglucose technique. Active neurons preferentially use glucose for oxidative metabolism. 2-Deoxy-D-[^{14}C]glucose is injected intravenously, the hypothalami are prepared and sectioned by conventional histologic techniques, and the concentration of grains in particular nuclei is evaluated microdensitometrically. *Autoradiography* can be used to localize neurotransmitter binding sites, trace axonal connections, locate sites of steroid receptors, and evaluate neuronal activity. The system to be studied is labeled, brain sections are prepared histologically, the slides are covered with a radiosensitive emulsion, and the emulsion is developed photographically. Just as there are immunocytochemical techniques to study the localization of neuropeptides, enzymes for catecholamine biosynthesis, and neurotransmitter or steroid receptors, there are also immunocytochemical techniques for studying the transcription of the message. The technique in widest use is called *in situ hybridization* histochemistry. Specific cDNA is used in place of a specific antibody (see Table 16-14). The hybridization of the specific cDNA occurs on the brain section that is then developed by autoradiographic techniques similar to those described earlier. Alternatively, nonradioactive probes allow visualization of a chromogenic reaction.

The great advantage of these approaches is that they can be used in combination. For example, a researcher may wish to identify hypothalamic neurons that have steroid receptors. Under these circumstances, it is possible to combine immunocytochemistry with steroid autoradiography.

☐ Creation of deficits of hypothalamic function

Surgical manipulations

Most surgical manipulations involving the hypothalamus and its targets, such as the pituitary, are performed to create deficit symptoms (Table 16-15).

Hypophysectomy. Hypophysectomy performed to remove the pituitary from the sella turcica by a parapharyngeal approach was first described in the rat by Philip Smith in 1927. Because the hypothalamus produces neurohormones that control the adenohypophysis and neurohormones that are released from the neurohypophysis, hypophysectomy creates many of the deficits of hypothalamic hypofunction. Although hypophysectomy creates deficits of all of the adenohypophysial and neurohypophysial hormones, a hypofunctional hypothalamus creates deficits of all of the pituitary hormones except prolactin. Prolactin levels in blood are virtually undistinguishable after hypophysectomy, but circulating prolactin

---

Table 16-14
NEUROANATOMIC METHODS FOR STUDYING HYPOTHALAMIC FUNCTION

Methods	Use
Immunocytochemistry	Visualize location of amines and peptides, activity of neurons
Tract tracing	Visualize axons
Autoradiography	Characterize transmitter binding sites, peptide message

Table 16-15
METHODS FOR CREATING DEFICITS OF HYPOTHALAMIC FUNCTION

Methods	Use
Hypophysectomy	Create deficits of neurohypophysial and adenohypophysial hormones and targets for adenohypophysial-releasing hormones
Stalk transactions	Create deficits of neurohypophysial hormones and releasing hormones at their targets
Surgical lesions	Destroy groups of cells or fibers suspected of involvement in hypothalamic function
Chemical lesions	Destroy groups of cells that synthesize and secrete specific neurotransmitters or neurohormones

levels increase as a result of a hypofunctioning hypothalamus. This response is the result of the removal of the pituitary lactotrope from the influence of dopamine, the physiologic prolactin-inhibiting hormone.

It is often difficult to verify the completeness of a hypophysectomy until postmortem inspection. Another problem associated with hypophysectomy is that it is difficult to prevent partial functional regeneration of the hypothalamohypophysial tract and permanently create deficits in oxytocin and vasopressin secretion. In some cases, the aspirated gland can be transplanted to sites distant from the hypothalamus. Deficit symptoms of all the hormones except prolactin persist. When transplanted to the sella turcica to allow revascularization by the hypophysial portal vessels, the deficit symptoms are reversed. These types of procedures allowed early anatomists to conclude that the critical link between the hypothalamus and adenohypophysis was neurovascular.

The stalk median eminence connecting the hypothalamus with the pituitary can be transected. This procedure causes disruption of the hypothalamohypophysial portal vasculature and the hypothalamohypophysial tract. This disruption is permanent if regrowth is prevented by placing a foil barrier in the transected area. Under these circumstances, the deficit symptoms are the same as hypophysectomy followed by transplantation to a site distant from the sella turcica.

Surgical lesions. To destroy deep-seated nuclei of the brain in areas such as the hypothalamus, the nuclei must be located with accuracy. This is achieved by placing the head of the subject in a stereotaxic apparatus, a device that holds it in a predefined, rigid position. With the aid of a map of the brain (i.e., stereotaxic atlas) and anatomic landmarks on the skull and the surface of the underlying brain, focal lesions that destroy discrete anatomic groupings of cell bodies can be placed by sending a current through an electrode. Similarly, fibers of passage may be destroyed by placement of a small knife that can be manipulated to deafferentate specific axons controlling the hypothalamus.

Chemical lesions

Selective lesions induced by neurotoxins have become a modern tool for studying the role of the hypothalamus (see Table 16-15). They can be site selective or transmitter selective. Monosodium glutamate or its more potent analog, kainic acid, are site selective for neuronal cell bodies and dendrites, sparing fibers of passage and axon terminals in areas outside of the blood-brain barrier such as the arcuate nucleus and median eminence, the OVLT, and the preoptic area. Glutamate, kainate, or its less toxic analog, ibotenic acid, may be injected systemically, and the substance lesions only those areas. Selectivity is further enhanced by local injections guided stereotaxically. Gold thioglucose selectively lesions the ventromedial nucleus of the hypothalamus.

Neurotoxic lesions can be made in catecholamine or indolamine neurons. 6-Hydroxydopamine or 6-hydroxydopa depletes catecholamines in the brain when injected in the third ventricle. These substances pass the blood-brain barrier, and their specificity may be restricted by stereotaxic injection in small volumes locally. The indoleamine neurotoxins are 5,6- or 5,7-dihydroxytryptamines that are injected intraventricularly or locally. Cysteamine, or β-mercaptoethylamine, depletes somatostatin in the central nervous system and the periphery. In the central nervous system, cysteamine depletes somatostatin in cell bodies and axons.

Certain plant lectins, such as ricin, are taken up and transported retrogradely along axons to cell bodies and ultimately kill the cell. Specificity is conferred by coupling an antiserum to the peptide made by the cell, and when the antiserum binds to the peptide, specific cells are killed. Alternatively, the cytotoxic plant lectins can be conjugated to hypophysiotropic peptides to kill target cells selectively.

☐ Stimulation of hypothalamic function

There are essentially three approaches to studying the excitation of the hypothalamus and its consequences (Table 16-16).

It is possible to study hypothalamic function in response to natural stimuli. For example, a researcher can record from specific areas of the hypothalamus or study the activities of specific neuropeptides or amine transmitters in the hypothalamus in response to exteroceptive

Table 16-16
METHODS FOR STIMULATING HYPOTHALAMIC FUNCTION

Methods	Use
Natural stimuli	To study the relation between physiologic stimuli and hypothalamic function
Chemical stimuli	To study the relation between neurotransmitters and hypothalamic function
Artificial stimuli	To excite groups of hypothalamic neurons and study the effect on a hypothalamically dependent endpoint

stimuli such as suckling, mating, volume expansion (e.g., hypertonic saline), volume depletion (e.g., hemorrhage), or alterations in temperature.

The activity of hypothalamic neurons can be studied in response to their natural chemical stimulators such as neurotransmitters and neurotransmitter agonists and antagonists. Moreover, the effects of endogenous peripheral factors such as hormones can be described. For example, the activity of estrogen on excitation of hypothalamic neurons involved in sexual behavior can be studied.

Hypothalamic neurons can be excited by artificial means and the consequence of their activity studied. The usual approach is to stimulate cells electrically or electrochemically. In the former case, neurons are excited by electrical depolarization, and in the latter case, depolarization is caused by the deposition of iron by the stimulating electrode. Using either of these approaches, the researcher can stimulate an area of the hypothalamus and measure a visceral endpoint.

☐ The direct application of active hypothalamic peptides or amines to physiologic targets reveals their physiologic role

The response of target cells to their physiologic affecters can be studied. For example, cells of the anterior pituitary can be enzymatically dissociated and placed in short-term culture. Hypothalamic peptides can be applied to the cultures by perifusion or by static incubation in monolayer cultures. The release of pituitary hormones into the media can be monitored, and intracellular transduction events can be studied with this approach. Using similar approaches, the receptors for the neuropeptides can be studied.

Physiologic Processes Controlled by the Hypothalamus

☐ The hypothalamus regulates pituitary hormone secretion

Characteristics of a neuroendocrine system

The key feature of a neuroendocrine system is the existence of the *neurohemal area*, which is the external zone of the median eminence at which neurosecretory axon terminals converge on a capillary bed that ultimately leads to and affects the secretions of the adenohypophysis. Similarly, neurosecretory axons comprising the hypothalamohypophysial tract terminate on sinusoids in the neurohypophysis and ultimately secrete their products to the peripheral circulation to affect visceral processes.

The axons terminating in the external zone of the median eminence secrete releasing hormones and release-inhibiting hormones into the hypothalamohypophysial portal plasma. To meet the definition of a releasing or release-inhibiting hormone, the substances in portal plasma must fulfill certain criteria (Table 16-17). The hormone must be extractable from hypothalamic or median eminence tissue. It must be present in the hypophysial portal blood in amounts greater than in the systemic circulation. Various concentrations of a particular releasing or release-inhibiting hormone in portal plasma must be correlated with the secretion rates of one or more of the anterior pituitary hormones in a variety of experimental conditions. The suspected hormone should stimulate or inhibit one or more pituitary hormones when administered in vivo or applied to pituitary cells or tissues in vitro, and the inhibitors that antagonize the actions of the suspected hormone should block or stimulate anterior pituitary hormone secretion. Target cells should have receptors for the releasing or release-inhibiting hormone.

Concept of feedback

The hypothalamic-pituitary-target axes can be characterized as a set of links over which information flows. As

Table 16-17
REQUIRED CHARACTERISTICS OF NEUROHORMONES

Activity must be extractable from whole hypothalamus or median eminence tissue.
Concentration in hypophysial portal blood must be greater than systemic circulation.
Dynamics in portal plasma must be correlated with dynamics of adenohypophysial hormone secretion.
Extracted material must be active in vivo and in vitro.
Inhibitors of neurohormones should affect physiologic endpoint.
Target cells should have receptors for neurohormones.

information is transmitted from link to link, it stimulates or depresses a biologic response in the next link, and it also influences the activity of an earlier link. This type of influence is referred to as a *feedback*.

There are two types of feedbacks. A *negative feedback* (Fig. 16-23) is one in which the activity of the downstream link inhibits the activity of one or more upstream link. The regulation of room temperature by a thermostatically driven furnace conceptually fits this model. The furnace, in response to regulation by a thermostat, raises the temperature of the room to a point preset on the thermostat. After that temperature is achieved, the thermostat senses that level and turns off the furnace. If the thermostat is inoperative, the thermostat runs extensively, and the temperature in the room rises to the limits of the furnace. In neuroendocrinology, an example of a negative feedback is the ability of adrenal corticosterone to inhibit CRH secretion into portal blood.

A *positive feedback* is one in which the downstream link enhances the activity of one or more upstream links. Conceptually, this is a more difficult mechanism to describe accurately. A voice-activated recording device is perhaps the best example of a positive feedback. The voice begins the recording device. When the voice ceases, the recording device stops. Unlike negative feedback, which is a dampening process and long term, positive feedback is relatively brief and inherently unstable. The feedback signal can be provided by a hormone itself or by a nonhumoral metabolite. The classic example of a positive feedback in neuroendocrinology is the ability of ovarian estradiol to stimulate GnRH secretion into portal blood.

Feedback systems can take several routes (Fig. 16-24). In the case of the hypothalamic-pituitary-target gland axis, a *long-loop feedback* would be achieved through blood borne from the peripheral target gland to affect the hypothalamus or pituitary. *Short-loop feedback* can be exemplified by a pituitary hormone influencing the secretion of its hypothalamic releasing or release-inhibiting factor. An *ultra–short-loop feedback* is represented by a pituitary hormone effecting its own secretion through an autocrine mechanism.

Regulation of the adenohypophysis by the hypothalamus

Gonadotropes. Overwhelming evidence favors the concept that the hypothalamic decapeptide known as GnRH is the peptide in hypophysial portal blood that is the physiologic humoral stimulator of LH and FSH secretion. Although FSH-releasing activities devoid of LH-releasing activity have been isolated from the hypothalamus, a distinct FSH releasing hormone has yet to be identified.

GnRH regulates LH and FSH secretion from the gonadotropes of the adenohypophysis in basal and ovulation-inducing "surge" states in female mammals. In rodents, the surge center is the medial preoptic area, and the basal center is in the medial basal hypothalamus. Males lack a functional surge center. The ovarian steroids, estrogen

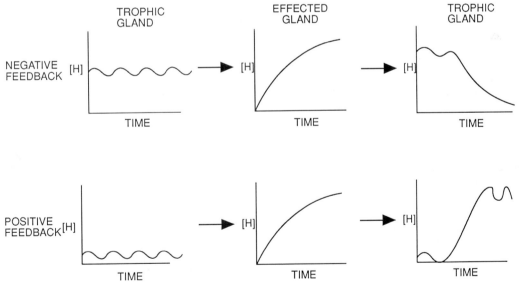

Figure 16-23
Negative feedback (*upper sequence*) and positive feedback (*lower sequence*) in the endocrine system. In a negative-feedback system, the trophic gland (e.g., adenohypophysis) secretes a signal (e.g., thyroid-stimulating hormone [TSH]), which stimulates the target gland (e.g., thyroid) to secrete its product (e.g., thyroid hormone), which feeds back to inhibit TSH secretion. In a positive feedback system, the trophic gland (e.g., adenohypophysis) secretes a signal (e.g., LH) at low rates that stimulates the target gland (e.g., ovary) to secrete its product (e.g., estradiol), whose increasing secretion rate allows the trophic gland to secrete luteinizing hormone in larger amounts. [*H*], hormone concentration in blood.

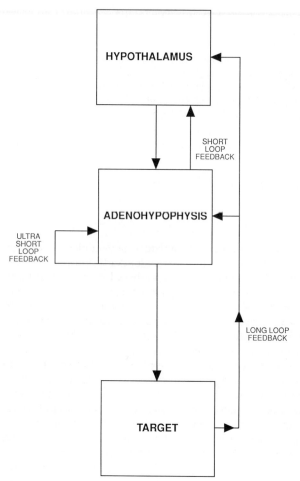

Figure 16-24
Feedback loops. In the hypothalamopituitary–target gland axis, a long-loop feedback system would be blood-borne from the peripheral target gland to actuate the hypothalamus or pituitary. A short-loop feedback system could be exemplified by a pituitary hormone influencing the secretion of its hypothalamic releasing or release-inhibiting hormone. An ultrashort-loop feedback system is characterized by a pituitary hormone influencing its own secretion through an autocrine mechanism.

and progesterone, inhibit LH and FSH secretion by acting directly at the gonadotrope and at the medial basal hypothalamus to inhibit GnRH release into portal blood. The ovarian steroids also stimulate a preovulatory surge of LH and FSH secretion by acting at the medial preoptic area to stimulate a surge of GnRH into portal blood. This is presumably achieved through a noradrenergic mechanism. In rodents, surgical isolation of the medial preoptic area from the medial basal hypothalamus prevents a steroid-induced surge of LH secretion. Neither ovarian or testicular steroids induce an LH surge in males, but the gonadal steroids inhibit LH and FSH secretion, regardless of the sex of the recipient. In primates, there is evidence that the surge center may reside in the basal hypothalamus and that ovarian steroids merely sensitize the gonadotropes to respond to unvarying pulses of GnRH. Monkeys bearing surgically isolated medial basal hypothalami

and infused with pulsatile GnRH respond to estradiol with an LH surge. This has led to the concept that the medial basal hypothalamus is a pulse generator for GnRH release into portal blood.

There is a clear sexual dimorphism of the medial preoptic area: males have a more intensely stained medial preoptic nucleus within this area than females. Such differentiation occurs perinatally. In rodents, castration of males within the first days of life prevents the appearance of this sexually dimorphic area in adults. Moreover, when adult, neonatally castrated males can respond to an estradiol challenge with a preovulatory-like LH surge. If testosterone is administered to phenotypic female rodents within the first days of life, they develop the male-type sexual dimorphic nucleus and the male pattern of gonadotropin secretion. Because they lack a surge center, they are anovulatory. Although GnRH is the physiologic regulator of LH and FSH secretion, other neuropeptides serve similar functions as neurohormones, neurotransmitters, or neuromodulators. These are listed in Table 16-18.

Lactotropes. The dominant hypothalamic control over pituitary prolactin secretion is inhibitory. Removal of hypothalamic influence over the adenohypophysis results in enhanced secretion of prolactin. Stalk transection, destruction of the medial basal hypothalamus, or placement of pituitary fragments or cells in culture results in the hypersecretion of prolactin. Moreover, in vivo treatment with dopamine antagonists results in the hypersecretion of prolactin. Conversely, in vitro treatment with dopamine agonists depresses pituitary prolactin secretion. These data, coupled with the inverse relation between dopamine levels in portal blood and peripheral blood levels of prolactin, suggest that dopamine is the prolactin release-inhibiting hormone. However, studies of the dynamics of prolactin release in response to lowered dopaminergic tone suggest that the lactotrope must also be under the influence of prolactin-releasing hormones of hypothalamic origin. Although TRH is one of the most widely studied candidates, others (see Table 16-18) have prolactin-releasing properties and may also play a role. Prolactin secretion in response to exteroceptive stimuli such as suckling may involve a lowering of dopamine levels in portal blood and an increase in the portal blood concentration of a putative prolactin-releasing hormone.

Thyrotropes. There is little doubt that the tripeptide pyro-Glu-His-Pro-NH$_2$ is TRH (see Table 16-18). TRH is the physiologic stimulator of TSH secretion from the adenohypophysis. The cell bodies whose axons terminate on the external zone of the median eminence are found primarily in the paraventricular nucleus. TSH stimulates thyroid hormone (i.e., thyroxine, triiodothyronine) secretion from the thyroid gland. The thyroid hormones diminish the release of TSH by lowering the response of the thyrotrope to TRH. This is a classic negative-feedback control system. Removal of the thyroid gland enhances the release of TSH into the peripheral circulation without affecting portal blood levels of TRH, suggesting that the primary control of TSH secretion by thyroid hormones does not reside at the hypothalamus. The hypothalamus

Table 16-18

PEPTIDE AND AMINES THAT ACT DIRECTLY ON ADENOHYPOPHYSIAL CELLS

Cell Type	Peptide or Amine*	Other Peptides or Amines*
Gonadotrope	GnRH	VIP, CCK, NPY, substance P, galanin, neurotensin
Lactotrope	Dopamine[†]	TRH, oxytocin, VIP, angiotensin II, somatostatin, LHRH
Thyrotrope	TRH	Somatostatin[†]
Corticotrope	CRH	AVP
Somatotrope	GHRH, somatostatin[†]	TRH

*AVP, arginine vasopressin; CCK, cholecystokinin; CRH, corticotropin–releasing hormone; GnRH, gonadotropin–releasing hormone; GHRH, growth hormone–releasing hormone; NPY, neuropeptide Y; TRH, throtropin–releasing hormone; VIP, vasoactive intestinal polypeptide.
†Inhibits function of the cell.

provides the drive (i.e., TRH), but the thyroid gland negatively regulates (i.e., thyroid hormone) the response of the thyrotrope (i.e., TSH) to that drive.

Corticotrope. ACTH secretion from the corticotrope is controlled primarily by CRH released into portal blood. ACTH stimulates the release of the steroid hormones of the adrenal cortex, which feed back negatively to inhibit ACTH release by acting at the hypothalamus and the adenohypophysis. AVP is also a potent ACTH-releasing hormone. Portal blood levels of AVP and CRH are positively correlated with ACTH-releasing stimuli such as stress. ACTH release is not caused by the action of a single peptide; it is the result of the actions of a hypothalamic complex.

Somatotrope. Although the somatotrope is predominantly under the stimulatory influence of hypothalamic growth hormone–releasing hormone (GHRH), it is also under the opposing influence of a growth hormone release-inhibiting hormone called somatostatin. Each neurohumoral peptide affects growth hormone secretion through distinct receptor sites on the somatotrope, and each plays a reciprocal neuromodulatory role for the other. The somatostatin neurons directly innervate GHRH neurons with the results of diminishing GHRH release into portal blood and reducing growth hormone release into the peripheral circulation. Stimuli known to release growth hormone also suppress the release of somatostatin into portal plasma.

The feedback control of growth hormone secretion does not fit the models of classic negative or positive feedback by target endocrine organs. For example, hypoglycemia stimulates growth hormone secretion by stimulating the release of GHRH into portal blood. Conversely, hyperglycemia inhibits growth hormone secretion by increasing somatostatin and decreasing GHRH levels in portal blood.

The hypothalamus and neurohypophysial function

Mechanism of secretion of neurohypophysial hormones. The hormone-neurophysin complex (i.e., vasopressin-neurophysin II or oxytocin-neurophysin I) is synthesized in cell bodies of the supraoptic or paraventricular nuclei. The complexes, still undergoing post-translational processing, are transported down the long axons comprising the hypothalamohypophysial tract to terminals adjacent to fenestrated capillaries in the neurohypophysis (Fig. 16-25). Adjacent to the fenestrated capillaries in the neurohypophysis are specialized glial-like cells known as *pituicytes* that regulate the microenvironment of the terminals. During this voyage, they are packaged in neurosecretory granules. At the axon terminal, the granule membrane fuses with the membrane of the axon terminal, and the hormone-neurophysin complex is exocytosed. The process coincides with the arrival of the action potential that depolarizes the membrane and allows the entry of sodium ions, opening the calcium channels, which plays an incompletely understood role in the exocytotic process. After exocytosis of the neurohormone, intracellular calcium is packaged into microvesicles and extruded and the membrane potential is restored by a sodium-potassium pump. The membranes of the evacuated neurosecretory granules are reformed from the surface of the axon terminal, where they are packaged into lysosomes and degraded or recycled, usually in areas of nonterminal swelling known as Herring bodies.

Stimuli for secretion of vasopressin. The two main drives for vasopressin secretion are an increase in osmolality of the plasma and a decrease in plasma volume (Table 16-19). These can be interrelated or independent stimuli.

Water deprivation causes an increase in plasma osmolality and a diminution of intracellular water. A change in plasma osmolality of as little as 1% is detected by osmoreceptive neurons that are distinct from the vasopressin magnocellular neurons in the hypothalamus. The osmoreceptive neurons stimulate vasopressin synthesis and release from the magnocellular neurons in the supraoptic area at a threshold of 280 mOsm/kg. The osmoreceptor neurons also stimulate thirst with a greater sensitivity of 290 mOsm/kg.

Vasopressin release is also stimulated by a 5% to 10% decrease in blood volume, blood pressure, or cardiac output. Hypovolemia is perceived by pressure receptors in the carotid and aortic arch and stretch receptors in the

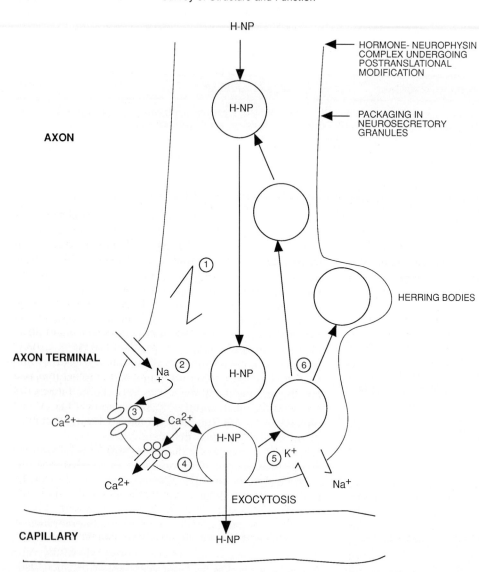

Figure 16-25
The mechanism by which the neurohypophysial peptide–neurophysin complex is axonally transported, processed, packaged, and secreted from the axon terminal. As the hormone neurophysin complex is transported down the axons in the hypothalamohypophysial tract, additional posttranslational processing is taking place. The mature complex is then packaged into neurosecretory granules, whose arrival at the axon terminal coincides with the arrival of the action potential (*1*). The membrane of the granule fuses with the axon terminal membrane and the product is exocytosed. The action potential is thought to play a role in the process by causing depolarization and entry of sodium (*2*), which allows entry of calcium through specific channels (*3*). Calcium plays an incompletely understood role in the exocytotic process. The intracellular calcium is then packaged into microvesicles (*4*) and extruded, and the membrane potential is restored by a sodium-potassium pump (*5*). The membranes of the evacuated neurosecretory granules are reformed (*6*) and packaged into lysosomes and degraded or recycled in areas of nonterminal swelling known as Herring bodies.

Table 16-19

PHYSIOLOGIC INPUTS FOR STIMULATION AND INHIBITION OF VASOPRESSIN AND OXYTOCIN SECRETION

Hormone	Stimulation*	Inhibition*
Vasopressin	↑ Plasma osmolality	↓ Plasma osmolality
	↓ Plasma volume	↑ Plasma volume
	↓ Blood pressure	↑ Blood pressure
	↓ Cardiac output	↑ Cardiac output
	↑ Noradrenergic tone	
Oxytocin	Suckling	Stress
	↑ Activation of cervicovaginal stretch receptors	
	↓ Placental progesterone	
	↑ Fetal cholesterol	
	↑ Sensory stimulation	

↑, increase; da, decrease.

walls of the left atrium, pulmonary veins, and the juxtaglomerular apparatus of the kidney. The afferent impulses of these sensors are carried through the ninth and tenth cranial nerves to the medulla and then through the midbrain over noradrenergic synapses to the magnocellular vasopressinergic neurons of the supraoptic nucleus. In the absence of any change in pressure, the receptors tonically inhibit vasopressin secretion. With acute volume depletion such as that caused by hemorrhage, noradrenergic inhibitory tone from the medulla to the hypothalamus is decreased, resulting in an increase in secretion of vasopressin. Volume depletion also stimulates central renin-dependent angiotensin release, which also stimulates vasopressin secretion and thirst.

Stimuli for secretion of oxytocin. Suckling is the best described stimulus for oxytocin secretion (see Table 16-19). The pathways are similar to those of vasopressin. The suckling stimulus is carried over afferent spinal pathways to the medulla and midbrain and then through cholinergic synapses to the paraventricular nucleus. Oxytocin release is pulsatile. Nipple suction by the young leads to synchronized activation of action potentials for 2 to 4 seconds in the paraventricular nucleus. From a resting "spontaneous" background of 1 to 10 spikes/second, these neurons generate a synchronized series of 70 to 80 action potentials within 3 to 4 seconds of application of the stimulus, resulting in the secretion of 0.5 to 1.0 mIU of oxytocin. This is followed by milk ejection from the mammary gland about 12 to 15 seconds later. These pulses of neuronal activity occur uniformly every 4 to 8 minutes. The characteristic of periodic bursting of action potentials at high frequency appears to be important for oxytocin secretion and consequent milk ejection.

The stimuli for oxytocin release during labor appear to be multiple. Activation of cervicovaginal stretch receptors by the growing conceptus is an important signal, and diminishing placental progesterone and elevated fetal free cortisol act in concert to stimulate oxytocin secretion sufficient to enhance contractions of the uterus.

Oxytocin secretion has a significant sensory component. For the human female, merely playing with the infant or sensing the cries of a hungry infant causes milk ejection, but emotional stress inhibits the secretion of oxytocin.

□ The hypothalamus regulates autonomic processes

Cardiovascular function

Independent of the control of blood pressure through the neuroendocrine function of the hypothalamus, the cardiovascular system is influenced by the hypothalamus through the autonomic nervous system (Table 16-20). These effects are mediated primarily by the sympathetic system through the vagus nerve. Stimulation of the posterior and lateral hypothalamus increases arterial pressure and heart rate, and stimulation of the preoptic area decreases heart rate and arterial pressure. The effects are mediated by the cardiovascular centers in the medulla and pons.

Cardiovascular response to alterations in environmental temperature or in defense reactions is also mediated by the hypothalamus. Dilation of blood vessels of the skin and constriction of deep visceral vessels occur in response to a hot environment. Cold exposure induces the converse responses. These responses are controlled by the preoptic and anterior hypothalamic areas. The defense reaction that is characterized by cutaneous vasoconstriction and muscular vasodilation is the consequence of discharge of sympathetic cholinergic vasodilators and sympathetic adrenergic excitation. These effects can be induced by selective stimulation of the anterior and posterior hypothalamus.

Thermoregulatory function

The control of body temperature by the hypothalamus is a classic example of integrative approach to the alteration of the internal milieu. The hypothalamus oversees autonomic compensations, such as alterations of blood flow and sweating; endocrine compensations, such as metabolism-regulating alterations of thyroid function; and musculoskeletal compensations, such as shivering, panting, and piloerection (Table 16-21). Temperature regulation by the hypothalamus is also a nonendocrine example of a feedback mechanism. The regulatory system actually collects temperature information from two sources: peripheral sources, such as the skin, visceral structures, and spinal cord, and central sources, such as thermosensors in the preoptic area and anterior hypothalamus, whose neurons are activated or inactivated by the temperature of the blood bathing them.

The hypothalamus bears dual mechanisms for controlling heat dissipation and heat conservation. The heat dissipation centers lie in the preoptic area and anterior hypothalamus, and the heat conservation centers lie in the posterior hypothalamus. Electrical stimulation of the preoptic area and anterior hypothalamus favors dilation of cutaneous blood vessels, panting, and suppression of shivering. All result in a drop in body temperature. Conversely, electrical stimulation of the posterior hypothalamus leads to cutaneous vasoconstriction, visceral vasodilation, shivering, and a suppression of panting. The metabolic response to temperature alteration also involves the hypothalamus. Exposure to cold enhances the animal's heat-generating metabolic rate by stimulating TRH-activated TSH secretion and subsequent thyroid hormone secretion.

It is clear from recordings of neurons in the preoptic area and anterior hypothalamus that there are warm-sensitive and cold-sensitive neurons. Warming of the skin or hypothalamus results in enhanced firing of warm-sensitive neurons and decreased firing of cold-sensitive neurons. Conversely, cooling of the skin or hypothalamus leads to opposite effects. These neurons integrate information from the periphery and the central nervous system.

The hypothalamus coordinates voluntary behavioral adjustments to extremes in environmental temperatures sensed at the hypothalamus and the skin. For example, in rats and monkeys trained to make behavioral adjustments to a hot environment, local warming of the hypothalamus in the face of normal ambient temperature results in the

Table 16-20
CONTROL OF CARDIOVASCULAR FUNCTION BY THE HYPOTHALAMUS

Area	Response*
Lateral and posterior hypothalamus	↑ Arterial pressure
	↑ Heart rate
Preoptic area	↓ Arterial pressure
	↓ Heart rate

* ↑ , increase; ↓ , decrease.

Table 16-21
THERMOREGULATORY FUNCTION OF THE HYPOTHALAMUS

Compensation	Area	Response
Autonomic	Preoptic area	Dilation of cutaneous blood vessels, sweating
	Posterior hypothalamus	Vasoconstriction
Musculoskeletal	Preoptic area	Panting, suppression of shivering
	Posterior hypothalamus	Shivering, suppression of panting, piloerection
Endocrine	Preoptic area	Increased thyroid function

appropriate behavioral adjustment to warmth. The hypothalamus also integrates a summation of the responses. When the hypothalamus and the environment are warmed, the behavioral response is greater than for either alone. In a hot environment, cooling of the hypothalamus completely suppresses the behavioral adjustment to elevation of environmental temperature. The hypothalamus assumes supremacy in the behavioral responses to alteration in temperature.

The hypothalamus mediates the response to pyrogens in pathologic states. Body temperature is regulated around a set-point. Substances called *pyrogens* that allow the temperature to deviate from that set-point can be produced by macrophages in disease states. The preoptic area appears to respond to one such pyrogen, interleukin-1. It has been suggested that the prostaglandins mediate the response to certain pyrogens and act at the preoptic area. Antipyretics such as indomethacin may act by blocking the synthesis of prostaglandins. The brain also contains a nearby antipyretic area within the septal nuclei. This area may use the peptide vasopressin. Injection of vasopressin directly into this area counteracts the effects of many known pyrogens. Antipyretics may also act by stimulating the release of vasopressin. Injection of a vasopressin antagonist prevents the antipyretic effects of indomethacin.

Defensive function

The hypothalamus is responsible for preparing the organism to respond to threatening or stressful situations. The *flight or fight response* represents an integrated constellation of responses to prepare for stressful situations (Table 16-22). Many of these responses are directly controlled by the hypothalamus, and others are indirectly influenced by the hypothalamus through its control of the endocrine system.

The hypothalamus stimulates a variety of cardiovascular compensations. In response to a perceived threat, blood pressure, heart rate, force of contraction, and rate of cardiac conduction velocity increase. The rate and depth of respiration increases. There is a shift of blood flow from the skin and splanchnic organs to the skeletal muscles, heart, and brain. Metabolic adjustments, including enhanced glycogenolysis and lipolysis, are made in anticipation of increased energy requirements.

Other autonomic alterations include mydriasis; ocular accommodation for far vision; contraction of the spleen

capsule, leading to increased hematocrit, piloerection, inhibition of gastric motility, and secretion; contraction of gastrointestinal sphincters; and sweating. Some of these responses are regulated by the autonomic nervous system directly, and others are controlled by hormones secreted in response to stressful stimuli.

Classically, epinephrine is secreted from the adrenal medulla in response to acute stressors. This catecholamine controls many of the metabolic demands of the flight or fight response. Glucocorticoids are secreted from the adrenal cortex in response to stressful stimuli. Secretion of glucocorticoids is controlled by ACTH secreted from the pituitary under the influence of hypothalamic CRH and AVP. In long-term stressful situations, this leads to suppression of the immune system. Other hormones whose secretion is stimulated in response to stress are β-endorphin (e.g., in pain perception), vasopressin (e.g., in renal function), glucagon (e.g., in carbohydrate mobilization), and prolactin (e.g., in immune responses). Growth hormone and insulin are typically inhibited during stressful circumstances. Many of these autonomic and hormonal responses are controlled by the anterior and ventromedial hypothalamus.

☐ The hypothalamus regulates behavioral processes

Ingestive behavior

It has been appreciated for some time that the hypothalamus bears two "centers" that exert opposing controls

Table 16-22
COMPONENTS OF THE FLIGHT OR FIGHT RESPONSE

Increase in blood pressure, heart rate, force of contraction, and rate of conduction velocity.
Increase in rate and depth of respiration.
Shift in blood flow from skin and splanchnic organs to skeletal muscles, heart, and brain.
Metabolic adjustments include enhanced glycogenolysis and lipolysis.
Other autonomic adjustments include mydriasis, accommodation for far vision, contraction of spleen capsule, piloerection, inhibition of gastric motility and secretion, contraction of gastrointestinal sphincters, and sweating.

over feeding behavior. Lesions restricted to the ventromedial nucleus of the hypothalamus produce hyperphagia and polydipsia, and lesions of the lateral hypothalamus produce aphagia. Stimulation of these areas leads to aphagia and hyperphagia, respectively. This understanding led to the concept that the ventromedial nucleus of the hypothalamus is a *satiety center* and the lateral hypothalamus is a *feeding center*. The ventromedial nucleus was originally thought to be a monitor of blood glucose. Adequate blood levels of glucose inhibit feeding by exciting the *glucostat* in this area. However, it is now clear that the limit of food intake is not determined by the volume or the quality of the food, but by the caloric value of the food. The ventromedial nucleus can be more accurately perceived as a *caloristat*.

Although these concepts have enjoyed long-standing acceptance, they are somewhat simplistic. The brain does not function in compartmentalized centers. Neural information instead is processed by way of circuits linking various areas of the brain.

There are several plausible explanations for the alterations of feeding behaviors. Feeding behaviors have a tremendous sensory component, and a lesion of the lateral hypothalamus may involve the trigeminal nerve and compromise the contribution of olfaction and taste. Similarly, lesions of the ventromedial hypothalamus have obliterated the response to foods that normally have aversive properties. The lesions may interfere with the set-point for maintenance of body weight. For example, force-fed obese rats do not become hyperphagic after they receive a lesion of the ventromedial nucleus. The lesions can also alter the balance of hormones that control the metabolism of foods. Lesions of the ventromedial nucleus can affect glucagon, insulin, ACTH, and growth hormone secretion. Animals bearing such lesions show a much greater release of insulin in response to food intake and a much greater deposition of fat. This is true even when the caloric value of food is controlled. Fibers of passage may be compromised by lesions or stimulation, and the observed effects may occur outside the site of manipulation.

Central and peripheral hormones and neurotransmitters influence feeding behavior. The hormone cholecystokinin (CCK) is released from the gut in response to food intake. In general, this hormone inhibits food intake and is perceived as a satiety hormone. Application of CCK directly to the paraventricular nucleus of the hypothalamus also inhibits feeding. Because CCK is also found within the hypothalamus, it is possible that this peptide is released from and acts within the hypothalamus as a satiety hormone. Glucagon, neurotensin, and calcitonin share this attribute of CCK. There is also evidence that CCK acts peripherally to inhibit feeding by causing nausea or similar gastric disturbance.

Among the stimulators of feeding behavior, NPY has been shown to play a major role. NPY levels well within the physiologic range evoke a feeding response in satiated male and female rats. NPY is found in the arcuate nucleus and median eminence, the medial preoptic area, and the paraventricular nuclei. However, NPY activity in the paraventricular nuclei is the only area in which the levels of the peptide are correlated with feeding behavior. Peptide or amine neurotransmitters may also selectively regulate the type of foods that are ingested. In animals given a choice of protein-, fat-, or carbohydrate-laden foods, application of norepinephrine directly to the paraventricular nucleus favors selection of the carbohydrate. In contrast, galanin stimulates the ingestion of fat, and opiates enhance the consumption of protein.

Thirst is regulated by tissue osmolality and vascular volume. These are controlled by AVP secreted from magnocellular neurons in the paraventricular nuclei and angiotensin II formed in the plasma and the brain. Although the drive for water ingestion is mediated through enhanced tissue osmolality or decreased vascular volume sensed by osmoreceptors in the brain and by baroreceptors in the brain and periphery, there appears to be a direct effect by hormones acting at the hypothalamus to mediate the behavioral response. The subfornical organ lies near the third ventricle and has fenestrated capillaries, permitting the entrance of blood-borne materials. The subfornical organ responds to low levels of angiotensin II in the blood and conveys information to the hypothalamus. It is possible that the communication is by way of a neuronally derived angiotensin II that affects the preoptic area. The preoptic area also receives information from peripheral baroreceptors. When water ingestion is required, the baroreceptors and angiotensin II stimulate the preoptic area, which activates other areas of the brain to begin drinking. The drive for termination of drinking is less well understood, but cessation of drinking is not merely the absence of the baroreceptor and osmoreceptor initiating signal.

Sexual behavior

The circumscribed behaviors leading to pregnancy and propagation of the species depend on the interaction of the gonads and the hypothalamus. In subprimate mammals, these events are driven by a heightened period of female sexual receptivity, called *estrus*; the estrous cycles coincide with the availability of a potentially fertilizable egg in the oviduct. A similar coincidence of gamete availability is not discretely defined in primates. Because primate cycles are overtly characterized by a period of breakdown of the lining and blood vessels of the uterus, called *menses*, these are referred to as menstrual cycles. Sexual receptivity is best studied in mammals who overtly display the behavior at discrete periods. For this reason, the rat is the most widely studied model of hypothalamic control of sexual behavior.

Hypothalamic control of sexual behavior in females. Sexual receptivity can be quantitated in female rats by a *lordosis quotient (LQ)*. Lordosis is the process by which the female arches her back, deflects her tail, and stands rigid to allow mounting and intromission by the male. The LQ is the number of times this event takes place divided by the number of attempts at mounting by the male and multiplied by 100.

Although the effects of various hormones on sexual behavior are species specific, the common hormone regulating most sexual behaviors is the ovarian hormone estrogen (Table 16-23). Estrogen receptors are present in the areas of the hypothalamus known to control sexual receptivity. Estrogen secretion is highest when sexual receptivity is increased. Although estrogen alone can enhance sexual receptivity, sexual receptivity is greatest when estrogen and progesterone levels are highest. Progesterone alone exerts little effect on sexual receptivity. Only in an estrogen-primed animal does progesterone further enhance sexual receptivity. Estrogens act by stimulating progesterone receptors in areas of the hypothalamus known to control sexual receptivity. Prolonged exposure to progesterone, as in pregnancy, causes a down-regulation of progesterone receptors in the hypothalamus and a decrease in sexual receptivity.

Four parts of the nervous system play a role in the control of female sexual behavior: the forebrain, ventromedial hypothalamic nucleus, midbrain central gray, and lower brain stem and spinal cord. Within the hypothalamus, lesions of the ventromedial hypothalamic nucleus depress sexual behavior in response to estrogen and progesterone. The ventromedial hypothalamic nucleus bears receptors for the ovarian steroids, and it is thought that this hypothalamic nucleus modulates the intensity or interpretation of sexually related sensory input. Estrogen receptors are also localized in the midbrain central gray. Neurons from the ventromedial hypothalamic nucleus and spinal cord project to the midbrain central gray. The spinal projection transmits tactile information provided by the male's mounting that is required for the induction of lordosis.

The neurotransmitter control of female sexual behavior is well described but cannot be reduced to participation by a single transmitter. Among the catecholamines, ascending noradrenergic fibers from the locus ceruleus regulate lordosis behavior by acting on α_1-noradrenergic receptors in the medial preoptic area and ventromedial hypothalamic nucleus. Norepinephrine may act in these areas by modulating progesterone receptors. Dopamine does not play a role in lordosis behavior but appears to

modulate proceptive behaviors such as ear wiggling, hopping, or darting.

Acetylcholine plays a role in the facilitation of lordosis behavior by estrogen. This steroid increases the activity of choline acetyltransferase and of acetylcholine receptors in the ventromedial hypothalamic nucleus. Moreover, acetylcholine applied directly to the medial preoptic area or ventromedial hypothalamic nucleus increases lordosis behavior, and acetylcholine antagonists applied to these same areas abolish or attenuate lordosis behavior.

Among the indolamines, serotonin appears to play an inhibitory role in sexual receptivity. The inhibition of serotonin synthesis excites lordosis behavior, and it has been suggested that serotonin is a sexual satiety neurotransmitter. A similar role has been proposed for γ-aminobutyric acid.

Some of the hypothalamic peptides that serve a neuroendocrine role in regulating the pituitary gland also serve a neurotransmitter role in regulating sexual behaviors in the female. The best example is GnRH. GnRH applied directly to the hypothalamus of ovariectomized female rats receiving an ineffective dose of estradiol demonstrate lordosis behavior. This suggests that GnRH stimulates LH secretion, ovulation, and subsequent sexual behavior in the female. Preoptic GnRH neurons project to the arcuate nucleus and median eminence area and subsequently release the peptide into the portal blood bathing the adenohypophysis, and they project to the midbrain central gray, the site mediating sexual receptivity. GnRH applied to the midbrain central gray stimulates sexual receptivity, and GnRH antisera applied to this area depresses sexual receptivity.

Pituitary hormones themselves may play a role in sexual receptivity. Prolactin applied directly to the midbrain central gray enhances sexual receptivity in rats receiving a low dose of estradiol. Conversely, pharmacologic depression of prolactin secretion at the time of anticipated onset of estrus depresses the magnitude of sexual receptivity. Pituitary hormones are released in response

Table 16-23
HYPOTHALAMIC CONTROL OF SEXUAL BEHAVIOR IN FEMALES

Chemical Mediator	Site of Action	Effect
Estrogen	Ventromedial nucleus, midbrain central gray	Increase LQ*
Progesterone	Ventromedial nucleus, preoptic area	Increase LQ response to estrogen
Norepinephrine	Ventromedial nucleus, preoptic area	Modulate progesterone receptors
Acetylcholine	Ventromedial nucleus, medial preoptic area	Modulate LQ response to steroids
Serotonin		Inhibit sexual behavior
Gonadotropin–releasing hormone	Midbrain central gray	Heighten response to estradiol
Prolactin	Midbrain central gray	Heighten response to estradiol

*LQ, lordosis quotient.

to the mating stimulus. It is established that prolactin is released from the adenohypophysis of rodents in response to excitation at the uterine cervix by the act of mating and transmitted to the hypothalamus through spinal pathways. The mating stimulus acts at the hypothalamus by lowering tuberoinfundibular dopaminergic tone, which leads to the release of prolactin. Prolactin activates the corpora lutea to maintain progesterone secretion, which maintains the subsequent pregnancy.

The other pituitary hormone released in response to the mating stimulus is oxytocin. The stimulus is transmitted over spinal pathways to enhance the activity of magnocellular neurons in the paraventricular nucleus. It has been suggested that after oxytocin is released in response to the mating stimulus it acts to enhance uterine contractions and favor sperm transport from the site of deposition at the mouth of the cervix to the site of fertilization at the oviduct. The lone weakness in this story is that, rather than enhancing contractions of the uterus from the cervical toward the ovarian direction, the contractions are enhanced in the ovarian-cervical direction, which would retard sperm transport.

Hypothalamic control of sexual behavior in males. The sexual behavior of male rodents has motivational and consummatory components (Table 16-24). Motivational behaviors are necessary to gain access to the female in heat. Consummatory behaviors are necessary for copulation. These behaviors include mounting, erection, intromission, and ejaculation.

Stereotypical male sexual behaviors are provoked by testosterone secreted from the Leydig cells of the testis. Testosterone controls male sexual behavior through two mechanisms. In peripheral tissues testosterone is converted to dihydrotestosterone (DHT). DHT is responsible for stimulating sensory receptors and may play a role in penile erection. Testosterone acts on the preoptic area of the hypothalamus to integrate the various consummatory components of male sexual behavior. The amygdala controls the motivational components of male sexual behavior. This appears to be a function of estrogen that has been aromatized from testosterone intraneuronally.

Among the catecholamines, dopamine from mesolimbic neurons appears to be the neurotransmitter controlling the motivational component of male sexual behavior.

The incertohypothalamic dopaminergic system appears to be responsible for the consummatory component. Neuropeptides modulate motivational and consummatory behaviors, and GnRH appears to act within the preoptic area to control consummatory behaviors. Endorphin neurons projecting from the amygdala to the preoptic area appear to have the opposite effects; they inhibit many consummatory behaviors. Other peptides that have been implicated in male sexual behaviors include substance P, NPY, α-melanophore–stimulating hormone, and oxytocin, but their physiologic significance has not been fully determined.

The preoptic area of the hypothalamus is sexually dimorphic, which is reflected in the pattern of LH secretion from the adenohypophysis. The dimorphic nature of the hypothalamus is also reflected in stereotypical male or female sexual behaviors. Males who are deprived of testosterone neonatally present a female cyclic pattern of LH secretion when challenged with estrogen as adults, and they present the typical female receptive lordotic pattern in response to estrogen and an aggressive male. Conversely, a female treated neonatally with testosterone shows the noncyclic pattern of LH secretion as an adult, and if treated with testosterone as adult, she will mount females in heat. Some of this sexual differentiation occurs in utero, but in rodents, most of it is determined neonatally. All hypothalami develop potentially as functionally female, and the differentiating event is the presence of androgen prenatally or neonatally.

Maternal behavior

The hypothalamus is intimately involved in mediating maternal behaviors stimulated by ovarian and pituitary hormones (Table 16-25). The rodent model is the most frequently studied. There are essentially four components of maternal behavior in the rat. Nestbuilding is the first behavioral sign. Late in pregnancy, the rat gathers bedding and any other materials available to it and prepares a nest in which she can deliver, nurse, and care for the young. She designs this to be the center of all her activities while the young are present. After the pups are born, the dam spends a large amount of time licking the neonates to clean them. The typical behavior rodents

Table 16-24
HYPOTHALAMIC CONTROL OF SEXUAL BEHAVIORS IN MALES

Chemical Mediator	Site of Action	Effect
Testosterone	Preoptic area	Control consummatory behaviors
Estradiol	Amygdala	Control motivational behaviors
Mesolimbic dopamine	Amygdala	Control motivational behaviors
Incertohypothalamic dopamine	Preoptic area	Control consummatory behaviors
Gonadotropin–releasing hormone	Preoptic area	Control consummatory behaviors
Endorphins	Preoptic area	Inhibit consummatory behaviors

Table 16-25
HYPOTHALAMIC CONTROL OF MATERNAL BEHAVIOR

Chemical Mediator	Site of Action	Effect
Estrogen after decline of progesterone	Medial preoptic area	Stimulate maternal behaviors
Prolactin	Preoptic area	Stimulate maternal behaviors in estrogen-primed rats
Oxytocin	Ventromedial nucleus	Stimulate maternal behaviors in estrogen-primed rats

share with all mammals is assumption of a nursing posture to allow the hungry young access to the mammary glands for retrieval of milk. As the nursing young mature, they tend to leave the convenience and safety of the mother's nest. The nursing mother then spends much time retrieving the young to the nest.

The hormonal drive for the onset of maternal behavior occurs during the prepartum period (see Table 16-25). By supplying foster pups late in pregnancy, the development of these behaviors can be characterized. The biochemical signal appears to be the gradual decline in progesterone secretion by the placenta coupled with the increase in ovarian estrogen secretion as the time of parturition approaches. The prepartum period can be envisioned as a period of hormonal priming. Maternal behaviors are the result of the combined actions of estradiol in the setting of decreasing progesterone, and they are influenced by adenohypophysial (or neural) prolactin. Oxytocin also may play a role. Estrogen affects maternal behavior largely through action in the medial preoptic area, mostly through the stimulation of estrogen receptors. Throughout pregnancy, there are more estrogen receptors in the preoptic area than in the entire hypothalamus. On the last day of pregnancy, the number of estrogen receptors in the rest of the hypothalamus rise to levels equivalent to those of the preoptic area. In addition to stimulating parental behaviors directly, estrogens also stimulate prolactin secretion.

It is the decrease of progesterone levels at the end of pregnancy that allows prolactin to exert its actions on the mammary gland to initiate and maintain lactation. Prolactin, secreted after parturition, has been implicated in the control of maternal behaviors. Hypophysectomy or treatment with the dopamine agonist bromocriptine can prevent many of the components of maternal behaviors, but infusion of prolactin directly into the preoptic area or into the CSF through the third ventricle stimulates maternal behaviors in estrogen-primed females rats. Prolactin can act on cells in the preoptic area and other circumventricular organs. Because prolactin is a large polypeptide, it is unlikely that it can cross the blood-brain barrier to affect neural structures. There are essentially three possibilities for the route prolactin may take to affect neural structures and maternal behavior. Pituitary prolactin may arrive at the hypothalamus by retrograde blood flow through the portal circulation. Alternatively, it has been shown that circulating prolactin has access to the CSF and brain through a receptor transport system located in the choroid plexes of the lateral, third, and fourth ventricles. However, the most recent evidence shows that specific areas of the hypothalamus contain prolactin mRNA, suggesting that prolactin is synthesized in these areas distinct from pituitary prolactin. Because estrogen can enhance brain and CSF levels of prolactin in hypophysectomized rats, it has been suggested that estrogen stimulates central prolactin synthesis and that the prolactin may enhance the sensitivity of estrogen-sensitive cells in the hypothalamus that regulate maternal behavior.

Among the hypothalamic peptides, oxytocin has been shown to promote maternal behavior when injected into the CSF of estrogen-treated rats. Oxytocin is ineffective when injected peripherally. Moreover, an oxytocin antagonist is effective in delaying maternal behavior when injected centrally. Destruction of the paraventricular nucleus, the source of oxytocin, also modifies maternal behaviors. The effects of oxytocin are correlated with the appearance of oxytocin cell membrane receptors in areas of the brain known to mediate maternal behavior, including the ventromedial hypothalamic nucleus of the hypothalamus, the bed nucleus of the stria terminalis, the anterior olfactory nucleus, and the central nucleus of the amygdala.

Emotional behaviors

The hypothalamus participates in emotional responses. For example, electrical stimulation of the lateral hypothalamus of cats results in many of the somatic and autonomic characteristics of anger such as piloerection, pupillary constriction, arching the back, raising the tail, and increased blood pressure. Similar rage-like responses can be elicited by decortication or by separating the hypothalamus from the cortex. The anger that is elicited is referred to as sham rage. Such animals respond to seemingly innocuous stimulation with a multitude of aggressive responses. The hypothalamus appears to act as an integrating center for these responses.

☐ The hypothalamus regulates rhythmic events

Types of rhythms

Rhythms (Fig. 16-26) are characterized by their *period* (i.e., time to complete one cycle), *frequency* (i.e., number of cycles per unit time), *phase* (i.e., points of reference

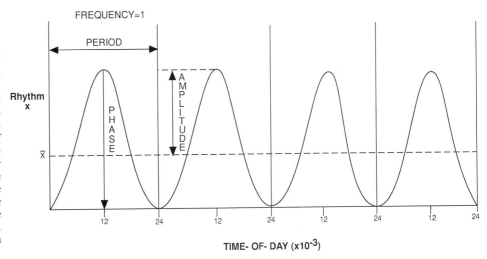

Figure 16-26
Parameters of a rhythm x over four complete cycles of any zeitgeber. Imagine that the zeitgeber is the lighting periodicity of an artificial 24-hour day, with daylight lasting from 7 A.M. to 7 P.M. The period (24 hours) is the time needed to complete one cycle of the rhythm. The frequency (once daily) of rhythm x is the number of cycles per unit of time. The phase is the maximum of the rhythm in reference to a time scale, such as that provided by the lighting periodicity or a clock. The amplitude is the deviation from the mean of the rhythm, $\bar{x}$.

on a time scale), and *amplitude* (i.e., magnitude of variation from the mean). Biologic rhythms are endogenous and self-sustaining. The only external cues may be provided by zeitgebers (i.e., time-givers such as the occurrence of light or dark) in some rhythms.

Biologic rhythms fall into one of four categories based on their period (Table 16-26): *circadian* (i.e., approximately 1 day and driven by zeitgeber), *ultradian* (i.e., <1 day and of much greater frequency, such as heart beat or respiration), *circannual* (i.e., >1 day and usually synchronized with seasonal events, such as seasonal fat deposition), and *infradian* (i.e., >1 day but shorter than 1 year, such as menstrual or estrous cycles).

Role of the hypothalamus in biologic rhythms

We have already described the role of hypothalamic neurohormones and neurotransmitters in infradian rhythms characterized by the menstrual and estrous cycles. The cyclic release of LH every 28 days in the human female involves participation of parts of the hypothalamus ranging from the most rostral to the most caudal boundaries.

In studying rodents, it is useful to appreciate the multifaceted roles of various areas of the hypothalamus regulating the various rhythms that comprise the estrous cycle. For example, GnRH neurons in the preoptic area respond to estrogen secreted every 4 to 5 days by secreting the peptide into portal blood and releasing a large bolus of LH into peripheral plasma. The preoptic area controls an infradian and a circadian rhythm. In rodents, ovariectomy and estrogen replacement results in an LH

surge at the same time on each day for the next several days. Shifting the lighting phase shifts the time of the occurrence of each surge by an equivalent amount. The lighting periodicity is the zeitgeber, and the biologic event is a true circadian rhythm.

Hypothalamic circadian rhythms require the participation of a timing device or clock within the hypothalamus. This role is served by the *suprachiasmatic nucleus* (SCN) of the hypothalamus. Destruction of the SCN results in the inability of the rat to transduce the lighting periodicity. A circadian rhythm that can be generalized to virtually all mammals is the adrenal corticosterone rhythm. In response to entrainment by lighting periodicity (rodents) or activity rhythms (man), corticosterone levels in the blood begin to increase and reach peak magnitudes at the same time each day. This pattern is driven by pituitary ACTH and hypothalamic CRH, as described previously. In rodents, the stimulus is the onset of darkness, but in humans, it is the onset of activity or wakefulness cycles. The secretion of most pituitary hormones is not just biphasic (basal and surge pattern) but is actually pulsatile, and blood levels of hormone at any time represent the summation of an ultradian pattern of hormone secretion from the cell. Such a rhythm is probably the consequence of the activity of a pulse generator within the hypothalamus regulating neurohormone secretion into hypophysial portal blood. A pulsatile ultradian rhythm of LH secretion is revealed in ovariectomized rats and monkeys. Measurement of multiple unit activity in the medial basal hypothalamus of monkeys reveals that the pulsatile pattern of LH secretion coincides with spikes

Table 16-26
CATEGORIES OF BIOLOGIC RHYTHMS

Rhythm	Approximate Period	Example
Ultradian	Much less than 24 h	Respiration, heart rate
Circadian	Approximately 24 h	Corticosterone rhythm
Infradian	Greater than 24 h but much less than 365 d	Menstrual cycles
Circannual	Seasonal, approximately 365 d	Hibernation

of multiple unit activity in the medial basal hypothalamus. This implies that the pulse generator for GnRH and subsequent LH secretion (at least in monkeys) resides within the medial basal hypothalamus.

Annual or circannual rhythms are not exclusively linked to the hypothalamus. Type I annual rhythms depend on the environment, and type II annual behavioral rhythms depend on an endogenous biologic clock. Type I rhythms are generally photoperiodically driven in that they require transduction of seasonal changes in day length. For example, as day length shortens during the late summer through early fall (i.e., short days), voles reduce their food intake, and their gonads involute. During the longer days of spring, food intake and gonadal weights return to normal. Type II annual rhythms are those that freerun; they require no environmental input and persist under constant environmental conditions. European starlings store fat before their demanding spring migration. Under constant light, temperature, and food availability, the rhythm of fat deposition persists. Each animal free runs with a period of 1 year and eventually become desynchronized under constant environmental conditions.

Photoperiodic measurement involves the SCN of the hypothalamus. Fibers from the retina of the eye terminate within the SCN. It is over this retinohypothalamic tract that the lighting periodicity is transduced. Efferent fibers from the SCN terminate in the paraventricular nucleus, which sends efferent fibers through the medial forebrain bundle to the spinal cord, which terminates on the intermediolateral cell column. Processes from these cells synapse in the superior cervical ganglion of the sympathetic chain. Postganglionic noradrenergic fibers from this area project to and innervate the pineal gland. By this pathway, the SCN generates a circadian rhythm in the pineal hormone melatonin, which is synchronized to the light-dark cycle. Melatonin is produced in greatest amounts during dark phases of the cycle and appears to be most important in mediating effects of annual rhythms.

The testes of male hamsters are most competent to produce sperm and normal levels of testosterone during the long days of summer. Pinealectomy prevents the loss of competency when animals are placed in the abbreviated illumination of short days. However, if pinealectomized hamsters receive long melatonin pulses, signaling short days or long nights, the testes regress independent of the environmental photoperiod. There appears to be a strain difference in the sensitivity to melatonin, and not all mammals depend on the pineal gland for the generation of biologic rhythms. For example, the rat, a photoperiodic mammal, has normally timed ovulation-inducing surges of LH release from the pituitary gland when pinealectomized. In animals responsive to melatonin, the sites in the central nervous system and periphery where melatonin acts to regulate biologic rhythms are not defined. Although there are marked species differences, the anterior hypothalamus, the SCN, and the adenohypophysis seem to be candidates.

Conclusion

The hypothalamus does not exert control over one process exclusive of other processes. The hypothalamus plays an integrative role in adapting the organism to meet the demands of its environment, and it can be appropriately characterized as an organ that uniquely ensures the perpetuation of the species.

SELECTED READINGS

Haymaker W, Anderson E, Nauta WJH (eds). The hypothalamus. Springfield, IL: Charles C. Thomas, 1969.

Jeffcoate SI, Hutchinson JSM (eds). The endocrine hypothalamus. New York: Academic Press, 1978.

Morgane PJ, Panksepp J (eds). Handbook of the hypothalamus. Anatomy of the hypothalamus, vol 1. New York: Marcel Dekker, 1979.

Neuroscience in Medicine, edited by P. Michael Conn.
J. B. Lippincott Company, Philadelphia © 1995

Chapter 17

*T*he Cerebral Cortex

MICHAEL W. MILLER
BRENT A. VOGT

The cerebral cortex forms a shell that covers most of the visible surface of the brain. Below its surface is a complex network of neurons and axons. The cerebral cortex is not uniform. It is composed of many structurally and functionally unique subunits that perform a wide range of sensory, motor, and mnemonic processes associated with cognition. The cerebral cortex organizes affective behaviors, including responses to painful stimuli, maternal and sexual behaviors, and the expression of rage and other emotions. Like other parts of the central nervous system, the cerebral cortex does not function in isolation, but it is part of an intricate plexus of overlapping circuits. In this chapter, we describe the morphology of cortical neurons and the way in which these neurons are assembled into clusters, columns, and areas. The intrinsic and extrinsic circuitries that enable the principal functions of the cerebral cortex are also elucidated.

Surface Features of the Cerebral Cortex

Magnetic resonance images of the cerebral cortex, like those in Figure 17-1, show that the cortex has a corrugated appearance. The crest of each fold is called a *gyrus*, and the depression between adjacent gyri is a *sulcus*. Very deep sulci are called *fissures*, and these include the space between the hemispheres, the interhemispheric fissure, and the lateral fissure (of Sylvius). The lateral surface of the cerebral cortex is composed of four lobes: the frontal, parietal, temporal, and occipital lobes. The medial surface contains extensions of each of these lobes and the limbic lobe.

The *frontal lobe* is the largest segment of the cerebral cortex. It extends from the rostral pole of each hemisphere to the central sulcus (of Rolando) and from the cingulate sulcus on the medial wall to the lateral fissure on the lateral surface. The frontal lobe is composed of five major gyri. The *precentral gyrus*, which contains motor cortex, runs parallel to the central sulcus and is bounded by the central and precentral sulci. Coursing perpendicular to the precentral gyrus on the lateral surface of the frontal lobe are the *superior, middle*, and *inferior frontal gyri*. The inferior frontal gyrus is further subdivided into the opercular, triangular, and orbital parts. An operculum is an extension of cortex that overlies or overhangs another region. In this instance, the operculum overlies cortex in the depths of the lateral fissure, the insular cortex (see Fig. 17-1*E,F*). The opercular and triangular divisions of the inferior frontal gyrus contain Broca's speech area. The *fronto-orbital gyri* form the ventral part of the frontal lobe and rest on the orbital plate of the frontal bone.

The *parietal lobe* is bounded rostrally by the central sulcus, medially by the cingulate sulcus, posteromedially by the parieto-occipital sulcus, posterolaterally by an imaginary line between the parieto-occipital sulcus and the preoccipital notch, and ventrally by the lateral fissure. Parietal cortex includes the postcentral gyrus, a strip of cortex that is posterior to the central sulcus and parallel to the precentral gyrus. The postcentral gyrus contains somatosensory cortex. Caudal to the postcentral gyrus are the superior and inferior parietal lobules, which are divided by the intraparietal sulcus. The inferior parietal lobule is composed of two gyri, the supramarginal gyrus, which caps the lateral fissure, and the angular gyrus, which straddles the border with occipital cortex and forms the banks of the end of the superior temporal sulcus. Two regions form the medial surface of the parietal lobe. One is the paracentral lobule, which surrounds the medial tip of the central sulcus. This lobule also includes a segment of frontal cortex. Caudal to the paracentral lobule is the precuneal cortex, which ventrally blends with the limbic lobe.

The *insula* is an island of cortex in the depths of the lateral fissure. It is bounded by cortex of the frontal, parietal, and temporal lobes. The dorsal border of the lateral fissure is composed of the opercular part of the inferior frontal gyrus and the parietal operculum, and the ventral border is composed of the temporal operculum. The rostral insula is part of the limbic system, although its specific functions are not fully understood. The posterior part of the insula is involved mainly in processing somatosensory information.

The *temporal lobe* is ventral to the lateral fissure. This lobe includes the superior, middle, and inferior temporal gyri that run parallel to the lateral fissure. The dorsal plane of the superior temporal gyrus contains the transverse gyri (of Heschl), where the primary representation of audition is located. The ventral part of the temporal lobe includes the occipitotemporal gyrus.

The *occipital lobe* includes cortex caudal to an imaginary line through the parieto-occipital sulcus and notch

Figure 17-1

Magnetic resonance imaging is used to examine the structure of the brain in vivo. (**A**) A lateral view of the cerebral cortex can be appreciated in a brain reconstructed from serial magnetic resonance images. The image is a compilation of a series of 1.5-mm optical sections. (**B**) The sulci and gyri can be identified on the lateral surface. The labels for the sulci are placed around the brain, and the labels for the gyri are placed at the appropriate site on the brain. (**C, D**) A parasagittal section reveals several features characteristic of the medial surface of the brain. (**E, F**) The horizontal and vertical lines shown in **C** identify the planes used in the horizontal and coronal sections shown in **E** and **F**, respectively. *cg*, cingulate gyrus; *co*, collateral sulcus; *cs*, cingulate sulcus; *hf*, hippocampal formation; *in*, insula; *lf*, lateral fissure. (MRI images courtesy of Nancy Andreasen, University of Iowa, Iowa City, IA)

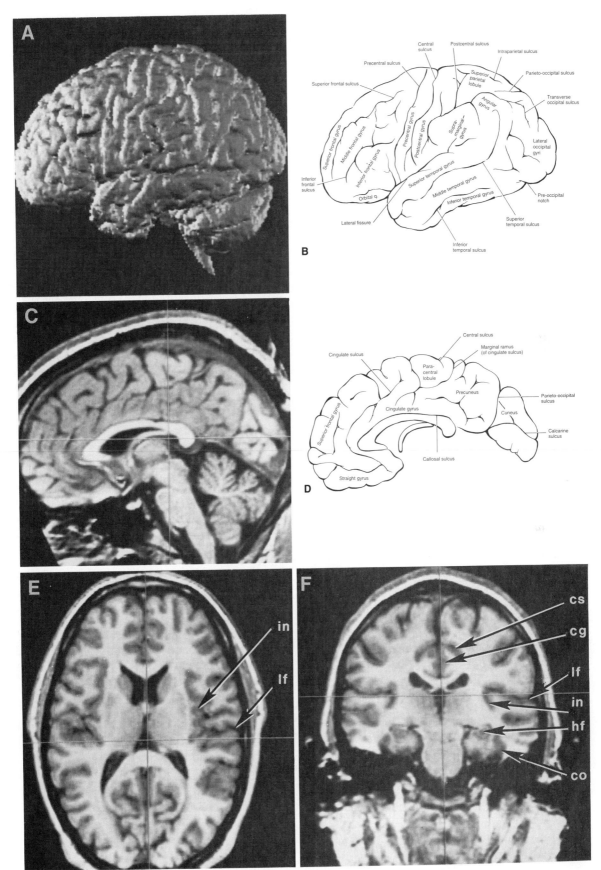

Figure 17-1

that demarcates it from the temporal lobe. The occipital lobe includes the calcarine fissure, within and around which is the visual cortex.

In 1868, the comparative neurologist Paul Broca defined the *limbic lobe*. This lobe is composed of the cingulate gyrus and hippocampal and parahippocampal gyri. The limbic lobe contains areas that are a major part of the anatomic limbic system (see Chap. 18). This cortical region is involved in olfaction, memory, and visceral, skeletal, and endocrine functions associated with emotional behaviors.

Cortical Cytology

☐ Projection neurons connect disparate sites

Most cortical neurons are projection neurons, and the most common cortical projection neuron is the pyramidal neuron (Table 17-1). A typical pyramidal neuron has a large pyramid-shaped cell body that gives rise to two sets of dendrites (Fig. 17-2). One set is a prominent dendrite that issues from the apex of the cell body. The apical dendrite often reaches layer I, where it arborizes in a tuft of dendrites. The apical dendrite also gives rise to smaller caliber collateral processes that often branch at 90° angles. The second set of dendrites is composed of processes that emanate from the base of the cell body. Unlike the apical dendrite, the branches of the basal dendrites bifurcate at acute angles. All of the dendrites, except the dendritic segments proximal to the cell body, are densely covered with small protuberances known as *spines*. A spine typically has the appearance of a lollipop, with a large rounded head that is attached to the dendritic shaft by a slender neck.

Apical dendrites of pyramidal neurons are often aggregated into clusters. Each cluster contains the apical dendrites of pyramidal neurons whose cell bodies are distributed in superficial and deep cortex. The apical dendrites of neurons whose cell bodies are in deep cortex form the core of the cluster, but the apical dendrites of neurons whose cell bodies are in superficial cortex are distributed at the periphery of the cluster. These clusters may define a functional unit or module.

With few exceptions, the axons of pyramidal neurons arise from the base of the cell body. In addition to emitting a local plexus of collaterals, the axons of pyramidal neurons have one of two projection patterns. They project from one region of cortex to another in the ipsilateral cortex (i.e., association projections) or contralateral hemisphere (i.e., callosal projections), or they descend to subcortical structures. The axons of projection neurons release the excitatory amino acids aspartate and/or glutamate as a neurotransmitter.

There are two subpopulations of projection neurons that do not have apical dendrites. Notable among these are the spinous stellate neurons in the intermediate zones of sensory cortices and the large, nonoriented neurons in superficial entorhinal cortex. Like the typical pyramidal neurons, these projection neurons have spinous dendrites, and they excite the postsynaptic neurons through the release of glutamate.

Most projection neurons respond to a stimulus with a series of spikes that are followed by prolonged after hyperpolarizations and after depolarizations (Fig. 17-3). These regular spiking neurons adapt to sustained stimuli.

☐ Local circuit neurons inhibit activity within the cortex

Based on their somatodendritic morphology, two groups of local circuit neurons can be discerned. One group is the stellate neurons. The stellate neurons have round cell bodies and an array of dendrites that radiate uniformly from their somata (Fig. 17-4). The second group of local circuit neurons has a polarized form. Their cell bodies are elongated and may be oriented radially or horizontally. Regardless of their orientation, the dendrites tend to arise from the attenuated poles of the cell body. The dendrites of all local circuit neurons are aspinous or sparsely spinous.

The pattern of the axonal arbors of local circuit neurons varies. The axon may arise from virtually any site on the cell body or from a proximal dendrite. The distribution of the axon arbors is restricted to the sphere of the dendritic arbors or extend beyond the dendritic field. Local circuit neurons use γ-aminobutyric acid (GABA) as their neurotransmitter (see Fig. 17-4A). Release of this

Table 17-1
FEATURES OF CORTICAL NEURONS

Feature	Projection Neurons	Local Circuit Neurons
Dendrites	Spinous	Aspinous
Axons	Local arbors and distant projections	Local arbors only
Synapses formed by		
Axosomatic afferents	Symmetric	Symmetric and asymmetric
Efferents	Asymmetric	Symmetric
Neurotransmitter	Glutamate and aspartate	GABA and neuropeptides
Discharge properties	Regular spiking	Fast spiking

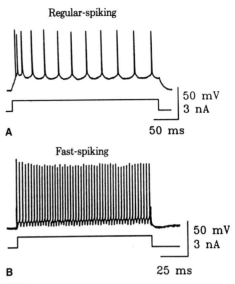

Regular-spiking

50 mV
3 nA

A   50 ms

Fast-spiking

50 mV
3 nA

B   25 ms

Figure 17-3
Intracellular recordings of neurons show that projection and local circuit neurons exhibit different firing patterns. (**A**) Projection neurons have a phasic, regular spiking pattern. (**B**) Stimulation of a local circuit neuron results in a continuous and maintained stream of fast spikes. (Connors BW, Gutnick MJ. Trends Neurosci 1990;13:99.)

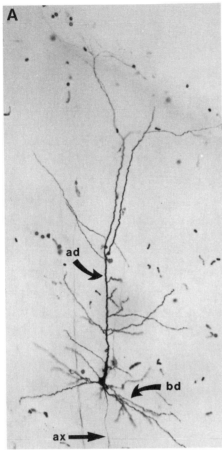

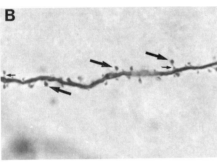

Figure 17-2
(**A**) This pyramidal neuron was intracellularly injected with the tracer, horseradish peroxidase, which was subsequently localized histochemically. This neuron has an apical dendrite (ad) that ascends from the apex of the pyramid-shaped cell body and reaches layer III. At this point, it branches to form an apical tuft that ramifies within layers I and II. The base of the cell body gives rise to a set of dendrites (bd) and to an axon (ax) that descends toward the white matter. (**B**) Each dendrite is invested with a coat of spines. A spine has a bulbous head (*large arrows*), which is attached to the dendritic shaft by a long, thin neck (*small arrows*).

neurotransmitter inhibits the activity of postsynaptic neurons. The local circuit neurons can release other neuroactive substances from their axonal terminals (see Fig. 17-4B,C). These peptides include acetylcholine, cholecystokinin, neuropeptide Y, somatostatin, substance P, and

vasoactive intestinal polypeptide. It is not yet understood how these other neuroactive substances interact with GABA on postsynaptic neurons to modulate neurotransmission, but it does appear that the release of the secondary substances is activity dependent. At low levels of excitation, local circuit neurons release only GABA, but at high levels of excitation, they release GABA and the other neuroactive compound.

Intracellular recordings of local circuit neurons reveal that cortical local circuit neurons have membrane and spiking properties that differentiate them from projections neurons. After suprathreshold stimulation, local circuit neurons discharge fast spikes lasting less than 0.5 milliseconds (see Fig. 17-3). Because these neurons exhibit little or no adaptation during a prolonged stimulation, the spike frequency remains the same, and the local circuit neurons transmit inhibitory information to postsynaptic targets with great fidelity.

☐ Cortical synaptology is a balance between excitatory and inhibitory information

Two structurally and functionally unique types of synapses are formed by cortical neurons. *Asymmetric synapses* are formed by axons that contain large vesicles and excitatory neurotransmitters, primarily glutamate. These synapses have postsynaptic densities that are composed of a protein kinase, and activation of the synapses leads to depolarizing potentials or excitatory responses in postsynaptic neurons. The *symmetric synapses* have presynaptic axons with small synaptic vesicles that contain

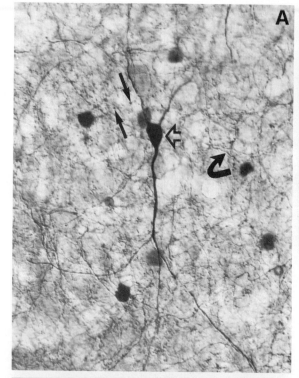

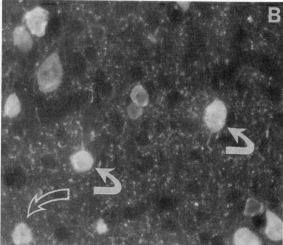

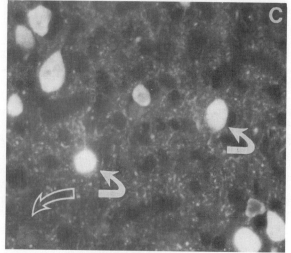

Figure 17-4

(**A**) Cortical local circuit neurons use γ-aminobutyric acid (GABA) as a neurotransmitter. The open arrow indicates the cell body of an aspinous stellate neuron, which was labeled immunohistochemically with an antibody directed against glutamic acid decarboxylase (GAD). GAD is the enzyme that catalyzes the rate-limiting step in GABA synthesis. GAD immunoreactivity is evident in the axonal processes (*curved solid arrow*) and in GABAergic axonal terminals (*straight solid arrows*). (**B**, **C**, **D**) Many GABAergic neurons colocalize with a peptide neurotransmitter. Immunofluorescence techniques were used to identify GABA-immunoreactive neurons (*solid arrows* in **B**), which were double-labeled with an anti-substance P antibody (*solid arrows* in **C**). Not all of the GABAergic neurons colocalize substance P (*open arrows* in **B** and **C**). (Courtesy of Steward Hendry, Johns Hopkins University, Baltimore, MD)

the inhibitory transmitter GABA. The presynaptic and postsynaptic densities of these synapses are approximately of equal thickness. Activation of symmetric synapses evokes hyperpolarizing potentials or inhibitory responses in postsynaptic neurons.

The distributions of asymmetric and symmetric synapses are among the many features that differentiate projection from local circuit neurons. The two types of synapses are largely segregated by projection neurons. The most common target of axons that form asymmetric synapses is the heads of dendritic spines. Most symmetric synapses are formed in the perisomatic region. This region includes the soma, the smooth surfaces of proximal dendrites, the axon hillock, and the initial segment of the axon. The result of this organization is that excitatory re-

sponses can be evoked over a large area of the pyramidal cell dendritic tree. The summed excitatory input is gated by inhibitory activity in the perisomatic region. Because the action potential initiation zone is in the initial segment of the axon, the symmetric synapses are strategically placed to modulate the discharge frequency of cortical pyramidal neurons. The asymmetric and symmetric synapses are intermingled along the smooth dendrites of local circuit neurons. The dispersion of inhibitory synapses in relation to excitatory ones results in a less pronounced gating of excitatory activity by local circuit neurons.

The efferents of the two classes of cortical neurons have different patterns of connectivity. The local arbors and projections of pyramidal neurons form excitatory, asymmetric synapses with postsynaptic targets. The axons of local circuit neurons form inhibitory, symmetric synapses. Hence, the two neuronal populations influence their targets in opposite ways.

The functional and connectional differences of the two populations of cortical neurons are evident during an epileptic seizure. In the absence of local circuit neuron-mediated inhibition, the activity of the projection neurons is unchecked. Projection neurons discharge without inhibitory modulation and produce depolarizing shifts that are composed of very large, excitatory, postsynaptic potentials. A common treatment for seizure activity is the administration of compounds such as valproic acid that have actions similar to GABA.

Structure and Distribution of Cortical Areas

The size and packing densities of neurons are not uniform in the cerebral cortex. Cytoarchitecture refers to the unique distributions of neurons in different parts of the cortex. The underlying tenet of this approach to neuroscience is that structural differences in the cerebral cortex are associated with functionally unique areas. Figure 17-5 has Nissl-stained sections through the somatosensory and motor cortices, areas that are caudal and rostral to the central sulcus, respectively. One of the striking features of cortical architecture is the horizontal alignment

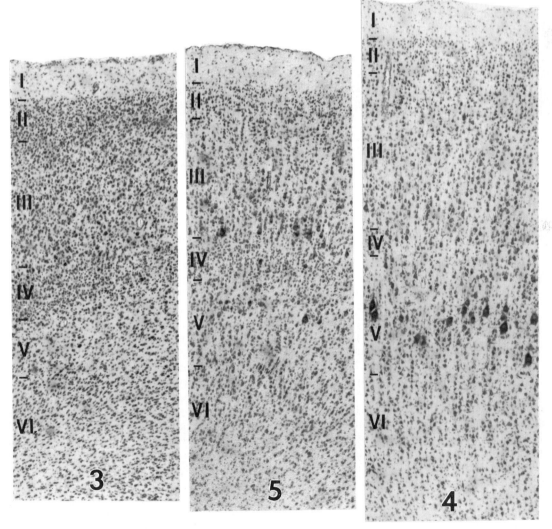

Figure 17-5
Photomicrographs of Nissl-stained sections through three sensorimotor areas to show differences in cytoarchitectural organization: primary somatosensory cortex (*area 3*); first somatosensory association cortex (*area 5*); motor cortex (*area 4*). Notice the highly graunular layers II and IV in area 3 and large pyramidal neurons in the deep part of layer III in area 5 and in layer V of area 4.

of its neurons into layers. It is generally accepted that neocortex or isocortex contains six layers.

The differentiation of cortical layers is largely based on the distinctive population of projection neurons in each layer. A neuron is considered to be in a layer according to the position of its cell body. A pyramidal neuron with a cell body in layer V may be referred to as a layer V neuron, or a layer may be described as having pyramidal neurons.

The most superficial cortical layer, the layer that abuts the pia mater, is layer I. Layer I is also referred to as the plexiform or molecular layer. Layer I is composed mainly of the apical tuft dendrites of pyramidal neurons and afferent axons. This layer is virtually devoid of neuronal cell bodies; it has only a few local circuit neurons, and projection neurons are absent from layer I. Layer II has a granular appearance and is densely populated by the cell bodies of small pyramidal neurons. Layer III is composed of medium and large pyramidal neurons. The sizes of their cell bodies increase with the depth, and the largest layer III neurons are deep, near the border with layer IV. Layer IV has a granular appearance. It is composed of the small, round cell bodies of stellate projection neurons and pyramidal neurons that do not have orienting apical dendrites. The cell packing density in layer IV often is the greatest of all the cortical layers. Layer V contains the largest pyramidal neurons. The cell packing density in this layer is the lowest of all cortical laminae. Layer VI has multiform projection neurons, imparting many different shapes to the cell bodies of layer VI neurons. Local circuit neurons are distributed in all cortical layers, but their distribution does not facilitate the cytoarchitectonic differentiation of cortical layers and areas.

A comparison of the somatosensory and motor cortices (see Fig. 17-5) provides an example of how cortical areas can be differentiated on the basis of their cytoarchitecture. The superficial layers I through IV of somatosensory cortex contain many small neurons that endow it with a granular appearance. This granularity is particularly evident in layers II and IV. Somatosensory cortex also is characterized by relatively small layer V pyramidal neurons. In contrast, motor cortex has almost no layer IV and relatively few small neurons in layers II and III. Instead, there are many large pyramidal neurons, particularly in layer V. The largest layer V pyramidal neurons are called Betz cells, named after the nineteenth century scientist who first described them. The Betz cells and other layer V projection neurons project axons to the spinal cord through the corticospinal tract. Although the pyramidal neurons in layer V of somatosensory cortex are smaller, some of them also contribute axons to the corticospinal tract.

Early in the twentieth century, Karl Brodmann produced one of the most thorough and enduring cytoarchitectural studies of the human cerebral cortex. Figure 17-6 is a copy of the classic cytoarchitectural map that summarizes his conclusions about the distributions of cortical areas. Each cytoarchitectural area is designated with an arabic numeral in the order in which he studied

them. The somatosensory cortex is area 3 and adjacent areas 1 and 2. These areas are in the postcentral gyrus of the parietal lobe. The motor cortex is area 4 and is in the precentral gyrus of the frontal lobe. The auditory cortex, or areas 41 and 42, is located in the transverse gyri on the dorsal aspect of the superior temporal gyrus. Visual cortex, or area 17, is found in the banks of the calcarine sulcus and the lateral wall of the occipital lobe. Other numbers used in this text refer to cortical areas designated by Brodmann.

Although most of the cytoarchitectonic areas on the lateral surface of the cerebral cortex are isocortical, many on the medial surface do not have six layers. The hippocampus, for example, has a pyramidal cell layer sandwiched between two plexiform layers. Cortical areas with fewer than six layers are referred to as *allocortex*. The allocortex is part of the "anatomic limbic system."

Allocortex is heterogeneous, and it is subdivided into three parts or moieties. *Archicortex* includes the hippocampal formation (i.e., hippocampus, dentate gyrus, and subiculum). *Paleocortex* includes the olfactory piriform area and the rostral insula. *Periarchicortex* includes many transitional or mesocortical areas, such as entorhinal, posterior parahippocampal, and orbital cortices, and much of cingulate cortex. An example of the cytoarchitecture of some allocortical moieties are represented by drawings of neuronal perikarya in Figure 17-7. The allocortical subiculum has a single pyramidal layer; the junction of this layer with the neuron-sparse molecular layer is indicated with an open arrow in this illustration. Entorhinal cortex or Brodmann's area 28 is an example of periarchicortex and forms a major part of the parahippocampal gyrus. One of its prominent features are the islands of large star neurons in layer II; these neurons are surrounded by a dashed line in Figure 17-7 to emphasize this "insular" arrangement. These neurons project into the hippocampal formation by way of the perforant pathway. The neocortical area 20 forms much of the inferior temporal gyrus. This area has six layers, including large layer III pyramidal neurons and granular layers II and IV.

One simple conceptualization of the cortical mantle is to view it as a sheet of layer V pyramidal neurons that is continuous from the hippocampus to isocortex. To this basic layer of neurons is added the superficial layer of pyramidal neurons, which is differentiated in periarchicortical and isocortical areas to form the many cytoarchitectural variations that compose the cerebral cortex.

The various components of the cerebral cortex are impacted differently in some neurologic and psychiatric diseases. This is exemplified by Alzheimer's disease. Figure 17-8 has examples of neurons from in three regions of a brain from a person afflicted with Alzheimer's disease. The tissue is stained for neurofibrillary tangles with the fluorescent dye thioflavin S. Observe the flame-shaped neurofibrillary tangles that fill the perikarya of projection neurons. The structure of neurofibrillary tangle–laden projection neurons is shown for the hippocampal formation, layer II of entorhinal cortex, and layer III in isocortex. Neurofibrillary tangles also accumulate in layer V

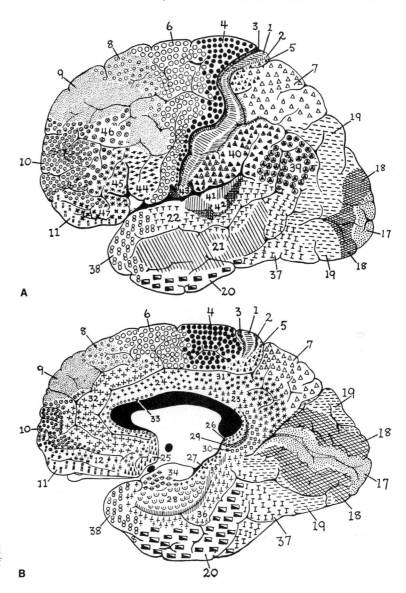

Figure 17-6
Brodmann's maps of the distribution of cortical areas on (**A**) the lateral and (**B**) medial surfaces of the cerebral cortex.

neurons of entorhinal and isocortical areas. This pathology in large projection neurons is indicative of ongoing degenerative processes. In many instances, the somata completely disappear, and only "ghost" neurofibrillary tangles remain. Because many of these large neurons are corticocortical projection neurons, there is a disruption of major efferent projection systems of the cerebral cortex in this disease. Degeneration of these neurons contributes to the cognitive and emotional impairments in Alzheimer's disease.

Functional Subdivisions of the Cerebral Cortex

Combined structural and functional studies led to systematic classification schemes of the principal functions of individual cortical areas. The main tools for function-

ally defining cortical areas are neuronal recording, electrical microstimulation, and positron emission tomography techniques.

☐ Sensory areas are defined by their thalamic input

Cortices involved in different aspects of sensation are subdivided into primary, first sensory association, second sensory association, and multimodal cortices. Primary sensory cortices contain neurons with spatially restricted receptive fields and properties that are dominated by thalamic afferents. For example, the lateral geniculate thalamic nucleus has neurons with receptive fields that are round and subtend as little as 1° of the visual field. Layer IV of primary visual area 17 (i.e., the lamina in which thalamic afferents principally arborize) has neurons with

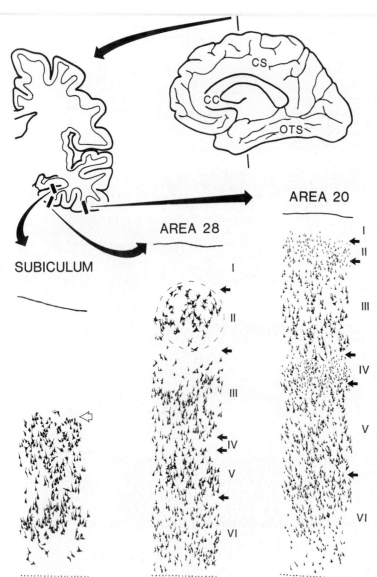

Figure 17-7
The cerebral cortex is not cytoarchitecturally uniform and has a progressive elaboration in its architecture in the medial to lateral direction. This is true for the dorsal and ventral surfaces, although it is illustrated here with areas from the ventral part of the cortex. The medial view of the brain shows the corpus callosum (CC), cingulate sulcus (CS), and occipitotemporal sulcus (OTS) and the plane through which a coronal section (*top arrow*) was taken for Nissl staining. The transverse section has three black rectangles indicating the subiculum, area 28, and area 20, in the medial to lateral direction. The three strips of cortex below are perikaryal drawings to show the distribution of neurons in each of these areas. The junction between the molecular and pyramidal cell layers of the subiculum is demarcated with an open arrow. In area 28, an island of layer II star pyramidal neurons is presented with a dashed line around it to emphasize that this is not a continuous layer of neurons as in most other cortical areas. Each area in the medial to lateral direction has a progressively more differentiated laminar architecture.

rectangular receptive fields, the size and orientation of which are the product of multiple thalamic inputs terminating on individual neurons.

Sensory spaces can have multiple representations in the primary sensory cortices. In primary somatosensory cortex, for example, there are four separate body representations. These separate sensory representations are contained within unique cytoarchitectural areas. The primary somatosensory cortex is composed of areas 3a, 3b, 1, and 2, and each area has its own representation of the body surface, called a *homunculus*. Each area is involved in the high-resolution localization and discrimination of somatic stimuli. Body regions with the highest density of receptors, such as the face and hands, also have the largest area of representation in the cortex, improving localization and discrimination of sensory stimuli in these areas. In visual cortex, a relatively large area of cortex is devoted to the central visual fields; this cortex responds to stimuli focused on the fovea. Each of the primary sen-

sory areas are indicated with vertical lines in Figure 17-9.

Parasensory association cortices receive their inputs mainly from the primary areas. Neuronal responses in these areas are more complex and involve the integration of a number of cortical inputs and those from the thalamus. As shown in Figure 17-9, each primary sensory area has a first parasensory association area: the first visual association area (VA1) for primary visual cortex includes areas 18 and 19, the first auditory association area (AA1) for primary auditory cortex includes part of area 22, and the first somatosensory association area (SA1) for primary somatosensory cortex includes area 5 and rostral area 40. In the left hemisphere, there is a specialization of the first auditory association area called *Wernicke's area*. Wernicke's area is a posterior part of AA1 and is involved in recognition of spoken language. Layer III pyramidal neurons are the main source and target of corticocortical projections, and the layer III neurons are par-

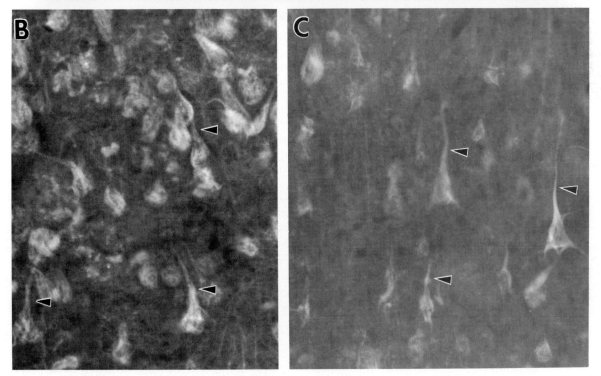

Figure 17-8
Micrographs of (**A**) thioflavin S–stained neurofibrillary tangles in layer II of area 28, (**B**) the subiculum, and (**C**) layer III of area 20 of a case of Alzheimer's disease. Notice in **B** and **C** that the apical dendrites of pyramidal neurons (*arrowheads*) are filled with these tangles, as are the somata. The border of layers I and II of the entorhinal cortex are marked with an arrow in **A**. In this island of layer II neurons, there are almost no primary (i.e., orienting) apical dendrites. Neurons in this layer instead have dendrites that radiate from around the soma (*arrowheads*), and these neurons are referred to as star cells.

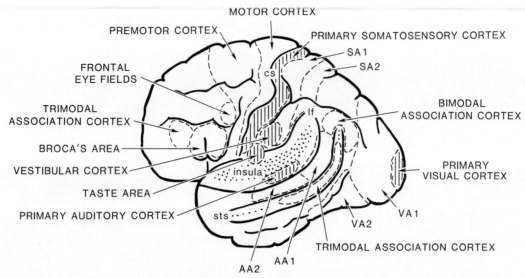

Figure 17-9
Topographic distribution of functional areas in the cerebral cortex. These regions were determined from a wide range of physiologic and anatomic studies of human and nonhuman primate brains and are highly schematic. They were plotted onto the same hemisphere as that used by Brodmann in Figure 17-6A so that functional regions can be compared with their associated cytoarchitectural areas. Primary sensory cortices are indicated with vertical hatching. A few sulci are labeled for orientation purposes and include the central sulcus (cs), lateral fissure (lf), and superior temporal sulcus (sts). The lateral fissure and superior temporal sulcus were "opened" to reveal the depths of these sulci. By lifting the frontal, parietal, and temporal opercula, the underlying insula can be seen. The taste area extends onto the upper bank of the parietal operculum, and the primary auditory cortex is on Heschl's gyri on the superior temporal plane (i.e., planum temporale). The superior temporal sulcus was "opened" to expose one of the trimodal association cortices.

ticularly well developed in sensory association areas. Figure 17-5 shows the architecture of area 5 with its large layer III pyramidal neurons.

Secondary sensory association cortices are characterized by their corticocortical connections. The second sensory association areas receive inputs from the first sensory association cortices and project to multimodal areas. There is a second sensory association cortex adjacent to each of the first association areas; the areas for visual cortex (VA2), for auditory cortex (AA2), and for somatosensory cortex (SA2) are shown in Figure 17-9.

Multimodal association areas receive inputs from more than one sensory modality and provide for intermodal associations among stimuli arriving in two or more second sensory association cortices. Thalamic nuclei that project to multimodal areas include the pulvinar, lateral posterior, and mediodorsal nuclei. These thalamic nuclei do not have a singular sensory function and are themselves likely sites for multimodal interactions. Multimodal areas can be classified as bimodal and trimodal cortices. Although there are many bimodal areas, Figure 17-9 presents only one as an example. It is on the angular gyrus dorsal to the tip of the superior temporal sulcus, and it receives inputs from the second association areas for the visual and somatosensory modalities. There are three major trimodal association areas (two are shown in Figure 17-9). One is in the ventral part of area 46 in prefrontal cortex, one is in the depths of the superior tem-

poral sulcus, and a third is in posterior parahippocampal cortex on the ventral surface of the cerebral cortex.

Each multimodal association area has projections to the cingulate and rostral parahippocampal cortices. These limbic areas are thought to be involved in monitoring the sensory environment, because they have neurons that respond to large sensory stimuli and lesions in cingulate cortex disrupt attention to sensory stimuli. Ablations in parahippocampal, hippocampal, and cingulate cortices disrupt memory formation and spatial orientation. The limbic cortices probably form an "end stage" in cortical processing of sensory inputs and provide mechanisms for memory that apply to complex patterns of sensory stimuli.

☐ The motor cortex projects to nuclei that innervate skeletal muscle

Activity in the various sensory spaces leads to environmentally adapted behaviors. The associations between sensory and motor cortices and the mechanisms by which particular motor sequences are generated are prominent issues in clinical neuroscience. Motor cortices have direct projections to the spinal cord or brain stem motor nuclei and are classified according to their contributions to movement, based on responses to electrical microstimulation and the length of time by which neuronal firing precedes movement. The primary motor cortex, which is Brodmann's area 4 in the precentral gyrus, has the lowest threshold for electrically evoked movements. Activation of neurons in the motor cortex engages a limited number of muscle groups, such as the lumbrical muscles for each finger. Neuronal activity in the primary motor cortex occurs just before movement. The finest control of motor activity is mediated by the primary motor cortex.

A dramatic example of the topographic representation of the musculature in motor cortex (i.e., the homunculus) is seen when seizure activity spreads through motor cortex in what is called the "Jacksonian march." After seizure activity is initiated in one part of the motor cortex, it induces convulsions in the associated part of the body. As the seizure activity spreads through the network of local axon collaterals of the projection neurons, the large depolarizing shifts in the projection neurons induces hyperexcitability in adjacent cortical areas. The result is convulsions that progressively include adjacent parts of the body. This seizure activity can spread across the entire motor cortex and by means of connections through the corpus callosum to the contralateral hemisphere. Convulsions that originally start in one focus can spread across the body surface and involve both sides of the body.

Classic accounts of motor cortex defined a secondary or supplementary motor area on the superior frontal gyrus medial to the primary motor cortex. The secondary motor area is one of several separate premotor areas. The neurons in premotor areas have direct projections to the primary motor cortex and to the spinal cord or brain stem motor nuclei. Electrical microstimulation in these areas can activate larger groups of muscles; the movements elicited have a longer latency than those resulting from stimulation of units in primary motor cortex. One of the premotor areas directs eye movements and is shown in Figure 17-5 as the frontal eye fields. Another area controls movements of the mouth during speech and is called *Broca's area*. Broca's area is directly connected to Wernicke's area, which is part of the first auditory association cortex in the left hemisphere. Language comprehension in Wernicke's area can control speech through Broca's area.

The classic studies of epileptic patients by the neurosurgeons Wilder Penfield and Herbert Jasper had a profound impact on our understanding of the functional organization of the cerebral cortex. They surgically exposed the cerebral cortex under general anesthesia and then allowed the patient to regain consciousness so that the patient's sensations could be reported after electrical stimulation or during seizure activity. In addition to observing sensory and motor phenomena after electrical stimulation of sensory and motor cortices, respectively, Penfield and Jasper observed that stimulation of limbic cortical areas evoked visceral and emotional sensations and memories. On patient, for example, had seizure activity that began with a "far-off" feeling. She smacked her

lips, swallowed, and complained of nausea. Borborygmi could be heard from her abdomen during the seizure, and these attacks were often associated with a feeling of panic. Electrical stimulation of the anterior temporal lobe and rostral insula evoked similar feelings and intestinal activity, and ablation of these cortices alleviated the seizure activity and associated emotions and autonomic activity. The limbic cortex contributes to visceromotor activity and emotion as well as operating as an end stage in sensory processing. These limbic areas include the orbital, insular, temporal pole, and anterior cingulate cortices.

Cortical Connectivity

☐ Thalamocortical associations define the cortical column

One of the principles of neuroscience is that the activity of sensory thalamic afferents defines the primary functions of a sensory cortical area. The receptive field characteristics of neurons in the lateral geniculate nucleus determine the visual properties of neurons in area 17. The thalamic projections terminate chiefly in layer IV, with minor projections to layers I and VI. As a result, the thalamic afferents synapse with any cell body or dendrite that compose these layers, regardless of where the cell body is located. It is also the case that projections from the ventrolateral and ventroanterior thalamic nuclei contribute to the motor functions of motor and premotor areas, respectively. The pivotal role of the thalamic input is confirmed by transplantation studies. After transplanting of fetal somatosensory cortex to visual cortex, the neurons from somatosensory cortex respond to visual stimuli.

The termination of thalamic afferent axons in the cerebral cortex determines functional cortical modules or columns. Vernon Mountcastle and his colleagues first showed that neurons in a vertical column of somatosensory cortex shared similar response properties. One column may contain neurons with preferential responses to stimulation of joints and deep tissues, and an adjacent column may contain neurons with preferential responses to light touch of the skin. The columns in the visual cortex represent ocular dominance and orientation specificity (Fig. 17-10). In sensory cortices, these columns are 500 to 1000 μm in diameter and reflect the distribution of thalamic axon terminals in the cortex.

Aggregates of neurons form columns in the motor cortex. Unlike the columns in sensory cortices, the columns of functionally similar neurons are assessed in terms of the output of motor cortex (i.e., by microstimulation techniques rather than receptive field mapping). The output columns in the motor cortex, which are larger than in the sensory cortices, are as large as 2 mm in diameter. These large sizes are the result of the size of dendritic fields of particularly large pyramidal neurons in the motor cortex and are specified according to the amount of

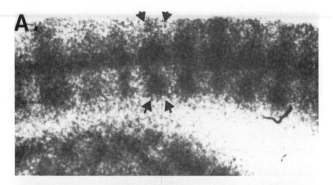

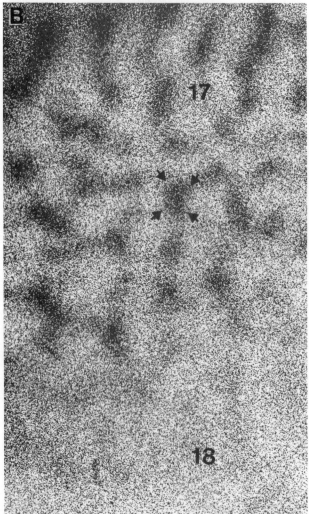

Figure 17-10
(**A**) Presentation of a bar of light with a particular orientation results in partial stimulation of visual cortex. The responsive neurons in area 17 are organized in vertical or radial columns (*arrows*). These columns were visualized autoradiographically using 2-[^{14}C]deoxyglucose, a radioactive tracer for glucose. This tracer was taken up and trapped in metabolically active neurons. The highest activity occurred in layer IV, the termination site of the thalamic afferents. (Hubel DH, Wiesel TN, Stryker MP. J Comp Neurol 1978;177:361.) (**B**) An autoradiograph of a tangential section through superficial cortex reveals a complex patchwork of activity. The orientation-specific columns (*arrows*) are present only in area 17 and not in the visual association cortex, area 18. (Gilbert CD, Wiesel TN. J Neurosci 1989;9:2432.)

cortex that must be electrically stimulated to evoke contraction in a small group of muscles.

Thalamocortical connections are reciprocal. A column of neurons defined by a particular thalamic input to layer IV contains pyramidal neurons in layer VI that project to the thalamus. These return projections are organized in a point-to-point fashion so that, for example, cortical points representing the central visual fields project to similar places in the lateral geniculate nucleus. These reciprocal connections are thought to be involved in modulating the receptive field properties of thalamic neurons.

☐ Monoaminergic afferents are involved in sleep patterns

The cerebral cortex is innervated by two ascending monoaminergic afferent systems. The noradrenergic and serotonergic systems arise from brain stem nuclei and project throughout the ipsilateral hemisphere. These projections are widely distributed and cross cytoarchitectonic borders.

Noradrenergic afferents project from neurons in the pontine and mesencephalic nuclei. The most notable of these nuclei is the locus ceruleus, so named because it appears as a blue site in the fresh brain. This is a small, pigmented nucleus in the floor of the fourth ventricle. The noradrenergic afferents terminate in all layers of cortex, but predominantly in layers I through IV. The cortical terminals of the noradrenergic afferents contain dense core vesicles. These vesicles package the norepinephrine before its exocytotic release into the synaptic cleft where the norepinephrine binds with α and β adrenoceptors. The laminar distribution of these receptors differs. The α_1 adrenoceptors are distributed in layers I through IV, but α_2 adrenoceptors are most common in layers I and IV. The β adrenoceptors are most densely distributed in layers I through III. The interaction of norepinephrine with these receptors produces different responses. For example, activation of α_2 adrenoceptors, which presumably are on the presynaptic membrane, opens potassium channels, inhibiting further release of norepinephrine. In contrast, binding of norepinephrine with the α_1 and β adrenoceptors closes potassium channels in postsynaptic neurons that are excited.

The serotonergic afferents arise from the dorsal and median raphe nuclei. These are small clusters of neurons along the midline of the pons and caudal midbrain. The raphe neurons also co-localize neuroactive peptides such as substance P, leu-enkephalin, and thyrotropin-releasing hormone. Serotonergic afferents innervate layers I, III and IV. The cortical serotonin receptors are largely confined to layers III and IV. The release of serotonin results in changes in the permeability of potassium channels, which leads to complex modulation of the activity of postsynaptic neurons.

The monoaminergic afferents to cortex are thought to be involved in the regulation of sleep and arousal. Serotonergic neurons in the dorsal raphe exhibit a slow, rhythmic activity. The pacemaker activity is modulated by the activity of noradrenergic afferents. This pacemaker activity is accelerated by the closing of potassium channels by norepinephrine. The activity of the pacemaker neurons changes with the state of arousal; it is highest when the person is awake, lower during periods of "slow-wave" sleep, and lowest during periods of dream or rapid eye movement (REM) sleep. The activity of noradrenergic raphe neurons rises during slow-wave sleep and falls during periods of REM. The phasic activity of a population of cholinergic neurons in the reticular formation (i.e., gigantocellular tegmental field) rises during REM sleep. These neurons are innervated by descending cortical projections that serve in a feedback capacity. It appears that an interaction of a number of neurotransmitter systems regulates the states of consciousness.

Cortical cholinergic afferents are involved in memory

Memory formation is a process that involves short-term and long-term events, many cortical areas, and a number of cortical afferents. It is beyond the scope of this chapter to describe what is known about the mechanisms of memory, although a few observations are offered. The hippocampus is critical for short-term memory, but long-term memories are probably stored in many of the multimodal association and limbic cortices. Because acetylcholine receptor antagonists interfere with memory formation, it is thought that cholinergic afferents are important for the operation of cortical circuits involved in memory formation. It is possible, for example, that cholinergic connections operate as an "enabling switch" that allows particular memories to be transferred from short-term to long-term storage.

There are two sources of acetylcholine in the cerebral cortex. About 70% originates in neurons with cell bodies in the magnocellular basal forebrain nuclei, and 30% arises from a subpopulation of cortical local circuit neurons. The basal forebrain nuclei include the medial septum, the basal nucleus of Meynert, the diagonal band of Broca, and the substantia innominata. Medial septal neurons project to the hippocampus. Projections from neurons in the basal nucleus of Meynert and medial segments of the diagonal band of Broca terminate in medial cortical areas, including entorhinal, medial prefrontal, and cingulate cortices. Neurons in the substantia innominata and lateral segments of the diagonal band of Broca project to many neocortical areas, including temporal, parietal, and occipital neocortical areas. Projections to the cerebral cortex generally terminate in layers V and VI, although there are projections to the superficial layers.

Disruption of cholinergic connections may contribute to the early signs of memory loss and spatial disorientation observed in patients with Alzheimer's disease. Neurofibrillary tangles and neuronal degeneration occur in the basal forebrain nuclei and are considered to be pathologic hallmarks of Alzheimer's disease. Memory deficits in Alzheimer's patients are associated with damage to the hippocampus that is engaged during short-term memory formation, to multimodal and periarchicortical limbic areas involved in long-term memory storage, and to the cholinergic system that may be involved in enabling these structures to convert short-term to long-term memories.

Interhemispheric connections are critical for the unification of cognitive activity

Each hemisphere contains the representation of the contralateral sensory field. Although there is a slight overlap at the midline, as in the region of central vision and the somatotopic representation of the trunk, the separation of the two perceptual spaces is rather clean. Despite this division, each person perceives a seamless representation of the world. The coordination of the processing that goes on in each hemisphere depends on axons that cross the corpus callosum.

Many cortical areas on one side of the brain are connected with areas in the contralateral hemisphere through callosal connections. Although the details about the callosal system vary among discrete cytoarchitectonic areas, some general patterns can be described. The axons of the callosal pathway arise from pyramidal neurons in layer III and to a lesser extent from layer V projection neurons. These axons descend into the white matter and pass across the corpus callosum. After entering the contralateral hemisphere, the axons follow a mirror image course and terminate in layers I, III, and IV of the homotopic or corresponding site in the contralateral hemisphere.

The function of callosal connections can be gleaned from the studies conducted by Roger Sperry and colleagues, who examined people whose corpora callosa and anterior commissures were transected. Overtly, such people with "split" brains operate perfectly well, but when challenged with certain tasks, it is clear that each hemisphere is not capable of executing all tasks. In one experiment, each split-brain patient was presented with an apple in his or her right or left visual fields. After presentation to the right visual field, a person was able to verbally describe the object. However, when the apple was placed in the left visual field, the person was unable to come up with the word apple but was able to point to an image of the apple when offered a choice of images. This experiment shows that both hemispheres receive sensory information but differ in their communicative capabilities.

The functional separation of abilities characterizes some of the hemispheric asymmetries. Although it is difficult to make generalizations about the functions of each hemisphere, each hemisphere is best able to mediate verbal or nonverbal communication. At the risk of oversimplifying the situation, it appears that the left hemisphere is expressive and best at analytic, rational thought, and the right hemisphere is perceptive and capable of emotional, intuitive thought.

The lateralization of function is expressed in normal, intact people as cerebral dominance in language and handedness. Virtually all right-handed people have dominant left hemispheres. As might be expected, a significant percentage (15%) of left-handed or mixed-handed people express right hemispheric dominance, but the dominant hemisphere for language in 70% of these people is the left hemisphere. Regardless of the outward expression of handedness, the left hemisphere is usually the dominant hemisphere for language. The brain dominance of an individual can be discerned from his or her writing posture. Right-handed persons who write with the standard grip and straight "stance" approach AND left-handed persons who write with a "hook" approach have dominant left brains. The few left-handed persons who write without a hook approach (i.e., hold the pencil straight toward the paper) and the rare right-handed persons who write with a hook have right-brain dominance.

☐ Cortex projects to various nonthalamic subcortical sites

In addition to the corticospinal (i.e., pyramidal) pathway described earlier, the axons of layer V neurons project to extrapyramidal targets, as summarized in Figure 17-11. These targets are sensory, motor, and integrative centers. Cortical afferents innervate sensory cranial nerve nuclei, such as the trigeminal brain stem nuclear complex and the solitary nucleus. These projections serve as feedback controls, modulating the output from second-order sensory nuclei. Other descending axons project to *motor centers*. These targets include the motor cranial nerve nuclei that have direct projections to skeletal muscle. These nuclei are the trigeminal motor, facial, supraspinal, and hypoglossal nuclei. These corticobulbar projections serve as upper motor neurons, analogous to the corticospinal projections.

The cerebral cortex also projects to three structures that project to the spinal cord. Projections from the frontal, parietal, occipital, and temporal lobes project to the superior colliculus. These afferents carry somatosensory, visual, and auditory information that enables the superior colliculus to execute its role in coordinating complex behaviors, such as tracking and attending to moving stimuli. *Cortical afferents*, primarily from the precentral gyrus, synapse with neurons throughout the red nuclei. Two nuclei in the reticular formation receive cortical input: the gigantocellular region of the medullary reticular formation and the oral area of the pontine reticular formation.

Extrapyramidal projections from layer V neurons also terminate in nuclei that are integrative centers. These targets include the caudate nucleus, the periaqueductal gray, and the pontine nuclei.

The cortical projections that innervate brain stem structures follow a pathway that is common with the corticospinal tract. The cortical projections pass through the subcortical white matter, the internal capsules, the crus cerebri, the base of the pons, and the pyramidal tracts.

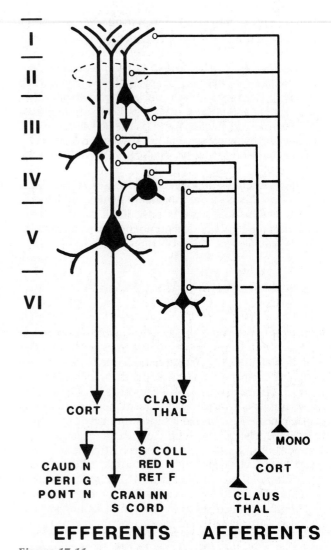

Figure 17-11
A summary diagram of the intrinsic organization and connections of a generic neocortical area. Three pyramidal neurons form a cluster (*dashed lines*). Excitatory connections are indicated with open axon terminals, and inhibitory connections are shown with black axon terminals. *caud n*, caudate nucleus; *claus*, claustrum; *cort*, corticocortical connections for ipsilateral and contralateral hemispheres; *cran nn*, sensory and motor cranial nerve nuclei; *mono*, monoamines including norepinepherine and serotonin; *peri g*, periaqueductal gray; *pont n*, pontine nuclei; *red n*, red nucleus; *ret f*, reticular formation; *s coll*, superior colliculus; *s cord*, spinal cord; *thal*, thalamus.

The only difference is that the axons exit from the common pathway at the level of the structure being innervated. For example, corticobulbar fibers that innervate the hypoglossal nuclei descend with the corticospinal neurons, but perpendicular arbors leave the pyramidal tract in the caudal medulla. The size of the descending common pathway decreases as the axons exit.

The cortex also is interconnected with the claustrum. The *claustrum* is a small sheet of neurons embedded in the subcortical white matter. The corticoclaustral interconnections largely parallel the corticothalamic interconnections. The claustrum sends cortical afferents to layer IV and, to a lesser extent, to layer VI. The cortex projects to claustrum; these projections originate in layer VI neurons in many cortical areas, including the primary auditory, somatosensory, and visual cortices. The function of the claustrocortical association largely remains a mystery, but it may play a role in the modulation of sensory receptive field properties. For example, claustral afferents to the visual cortex contribute to the property of end inhibition, whereby the sizes of certain receptive fields are limited by antagonistic effects. The corticoclaustral connections, like the corticothalamic projections, probably feed back on the claustrum to modulate the activity of the claustral neurons. These reciprocal projections may also provide a system for the integration of the three sensory spaces within the claustrum.

Organization of the Cerebral Cortex

The organization of the cerebral cortex from the perspectives of intrinsic circuitry and afferent and efferent connections is summarized in Figure 17-11. It is built around a small group of layer III and V neurons whose apical dendrites are aggregated into a cluster. The afferent connections shown include excitatory thalamic afferents to layer IV, excitatory cortical connections from ipsilateral and contralateral cortices to layer III, and a diffuse distribution of serotonergic and noradrenergic projections that innervate elements in all cortical laminae. A single local circuit neuron is shown in layer IV with its inhibitory connections with pyramidal neurons, but these neurons are diffusely distributed throughout all layers of the cortex.

Cortical neurons project to several extrinsic sites. The actual density of any one of these projections depends on the specific cortical area under consideration. All cortical areas have ipsilateral cortical projections that originate in layer III and projections to the thalamus, pontine, and caudate nuclei that originate from layer V neurons. Figure 17-11 does not intend to imply that each layer V pyramidal neuron makes all of these projections. It is likely that a single pyramidal neuron in layer V has axon collaterals that project to between one and three subcortical sites. Specialization in projection systems for each cortical area occurs in terms of inputs to the superior colliculus, reticular formation, sensory and motor cranial nerve nuclei, and the red nucleus. Many of these connections are unique to parts of particular motor and premotor areas.

Sensory and motor information interact in the cortex; visual, auditory, and somatosensory information can be integrated, stored, and recalled so that the appropriate motor system(s) can be activated. All of these processes are essential for mnemonic mechanisms and cognition. The cortex is pivotal for all of the processing that underlies the complex thought and communication considered to be uniquely human. Nevertheless, the cerebral cortex does not act in isolation. The cortex can function only because of the diverse interconnections with other central nervous system structures.

SELECTED READINGS

Peters A, Jones EG (eds). Cerebral cortex, vol 1–10. New York: Plenum, 1984-1994.

Vogt BA, Gabriel M. Neurobiology of cingulate cortex and limbic thalamus. Boston: Birkhäuser, 1993.

Clinical Correlation
DEMENTIA AND ABNORMALITIES OF COGNITION

ROBERT L. RODNITZKY

Dementia is a state of cognitive impairment resulting from the progressive loss of previously acquired mental abilities. There is a mild decline in cognitive function that occurs with normal aging, but it seldom results in serious functional disability. In dementia, cognitive dysfunction is sufficient to prevent the normal accomplishment of standard activities of daily living such as dressing, cooking a meal, or balancing a checkbook.

Cortical or Subcortical Brain Pathology

The clinical characteristics of dementia occurring in diseases affecting the cerebral cortex differ from those of subcortical dementia. In cortical disorders such as Alzheimer's disease, the symptoms specific to cortical dysfunction include aphasia, apraxia (i.e., inability to organize and perform a motor task despite normal strength and coordination), and agnosia (i.e., inability to recognize). In dementias primarily related to subcortical pathology, such as that occurring in Parkinson's disease, these features are usually absent. In both forms, memory loss occurs, but it is typically less severe in subcortical dementia. The memory loss associated with subcortical dementia is frequently amenable to cuing, and a hint often suffices to bring the forgotten thought to mind. It has been said that those with cortical dementias forget, and those with subcortical dementia, who can recall after being cued, forget to remember.

Alzheimer's Disease

☐ Alzheimer's disease is the most common cause of severe dementia

Although more than 70 causes of dementia have been identified, Alzheimer's disease is responsible for more than 60% of cases of dementia in persons older than 65 years of age. This condition is characterized histologically by the presence in the brain of abnormal aggregations of cytoskeletal filaments known as *neurofibrillary tangles* and by abnormal structures referred to as *senile plaques*. Both of these microscopic abnormalities are found predominantly in neocortical areas and the hippocampus. Senile plaques are extracellular deposits of β-amyloid protein, which is derived from the transmembrane portion of a more complex protein, the amyloid precursor protein (APP). Recent evidence suggests that, in familial cases of Alzheimer's disease (a relatively rare form), there is a point mutation in the gene on chromosome 21 which encodes the APP. The manner in which senile plaques and β-amyloid protein relate to the pathogenesis of Alzheimer's disease is a matter of some debate. Some researchers consider amyloid to be a cause, and others think it is the effect of neuronal death in Alzheimer's disease.

☐ Memory loss is the most constant feature of Alzheimer's disease

The clinical hallmark of Alzheimer's disease is memory loss. In some cases, remote memory is less affected than immediate or intermediate-length memory, but most commonly, all forms are involved to some extent. At first, memory loss is mild and only slightly impairs routine daily activities, but with progression, there are more serious manifestations of memory loss, such as inability to recognize relatives or a tendency to become lost in one's own home. In Alzheimer's disease, memory impairment appears to be related to neuronal loss in the hippocampus, and a reduction in the number of cholinergic neurons in the basal nucleus of the forebrain. The importance of neurons in the pathogenesis of memory loss is suggested by the fact that pharmacologic enhancement of cholinergic function appears to improve memory. Whether damage to cholinergic neurons is a primary event in Alzheimer's disease or merely the result of retrograde degeneration secondary to cortical neurodegeneration is not understood.

Sudden Abnormalities of Memory and Cognition

Sudden abnormalities of memory and cognition are more likely caused by brain tumors or stroke than degenerative conditions. Symptoms similar to those seen in Alzheimer's disease can occur as a result of localized damage to specific regions of the brain from a brain tumor or cerebral ischemia. *Transient global amnesia* is the sudden but temporary inability to encode new memories. This syndrome is usually caused by temporary dysfunction of both hippocampi due to causes such as epileptic discharges or ischemia. Affected patients remain alert but repeatedly ask the same orienting questions, because they cannot retain the information provided in previous answers. Within minutes to hours, the episode passes, but there is little or no recall of the total event.

Brain tumors or stroke involving the right parietal lobe commonly result in *agnosia*, an inability to recognize. *Autotopagnosia* refers to the inability to recognize a part of one's own body. After a right parietal stroke, a person with autotopagnosia may identify his or her own left hand as belonging to the examiner instead. *Prosopagnosia* is a specific agnosia that can result from highly localized cortical damage. It is characterized by the inability to recognize and identify familiar faces. This condition occurs in persons who have suffered damage to the visual association cortices in both hemispheres.

Apraxia is the inability to perform skilled motor movements despite normal coordination and preserved motor, sensory, and cognitive function. The apractic patient may properly grasp and lift a spoon, but he cannot demonstrate how it is used. Apraxia most commonly occurs after damage to the inferior parietal region of the dominant hemisphere. Certain forms of apraxia are more related to impaired perception of spatial relations and are caused by lesions of the nondominant hemisphere. Dressing apraxia is an example of this phenomenon. The affected person cannot correctly perform acts such as putting an arm into the sleeve of a coat, buttoning buttons, or donning a garment on the correct part of the body.

SELECTED READINGS

Katzman R, Jackson JE. Alzheimer's disease: basic and clinical advances. J Am Geriatr Soc 1991;39:516.

Kosik K. Alzheimer's disease: a cell biologic perspective. Science 1992;256:780.

Neuroscience in Medicine, edited by P. Michael Conn.
J. B. Lippincott Company, Philadelphia © 1995

Chapter 18

*T*he Limbic System

J. MICHAEL WYSS
THOMAS VAN GROEN

Neuroanatomy of the Limbic System

The limbic system is a relatively new concept that has undergone considerable analysis and redefinition since its introduction in 1952. The areas included under the umbrella of the limbic system include the hippocampal formation, septal nuclei, cingulate cortex, and amygdala. All these regions are telencephalic, but they are structurally more primitive than cerebral neocortex, and they are highly interconnected by direct projections between the limbic regions and by indirect projections through the diencephalon. These interrelated areas of the telencephalon and diencephalon form the limbic system, and although the individual limbic areas are functionally diverse, serving emotional and motivational behavior on the one hand and learning and memory on the other, the high degree of interconnection suggests that these areas have an underlying unity.

In contrast to the neocortex, which can be divided into six layers (i.e., five neuronal cell layers and one superficial molecular layer), the limbic cortical areas tend to be structurally simpler, and they arise earlier in mammalian phylogeny. These areas have often been called the ancient (i.e., archicortex), old (i.e., paleocortex), and middle (i.e., mesocortex) cortices on the basis of the development of each individual area. Despite their phylogenetically immature appearance, many limbic areas of cortex reach their greatest development in humans, and compared with primary sensory and motor cortex, they expand more rapidly in size from lower primates to humans (Fig. 18-1). These areas are not antiquated remnants of earlier evolution but are brain regions that have continued to develop structurally and functionally throughout phylogeny.

Although Willis (1664) was the first to refer to the cortical regions that form the medial edge of the telencephalon as *limbus* (i.e., Latin meaning border), Broca (1878) was the first to popularize the designation of this cortex as *limbic*. "Le grand lobe limbique" was envisioned by Broca as comprising the entire gray mantle that lay as a transition between the neocortex and the diencephalon and formed a circle around the interventricular foramen (see Fig. 18-1). The limbic lobe is rostrally connected to the olfactory bulb and olfactory cortex and is smaller in microsmatic animals and bigger in macrosmatic animals. These facts led Broca to suggest that the function of the limbic lobe was related to olfaction. Several researchers expanded on this view, even to the point of suggesting that the limbic lobe was a smell brain (i.e., rhinencephalon), but by 1940, several lines of evidence suggested that these regions of cortex were involved in other functions, especially emotion.

In 1937, Papez hypothesized that there was an anatomic basis for the emotional disturbances of some of his psychiatric patients: damage to the circuit called the *Pa-*

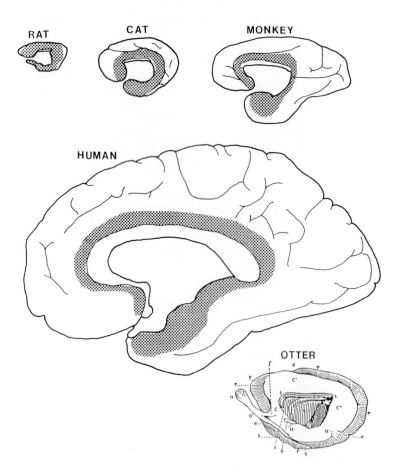

RAT   CAT   MONKEY

HUMAN

OTTER

Figure 18-1
Compared with other animals, the limbic cortex reaches its largest extent in humans. These drawings illustrate the size and position of the limbic cortex (*stipple*) in the three animals in which it most often has been studied and in the human. Broca's original concept of the unity of "le grand lobe limbique" is shown (*lower right*) in the otter, the animal in which he originally identified it. Broca, like many neurologists of his day, often used animals with a highly developed sense of smell, such as the otter, for anatomic studies. In these animals, the continuity of the limbic cortex with the olfactory bulb is prominent, but in the primate, this attachment is less striking. All figures are drawn approximately to the same scale.

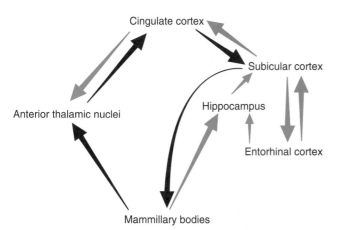

Figure 18-2
The Papez circuit as originally proposed (*black lines*) and with the later observed connections (*shaded*). In Papez's original work, the hippocampal formation was viewed as a single area.

pez circuit (Fig. 18-2) that normally interrelates limbic cortical areas with the diencephalon. Two years later, Klüver and Bucy demonstrated that extensive lesions to the temporal lobe that primarily damage two parts of the limbic system (i.e., amygdala and hippocampal formation) profoundly influenced the affective behavior of subhuman primates. In 1952, on the basis of these and other studies, MacLean suggested that the term *limbic system* (see Fig. 18-1) should be applied to the limbic areas, and he emphasized that the limbic system elements were at the interface between somatic and visceral areas of the brain and could relate these systems to each other and to the ongoing behavior of the organism. Nauta later expanded on the definition of the limbic system by showing that areas of the limbic cortex (but not neocortex) were connected directly to many areas of the hypothalamus and brain stem.

Although researchers still differ about the exact number of areas that comprise the limbic system, most agree

that it includes the *hippocampal formation, septal nuclei, mammillary bodies, anterior thalamic nuclei, cingulate (midline) cortex, and amygdala.*

□ The hippocampal formation is divided into three major regions

Overall structure

Within the cerebral cortex, the hippocampal formation is the most easily differentiated region. Most of the area (i.e., the hippocampus which includes the hippocampus proper and the dentate gyrus) has only a single neuronal cell layer and contrasts sharply with the six layers of neocortex. This relative simplicity has led many researchers to use these regions of the hippocampal formation as a model system to illuminate the structure and function of the more complex neocortex.

Although some differences in nomenclature continue to confuse the hippocampal literature, the following terms are most widely used and are best defined. *Hippocampal formation* typically refers to the *dentate gyrus,* the *hippocampus proper* (i.e., Ammon's horn or cornu Ammonis), the *subicular cortex,* and the *entorhinal cortex* (Fig. 18-3). The entorhinal and subicular cortices are the primary areas responsible for the extrinsic connections of the hippocampal formation. The hippocampus proper and the dentate gyrus are primarily concerned with processing the information that passes through them. The most common nomenclature for the areas of the hippocampus proper is that of Lorente de Nó who divided it into CA (cornu Ammonis) regions CA_1 through CA_3 (see Fig. 18-3). M. Rose divided the hippocampus proper into five fields, designated h_1 through h_5, and his terminology remains popular among some neuropathologists.

The hippocampal formation lies in the temporal lobe of the cerebral cortex, just medial to the inferior horn of the lateral ventricle (Fig. 18-4), and to some early anato-

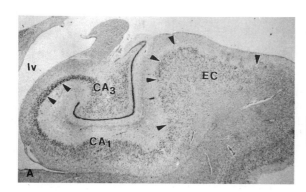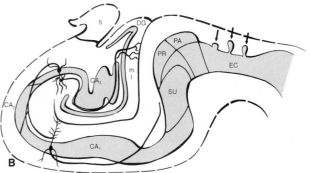

Figure 18-3
(**A**) Nissl-stained coronal section of a human hippocampal formation, taken from the midanteroposterior level. (**B**) The line drawing illustrates the divisions of the hippocampal formation and the relative appearance (*red*) of the granule cells and pyramidal neurons in the dentate gyrus and CA fields, respectively. *CA,* the fields of the hippocampus proper; *DG,* dentate gyrus granule cell layer; *EC,* entorhinal cortex; *fi,* fimbria; *lv,* lateral ventricle; *ml,* dentate gyrus molecular layer; *PA,* parasubiculum; *PR,* presubiculum; *SU,* subiculum.

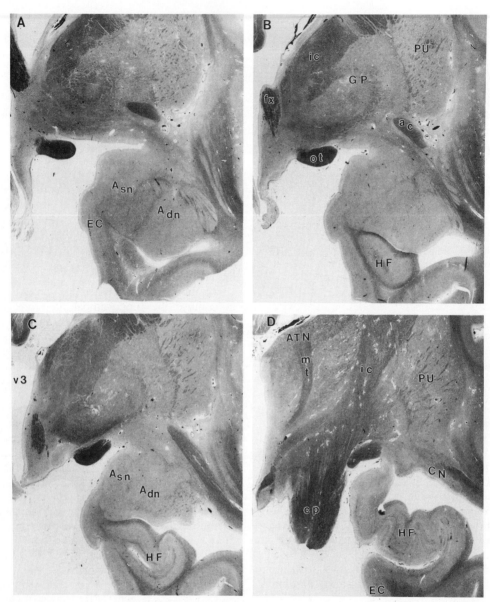

Figure 18-4

(**A–D**) Four photomicrographs of Weil-stained coronal sections through the human amygdala and rostral hippocampal formation. A_{dn}, deep nuclei of the amygdala; A_{sn}, superficial nuclei of the amygdala; *cp*, cerebral peduncle; *EC*, entorhinal cortex; *GP*, globus pallidus; *HF*, hippocampal formation; *PU*, putamen.

mists, its gross appearance looked like a seahorse (i.e., hippocampus in Greek). Although the hippocampal formation stretches from the amygdala to the splenium of the corpus callosum, the intrinsic structure of the hippocampus proper is best appreciated in its middle one third. In this region, the cortical sheet of hippocampal tissue is folded over itself in the characteristic S shape (see Fig. 18-3). In more rostral regions of the hippocampal formation, the entire S-shaped hippocampal cortex bends back on itself, and the CA fields become much more difficult to differentiate with precision (see Fig. 18-4D). Similarly, at caudal levels the simple "S" shape is obscured as the formation bends near the splenium of

the corpus callosum. The hippocampal formation continues as a small band of neurons (i.e., the indusium griseum) above the entire length of the corpus callosum that extends around the genu of the corpus callosum and descends into the septal area, where it is known as the taenia tecta.

Regions of the hippocampus

The *dentate gyrus* is a trilaminar structure that caps the distal tip of the CA fields (see Fig. 18-3). The main cell layer is composed of tightly packed granule cell neurons that have "apical" dendrites that project primarily toward the pial surface, but the granule cell neurons also give rise to less conspicuous basal dendrites that extend toward the hilus. These granule cells are generated rather late in embryonic development, and some of these neurons continue to divide into adulthood, albeit at an extremely slow rate.

Superficial to the granule cell layer lies the molecular layer, which contains most of the dendrites of the granule cells, and deep to the granule cell layer lies the polymorphic layer or hilus. Axons of the polymorphic neurons in this region project to the molecular layer of the dentate gyrus, but they do not appear to project to the hippocampus proper.

The hippocampus proper is made up of a single layer of pyramidal neurons. On the basis of cellular morphology, the hippocampus proper has been divided into an area that lies proximal to the dentate gyrus (i.e., *area CA₃*), an area that lies distal to the dentate gyrus and contains neurons that are slightly smaller in size and more widely scattered (i.e., *area CA₁*), and a short, mixed-cell region (i.e., *area CA₂*) that lies between CA₃ and CA₁ (see Fig. 18-3). Of the seven million neurons in the CA fields, about 67% reside in CA₁, 30% lie in CA₃, and 3% are in CA₂.

Area CA₃ has a relatively compact and cohesive pyramidal cell layer. Within the dentate hilar area, CA₃ bends toward one blade of the dentate gyrus and then bends back toward the opposite blade (see Fig. 18-3). In addition to the relatively large size of the CA₃ pyramidal neurons, the structure of the apical dendrites differentiates these neurons from those in CA₂ and CA₁. The apical dendrites of the CA₃ pyramidal neurons bifurcate close to their somata, but the apical dendrites of the CA₁ and CA₂ pyramidal neurons have smaller branches along most of their length (see Fig. 18-3*B*). The granule cell neurons of the dentate gyrus send their axons, the *mossy fibers* (so called because of the large terminals that stud these axons), to the CA₃ neurons, and these axons terminate massively on the proximal shaft of the apical dendrites in a layer that is named the *stratum lucidum*. In this layer, the postsynaptic specializations on the CA₃ dendrites resemble large thorns, a characteristic not seen on the CA₁ and CA₂ apical dendrites. The apical CA₃ dendrites extend from the stratum lucidum into the *stratum radiatum* and *stratum lacunosum or moleculare*; each of these layers contains distinct afferent inputs to the pyramidal neurons.

Above the molecular layer of the CA fields is the obliterated *hippocampal fissure*. The CA₃ basal dendrites extend into the *stratum oriens*, where scattered basket and polymorphic neurons are also located. Between the stratum oriens and the lateral ventricle lies the *alveus*, a major fiber bundle that carries information to and from the hippocampus. Many axons in the alveus course through the *fimbria*, a large fiber bundle that lies medial and dorsal to the CA fields (see Fig. 18-3), and fornix en route to subcortical and diencephalic sites.

The borders of area CA₂ are not easily discriminated in Nissl-stained material because some neurons of CA₃ tend to lie deep to the CA₂ pyramidal cells. Other features differentiate the CA₂ area. CA₂ contains many large pyramidal neurons that are similar in size to those in CA₃, but CA₂ has a more compact pyramidal cell layer. In contrast, CA₁ has slightly smaller and more widespread pyramidal neurons (see Fig. 18-3*A*). The mossy fibers do not extend to the CA₂ neurons, and the CA₂ neurons can be selectively labeled for some neurotransmitters. Because the mossy fibers do not extend to area CA₂ or CA₁, these regions lack a stratum lucidum, and the stratum radiatum is adjacent to the pyramidal cell layer.

Subicular cortex

Researchers have identified as many as six subregions in the subicular cortex, but it is useful to consider three major divisions in the human subicular cortex (see Fig. 18-3). Adjacent to CA₁ is the *subiculum proper*, which has a single pyramidal layer that is wide and can be subdivided into a superficial, large cell sublayer and a deep, small cell sublayer. Several researchers suggest that there is a prosubicular division between CA₁ and subiculum proper, but the differences between this area and the subiculum are equivocal. The border of the subiculum with the *presubiculum* (i.e., area 27 of Brodmann) is marked by the appearance of a tightly packed layer of small pyramidal neurons (i.e., lamina principalis externa) that caps the inner layer (i.e., lamina principalis interna), which appears in the same position as the subiculum's pyramidal cell layer. As in all areas of the hippocampal formation, adjacent areas tend to slide over or under each other in transition regions (see Fig. 18-3). The *parasubiculum* (i.e., area 48 of Brodmann) lies between the presubiculum and the entorhinal cortex. Similar to the presubiculum, the parasubicular has two major neuronal cell layers, but the neurons in the lamina principalis externa are larger than those in the corresponding layer of the presubiculum.

Entorhinal cortex

The human entorhinal cortex (i.e., area 28 of Brodmann) is ventral to the rostral half of the hippocampal formation and the entire amygdaloid complex (see Figs. 18-3 and 18-4). The entorhinal cortex has been divided into as many as 23 separate areas, but all these areas are similar in morphology, and the cortical area is not subdivided in this chapter. The differentiating elements of the entorhinal cortex are in layers II and IV. The neuronal cell bodies in layer II form prominent, large cell islands, especially at the rostral levels of the entorhinal cortex. In contrast, layer IV consists of a dense fiber plexus (i.e., the lamina dissecans) that clearly separates the small pyramidal neurons in layer III from the larger pyramidal neurons in layer V. The entorhinal cortex is bordered by area 35 (i.e., part of the perirhinal cortex). Although the entorhinal cortex was named for its proximity to the rhinal sulcus in subprimates and monkeys, the rhinal sulcus in humans is rostral to the entorhinal cortex. In humans, the collateral sulcus forms the border of the parahippocampal gyrus, but it does not faithfully mark the lateral border of the entorhinal cortex, which lies in the medial part of the parahippocampal gyrus.

☐ The septal nuclei are a limbic subcortical area

The *septal region* is a subcortical telencephalic region that is highly interconnected with the hippocampus through the fornix, and it lies near the genu of the corpus callosum and medial to the lateral ventricles (Fig. 18-5A). Few or no neuronal cell bodies are within the septum pellucidum, which is a thin sheet of tissue that is dorsal and caudal to the anterior commissure and separates the lateral ventricles; conversely, most septal neurons lie ventral to the septum pellucidum and rostral to the anterior commissure. The septal nuclei are separated into two major divisions, the *medial and lateral septal nuclei*. The

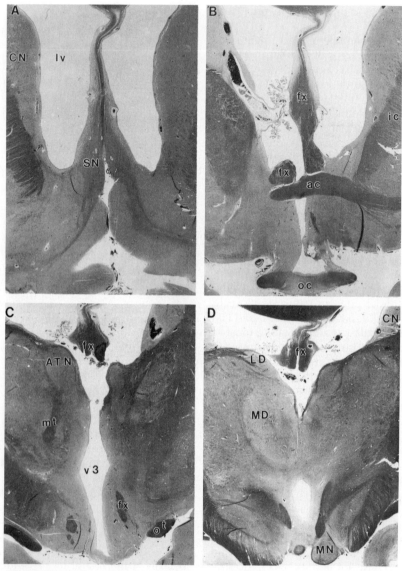

Figure 18-5
Four photomicrographs of the Weil-stained (for myelinated axons) coronal sections through successively more caudal areas of the human brain. (**A**) The septal nuclei. (**B**) The descending fornix in relation to the anterior commissure. (**C**) The rostral anterior thalamic nuclei and midhypothalamus. (**D**) The posterior and anterior thalamic nuclei, the lateral dorsal nucleus, and the mammillary complex. Each of these figures is cut asymmetrically; the left side of the brain is further caudal than the right side. *ac,* anterior commissure; *ATN,* anterior thalamic nuclei; *CN,* caudate nucleus; *fx,* fornix; *ic,* internal capsule; *LD,* lateral dorsal nucleus of the thalamus; *lv,* lateral ventricle; *MD,* medial dorsal nucleus of the thalamus; *MN,* mammillary nuclei; *mt,* mammillothalamic tract; *oc,* optic chiasm; *ot,* optic tract; *SN,* septal nuclei; *VA,* ventral anterior thalamic nucleus; *v3,* third ventricle.

lateral septal nucleus has relatively small neuronal cell bodies and is primarily an area for receipt of information. In contrast, the medial septal nucleus has larger neuronal cell bodies and is the major origin of the hippocampal projections of the septal nuclei. In addition to the medial and lateral nuclei, the septal complex includes the septohippocampal, the septofimbrial, and the triangular septal nuclei, the bed nucleus of the stria terminalis, and the *diagonal band of Broca*.

☐ The mammillary bodies are the major limbic hypothalamic area

Although several areas of the hypothalamus are interconnected with limbic cortical regions, the mammillary bodies have the most prominent connections to this cortex. The *medial mammillary nuclei* in human are so large that they bulge out from the base of the hypothalamus, giving rise to their suggestive name (Fig. 18-5*D*). The mammillary complex is typically divided into four nuclei. The largest nucleus is the *medial mammillary nucleus*, lateral to which is the *lateral mammillary nucleus*, and further lateral lies the *tuberomammillary nucleus*. The neurons in the latter two nuclei are much larger than those in the medial mammillary nucleus. Above the medial mammillary nucleus is the *supramammillary nucleus*. Each of these four regions has distinct connections and cytoarchitecture. The major projection from the mammillary bodies is through the *mammillothalamic tract*, which terminates in the anterior thalamic nuclei.

☐ The anterior thalamic nuclei are the primary limbic thalamic area

The anterior nuclei of the thalamus are dorsomedial to the internal medullary lamina and classically are divided into three nuclei: the *anterior dorsal, anterior medial, and anterior ventral thalamic nuclei* (Fig. 18-5*C*). In lower primate and subprimate species the three nuclei are distinct, but in the human brain, the anterior medial nucleus is difficult to discriminate from the anterior ventral nucleus, and the two have often been grouped together and called the principal anterior nucleus, but the connections of these areas are very different. The anterior dorsal nucleus is by far the smallest of the three anterior thalamic nuclei, and its relative volume does not increase substantially in humans compared with lower primates. In all species, the cytoarchitecture of the anterior dorsal nucleus is distinct from that of the anterior ventral and the anterior medial nuclei, because its neurons are more tightly packed and stain darker than the others.

The *lateral dorsal nucleus* lies on the dorsal surface of the thalamus and is surrounded by a fiber capsule, giving an appearance similar to the anterior nuclei that it borders (see Fig. 18-5*D*). Although its Nissl-stained appearance and location are distinct from those of the anterior nuclei proper, in fiber stains, the lateral dorsal nucleus appears to be a caudal continuation of the anterior group,

and many of its connection are similar; it is often considered as part of the anterior thalamic group.

Several other thalamic nuclei have major projections to the limbic cortex. These include the *medial dorsal nucleus*, the *midline nuclei* (especially the parataenial, paraventricular, and reuniens nuclei), the *intralaminar nuclei,* and the *lateral thalamic nuclei* (i.e., lateral posterior and medial pulvinar nuclei). Modern tract tracing studies show that only distinct regions of these nuclei are limbic and that other regions are related primarily to nonlimbic areas of neocortex.

☐ The cingulate cortex comprises the limbic midline cortex

The major projection of the anterior thalamic nuclei is to the cingulate cortex, an area that forms a major component of the circuit of Papez. This cortical region lies below the cingulate sulcus and surrounds the corpus callosum from its rostrum to its splenium. This cortex typically is designated according to the numbering system of Brodmann (1909) or the names suggested by Rose (1928). Both schemes recognize that a major division of the cingulate cortex lies immediately behind the splenium of the corpus callosum, deep within the sulcus of the corpus callosum. This part of cingulate cortex includes areas 26, 29, and 30 of Brodmann and is the retrosplenial cortex of Rose. The part of cingulate cortex that is rostral to the splenium of the corpus callosum includes *areas 23, 24, 25, 32, and 33 of Brodmann*. In humans, the Brodmann numbering system of the rostral cingulate cortex is preferred to Rose's six divisions of *infraradiata cortex*.

☐ The amygdala is divided into two major regions

The amygdala, whose name reflects its almond-like appearance, is a mass of gray matter that lies within the temporal lobe, immediately rostral to the inferior horn of the lateral ventricle (see Fig. 18-4). In Weigert-stained sections the amygdala looks similar to the striatum, and it appears continuous with the lentiform nucleus dorsally and the tail of the caudate nucleus caudally. Together with the fact that the amygdala and the striatum develop from the embryonic striatal ridge, this suggested to early anatomists that the amygdala was part of the basal ganglia, but later connectional and immunohistochemical studies demonstrated that it is connected with and functions as part of the limbic system.

The amygdala usually is divided into two major regions: the *corticomedial* (or superficial) and the *basolateral* (or deep) *nuclear groups*. The deep nuclear group, which has the greater volume in the amygdala, contains four nuclei: the lateral, the basal lateral and the basal medial nuclei, and the amygdaloclaustral area. The superficial group of nuclei includes the anterior, the medial and the cortical amygdaloid nuclei, the periamygdaloid cortex, and the nucleus of the lateral olfactory tract, a nucleus that is poorly developed in human brain. On the basis of its structure and connections, the central nucleus is considered as a separate region from the two large nuclear groups of the amygdala. At its caudal end, the superficial group is continuous with the amygdalohippocampal area, a transition zone between the amygdala to the hippocampal formation.

Major Connections of the Limbic System

☐ The hippocampal formation projects mainly via the fornix and cortical associational axons

Intrinsic connections

The intrinsic connections of the hippocampal formation form a serial pathway with several collateral and feedback projections added onto the serial path. The serial connections of the hippocampal formation, called the trisynaptic pathway (Fig. 18-6), begin with the entorhinal cortex projections to the dentate gyrus granule cells, but the entorhinal cortex axons have collaterals to the CA fields and the subiculum. The largest input to the dentate granule neurons originates in *layer II and III neurons of the entorhinal cortex*; the axons terminate in the molecular layer of the dentate gyrus on the outer two thirds of the granule cell dendrites. The position of the cells of origin within the entorhinal cortex dictates the radial position of the terminals on the dentate granule cell dendrites. In monkeys, axons of neurons in the "lateral" entorhinal cortex terminate in the outer one third of the molecular layer, but axons originating in the "medial" entorhinal cortex terminate in the middle one third of the

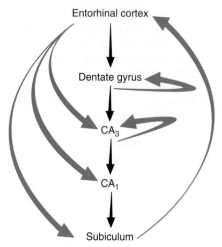

Figure 18-6
The primary intrinsic circuit of the hippocampal formation is shown by the heavy black lines, and the feedback and feedforward collaterals are shaded.

dentate molecular layer. The polymorphic neurons of the dentate hilus project to the proximal one third of the dentate molecular layer.

The granule cells of the dentate gyrus project to the pyramidal neurons in area CA_3 through their mossy fibers (see Fig. 18-3B). The CA_3 pyramidal neurons send out a projection, known as the *Schaffer collaterals*, to the apical and basal dendrites of distal parts of CA_2 and CA_1. CA_1 has a massive projection to the subiculum, which projects back to the entorhinal cortex (see Fig. 18-6). The other two areas of the subicular cortex, the presubiculum and the parasubiculum, do not receive direct input from the hippocampal formation, but they give rise to dense projections to the entorhinal cortex, with presubicular axons terminating in layers I and III and parasubicular axons terminating in layer II of the entorhinal cortex. All three areas of the subicular cortex have connections with the retrosplenial cortex.

Several feedback loops modify the information flow in the hippocampal formation. The dentate hilar, polymorphic neurons project back to the proximal dendrites and cell bodies of the granule cells and potently influence information flow through the granule cells. The polymorphic neurons underlying the CA fields and the basket cells in the CA pyramidal cell layer also provide local feedback to the CA pyramidal neurons. The CA_3 neurons have intrinsic connections within the CA_3 field. The primary cell layer of all CA fields and the dentate gyrus contains basket cell neurons that project to and potently inhibit neurons in the layer in which they reside.

Afferent connections

Papez's original hypothesis was that the hippocampal formation received a major input from sensory areas of the cerebral cortex and that the hippocampal formation processed this information and, through the fornix, projected the processed information to the mammillary bodies from which the appropriate emotional response could be coordinated (see Figs. 18-2 and 18-7). During the past 2 decades, this circuitry has been elucidated, and it has also become clear that the afferent and efferent connections of the hippocampal formation are far more varied and complex than Papez imagined.

Most extrinsic inputs to the hippocampal formation terminate in the *entorhinal or subicular cortices* (Fig. 18-7). The entorhinal cortex receives inputs from several areas of the temporal, frontal, and midline cortices. The perirhinal cortex and the temporal polar cortex project to the lateral portion of the entorhinal cortex. Researchers have shown that dorsal temporal, insular, orbitofrontal, infralimbic, prelimbic, cingulate, and retrosplenial cortices have significant projections to the entorhinal cortex. Several subcortical areas also have significant direct projections to the entorhinal cortex. The lateral nucleus of the amygdala has a dense projection to the lateral entorhinal cortex, and other areas of the amygdala have less prominent projections to the entorhinal cortex. The claustrum also projects to the entorhinal cortex, as do the paraventricular and reuniens nuclei of the thalamus.

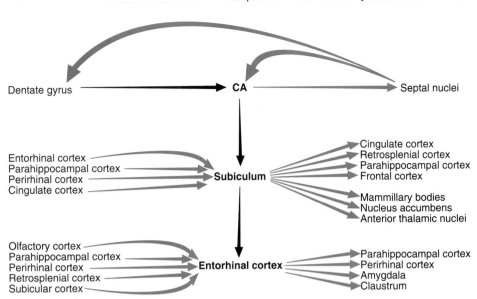

Figure 18-7
The primary extrinsic connections of the hippocampal formation are shaded. The black lines illustrate the intrinsic projections.

The subicular cortex is the second major area of the hippocampal formation receiving extrinsic inputs. Temporal, frontal, and midline cortices project rather densely to distinct divisions of the subicular cortex, and the dorsolateral parietal cortex also projects to these areas. The visual cortex (i.e., area 18b) of the rat also projects to the subicular cortex. Most of the thalamic input to the hippocampal formation terminates in the subicular cortex. Each of the anterior nuclei of the thalamus projects to selective areas of the presubicular and parasubicular cortex, and each nucleus has a distinctive terminal pattern in these cortical areas. The midline thalamic nuclei (i.e., parataenial and reuniens nuclei) also have prominent projections to the subicular cortex.

The inputs to the entorhinal cortex, including those from the subicular cortex, probably provide most of the important sensory and motor information to the dentate gyrus and hippocampus, but a few extrinsic inputs directly project onto the dentate granule cells and the CA pyramidal neurons. The most prominent of these projections originates in the medial septal nucleus and nucleus of the diagonal band of Broca and reaches the hippocampus through the fornix. Axons from the septal nuclei terminate throughout the hippocampal formation, but the major terminal field is in the molecular layer of the dentate gyrus and in area CA_3 (see Fig. 18-3). Approximately 50% of the axons in this projection use acetylcholine as a neurotransmitter. A second direct projection to the dentate gyrus and CA fields originates in the supramammillary nucleus and terminates prominently on the proximal dendrites of the dentate granule cells and in the CA fields. The nucleus reuniens of the thalamus has a prominent projection to the distal part of the CA_1 apical dendrites and a weaker projection to other areas of the hippocampal formation.

Several brain stem nuclei that project to most of the telencephalon also project to the hippocampal formation. These include a noradrenergic projection from the *locus ceruleus*, a serotonergic projection from the *raphe nuclei* (i.e., dorsal raphe and central superior nuclei), and a smaller, dopaminergic projection from the *ventral tegmental area of Tsai*. The *periaqueductal gray*, the lateral and dorsal tegmental nuclei, and reticular nuclei of the brain stem also project to the hippocampal formation.

Efferent connections

The major projections from the hippocampal formation originate in the subicular and entorhinal cortices, and the axons of the CA_1 through CA_3 neurons give rise to a significant projection to the septal nuclei (primarily the lateral septal nucleus) through the precommissural fornix (see Fig. 18-7). The subicular and entorhinal cortices also project to the lateral septal nucleus, albeit to a lesser extent. The subiculum has a very dense projection to the *nucleus accumbens*, which is considered by many to be part of the limbic "striatum."

The *postcommissural fornix* originates in the subicular cortex. These subicular axons innervate the anterior and lateral dorsal thalamic nuclei, the ventromedial hypothalamic nucleus, and the mammillary complex. The mammillary complex receives the densest of these diencephalic projections. Each component of the subicular cortex projects in a distinct pattern. For example, only the subiculum innervates the medial mammillary nucleus, but the presubiculum and parasubiculum densely innervate the lateral mammillary nucleus, and the subiculum has a much less dense projection to this nucleus.

The subicular and entorhinal cortices are also the origin of projections to the cerebral cortex. The subicular complex projects to the cingulate, retrosplenial, and medial orbital cortices and to neurons in the parahippocampal gyrus, including the perirhinal cortex. The entorhinal cortex provides a dense projection to the cerebral cortex, where its axons terminate prominently in perirhinal, caudal parahippocampal, retrosplenial, and temporal polar cortices, and it probably sends reciprocal projections to all other areas that project to it. The cortical areas to which subicular and entorhinal cortices project have widespread connections with almost all associational cortices. The hippocampal formation also projects to the amygdala. The entorhinal cortex has a small projection to the lateral and basal nuclei, but most of the entorhinal axons in these nuclei are destined for the more rostral substantia innominata. A small area at the border of CA₁ and subiculum has a substantial amygdaloid projection that terminates in the basal nucleus and the periamygdaloid nucleus. The subicular and entorhinal cortices are interconnected through direct reciprocal projections.

The constellation of these connections suggests that information flow in the hippocampal formation is primarily serial, but as shown in Figure 18-6, there are many feedback loops in this circuit.

☐ Axons to and from the septal nuclei course in the fornix and medial forebrain bundle

The best known of the connections of the septal nuclei are those with the hippocampus proper, which were previously described. In addition, the septal nuclei receive afferents from the amygdaloid complex, the hypothalamus, and the medial midbrain reticular region through the medial forebrain bundle and a projection from olfactory regions.

Efferent axons of the septal nuclei course through the *stria medullaris* and the *medial forebrain bundle* to terminate in the habenula and hypothalamus, respectively. Some medial forebrain bundle axons from the septal nuclei extend to midbrain tegmentum, and other extend to brain stem nuclei.

☐ The mammillary bodies relay hippocampal information to the anterior thalamic nuclei

The mammillary complex connections of the limbic system were important to Papez's hypothesized circuit of emotion. The *fornix* splits into two segments at the anterior commissure, and the postcommissural complement of axons, which originate in the subicular cortex, traverses the hypothalamus and terminates in the mammillary complex. The mammillary complex is also innervated by the dorsal and ventral tegmental nuclei and the anterior hypothalamus.

Most efferent axons leave the mammillary nucleus by way of the *mammillothalamic tract* or the *mammillotegmental tract*. Through the mammillothalamic tract, the medial mammillary nucleus projects to anterior ventral and anterior medial thalamic nuclei, and the lateral mammillary nucleus projects primarily to the anterior dorsal nucleus. The lateral dorsal nucleus also receives a small projection from the mammillary complex. The mammillotegmental tract carries axons to dorsal and ventral tegmental nuclei in the midbrain. The supramammillary nuclei have a significant projection to the dentate gyrus.

☐ The anterior thalamic nuclei project primarily to the posterior cingulate cortex and the hippocampal formation

The anterior thalamic nuclei are innervated by three major pathways. The mammillothalamic tract and the fornix from the subicular cortex supply approximately equal numbers of axons to all three nuclei. Corticothalamic fibers from the cingulate, the retrosplenial, and the subicular cortices provide the third major afferent. One other notable input is a small projection from the visual system. A direct projection from the retina to the anterior nuclei has been reported to exist in the rat. The visual cortex projects to this area, although most of these projections terminate in the lateral dorsal nucleus. This nucleus also receives prominent inputs from retrosplenial and subicular cortices and a small input from the mammillary complex.

The thalamocortical projections from the anterior thalamic nuclei supply the entire cingulate, retrosplenial, and subicular cortices with a thalamic innervation, but each of the thalamic nuclei ends in distinct regional and laminar patterns (Fig. 18-8). Data from subprimates indicate that, unlike most thalamic projections, which terminate primarily in layer IV of the cortex, the projections from the anterior and lateral dorsal nuclei primarily terminate in layers I and IV of the cortex. The anterior dorsal and anterior ventral projections to the retrosplenial (i.e., area 29) and subicular cortices are largely overlapping in their regional spread but are distinct in their laminar, terminal patterns, with the anterior ventral nucleus projection largely confined to the outer third of layer I of the cortex and the anterior dorsal nucleus projection terminals spread throughout layer I. The anterior medial thalamic nucleus projects to the posterior cingulate cortex (i.e., area 23) but provides a sparse projection to anterior cingulate cortex (i.e., area 24), an area that is densely innervated by the parataenial, paraventricular, and intralaminar nuclei. The anterior medial nucleus also projects to the prefrontal and orbitofrontal region of cortex. Studies in the rat suggest that the projections to the posterior limbic cortex from the anterior ventral thalamic nucleus form a precise map in which individual thalamic neurons project to restricted groups of cortical neurons.

The lateral and medial thalamic nuclear groups also project to the limbic cortex. The projection from the lateral dorsal nucleus largely overlaps the projections of anterior dorsal and anterior ventral thalamic nuclei to the posterior limbic cortex and extends into the medial parietal cortex. The medial pulvinar thalamic nucleus projects to the posterior cingulate and retrosplenial cortex in primates. Two different parts of the medial dorsal nucleus, the densicellularis and parvicellularis, project to much of the cingulate cortex, including areas 25 and 32. The medial subnucleus of the ventral anterior thalamic nucleus projects to the cingulate cortex. Although these latter thalamic areas are not usually considered part of the limbic system, some data demonstrate that the portions of these nuclei that project to the limbic cortex do

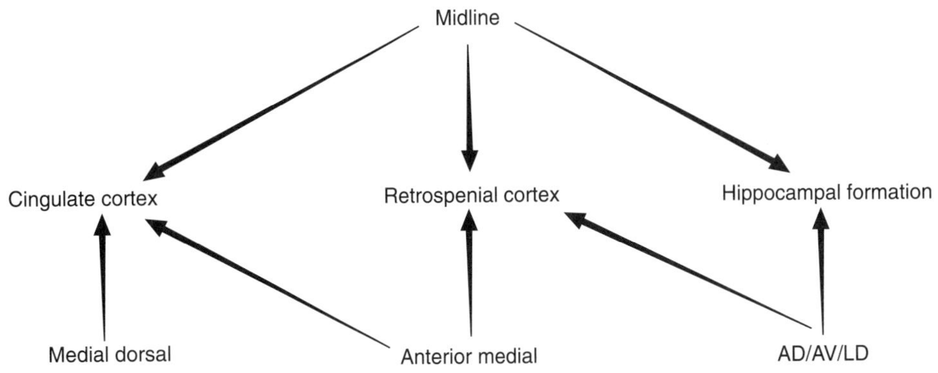

Figure 18-8
Each of the thalamic nuclei that project to the cingulate cortex and hippocampal formation has a distinct terminal field (*arrows*). The midline group includes the parataenial and reuniens nuclei.

not project widely to other cortical targets, and these can be considered limbic subnuclei.

In contrast to the anterior cingulate cortex, the posterior cingulate cortex is connected primarily to the hippocampal formation

Afferent connections

The thalamic projections to the cingulate cortex are extensive, but the cortical connections are equally important. The entire cingulate cortex, including the retrosplenial cortex, is interconnected by extensive commissural and associational projections, with the latter linking the posterior and anterior segments. Much of the caudal cingulate cortex also receives extensive projections from the subicular cortex and to a lesser extent from the entorhinal cortex. Several neocortical areas, including the somatosensory and visual association cortices, prefrontal cortex, parietal lobe, and superior temporal sulcus, project to the cingulate cortex.

As in other parts of the cortex, the cingulate cortex is innervated by cholinergic (i.e., from the diagonal band of Broca), noradrenergic (i.e., from locus ceruleus), and serotonergic (i.e., from the dorsal raphe and central superior nuclei) axons. The ventral tegmental area of Tsai sends a dense dopaminergic projection to the anterior cingulate cortex but only an extremely sparse dopaminergic projection to the posterior cingulate and the retrosplenial cortices. Each of these transmitter-specific systems terminates in a distinct laminar pattern.

Efferent connections

The projections of the cingulate and retrosplenial cortices to the subicular and entorhinal cortices, which course through the cingulum bundle, close the hypothesized circuit of Papez (see Fig. 18-2). This projection is relatively dense and arises from many regions of the cingulate cortex, but it originates primarily in the retrosplenial cortex. Cingulate cortex is reciprocally connected with areas of associational neocortex. All areas of

the cingulate cortex project to the medial striatum, and all areas are connected to the anterior thalamic nuclei.

The anterior cingulate cortex has substantial projections to the hypothalamus and brain stem, especially to areas related to visceral regulation. The cingulohypothalamic projection arises primarily from areas 23, 24, 25, and 32 and terminates primarily in the lateral hypothalamic area. Midbrain projections from the cingulate cortex arise primarily in areas 24 and 25, and these terminate in the periaqueductal gray matter, the raphe nuclei, and the deep layers of the superior colliculus. Areas 23, 24, 25, 29, and 32 project to the ventral pontine nuclei in a topographic manner. Sparse, medullary projections arise primarily from area 25 and end in the nucleus of the solitary tract. A few axons from the cingulate cortex, primarily from the dorsal areas 23 and 24 near the motor cortices, pass into the spinal cord, although their precise termination is unknown.

The amygdala is connected to both the cerebral cortex and the hypothalamus

Intrinsic connections

The intrinsic connections of the amygdala have been difficult to resolve with precision because of the small, irregular size of many amygdaloid nuclei and the fact that many axons pass through nuclei in which they do not terminate. These axons could be inadvertently labeled by tracers placed in a nucleus. With the advent of several modern techniques that minimize these problems, the intrinsic circuitry of the amygdala has begun grudgingly to give way. The lateral nucleus, which receives the most direct sensory inputs, projects to all subdivisions of the amygdala. A connection that is particularly of interest is the projection from the lateral nucleus to the basal nuclei, which are the chief sources of the reciprocal projections to the sensory cortices. The basal nuclei project to all other amygdaloid nuclei except the lateral amygdaloid nucleus. The central and superficial nuclei have many intraamygdaloid connections, but few of these projections are to the deep nuclei. These connections suggest that processing through the amygdala is achieved primarily in series (Fig. 18-9).

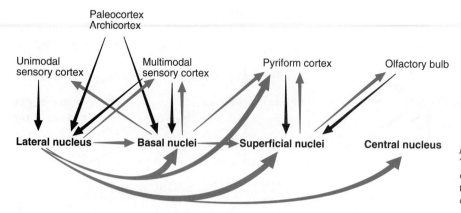

Figure 18-9
The primary cortical inputs to the amygdala are shown in black, and the main telecephalic projections from the amygdala are shaded.

Afferents connections

The major afferent connections of the amygdala differentiate the two components of the complex. The *olfactory bulb* projects directly to the superficial amygdaloid nuclei through the lateral olfactory tract, but few or no olfactory bulb axons directly innervate the deep group of nuclei. Within the superficial nuclear group, dense olfactory bulb and anterior olfactory nucleus projections terminate in the anterior cortical nucleus, the nucleus of the olfactory tract, and the periamygdaloid cortex. Most areas of the deep nuclear group receive an olfactory related input, but this originates in the piriform cortex and from intrinsic amygdaloid projections from the superficial nuclear group. The only major area of the amygdala that appears to lack an olfactory bulb or a direct olfactory cortex input is the *central nucleus*. The olfactory bulb projections to the amygdala are reciprocated from the superficial amygdala (see Fig. 18-9).

The amygdala, especially the lateral nucleus, is innervated directly by several other *unimodal, sensory areas* of the cerebral cortex, but it does not receive significant projections from primary sensory cortices except for the olfactory input. In addition to these single-modality sensory inputs, most other parts of the temporal lobe supply polysensory inputs to the amygdala. Most of these projections are directed at the lateral and basal nuclei.

Somatosensory information reaches the amygdala through projections from the *insular and orbital cortices*. The most direct pathway for this information is from the secondary somatosensory cortex to the posterior insular cortex, which projects heavily to the lateral nucleus of the amygdala. The caudal orbital cortex projects to the basal nucleus, and this region may also project to the central nucleus. The anterior cingulate cortex (i.e., areas 24, 25, and 32) projects primarily to the basal nucleus.

Several *thalamic nuclei* project to the amygdala. Auditory input is directed to the lateral and central nuclei through projections from the medial geniculate nucleus. Gustatory projections reach the lateral nucleus from the ventral posterior medial nucleus (i.e., pars medialis). Other thalamic projections to the amygdala originate in the midline, intralaminar, and the medial pulvinar nuclei.

The *hypothalamus* sends a moderately dense projection to the amygdala, which reciprocates the amygdalohypothalamic projections. The most prominent projections originate in the ventromedial and lateral hypothalamic nuclei and terminate in the deep nuclei of the amygdala.

A significant cholinergic projection to the amygdala originates from the magnocellular neurons of the basal forebrain. Specifically, the nucleus basalis of Meynert and the horizontal and vertical limbs of the diagonal band of Broca project to the deep and superficial nuclei, and the substantia innominata project to the central amygdaloid nucleus. Brain stem projections to the amygdala arise in the parabrachial nucleus (to the central nucleus), the pedunculopontine nucleus, the ventral tegmental area of Tsai (i.e., dopaminergic), the subceruleus (i.e., noradrenergic), the locus ceruleus (i.e., noradrenergic), and, perhaps the noradrenergic and dopaminergic neurons in the medulla.

Efferent projections

The amygdala projects directly to many cortical areas. The entorhinal cortex and the subicular cortex receive a substantial projection from the amygdala. Evidence suggests that some of these efferents may terminate on the distal dendrites of the hippocampus proper. The basal amygdaloid nuclei project densely to several areas of the unimodal sensory cortex that project primarily to the lateral nucleus of the amygdala (see Fig. 18-9). The amygdala projects to many polysensory regions of the cortex, and many of these projections arise from the lateral nucleus. Within the temporal lobe, the rostral one third of the temporal cortex receives the densest amygdaloid projections, but most of the temporal lobe and much of the occipital lobe are innervated to some extent. The amygdala also projects densely to the anterior cingulate cortex (i.e., areas 23, 24, 25, and 32) and the frontal and insular cortices.

Two major fiber tracts connect the amygdala with the diencephalon and brain stem (Fig. 18-10). The *stria terminalis* is the more obvious of the two, because it arches along medial to the entire extent of the body and tail of the caudate nucleus and descends lateral to the fornix at the level of the anterior commissure. Most axons in the stria terminalis terminate in the bed nucleus of the stria

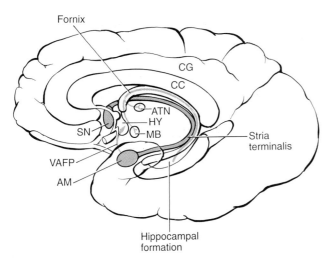

Figure 18-10
This figure depicts the three major fiber bundles that carry the primary output axons from the amygdala (*AM; VAFP,* ventral amygdalofugal pathway) and the hippocampal formation to subcortical and diencephalic sites. *CC,* corpus callosum; *CG,* cingulate gyrus; *HY,* hypothalamus; *NBF,* basal forebrain including nuclei accumbens, diagonal band of Broca, and basalis; *SN,* septal nuclei.

terminalis, which is located dorsal to the anterior commissure. This nucleus, which is considered by some to be a rostral extension of the amygdala, has dense projections to the hypothalamus and brain stem. Axons in the postcommissural stria terminalis project to the anterior hypothalamus, and some of these axons probably course further caudally to the brain stem through the medial forebrain bundle. Other fibers terminate in the preoptic region, in the cell-poor region surrounding the ventromedial hypothalamic nucleus (i.e., axons derived from the amygdalohippocampal area), and the core of the ventromedial hypothalamic nucleus (i.e., from the medial nucleus of the amygdala).

The *ventral amygdalofugal tract* provides a second pathway for amygdaloid axons that are destined for subcortical, diencephalic, and brain stem areas. This projection passes rostrally from the amygdala, courses beneath the lenticular nucleus through the substantia innominata,

and terminates in the lateral preoptic nucleus and the full extent of the lateral hypothalamic area. Caudal to the hypothalamus, axons from the central nucleus of the amygdala terminate in several brain stem nuclei (Fig. 18-11). A few axons appear to continue into the cervical spinal cord.

The amygdalofugal fibers also terminate in the medial dorsal thalamic nucleus. These axons originate from most areas of the amygdala (the medial and the central nuclei appear to be exceptions) but especially from the deep nuclei and the periamygdaloid cortex. There is also a reciprocal connection to the midline thalamic nuclei, primarily from the central and medial amygdaloid nuclei. The central and basal amygdaloid nuclei have a reciprocal connection with the basal forebrain. One of the densest amygdalofugal projections is that from the basal nuclei to the striatum. Although much of this terminates in the "limbic" striatum (i.e., the nucleus accumbens), there is a substantial projection to the ventromedial caudate and putamen.

Functional Considerations

☐ The limbic system is involved in affective behavior

A major insight into the neurologic mechanisms controlling affective state (i.e., personality, emotion, and social behavior) was Papez's 1937 paper in which he proposed that the "limbic lobe" of the cortex formed the anatomic circuit that coordinates *emotion*. Although this paper was very speculative and lacked rigorous experimental support, it quickly gained considerable acceptance, partially because of the dominant trend in psychology and psychiatry at that time. Freudian psychoanalytic thinking readily welcomed the idea that the old and ancient parts of the brain were responsible for emotional and instinctive behavior and that the newer areas of the brain, such as the neocortex, were primarily attentive to conscious tasks and controlled behavior.

In a study published in 1939, Klüver and Bucy reported that bilateral temporal lobectomy in monkeys produced dramatic behavioral changes, most of which could be de-

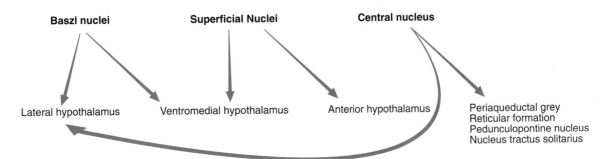

Figure 18-11
The major subcortical, diencephalic, and brain stem connections of the amygdala are shown by the arrows.

fined as affective disorders. The Klüver-Bucy syndrome was characterized by a markedly increased sexual activity that was often inappropriate (e.g., monkeys mounting other species or chairs), a loss of fear and a resulting flattening of emotions, an enormous increase in oral behavior (e.g., the animal puts almost anything in its mouth), and indiscriminate dietary behavior. Subsequent studies have demonstrated that damage to the amygdala causes the sexual, appetitive, and affective dysfunctions in these animals. The visual and memory losses that Klüver and Bucy reported in their original primate experiments appear to be the result of damage to other areas of the temporal cortex. The Klüver-Bucy syndrome has been shown to occur in humans who have selective damage to the amygdala.

At around the time that Klüver and Bucy demonstrated the effects of amygdalectomy on emotional behavior, Jacobson reported his neurosurgical studies of chimpanzees, and in his presentation, he parenthetically alluded to the calming effect that a prefrontal lesion had on the behavior of one particularly neurotic female chimp. This led Egas Moniz, the young Portuguese neuropsychiatrist, to return home immediately and begin treating severe mental disorders using prefrontal lobotomy. Moniz launched modern psychosurgery, for which he was awarded the 1949 Nobel Prize in Physiology and Medicine.

During the 1940s through the 1970s, many clinicians carried out psychosurgery on various limbic regions, especially the amygdala and the cingulate and the prefrontal cortices, and for a time, these lesion techniques became the method of choice to relieve severe emotional disorders; however, except for the study of Klüver and Bucy and the work of a few other behaviorists, the method lacked an empiric basis until the late 1960s, when critical experiments on the connections and functions of these areas were undertaken.

Although the questionable effectiveness of psychosurgery as a treatment for psychiatric diseases has considerably dampened enthusiasm for the treatment, studies of patients who received these surgical interventions and the literature in this field have demonstrated that the circuit of Papez is not a major contributor to emotional control. The areas that Papez did not include in his circuit (e.g., the amygdala, medial dorsal thalamic nucleus, anterior hypothalamus) or that he thought were peripheral contributors to the circuit (e.g., prefrontal cortex, rostral cingulate cortex) appear to play a major role in emotional control and the conscious perception of emotional experience.

Disorders in the limbic cortex have been linked to several psychiatric diseases, (e.g., *autism, schizophrenia*). Autistic children display abnormalities in temporal lobe electroencephalographic (EEG) patterns and enlargement of the temporal horn of the lateral ventricles along with some Klüver-Bucy–type affective disorders. In monkeys, the behavioral signs of autism can be mimicked by lesions of the temporal lobe, but this occurs only if the lesions are made very early in the monkey's life. The linkage between schizophrenia and limbic system dys-

function is suggested by several clinical and pathologic studies.

The frontal and temporal cortices appear to be somewhat smaller in schizophrenic patients, and pyramidal neurons in the CA fields of the hippocampus are often abnormally oriented in these patients. In schizophrenic patients, there appear to be several imbalances in neurotransmitters that are selective for the limbic system. Most prominent among these is the dopamine system, which has a particularly dense projection to anterior limbic system regions. Many effective therapies for schizophrenia target the dopamine system. In schizophrenics, the areas of the limbic system that receive a dense dopamine projection display selective imbalances, including decreased glucose uptake. Lesions of the cingulate cortex can result in akinetic mutism.

☐ The limbic system contributes to normal learning and memory

Although the relation between emotional dysfunction and the limbic system appear primarily to involve the amygdala and anterior cingulate cortex, learning and memory deficits that accompany affective disorders appear to be the product of damage to more caudal limbic regions, especially the hippocampal formation and the posterior cingulate and retrosplenial cortices.

Before discussing the relation between the limbic system and learning and memory, it is useful to clarify certain terms that define memory. Memory can be divided into *short-term memory*, which endures for a very brief time, and *long-term memory*, which is retained long after the conscious perception and consideration of an event. Long-term memories are further divided into *declarative memory*, which refers to the memory of facts that can be recalled to consciousness, and *procedural memory*, which refers to patterns of behavior that are learned (e.g., motor skills, procedural skills). The limbic system is primarily involved in the transfer of declarative memory from short-term to long-term traces.

Much of our early understanding of the relation between learning and memory and the limbic system came from studies of patients with damage to the brain. At the turn of the century, Bekhterev suggested that damage to the temporal lobe was related to a severe memory impairment in one of his patients, but it was not until the 1950s that the association was widely accepted. In 1953, William Scoville operated on a patient who presented with intractable, severe generalized epileptic seizures. The bilateral temporal lobe resection was a successful treatment, but it left the patient with a severe anterograde memory deficit (i.e., he was unable to convert short-term to long-term memory). The patient displayed little retrograde memory deficit (i.e., he was able to recall events that occurred before his operation).

During the subsequent half century, this patient, designated H.M., has been studied intensively. A flavor of his dysfunctional state comes from the detailed 1968 report of Milner and colleagues, in which H.M. is shown to fail to recall almost any event, even very traumatic events such as the death of his father. Only those events that were continuously repeated were remembered, and they were remembered only vaguely. For instance, when riding home with Dr. Milner in 1966, H.M. was asked to provide directions to his new home, in which he had lived for more than 8 years. Instead, he led Dr. Milner to the house he lived in before his surgery. He recognized this was not the correct location, but he could not provide any directions to the new home. In contrast to his obvious deficit in declarative memory, H.M. is quite accomplished in tasks requiring procedural learning skills, such as the mirror-draw task in which he displayed normal memory. H.M. demonstrates the importance of the temporal lobe in declarative memory, and his case shows that there is a dissociation between the areas of the brain that serve declarative compared or procedural memory.

Deficits in many other patients have confirmed the role of the posterior limbic cortex in mnemonic tasks, but patient R.B. provides the clearest clinical picture of the relation, because rigorous testing of this patient has been followed by clear histologic findings. While undergoing coronary bypass surgery, R.B. had a transient ischemic episode, after which he displayed severe anterograde amnesia but no significant retrograde amnesia. His memory for past events appeared somewhat improved compared with normal subjects. Postmortem examination 5 years after the ischemic episode revealed a very selective bilateral lesion of the CA_1 fields of the hippocampus, and no apparent damage to any other limbic region of the brain (Zola-Morgan et al, 1991). Remarkably, the severe deterioration of declarative memory ability in R.B. was not accompanied by any apparent alteration in emotion or cognitive function.

Animal studies have supported the role of the hippocampus and posterior limbic cortex in mnemonic tasks. In monkeys, damage to the hippocampal formation impairs memory but has little effect on emotional behavior. In contrast, selective damage to the amygdala causes affective alterations but few or no alterations in learning and memory.

Several other areas of limbic cortex appear to contribute to learning and memory. As would be expected from the connections, damage to subicular and entorhinal cortices effects learning and memory. The retrosplenial cortex, with which the hippocampal formation is highly interconnected, has been shown to participate in learning and memory. Animal studies suggest that the retrosplenial cortex contains areas that are involved in mapping out spatial relations, and lesion of this region in rat cause a significant decline in learning. Valenstein reported a case in which severe anterograde memory impairment was diagnosed in a patient with very selective neurologic damage that was confined to the retrosplenial cortex. Animal studies suggest that the cingulate cortex, like the retrosplenial cortex, may play a role in learning and memory. Damage to the cingulate cortex causes a decrease in learning ability in monkeys, but these deficits may relate more closely to an inability of the animal to correctly weigh the importance of events.

Two subcortical limbic regions should be mentioned in relation to learning and memory. Some patients with relatively selective vascular lesions of the basal forebrain and patients with *Alzheimer's disease* display dementia and loss of the cholinergic projection to the limbic cortex. Animal studies suggest that selective lesions of the basal forebrain projection to the limbic cortex reduce learning and memory abilities, and reestablishing this pathway by means of transplantation of embryonic basal forebrain tissue ameliorates the lesion-induced deficits. These data suggest that the basal forebrain projection to the limbic system is a contributor to learning and memory.

The diencephalon appears to play a role in learning and memory. In 1887, Korsakoff characterized a syndrome in which a severe loss of memory was present in chronic alcoholics. Patients with *Korsakoff's syndrome* display anterograde amnesia, especially for paired-association tasks, extensive retrograde amnesia for events that occurred throughout their adult life, confabulation (i.e., making up stories that are relatively plausible to cover up their memory impairments), reduced frequency of speech, little perception of their memory loss, and generalized apathy. The cause of Korsakoff's disease in many patients appears to be a lack of dietary thiamine (i.e., vitamin B_1) which is caused by the malnutrition that accompanies the long drinking bouts in chronic alcoholics. The syndrome consists of variable damage primarily to the medial dorsal nucleus of the thalamus, the frontal cerebral cortex, and the mammillary bodies. The pivotal position of the mammillary bodies in the posterior limbic circuitry suggests that this damage is critical to Korsakoff's syndrome, but the anatomic mechanisms underlying the disease remain unresolved.

The severe retrograde amnesia and affective disorders in Korsakoff's syndrome are not observed in amnesic patients with primary temporal lobe dysfunction. Although amnesia and Korsakoff's syndrome involve memory deficits that are related to limbic cortex dysfunction, the two diseases have different neuropathologic and functional attributes.

☐ The hippocampal formation plays a role in epilepsy

The hypothesized relation between hippocampal formation damage and epileptic seizures dates to the early 19th century, and by 1880, Sommer had clearly identified an area of the hippocampus proper (i.e., *Sommer's sector* or area CA_1), which was consistently damaged in the epileptic patients he studied. Several subsequent reports supported the idea that, in patients with temporal lobe epilepsy, the epileptogenic focus is often in the hippocampus proper (Fig. 18-12), and that limited resection of the hippocampal focus can abolish subsequent seizure activity. In many patients, analysis of the degree of hippocampal sclerosis compared with the frequency and severity of epileptic activity suggests that the damage to the hippocampus was present before the initial epileptic activity.

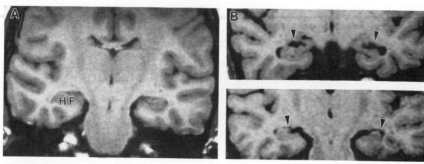

Figure 18-12
A magnetic resonance coronal section through the temporal lobes of (**A**) a normal adult
and (**B**) an adult with severe, long-standing epilepsy. Notice the marked bilateral atrophy
of the hippocampus (*arrowheads*) in the epileptic patient.

The pyramidal neurons in the hippocampus are contributors to the initiation of epileptic seizure activity and
are damaged by recurring seizures. This pattern occurs
in children and adults and could create enlarged sclerotic
foci that would increase the probability and or intensity
of additional seizure activity.

□ Limbic system abnormalities are present in Alzheimer's disease

In the normal course of aging, some neurons in the limbic cortex degenerate, potentially compromising limbic
memory circuits. In contrast, aging persons with Alzheimer's disease display extensive neuronal damage (i.e.,
neurofibrillary tangles and β-amyloid plaques) and cell
loss, especially in the limbic and olfactory cortical regions. In the hippocampal formation, area CA_1 and the
subicular and entorhinal cortices are most severely effected. Damage to subicular neurons would significantly
affect the major outputs of the hippocampal formation,
the fornix projection to septal and diencephalic nuclei,
and the innervation of entorhinal cortex (see Fig. 18-10).
The prominent damage to the layer II pyramidal neurons
of the entorhinal cortex compromises the entorhinal
projection to the hippocampus. These combined lesions
isolate the hippocampal formation and probably produce
memory impairment.

Damage to other limbic areas probably contributes to
many of the affective changes in Alzheimer's patients. The
olfactory regions of telencephalon, the amygdala, the cingulate cortex, and the hypothalamus are consistently
compromised in Alzheimer's patients, and each of these
areas plays an important role in the regulation of emotional stability. The cholinergic neurons in the diagonal
band of Broca and nucleus basalis of Meynert also are
damaged in Alzheimer's disease, significantly reducing
the cholinergic input to the limbic cortex. The pathogenesis of Alzheimer's disease is discussed in detail in Chapter 17.

□ The hippocampus as a model system has provided important insights into the functioning of the cerebral cortex

Because of its relative simplicity, the limbic cortex, especially the hippocampal formation, has often been used
as a model system for investigating neuronal plasticity.
Most studies have shown that the hippocampal formation
displays a high degree of active reorganization, an attribute that may account for its central role in learning and
memory and its contribution to epilepsy. The plasticity
displayed is of several types. Many studies have investigated the ability of axons to reinnervate areas of the hippocampal formation that have been denervated. These
studies have produced important insights into the role of
trophic factors in neuronal reorganization. The hippocampal system also has provided an excellent model in
which to characterize the changes that occur in postsynaptic neurons after denervation.

At the synaptic and molecular level, the hippocampal
system has been the major model in which *long-term potentiation* (LTP) has been studied. LTP is a long-lasting
(i.e., 15 minutes to several days) increase in the efficiency
of synaptic transmission that can be induced by short-
lasting high-frequency stimulation. This phenomenon,
which was first identified in the entorhinal to dentate
projection, was subsequently demonstrated in many
other areas of the nervous system. Many tout LTP as an
important experimental model of memory.

Limbic System Overview

The limbic system is a highly interconnected group of
regions that receive diverse multimodal sensory information. The anterior portions of the system (e.g., the
amygdala, anterior cingulate cortex) primarily regulate
affective behavior and visceromotor function, and the
posterior portions (e.g., the hippocampal formation, retrosplenial cortex) are predominantly involved in the

should quickly begin to illuminate the unity and diversity of the limbic system (Fig. 18-13).

SELECTED READINGS

Aggleton JP. The amygdala: neurobiological aspects of emotion, memory, and mental dysfunction. New York: Wiley-Liss, 1992.

Broca P. Anatomie comparée des circonvolutions cérébrales. Le grand lobe limbique et la scissure limbique dans la série des mammifères. Rev Anthrop 1878;2:285.

Chan-Palay V, Köhler C. The hippocampus. Neurology and neurobiology, vol 52. New York: Alan R. Liss, 1989.

Isaacson RL. The limbic system, 2nd ed. New York: Plenum Press, 1982.

Klüver H, Bucy PC. Preliminary analysis of function of the temporal lobes in monkeys. Arch Neurol Psych 1939;42:979.

Kolb B, Whishaw IO. Fundamentals of human neuropsychology. New York: WH Freeman, 1990.

Lopes Da Silva F, Witter MP, Boeijinga PH, Lohman A. Anatomical organization and physiology of the limbic cortex. Physiol Rev 1990;70:453.

MacLean PD. Some psychiatric implications of physiological studies on the frontotemporal portion of the limbic system (visceral brain). Electroencephalogr Clin Neurophysiol 1952;4:407.

Rosene DL, Van Hoesen G. The hippocampal formation of the primate brain: a review of some comparative aspects of cytoarchitecture and connections. In Peters A, Jones EG. Cerebral cortex. New York: Plenum Press, 1987;6:345.

Scoville WB, Milner B. Loss of recent memory after bilateral hippocampal lesions. J Neurol Neurosurg Psychiatry 1957;20:11.

Valenstein E, Bowers D, Verfaellie M, Heilman KM, Day A, Watson RT. Retrosplenial amnesia. Brain 1987;110:1631.

Valenstein ES. Brain control: a critical examination of brain stimulation and psychosurgery. New York: John Wiley & Sons, 1973.

Vogt BA, Gabriel M. The neurobiology of cingulate cortex and limbic thalamus. New York: Springer-Verlag, 1992.

Zola-Morgan S, Squire LR, Alvarez-Royo P, Clower RP. Independence of memory functions and emotional behavior: separate contributions of the hippocampal formation and the amygdala. Hippocampus 1991;1:207.

Figure 18-13
The sixteenth century Italian physician Aranzi had a vivid imagination when he named the hippocampus for its resemblance to the mythic seahorse, shown here being ridden by Triton in a seventeenth century drawing.

temporary storage of information, including the encoding of spatial relations. In contrast to this diversity, the limbic cortical areas are highly interconnected and have prominent connections with limbic subcortical, diencephalic, and brain stem nuclei. These areas are functionally interrelated.

Emotional state can significantly influence learning and memory, and elements of learning and memory (e.g., habituation and orientation) can importantly alter emotional status and arousal. Riding the wave of the rapidly progressing neurobiologic techniques, future studies

Clinical Correlation
MENTAL ILLNESSES

NANCY C. ANDREASEN

Mental illnesses are the diseases within medicine that affect the core elements that define humanity and differentiate human beings from other animals: personality, goal-directed behavior, language, creativity, abstract thinking, emotion and mood, and social organization. When these various functions are working well in normal human beings, we can watch them marry and parent their children effectively, design roads or buildings, create business ventures, play chess, or write plays, novels, or music. When these normal functions are disrupted, we see the often-tragic consequences of mental illness: despondency and despair, suicide, terrifying distortions of reality, or loss of the capacity to think well and remember.

The various symptoms of mental illness, such as hearing voices when no one is around (i.e., hallucinations) or feeling persecuted (i.e., delusions), almost certainly represent disruptions in brain structure, circuitry, and chemistry. Unlike many of the neurologic disorders, which are usually caused by disruptions of a single system or clearly visible lesions, the neural disruptions that produce mental illness are probably complex, multiple, and not readily observable, because they occur primarily at the biochemical or molecular level. The identification of the underlying mechanisms of major mental illnesses is one of the most exciting "search and destroy" missions in modern medicine. The availability of techniques such as neuroimaging (see Chap. 33) or the various applications of molecular biology have at last given psychiatrists the tools that can be used to study the brain in the fine-grained manner necessary to map the mechanisms of mental illness.

Mental illnesses range from mild to severe. Milder mental illnesses include personality disorders (e.g., obsessive-compulsive personality, antisocial personality) and various adjustment disorders. Although these disorders are often seen as primarily psychological, all psychological experiences ultimately must be defined at the neural level. Much evidence indicates that even disorders such as obsessive-compulsive personality have identifiable abnormalities on positron emission tomography

(PET) scans (i.e., increased perfusion in the prefrontal cortex and basal ganglia) and that there may be a strong genetic component to antisocial behavior.

There is little controversy about the importance of neurobiologic components for the more severe mental illnesses. These include the dementias (see Chap. 17), schizophrenia, bipolar disorder, major depression, and severe forms of anxiety disorders, especially panic disorder. This chapter provides an overview of these severe forms of mental illness, briefly describing the characteristic symptoms and summarizing current knowledge about their neurobiologic substrates.

Schizophrenia

Schizophrenia is typically a catastrophic illness that begins in adolescence or early adulthood. Because its symptoms often produce severe incapacity, its original name was "dementia praecox." It was initially identified and defined in the early 20th century by a team of psychiatrists that included Alois Alzheimer, and was differentiated from dementia in the elderly (later renamed Alzheimer's disease) on the basis of age of onset. Dementia praecox was later renamed "schizophrenia", which means "shattered mind".

Schizophrenia is characterized by a mixture of signs and symptoms, no single one of which is necessarily present. The absence of a single defining feature sometimes makes this disease difficult for students to conceptualize. The disorder is essentially defined by the presence of several from a group of characteristic symptoms, accompanied by significant deterioration in functioning and a relatively chronic course (i.e., presence of some symptoms for at least 6 months). Typically, some symptoms of the disorder persist for the remainder of the person's life. The combination of significant incapacity, onset early in life, and chronicity of illness makes schizophrenia a particularly tragic disorder, and it is quite common, occurring in between 0.5% and 1% of the population.

The symptoms of schizophrenia are often divided into two broad groups: positive and negative. *Positive symptoms* include delusions, hallucinations, disorganized speech, and disorganized or bizarre behavior. These symptoms are referred to as positive because they represent distortions or exaggerations of normal cognitive

This research was supported in part by NIMH Grants MH31593, MH40856 and MHCRC 43271, The Nellie Ball Trust Fund, Iowa State Bank and Trust Company, Trustee; and a Research Scientist Award, MH00625.

or emotional functions. *Hallucinations* are abnormalities in perception (i.e., hearing voices without a stimulus); *delusions* are distortions in inferential thinking (i.e., misinterpreting information, usually in a way that suggests danger or harm); *disorganized speech* is a disruption in language; and *disorganized behavior* is a disruption in motor or behavioral monitoring and control. Some positive symptoms are always present in patients with schizophrenia, but the pattern may vary. One particular patient may be especially psychotic and experience hearing voices and feeling persecuted. Another patient may be more disorganized, speaking in an incoherent manner and displaying abnormalities in behavior such as agitation, wearing strange clothing, or making clearly inappropriate sexual overtures.

The *negative symptoms* of schizophrenia reflect a loss or diminution of functions that are normally present. Negative symptoms include *alogia* (marked poverty of speech or speech that is empty of content); *affective flattening* (a diminution in the ability to display expression of emotion); *anhedonia* (an inability to experience pleasure or loss of interest in social interaction); *avolition* (an inability to initiate or persist in goal-directed behavior); and *attentional impairment*. Although the positive symptoms of schizophrenia are often colorful and draw attention to the patient's illness, the negative symptoms tend to impair the person's ability to function in normal daily life. These are the symptoms that prevent patients with schizophrenia from having normal family relationships, attending school, holding a job, or forming friendships and intimate relationships.

The positive and negative symptoms and the corresponding functions that are distorted or diminished are summarized in Table 18-1.

The profound and pervasive cognitive and emotional disturbances that characterize schizophrenia suggest that it is a serious brain disease affecting multiple functions and systems. Identification of the neurobiologic substrates of schizophrenia is one of the more hotly pursued areas in contemporary neuroscience. The various symptoms suggest possible involvement of a variety of cortical and subcortical brain regions. Auditory hallucinations

and disruptions in linguistic expression suggest involvement of auditory cortex and perisylvian language regions. Other positive symptoms, such as delusions or disorganized behavior, are more difficult to localize to specific regions and imply the involvement of larger systems and circuits. Negative symptoms are possibly related to the prefrontal cortex, which mediates goal-directed behavior, fluency of thought and speech, and other functions that are lost or diminished. Some support for abnormalities of these various brain regions has been provided by studies using magnetic resonance (MR), single photon emission computed tomography, and PET. Regions in which abnormalities have been found rather consistently include the prefrontal cortex, the superior temporal gyrus, the hippocampus, and the basal ganglia.

In addition to specific localizations, a large mass of evidence suggests the presence of less specific abnormalities in brain structure in schizophrenia. The bulk of this work has been done with computed tomography (CT), a neuroimaging technique that has been available since the mid-1970s. More than 75 studies have been completed using CT scanning; approximately three fourths of them have indicated that patients with schizophrenia (evaluated as a group) have enlargement of the ventricular system compared with normal volunteers. An additional abnormality includes prominent sulci, originally assumed to be due to cortical atrophy but more recently considered to be due to abnormal brain development (i.e., dysgenesis).

The pathophysiologic significance of these abnormalities is uncertain. Several studies using CT and MR have evaluated patients early in the illness and have found these abnormalities to be present at onset. CT and MR studies have been done of discordant monozygotic twins, consistently indicating more prominent structural brain abnormalities in the ill twin. These findings imply that schizophrenia may be a neurodevelopmental disorder that is produced by a mixture of genetic and environmental causes. Twin studies have been used to estimate the degree of genetic contribution that occurs in schizophrenia. When concordance rates are compared in monozygotic and dizygotic twins, the concordance rate in the former is approximately 40% to 50%, and the rate in the latter is approximately 10%, suggesting that there is a prominent genetic contribution to schizophrenia, but also that nongenetic factors must play a role. If the etiology were totally genetic, the concordance rate in monozygotic twins would be 100%.

The cerebral neurochemistry of schizophrenia has been partially illuminated during recent years. The dopamine hypothesis remains the primary working theory, although it is almost certainly an oversimplification. The dopamine hypothesis suggests that the symptoms of schizophrenia are produced by a functional excess of dopamine at crucial brain regions (e.g., limbic regions, language regions). Evidence to support the dopamine hypothesis is derived primarily from studying the action of neuroleptic medications; the therapeutic efficacy of these antipsychotic agents has been shown to be closely correlated with their ability to block dopamine type 2 recep-

Table 18-1
SYMPTOMS OF SCHIZOPHRENIA

Symptom	Function Distorted or Lost
Positive	
Hallucinations	Perception
Delusions	Inferential thinking
Disorganized speech	Language
Bizarre behavior	Behavior monitoring
Negative	
Alogia	Fluency of speech
Affective blunting	Fluency of emotional expression
Anhedonia	Hedonic capacity
Avolition	Volition, drive

tors. In addition, postmortem studies have identified an increase in dopamine receptors in the brains of persons with schizophrenia compared with normal controls, and some in vivo studies have found an increase in the levels of dopamine metabolites. Because some newer neuroleptics have potent D_1 or 5-HT$_2$ receptor blockade, simple versions of the dopamine hypothesis are being reappraised.

Mood Disorders

Unlike schizophrenia, which has no single characteristic symptom, the mood (i.e., affective) disorders can be defined relatively simply. Mood disorders are characterized by abnormalities in the aspect of emotional experience and expression that we refer to as mood. Normally, a person's mood hovers around neutrality, with mild feelings of elation being experienced as happiness and mild feelings of sadness being experienced as unhappiness. When mood swings outside the normal range, the normal fluctuations in emotion become pathologic. The pathologic extreme of elation is known as *manic disorder*, and the pathologic extreme of depression is known as *major depressive disorder*. A person who experiences both of these poles of mood abnormality within a single episode or at different times during his or her life is said to have *bipolar disorder*.

Mood disorders are among the most common of mental illnesses. Bipolar disorder occurs in approximately 1% of the population, but the lifetime prevalence of major depression may be 10% or greater. Mood disorders appear to have a particularly close association with creativity and high levels of achievement. Van Gogh, Lincoln, Tolstoy, and William James are just a few of the notable individuals who have experienced mood disorders.

Mania is defined by the presence of an abnormally elevated, expansive, or irritable mood, plus at least some of a variety of other symptoms that characterize the overall syndrome. The related symptoms include grandiosity, decreased need for sleep, increased talkativeness, racing thoughts, distractibility, increased goal-directed activity, and excessive involvement in pleasurable activities that have a high potential for painful consequences (e.g., buying sprees, sexual indiscretions, foolish business ventures). A patient with mania is usually cheerful and enthusiastic, although he or she may become irritable if thwarted. Because mania is often subjectively pleasurable, manic patients may be reluctant to seek treatment, at least early in the course of illness. However, family and friends recognize the person's behavior as unusual, potentially dangerous, and often annoying and intrusive. Because judgment tends to be impaired during mania, and because people with mania tend to be grandiose and expansive, they often overextend themselves and get into difficulties at work, at home, or financially.

Major depression presents as the opposite extreme from mania. A person with major depression feels despondent, blue, and down in the dumps. The depressed mood is typically accompanied by a large number of associated symptoms that characterize the disorder, including diminished interest or pleasure, sleep disturbance, changes in appetite and weight, psychomotor agitation or retardation, decreased energy, feelings of worthlessness or excessive guilt, difficulty concentrating, indecisiveness, and recurrent thoughts of death or suicide. Although mania is usually subjectively pleasant, depression is characterized by intense psychological pain. Because the experience of depression is so painful, people with this disorder have a high suicidal risk; approximately 15% of patients hospitalized for depression kill themselves. Although suicide is the most serious complication of depression, other social and personal complications also occur as a consequence of the diminished interest, withdrawal, decreased energy, and sleep and appetite difficulties that occur during a depressive episode.

The neurobiology of the mood disorders has been extensively studied. These disorders have a prominent familial component. The morbid risk for mood disorder is at least doubled over the general population rate in the first-degree relatives of people who suffer from mood disorder. This evidence suggesting that mood disorders tend to run in families does not necessarily indicate genetic transmission; role modeling, learned behavior, and social factors such as economic deprivation may also play a role. Twin studies have complemented these family studies and suggested that mood disorders are genetic in additional to familial. The overall monozygotic to dizygotic twin ratio is approximately 4 : 1, quite similar to that observed for schizophrenia.

Mood disorders were also among the first major mental illnesses to be studied with molecular biology using the linkage method. Linkages have been reported on the short arm of chromosome 11 and the X chromosome, but these linkages have not been consistently replicated. It is likely that the mood disorders are etiologically heterogeneous and multifactorial, making linkage to a single major locus unlikely.

The cerebral neurochemistry of the mood disorders has been extensively explored. Two major transmitter systems have been implicated: the norepinephrine system and the serotonin system. Much of the evidence for the role of these neurotransmitters and mood disorders derives from psychopharmacology. Most medications that have been efficacious for the treatment of depression have increased the amount of norepinephrine or serotonin functionally available at nerve terminals. The primary mechanism for most medications is inhibition of reuptake. Monoamine oxidase (MAO) inhibitors are also effective antidepressants, and they act by inhibiting the breakdown of catecholamines. Reserpine, which depletes monoamines, tends to worsen depression.

The chemical mechanisms that explain the occurrence of mania are less well understood; the primary medications used to diminish manic symptoms include lithium and the neuroleptics. The mechanism of action of lithium is uncertain, but the antipsychotics tend to antagonize the actions of dopamine or serotonin.

A variety of neuroendocrine and neurophysiologic abnormalities have also been observed in the mood disorders. The disturbances in sleep, appetite, and the normal diurnal variation that are common in depression suggest abnormalities in cortisol regulation. Extensive studies of the hypothalamic-pituitary-adrenal axis have been completed; most evidence suggests that the observed abnormalities are probably occurring at the high end of this axis, i.e., in the cortex or hypothalamus. Neurophysiologic studies, especially those using sleep EEG, also support the likelihood that general cerebral dysregulation is occurring in the mood disorders, because patients suffering from depression have a variety of EEG abnormalities during sleep, including decreased slow wave sleep, shortened rapid eye movement (REM) latency, and abnormally prolonged periods of REM sleep. All these abnormalities in sleep EEG are consistent with the subjective complaints of insomnia that depressed patients express.

Anxiety Disorders

Anxiety disorders are a somewhat heterogeneous group of conditions that are often considered to be at the far end of an underlying continuum of mental illnesses, one end of which comprises the very severe and clearly biologically produced conditions (e.g., dementia, schizophrenia) and the other end of which is defined by disorders that are predominantly psychological (e.g., mild anxiety, other neuroses, personality disorders). Within this continuum, the mood disorders reside in the middle, with severe forms having a clear biologic component and milder forms having a prominent psychological substrate. Evidence suggests that many anxiety disorders also have prominent neurobiologic substrates. Two important examples are panic disorder and obsessive-compulsive disorder.

☐ Panic disorder

Panic disorder is defined by recurrent *panic attacks,* also referred to as anxiety attacks. A panic attack is experienced as a subjectively catastrophic sensation of terror or fear, accompanied by a variety of additional symptoms, such as shortness of breath, dizziness, palpitations or tachycardia, trembling, sweating, choking, nausea, paresthesias, flushing or chilling, or chest pain or discomfort. The person experiencing a panic attack often feels as if he or she is dying, going crazy, or losing control. Although many people have experienced a panic attack sometime during their lives (e.g., when exposed to some genuinely frightening situation such as turbulence on an airplane or some psychologically threatening situation such as being called on in class or asked to give a speech), panic disorder is a pathologic condition by vir-

tue of the frequency of the panic attacks and the absence of any clear stimulus. Although the first panic attack of some persons with this condition is provoked, the panic attacks assume a life of their own and occur without any obvious reason in patients with the full-blown disease.

Panic disorder is sometimes accompanied by a related condition, *agoraphobia,* which is a fear of public places or leaving home. Patients with a combination of panic disorder and agoraphobia essentially become housebound.

Panic disorder and agoraphobia occur in 0.5% to 3% of the population. When panic attacks are frequent, the condition can be very disabling.

Patients with panic disorder typically consult general practitioners, cardiologists, or gastroenterologists. After a thorough physical and laboratory workup, no specific somatic cause is identified. However, the disorder can usually be diagnosed clinically simply on the basis of a careful history.

The neurobiologic substrates of panic have been extensively explored. No single specific abnormality has been identified, but a variety of abnormalities have been suggested: carbon dioxide hypersensitivity, disturbances in lactate metabolism, increased catecholamine levels, abnormalities in γ-aminobutyric acid (GABA), and abnormalities in the locus coeruleus. Some of the evidence for these possible abnormalities is based on the ability of different substances to induce panic attacks, including isoproterenol, carbon dioxide, or sodium lactate. Many of the drugs used to treat panic activate GABA, providing some evidence for the role of this neurotransmitter. Other drugs that are effective in treating panic diminish norepinephrine transmission.

☐ Obsessive-compulsive disorder

Obsessive-compulsive disorder was once considered a classic example of a psychological condition, but more recently it has been observed to be explicable primarily on neural levels. This illness is defined by the presence of obsessions or compulsions. Many people have experienced mild obsessions or compulsions, just as many have experienced one or two panic attacks or transient periods of unhappiness and depression. Obsessive-compulsive disorder becomes an illness when the obsessions or compulsions significantly interfere with a person's normal routine, occupational functioning, or social activities.

Obsessions are recurrent thoughts or impulses that are experienced as intrusive or senseless; obsessions include the fear of harming someone, sexual images, or thoughts perceived as sinful or blasphemous. Compulsions are repetitive behaviors, such as hand-washing or checking rituals, that are often performed in response to obsessions and may be perceived as neutralizing or preventing a dreaded situation. In full-blown obsessive-compulsive disorder, these symptoms produce marked distress and consume an unpleasantly large proportion of a person's day. Although obsessive-compulsive disorder traditionally was considered to be rare, epidemiologic studies have suggested that it may be rather common, with a lifetime prevalence as high as 1% or 2% of the population.

Current thinking about obsessive-compulsive disorder links it to a variety of other conditions that are also characterized by repetitive behaviors, such as Tourette's syndrome. These two syndromes are linked in twin and family studies, suggesting that they may have a genetic relationship. Neuroimaging studies have helped to indicate which brain regions may be involved in obsessive-compulsive disorder. Two groups of investigators, using PET, showed that patients with obsessive-compulsive disorder have increased metabolic rates in the caudate nuclei and in the prefrontal cortex. The increased prefrontal activity is consistent with the tendency of obsessionals to ruminate, plan excessively, and think in an overabstract way. The prefrontal cortex and the basal ganglia have important reciprocal interconnections, and the involvement of the basal ganglia may also explain the motor components of obsessive-compulsive disorder. The most widely accepted model for the cerebral biochemistry of obsessive-compulsive disorder implicates serotonin. Drugs that have a potent antiobsessional effect, such as clomipramine or fluoxetine, are relatively potent serotonin reuptake inhibitors.

REFERENCES

Andreasen NC, Black DW. Introductory textbook of psychiatry. Washington, DC: American Psychiatric Press, 1991.

Andreasen NC, Olsen S. Negative vs positive schizophrenia. Arch Gen Psychiatry 1982;39:789.

Barlow DH. Anxiety and its disorders—the nature of treatment of anxiety and panic. New York: Guilford, 1988.

Carlson A. The current status of the dopamine hypothesis of schizophrenia. Neuropsychopharmacology 1988;1:179.

Casanova MF, Kleinman JE. The neuropathology of schizophrenia: a critical assessment of research methodologies. Biol Psychiatry 1990;27:353.

Goodwin FK, Jamison KR. Manic-depressive illness. New York: Oxford University Press, 1990.

Gorman JM, Liebowitz MR, Fyer AJ, Stein J. A neuroanatomical hypothesis for panic disorder. Am J Psychiatry 1989;146:148.

Pakel ES (ed). Handbook of affective disorder. New York: Guilford, 1982.

Roy-Byrne P (ed). Anxiety: new findings for the clinician. Washington, DC: American Psychiatric Press, 1989.

Neuroscience in Medicine, edited by P. Michael Conn.
J. B. Lippincott Company, Philadelphia © 1995

Chapter 19

The Basal Ganglia and the Thalamus

IGOR A. ILINSKY
KRISTY KULTAS-ILINSKY

THE BASAL GANGLIA

Components and Historic Perspective

The definition of the term basal ganglia has evolved over the years from a purely descriptive anatomic name for a group of large subcortical cellular masses in the telencephalon to a functional concept of a system of intricately interrelated subcortical nuclei that play an important role in controlling movements. In addition to telencephalic structures such as the *striatum* and *globus pallidus* (Figs. 19-1 and 19-2), two other structures, the *substantia nigra* and *subthalamic nucleus*, are considered part of the basal ganglia system. Many of these structures are subdivided further into subnuclei (Table 19-1).

Lesions in these nuclei cause different movement disorders. The features of the disorders are site specific; the symptoms depend on the nucleus or nuclei primarily involved in the lesion. However, a unifying feature of the pathologic manifestations of the lesions in the basal ganglia is the absence of muscle paralysis. This sharply contrasts with lesions in the motor cortex or at any other level of the corticospinal (i.e., pyramidal) tract, which usually result in muscle paralysis. This and other differences between the two systems were emphasized in the term previously used for the basal ganglia, the extrapyramidal system. Although this term is still being used in the clinical setting, the concept of two functionally different systems involved in the motor control implied by this terminology does not agree with our understanding of movement mechanisms. According to the modern concept, there is one motor control system, in which basal ganglia, cerebellum, and many other brain centers perform specific functions. The function of the basal ganglia is not limited to movement control. The role of these nuclei in cognitive functions has become more obvious in experimental and clinical findings.

In this section, we describe the major intrinsic and extrinsic connections of the basal ganglia and specific functional and anatomic features of their components. This description of basal ganglia nuclei is accompanied by a comment on the pathologic conditions associated with some of them. For an extended discussion about the causes and clinical aspects of basal ganglia diseases, see the Clinical Correlation at the end of this chapter.

To more easily understand the circuitry of the basal ganglia, it is helpful to consider it as consisting of three parts: the *input structures* that receive the afferents from outside structures; *internal circuits*, with little connection to structures outside the basal ganglia; and *output structures* that deliver the resulting activity to other brain centers. The striatum, which consists of the *caudate, putamen*, and *nucleus* accumbens (see Table 19-1 for synonyms), represents the major input structure of the basal ganglia. Afferents from several brain regions converge on its cells. The *medial part of the globus pallidus,* GPm, and the *pars reticularis of substantia nigra,* SNr, and the region known as a *ventral pallidum* are the output structures of the basal ganglia circuitry. Other basal ganglia structures listed in Table 19-1, such as the pars compacta of substantia nigra (SNc), lateral globus pallidus, and subthalamic nucleus, are interconnected with one another or

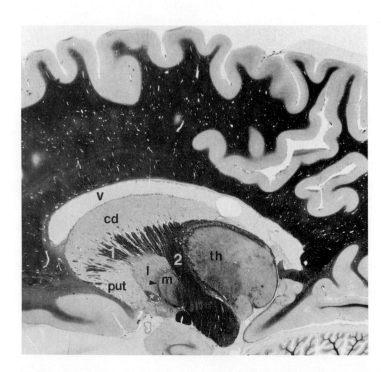

Figure 19-1
Photomicrograph of a part of a sagittal section through the human brain stained for myelin. Notice the position of the caudate (cd) underneath the subcortical white matter adjacent to the lateral ventricle (v). Two limbs of the internal capsule, the anterior (1) and posterior (2), separate the caudate from putamen (put) and globus pallidus from thalamus (th), respectively. Two parts of the globus pallidus, the lateral (l) and medial (m), are separated by a lamina (*arrowhead*). The white asterisk marks the anterior commissure. (Courtesy of the Yakovlev Collection, Armed Forces Institute of Pathology, Washington, DC.)

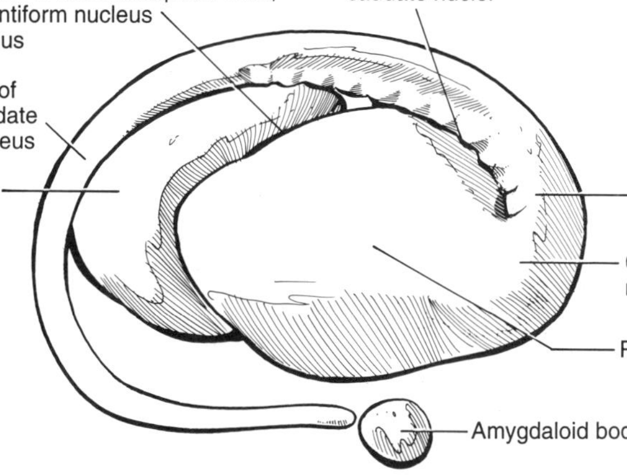

Figure 19-2
Drawing of an isolated striatum, illustrating its spatial relation to the thalamus and amygdala. The internal capsule is removed, and the cleft where the fibers of the internal capsule run is indicated. (Carpenter M. Human neuroanatomy, 7th ed, Baltimore: Williams & Wilkins, 1976.)

with the input and output nuclei previously described, and they can be considered as components of the internal circuits. We first discuss the organization of the two key structures, the SNc and the striatum.

Substantia Nigra

The SNc occupies the dorsal tier of the substantia nigra and is composed of densely packed (hence the name *pars compacta*, meaning a compact part), triangular or fusiform neurons with a moderate number of processes. These cells use dopamine as a neurotransmitter and are rich in a black pigment called melanin. In healthy persons, the concentration of the melanin is so high that it is visible in gross autopsy specimens, which gave the name to the whole structure; substantia nigra means black substance. The axons of SNc neurons form the nigrostriatal pathway, which supplies dopamine to the striatum. SNc neurons fire tonically at a low frequency. The resulting synaptic activity at the terminals of the nigro-

striatal pathway is thought to exert a tonic, modulatory (mostly inhibitory) effect on striatal neurons. This pathway does not seem to carry highly specific coded information; instead, the dopamine released from its terminals regulates the overall excitability of striatal neurons. Another function of the dopamine in the striatum is to regulate the release of other neurotransmitters from axon terminals by a mechanism that is briefly described later in this chapter.

For reasons that are not entirely understood, the SNc neurons are vulnerable to a variety of adverse factors that cause selective death of these cells. After as much as 80% of the neuronal population in the SNc has degenerated, symptoms of Parkinson's disease start to manifest. The most obvious of symptoms are akinesia or bradykinesia (i.e., absence or slowness of movement), muscle rigidity, and tremor at rest. Dopamine is released at the axon terminals of SNc cells in striatum and from the dendrites of the SNc cells locally. The released dopamine controls the excitability of the dopaminergic cells themselves through autoreceptors located on their soma and dendrites and influences the activity of cells in the SNr (Fig. 19-3). Dur-

Table 19-1		
BASAL GANGLIA NOMENCLATURE		
Term	**Synonym**	**Components**
Corpus striatum		Caudate, putamen, globus pallidus, nucleus accumbens
Neostriatum	Striatum	Caudate, putamen, nucleus accumbens
Paleostriatum	Pallidum or globus pallidus (including ventral pallidum)	Medial part, lateral part
Lentiform nucleus	Lenticular nucleus	Globus pallidus and putamen
Subthalamic nucleus	Body of Luys or corpus Luysi	
Substantia nigra	Substance of Sömmering	Pars compacta and pars reticularis
Ventral tegmental area	Area of Tsai	

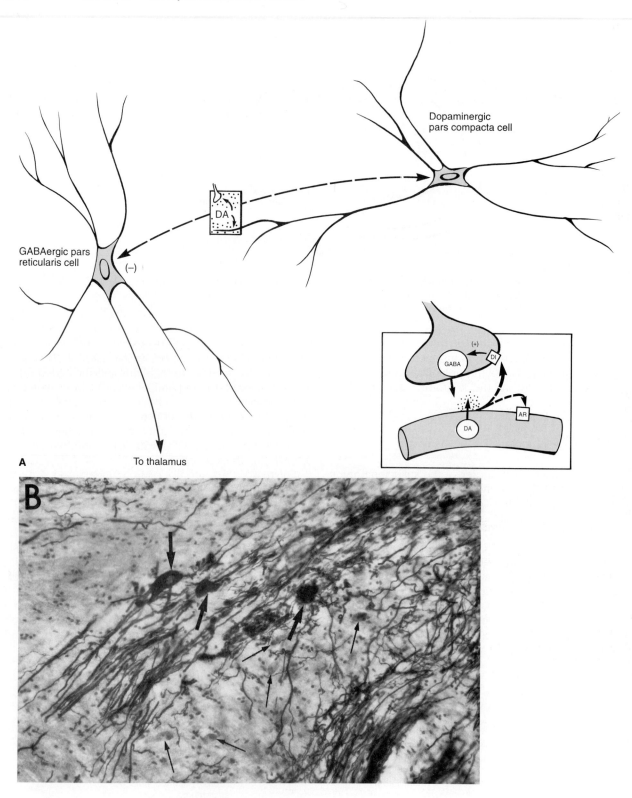

Dopaminergic
pars compacta cell

GABAergic pars
reticularis cell

(−)

DA

(+)

GABA

DI

DA

AR

A

To thalamus

B

ing the course of Parkinson's disease, as the dopaminergic SNc cells continue to die, dopamine deficiency develops at both release sites. This sequence of events accounts for many of the symptoms, especially bradykinesia. Treatment with L-dopa, (3,4-dihydroxyphenylalanine) an amino acid precursor of dopamine, relieves many of these symptoms.

Medially, the SNc is continuous with the *ventral tegmental area of Tsai*. Although technically a separate entity, the area of Tsai is similar to SNc in several respects: its cells are also dopaminergic, and it projects to the striatum. The main difference between the two nuclei is in the topography of the terminal fields of their axons in the striatum.

Striatum

☐ The striatum receives the bulk of the input to basal ganglia from extrinsic sources

The striatum is the largest and most complex basal ganglia structure. Functionally, it is the most important structure, because it receives almost all afferents to the basal ganglia circuitry from the outside sources. The access of information from other brain regions to the basal ganglia is mostly achieved through the striatum. Different levels of organization are recognized in the striatum. We start with the most obvious at the gross level, the major nuclear subdivisions and their afferent-efferent connections, and then proceed with the microscopic and neurochemical organization of the striatum.

☐ The striatum consists of three subdivisions

The three major subdivisions of the striatum are the caudate, putamen, and nucleus accumbens. Some of their topographic relations are illustrated in Figure 19-2. The caudate is further subdivided into three parts: the head, the body, and the tail. The head is the most anterior part that has the largest dorsoventral extent; its medial wall borders the anterior horn of the lateral ventricle. The body is the part that is located above the thalamus. The tail forms when the body acquires the position posterior to the thalamus before turning anteriorly at its most ventral aspect. The most medioventral part of the head of the caudate is the nucleus accumbens. The putamen is the largest region in the striatum. It is lateral and ventral to the caudate and is separated from it by the internal capsule (see Fig. 19-1).

☐ The cerebral cortex is the major source of afferents to the striatum

Cortical afferents

Virtually all areas of the cortex project to the striatum. The corticostriatal projections are topographically organized, but the details of this topography have not been entirely worked out. Nonetheless, the association cortices from different lobes of cerebral hemispheres appear to project predominantly to the caudate, and the motor and somatosensory cortices project predominantly to the putamen (Fig. 19-4). These cortical afferents overlap with the dopaminergic input from the SNc. The limbic cortices and hippocampus project predominantly to the nucleus accumbens and to ventral regions of the putamen, where they overlap with the dopaminergic input from the ventral tegmental area of Tsai. These ventral striatal regions are often considered as the functionally distinct "limbic" striatum in contrast to the more dorsal "motor" regions of striatum.

Other striatal afferents

Another extrinsic afferent system to the striatum originates in the thalamus. The thalamic nuclei that give rise to these projections belong to the intralaminar and midline groups. There is also a certain topographic organization of thalamostriatal projections: the centromedian nucleus is connected predominantly with the putamen, other intralaminar nuclei project predominantly to the

◄*Figure 19-3*
(**A**) The targets of action of dopamine in the substantia nigra. Dopamine (DA) released from the dendrites of pars compacta cells acts on the GABAergic cells of the pars reticularis of the substantia nigra (SNr) by inhibiting their activity. It also increases the release of GABA from striatonigral terminals and controls the discharge rate of nigrostriatal cells themselves by interaction with autoreceptors (AR). *DI,* type I dopamine receptor. (**B**) Photomicrograph of substantia nigra in the Rhesus monkey stained with antibody to tyrosine hydroxylase (i.e., enzyme producing dopamine) and lightly counterstained with thionin. The photomicrograph illustrates dopaminergic cell bodies (*thick arrows*) and bundles of dopaminergic dendrites extending into SNr. The nondopaminergic SNr cells are at the tips of the thin arrows. Small, round, darkly stained structures are nuclei of glial cells. In primates, the two parts of the substantia nigra interdigitate, and dopamine released from the dendrites of the pars compacta is available to the pars reticularis, where it interacts with presynaptic dopamine receptors on striatonigral GABAergic terminals (inset in **A**).

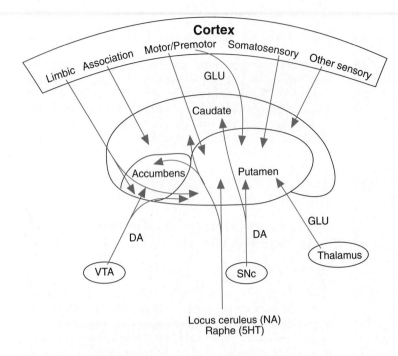

caudate, and the midline nuclei are connected with the limbic striatum. Corticostriatal and thalamostriatal afferent fibers use glutamate as the neurotransmitter at their terminals, and both are excitatory to striatal neurons.

Additional excitatory afferent input to the striatum is derived from the basal nucleus of amygdala. These afferents are distributed mainly to the nucleus accumbens and ventral regions of the putamen (i.e., limbic part of the striatum). Aspartate has been suggested as a putative neurotransmitter in this pathway.

The striatum receives diffusely distributed afferents from the noradrenergic cells of the locus ceruleus and serotoninergic cells of the dorsal raphe nucleus.

□ How are different inputs organized in relation to striatal neurons?

The organization of inputs to the striatum is such that they all converge on striatal cells, but the density of each input on a single cell is low. Certain combinations of cortical areas converge on specific pools of striatal neurons. For example, cortical areas controlling the muscles around one joint and those receiving somatosensory input from the related body region may converge in one striatal locus. The adjacent locus may receive similar inputs that represent a different part of the body. Such an arrangement has been compared with a combinational logic circuit in which the firing of a striatal cell depends on the right combination of inputs activated. Future studies may show that the same principle applies to the organization of other striatal afferents.

□ Striatum contains morphologically and neurochemically distinct cell types

Projection neurons and regulation of striatal output

There are different cell types in the striatum; the exact number of types depends on the species and perhaps on the techniques used in their identification. However, everyone agrees that the most numerous cell type is the *medium spiny neuron.* The name of these cells is derived from the fact that their dendrites are densely covered with spines (Fig. 19-5). The spines significantly increase the size of the receptive fields of these cells. They serve as the postsynaptic sites for cortical and thalamic afferents to striatum. Nigral and raphe inputs terminate on the spine necks or dendritic shafts of medium spiny neurons.

The medium spiny cells are the projection neurons of the striatum. The axons of these cells travel out of the striatum to contact neurons in the globus pallidus and substantia nigra. The neurotransmitter GABA is released by the axon terminals of medium spiny neurons. The organization of striatal output is illustrated Figure 19-6.

Although morphologically similar and using the same neurotransmitter, the medium spiny neurons are a neurochemically heterogeneous cell population considering their content of neuroactive peptides. Subsets of the medium spiny neurons with different peptide contents project to different targets outside the striatum. Cells that project to the SNr and the medial globus pallidus express substance P and dynorphin, and the cells that project to the lateral globus pallidus contain enkephalin (see Fig. 19-6). Differences in peptide expression provide a mechanism for differential regulation of activity of medium

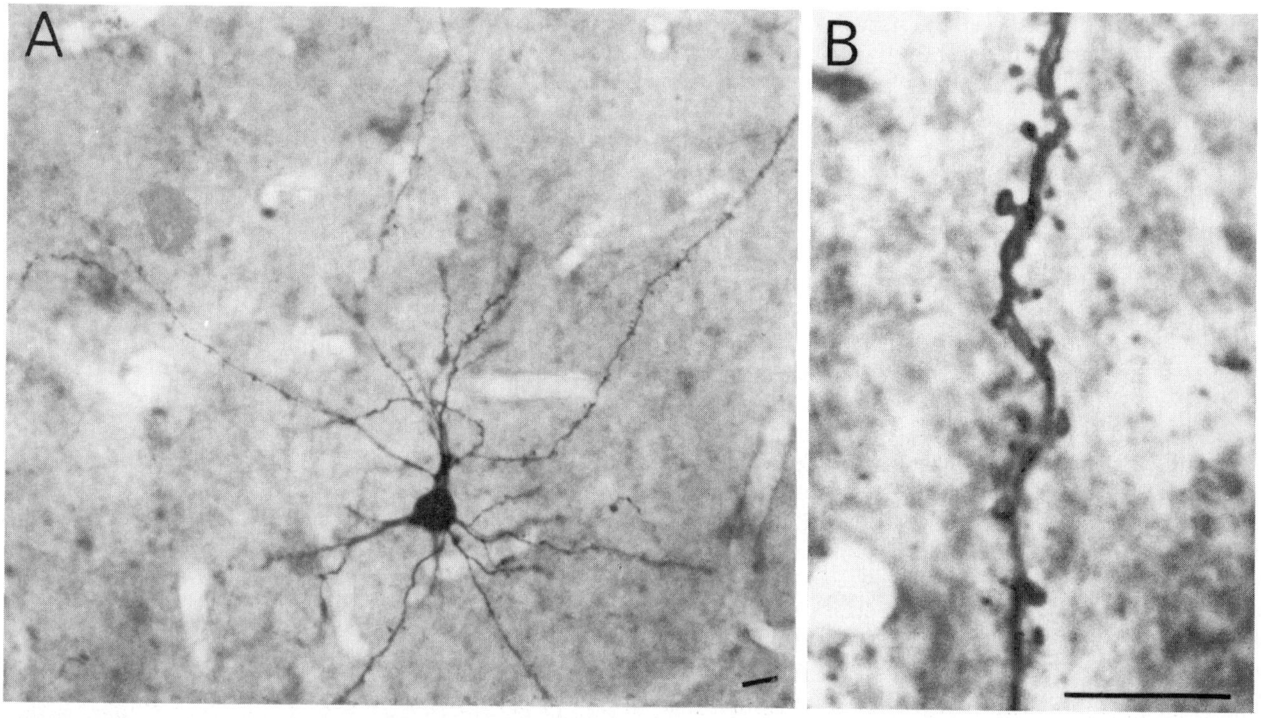

Figure 19-5
(**A**) Example of a medium spiny neuron in the rhesus monkey caudate nucleus. Bar = 10 μm.
(**B**) A spine-covered dendrite at higher magnification. Notice the enlarged spine heads and thin necks
connecting the spines to the dendritic trunk. Bar = 10 μm.

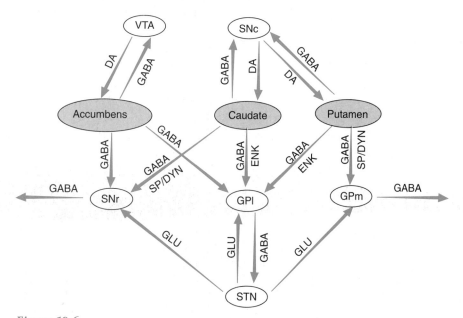

Figure 19-6
Topography of the striatal output and major internal loops of the basal ganglia. The direct
output from the caudate travels mostly to the pars reticularis of the substantia nigra (SNr),
and the direct output from the putamen targets the medial globus pallidus (GPm). The indirect output from the striatum is directed through the lateral globus pallidus (GPl) and
subthalamic nucleus (STN) to the same final targets. Neurotransmitters and peptides involved
in each of these connections are shown along the arrows. *ENK*, enkephalin; *DA*, dopamine;
DYN, dynorphin; *5HT*, serotonin; *GABA*, γ-aminobutyric acid; *GLU*, glutamate; *NE*, norepinephrine; *SNc*, pars compacta of the substantia nigra; *SP*, substance P; *VTA*, ventral tegmental
area.

spiny neuron subpopulations. The striatal cells projecting to different targets can function independently, because they are controlled by different genes and are influenced by different factors. Moreover, study findings suggest that D_1 and D_2 types of dopamine receptors may be differentially localized to the dynorphin- and substance P–containing cells and to enkephalin-containing cells, respectively.

There is a differential distribution of presynaptic dopamine receptors in the striatum and striatal targets. In the striatum, the presynaptic location of D_2 receptors (Fig. 19-7) has been indicated on the dopaminergic terminals of the nigrostriatal pathway, glutamatergic buttons of cortical origin, and terminals of axon collaterals of medium spiny neurons. D_1 receptors are located on the surface of the GABAergic terminals of striatonigral afferents in the SNr. Activation of presynaptic D_2 receptors results in reduction of neurotransmitter release, but the activation of presynaptic D_1 receptors leads to the increase of neurotransmitter release. Thus, the firing of striatal cells and the release of GABA at their terminals in the substantia nigra are controlled by the local levels of dopamine. Such differential distribution has inspired the use of a combination of D_1 and D_2 receptor agonists in treating parkinsonism.

Interneurons

Other cell populations of the striatum are interneurons or local circuit neurons, LCNs. They make up 4% to 23% of the total striatal neurons, depending on the species. A remarkable and rare cell (i.e., only 2% of the total cell population) is the *giant aspiny neuron*. These cells have elongated somata and few, but very long, spineless dendrites. Their axons bifurcate repeatedly to form dense plexuses within the confines of their dendritic arbors, with a radius of as much as 1 mm from the soma. These are the cholinergic cells of the striatum.

Another type of interneuron is represented by small to *medium-sized aspiny cells* that are GABAergic. Although these interneurons use the same neurotransmitter as striatal projection neurons, the two cell types differ with respect to the levels of expression of isoforms of the GABA-producing enzyme, glutamic acid decarboxylase (GAD). The medium spiny neurons express predominantly the GAD isoform with a molecular weight of 65,000, but the GABAergic interneurons express the GAD isoform with a molecular weight of 67,000. The two isozymes have different functional properties and are the products of different genes. This difference permits differential regulation of GABA synthesis in the two types of GABAergic cells of the striatum.

A third type, the *aspiny interneuron*, is also recognized in the striatum. Morphologically, these cells are similar to cholinergic interneurons, but they display no cholinergic markers. Instead, they contain a high concentration of somatostatin. These cells also express neuropeptide Y and contain high levels of nitric oxide synthase, the enzyme involved in the synthesis of nitric oxide, which is a free radical that appears to play a role in neurotransmission.

The striatal medium spiny neurons degenerate in the course of a hereditary movement disorder known as Huntington's disease or Huntington's chorea. This syn-

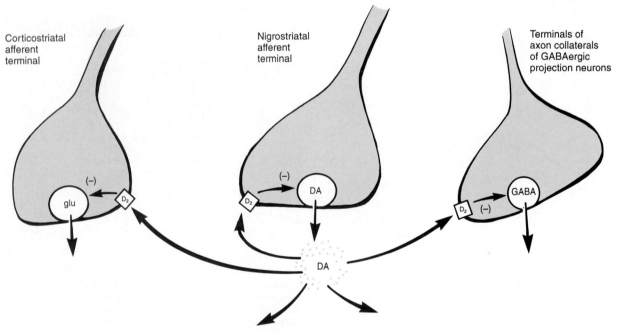

Figure 19-7
The drawing illustrates the effects of dopamine (DA) on the release of neurotransmitters in the striatum resulting from its interaction with type 2 dopamine receptors (D_2). A presynaptic location of D_2 receptors has been reported for glutamatergic, dopaminergic, and GABAergic terminals. Activation of these receptors supposedly leads to a reduction in neurotransmitter release. *glu*, glutamate.

drome is characterized by involuntary movements of distal parts of hands and legs, uncontrollable contractions of muscles of facial expression and tongue, and some cognitive deficits. At the onset of the disease, the subpopulation of striatal cells projecting to the lateral globus pallidus (i.e., enkephalin-containing cells) are thought to be affected first. At advanced stages, all medium spiny neuron populations are involved, and the striatal atrophy becomes noticeable. The disease is accompanied by increased GABA receptor sensitivity at striatal projection sites, but attempts to treat this disease with GABA agonists have been unsuccessful.

☐ Striatal compartments represent another level of organization

In routine histologic preparations, areas of different cell density can be recognized in the striatum of primates. The regions of higher cell density have been known as cell islands, but in the modern literature, they are called *striosomes or patches*. The lower cell density areas

around the striosomes are called *matrix*. These two distinct compartments also differ neurochemically. The distribution of most of the neurochemical markers in the striatum is patchy; there is a mosaic of areas of higher and lower concentrations of different substances (Fig. 19-8). The boundaries of the neurochemically distinct compartments coincide with those of the striosomes and matrix. For example, the degree of acetylcholinesterase staining is greater in matrix than in striosomes, although the latter display a higher density of mu opiate receptors and many other substances. In lower species such as carnivores and rodents, the existence of the two striatal compartments can be revealed only neurochemically, and differences in the cell densities are not observed.

The functional significance of striatal compartments has not been entirely explained, but some insights have been obtained from anatomic correlations between the two compartments and their associations with striatal afferents and targets of its projections. For example, the dendritic arbors of medium spiny neurons respect the compartmental boundaries; they are confined to the compartments that contain their cell bodies. However,

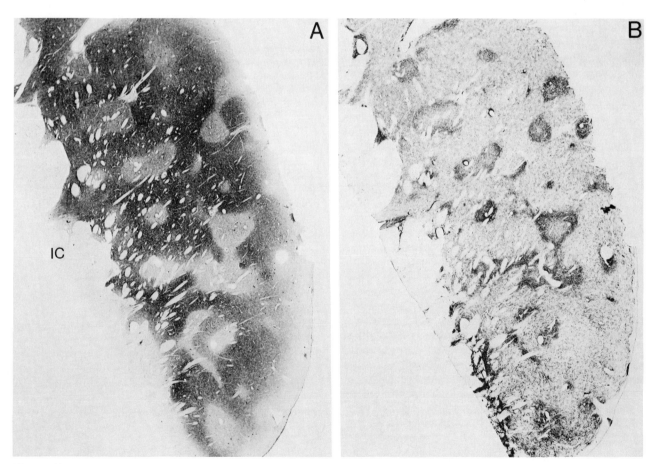

Figure 19-8
Photomicrographs of striatal compartments as seen in adjacent coronal sections through the head of the caudate nucleus in the human brain. (**A**) Staining for acetylcholinesterase (AChE). The AChE-poor areas correspond to striosomes; the AChE-rich areas represent the matrix. (**B**) Distribution of enkephalin-like immunoreactivity. The staining is more intense in striosomes than in the matrix. *IC,* internal capsule. (Modified from Graybiel A. Functions of the basal ganglia. London: Pitman, 1984.)

the dendrites of some types of interneurons cross the compartmental boundaries. The axonal plexuses formed by the cholinergic aspiny neurons appear to be more dense in the matrix, which may account for its higher acetylcholinesterase content. Some data indicate that certain sets of cortical afferents prefer specific striatal compartments. Similarly, the output targets of medium spiny neurons located in the various compartments may differ. In the rat, for example, striosomes project preferentially to the SNc, and the matrix cells project preferentially to the SNr.

□ Output pathways of the striatum have different targets and functional properties

The previous description illustrates the structural and neurochemical diversity of the striatum. Complex mechanisms have evolved for processing multimodality information arriving to this nucleus from different brain regions, especially from the vast cortical areas. All this activity converges on striatal neurons.

The striatum gives rise to several distinct inhibitory pathways that were mentioned in relation to the subpopulations of medium spiny neurons. Two pathways target the intrinsic basal ganglia nuclei. These are the *striatonigral pathway to the SNc* and the *striatopallidal pathway to the lateral globus pallidus* (see Fig. 19-6). Projecting to the output nuclei of the basal ganglia circuitry are the *striatonigral pathway to the SNr* and the *striatopallidal pathway to the medial globus pallidus* (GPm). Although the names of the pathways are similar, the distinction between the striatal efferents to intrinsic basal ganglia nuclei and to the output structures of the basal ganglia is important. These pathways originate from different cell populations, are regulated by different mechanisms, feed into different circuits, and are differentially involved in basal ganglia diseases. Additional distinctions exist between the outputs to SNr and GPm: the former originates predominantly in the caudate, and the latter is derived from the putamen (see Fig. 19-6).

The organization of the output from the third striatal component, called the limbic striatum, is not worked out as completely. It is thought that the output from the ventral striatum is directed to the ventral pallidum, which is an ill-defined area ventral to the lateral globus pallidus underneath the anterior commissure. The differential distribution of striatal efferents to different output nuclei provides the basis for the concept of parallel information-processing channels within the basal ganglia. This parallel arrangement of the striatal output pathways is maintained to a large extent at the level of the thalamus and cortex.

All striatal targets are substantially smaller than the striatum itself, and they contain significantly fewer cells. Starting with the corticostriatal afferents, there is a sequential convergence of information in the striatum and then even more on the targets of striatal output. GPm and SNr, the nuclei that give rise to the basal ganglia outputs,

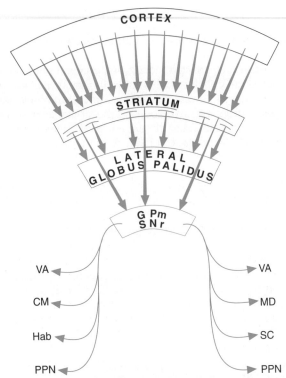

Figure 19-9
Convergence and divergence of basal ganglia connections. The cortical cell pools converging on striatal cells are large, because the striatum contains fewer neurons than the cortex. The pars reticularis of the substantia nigra (SNr) and globus pallidus (GP) contain fewer neurons than the striatum. The striatal cell pools converging on single lateral globus pallidus (GPl) cells are smaller than those on medial globus pallidus (GPm) and SNr cells, because there are more neurons in the GPl than in the GPm and SNr. The number of output cells in the GPm and SNr is several orders of magnitude smaller than the number of cells providing information to the striatum, but the output from this small number of cells diverges to a large number of targets through axon collaterals. The dashed line illustrates the indirect GPl output through the subthalamic nucleus. *CM*, centromedian nucleus; Hab, lateral habenular nucleus; *MD*, mediodorsal nucleus; *PPN*, pedunculopontine nucleus; *SC*, superior colliculus; *VA*, ventroanterior nucleus.

contain the smallest number of neurons. However, the efferents of these nuclei are highly divergent. These associations are illustrated in Figure 19-9.

Despite the fact that the medium spiny neurons have the highest density of excitatory inputs compared with cells in all other basal ganglia nuclei, these cells are silent most of the time. The medium spiny neurons fire only episodically, generating a train of action potentials at irregular interspike intervals within each train. In many instances, the firing episodes of striatal neurons are related to the onset of movements.

Pallidosubthalamic Loop

We next consider the organization of the two intrinsic basal ganglia nuclei: the *lateral globus pallidus* (GPl) and the *subthalamic nucleus*.

The GPl extends along the medial aspect of the putamen but is shorter than the putamen in its anteroposterior extent and narrower in its mediolateral extent. The dorsal part of the GPl borders the internal capsule. The cells in this nucleus are larger than striatal neurons and do not have spines. The striatal efferents establish numerous synapses on the dendrites of GPl cells, resulting in dense coverage of the dendritic membrane with GABAergic terminals. Another less-dense synaptic input to these cells is provided by small terminals of the axons from the subthalamic nucleus. These axon terminals are excitatory for GPl neurons. The GPl cells themselves are GABAergic and exert an inhibitory effect on their target, the subthalamic nucleus. This GPl to subthalamic nucleus to GPl connection represents a closed loop within the basal ganglia circuitry (see Fig. 19-6).

The subthalamic nucleus lies ventral to the zona incerta and dorsal to substantia nigra in the subthalamus from which it derives its name (see Figs. 19-12 and 19-13). It is an almond-shaped, compact nucleus that is densely packed with small, lightly staining cells. In addition to the inhibitory input from GPl, these cells receive direct excitatory input from the premotor cortex. This cortically derived excitation is thought to be instrumental in making it possible for subthalamic nucleus neurons to fire by overriding the inhibition derived from the globus pallidus. The evidence has been consistent that the subthalamic nucleus neurons release glutamate at their terminals and exert an excitatory influence on their targets. The subthalamic nucleus cells themselves represent a rare example of an excitatory neuron within the basal ganglia circuitry, in which the overwhelming majority of other cell types exert inhibitory influences on their targets (see Fig. 19-6). Besides the GPl, the two other targets of subthalamic nucleus projections are the output nuclei from the basal ganglia circuitry, SNr and GPm.

Lesions of the subthalamic nucleus are rare. However, if as much as two thirds of this nucleus is destroyed, a characteristic movement disorder known as hemiballismus occurs. This is manifested by large-amplitude, uncontrollable, ballistic movements in proximal joints.

Basal Ganglia Output Patterns

There are two major features of the organization of basal ganglia outputs. First, the output cells are almost continuously active, and by releasing the inhibitory neurotransmitter GABA, they appear to suppress the activity of the target cells for most of the time. There is speculation that the target cells are allowed to fire only during short episodes when the striatal neurons are active and suppress the activity of the GPm and SNr cells. This occurs at the onset of or during movements in the context of complex behavioral patterns. Second, the effects of the basal ganglia system on its targets may be synchronized to some extent. This may occur because the GPm and SNr projections are organized so that a single nigral or pallidal neuron may project to different targets through its axon collaterals. In this way, several distant targets of the basal ganglia output may receive the same information at about the same time. Unlike the high degree of convergence within the nuclei of the basal ganglia, the output from the basal ganglia circuitry is highly divergent (see Fig. 19-9), which is consistent with the important role of the basal ganglia in complex behaviors.

☐ Anatomic and functional organization of the output nuclei

The two best-studied output nuclei of the basal ganglia circuitry are the GPm and SNr. These two structures have many features in common and are considered together. Both nuclei are composed of sparsely distributed, relatively large multipolar neurons with a few, long, sparsely branching dendrites. The SNr and GPm neurons use GABA as a neurotransmitter and inhibit their targets. The main input to the GPm and SNr is derived from the striatum; another is derived from the subthalamic nucleus. The synaptic arrangement on the dendrites of GPm and SNr is therefore similar to that in the GPl, which receives similar afferents. The dendritic shafts of the GPm and SNr neurons are densely covered with GABAergic synapses, and the excitatory synapses formed by subthalamic nucleus axons are less numerous. Despite this high density of inhibitory input, the SNr and GPm cells discharge tonically at a high frequency. This firing pattern seems to be determined by inherent membrane properties of these cells, but it is not clear how the afferent activity regulates it. However, the firing patterns of these basal ganglia output nuclei change in relation to different movement parameters, such as direction of movement, velocity, or amplitude. An increase in the discharge rate is the most frequently observed change, although some cells also display a decrease in the firing rate.

Despite numerous similarities between the properties of the two basal ganglia output structures, there is also a significant difference: the functions of the pallidal and nigral cells are related to two different kinds of movements. GPm neurons are predominantly involved with limb movements. SNr neurons are predominantly associated with the orofacial, eye, head, and neck movements. More importantly, changes in the basal ganglia output activity are not associated with simple stereotyped movements but are associated with movements within the context of complex behaviors with cognitive, motivational, and sensory elements.

☐ Targets of the basal ganglia output

The basal ganglia output has ascending and descending components (Fig. 19-10). The ascending output is organized in a form of parallel channels. The major ascending output from the GPm is directed to a large region in the lateral part of the *thalamic ventroanterior nucleus*. The smaller component of the pallidothalamic pathway ter-

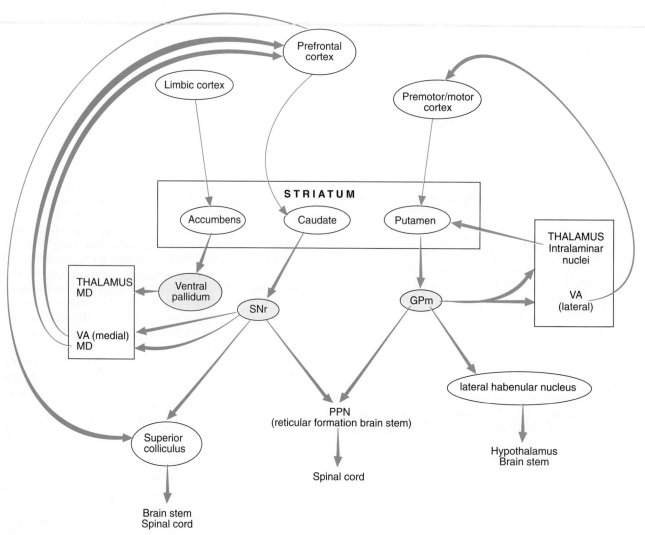

Figure 19-10
Output pathways of the basal ganglia. Two ascending parallel output channels originate in the caudate and putamen, respectively, and reach target areas of the cortex through different thalamic nuclei. The third, limbic-related output is channeled through the pars reticularis of the substantia nigra (SNr) and ventral pallidum eventually to reach large areas of prefrontal and limbic cortices. *GPm,* medial globus pallidus; *MD,* mediodorsal nucleus; *PPN,* pedunculopontine nucleus; *VA,* ventroanterior nucleus.

minates in the intralaminar nuclei, mainly in the *centro-median nucleus.*

The output from the GPm is organized in the form of distinct fiber bundles called the *ansa lenticularis* and *lenticular fasciculus.* The former leaves GPm as a compact fiber bundle and courses around the internal capsule to enter the subthalamus. The latter starts in a form of small fascicles that individually cross the internal capsule. Directly above the subthalamic nucleus, these fascicles combine to form the lenticular fasciculus. As the fibers of the ansa lenticularis continue anteriorly in the subthalamus, they combine with the lenticular fasciculus. Together they become the anterior component of the thalamic fasciculus. This component then enters the lateral part of the ventroanterior nucleus.

The pallidal afferent territory in the ventroanterior nucleus projects to the premotor cortex (Brodmann's area 6), including its part on the medial surface of the hemisphere known as the supplementary motor area. It also projects to the primary motor cortex. These cortical regions contribute to the corticospinal, corticobulbar, and corticopontocerebellar pathways. The basal ganglia output through the GPm, ventroanterior nucleus, and thalamocortical connections can influence the lower motor neurons and cerebellum by means of various corticifugal systems. Functionally, this basal ganglia output is primarily involved with limb movements.

The major ascending output from the SNr is also directed to the ventroanterior nucleus of the thalamus. However, it is restricted to its medial region, which is distinct from that receiving pallidal input. A substantial component of the nigrothalamic projections terminates in the *mediodorsal nucleus* of the thalamus. Unlike the pallidothalamic pathway, the nigrothalamic does not form distinct fiber bundles. Instead, the pathway reaches the ventroanterior nucleus in the form of diffuse fibers through the subthalamus.

The ventroanterior nucleus and mediodorsal nucleus regions receiving the nigral afferents are connected with the large areas of prefrontal cortex, including the frontal eye fields (Brodmann's area 8) and probably Broca's area. These cortical areas contribute to the corticobulbar and corticopontocerebellar projections. The frontal eye fields also give rise to a massive projection to the superior colliculus.

Based on its connections and the results of electrophysiologic studies, the basal ganglia output through SNr is known to be involved with orofacial, eye, head, and neck movements.

The descending output from the GPm and SNr is directed to the midbrain. One of the targets is the region in the midbrain known as the pedunculopontine nucleus. This nucleus is a part of the brain stem reticular formation and is connected with the cells that give rise to the reticulospinal tract. It has not been established how significant this basal ganglia output is in primates, but if proven important, it may represent a more direct route than the ascending pathways through the thalamus for the basal ganglia output to reach the lower motor neurons.

Another functionally significant descending output derived primarily from the SNr is directed to the deep layers of the superior colliculus, which also receives input from the frontal eye fields. In a way, the superior colliculus receives a dual input from SNr—one that is direct and one indirect through the nigral-thalamic-cortical-col-licular connection. The superior colliculus is an integrative brain center controlling vestibular-ocular reflexes, saccadic eye movements, and other oculomotor functions. The basal ganglia output through the SNr is involved in the control of voluntary and reflex eye movements.

The topographic segregation of the basal ganglia pathways related to movements of different parts of the body is channeled through motor-related regions of the striatum and thalamus. These regions include the dorsal caudate and putamen, ventroanterior nucleus, and a distinct part of the mediodorsal nucleus.

The topography of the basal ganglia outputs originating in the limbic striatum is more enigmatic. It is known that the output from the nucleus accumbens and ventral striatum is channeled through the region known as the ventral pallidum and probably through the SNr, reaching the mediodorsal nucleus of the thalamus. The mediodorsal nucleus regions receiving this input are connected with the orbital, medial prefrontal and limbic cortices. Output from the GPm also reaches the lateral habenular nucleus of the epithalamus, which has strong connections with the hypothalamus and some other limbic structures. The limbic component of the basal ganglia system has been implicated in emotional behavior and psychiatric disorders.

The areas of the cortex that receive multisynaptic input from the basal ganglia system give rise to topographically organized feedback projections to the regions of striatum, where the basal ganglial-thalamic-cortical pathways originated. These feedback projections represent only a small component of the overall corticostriatal afferents.

THE THALAMUS

The thalamus forms the most rostral part of the brain stem and is the major component of a larger entity known as the diencephalon (Fig. 19-11). In addition to the thalamus, the diencephalon contains the hypothalamus, subthalamus, and epithalamus. This section discusses the thalamus; other components of the diencephalon are covered in chapters on the limbic system and hypothalamus.

The term *thalamus* is a Greek word meaning inner chamber. The origin of the term is ascribed to Galen (2nd century A.D.), who traced the optic nerve fibers to an oval mass closely associated with the brain ventricles. After Galen's description, this part of the brain became known as an optic thalamus or "a chamber of vision." Although it was later realized that the thalamus was not a cavity but a conglomerate of gray matter, the term remained with only a slight modification; "optic" was dropped when it was discovered that all other sensory modalities, except olfaction, are processed in the thalamus.

Phylogenetic development of the thalamus is closely related to the expansion of the neocortex, which reaches its peak in primates. Evolutionary specialization involves different thalamic regions to a different degree; some areas undergo major qualitative changes, and others change mainly in size. Examples of both types are described in this chapter.

Coordination of Thalamic Anatomy and Physiology

☐ External and internal landmarks help gross delineations in the thalamus

For an understanding of thalamic topography, it is helpful to be familiar with some external (see Fig. 19-11) and internal landmarks (Fig. 19-12). As seen in Figure 19-12, the thalamus extends from the wall of the third ventricle to the *internal capsule*. In the middle, it is transversed in the anteroposterior direction by a thin sheet of fibers called the *internal medullary lamina*. Anteriorly, the

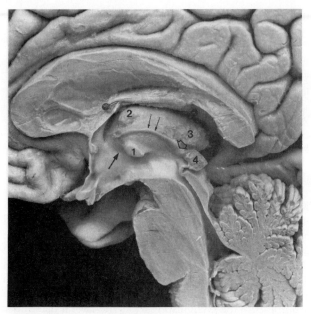

Figure 19-11
View of the human thalamus from the third ventricle, with the fornix pinned away to reveal its dorsal surface. The external landmarks of the thalamus are the hypothalamic sulcus (*thick arrow*); stria medullaris (*two thin arrows*), a fiber bundle connecting the habenula with hypothalamus and basal ganglia; massa intermedia or interthalamic adhesion (1), the anterior nucleus forming an eminence anterodorsally (2), habenula (*open arrowhead*) with the pineal (4) posterior to it, and the pulvinar (3) extending posterolaterally.

lamina splits in half to surround the anterior thalamic nucleus; posteriorly, it encompasses another large subdivision called the *centromedian nucleus*. Sensory and motor nuclei are lateral to this lamina, and the *mediodorsal nucleus* is medial to it.

Another sheet of fibers called the *external medullary lamina* surrounds the thalamus anteriorly and laterally. This lamina is separated from the internal capsule by a thin sheet of gray matter that forms the *reticular nucleus* of the thalamus. In sections through the thalamus this nucleus appears as a narrow band of gray matter squeezed between the internal capsule and external medullary lamina. This gray matter continues ventrally into the subthalamus to become the *zona incerta* (Fig. 19-13).

What does the thalamus do?

The role of the thalamus can be defined as the last station before the cortex in processing ascending subcortical information. It has also been called a gatekeeper to the cortex or to the conscious brain. The thalamus is composed of a large number of nuclei or subdivisions, each consisting of a group of cells with similar morphology and function. The function of each thalamic nucleus is largely defined by the source of its subcortical afferents and the major cortical target of its projections (Fig. 19-14).

The cortical connections of thalamic nuclei are mostly reciprocal in that each nucleus projects to and receives afferents from the same cortical regions. Corticothalamic afferents arise from layer VI cells. Thalamocortical axons terminate mostly in middle cortical layers III or IV, depending on the area. In some instances they form additional terminal plexuses in layer I.

One thalamic nucleus represents an exception to the general topographic pattern. The *reticular nucleus* of the thalamus does not project to the cortex. Instead, its axons project to all other thalamic nuclei. However, the reticular nucleus, positioned between the thalamus and the internal capsule (see Figs. 19-12 and 19-13), is on the path of the corticothalamic and thalamocortical fibers. When traversing this nucleus, both types of fibers give off collaterals that form synapses on its neurons. In this way, the reticular nucleus occupies a strategic position that allows it to control the flow of information between the thalamus and the cortex.

☐ Internal thalamic circuits process information destined for the cortex

Understanding the details of the overall thalamic organization mandates a review of some principles of organization of neural circuits within thalamic nuclei and the association of these circuits with the thalamic reticular nucleus. A simplified diagram of a basic thalamic circuit is illustrated in Figure 19-15. This diagram is applicable to all thalamic nuclei except the ventroanterior region. The predominant cell type in the thalamus is a thalamocortical *projection neuron* (Fig. 19-16). These are usually medium- to large-sized multipolar neurons that have dendrites radiating in all directions.

Another cell type is represented by very small cells that have little perikaryal cytoplasm and few dendrites (see Fig. 19-16). These are inhibitory interneurons, called *local circuit neurons* (LCN), which use γ-aminobutyric acid (GABA) as a neurotransmitter. In addition to conventional synapses formed by their axon terminals on as yet unidentified locations on thalamocortical projection neurons, these cells establish unconventional synapses (Fig. 19-17). In the latter, the vesicle-containing dendrites of GABAergic LCN are presynaptic, but the dendrites or soma of thalamocortical projection neurons are postsynaptic. These synapses are called *dendrodendritic* or *dendrosomatic*.

The unconventional synapses are commonly part of even more complex synaptic arrangements, of which the best understood are serial synapses, triads, and glomeruli. Schematic drawings of some examples of these synaptic relations are shown in Figure 19-18. Terminals of the major subcortical or cortical afferents are the main components of these complex synapses in each thalamic nucleus. The essence of these arrangements is that, in addition to termination on a projection neuron, an afferent terminal establishes a synapse on a dendrite of a LCN. This LCN contacts projection neuron directly or through

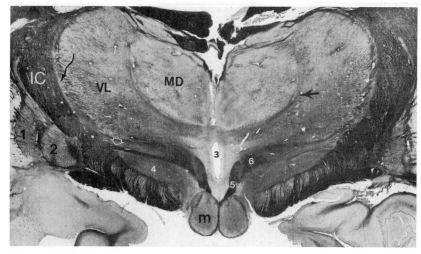

Figure 19-12

Coronal section through the human thalamus stained for myelin. Notice the location of two medullary laminae: external (*open arrowhead*) and internal (arrow). *IC,* internal capsule; *M,* mammillary bodies; *MD,* mediodorsal thalamic nucleus; *block white arrow,* lenticular fasciculus; *wavy arrow,* reticular nucleus of the thalamus, *1,* lateral globus pallidus; *2,* medial globus pallidus; *3,* third ventricle; *4,* subthalamic nucleus; *5,* mammillothalamic tract; *6,* thalamic fasciculus.

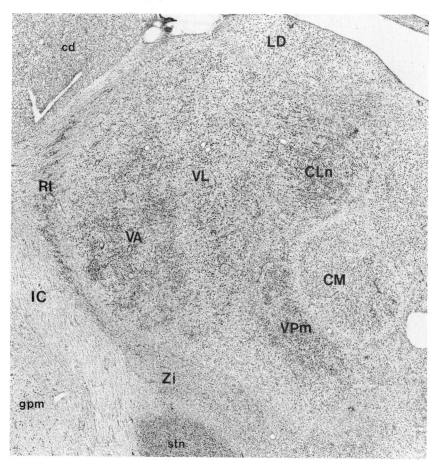

Figure 19-13

Photomicrograph of the cytoarchitectural distinctions of a Nissl-stained sagittal section through the monkey thalamus. Anterior is to the left, marked by internal capsule (IC). The reticular nucleus (Rt) occupies the most anterior position and is ventrally continuous with the zona incerta (Zi) in the subthalamus. Thalamic nuclei: *CLn,* centrolateral; *CM,* centromedian, which is incapsulated by split fibers of the internal medullary lamina; *LD,* laterodorsal; *VA,* ventroanterior; *VL,* ventrolateral; *VPm,* ventroposteromedial. Parts of three basal ganglia nuclei are also seen in this photo: *cd,* caudate; *gpm,* medial globus pallidus; *stn,* subthalamic nucleus.

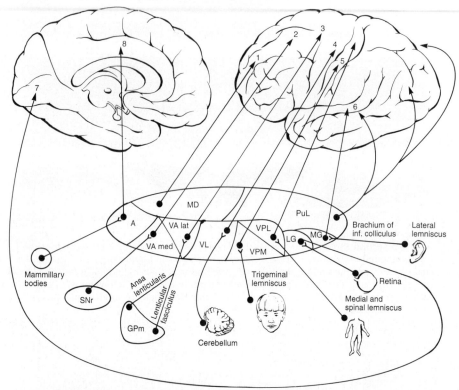

Figure 19-14
Diagram of the input-output relations of thalamic nuclei. Thalamic nuclei are outlined from left to right in their anteroposterior sequence. The major cortical projection targets of each nucleus are numbered: *1*, prefrontal; *2*, frontal eye fields; *3*, premotor; *4*, primary motor; *5*, primary somatosensory; *6*, primary auditory; *7*, primary visual; *8*, cingulate. *A*, anterior nucleus; *GPM*, medial globus pallidus; *LG*, lateral geniculate; *MD*, mediodorsal nucleus; *MG*, medial geniculate; *PuL*, pulvinar nucleus; *SNr*, reticularis of substantia nigra; *VA med*, ventroanterior nucleus medial part; *VA lat*, ventroanterior nucleus lateral part; *VL*, ventrolateral nucleus; *VPL*, ventroposterolateral nucleus; *VPM*, ventroposteromedial nucleus.

a cascade of interposed LCN dendrites. In this way, the participating afferent input has a dual influence on thalamocortical neurons. One influence is direct, which is usually excitatory; another is indirect and inhibitory and is mediated by GABAergic synapses formed by LCN dendrites. The inhibitory interneurons modulate the afferent inputs to projection neurons in the thalamus.

It has been suggested that the unconventional dendrodendritic synapses established by LCN may change the conducting properties of projection neuron dendrites locally, without much impact on their trigger zone. What makes this system more complex is the dual relation of the LCN with projection neurons. In addition to the synapses formed by presynaptic dendrites, there are synapses formed by LCN axons. These synapses may control the overall firing pattern of projection neurons.

The neurons of the reticular thalamic nucleus, like LCN, use GABA as a neurotransmitter and have an inhibitory influence on other thalamic nuclei. The reticular nucleus cells have complex firing patterns that depend on the subject's level of vigilance. In an awake state, reticular neurons are depolarized and generate tonic dis-

charges. Under these conditions, thalamocortical neurons function in a relay-type mode; they actively process the arriving subcortical information and transmit it to the cortex. During deep sleep, the thalamic reticular nucleus neurons fire in rhythmic bursts (Fig. 19-19) that increase the hyperpolarization of thalamocortical neurons. However, by a phenomenon known as a *postinhibitory rebound*, which is caused by changes in the membrane calcium conductance, the thalamic projection neurons also begin to generate rhythmic bursts. This synchronized thalamic bursting pattern is reflected in the spindle waves that are recorded in the electroencephalogram during sleep and drowsiness. In this condition, the transfer of information from the thalamus to the cortex is completely blocked.

Thalamic mechanisms and the role of individual neuronal elements in them are not completely understood. Numerous features of thalamic organization have not received proper explanation. For example, the number and type of complex synaptic arrangements and the density of interneurons vary among thalamic nuclei. The local inhibitory circuits formed by LCN appear to be more de-

CORTEX

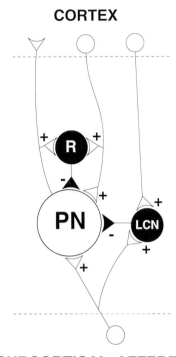

SUBCORTICAL AFFERENT

Figure 19-15
Simplified circuit diagram of a typical thalamic nucleus. Open triangular profiles mark excitatory (+) inputs; black triangular profiles mark inhibitory (−) GABAergic inputs. *LCN,* local circuit neuron; *PN,* projection neuron; *R,* reticular thalamic nucleus neuron.

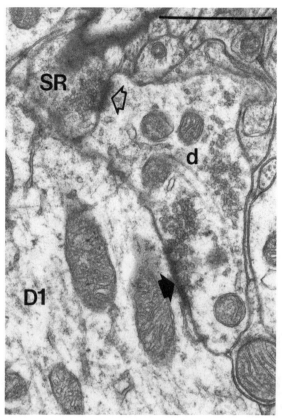

Figure 19-17
Electronmicrograph of a typical dendrodendritic synapse (*black arrow*) formed by vesicle-containing dendrite (d) of a local circuit neuron on a primary dendrite (D1) of a projection neuron in the monkey ventrolateral nucleus. This is a part of a serial arrangement because d receives a synaptic contact (*open arrow*) from a small button with round vesicles (SR). Such buttons are usually terminals of excitatory corticothalamic afferents, and dendrodendritic contacts are usually inhibitory. (Kultas-Ilinsky K, Ilinsky IA. J Comp Neurol 1991;314:319.)

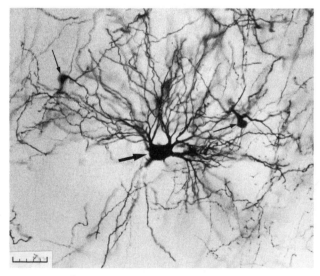

Figure 19-16
Examples of two nerve cell types in the monkey ventroanterior nucleus. Notice the elaborate dendritic arbor of the projection neuron (*thick arrow*) that points to its soma and the small bodies (*thin arrows*) and few dendrites of the local circuit neurons. (Courtesy of Dr. Clement Fox Collection at Wayne State University, Detroit, MI).

veloped in some nuclei compared with others. There are significant differences between species, with a trend toward an increase of complexity of local neuronal circuits in primates.

Subdivisions of the Thalamus

Thalamic nuclei can be grouped roughly into four general categories: modality-specific nuclei; multimodal or association nuclei; nonspecific or diffusely projecting nuclei; and the reticular nucleus, which for the reasons given previously can be considered as a separate entity. This grouping is based on similarities of anatomic and functional features within each group. There are several nomenclature systems for thalamic nuclei, and some of them are rather complex. A simplified classification of the nuclei in the primate thalamus is used in this chapter as we consider each of the nuclear groups in more detail.

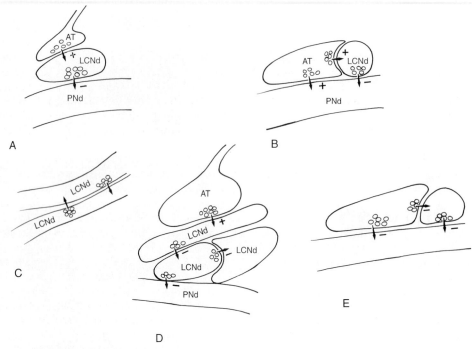

Figure 19-18
Examples of complex synaptic arrangements in thalamic nuclei. (**A**) Simple and (**D**) complex serial synapses. (**C**) reciprocal synapse. (**B, E**) Examples of triads, **E** is an all inhibitory triad. Glomeruli (not shown) contain several serial and triad-type complexes. *AT,* afferent terminal; *LCNd,* local circuit neuron dendrite; *PNd,* projection neuron dendrite. Arrows indicate the direction of synaptic transmission.

☐ Modality-specific nuclei

The nuclei combined in the modality-specific group share several common features. In a broad sense, they are all modality specific. The core of the group is composed of the nuclei that process somatosensory, visual, and auditory information. Additional members of the group are the subdivisions that process subcortical information derived from the basal ganglia, cerebellum, and limbic system. Another characteristic feature of the nuclei in this group is that the major output from each of them is directed toward one well-defined cortical region, such as a primary sensory cortex of corresponding modality or a primary motor cortex. Within each of these thalamic nuclei are cell populations that have additional cortical projections.

Modality-specific nuclei are represented by six thalamic subdivisions: *anterior, ventroanterior, ventrolateral, ventroposterior, medial geniculate,* and *lateral geniculate nuclei.* With the exception of anterior nuclei, all others are lateral to the internal medullary lamina forming a ventral tier. Medial and lateral geniculate bodies are also considered components of a distinct entity known as the metathalamus.

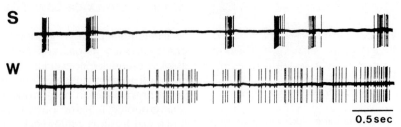

0.5 sec

Figure 19-19
Single-cell recording of activity of a reticular nucleus neuron during sleep (S) and wakefulness (W). Notice the two distinct firing patterns: short, high-frequency bursts during sleep and tonic, lower-frequency discharge in the awake state. (Steriade M, J Neurosci 1986;6:68).

Anterior nuclei

By virtue of their connections with the cingulate cortex, a component of the limbic lobe, the anterior nuclei are a part of the limbic system. This also determines their function in emotional behavior, learning, and memory processes.

In the human brain, the largest nucleus of the anterior group is the principal anterior nucleus. A small, distinct cytoarchitectonic entity in the group is the anterodorsal nucleus. In other species, instead of the principal nucleus, two distinct subdivisions are found: anteromedial and anteroventral nuclei. The laterodorsal nucleus is in the most dorsal aspect of the thalamus (see Fig. 19-13) in all species. Although it is not a part of the anterior nuclear group topographically, it shares similar anatomic, connectional, and functional features with the group; it is also a part of the "limbic thalamus." All nuclei of the limbic thalamus have reciprocal connections with the cingulate cortex.

Thalamocortical projections from the anterior nuclei travel in the *anterior limb* of the internal capsule and *cingulum*. Their terminal zones in the cingulate cortex are topographically organized. Fibers originating in the medial aspects of the principal anterior nucleus project predominantly to the anterior regions of the cingulate cortex. Those from the lateral parts of the principal nucleus, anterodorsal and laterodorsal nuclei, form denser terminal fields in the posterior cingulate cortex.

The projections from the limbic thalamus, although providing the bulk of the subcortical input to the cingulate cortex, are not the only thalamocortical fibers that reach this cortex. Some other thalamic nuclei contribute to thalamocingulate projections but to a lesser extent. Among them are the *ventroanterior, ventrolateral, intralaminar,* and *midline nuclei* and parts of the *pulvinar.* In this way, the cingulate cortex receives information from almost all sensory and motor systems of the brain.

Afferents to the limbic nuclei of the thalamus originate in several structures. The major subcortical input is derived from the mammillary bodies of the hypothalamus (see Fig. 19-14). These fibers form a compact bundle, the *mammillothalamic tract* (see Fig. 19-12), which ascends through the thalamus and terminates in the principal anterior and anterodorsal nuclei. Axons of cells located in the medial part of the mammillary body, the medial mammillary nucleus, terminate in the ipsilateral principal anterior nucleus. Those originating from the lateral mammillary nuclei terminate in the anterodorsal nucleus bilaterally. It is not known whether fibers from mammillary bodies terminate in the laterodorsal nucleus.

Another substantial afferent input to the limbic thalamus is derived from the hippocampal formation. This projection originates from the cells of the subiculum and presubiculum. The hippocampal afferents terminating in the principal nucleus and anterodorsal nucleus travel in the fornix; those terminating in the laterodorsal nucleus travel in the fornix and the retrolenticular limb of the internal capsule.

Ventroanterior nucleus

The ventroanterior nucleus is the most anterior nucleus in the ventral tier. It is lateral to the principal anterior nucleus in the same coronal plane. Anteriorly, it conforms to the anterior thalamic curvature bordering the external medullary lamina directly behind the reticular thalamic nucleus (see Fig. 19-13). In primates, including humans, the ventroanterior nucleus is divided into several subdivisions. In addition to cytoarchitectonic distinctions, these subdivisions differ with respect to their cortical and subcortical connections.

The ventroanterior nucleus represents the major basal ganglia afferent territory in the thalamus, because the two ascending basal ganglia outputs, the nigrothalamic and pallidothalamic pathways, terminate in it (see Fig. 19-14). It participates in movement control. This nucleus has undergone significant evolutionary specialization. In primates, it has acquired a significant size, and the terminal zones of nigral and pallidal projections, which overlap in lower species, have become entirely segregated. The nigrothalamic pathway, which originates in the substantia nigra pars reticularis (SNr), terminates in the medial part of the ventroanterior nucleus that contains large neurons and is known as the *magnocellular nucleus.* The pallidothalamic pathway, formed by the combined fibers of *ansa lenticularis* and *lenticular fasciculus* (see Fig. 19-12) that enter the thalamus as a part of the *thalamic fasciculus,* terminates more laterally in the ventroanterior nucleus. This part of the nucleus occupies a rather large thalamic territory composed of medium-sized neurons arranged in clusters (see Fig. 19-13). The basal ganglia–thalamic connection is ipsilateral.

The two regions of the ventroanterior nucleus have different cortical connections. The medial magnocellular ventroanterior nucleus is reciprocally connected with the prefrontal cortex in Brodmann's area 8 or the frontal eye fields. The pallidal afferent territory of the ventroanterior nucleus is connected with the premotor cortex or Brodmann's area 6, including the supplementary motor area. It also has some connections with the primary motor and cingulate cortices. As discussed in the preceding section of this chapter, basal ganglia output channeled through the medial globus pallidus is involved preferentially with limb movement. Output through the SNr is related to eye, orofacial, head, and neck movements. There is some degree of topographic organization in the basal ganglia output. This organization is maintained at the thalamic level and preserved at the level of the neocortex.

The regions of the cortex receiving multisynaptic input from the basal ganglia give rise to a number of corticifugal pathways. Among the most important pathways functionally are the projections from the frontal eye fields to the superior colliculus and oculomotor nuclei of the brain stem and the corticopontocerebellar connection.

The ventroanterior nucleus has unique anatomic and functional features. Unlike other thalamic nuclei, the major subcortical afferents to the ventroanterior nucleus, the nigrothalamic and pallidothalamic pathways, are inhibitory. They use GABA as a neurotransmitter and terminate in large numbers at proximal parts of projection neurons. These extrinsic GABAergic inputs to the ventroanterior nucleus add to the synapses formed by intrinsic GABAergic systems (i.e., LCN and reticular nucleus axons), which are found in all thalamic nuclei. The only well-established excitatory input to the ventroanterior nucleus is provided by corticothalamic fibers.

The ventrolateral nucleus

The cytoarchitecturally rather homogeneous ventrolateral nucleus lies just posterior to the ventroanterior nucleus. The ventrolateral nucleus neurons are larger than ventroanterior nucleus cells and more sparsely distributed (see Fig. 19-13). The major subcortical afferent system terminating in the ventrolateral nucleus is the cerebellothalamic pathway (see Fig. 19-14), and its major fiber component is derived from the contralateral dentate nucleus. The cerebellothalamic fibers enter the ventrolateral nucleus from its ventral aspect as a component of the *thalamic fasciculus* (see Fig. 19-12). The fact that this nucleus is the major target of the ascending cerebellar output determines its function, which is related to the control of coordinated skilled movements and motor learning. The ventrolateral nucleus and the ventroanterior nucleus are together called the motor thalamus. In contrast to the fine, point-to-point somatotopic maps in somatosensory thalamic nuclei, the multiple body representations in the motor thalamus are rather crude as they relate to large groups of muscles (Fig. 19-20).

The major target of the thalamocortical projections from the ventrolateral nucleus is the primary motor cortex (Brodmann's area 4) in the precentral gyrus (see Fig. 19-14). The ventrolateral nucleus also is connected with nonprimary somatosensory cortices (Brodmann's areas 5 and 7) in the parietal lobe. This posterior parietal cortex plays a role in decoding sensory stimuli that provides spatial information for targeted movements. The population of cells that gives rise to the parietal cortex projection is distinct from that projecting to area 4. The former occupies the dorsal part of the ventrolateral nucleus, and the latter occupies its ventral aspect. On this basis, the ventrolateral nucleus can be divided into two subnuclei that both receive afferents from the cerebellum but are interconnected with different cortical regions. Moreover, a limited neuronal population of the ventrolateral nucleus projects to the limbic and premotor cortices. Through the ventrolateral nucleus and its connections, cerebellar output that carries information related to skilled movements has direct access to the final motor pathway (i.e., corticobulbar, corticospinal) and to limbic and somatosensory association cortices.

Of special significance among other subcortical inputs to the ventrolateral nucleus are the sparse direct and indirect (i.e., through some brain stem nuclei) spinothalamic inputs that have been demonstrated in anatomic and physiologic studies. These inputs interact with the cerebellum-derived activity on ventrolateral nucleus neurons.

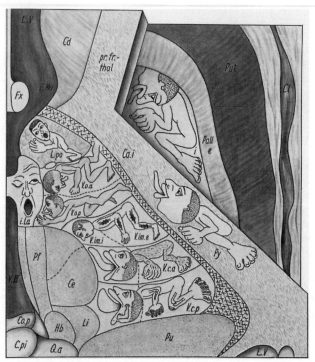

Figure 19-20

Multiple somatotopic representations in the human thalamus in a horizontal section as deduced from electrophysiologic recordings from patients during stereotactic surgeries. The homunculi illustrate different sensory representations of the ventral posterior thalamus (V.c.p., V.c.a.) and different muscle groups in the motor thalamus (L.p.o., V.o.a., V.o.p., V.i.m). Additional homunculi are shown in the internal capsule (Ca.i) and medial globus pallidus (Pall. i). The nuclear nomenclature is that of Hassler. (Hassler R, Mundinger F, Riechert T. Stereotaxis in parkinson syndrome. New York: Springer-Verlag, 1979).

According to available data, there seems to be no convergence at the thalamic level of motor-related information arriving from the two basal ganglia outputs and cerebellum. Each of these systems has its own channel for delivering information to discrete cortical areas. The final integration of information arriving from the basal ganglia and cerebellum is thought to take place at the level of neocortex, with corticocortical connections playing an important role in this process. However, the accurate identification of the areas involved and the actual mechanism of such interactions are unknown. The motor thalamus appears to have reached the highest degree of complexity of its neuronal circuits in primates, which may be directly related to the increased complexity of motor behavior in primate species.

The ventroposterior nucleus

This large, thalamic area posterior to the ventrolateral nucleus is involved in processing somatosensory information. The characteristic feature of the ventroposterior nucleus, common to all thalamic nuclei processing sensory modalities, is its precise somatotopic organization. Based on topographic and cytoarchitectonic distinctions, two major subnuclei are recognized within the ventroposterior nucleus: the *ventroposteromedial* and *ventroposterolateral subnuclei*. A branch of the internal medullary lamina forms a distinct border between the two nuclei in primates. The ventroposteromedial subnucleus is composed of small, densely packed cells (see Fig. 19-13), but the ventroposterolateral subnucleus cells are larger and less densely packed. In both subnuclei, the neurons are arranged in clusters. The clusters are formed by cells involved with the same modality and receptive field.

The ventroposterolateral subnucleus receives its major inputs from the *medial lemniscus* and *spinal lemniscus* (see Fig. 19-14). The extremities and the trunk are represented in the ventroposterolateral nucleus; all sensory modalities from these body regions are processed in this nucleus. In contrast, the ventroposteromedial subnucleus receives its major input from the *trigeminal lemniscus*, representing ventral and dorsal trigeminothalamic pathways that carry all somatic sensory modalities from the face. The gustatory sense is also thought to be represented in the primate ventroposteromedial subnucleus. Both ventroposterior subnuclei contain precise somatotopic maps of corresponding body parts (see Fig. 19-20).

After entering the ventroposterior nucleus, lemniscal axons arborize profusely and terminate in a rod-like space that encompasses a cluster of thalamic neurons. Because the terminals within each rod carry the information about a specific somatosensory modality and place on the body, they convey these same properties to the cells in clusters. The thalamic cells then convey this information to neurons within a corresponding cortical column of the primary somatosensory cortex. These connections provide the basis for preserving the somatotopic organization within the somatosensory system.

The major thalamocortical projection from the ventroposterior nucleus terminates in the primary somatosensory cortex, which includes Brodmann's areas 1, 2, 3a and 3b in the postcentral gyrus. These same cortices generate the bulk of corticothalamic afferents to the ventroposterior nucleus. Additional cortical connections of the ventroposterior nucleus are made with the nonprimary somatosensory cortices in the parietal lobe.

The ventroposterior nucleus contains numerous LCN and a great variety and number of complex synaptic arrangements. A difference exists in the mode of termination of medial lemniscal and spinothalamic fibers. The former establish synapses on projection neurons and LCN and represent major components in complex synapses. The spinothalamic input targets predominantly projection neurons and avoids complex synaptic arrangements.

Lateral geniculate nucleus

The lateral geniculate nucleus is the thalamic center for processing visual information. Unlike other thalamic nuclei, the lateral geniculate nucleus is a laminated structure. The cells of this nucleus are organized into six distinct layers. The most ventral two layers are composed of large cells and called magnocellular layers. The dorsal layers (3 to 6) are composed of small cells and called parvicellular layers.

The major afferent inputs to the lateral geniculate nucleus are derived from the retina (see Fig. 19-14). Retinogeniculate fibers travel in the optic tract and terminate in the lateral geniculate nucleus layers in an orderly fashion preserving a point-to-point map of the visual space. Layers 2, 3, and 5 receive retinofugal axons from the ipsilateral eye, and the fibers from the contralateral eye terminate in layers 1, 4, and 6. Different functional classes of the retinal cells are thought to terminate on different lateral geniculate nucleus layers. In this way, corresponding functional classes of thalamic neurons are established in the lateral geniculate nucleus layers (see Chap. 23).

Additional subcortical afferents to the lateral geniculate nucleus arise from the superior colliculus, the pretectal nucleus of the optic tract, and some other brain stem centers.

The axons of geniculocortical neurons terminate in the primary visual cortex, Brodmann's area 17. Additional connections are established with areas 18 and 19. Corticothalamic projections to the lateral geniculate nucleus originate from all three of these cortical areas.

Synaptic connections in the lateral geniculate nucleus are understood better than those in other thalamic nuclei. The two functional classes of geniculate neurons known as X and Y cells differ in the distribution on them of synaptic inputs from different sources. The manner in which the retinal axons terminate on X and Y cells also varies. The X-type retinal axons are involved in complex synaptic associations with projection neurons and LCN that are known as glomeruli. The Y axons terminate directly on corresponding lateral geniculate nucleus projection neurons and seem to have less interaction with interneurons. The corticothalamic afferents terminate on the distal parts of dendritic arbors of projection neurons. When activated, they facilitate transmission at more proximally located retinothalamic synapses. Less is known about the interaction between retinal and other subcortical inputs to the lateral geniculate nucleus.

Medial geniculate nucleus

The medial geniculate nucleus is posterior and medial to the lateral geniculate nucleus. Although several cytoarchitectonic subdivisions are thought to exist in this nucleus, they are not easily differentiated. The medial geniculate nucleus can be roughly divided into ventral and dorsal parts. The ventral subdivision has a laminar organization similar to the lateral geniculate nucleus but less distinct. The dorsal subdivision is not laminated and consists mainly of medium-sized neurons with groups of large cells concentrated medially.

The medial geniculate nucleus is the thalamic center for processing auditory information. The major subcortical afferent input to the medial geniculate nucleus is derived from the auditory centers of the brain stem. The bulk of these fibers arise from the inferior colliculus and reach the medial geniculate nucleus through the *brachium of the inferior colliculus* (see Fig. 19-14). Some *lateral lemniscus* fibers also reach the medial geniculate nucleus directly. There is a point-to-point representation, called a tonotopic map, of the cochlea in the medial geniculate nucleus. However, the details of this organization are not worked out as well as those for the retinotopic map in lateral geniculate nucleus or the somatotopic map in ventroposterior nucleus.

Additional subcortical afferents to the medial geniculate nucleus are provided by collaterals of fibers from the *brachium of the superior colliculus* derived from the optic nerve and by collaterals of spinothalamic and medial lemniscal fibers.

The medial geniculate nucleus has reciprocal cortical connections with the primary auditory cortex (Brodmann's areas 41 and 42) in Heschl's gyri. Some subdivisions of the medial geniculate nucleus have additional connections with association auditory cortices in the superior temporal gyrus and insula.

☐ Multimodal association nuclei

The multimodal association nuclei occupy a large, poorly delineated area of the dorsal posterior thalamus. Several features of nuclei in this group differentiate them from the modality-specific nuclei. First, they have widespread cortical connections with association areas in the frontal, parietal, and temporal lobes. Second, unlike specific nuclei, thalamic association nuclei do not have one dominant subcortical afferent input; afferents to these nuclei originate from many different brain regions and have equal weight. None of these afferents project to the entire territory of the nucleus but instead tend to terminate in patches. Functions of association nuclei cannot be defined in precise, modality-specific terms, but they are all related to higher brain functions such as language and learning. Association nuclei reach their largest size and diversity in primates, especially humans.

Two major association nuclei considered in this chapter are the mediodorsal nucleus and lateroposterior nucleus–pulvinar complex.

Mediodorsal nucleus

The mediodorsal nucleus is represented by a large, spherical cell mass just medial to the internal medullary lamina (see Fig. 19-12). It can be seen at the same coronal levels as the ventrolateral nucleus. The mediodorsal nucleus develops in parallel with the prefrontal cortex, with which it has extensive connections. The function of the mediodorsal nucleus is related to motivational behavior in general and is specifically pertinent to complex behavior involving eye, head, and neck movements.

Different regions of the prefrontal cortex project to different mediodorsal nucleus subdivisions. These subdivisions are not always easily recognizable because of subtle cytoarchitectonic distinctions. Olfactory cortex and orbital gyri in the frontal lobe have strong connections with one of these subdivisions. Subcortical afferents to the mediodorsal nucleus originate from basolateral amygdala,

SNr, ventral pallidum, mesencephalic raphe nuclei, and locus ceruleus. The topographic and functional relations of the terminal zones of different subcortical afferents in the mediodorsal nucleus are poorly understood.

The conspicuous feature of the mediodorsal nucleus is the large number of LCN, which are larger in size than those in other thalamic nuclei and display rather elaborate dendritic arbors. Another remarkable feature of the synaptic organization of the mediodorsal nucleus is a large number of glomeruli. The functional significance of the abundance of these complex synaptic arrangements is not clear.

Lateral posterior nucleus and pulvinar complex

The lateral posterior nucleus–pulvinar complex is located in the dorsoposterior aspect of the thalamus and is the largest thalamic subdivision (see Fig. 19-11). Although the two nuclei display some subtle cytoarchitectonic distinctions, they are commonly considered together because of connectional and functional similarities. This cellular mass has extensive connections with association cortices in parietal, occipital, and temporal lobes. The medial pulvinar is connected with the cingulate cortex. The subcortical inputs to the complex are derived from a variety of sources, including the retina, superior colliculus, and pretectum. The afferents from the superior colliculus originate in the superficial and deep layers. The superficial layers receive direct input from the retina, and the deep layers represent the center of convergence for motor-related inputs from the basal ganglia, cerebellum, and brain stem centers processing various sensory modalities. Based on these connections, lateral posterior nucleus–pulvinar complex is considered to be an integrative center for sensory and motor information related primarily to vision. It is also thought to play a role in the learning and memory associated with sensory experiences.

☐ Nonspecific or diffusely projecting nuclei

The nonspecific or diffusely projecting nuclei are grouped together on the basis of their widespread cortical projections and because their main projection target is to the striatum, rather than the cortex. These nuclei derive their names from their location in the thalamus. Those within the internal medullary lamina are called intralaminar nuclei, and those along the medial wall of the thalamus bordering the third ventricle are called midline nuclei.

Intralaminar nuclei

The nuclei of the internal medullary lamina are divided into anterior and posterior groups. Two main nuclei are recognized in the anterior group and are called the *paracentral* and *centrolateral nuclei*. In the posterior group, the two main nuclei are the *parafascicular* and *centro-*

median nuclei (see Fig. 19-13), which are located at the same anteroposterior level of the thalamus as the somatosensory nuclei. The distinguishing feature of the intralaminar nuclei, in addition to sparse and diffuse cortical projections, is that the bulk of their output is directed to the neostriatum. Moreover, the thalamocortical projections from the intralaminar nuclei are represented in part by axon collaterals of the thalamostriatal projection neurons.

The two nuclei in the posterior intralaminar group, although close to one another, differ cytoarchitecturally. The centromedian cells are lightly stained, medium sized, and oval. The parafascicular neurons are smaller, darker, more elongated, and densely packed. The centromedian nucleus achieves its largest size in humans, and the parafascicular nucleus is less distinct in primates than in rodents. The posterior intralaminar nuclei receive their major input from the medial globus pallidus. These fibers are actually collaterals of the pallidothalamic fibers terminating in the ventroanterior nucleus. The anterior intralaminar nuclei receive their major input from the cerebellum and spinal cord. The anterior and posterior nuclear groups also receive afferents from the deep layers of the superior colliculus and pretectum. The centromedian nucleus also receives input from the periaqueductal gray and dorsal raphe nuclei.

Although the intralaminar nuclei receive major inputs from the same subcortical structures as the modality-specific nuclei of the thalamus, the intralaminar nuclei also receive multiple inputs from other subcortical structures, most of which are part of the brain stem reticular formation. In addition to the direct brain stem inputs, the intralaminar nuclei receive indirect input from the brain stem reticular formation through the reticular thalamic nucleus.

The functional role of the intralaminar nuclei cannot be defined simply. By virtue of their strong connections and strong association with the reticular formation, they are regarded as a part of an "ascending activating system" that regulates the mechanisms of cortical arousal. As a part of the basal ganglia circuitry, they participate in the motor control mechanisms. The intralaminar nuclei may participate in the regulation of tolerance to pain, a function apparently subserved by their direct spinothalamic input.

Midline nuclei

The midline region is filled with small to medium-sized, darkly stained cells that form several cytoarchitectonic entities. The most prominent cell groups are *paraventricular, rhomboid or central,* and *reuniens nuclei.* In primates, these nuclei are relatively inconspicuous compared with their counterparts in nonprimate species.

The midline nuclei project to the prefrontal and limbic cortices, the amygdala, and the ventral striatum. Subcortical afferents to the midline nuclei originate in the hypothalamus, amygdala, and several midbrain and medullary structures, including the nucleus of solitary tract, periaqueductal gray, parabrachial and raphe nuclei, and

locus ceruleus. Spinothalamic fibers also reach the midline region. Because all these fiber systems contain a variety of peptides and biogenic amines, the midline region is the richest area in the thalamus in these substances and associated receptors. The function of the midline nuclei is not clearly defined, but because they are intimately related to the ventral striatum and amygdala, they are probably involved with emotional aspects of motor behavior and with autonomic functions and memory.

Transmitters and Related Substances in the Thalamus

Unlike many other brain regions, the thalamus has relatively few neuroactive peptides and biogenic amines. The peptides and amines are contained within the incoming afferent fibers that originate mostly from the spinal and lower brain stem regions. There are no aminergic cells in the thalamus and few peptidergic neurons.

The major neurotransmitter-related substances in the thalamus are excitatory amino acids, mainly glutamate and glutamate receptors and GABA and GABA-benzodiazepine receptors. Glutamate is a suspected neurotransmitter in corticothalamic afferents and thalamocortical projection neurons. Many glutamate receptor subtypes have been found in the thalamus, including NMDA (*N*-methyl-D-aspartate), AMPA (amino-3-hydroxy-5-methyl-4-isoxazolepropionate), and metabotropic receptors. The kainic acid–sensitive glutamate receptor subtypes are found only in the reticular thalamic nucleus. GABA is the neurotransmitter in several synapses formed by thalamic LCN and reticular nucleus neurons. It is also the neurotransmitter at the terminals of extrinsic afferents from the basal ganglia. In view of the important role of GABA in different thalamic circuits, it is not surprising that the thalamus also contains high concentrations of GABA and benzodiazepine receptors. Benzodiazepine receptors are especially dense in the limbic and midline nuclei.

Clinical Expressions of Thalamic Lesions

Although uncommon, thalamic lesions represent a serious clinical problem when they occur. Neurologic conditions associated with these lesions can be devastating and are often incurable.

Intracerebral tumors located in the diencephalon manifest with a mixture of general and local symptoms. Local manifestations of diencephalic lesions are usually complex. They can include sensory, motor, and endocrine components because of the functional diversity of the thalamus itself and because of involvement of surrounding structures, such as the hypothalamus. The one indicative sign of a thalamic lesion is severe impairment or loss of somatic sensations from the contralateral side. For unknown reasons, the most affected somatic sensations are deep sensations and two-point discrimination.

Stroke frequently occurs in the thalamogeniculate artery, which is a branch of the posterior cerebral artery supplying the lateroposterior half of the thalamus. Occlusion of this artery is often accompanied by devastating pain in addition to sensory loss, although in some cases, severe facial pain is experienced without any sensory loss. The latter condition is known as *thalamic pain syndrome* or *Dejerine-Roussy syndrome*. The devastating pain that accompanies this condition is intractable to analgesics, and its mechanism is not understood. Vascular strokes at more anterior thalamic locations are usually accompanied by temporary hemiparesis and aphasia if a dominant hemisphere is involved.

A few words should be said about surgical interventions in the thalamus for treatment of some neurologic conditions. In the early 1950s, with the introduction of stereotactic techniques for clinical applications, surgical lesions often were created in the thalamus in patients with movement or sensory disorders. The most successful have been lesions of the ventrolateral nucleus for treatment of symptoms of parkinsonism, especially tremor, and lesions in somatosensory and centromedian nuclei for pain relief. The former technique lost its significance after the introduction of L-DOPA therapy, although it is still used in medically resistant forms of parkinsonian tremor, ballism, and choreic movements. Thalamic lesions for treatment of incurable pain, including thalamic pain syndrome and phantom pains, i.e., pain associated with amputated extremities, can also be effective.

Lesioning in some instances has been replaced by intracerebral stimulation of appropriate regions with chronically implanted electrodes. The major limitations of these approaches are that the results are not predictable and that there is no rational explanation for the positive effects or ineffectiveness of these manipulations. Interventions at the level of the thalamus have the major disadvantage of being empiric, which reflects our lack of fundamental knowledge about thalamic mechanisms.

SELECTED READINGS

Delong M, Crutcher M, Georgopoulus A. Relations between movement and single cell discharge in the substantia nigra of the behaving monkey. J Neurosci 1983;3:1586.

Georgopoulus A, Delong M, Crutcher M. Relations between parameters of step-tracking movements and single cell discharge in the globus pallidus and subthalamic nucleus of the behaving monkey. J Neurosci 1983;3:1599.

Gerfen C. The neostriatal mosaic: multiple levels of compartmental organization. TINS 1992;15:133.

Hassler R, Mundinger F, Riechert T. Stereotaxis in parkinson syndrome. New York: Springer-Verlag, 1979.

Jones E. The thalamus. New York: Plenum Press, 1985.

Shepherd GS (ed). The synaptic organization of the brain, 3d ed. New York: Oxford University Press, 1990.

Steriade M, Jones EG, Ilinas R. Thalamic oscillations and signaling. New York: John Wiley & Sons, 1990.

Clinical Correlation
DISORDERS OF THE BASAL GANGLIA

ROBERT L. RODNITZKY

The classic clinical syndromes resulting from abnormalities of the basal ganglia are disorders of movement. These may take the form of excessive involuntary movements (i.e., hyperkinesia) or decreased movement (i.e., hypokinesia). Hypokinesias such as bradykinesia (i.e., slow movement) or akinesia (i.e., absence or difficult initiation of movement) are largely seen in Parkinson's disease and a few conditions that mimic this disorder. There are several forms of hyperkinesia and many different disease states that cause these symptoms. Among the most common forms of hyperkinesia are chorea, dystonia, tremor, and tics. The complex anatomic connections and physiologic associations of the basal ganglia often make it difficult to discern the anatomic locus of disease responsible for these abnormal patterns of movement. Similarly, with the exception of Parkinson's disease, the neurotransmitter aberrations responsible for many movement disorders are unknown and are often defined only by the mode of action of the drugs used to treat them.

Chorea

Chorea is an involuntary movement with a variety of causes. It is characterized by arrhythmic, rapid, involuntary movement that flows from one part of the body to another in nonstereotypic fashion. When chorea is severe, of high amplitude, and prominent in proximal parts of an extremity, it is referred to as *ballism*. There are many causes of chorea, including rheumatic fever, metabolic imbalance (e.g., hyperthyroidism), and the use of certain drugs (e.g., amphetamines, levodopa). The form associated with rheumatic fever is known as *Sydenham's chorea*. One of the most common non–drug-related causes of chorea is *Huntington's disease*, an autosomal dominant condition which invariably progresses to severe disability and death.

It had long been known that examination of the brains of patients with Huntington's disease at autopsy revealed remarkable gross atrophy of the head of caudate nucleus. Modern neuroimaging techniques such as computerized tomography or magnetic resonance imaging can demonstrate this finding during the patient's life. At the cellular level, the chorea of Huntington's disease has been related to a loss of striatal GABA-enkephalin neurons that are the origin of the indirect inhibitory pathway to the external segment of the globus pallidus. It is theorized that when these pallidal cells are not adequately inhibited, there is a resultant increase in inhibition of the subthalamic nucleus, reducing the excitatory drive on neurons in the internal segment of the globus pallidus. The decrease in output of the internal segment of globus pallidus reduces inhibition of the thalamus, leading to an increase in cortical activation and chorea.

This explanation of the origin of chorea in Huntington's disease is consistent with the fact that the most severe and acute form of chorea, ballism, is usually associated with a direct lesion of the subthalamic nucleus. The most common of these direct lesions is an infarction of the subthalamic nucleus resulting in contralateral ballism or *hemiballism*.

In the chorea of Huntington's disease and the ballism of subthalamic nucleus infarction, dopamine-blocking (e.g., haloperidol) or dopamine-depleting (e.g., reserpine) drugs are effective in reducing the abnormal movements, and dopaminergic drugs worsen them. Although dopaminergic cells are not the focus of pathology in either of these conditions, dopamine blockade is effective in reversing the chorea associated with them. These drugs, by reducing the inhibitory effect of dopamine on striatal GABA-enkephalin neurons, enhance the inhibiting influence of these cells, resulting in less inhibition of the subthalamic nucleus. The often remarkable effect of antidopamine drugs on chorea is a prime example of the fact that the researcher cannot always infer from a salutary pharmacologic response which cell group or neurotransmitter is primarily affected by the disease process in basal ganglia disorders. One obvious exception to this observation is *Parkinson's disease*, for which replenishment of a deficient neurotransmitter, dopamine, does reflect the primary neurochemical abnormality.

Dystonia

Dystonia is an abnormal hyperkinetic movement distinct from chorea. Dystonia is an involuntary movement that is twisting, somewhat sustained, and often repetitive. With

367

time, the body part affected by this abnormal movement may develop a fixed, abnormal posture.

Dystonia can be described according to its distribution within the body as focal, segmental, or generalized. Focal dystonia implies involvement of a single part of the body such as the hand, and segmental dystonia involves two or more adjacent areas of the body such as the neck and arm. Generalized dystonia is defined as involvement of the lower extremities plus any other body part. Examples of focal dystonia are writer's cramp, an involuntary contraction of hand or finger muscles while writing, and torticollis, an involuntary turning or tilting of the head. Torticollis plus facial or eyelid dystonia constitute segmental dystonia.

Idiopathic torsion dystonia is the most common condition resulting in generalized dystonia. It is an autosomal dominant disorder that typically begins in childhood or adolescence. The gene for this condition has been localized to the long arm of chromosome 9. In this disorder, like most spontaneously occurring dystonic disorders, there are no apparent structural abnormalities of the basal ganglia and no abnormalities of neurotransmitter function that have been consistently demonstrated. However, neurotransmitter manipulation sometimes results in marked reduction of dystonia. The most effective results are obtained with anticholinergic agents. A smaller subset of patients respond remarkably to the dopamine precursor L-dopa.

Although there is no obvious structural pathology of the basal ganglia in the idiopathic dystonias, discrete structural lesions of these structures caused by tumor, trauma, or stroke can result in an identical movement disorder. In these cases, the brain abnormality leading to dystonia is most likely to be located in the putamen, but it can also be found in areas related to the basal ganglia through efferent or afferent pathways, such as the thalamus or cerebral cortex. Typically, dystonia due to unilateral structural abnormalities of the basal ganglia or their connections is confined to the extremities on the contralateral side of the body and is referred to as *hemidystonia*.

Pending the development of better pharmacologic therapies for dystonia based on a fuller understanding of its etiopathogenesis, the use of botulinum toxin has emerged as one of the most effective treatments for this condition. Partial weakening of the muscles responsible for the dystonic movement can be accomplished by directly injecting them with small amounts of botulinum toxin. This substance works by inhibiting release of acetylcholine from nerve terminals, effectively producing chemical denervation of muscle with resultant weakness and atrophy.

Tics

Tics are sudden, brief, stereotyped movements. The movement is involuntary and complex, such as blinking of the eye or shrugging of the shoulder. Movements such as these are known as *motor tics*. When the vocal apparatus is involved, a tic may consist of a nonspecific vocalization, such as a grunt or a distinct verbalization including recognizable words or phrases. *Tourette's syndrome* is a condition in which motor and vocal tics occur. Tourette's syndrome, thought to be a hereditary disorder, begins in childhood and is more common in boys. It is often associated with certain behavioral abnormalities such as attention deficit disorder or a variety of compulsions or ritualistic behaviors. Like idiopathic torsion dystonia, its anatomic and neurochemical basis are not fully understood. Tics, like chorea, are effectively suppressed by dopamine-blocking drugs, but no definite abnormalities of dopaminergic cells or pathways have been demonstrated in the brains of affected persons.

Basal Ganglia Disorders Causing Multiple Forms of Abnormal Movement

Some basal ganglia disorders can result in several different forms of abnormal movement. Several distinct diseases of the basal ganglia can result in several or all of the previously described abnormal movements. *Hallervorden-Spatz syndrome*, a condition associated with neuronal loss and increased iron storage primarily in the SNr and internal globus pallidus can result in chorea, bradykinesia, or dystonia in any combination. This same spectrum of disordered movements plus tremor can occur in *Wilson's disease*, a condition characterized by neuronal loss and excessive copper storage in the brain, especially the putamen. In this condition, symptoms can be reversed by decoppering the central nervous system with medications.

Tardive dyskinesia is a syndrome in which abnormal movements occur after prolonged administration of dopamine-blocking drugs such as chlorpromazine or metoclopramide. In this condition, the abnormal movement is usually chorea but occasionally is dystonia. The abnormal movements of tardive dyskinesia can affect any part of the body, but they are most often focused on the mouth and tongue. A common, but unproven explanation for the pathogenesis of this condition is that dopamine receptors develop compensatory supersensitivity as a result of being subject to chronic pharmacologic blockade.

The class of dopamine receptor that is blocked seems to be important in the pathogenesis of this syndrome, because dopamine-blocking drugs that preferentially block the D_3 rather than the D_2 dopamine receptor are much less prone to cause tardive dyskinesia.

Neuroscience in Medicine, edited by P. Michael Conn.
J. B. Lippincott Company, Philadelphia © 1995

Chapter 20

Spinal Mechanisms for Control of Muscle Length and Force

WILLIAM Z. RYMER

Mechanical Properties of Muscle Tissue

All motor commands from the central nervous system are expressed through changes in the magnitude of neural excitation of skeletal muscle. The changes in muscle excitation generate force and motion, the magnitude of which depends on the properties of the muscle tissue and the mechanical loads experienced by the muscle. This chapter begins with a description of the neural regulation of movement and the relevant mechanical properties of muscle, with emphasis on the dual roles of skeletal muscle as a force generator and a mechanical impedance with elastic (i.e., spring-like) and viscous (i.e., frictional) properties.

☐ Muscle acts as a force generator

Skeletal muscle acts as a machine that transforms chemical energy, stored in the form of high-energy phosphate bonds in the ATP molecule, into mechanical energy that is elaborated as force or motion. The cellular mechanisms that transduce energy into the mechanical actions of muscle ultimately regulate movement. Many aspects of neural activity are related to the special stimulus requirements of muscle.

Types of fibers

Striated muscle is composed of fascicles of *myofibers*, which are the cells of muscle tissue (Fig. 20-1A). Each multinucleated myofiber contains many bundles of slender filaments, called *myofibrils*.

With light or electron microscopy, the regular structure of striated muscle is apparent. Dark bands alternate with light bands, which are bisected by a narrow, dark line, called the *Z line*. The segment from one Z line to another is called a *sarcomere*, which is the unit of contraction because it is the force-generating element of the myofibril is the sarcomere (Fig. 20-1B). Each myofibril is constructed of a repeating array of sarcomeres. It is this recurring structure that produces the striated pattern of the fibril and of the whole fiber. The sarcomere contains thick and thin filaments, which bear the chemical moieties responsible for generating force.

The *thick filament* is made up primarily of *myosin*, which is a long-tailed molecule with a globular head and a flexible neck. The myosin molecule is made up of two identical heavy chains, which form the body of the molecule, and two pairs of light chains, which are seques-

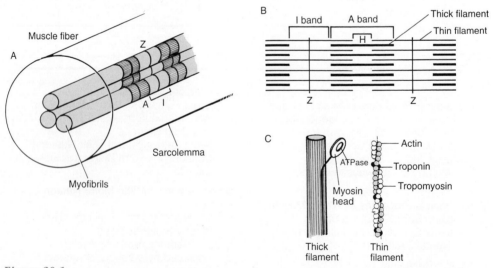

Figure 20-1

(**A**) The constituent elements of the muscle include the muscle fiber, the membrane or sarcolemma, and the myofibrils, which appear striated because of the A and I bands, which have different optical properties. The myofibrils are the bundles of filaments that comprise the sarcomeres, which are the segments between adjacent Z lines. (**B**) Components of the sarcomere. Each sarcomere contains thick and thin filaments. Thick filaments reside in the center of the sarcomere, and the thin filaments traverse the area from one sarcomere to another through the boundary, called the Z line. The thick filaments carry the myosin crossbridges, which interact with the thin filaments. The thin-filament actin carries a receptor that binds the myosin crossbridge. (**C**) Molecular constituents of the sarcomere. The thick filament is composed primarily of myosin, which has a long tail and a protruding head forming the crossbridge. The head incorporates an ATPase binding location, which is needed for establishing the high-energy mechanical state of the crossbridge. The thin filament is made up of actin monomers, which are organized in a helical chain, forming F actin. Regulatory proteins lie in the grooves of the F actin strands. These are tropomyosin, an enlongated molecule that lies in each of the two grooves of the actin helix, and troponin, a peptide that occurs at periodic locations along the thin filament.

tered in the head of the myosin molecule. Different types of muscle tissues (e.g., skeletal, cardiac, smooth), muscle at different stages of development (e.g., embryonic, neonatal), and muscles from different animal species show differences in the myosin heavy chain isoforms. The myosin molecules are laid out in the thick filament so that their tails are packed in parallel, oriented along the long axis of the filament, and the heads protrude at regular intervals around the circumference of the thick filament (Fig. 20-1C).

The myosin head is a key locus at which the transduction from chemical to mechanical energy takes place (see Fig. 20-1C). Transduction requires binding of the myosin head to receptor sites on the thin filament, after which conformational changes occur in the crossbridge that are instrumental in producing muscle contraction. The myosin head is also the locus for an ATPase that facilitates the process of muscle contraction.

The *thin filament* is composed of several different proteins. The major protein, *actin*, is a globular monomer that is packed in the form of a twisted chain, called F actin (see Fig. 20-1C). The thin filament also contains regulatory proteins, including *tropomyosin*, an elongated, rod-shaped molecule lying along the actin molecular helix, and *troponin*, which is a globular molecule found at periodic intervals along the thin filament. The troponin consists of various peptide subunits, including troponin C, which is structurally similar to calmodulin and binds four Ca^{2+} ions on each subunit; troponin I, which inhibits the binding of actin to myosin; and troponin T, which binds the troponin complex to tropomyosin. Troponin and tropomyosin regulate contraction by controlling access of the myosin head to the actin receptor sites on the thin filament. Force is generated when myosin heads bind to exposed actin receptor sites on the thin filament.

Sliding filament theory

In 1954, A.F. Huxley and R. Neidergeke together with H.E. Huxley and E.J. Hanson proposed that force generation takes place in the sarcomeres of skeletal muscle when the myosin crossbridges on the thick filament bind to actin receptor sites on the thin filament, causing the thin filaments to slide between thick filaments toward the center of the sarcomere (see Fig. 20-1B). This is called the *sliding filament theory*, because the force is generated by filament motion, rather than by length or force changes within one or other type of filament. It is widely accepted as the most plausible mechanism for muscle force generation. In the process of binding the myosin globular heads to receptor sites of the adjacent actin filaments, there is a degradation of chemical energy (e.g., hydrolysis of ATP to ADP plus inorganic phosphate) and an associated conformational change in the crossbridge, pulling the thin filament toward the center of the sarcomere. This sequence usually results in a reduction of sarcomere length.

Contraction is accomplished by the cyclic formation and disassociation of bridges between myosin heads and actin filaments. A crossbridge may bind, undergo confor-

mational change, detach, and then rebind to a different actin receptor site. This process can continue as long as the level of free calcium ions (Ca^{2+}) in the sarcoplasm remains high and as long as ATP synthesis keeps pace with the need for high-energy phosphate bonds.

☐ How is muscle contraction initiated?

Calcium ions regulate many cellular processes, including cell motility, secretion, and synaptic transmission (Fig. 20-2). Calcium also appears to be important for muscle contraction in at least two ways. Initially, increases in the concentration of Ca^{2+} appear to control directly the sequence of events of contraction by regulating the inter-

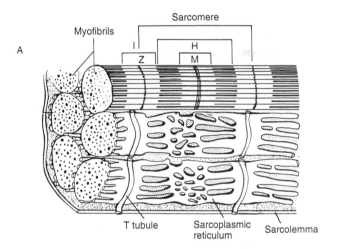

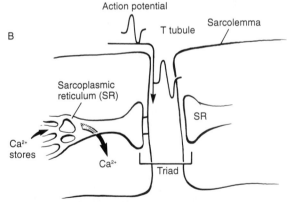

Figure 20-2

(**A**) Components responsible for excitation-contraction coupling. The muscle contraction is initiated by Ca^{2+} release from stores within the sarcoplasmic reticulum (SR). An action potential, initiated in the end-plate region, traverses the sarcolemma and is propagated inward toward the fiber center by conduction of the action potential into the T tubule. (**B**) The structures responsible for excitation-contraction coupling are seen in more detail. The terminal cisternae of the sarcoplasmic reticulum are arranged symmetrically around the sarcolemma of the T tubule, forming a triad. Depolarization of the plasma membrane of the muscle cell induces calcium release from the adjacent sarcoplasmic reticulum by chemical or voltage signals.

action of myosin with actin. Subsequently, reductions in Ca^{2+} levels terminate contraction.

The processes of excitation and contraction coupling rely on specialized structures called the *transverse tubule* (T tubule) and the sarcoplasmic reticulum (see Fig. 20-2). The T tubules are tubular invaginations of the myofiber's plasma membrane that communicate with the extracellular medium and extend into the cell's interior at the level of the Z line. The T tubules are in close contact with the sarcoplasmic reticulum and can support action potential propagation well into the cell interior. The myofibrils of each myofiber are surrounded by a network called the *sarcoplasmic reticulum*, which is a closed set of tubules and cisterns, and although it does not communicate with the cell surface, it approaches the T tubule, forming specialized structures called *triads*. These regions of apposition form the locus for chemical or perhaps voltage signaling between electrical and chemical intracellular events. The tubes are used to regulate signal transmission from the cell surface to the interior and to control the Ca^{2+} flux within the cell. Calcium ions probably are released from storage sites in the sarcoplasmic reticulum in response to electrical or chemical cues that are emitted after depolarization of the surface membrane, called the *sarcolemma*, and the T tubule.

Under normal resting conditions, the concentration of Ca^{2+} in the sarcoplasm is held at barely detectable levels ($<10^{-12}$ mmol/L). In response to neurally initiated depolarization of the sarcolemma, Ca^{2+} release develops swiftly, reaching concentrations of 10^{-7} mmol/L or higher in some types of muscle cells. The effect of the increase in Ca^{2+} is to change the conformation of the regulatory proteins on the thin filament, allowing nearby myosin crossbridges to bind to exposed actin receptors sites.

The binding of crossbridges to actin and the subsequent conformational change in an energy-releasing process do not initially require ATP degradation, although it is required for disassociation of the bridges. This helps to explain the phenomenon of muscle *rigor*, in which muscle becomes quite stiff after death. The loss of ATP production causes Ca^{2+} accumulation in sarcoplasm, exposing actin receptor sites to myosin crossbridge interactions. The crossbridges then bind, and without a source of energy to move the cycle along, the actin and myosin filaments remain tightly crosslinked, no longer able to easily slide past each other. The static crosslinking produces muscle rigor.

Energy is required to detach the myosin from the actin receptor site and to allow the crossbridge to be reconfigured to its original high-energy state. This detachment is necessary for crossbridge cycling and for force generation or muscle shortening to continue.

☐ How does muscle contraction end?

An increase in Ca^{2+} concentration is the primary event that initiates contraction, and Ca^{2+} reductions terminate contraction. After Ca^{2+} release is terminated, governed by the recovery of the muscle fiber membrane potential

to normal levels, Ca^{2+} is reabsorbed swiftly by active transport from the sarcoplasm into the sarcoplasmic reticulum. This reabsorption process is energy dependent, because the Ca^{2+} has to be reabsorbed against a substantial concentration gradient. One molecule of ATP is hydrolyzed for every two Ca^{2+} ions that are reabsorbed.

☐ Muscle fibers utilize different metabolic pathways to generate energy

ATP is used to promote crossbridge detachment, produce conformational change in the myofibril, and support active transport of Ca^{2+} into the sarcoplasmic reticulum. ATP therefore plays a pivotal role in many processes that are responsible for muscle contraction.

ATP is generated as a result of several biochemical processes in the muscle fiber, including degradation of free fatty acids, which are absorbed from the bloodstream, and the degradation of glucose. The glucose may originate directly from capillary absorption, or it may be generated by the degradation of intracellular glycogen stores. The breakdown of glucose or free fatty acids can proceed using the machinery of oxidative metabolism or of glycolytic metabolism when oxygen-dependent mechanisms are less readily available.

Fibers in different skeletal muscles and fibers in the same muscle often have different metabolic properties, allowing the functional specialization of mechanical muscle performance. The specializations include the speed of muscle contraction and the degree of muscle fatigability. Some muscle fibers contain high glycogen concentrations, and these appear to rely primarily on glycolytic pathways to generate the ATP necessary for contraction. Glucose is released from stored glycogen by a phosphorylase enzyme and is degraded to lactate, generating a few molecules of ATP per molecule of glucose degraded. ATP may also be generated on a short-term basis by the transfer of high-energy phosphate from its storage location on creatine. Other muscle fibers rely on oxidative metabolism, which is a far more efficient means to generate the high-energy phosphate bond of ATP. These fibers typically use absorbed glucose or free fatty acids and oxygen and have highly specialized metabolic apparatus for oxidative phosphorylation, including large numbers of mitochondria, substantial intracellular myoglobin (i.e., a binding compound that facilitates oxygen storage), and relatively modest or absent glycogen stores. These oxidative-type fibers are usually surrounded by a dense capillary network, presumably to facilitate oxygen and free fatty acid delivery to the fibers.

Mechanical Properties of Whole Muscle

Neurally activated muscle has several important mechanical properties that are instrumental in the performance of movement.

☐ Muscle behaves like a spring

When an active muscle is stretched, the muscle force output increases progressively with increasing muscle length (Fig. 20-3A). This proportional correlation between force and length is the defining property of a spring. For a given level of neural excitation, the force increases with increasing muscle extension, until the maximal physiologic length is reached (see Fig. 20-3A). If the maximal length is exceeded, the muscle force begins to decline, although this extended length is not normally achieved in intact limbs. The form of the length-tension relation

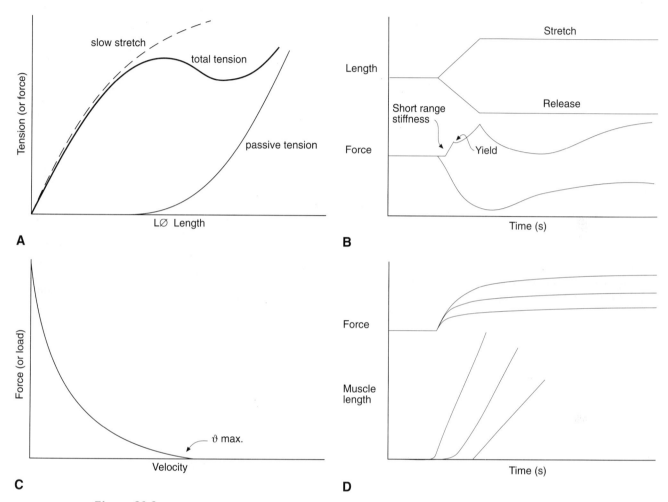

Figure 20-3

Mechanical properties of skeletal muscle (**A**) Length of tension properties. The isometric length-tension relation is derived by stimulating the muscle nerve at a constant intensity and frequency while holding the muscle rigidly at a designated length. If this length is progressively increased, muscle force increases progressively until reaching a maximal value at the maximal physiologic length in the body (L_0). If extension is continued beyond this maximum, the muscle force begins to fall, before ultimately increasing again at extreme lengths. This decline and subsequent increase in force at longer isometric lengths is attributable to the contribution of passive tissues surrounding the fibers. (**B**) Mechanical properties of active muscle subjected to symmetric stretch and release. When electrically stimulated muscle or deafferented active muscle is subjected to symmetric stretch and release of comparable velocity and amplitude, the force responses are grossly asymmetric. During stretch, the muscle initially has a region of elevated stiffness called the short-range stiffness, which is followed by an abrupt decline in stiffness, called the yield. At the end of the stretch, there is a secondary decline in force, which gradually recovers, with the muscle reaching a steady state only after several hundred milliseconds. During shortening, there is a relatively smooth decline in force, which settles to an isometric value. (**C**) Force-velocity relations. When muscle is allowed to shorten against a constant load, the magnitude of the shortening velocity, when measured at a constant length, is related to the load magnitude as a hyperbolic function. (**D**) Correlations between load and shortening velocity. If muscle force is generated with the muscle attached to the load, no shortening can occur until muscle force exceeds the load magnitude. When the load is minimal, this level is reached quickly, and muscle shortening begins relatively early and takes place at a relatively high velocity. When the load is maximal, the shortening begins after a longer period and takes place at a relatively slow velocity.

varies somewhat with the rate of neural excitation; the length at which the force begins to increase steeply depends on the stimulus rate, and the peak tension occurs at different lengths. The reasons for these length-dependent changes in force are not clear, but changes in the available population of bound crossbridges and length-related increases in Ca^{2+} release from the sarcoplasmic reticulum are potentially significant factors.

This spring-like description of muscle mechanics is broadly accurate for *isometric* length conditions, in which the length of the muscle is clamped at each measurement point, but it becomes even more precise for slow stretches, which generate a near-linear force-length relation, resembling a simple spring even more closely. The significance of these spring-like responses will become clearer after the effects of reflex action are explained, but one benefit of spring-like properties is that muscle forms a compliant interface with the external world and acts somewhat like a shock absorber.

Deviations from spring-like behavior

If active muscle devoid of reflex control is stretched rapidly from an initial isometric state, the spring-like behavior is disrupted, and muscle stiffness can be seen to change sharply with increasing stretch. After the stretch exceeds a fraction of a millimeter (typically 300 to 400 μm), the initial steep rise in force is interrupted, and muscle force declines sharply, sometimes falling below the prestretch level (Fig. 20-3*B*). The initial high-tension region is called the *short-range stiffness*, and the subsequent sharp decline in the force of the muscle is called the *yield*. Although the short-range stiffness and yield are most distinct in slow-twitch muscles such as the soleus, there is routinely a change in stiffness during stretch even in fast-twitch muscle after the length change exceeds a fraction of a millimeter. The initial high degree of stiffness is attributable to the stiffness of a population of attached myosin crossbridges, and the steep decline in force and stiffness is a result of stretch-induced crossbridge rupture.

Asymmetry of muscle mechanical response to stretch and release

There is a profound asymmetry of the force response to symmetric stretch and release, which represents a substantial deviation from classic spring-like behavior (see Fig. 20-3*B*). Although the force changes after a few hundred micrometers for stretch or release are usually symmetric, the two responses then depart substantially from this pattern. Muscle remains quite stiff during release, but it often shows a substantial decline in overall stiffness during stretch.

These mechanical characteristics represent a significant departure from the spring-like behavior of muscle, and they pose substantial difficulties for any neural control mechanism, because the profound changes in force develop quickly.

Force-velocity relations

Although muscle shows spring-like behavior for slow extensions and slow shortening, the magnitude of the force generated during the shortening of muscle is determined primarily by the speed with which the shortening occurs (Fig. 20-3*C,D*). There is a well-defined correlation between the speed with which a muscle can shorten and the load it can carry. For example, when a muscle is activated electrically and allowed to shorten against a load, a characteristic sequence of length and force changes occurs, in which the shortening speed declines with increasing load magnitude.

A typical shortening experiment is diagrammed in Figure 20-3*C* and 20-3*D*. Initially, as the muscle force develops, it may not be sufficient to overcome the opposing load, and no motion takes place. This is the *isometric phase*, in which force is increasing without an accompanying reduction in muscle length. After the force generated exceeds the magnitude of the opposing load, the muscle begins to shorten progressively at a constant velocity. The shortening velocity is rapid when loads are small, and the shortening velocity declines when the applied load is increased. Ultimately, shortening velocities become very slow when loads are very large, when measured with respect to the maximal force-generating capacity of the muscle.

The relation between the shortening velocity and applied load is well characterized. Figure 20-3*C* illustrates a typical force-velocity relation for mammalian skeletal muscle, showing that the decline in force, relative to the isometric state, is steep, even at modest shortening velocities, and indicating that the effect of movement on muscle force generation is profound.

The form of the force-velocity relation is given by the classic Hill equation:

$$(P + a)(V + b) = c$$

in which P is the load in newtons, V is the shortening velocity in millimeters per second, and a, b, and c are constants. Although the classic force-velocity relations were described using muscle shortening against controlled loads, the converse relation also applies; the maximal force generated falls steeply when muscle velocity is regulated by the experimenter. Although these force-velocity relations may seem to be somewhat arcane, they are important in regulating the speed of human movement, and they ultimately limit motor performance.

☐ Muscle contraction is initiated by neural excitation

The beginning of this chapter described the sequence of events between the neural excitation of muscle and the resulting force generation. This section describes the relations between the frequency or rate of neural excitation and the resulting muscle force.

When the motor axon is electrically activated by a single, short pulse or naturally stimulated by synaptic excitation of the motor neuron, a single action potential is transmitted from nerve to muscle, producing a transient increase in muscle force, which is called *muscle twitch* (Fig. 20-4). There is a substantial delay between the arrival of the excitatory potential in the muscle and the beginning of muscle force generation. This delay is described as the *excitation-contraction delay*, and it may be 3 to 5 milliseconds or longer, depending on the type of muscle. In all mammalian skeletal muscles, the twitch has a characteristic form, in which there is a relatively rapid rise from the onset to peak force, followed by a gradual decay. The time to peak force and time to one-half peak force during the declining phase vary greatly in different types of muscles, but the twitch is routinely asymmetric, with a prolonged declining phase.

If muscle is activated repeatedly by a train of action potentials, the mean force level generated varies greatly with the rate of neural activation (see Fig. 20-4A). If the excitation rate is sufficiently low, so that the force generated by the twitch returns to baseline between each twitch, there is no net increase of the force, and the maximal force reached is that generated at the peak of each twitch. However, as the rate of stimulation is increased, the new nerve impulse arrives before the force generated by the previous twitch has completely dissipated. Because of force summation, successive twitches generate a progressively increasing force level, which increases over the first few impulses and reaches a plateau level that

remains relatively constant for several seconds. If the individual twitches are still discernible, the force plateau is described as a *partially fused tetanus*. If no individual twitch transients are evident and the force trace is smooth, the response is described as a *fused tetanus*.

The practical implications of this force-frequency association are twofold. As shown in Figure 20-4B, there is a nonlinear relation between stimulus rate at which motor axons are activated and the resulting mean muscle force, although the effects of such rate increases vary in different kinds of muscles. The plot of this relation is sigmoid. The example shows that significant relative increases in rate produce a negligible increase in mean force at low rates, usually between 3 and 6 impulses/second. However, after rates of 8 to 10 impulses/second are reached, further increases in the rate produce substantial increases in the mean force, and this high sensitivity to increasing the stimulus rate continues until relatively high rates are achieved. For one of the muscles illustrated (see Fig. 20-4B), the stimulus rate may ultimately reach 30 to 40 impulses/second.

The nonlinear relation between stimulus rate and force requires that the motor units must be activated at particular rates for the muscle to be an optimally effective force generator. In most instances, the nervous system generates motor neuron discharge rates that lie within the steeply rising portion of the sigmoid plot of the relation for each motor unit. Although the plotted form of

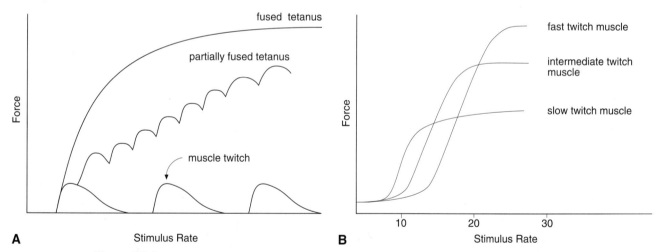

Figure 20-4
Relations between stimulus rate and muscle force. (**A**) Affect of stimulus frequency on muscle force output. When muscle is stimulated with a single, short stimulus to the muscle nerve, the resulting force change is described as a muscle twitch in which the force rises quickly to a maximum and then decays slowly. If stimuli are applied at low rates, there is no net force accumulation. If the stimulus rate is increased so that the mean force level accumulates, the force rises progressively, and it is described as a tetanus. If the individual twitch transients are still visible, the tetanus is called an unfused tetanus, and if the force trace is entirely smooth, the tetanus is described as a fused tetanus. (**B**) Force rate relations for different muscles. When a slowly contracting muscle, such as the soleus, is stimulated at different rates, the force begins to rise at relatively low rates, and the plot is a sigmoidal curve, demonstrating a maximal force arising at 20 to 30 pulses/second. For a fast-twitch muscle, the steep portion of the sigmoidal curve is moved significantly to the right of the plot and may not be reached until rates of more than 10 pulses/second are applied. Maximal force values may not be reached until rates exceed 50 or 60 pulses/second.

the rate-force relation is routinely sigmoid, the exact magnitude and shape of the sigmoid curve varies greatly for different types of muscle fibers. Slowly contracting muscle fibers reach the steep portion of their sigmoid curve at relatively low discharge rates, and rapidly contracting fibers need more frequent neural activation to achieve a full tetanus. In rapidly contracting muscles, the plot of the sigmoid relation is moved to the right. The form of the force-rate relation changes for muscle fibers as they become fatigued or as they are activated repeatedly and achieve a state called *potentiation*, in which twitch size increases during repetitive activation.

The means by which the motor neuron rate is tuned to match the contractile properties of the associated muscle fibers is not entirely understood, although the match must be achieved quickly to accommodate the changing contractile properties of muscle fibers during repetitive activation. For example, in fatigue, the muscle fiber contraction and relaxation times slow substantially, meaning that lower motor neuron discharge rates are required to achieve maximal force. When other fibers are activated repeatedly, twitch size may increase, and contraction times may change because of potentiation.

Muscle Receptors

Skeletal muscle contains several types of specialized receptors, whose properties, structure, and functional contributions to movement regulation are relatively well understood. Figure 20-5 illustrates the three main classes of muscle receptors in muscle: muscle spindle receptors, Golgi tendon organs, and free nerve endings.

☐ Muscle spindle receptors respond to changes in muscle length

The *muscle spindle* consists of a cluster of slender muscle fibers, called *intrafusal fibers*, contained within a fluid-filled capsule (see Fig. 20-6A). The muscle spindle lies adjacent to other regular muscle fibers and traverses most or all of the length of the muscle from the tendon of origin to the tendon of insertion. Because of this arrangement, the muscle spindle is said to lie in parallel with regular skeletal muscle fibers. Under isometric conditions, increases in the active muscle force, such as those mediated by neural excitation of muscle fibers, elongate the elastic elements, including the tendon, and reduce tension on the spindle and spindle receptors, causing a reduction in afferent discharge rate. This force-induced reduction of discharge, called *unloading*, is often used as a test to differentiate muscle spindle receptors from the in-series receptors, the tendon organs, which increase their discharge during an increase in active muscle force.

The muscle spindle carries two kinds of specialized sensory terminals in a dense and highly specialized efferent innervation. The structure of these receptor terminals is shown in Figure 20-5B. There is a large, typically annulospiral ending wrapped around the central portion of all intrafusal fibers in the spindle, and a smaller, often branching sensory terminal located more peripherally, toward the polar regions of the spindle. Based on their responses an increase in muscle length, the first type is called the *primary ending*, and the second type is called the *secondary ending* (Fig. 20-6).

Intrafusal fibers display specialized anatomic features that separate them into two categories. A small fraction of the intrafusal fibers, usually only one or two in each spindle, have clusters of nuclei in the central or equatorial region of the fiber and are called *nuclear bag fibers*. Most of the intrafusal fibers are slender and elongated, with nuclei arranged in chains. These are called *nuclear chain intrafusal fibers*. There are subspecializations of these intrafusal fibers (e.g., bag$_1$, bag$_2$, long and short chains), but these distinctions are not vital for understanding spindle receptor function.

The behavior of the muscle spindle receptor appears to be governed primarily by the mechanical properties of the supporting intrafusal muscle fiber, although the intrinsic biophysical properties of the primary and secondary receptor terminal areas may also be somewhat different. Nuclear bag fibers have little or no contractile material in the equatorial regions where the nuclei are located. Instead, the contractile regions are confined to the intrafusal fiber poles. If intrafusal fibers are stretched, the mechanical resistance exerted by the pole is likely to be different from that of the equatorial region, even at rest. Because the poles are more viscous, more rapid stretches extend the equatorial regions disproportionately, because the poles are more resistant to stretch under these conditions. Because the primary ending terminals are located mainly around the equatorial regions, more rapid muscle stretches affect the receptor terminals to a greater extent.

Another way in which intrafusal structure affects spindle receptor behavior is that efferent spindle innervation, which activates the polar regions of the muscle spindle, produces increased force and shortens the polar regions at the expense of the more elastic equatorial zone. This is especially likely in the bag fibers, which appear to have the greatest difference between the mechanical properties of the poles and equator. Although the nuclear chain fiber shows less structural inhomogeneity, mechanical observations indicate that their polar regions may also be somewhat less stiff than the equatorial zones. These differences in the structural and mechanical properties of the polar region may help to explain the different responses of primary and secondary endings to muscle stretch.

Figure 20-6 illustrates the different responses of primary and secondary spindle receptor afferent fibers to constant velocity stretches of the receptor-bearing muscle, beginning at a relatively short length. The primary spindle afferent shows a substantial increase of discharge during the dynamic phase of stretch, and this discharge rate drops to a much lower level after a new constant length is achieved. This appearance is broadly comparable to that of a velocity sensor, which would be expected

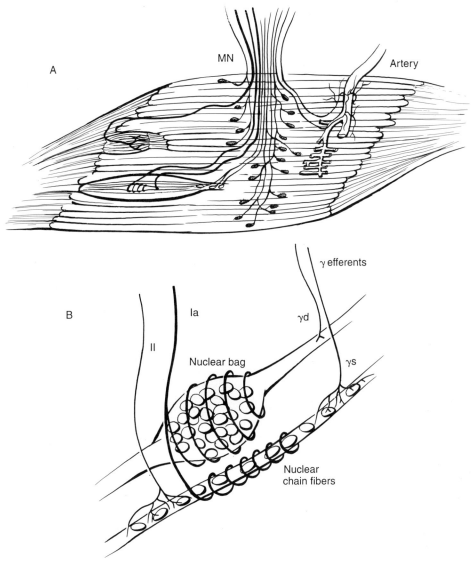

Figure 20-5
Organization of muscle receptors. (**A**) Extrafusal muscle, with fibers reaching from tendon at one end to tendon of the other end. The motor innervation terminates as a motor point in which end-plates are distributed across the muscle. Two of the key encapsulated muscle receptor organs are the muscle spindle and the tendon organ. The muscle spindle consists of an encapsulated structure in which small fibers, the intrafusal fibers, reach from tendon at one end to tendon at the other. Around the central portion of the spindle, there is a receptor terminal with an annulospiral structure called the primary ending. In the polar regions of the spindle, the branched or coiling terminal is the secondary ending. At the proximal end of the spindle, there is a separate spindle innervation, called the fusimotor innervation. (**B**) Expanded view of the intrafusal fibers and receptor terminals of the muscle spindle. The spindle contains large nuclear bag fibers, which are characterized by a cluster of nuclei in the central or equatorial region. Other intrafusal fibers, called nuclear chain fibers, have the nuclei arranged serially. The primary ending usually has an annulospiral appearance, in which the receptor is coiled around all of the intrafusal fibers. The secondary ending has a branched or coiled structure and is located more peripherally. Gamma efferent innervation (γ) is illustrated with gamma plates on the bag fiber and branching terminals on the chain.

to increase its output substantially during the dynamic phase of stretch if the velocity reaches a constant level and to drop the output substantially if the velocity falls to zero. The secondary spindle afferent is much less influenced by the speed of the stretch and appears to follow the length changes more closely, displaying relatively little dynamic overshoot during the ramp stretch.

When characterized in this fashion, it is possible to attribute the predominant *velocity* sensitivity to muscle spindle primary receptors and the predominant *length*

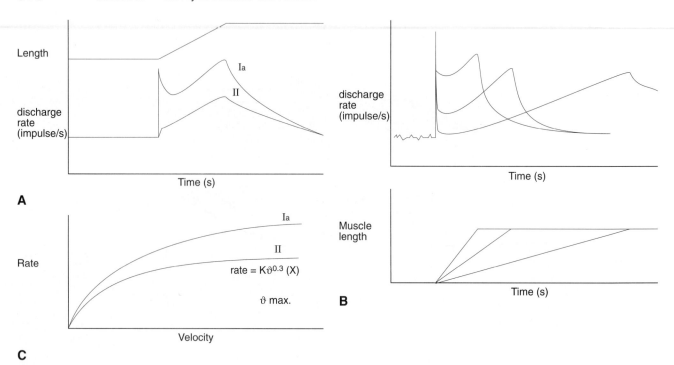

A

B

C

Figure 20-6
(**A**) Responses of primary and secondary endings to constant-velocity stretch of a given amplitude. The primary ending shows a very steep step-like increment in the discharge rate at the beginning of the stretch, terminating in a brief cluster of action potentials, emitted at a very high frequency. This is called the initial burst. The rate increases progressively with increasing length and then falls steeply at the end of the ramp of the plotted curve, returning to a substantially lower level. The secondary ending has a much smaller incremental increase at the beginning of the ramp of the curve and may have a limited dynamic overshoot. (**B**) Velocity dependence of the primary ending discharge rate. Three velocities are illustrated, but the discharge rate increment changes modestly. (**C**) The rate increment changes modestly with increasing velocity. A 100-fold increase in stretch velocity produces a twofold to threefold increase in the discharge rate. The form of this correlation is that of a power function, with the velocity exponent ranging between 0.2 and 0.3. The graph also demonstrates that primary and secondary endings are scaled versions of each other.

sensitivity to the secondary endings. A more extensive comparison of the impact of different stretch velocities on various receptor types indicates that neither primary or secondary endings are especially sensitive to stretch velocity, because a 100-fold increase in stretch velocity induces only a twofold to threefold increase in the discharge rate in either kind of receptor (see Fig. 20-6*B,C*). The relative increase in discharge with increasing velocity is comparable between the two classes of receptors, presumably reflecting the similarities in mechanical properties of intrafusal fibers supporting both receptor types.

Dependence of spindle receptor afferent discharge on stretch amplitude

The response of the spindle receptor is also strongly influenced by the amplitude of stretch. Stretches of 100 to 200 μm in amplitude produce disproportionately large increases in the discharge rate, which would be unsustainable if the change in length continued over many millimeters. This highly sensitive region is called the *small*

signal region, and it is responsible for the *initial burst*, which is visible at the beginning of a large-amplitude, constant-velocity stretch (see Fig. 20-6*A*). This highly sensitive small signal response is also visible during other kinds of length stimuli, such as sinusoidal or square wave length change, and it is noteworthy in that the behavior of spindles in this region is essentially linear, which means that spindle afferent discharge rate adjusts with increasing stretch amplitude and velocity.

Efferent innervation of the muscle spindle

As illustrated in Figure 20-7, all muscle spindles receive several types of efferent or motor innervation. This efferent innervation, which is broadly described as *fusimotor* innervation, arises from small motor neurons in the ventral horn through small-diameter, slow-conducting myelinated fibers, called *gamma fibers*. Alternatively, the efferent innervation may come from branches of regular, large-diameter skeletomotor fibers, displaying more rapid fiber conduction velocity. These larger fusimotor

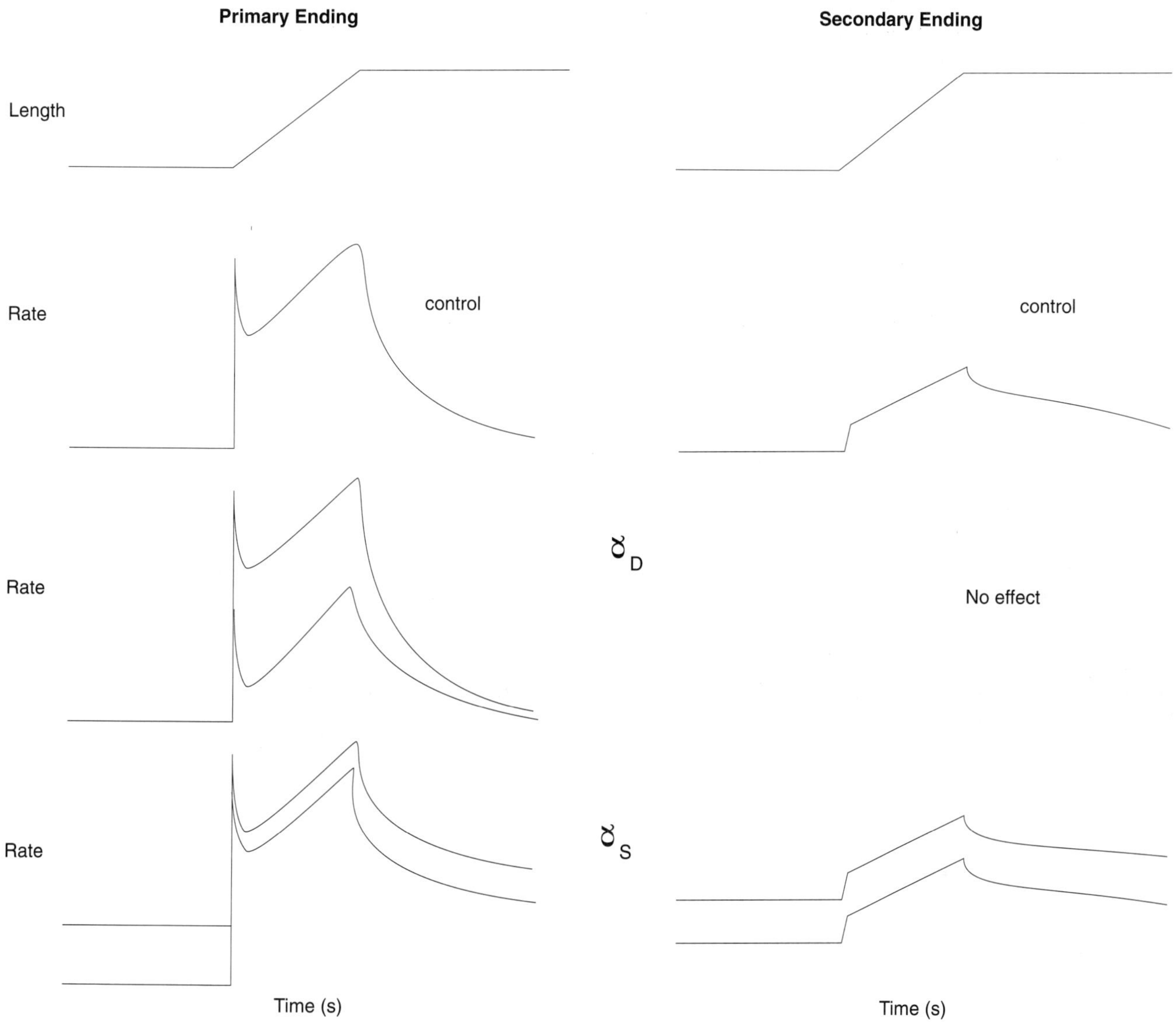

Primary Ending   **Secondary Ending**

Figure 20-7

Fusimotor effects on spindle afferents. This figure compares the effects of dynamic and static gamma motor neuron stimulation on primary and secondary muscle spindle afferents exposed to similar stretches. The second panel (from the top) on each side illustrates the control responses of primary and secondary endings, without added fusimotor input. Stimulation by gamma efferents of bag fibers produces a substantial increase in the dynamic response without a significant change in the initial and final firing rates. The dynamic motor neurons are called gamma$_d$ fibers. The efferent innervation of the chain fibers enhances the length sensitivity and discharge rate of both the primary and secondary endings. Because both initial and final or plateau rates are increased by gamma neuron stimulation, with little or no effect during the dynamic phase, this static efferent innervation is called gamma$_s$.

fibers are referred to as *beta fibers* or *skeletofusimotor fibers*, reflecting the fact that they originate from branches of regular motor neuron efferents.

The actions of gamma and beta innervation on the muscle spindle are broadly comparable and are not be discussed further here, although the relative use of these pathways is undoubtedly quite different, given their differing cellular origins and the probable difference in recruitment and rate behavior of the originating spinal neurons.

The effect of the fusimotor innervation on spindle afferent discharge depends on the particular intrafusal fiber that is innervated. The fusimotor efferents that innervate bag fibers enhance the dynamic responsiveness of the primary endings, because these endings are present on the bag fiber; they are described as gamma$_d$ or beta$_d$ dynamic fibers. The fusimotor fibers that innervate nuclear chain fibers enhance the length sensitivity and static background discharge rate of primary and secondary

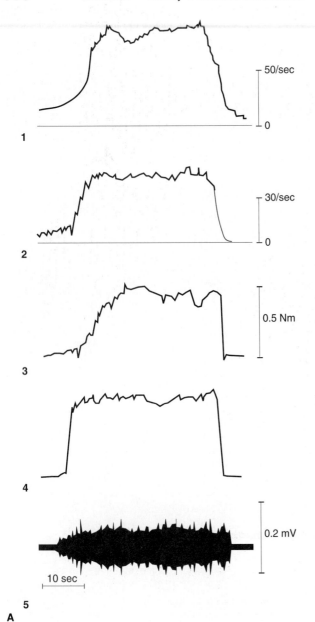

Figure 20-8

Response of human muscle spindle afferents during voluntary contraction. (**A**) Response of a primary ending in a finger flexor during isometric contraction. Panel 1 shows the afferent firing rate; panels 2 through 4 show the torque increase; and panel 5 shows the associated electromyographic (EMG) response. The afferent rate increases steeply, even before there has been a significant increase in torque or EMG activity. The fact that the afferent rate increases sharply and is sustained even with progressive force increase indicates that a substantial increase in fusimotor activity must have taken place to offset tendon elongation and the associated internal shortening of the spindle. (**B**) Torque rate relations. The responses of one primary and two secondary endings illustrate that the discharge rate increases modestly but significantly with increasing isometric torque. (**C**) Response of muscle spindle afferent (nerve) from a finger flexor during a slow voluntary shortening. The visual target for movement is the command trace. The next series of traces (joint angle) depicts the sequence of actual movements. The plots labelled nerve are raster plots, in which the occurrence of an action potential appears as a dot. The lowest panel is a histogram, summarizing mean firing of the afferent during each phase of movement. Under conditions of slow voluntary motion, the fusimotor input to a muscle spindle may be sufficient to offset the muscle length change. In this sequence of voluntary movements, the recording of muscle spindle primary afferent responses is shown as a raster diagram. The response during the shortening and plateau phases is somewhat higher than that during the initial static or hold phase. However, the rate changes are modest, indicating the efferent innervation is able to compensate almost exactly for the change in muscle length.

continued

endings. For this reason they are described as static fusimotor fibers or gamma$_s$ and beta$_s$ fibers.

Figure 20-7, illustrates the different responses elicited by activating one or the other class of fusimotor efferent fibers individually. This situation is unlikely to take place in living systems, because both types of efferents are normally activated together. However, activation of individual gamma$_d$ fibers showed that spindle primary ending responses increase substantially, and gamma$_s$ activation produced changes primarily during the constant length or isometric phase of muscle extension.

Muscle afferent recordings in intact subjects

Several methods can be used to record afferent discharge in essentially intact, unanesthetized animal and human subjects. The approaches include recordings from dorsal root afferents in the cat using fine wire electrodes, dorsal root ganglion recordings in the cat and monkey using sharp, pin-like electrodes, and intraneural or microneurographic recordings in human subjects with insulated tungsten electrodes. The dorsal root and dorsal ganglion recordings in the cat have revealed the patterns of afferent discharge in a range of motor behaviors, including stance and locomotion, and the human studies have allowed accurate quantification of spindle afferent discharge during changing voluntary force and during voluntary movements of controlled velocity.

Many of these studies have found a consistent pattern of spindle afferent excitation, in which motor neurons are activated broadly in concert with spindle afferent rate increases, suggesting that fusimotor activity increases

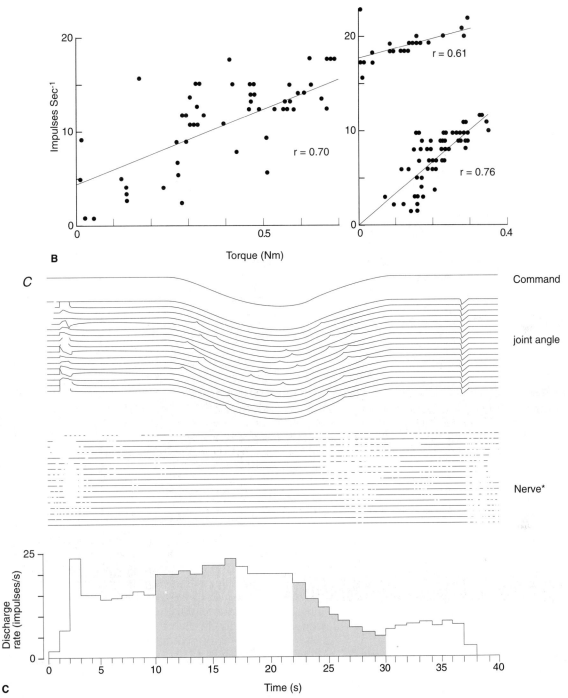

B

C

Command

joint angle

Nerve*

C

Figure 20-8 Continued

with increasing skeletomotor activity. In many experiments in which alpha and gamma discharges are recorded simultaneously, gamma neurons appear to be largely activated before skeletomotor (i.e.. alpha) activity begins, and by the time extrafusal muscle fiber activation takes place, there has already been large-scale activation of gamma fibers, producing *alpha-gamma co-activation*. The rules governing the activation of fusimotor neurons are not clear, although it has been demonstrated that gamma neurons may sometimes be activated without concurrent activation of alpha motor neurons. Whether there is the capacity to independently control fusimotor and skeletomotor (i.e., spinal motor neurons) remains to be verified.

Studies from intact human subjects have also been revealing in relation to afferent discharge during voluntary muscle contraction in isometric contractions or during slow movements. Such studies, which have been performed in muscles of upper and lower limbs of human subjects, have the capacity to define the relation between fusimotor and skeletomotor activation more thoroughly than is possible in animal models because of the voluntary cooperation provided by the human subjects. The studies of human volunteers have shown that there is substantial variation in the level of fusimotor input during most naturally occurring movements and that this fusimotor input is powerful in its effects on spindle afferent discharge.

During voluntary isometric contraction, the discharge rate of muscle spindle afferents increases substantially, reaching a level much greater than that recorded at rest (Fig. 20-8). In most of these studies, the impact of gamma efferent innervation to the spindle is difficult to deduce, because increasing the force induces elongation of tendon and other series of elastic components, allowing internal spindle shortening to occur. This elongation offsets fusimotor action to some degree. However, the finding (see Fig. 20-8A) that the afferent discharge rate increases with increased force indicates that fusimotor input is more than able to offset the change in length of a series of elastic elements. When muscle activation produces voluntary shortening, muscle spindle discharge rates may fall only slightly if the rate of muscle shortening is modest (see Fig. 20-8C).

The overall results of these studies have shown that muscle primary spindle afferents are strongly activated during voluntary contraction, that this activation is very strong at the lowest force levels, and that it may then increase relatively little with increasing force. This type of fusimotor activation can compensate for muscle fiber shortening to a large degree, except if the muscle shortening occurs rapidly.

Studies of humans and animals suggest that gamma$_d$ and gamma$_s$ fusimotor activation occurs together, although the possibility of independent activation of gamma$_d$ and gamma$_s$ fibers under some naturally occurring conditions has not been excluded.

Functions of fusimotor input

The mode of operation of fusimotor input to muscle spindles and its functional role is a matter of continuing debate, but it is likely that at least some of the functions described here are fulfilled.

The gamma$_s$ activation takes up the slack in muscle spindles, allowing primary and secondary endings to respond sensitively to added small changes in length. This raises the possibility that, with the help of fusimotor input, spindle receptors are able to maintain a broad dynamic range but are still able to respond sensitively to small perturbations of length.

Fusimotor innervation (i.e., gamma or beta) may match the muscle spindles to the changing mechanical properties of muscle. For example, muscle becomes stiffer and more viscous as its level of activation increases. Efferent innervation to the spindle may adjust the mechanical properties of the intrafusal fibers to optimize the pattern of spindle response to compensate for the altered mechanical muscle properties.

The gamma$_d$ input induces a substantial increase in the dynamic spindle response during muscle stretch, but it has much less effect during muscle shortening. These effects may be important in providing the appropriate asymmetry of reflex action in stretch and release of muscle.

☐ Golgi tendon organs act as force transducers

Structure-function relations of the tendon organ

The *tendon organ* is a receptor encapsulated by connective tissue. It consists of a branching nerve terminal that is interwoven with collagen and elastic fibers lying between a group of muscle fibers and the tendon (Fig. 20-9A). The tendon organ lies "in series" with this small cluster of muscle fibers, and it is subjected to mechanical strain when the muscle fibers are active.

Encoder mechanisms

The tendon organ acts like a strain gauge. Although the exact means by which muscle force is transduced by the tendon organ receptor is unknown, it is probably mediated by regional strain on nerve terminals as they are compressed among the tendinous fascicles. Few tendon organs are located within the body of the tendon itself; they usually reside within the muscle, at the muscle fiber-tendon boundary. Tendon organs may be scattered through many muscles in regions quite distant from the tendon, although they are often near collagenous tissue planes.

Tendon organ responses to muscle force change

Tendon organs are excited most readily by active muscle force increases, as are produced by neural activation of muscle, rather than by passive muscle force increases, as induced by muscle stretch. Typically, under conditions of physiologic activation, tendon organ discharge increases more or less proportionately with increasing force, and the response of the tendon organ follows the changes in force closely. Figure 20-9B shows a typical sequence of excitation of two different tendon organs, showing that each begins to discharge at a particular threshold force and their discharge rate increases irregularly at low forces, depending on the recruitment of particular muscle fibers belonging to the subset of fibers attached to the tendon organ receptor. After many muscle fibers are active, the impact of added muscle fiber activity on tendon organ discharge is diminished, and the discharge rate of the tendon organs follows the changing force quite accurately.

Although recordings of tendon organ afferents in intact conscious animal or human models are relatively rare, several studies have shown that tendon organ activity in physiologically activated muscle closely reflects the instantaneous muscle force recorded at the tendon, with relatively little sensitivity to the rate of change of the muscle force. This pattern of response is summarized in Figure 20-9C, which shows that rate of tendon organ dis-

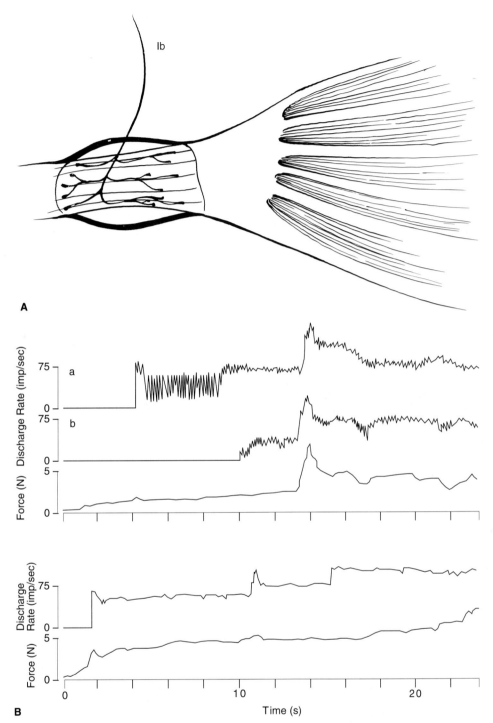

Figure 20-9

(**A**) Expanded view of a Golgi tendon organ, showing the encapsulated receptor area around the tendon fascicles and the limited number of muscle fibers attaching to those collagen fascicles. (**B**) The sequence of activation of the two tendon organs was recorded from the same soleus muscle in a decerebrated cat preparation. Although the thresholds of the two receptors are somewhat different and the initial discharge rates are somewhat variable, the rates become quite smooth and follow accurately the minor fluctuations in force after a significant force level is achieved. (**C**) There is a straight line relation between reflexively generated static isometric force and the discharge rate. The relation is not strictly linear, because there is a non-zero intercept on the ordinate. (**D**) The graph summarizes the relations for a population of tendon organ afferents; the data was drawn from the soleus muscle in different preparations.

continued

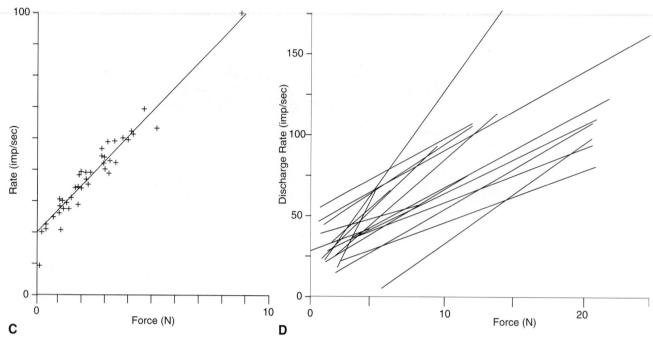

Figure 20-9 *Continued*

charge increases linearly with increasing force over much of the force range, although the force-rate relation often displays an upward convexity at higher forces. This relative linearity and sensitivity of the tendon organ response is unexpected because the numbers of muscle fibers sampled by any tendon organ is relatively small, typically 12 to 16 muscle fibers per tendon organ in the large limb muscles of the cat (see Fig. 20-9*A*). Despite this limited sample, the response patterns of individual tendon organ are surprisingly close to the force variations of the whole muscle, as registered at the tendon (see Fig. 20-9*B*), except at very low forces, at which only one or two of the muscle fibers attached to the tendon organ may be activated.

☐ Muscle contains other types of mechanoreceptors

The major muscle spindle and tendon organ afferents usually constitute less than 50% of the total sensory innervation from a typical muscle, as estimated from the total number of afferent fibers in the muscle nerve. There is also a substantial population of unmyelinated fibers in the nerve, whose receptor terminals have the appearance of *free nerve endings*, which are distributed throughout the muscle, including the muscle and tendon surfaces. These afferent fibers have different conduction velocities, including a few belonging to the most rapidly conducting afferent fiber group (group I), but in most cases, the free nerve ending afferents exhibit conduction velocities of small myelinated group III fibers and group IV fibers, most of which are unmyelinated. Although some of these

afferent fibers arise from specialized nerve terminals, such as pacinian corpuscles, most of them originate in free nerve endings, which have no discernible specialized terminal structure using light microscopy.

Although the nerve terminals of free nerve endings do not appear to be specialized structurally, they do exhibit a range of functional specialization. Some terminals are responsive to nociceptive input, and others respond primarily to mechanical stimuli, such as pressure or tension, although with much less sensitivity than the encapsulated receptors (i.e., spindle and tendon organs) described earlier. Other small afferents arise from receptors that respond to thermal and metabolic changes in the muscle, which may indicate that they have a role in cardiovascular and neuromuscular responses to exercise and in neuromuscular compensation for fatigue.

Afferent Pathways to the Spinal Cord

☐ Afferents are classified according to their conduction velocity and diameter

As summarized in Table 20-1, muscle afferents display a range of diameters and conduction velocities. The most rapidly conducting fibers, which are also those with largest diameter (group I) typically conduct at velocities of 100 m/second or more. Conduction velocities and diameters are well established for the cat, and primary ending afferent conduction velocities may reach 120 m/second.

Table 20-1

MUSCLE AFFERENT NERVE FIBER DIAMETERS AND CONDUCTION VELOCITIES

Type of Afferent	Mean Nerve Fiber Diameter (μm)	Mean Nerve Fiber Conduction Velocity (m/s)	Functional Roles of Afferents
Group I	15	90	Primary muscle spindle endings (Ia) Golgi tendon organs (Ib)
Group II	8	48	Secondary spindle endings
Group III	4	24	Free nerve endings, myelinated
Group IV	<1	1	Free nerve endings; unmyelinated

Secondary spindle afferents conduct at velocities ranging from 24 to 72 m/second, and small myelinated and unmyelinated fibers from skin and muscle (i.e., groups III and IV) conduct at velocities of less than 1 to 24 m/second. Conduction boundaries for the different fiber populations are not as well defined for human, and there is not a clearly defined conduction velocity boundary between primary and secondary spindle afferents.

Afferent pathways make connections with interneurons, motor neurons, and tract neurons

Sensory fibers enter the spinal cord primarily through the spinal dorsal roots, although there are a small number of unmyelinated nerve fibers and a few myelinated fibers that enter through the ventral roots. The anatomic distribution of the major afferent pathways within the spinal cord is well known (see Chap. 11).

Large myelinated muscle afferents (i.e., groups I and II) entering through the dorsal roots tend to segregate medially as they enter the spinal cord and then travel ventrally through the medial portions of the dorsal spinal gray matter before diverging to make synaptic connections with neurons in the intermediate spinal gray matter and ventral horn (Fig. 20-10). Some fibers terminate by making synaptic connections on regional interneurons, and others continue on to make synaptic connections directly with motor neurons in the ventral horn.

Small myelinated (Aδ) and unmyelinated (C) fibers make synaptic connections with neurons in dorsal gray matter, including the most superficial regions of spinal gray, called lamina I, which is largely concerned with the processing of pain-related or nociceptive information (see Fig. 20-10A). Other mechanoreceptor group III and IV afferents project to deeper regions of the dorsal gray area (i.e., laminae IV and V) before the signals are propagated to other spinal and supraspinal sensory systems.

Afferent pathways and their neuronal connections form distinct spinal circuits

Muscle afferent information is relayed to interneurons in the dorsal horn or intermediate gray matter or relayed to motor neurons in the ventral gray matter with monosynaptic or oligosynaptic connections in each location. Most interneuronal elements of the spinal cord have not been fully identified in mammalian preparations. A few types of interneurons have been characterized, primarily because of the availability of specific and practical diagnostic electrophysiologic tests.

Spinal information processing

Cutaneous and muscle afferent information follows several routes in the nervous system. Large myelinated afferents enter the spinal cord through the dorsal roots and then may follow one of two distinct paths. Large myelinated afferents from skin, muscle, or joint may branch and send the fibers into the white matter of the dorsal column and dorsolateral column, where they may travel rostrally for many centimeters. Some of these large afferents may reenter dorsal gray matter in proximal segments, where they may synapse with regional neurons, with new postsynaptic fibers reentering dorsal and dorsolateral columns. A few fibers may traverse dorsal white matter without relay all the way to the dorsal column nuclei (i.e., gracile and cuneate). Muscle afferents may circumvent the gracile and cuneate relays and may make a separate relay in the brain stem, although this relay has not been shown to be important in primates. The afferent fibers from the gracile, cuneate, and brain stem muscle relay nuclei pass signals to higher centers, including the thalamus, cerebellum, and cortex. The alternative afferent destination is directly to interneuronal and motor neuronal circuits, as described previously.

Muscle Ia afferents make extensive monosynaptic connections with virtually all motor neurons innervating the

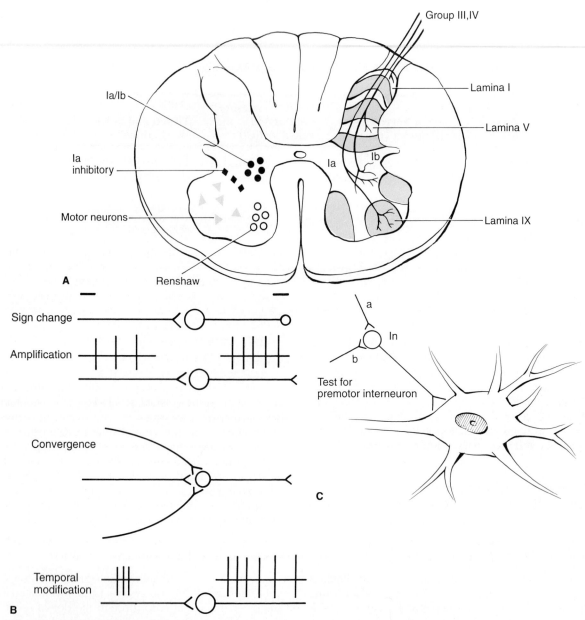

Figure 20-10

(**A**) Interneuronal circuitry and afferent connections of the spinal cord. The spinal cord gray matter is composed of a variety of neurons that have specialized features demonstrated on light microscopy. These features have been used to classify the regions of the gray matter into the twelve laminae, with lamina I located most superficially in the dorsal gray matter. These laminae are useful for describing the preferred destination for particular classes of muscle afferent fibers. Large-diameter muscle afferents (i.e., Ia and Ib) travel medially through dorsal gray matter. Ib fibers terminate in the intermediate gray matter in laminae V through VII. Group Ia fibers may emit collaterals before proceeding into the ventral horn toward laminae IX, the location of spinal motor neurons. The cross-sectional drawing shows the major neuronal elements of the spinal cord, including spinal motor neurons, Renshaw interneurons located medial to the motor neuron core, Ia inhibitory interneurons, and Ia and Ib inhibitory interneuronal groups located dorsal or medial to the motor neuron pool. (**B**) Types of computational operations performed by spinal interneurons. The most common operation is a sign change, in which an excitatory input, such as an afferent, activates an inhibitory interneuron, which induces inhibition at its postsynaptic sites. Interneurons can amplify their input by transforming a low-frequency input train to a high-frequency output. Interneurons may also integrate spatially by means of a convergence of afferent inflow from different sources onto a particular interneuron. They may also induce changes in the temporal pattern, such as changing a transient input consisting of a few impulses into long-lasting discharge. (**C**) Key factors underlying a commonly used test for a premotor interneuron. Activation of afferent a or b individually may not be sufficient to cause the interneuron to reach the threshold, and the subsequent synapse at the neuron reveals no synaptic potential. However, if a and b are activated simultaneously, the interneuron is recruited and postsynaptic potentials are manifested in the spinal motor neuron.

muscle from which the spindle afferents originate (i.e., homonymous motor neuron pool) and with many motor neurons from nearby synergists. The result is extensive divergence of afferents to many motor neurons and extensive convergence of Ia afferents onto individual motor neurons.

Spinal interneuronal systems

Spinal *interneurons* are defined largely on anatomic grounds. Interneurons are neurons whose axons extend relatively short distances within the cord, usually no more than a few spinal segments. The cell body diameters are usually less than 50 μm, and many neurons are even smaller. Although muscle afferents from muscle spindles (i.e., primary and secondary spindle afferents) make some direct or *monosynaptic connections* with spinal motor neurons in the ventral horn, almost all afferents, including those from primary and secondary endings, make their first synaptic connections with neurons in dorsal or intermediate spinal gray matter.

Interneuronal information processing

Interneurons are an important component of spinal information processing, because they perform several important computational operations (see Fig. 20-10*B*). The most common computational operation is a sign change (i.e., inhibition), in which, for example, an afferent originating from muscle or skin produces synaptic excitation at the first synaptic relay and then activates an inhibitory interneuron, producing inhibition at subsequent postsynaptic sites.

Excitatory interneurons may amplify information received from the periphery, or they may change the spatial distribution of that information by virtue of the divergent pattern of their nerve terminals. Interneurons may also change the time course and the frequency content of incoming signals by filtering out high frequencies or by providing integrator kinds of operations. Interneurons may provide other signal processing, such as reshaping or changing the transmitted signal from a step to a more transient or pulse-like response.

Because the input from a single afferent is sometimes modest, it may be necessary to sum the input from several sources to reach the interneuron threshold. A single afferent fiber input may not elicit a interneuron response by itself, but the neuron threshold may be reached if two or more afferent inputs are excited simultaneously. If these afferent sources are widely dispersed, the interneuron also may integrate the spatial information.

The subliminal effect of single afferents on interneurons is useful in detecting the existence of an interneuron in a projection pathway from an afferent to a motor neuron (see Fig. 20-10*C*). For example, an interneuron is thought to be present when activation of an individual afferent pathway (e.g., a peripheral nerve) gives rise to no synaptic potential in a motor neuron, but simultaneous activation of two afferent inputs does induce a visible excitatory or inhibitory postsynaptic potential.

Some interneurons, called *propriospinal neurons*, are specialized for spinal relay. Others may promote *rhythm generation*, especially in the more proximal lumbar spinal segments, where they may contribute to the cyclic motor neuron excitation that takes place as part of locomotion.

The primary role of interneurons appears to be as summing or integrating elements, in which convergent input from a variety of sources, sometimes including different sensory modalities, is integrated and passed on to the next step of information processing.

Interneuron transmitter systems

Most interneurons are inhibitory in their postsynaptic effects, and they release inhibitory transmitters such as glycine or γ-aminobutyric acid (GABA) from their presynaptic terminals. Others are excitatory, presumably releasing glutamate, aspartate, or other excitatory amino acids from their presynaptic terminals.

Functions of identified interneurons

Most of the interneuronal systems of the mammalian spinal cord are not fully identified, and the interneurons that are named are not necessarily the most important, because interneuron naming is often arbitrary and frequently depends on some distinctive electrophysiologic feature, such as the neuron response to synchronous antidromic ventral root stimulation, which may have little relation to the interneuron's functional roles. Some of the named spinal interneurons are Ia inhibitory, Renshaw, Ia-Ib inhibitory, II excitatory, and presynaptic inhibitory.

Renshaw neurons. Identification of interneurons has been based on a variety of circumstantial electrophysiologic findings, which may be unrelated or only obliquely related to the functional role of these interneuronal systems (Fig. 20-11*A* and see Fig. 20-10*A*). For example, Renshaw interneurons, which are small interneurons located in the ventral horn medial to the spinal motor nuclei, display an unique high-frequency bursting discharge after antidromic excitation of spinal motor axons in the ventral root. These interneurons receive excitatory input from motor axon collaterals and synapse on regional spinal motor neurons.

The transmitter released at the motor neuron axon collaterals to Renshaw neurons is acetylcholine, which is to be expected, because this transmitter is also released at the other terminal branches of the motor axon at the neuromuscular junction. The cholinergic postsynaptic receptors on the Renshaw neuron are *muscarinic* and *nicotinic receptors*. Antidromic excitation of the Renshaw neuron is only partly impaired by separate administration of nicotinic or muscarinic acetylcholine antagonists, but it is essentially eliminated by the simultaneous administration of both cholinergic blockers. For example, combinations of cholinergic nicotinic antagonists, such as mecamylamine and dihydro-β-erythroidine, and muscarinic antagonists, such as atropine, severely reduce anti-

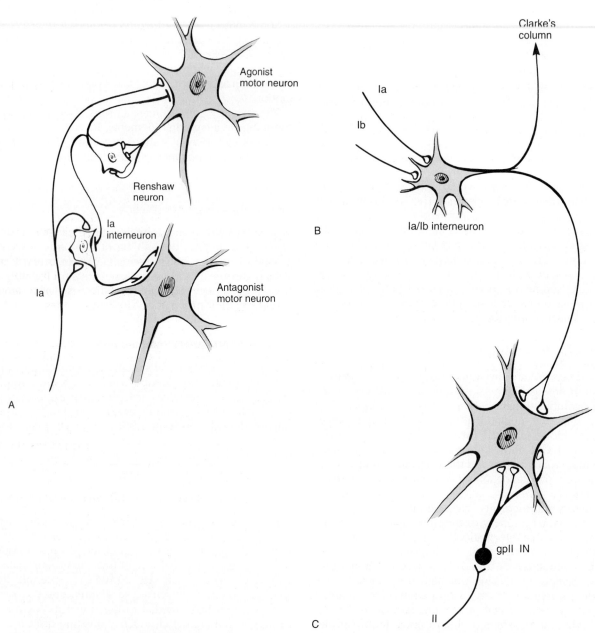

Figure 20-11

Types of spinal interneurons. (**A**) This important circuit involves linkages among motor neurons and flexor and extensor muscle groups associated in interneuron pools. Flexor motor neurons produce collaterals that innervate and excite Renshaw interneurons. Renshaw axons may inhibit the originating motor neuron, the flexor, and the Ia inhibitory interneuron. The Ia inhibitory interneuron receives Ia afferent input from a flexor, which induces excitation of that flexor motor neuron. Activation of the Ia inhibitory interneuron inhibits the antagonist extensor motor neuron. Excitatory synaptic terminals are drawn as enclosed circles, and inhibitory terminals are presented as bars. (**B**) The convergence of Ia and Ib afferents on an inhibitory interneuron inhibits the activity of a typical extensor motor neuron. The Ia and Ib interneurons also project to Clarke's column, which is the nucleus of origin of the dorsal spinocerebellar tract. (**C**) Group II afferents excite an interneuron, located in the proximal lumbar segments of the cord, which induces excitation of motor neurons in lower lumbosacral segments.

dromically elicited Renshaw activity, confirming the diversity of postsynaptic cholinergic receptor types on the Renshaw neuron.

Axons of Renshaw neurons make synaptic connections with regional motor neurons and other interneurons, including Renshaw neurons of other motor neuron pools and regional Ia inhibitory interneurons. The transmitter released by Renshaw axon terminals is thought to be glycine, a naturally occurring amino acid. GABA may also be released at Renshaw terminals, but this has not been independently verified.

Ia inhibitory interneuron

The Ia inhibitory interneuron (see Figs. 20-10 and 20-11) receives monosynaptic excitatory input from Ia afferents in one muscle and then makes inhibitory synaptic connections to motor neurons of opposing or antagonist muscles. By virtue of these connections, the Ia interneuron is thought to promote *reciprocal innervation*, in which agonists and antagonists acting about a joint are prevented from being active simultaneously.

Although the functional role and utility of Ia inhibitory interneurons is relatively easy to comprehend, the identification of these interneurons is circumstantial and depends on the fact that they are inhibited by antidromic ventral root excitation, apparently through Renshaw neurons. An interneuron activated by Ia afferents of a particular muscle but silenced by ventral root stimulation meets the necessary defining criteria.

The transmitter released at Ia interneuron axon terminals is thought to be *glycine,* for which no highly specific antagonists have been identified. Nonspecific antagonist effects are mediated by the agent strychnine, but the strychnine effects are not localized, instead impacting many synaptic locations within the central nervous system.

Ia-Ib interneurons. Golgi tendon organ afferents make synaptic relays in the intermediate gray matter of the spinal cord, producing autogenic inhibition of homonymous or synergist motor neurons (see Figs. 20-10 and 20-11). These interneurons, which were originally called Ib afferents, also receive muscle afferent input from Ia spindle receptor afferents, making their integrative function more complex than originally perceived. The major driving input is primarily from Golgi tendon organs, which raises the possibility of a *force regulatory action* by this interneuronal system.

These interneurons also project to Clark's column neurons, which lie in proximal lumbar segments and give rise to axons traveling as the dorsal spinocerebellar tract. The existence of an inhibitory connection from muscle Ib afferents through the Ib inhibitory interneuron to homonymous and other synergist motor neurons raises the possibility of a closed-loop force regulator, in which force increases sensed by tendon organs inhibit homonymous motor neurons. This force regulation has been difficult to demonstrate, because the animal preparations that allow study of intact reflex spinal pathways, such as the decerebrate cat model, show substantial suppression

of many segmental inhibitory interneuronal systems, including the Ia-Ib inhibitory pathway.

Several studies of humans have provided indirect evidence for the operation of a force regulator. These studies have used fatigue as the test probe to evaluate force feedback compensation, relying on the fact that fatigue induces a substantial short-term loss of muscle contractile force. However, although fatigue is the most common source of a loss of force-generating capacity, there are also probably substantial changes in afferent inflow from several types of muscle receptors, which may exaggerate the degree of responsiveness of the Ib pathway compared with the normal path. Muscle fatigue induces changes in muscle temperature, pH, and metabolic state, all of which may change the spontaneous discharge patterns and the mechanical responsiveness of group III and IV muscle afferents. Because free nerve ending afferents and Ia and Ib afferents may converge on the same set of Ib interneurons, alterations in the baseline discharge of group III and IV afferents may alter the responsiveness of the Ib pathway.

Group II excitatory interneurons. An extended series of studies has demonstrated that there exist interneurons that receive selective input from secondary spindle afferents, which make excitatory synaptic connections to lumbosacral spinal motor neurons and which are in the proximal lumbar spinal cord segments of the cat (see Figs. 20-10 and 20-11). The functional role of this interneuronal system is not understood, although interneuronal structural and electrophysiologic properties and connections appear to be clearly defined.

Flexion reflex pathways. There is no uniquely identified set of interneurons involved in flexion reflexes. However, many of the interneurons located in the intermediate gray matter of the spinal cord and in deeper laminae are involved in processing information from cutaneous and subcutaneous sensory endings and from high-threshold muscle afferents, especially those with smaller diameters (e.g., groups II, III, IV). Excitation of any of these afferents often results in a coordinated pattern of flexor muscle activation. Although the range of afferent input eliciting flexion reflexes is diverse, the ultimate effect is relatively stereotyped. The activation of many of these afferent systems induces consistent excitation of flexors and inhibition of extensor muscles, primarily by means of the Ia inhibitory interneuronal system.

The individual neuronal elements of the flexion withdrawal reflex system are not clearly separable, and there appears to be an extensive polysynaptic chain involving multiple relay sites. The transmitters of such systems are not fully identified, although the excitatory synaptic connections are probably mediated by excitatory amino acids, such as glutamate and aspartate, and it is likely that GABA plays an important role in mediating inhibition.

Presynaptic inhibitory interneurons. Some interneurons in the dorsal gray matter of the spinal cord terminate on presynaptic terminals of afferents and of other interneurons. These interneurons release GABA, which depolarizes presynaptic terminals and reduces the amount of transmitter released by each incoming action potential. This reduction of transmitter release is mediated by reducing the amount of Ca^{2+} that enters the terminal with the arrival of each action potential.

Motor Neurons and the Motor Neuron Pool

The physiology of CNS neurons began with the study of spinal motor neurons, primarily in the 1950s and 1960s by John Eccles and his colleagues in Canberra, Australia. For many years, the motor neuron was the prototype for the study of central neurons, and it remains one of the most extensively studied neurons in the mammalian nervous system.

A spinal motor neuron is the neuron directly innervating skeletal muscle, primarily through axons traversing the ventral roots (see Fig. 20-11). All of the motor neurons innervating a given muscle are together called the *motor neuron pool* (see Fig. 20-11). Each motor axon innervates a group of muscle fibers in the muscle. The motor neuron with the innervated muscle fibers is called the *motor unit*. The group of muscle fibers innervated by one motor axon is called the *muscle unit*.

☐ Motor neurons have distinct biophysical properties

One type of spinal motor neuron is characterized by a relatively large cell body (i.e., soma) and innervates muscle fibers in skeletal muscle. This type of motor neuron is referred to as an *alpha or skeletomotor neuron*. A second group of smaller motor neurons, called *gamma or fusimotor neurons*, innervates muscle spindles. The alpha and gamma motor neurons are interspersed throughout the ventrolateral portion of spinal gray matter. An additional class of neurons called beta or skeletofusimotor motor neurons is also present in many vertebrate systems, but these neurons have no features distinguishing them structurally or physiologically from alpha motor neurons. Beta motor neurons are characterized solely by the fact that they jointly innervate skeletal muscle and intrafusal fibers of muscle spindles in the same muscle.

Although the soma of a spinal motor neuron may be quite large, approaching 100 μm or more, the dendritic arbor is even larger and may extend for several millimeters radially from the cell soma out into white matter and far up into the dorsal gray matter.

Chemical conductances

Spinal motor neurons are characterized by a range of specific chemical- or voltage-activated conductances.

Activation of excitatory synaptic projections to spinal motor neurons, such as those originating from Ia afferents, produces an excitatory postsynaptic potential. As shown in Figure 20-12, the synaptic potential begins quickly after the arrival of the action potential in the presynaptic nerve terminal. This small delay is important, because it indicates that the effects are not mediated by direct electrical transmission. After an interval of less than 1 millisecond, there is a rapidly depolarizing voltage change, which may reach a peak in 3 to 5 milliseconds and then decay gradually over the ensuing 15 to 20 milliseconds.

From a variety of biophysical studies of motor neurons, it is evident that the time course of the synaptic potential depends jointly on the magnitude and time course of synaptic current injection, which may last only 100 to 300 microseconds, and on the electrical properties of the cell membrane. The latter is referred to as the resistance-capacitance properties of the membrane. The excitatory synaptic potential follows its particular time course, primarily because of charge storage and distribution on the membrane "capacitor." This membrane charge is slowly dissipated over the subsequent 10 to 20 milliseconds, with a time course that depends on the electrical properties of the membrane and on the physical location of the synaptic terminals on the cell surface.

The effective current flow seen at the axon hillock, the preferred site of action potential initiation of the motor neuron, is strongly influenced by membrane properties and by the cell geometry, especially the pattern of dendritic branching. The excitatory synaptic potentials that originate on distal dendrites are greatly attenuated and filtered by the electrical properties of the cell.

When some inhibitory inputs are activated, such as those from Ia inhibitory interneurons or from Renshaw cells, the inhibitory postsynaptic potentials begin 1 to 2 milliseconds later because of the time needed to activate one or more intervening interneurons, but the synaptic potentials have a broadly comparable time course, although with hyperpolarizing voltage changes. A typical sequence and time course of these synaptic potentials is diagrammed in Figure 20-12.

Voltage-gated conductances

In addition to the chemically activated conductances, there are voltage-gated conductances, several of which are closely comparable to those described in peripheral myelinated axons. Some conductances are mediated by voltage-sensitive Na^+ and K^+ channels, which have an uneven distribution over the surface of the motor neuron and proximal axon. The excitable Na^+ channels are highly concentrated in the first node of Ranvier and on the regional membrane close to the axon (i.e., axon hillock), but they are relatively less concentrated over the cell soma and dendrites. When current is injected into a cell, the first point at which action potential generation is initiated is at the axon hillock and first node, and this is followed by the retrograde invasion of the action potential into the soma and dendrites, as well as the standard orthodromic invasion of the motor axon. This discontinuity in the activation sequence is visible during standard

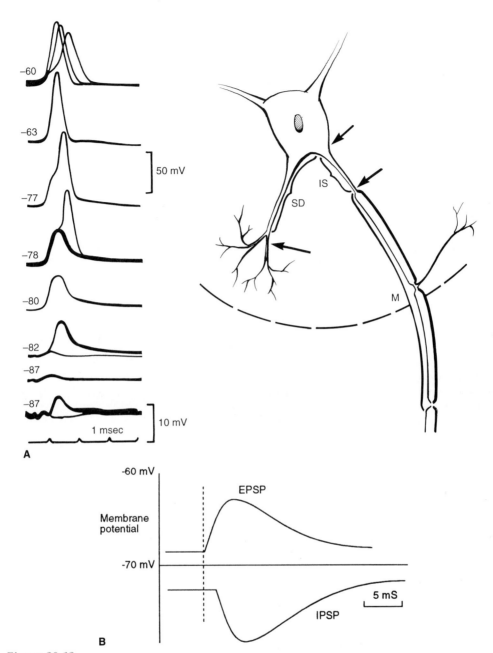

Figure 20-12
(**A**) Diagram of a motor neuron, showing the cell soma and dendrites (SD segment), the initial segment or axon hillock, and the first node of Ranvier. The changes in action potential formation result as a motor neuron is stimulated antidromically. At −87 mV, only a small potential, called the M potential, is visible, as the potential reaches −80 mV, a partial action potential, called the A spike, is visible, indicating invasion of the initial segment of the motor neuron. As the motor neuron potential is further depolarized to −78 mV, a full action potential is visible, with a clear inflection still evident at the transition between and A and B spikes. (**B**) The diagram plots the time course of an excitatory postsynaptic potential (EPSP) and inhibitory postsynaptic potential (IPSP). Both have a relatively similar rise time, reaching a peak in 3 to 5 milliseconds, and a prolonged decay. The inhibitory postsynaptic potential begins a few milliseconds after the excitatory potential because of the intervening inhibitory interneuron.

intracellular recordings as a minor inflection on the rising edge of the voltage recording of the action potential.

In addition to the standard Na^+ and K^+ channels, there are slower-speed K^+ channels that are activated during and after the repolarization phase of the action potential. There are also Ca^{2+} conductances that may change membrane voltage directly by generating inward current flow, or they may mediate secondary changes in other conductances, such as Ca^{2+}-sensitive K^+ conductances.

Motor neuron stability and plateau potentials

The use of unanesthetized preparations, such as the decerebrate cat model, the spinal cord, and brain stem slice, and culture preparations of spinal motor neurons has revealed that other kinds of membrane potential responses, not reported in the standard anesthetized preparations, can appear under particular conditions.

For example, in the course of classic intracellular motor neuronal recordings in anesthetized cats, a brief square wave current injection gives rise to a square wave voltage transient, filtered by the resistance-capacitance properties of the cell. In the unanesthetized preparation, a similar intracellular current injection can produce a sustained depolarization, which may continue for tens of milliseconds, long after the transient depolarizing stimulus has been completed. These sustained voltage changes, or *plateau potentials* may be the consequence of inward Ca^{2+} flow, or they may be a result of a secondary reduction of K^+ outflow, mediated by changes in free intracellular Ca^{2+} or K^+ conductances. In either case, they appear to depend on the presence of compounds called *neural modulators*, such as serotonin or norepinephrine, which are released from descending brain stem pathways.

The presence of serotonin in the neuronal environment changes the form of the input-output relations of spinal motor neurons. A transient input signal is transformed into a sustained discharge. We do not know if this mode of operation is commonly used or if it is used only in selected circumstances. If, for example, it is widely used in normal posture and gait, brain stem or spinal cord injuries that interrupt these descending systems may adversely impact motor neuronal responses, promoting relatively ineffectual motor neuronal discharge and exacerbating muscle weakness and increased fatigability.

☐ Motor neurons are reflexively activated by a variety of afferent pathways

Spinal cord regulation of muscle force

There are two broadly different ways in which activation of the motor neuron pool by afferent inflow, by segmental interneuronal input, or by descending spinal pathways can increase muscle force (Fig. 20-13). The first mode is *recruitment* of motor neurons, which is the transition from a passive state, in which the motor neuron is quiescent, to an excited state, in which the motor neuron is emitting action potentials. The second mode of force regulation is achieved by increasing the rate of discharge of individual motor neurons, a process called *rate modulation*. The first option progressively activates more motor neurons and muscle fibers, filling in all the elements in the pool until the pool is completely recruited. The second option, rate modulation alters the individual force output of single motor units by virtue of their capacity to produce a partially fused tetanus, in which the mean force output becomes progressively greater with the increasing rate of motor neuron discharge.

In living systems, these two regulatory mechanisms are closely interwoven, with recruitment being the dominant source of force increase at low levels of motor neuron pool excitation and rate modulation generating a progressively greater impact on muscle force output as more motor neurons are activated.

Characteristics of motor neuron recruitment

When a motor neuron belonging to a particular motor neuron pool is subjected to increasing excitatory synaptic input, such as from muscle spindle Ia afferent fibers, the resulting progression of excitation and recruitment in different motor neurons is orderly and virtually stereotypic, obeying the *size principle* (Fig. 20-14). The size principle, enunciated first by Henneman and colleagues in 1965, states that motor neurons are recruited in a defined rank order, in which small-sized motor neurons are recruited first, followed by larger and larger motor neurons. Conversely, if motor neurons in a pool are already active, the application of an increasing inhibitory input, such as from antagonist muscle Ia afferents, causes a derecruitment of motor neurons, in which the largest motor neurons are the first to drop out. With increasing inhibition, derecruitment continues progressively until the smallest motor neurons are silenced.

Size of motor neurons. The original description by Henneman of the relation between recruitment rank order and size was derived from studies using ventral root recordings of motor axons, in which the size of the extracellular action potential recorded from the ventral root was used as an indication of the size of the motor axon and, by inference, as an indication of the size of the associated motor neuron. A typical pattern of activation is illustrated in Figure 20-13C, which shows that increasing muscle extension gives rise to recruitment of ventral root action potentials of progressively greater amplitude. This correlation between action potential size and motor neuron somatic size is broadly applicable to the bulk of motor neuron axons recorded in the ventral root, but it may not hold in fine detail, partly because unrelated technical factors may influence the size of the recorded action potential, but mostly because motor neuron size itself may not be the primary factor governing motor neuron recruitment order.

When motor neuron recruitment order is plotted against tetanic tension of the associated motor unit, the

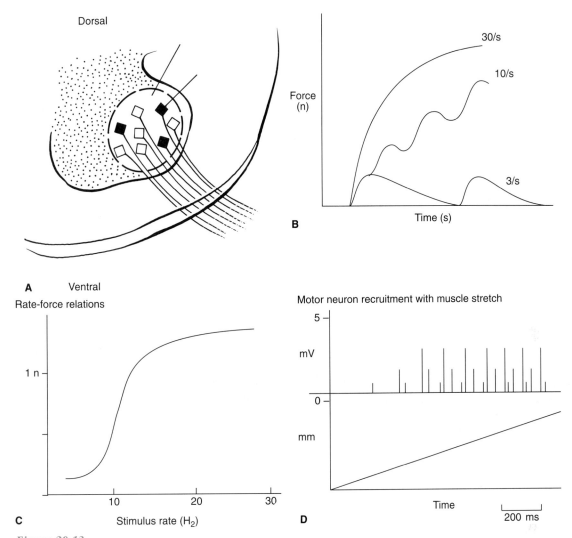

Figure 20-13
Regulation of muscle force. (**A**) Recruitment of motor neurons is manifested as activation of the elements in the motor neuron pool. Recruitment can augment muscle force only if motor neurons are incompletely activated. (**B**) Rate modulation is available at all levels of recruitment. The discharge rate of the motor neuron strongly influences the net force output of the motor unit. At very low discharge rates, individual twitches generated by motor neuron excitation do not summate significantly, but at rates exceeding 6 or 8 pulses/second, a partially fused tetanus develops. Under normal conditions, motor units operate in a response region in which their tension output is partially fused. (**C**) The characteristic force and stimulus rate relation for an individual motor unit, demonstrating the sigmoidal curve describing the operating region in which small changes in the impulse rate (10 to 20 pulses/second) induce substantial changes in force. (**D**) Typical outcome of an experiment in which motor units are recruited progressively during slow muscle stretch. The first motor units to be activated have small action potentials, and as the excitation is increased, progressively larger action potentials are introduced, reflecting the recruitment of larger and larger motor neurons.

relation becomes much clearer and is more linear than the plotted relation between recruitment rank order and axon action potential size, and recruitment reversals, in which the larger motor neuron appears to be the first activated, are much less common. Although tetanic motor unit force has no obvious causal relation with the neuronal factors governing recruitment, it presumably provides a more sensitive marker of net motor unit properties than is available by relying on action potential size.

Physiologic mechanisms of the size principle. The physiologic factors responsible for the orderly recruitment of motor neurons by size are not fully understood, despite intensive and continuing investigation. Henneman originally proposed that motor neuron size itself could be the governing factor. The idea was that, for a given synaptic current applied to all motor neurons in the pool, the magnitude of the resulting excitatory synaptic potentials in motor neurons would be determined

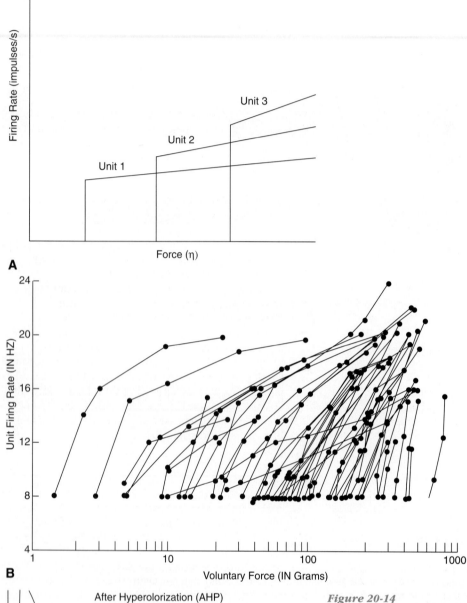

A

B

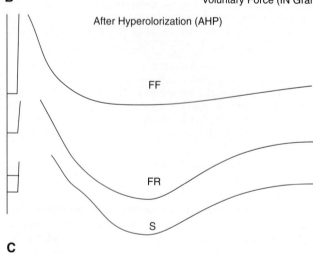

C

After Hyperolorization (AHP)

FF

FR

S

Figure 20-14

Relation between recruitment and rate modulation for different motor units recruited at different motor forces. (**A**) Three motor units were recruited during progressively increasing isometric force (*abscissa*). Unit 1 is recruited at a low force at a rate of 8 pulses/second, and subsequent units are recruited at slightly higher rates. The slope of the relation between firing rate and force is somewhat steeper for the later units, indicating an increasingly important contribution from rate modulation. (**B**) The range of rate modulation and recruitment forces were recorded from motor units in the human extensor digitorum communis muscle. Voluntary force is depicted on the abscissa in a logarithmic scale, and the unit firing rate is shown on the ordinate. There are units recruited across a broad range of force, with units recruited at the lowest forces activated routinely at 8 pulses/second, but these lower threshold units often increase their rate over a limited range and then saturate. Units recruited at higher forces showed a steeper force-rate relation. (**C**) Relation between after-hyperpolarization (AHP) and motor neuron type. S-type motor neurons show the largest and most long-lasting AHP, and FF-type motor neurons show the smallest and the shortest AHP. FR motor neurons show intermediate AHP depth and duration. The magnitude of the AHP is closely correlated with the maximal firing rates of these different units.

by the effective electrical "input resistance" of the motor neuron, which is inversely related to the surface area of the cell. With all other factors held constant, a small neuron would present a higher input resistance than a large neuron because of the reduced membrane area available to transmit the current, and the resulting excitatory potentials in smaller motor neurons would be expected to be larger. Because the neuron voltage threshold is essentially constant among different motor neurons, the differences in excitatory postsynaptic potential size would dictate the order of recruitment.

Evidence indicates that it is highly unlikely that the synaptic current to each motor neuron remains constant across all motor neurons of the pool. It appears that the net synaptic current varies systematically with the number of synaptic terminals distributed over the motor neuron surface and is likely to be greater for larger motor neurons. It is also possible that presynaptic transmitter release, postsynaptic receptor characteristics, and intrinsic membrane properties may vary systematically in different kinds of motor neurons. The relative importance of these factors has not been determined.

Factors governing motor neuron discharge rate. Motor neuron discharge rates vary systematically with recruitment rank order in ways that appear to be well matched to the mechanical properties of the associated muscle fibers. For a given motor neuron pool (see Fig. 20-14), the firing rate at motor neuron recruitment increases with increasing recruitment rank order in larger motor neurons, although this rate increase is relatively modest. A potent source of force increase is the increase in motor neuron firing rate with increasing rank order, and later recruited motor neurons display typically steeper rate increases with increasing force.

The mechanisms regulating motor neuron firing rates are not entirely understood because intracellular current injections can induce motor neurons to fire at extremely high rates, much higher than are seen under physiologic conditions. The most obvious factor limiting motor neuron discharge rate is the after-hyperpolarization (AHP), which is the hyperpolarizing voltage swing that follows the occurrence of the action potential. Although motor neurons display differences in the magnitude of the AHP (see Fig. 20-14C), which are related to their different firing rates, it is not clear that the AHP is the sole influence governing discharge rate under normal conditions. For example, rate variations arising during locomotion may be profound but may not be connected to the magnitude of AHP. It is therefore likely that synaptic current is sometimes limited by some other means. Whatever the mechanism of rate regulation, there is a systematic trend in which initial motor neuron firing rate at recruitment increases with rank order of recruitment, and the impact of increasing discharge rate on net muscle force is progressively important in higher-threshold motor units, primarily because the force generated is greater in the larger motor units.

Regulation of Movement

☐ Motor units are the functional elements of motor control

When a motor axon is excited directly, such as by electrical stimulation of a small ventral root fascicle, or if the related motor neuron is activated electrically by intracellular current injection, the muscle fibers innervated by this motor axon are excited and show a *twitch*, whose mechanical features are related to the number and mechanical properties of the innervated muscle fibers. There has been extensive study of these motor unit properties and of the relation between the mechanical behavior of the muscle fibers and the electrophysiologic properties of the innervating motor neuron. Several of these findings are summarized in Figure 20-15.

Motor units vary in the size and time course of the twitch elicited by a single stimulus and in the resistance to fatigue manifested during repetitive activation. The magnitude of the motor unit twitch varies greatly among mammalian muscles and different motor units in the same muscle. Much of our knowledge is drawn from the hind limb muscles of the cat, especially the medial gastrocnemius. These studies have shown that many muscles, such as the medial gastrocnemius, are heterogeneous, possessing units with a spectrum of mechanical and biochemical properties.

Twitch size

When the twitch responses of different units in a given muscle are compared, the twitch tension or force amplitude varies from several grams to a few milligrams (see Fig. 20-15). Several factors contribute to the differences in twitch tension.

First, large motor units may have more muscle fibers innervated by a motor axon than smaller units. The number of muscle fibers innervated by a single motor axon, which is called the *innervation ratio*, may be larger in the units generating larger tensions. In the case of the medial gastrocnemius muscle, the innervation ratio for large twitch units is greater than that for small units, but the difference is not enough to explain the difference in twitch force. For example, the innervation ratio may vary from 250 to 700 muscle fibers per single axon, a threefold range, but the differences in force may vary by 50- or 100-fold.

A second factor contributing to the generation of muscle force is the *cross-sectional diameter of the muscle fiber*. The cross-sectional diameter is important because it reflects the number of myofibrils that can contribute to force generation. These myofibrils are arranged in parallel and can add to net fiber force independently. The average cross-sectional area of individual muscle fibers in large-twitch motor units is substantially greater than that of small-twitch units, although the difference in area is not quite enough to explain the differences in twitch and tetanic force.

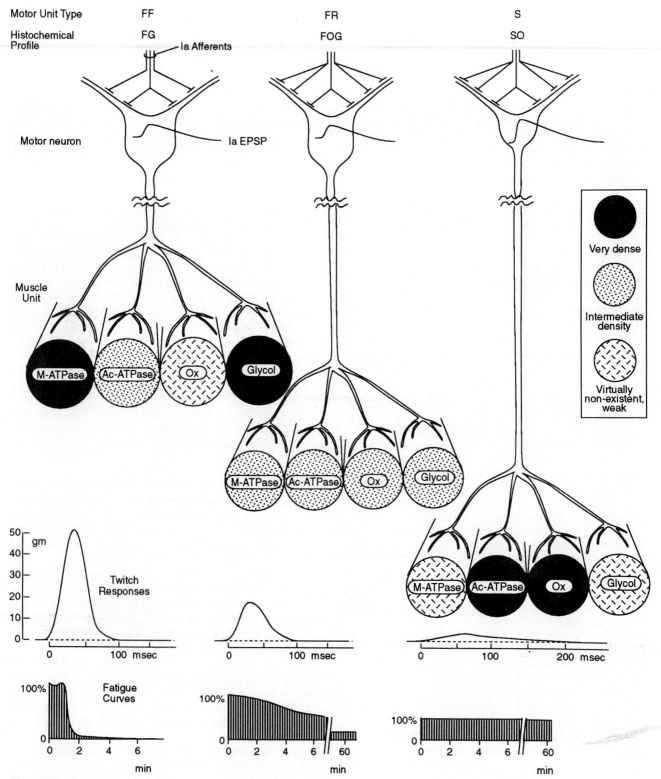

Figure 20-15

Relations between motor neurons and the muscle fibers they innervate. Three different-sized motor neurons innervate muscle fibers of varying diameters (*upper section*). The twitch induced by transient motor neuron excitation reaches the highest peak force in the shortest time for the largest motor neuron, and has the lowest peak force and slowest rise time for the smallest motor neuron. Repetitive activation of the large neuron (*left*) results in a force tetanus which declines very rapidly, decaying almost completely within 2 minutes, whereas the smallest neuron (*right*) generates a smaller tetanus, which does not decline over many minutes; the intermediate unit shows a partial decline in tetanic force. These differing degrees of fatigue sensitivity are used in conjunction with differences in twitch contraction time to classify motor units (*see upper table and text*). Metabolic correlates of the physiological differences are also shown (*central portion*). S-type units show dense concentrations of oxidative (Ox) and mitochondrial (Ac-ATPase) enzymes. FF units stain strongly for glycolytic enzymes and have high concentrations of myosin bound ATPase. FR units have a mixture of both oxidative and glycolytic enzymes. These staining differences support an alternative histochemical classification scheme (*top*): fast glycolytic (FG), fast oxidative glycolytic (FOG), and slow oxidative (SO). The largest neurons receive the most extensive Ia afferent input but with low terminal density, while the smallest motor neurons receive the highest density of Ia afferent input.

The third factor is the specific force-generating capacity of each fiber type, called the *specific tension*. It is possible that fibers belonging to different types of motor units are capable of generating different forces, even if the effects of innervation ratio and of cross-sectional area are eliminated. The *force per unit area*, which is the measure of specific tension, is potentially greater in fast-twitch than in slow-twitch fibers. Attempts to calculate the specific tension of different muscle fibers show at most a threefold difference between fast and slow twitch units, but the accuracy of such estimates is open to question, and the contributions of specific tension are probably even more modest than the aforementioned twofold difference.

Innervation ratios and cross-sectional areas are probably the more important sources of difference in twitch tension for different motor unit types, although specific tension differences may also contribute.

Twitch time course

Different motor units vary greatly in the speed of their contraction, which is measured from the onset of the twitch transient to the peak of the twitch (see Fig. 20-15). This speed of contraction turns out to be an important marker of unit specialization and an excellent predictor of the fatigability of the unit during repetitive activation.

For the case of the cat medial gastrocnemius muscle, motor units are broadly divisible into *fast-twitch and slow-twitch units*, with a boundary value of 55 milliseconds. Units with twitch contraction times longer than 55 milliseconds are called slow-twitch or S-type units. They are able to generate only small twitches and modest levels of tetanic force during repetitive activation. The fast-twitch or F-type units have contraction speeds less than 55 milliseconds, are usually able to generate much larger twitches, and generate correspondingly greater tetanic forces. However, they are not able to sustain this tetanic force without change during prolonged repetitive activation.

The physiologic basis of the differences in twitch contraction times is not fully established, but histochemical techniques have revealed differences in the staining of muscle fibers for a particular myosin ATPase. This ATPase is located on the myosin head and appears to regulate the speed of crossbridge recycling during muscle fiber excitation. This ATPase is not entirely responsible for regulating the speed of twitch contraction, but it may be implicated as one component of the regulatory process.

Fatigability of motor units

When a motor unit is subjected to repetitive activation at high frequency (e.g., 30 pulses/second for 300-millisecond bursts), the unit generates a sustained tetanus, whose force magnitude may be several fold greater than that of the twitch (see Fig. 20-15). This force response to prolonged, repetitive activation has proven to be a helpful classification tool. When an S-type motor unit is activated repetitively, the force response reaches a sustained tetanus and remains at the same level for prolonged periods of time, often as long as several hours. When measured at 2 minutes, the time chosen arbitrarily for evaluation, the force has dropped by less than 25%.

When these motor unit fibers are examined with histochemical stains, the S-type fibers contain many mitochondria and stain heavily for mitochondrial enzymes, such as succinic dehydrogenase, and for mitochondrial ATPases. The S fibers also contain substantial concentrations of myoglobin and are surrounded by a dense capillary network. There is usually relatively little intracellular glycogen stored. The enzymatic profile indicates that these fibers are specialized for oxidative phosphorylation and depend on transported blood glucose or free fatty acids to generate the necessary ATP. Because oxidative phosphorylation is an efficient means of generating ATP, muscle contraction can continue without decrement for prolonged periods if blood flow is sufficient to deliver the needed metabolic substrates.

In contrast, F-type units show a broad range of mechanical behaviors during repetitive activation and a correspondingly broad range of metabolic and enzymatic features on histochemical analysis. During repetitive excitation, some motor units fail to sustain the tetanic force for even 2 minutes, falling to less than 25% of the initial level. These are classified as fast-twitch, fatigable (FF) units. Other F-type units are able to sustain their tetanic force more readily, falling to between 25% and 75% of the initial tetanic force at the 2 minute mark. These are called fast-twitch, fatigue-resistant (FR) units.

Histochemical analyses of the fast-twitch units shows an array of histochemical profiles. The most fatigable FF units have high concentrations of myosin ATPase, few mitochondria, little or no myoglobin, a meager capillary network, and substantial glycogen stores. These units appear to rely on glycolysis to generate the necessary ATP for sustaining contraction and are therefore called FG units in current metabolic classification schemes.

The units showing intermediate degrees of fatigability reveal some residual oxidative machinery on histochemical analysis, including mitochondria and associated mitochondrial enzymatic staining. Because the histochemical profile is mixed, these units are classified as FOG in metabolically based schemes.

It appears that the degree of fatigue, estimated from the loss of force, is related to the capacity to sustain high levels of ATP production. In FF or FG fibers, ATP synthesis declines when glycogen is lost and muscle contraction declines, but in FR or FOG fibers, the residual oxidative contributions delays fatigue onset.

☐ Reflexes are a major component of the neural regulation of movement

Previous sections of this chapter described the spinal elements implicated in the control of muscle force, including the neurons of the spinal cord, the muscle receptors, and the muscle itself. The sections that follow address the functional contribution of these various elements to the control of movement.

Broadly speaking, the spinal cord is engaged in three aspects of movement regulation. In *information transmission*, the spinal cord relays afferent information to higher centers in spinal cord, brain stem, and beyond and *transmits efferent commands* from higher centers to the motor nuclei of spinal cord. In *reflex action*, spinal cord neurons and their connections form the substrates for a variety of *sensory-motor reflexes*. In *pattern generation*, spinal and brain stem interneurons form the basis for *oscillatory neuronal discharge,* which underlies *rhythmic behaviors* such as locomotion, respiration, and mastication.

Types of reflexes

A *reflex* is defined as a stereotypic motor response to a particular sensory input. Reflexes vary broadly in the complexity of their motor responses and in the number and diversity of neural elements that are used. Reflexes may be relatively simple, involving sensory afferents such as Ia afferents from one muscle inducing activation of motor neurons innervating the same muscle. Such reflexes, which include the *tendon jerk* and the *tonic stretch reflex*, are often referred to as *autogenetic* or *homonymous*. Reflexes elicited by muscle afferents of one muscle acting on on motor neurons of a neighboring muscle are often described as *heteronymous*. When such reflexes result in coordinated responses between two or more muscles with similar mechanical actions, the muscles are said to be acting as *synergists*.

At a somewhat higher level of reflex organization, there is a *reciprocal pattern of activation*, in which inhibition is exerted by Ia afferents of one muscle on motor neurons of the antagonist muscle through the Ia inhibitory interneuron.

At the next level, there is a more complex array of reflexes in which the response to a sensory input may be relatively complex and even repetitive or rhythmic. These reflexes include responses such as the *flexion withdrawal reflex*, in which a noxious or nonnoxious stimulus to the skin or other deep tissues evokes a broad-scale coordinated withdrawal of a limb by systematic activation of flexor muscles at several joints.

Beyond the flexion withdrawal reflex, there is an array of complex, goal-directed reflexes, such as wipe and scratch reflexes, in which local excitation of skin surface gives rise to a coordinated and often repetitive motion by the animal to remove the irritant focus from the skin. In mammalian quadrupeds, this is usually called the *scratch reflex*, and in amphibians, it may be referred to as the *wipe reflex*. The fact that the action attempting to eradicate the irritant focus is repetitive and rhythmic in character suggests that the response may also involve an *oscillator* or *pattern generator* that rhythmically excites spinal circuits.

Spring-like properties of the stretch reflex

Since the work of Sherrington, it has been known that stretch of a muscle in a physiologically active animal, such as an unanesthetized decerebrate preparation, gives rise to a substantial increase in motor output, which is reflected as an increase in muscle force. This increase in force is induced by orderly recruitment and rate modulation of motor neurons, in a pattern characterized in earlier sections of this chapter. This increase in motor output results in a systematic increase in muscle force, in which the force rises smoothly and approximately in proportion to the degree of muscle extension. Conversely, when the muscle is allowed to shorten, the reduction in force is essentially proportional to the reduction in muscle length.

These features of the stretch reflex response can be characterized as spring-like, because a change of length is accompanied by a proportional change of force, the defining feature of a simple spring mechanism. When the muscle stretch is maintained, the increase in force is also maintained, verifying the spring-like property, because a spring continues to resist when extended. However, when muscle is stretched at a constant velocity, it also shows a force overshoot at the end of the stretch, suggesting that there is some degree of dynamic or velocity sensitivity.

Comparison of the forces generated over a broad range of velocities indicates that the effect of increasing velocity is relatively modest. A 100-fold increase in stretch velocity induces less than a twofold change in muscle force. When the force increment induced by stretch, measured at a constant length increment, is plotted against stretch velocity, the relation is nonlinear and is well described by a power function in which force increases with velocity raised to the power of 0.2. (i.e., $F = K\ v^{0.2}$). It appears that muscle is a relatively weak, viscous element, because the increase in force is much less than proportional to the increase in velocity. However, the force generated increases relatively steeply at very low velocities, so that muscle emulates a frictional device, resisting motion most powerfully at the onset of motion, where velocities are small.

Comparison of reflex and areflexive behavior of muscle

When active muscle is removed from reflex control, such as by dorsal root section, muscle displays an asymmetric response to stretch; stiffness is modest during stretch but large in shortening. When reflex mechanisms are intact, the asymmetric mechanical properties of muscle are fully compensated, preventing muscle yield from being manifested.

As shown in Figure 20-16, if the mechanical or areflexive response to stretch and release is superimposed on the reflex response (matched for the same initial force), the early phase of the force response to muscle stretch in both cases is virtually identical, indicating that the re-

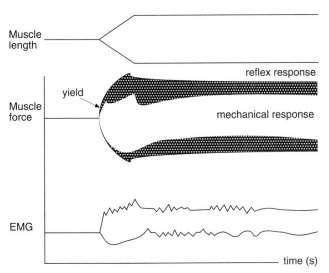

Figure 20-16

Comparison of mechanical and reflex responses of reflexively active soleus muscle of the cat. Muscle is exposed to symmetric stretch and release of constant velocity. In the central part of the figure, the mechanical response is replicated, showing the short-range stiffness, yield, and asymmetric force characteristics. The overall reflex response (*top*) is much more spring-like, because stretch and release produce a smooth and progressive change in the force, which is essentially symmetric. Force changes are sustained even during the plateau phase of the length change. The electromyographic (EMG) responses are also asymmetric, reflecting the different reflex actions during stretch and release. During stretch, the EMG response increases substantially and shows a sustained level of activation during the plateau. During release, there is a modest EMG reduction that is relatively smaller in magnitude, although with a sustained component during the hold phase.

sponse to stretch is initially governed by the intrinsic mechanical properties of active muscle. However, at the point that yielding should have taken place, the reflex response continues smoothly without discontinuity, indicating that effective compensatory mechanisms must have been operating.

The response to muscle shortening is much more similar for reflex and mechanical or areflexive responses, indicating that there is less requirement for neurally mediated compensation during the shortening phase of motion. These differences in the mechanical behavior are mirrored in the electromyographic (EMG) responses, which show substantial EMG increases during stretch and relatively modest EMG reductions during shortening.

The finding that the more complex mechanical properties of muscle, such as the muscle yield, are obscured in the presence of reflex action indicates that the *reflexes linearize and smooth the mechanical behavior of muscle*. In the absence of reflex action, the onset of muscle yield occurs very early in relation to stretch onset, and the mechanical properties of muscle would normally be expected to change abruptly, within 20 to 50 milliseconds of stretch onset. This rapid change is likely to impose severe time constraints on compensatory responses, which may take an additional 25 to 50 milliseconds to

elicit an appropriate mechanical response. A second issue is that the mechanical response is quite asymmetric, requiring substantial reflex compensatory responses in stretch but relatively modest responses in shortening.

These dual *constraints of timing and asymmetry* indicate that straightforward feedback control mechanisms are unlikely to be responsible for compensating muscle yield and asymmetric muscle mechanical behavior, because the time constraints are too severe to allow errors to be detected, transmitted to the cord, and the corrective neural command relayed to and implemented in the muscle. Although the debate about the nature and consequences of reflex action continues, investigators think that much of the compensatory response for muscle yield is built into the response characteristics of the muscle spindle receptors.

This is an unexpected solution to the problem of controlling muscle force, because muscle spindle receptors are usually designated as primary length sensors, rather than as sensors regulating muscle force. In most published experiments, muscle length is controlled by the stretcher, and muscle spindle responses would be expected to be essentially invariant and independent of muscle force levels. However, this receptor is the only candidate with a short onset time for activation, and it has the requisite pattern of asymmetric response to stretch and release. It appears that the neural mechanisms mediating stretch reflex action act predictively, at least initially, when the muscle spindle receptor issues a response that is appropriate for correcting impending changes in muscle properties, such as muscle yielding and asymmetric muscle stiffness.

Although tendon organ responses could contribute to and promote the improvement of muscle mechanical properties, the speed of tendon organ–mediated feedback is too slow to prevent the manifestation of rapid-onset mechanical changes, such as muscle yielding. Moreover, the appropriate compensatory muscle mechanical responses occurs even when Ib interneuron responses are modest or absent.

Assessment of the hypothesis of stiffness regulation

The dual and competing actions of muscle length feedback, which can promote increased stiffness of muscle, and force feedback, which can reduce muscle stiffness, are simultaneously active in many conditions, suggesting that the regulation of stiffness to some predetermined level may be a function of stretch reflex mechanisms (Fig. 20-17). This view was advanced by Nichols and Houk (1976), who showed that the stiffness of muscle was more constant in the presence of reflex action, although the actual magnitude was not held closely to a specific value.

Most researchers think that the evidence in support of a primary role of the stretch reflex as a stiffness regulator is limited, because muscle stiffness does not remain constant or even approximately constant under most operating conditions. However, the presence of force and

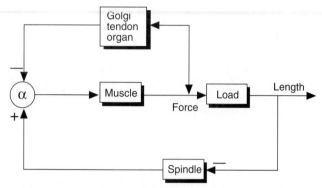

Figure 20-17
Servocontrol of the spinal cord and the regulation of muscle length and force. The diagram documents the classic associations among the segmental motor apparatus and muscle sensors and the central connections in regulating force. The alpha motor neuron serves as summing junction for the afferent inflow. Alpha motor neurons innervate muscle, which exerts force against the load of the limb or against additional superimposed loads. After muscle has acted on the load, there may be a resulting change in muscle length, which alters the spindle afferent discharge. Muscle tendon organs are also activated by increasing muscle force. Muscle spindle receptors project directly to the spinal neurons. Increasing muscle length induces a force that opposes the length change in this negative-feedback loop. Muscle force sensors, such as tendon organs, also inhibit homonymous motor neurons, closing a force-regulating loop. In combination, simultaneous force and length regulation achieves net regulation of muscle stiffness.

length sensors, together with their reflex connections, produces a much more spring-like behavior of muscle, and it is this characteristic of muscle that is likely to be important in the control of movement. For example, the presence of predictive reflex compensation means that the sharp extension of ankle extensors, such as the soleus extension that takes place during the stance phase of locomotion, would induce a smooth force increase and enhance effective damping of the contact point.

Pattern generation

Virtually all nervous systems contain groups of neurons that have the capacity to generate rhythmic bursting behavior, even if isolated from other neuronal systems. These neurons may show spontaneous bursting, or they may need a gating signal, as is provided by monoaminergic agents like norepinephrine, but the pattern of discharge does not depend on incoming afferent or descending signals; these are called *pattern generator* neurons.

On the basis of extensive studies from acute and chronic spinally transected preparations, including cat and turtle, it is evident that such neuronal clusters can induce repetitive discharge sufficient to drive locomotion. The finding that such locomotion-like patterns of motor neuronal discharge do not depend on descending, ascending, or peripheral afferent inputs supports the view that locomotion emerges from the activities of a discrete oscillator, located primarily within the interneurons of the spinal cord.

In the presence of appropriate tonic pharmacologic drive, provided by norepinephrine or by other α-adrenergic agonists (e.g., clonidine), the interneuronal systems continue to discharge, showing the essential features responsible for rhythmic bursting. The pattern generator appears to be activated by descending inputs from locomotion control regions in the mesencephalon and other areas of the brain stem. Some of these brain stem regions release of norepinephrine in the spinal cord; others may trigger norepinephrine release indirectly.

The discharge of the spinal pattern generators is subject to peripheral modulation by appropriate segmental afferent inflow. Cutaneous stimulation of a foot in a quadruped during the swing phase of locomotion gives rise to avoidance behavior in which the leg is caused to circumvent the obstruction, but when the leg is still weight bearing, similar cutaneous stimulation is ineffectual, indicating that there is a substantial phase-dependent modulation of the effects of sensory input.

SELECTED READINGS

Burke RE, Levin DN, Tsairis P, Zajac FE. Physiological types and histochemical profiles in motor units of the cat gastrocnemius. J Physiol (Lond) 1973;234:723.

Burke RE. Motor unit types of cat triceps surae muscles. J Physiol (Lond) 1967;194:141.

Burke RE. Motor units: anatomy physiology and functional organization. In Brooks V (ed). Handbook of physiology, sect I. Washington, DC: American Physiological Society, 1981:345.

Crago PE, Houk JC, Rymer WZ. Sampling of of total muscle force by tendon organs. J Neurophysiol 1982;47:1069.

Henneman E, Mendell LM. Functional organization of the motoneuron pool and its inputs. In Brooks V (ed). Handbook of physiology, sect I. Washington, DC: American Physiological Society, 1981:423.

Houk JC, Rymer WZ. Neural control of muscle length and tension. In Brooks V (ed). Handbook of physiology, sect I. Washington, DC: American Physiological Society, 1981:257.

Huxley AF, Niedergerke R. Structural changes in muscle during contraction. Nature 1954;173:971.

Huxley HE, Hanson EJ. Changes in the cross-striation of muscle during contraction and stretch, and their structural interpretation. Nature 1954;173:973.

Mathews PBC. Muscle spindles, their messages and their fusimotor supply. In Brooks V (ed). Handbook of physiology, sect I. Washington, DC: American Physiological Society, 1981:257.

Nichols TR, Houk JC. Improvement in linearity and regulation of stiffness that results from actions of the stretch reflex J Neurophysiol 1976;39:119.

Stein RB. What muscle variable does the nervous system control in limb movements. Behav Brain Sci 1992;5:535.

Vallbo AB. Discharge patterns in human muscle spindle afferents during isometric voluntary contractions. Acta Physiol Scand 1970; 80:552.

CHEMICAL MESSENGER SYSTEMS

Neuroscience in Medicine, edited by P. Michael Conn.
J. B. Lippincott Company, Philadelphia © 1995

Chapter *21*

*C*hemical Messenger Systems

DAVID K. SUNDBERG

Monoamine Oxidase Inhibition May
Benefit Patients With Parkinson's
Disease
The Lewy Body Is the Pathologic
Hallmark of Parkinson's Disease
Dopaminergic Preparations Are the
Most Effective Therapy for
Parkinson's Disease

Severe Complications Can Occur After
Years of Levodopa Therapy
Tissue Transplantation May Improve
the Symptoms of Parkinson's
Disease

Neurotransmitters, Neurohormones, and Neuromodulators

Much of the information about brain function was originally derived from studies of the peripheral nervous system. These studies, which started in the late 1800s and the beginning of this century, focused mainly on the somatic (e.g., neuromuscular junction) and autonomic (i.e., sympathetic and parasympathetic) nervous systems. These areas offered an accessible, discrete, and less complex model of interactions between neurons and target cells. The science of grinding up specific organs in an effort to extract their "vital essence" began in earnest in the middle 1800's and led to the concept of chemical messenger systems. Neurotransmitters, neurohormones, and neuromodulators are the chemical messengers that allow one cell to communicate with another cell or with itself.

☐ Neurotransmitters are small organic molecules that carry a chemical message from a neuronal axon or dendrite to another cell or nerve

Even in ancient times, remarkably accurate concepts existed suggesting that the large nerves of the body, which could be easily visualized during dissection, carried a substance or substances that coordinated and activated the body. The Greeks and Romans called this substance *psychic pneuma*, a product of vital and animal pneuma from the lungs and heart. Early Hindus suggested that the spinal cord and the sympathetic chain ganglia were channels, called *chakra*. These chakra carried a substance, called *prana*, the flow of which could be augmented by the practice of Yoga. As early as 3000 B.C., the Chinese taught about the flow of *chi* or an energy that flowed through the body in channels that could be helped by the medical practice of acupuncture and by an exercise called tai chi chuan.

The classic neurotransmitters are now thought of as small molecules, synthesized and stored in secretory granules of the axons of nerves. Chemically, these transmitters can be biogenic amines (e.g., the catecholamines norepinephrine and dopamine and the indolamines, ser-

---

Table 21-1
STEPS TO IDENTIFY A SUBSTANCE AS A NEUROTRANSMITTER

1. Anatomic: The substance must be present in appropriate amounts in the presynaptic process.
2. Biochemical: The enzymes that synthesize the substance must be present and active in the terminal, and enzymes that degrade it must be present in the synapse.
3. Physiologic: Stimulation of the presynaptic axon should cause the release of the substance, and application of the specific substance should mimic stimulation of the nerve.
4. Pharmacologic: Drugs that affect specific enzymatic or receptor-mediated effects of the proposed transmitter substance should alter nerve stimulation through changes in synthesis, storage, release, uptake, or stimulation or blockade of the receptor.

---

otonin, and histamine), acetylcholine or amino acids (e.g., γ-aminobutyric acid [GABA], glutamate, glycine), or small proteins (i.e., peptides) that are synthesized in the cell body and carried to the axon by axoplasmic transport. These transmitters inhibit or stimulate the function of the target cell by combining with receptors and initiating a biochemical change in the membrane, cytoplasm, or nucleus of that cell.

The process is called *nerve transmission* and refers to the passage of biochemical information by means of a chemical messenger from a nerve across a specific junction to another cell. This should not be confused with *nerve conduction*, which is the passage of an electrochemical current down a nerve axon. Table 21-1 lists some proposed criteria for classifying a substance as a neurotransmitter.

☐ Neurohormones are chemical messengers that are secreted by the brain into the circulatory system and alter cellular function at a distance

The word *hormone* was originally coined by E.H. Starling in 1905 in reference to a "chemical messenger which, speeding from cell to cell . . . coordinates the activities and growth of different parts of the body." A neurotransmitter, such as epinephrine from the adrenal medulla, can also be a neurohormone. In the brain, certain hypothalamic neurons make small proteins, the *peptide neurohormones*. These are secreted from axons directly onto a portal vasculature system, in which they are carried to the anterior pituitary to influence endocrine function. The early Greeks and the Romans ascribed a similar function to this part of the brain. Galen and later Vesalius suggested that the hypothalamus produced *pituita* (Latin for phlegm), which was distilled from the ventricular system and secreted through the hypothalamus into the pituitary and then to the nose.

Just as a neurotransmitter can be a neurohormone, the opposite can also occur. Many of the neurohormone-producing cells have axons that project into other areas of the brain, the brain stem, and spinal cord to influence other somatic and behavioral functions. A full discussion of this subject is found in Chapter 16.

☐ Neuromodulators are transmitters or neuropeptides that alter the endogenous activity of the target cell

Excitable cells are unique in having a specific complement of ion channels that dictate how that cell behaves. The spontaneous behavior of cells is sometimes referred to as *endogenous activity*. This activity is best identified when the cell or nerve is removed from its normal environment and studied in vitro. One of the best examples is the rhythmic activity of the viscera that is modulated by the autonomic nervous system. When removed from a frog, the heart continues to beat if incubated in an ionic buffer containing calcium. The rate of contraction of the heart depends on the rhythmic action of ion pumps that maintain a gradient of sodium and potassium across a relatively leaky membrane. Sympathetic nerves alter the intrinsic strength and rate of contraction of the heart through cardiac β receptors that modulate the activity of intracellular messengers and membrane ion channels.

Neurons in the brain also have ionic pumps that rhythmically maintain an endogenous activity or "firing pattern." Neuromodulation allows a cell or a nerve to adapt its activity to changes that occur in its environment.

Although earlier investigators assumed that each neuron contained only one transmitter, many nerves contain several transmitters. Studies have shown that the main transmitter and the co-transmitter can be specifically released from the same nerve at different frequencies of nerve activity because of the distribution of calcium channels within the terminal. As neuromodulators, co-transmitters serve a feedback role and are referred to as *autocoids*. By combining with the autoreceptors on its own presynaptic membrane, a transmitter can augment or inhibit further nerve activity. In some nerves, such as the peripheral sympathetic nerves, a substance can have transmitter and modulator functions. For example, norepinephrine can activate a postsynaptic receptor and influence target tissue function (e.g., heart rate) or stimulate a presynaptic receptor (e.g., α_2) to inhibit further norepinephrine release.

An experimental technique called *immunocytochemistry* is used to visualize a transmitter or the specific enzymes needed to make a neurotransmitter within a nerve cell or axon. By using two antibodies to the specific substances and stains that bind to the antibodies, the investigator can visualize the transmitter and the co-transmitter within the same axon or nerve cell body. In many cases, a typical small neurotransmitter and a neuropeptide coexist in nerves as the pair of co-transmitters. Either of the two substances may serve the neuromodulator function. For example, although norepinephrine is the transmitter and neuropeptide Y (NPY) the modulator in sympathetic nerve endings, it may be the reverse in the brain. A list of transmitters with documented co-transmitters is given in Table 21-2.

The classic smaller neurotransmitters are stored in synaptic vesicles (30 nm in diameter) that are manufactured and packaged within the axon terminal itself. The larger peptide co-transmitters are stored in large synaptic vesicles that are synthesized and assembled in the neuron cell body (i.e., soma) and reach the terminal by axonal transport, a process that can take hours or days.

The small transmitter vesicles are concentrated in an active zone of the terminal, associated with an electron-dense area of membrane that contains many calcium channels. Low-frequency nerve stimulation causes a local increase in calcium in the active zone and results in the

Table 21-2
COEXISTENCE OF NEUROTRANSMITTERS AND MODULATORS

Neurotransmitter	Co-Transmitter
Catecholamines	
Dopamine	Cholecystokinin, enkephalin
Norepinephrine	Neuropeptide Y, neurotensin, enkephalin
Epinephrine	Enkephalin, neuropeptide Y
Serotonin	CCK, enkephalin, substance P
Acetylcholine	VIP, substance P, enkephalin, CGRP
GABA	Enkephalin, neuropeptide Y, CCK
Glutamate	Substance P
Glycine	Neurotensin

CCK, cholecystokinin; *GABA*, γ-aminobutyric acid; *VIP*, vasoactive intestinal polypeptide.

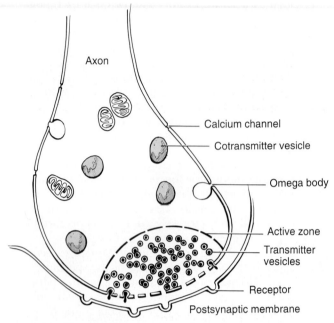

Figure 21-1

The diagram of an axon terminal shows the active zone with many transmitter vesicles and calcium ion (Ca²⁺) channel. The co-transmitters are localized in larger granules in the periphery of the terminal where there are fewer Ca²⁺ channels. To achieve the necessary Ca²⁺ concentrations for co-transmitter release, a higher frequency of nerve stimulation is necessary.

release of the small transmitter by fusion of the anchored vesicle with the membrane. The intracellular calcium is rapidly sequestered by binding to neuronal proteins and being pumped out of the axon. In this way, calcium levels only reach a threshold value in the active zone during low-frequency stimulation. For the peptide co-transmitters, the larger vesicles are found in other parts of the axon terminal, distant from the active zone. Higher-frequency nerve stimulation results in increased calcium concentrations in these areas of the terminal with co-transmitter release. Low-frequency nerve stimulation can specifically release the small transmitter in the active zone, and higher frequencies effect co-transmitter release. These interactions are shown in Figure 21-1.

☐ Transmitters can be classified as fast or slow

Synaptic transmission can be mediated in a rapid or a slow fashion, depending on whether the action persists for milliseconds, seconds, minutes, hours, or days. This property is a function of the specific transmitter and the type of receptor used to mediate the postsynaptic event. For example, certain amino acids and acetylcholine at the nicotinic receptor cause a rapid alteration in postsynaptic function that can be measured in milliseconds. The molecular mechanism is mediated through the opening or closing of ionic channels. The acetylcholine and gluta-

mate receptors are composed of five protein subunits that form a channel through the membrane for sodium or calcium. Activation of these receptors induces a rapid change in the ionic movement across the membrane.

Other transmitters induce slower cellular events. Acetylcholine at the muscarinic receptors and the catecholamines induce a slow change in the target cell that can take several seconds to minutes. These receptors are represented in the membrane by a single subunit, and they initiate a biochemical event within the cell by activation of a G protein. Still other transmitters, exemplified by the neuropeptides and many growth factors, have effects that occur over periods of hours or days.

The site of synthesis and axonal transport of transmitters and the enzymes needed for their metabolism is an important factor in the rate of transmission. Axonal transport occurs at a rate of 1 to 400 mm/day. The slow peptide transmitters, which are synthesized in the cell body and require transport to the terminal before release, are extremely potent agents that require some time for their actions. The fast transmitters would be easily depleted if they had to rely on axoplasmic transport to supply the needed stores in the axon. These transmitters are all synthesized and stored in the axon, and conservation of the released transmitters is achieved by re-uptake and storage into the nerve axon. Although fast transmitters usually act by changing the distribution of ions across a membrane and altering electrical potentials, the slow transmitters cause biochemical changes within the cell. These changes may include phosphorylation of proteins, activation of enzymes and second messengers, or even activation of genetic transcription and production of new proteins.

☐ Nerve activity can be measured by determining the firing rate or rate of neurotransmitter release or turnover

One of the most important issues in neurobiology is determining the activity of a specific nerve or type of nerve under various physiologic or behavioral states. These challenges have been primarily addressed by electrophysiologic or neurochemical methods.

Nerve activity estimated by electrical behavior

Nerve activity can be estimated by measuring the electrical behavior. This is done by inserting an electrode into the brain and measuring electrical activity during different conditions. This technique can be extremely difficult in an awake, freely moving animal, in which movement of the electrode by as much as a few microns can confound results. Refinements of the electrophysiologic methods have progressed from measurement of nerve bundles to measurements of single neurons and to the new technique of *patch clamping*. In the patch clamp technique, the properties of a little patch of membrane from an individual nerve is studied after being sucked onto the end of a micropipette. This can demonstrate the actual types of ionic channels on a small, isolated piece of membrane.

Nerve activity determined by the rate of transmitter release or turnover

Various techniques have been developed to measure the rate of release of a transmitter. These include monitoring the release from isolated nerves in tissue culture (in vitro), placing a small cannula (i.e., push-pull cannula) within a brain area of an anesthetized or awake animal and perfusing that area of the brain with a physiologic media, or implanting a similar perfusion cannula attached to a dialysis membrane (i.e., microdialysis) that restricts the recovery of brain elements to only specific molecular weights. Transmitter release can then be determined in the media, perfusate, or dialysate.

Several turnover techniques have been used to measure neuronal activity. The rate of use of a transmitter can be estimated by measuring its disappearance after chemically inhibiting its synthesis. An example of this technique is injecting α-methyl-*p*-tyrosine (250 mg/kg), which inhibits tyrosine hydroxylase, into an animal and measuring the rate of disappearance of norepinephrine or dopamine. Turnover can also be estimated by measuring the rate of accumulation of the transmitter or disappearance of its metabolite after chemically inhibiting its metabolic breakdown. An example is measuring the appearance of serotonin or the disappearance of its primary metabolite, 5-hydroxyindoleacetic acid, after inhibition of the enzyme monoamine oxidase (MAO) with the drug pargyline or iproniazid.

All of these techniques have provided valuable information about the activity of specific neurotransmitter pathways under different physiologic conditions and behavioral states. They have allowed the development of many behavioral models, such as the concept that the transmitter dopamine is involved in feelings of reward and reinforcement.

Small-Molecular-Weight Transmitters

The small-molecular-weight transmitters are made locally within the nerve cell axon or dendrite. The well-established small-molecular-weight transmitters and modulators include three amino acids, five biogenic amines (essentially decarboxylate derivatives of amino acids), and acetylcholine. All of these molecules are synthesized and stored in the axon terminal, from which they are released after the arrival of an action potential. These transmitters can act fast or slowly, depending on their postsynaptic receptor type. If their synthesis required axoplasmic transport from the cell body, they would soon be depleted. The enzymes responsible for making these compounds are synthesized in the cell body and transported to the axon, where the actual transmitter synthesis takes place.

The discussion of these transmitters focuses on the history of the discovery of the transmitter, the mechanism of their biosynthesis, secretion and catabolism (i.e., degradation), their anatomic distribution, and their functions and physiologic actions. The function of most of these chemical messenger systems is still poorly understood. Some of these functions are assessed in cases of behavioral or physiologic alterations due to disease states, specific lesions, or the effects of pharmacologic intervention.

☐ Acetylcholine is the prototypical neurotransmitter

The first neurotransmitter to be isolated and identified was the small ester of the lipid choline, *acetylcholine*. The entire concept of neurotransmission developed from its discovery. In the early 1900s in Germany, Otto Lowe showed that a chemical substance, released on stimulation of the vagus nerve of an isolated frog's heart, would slow the beating of another frog's heart bathed in the same media. He called this substance *vagus-stuff* because it was released by stimulation of the vagus nerve. Nerve systems that use acetylcholine as a transmitter are called *cholinergic nerves*.

Anatomy and function of cholinergic neurons

Cholinergic neurons are found in the central and peripheral nervous system. Peripheral nerves that use acetylcholine include several important pathways, including the motor neurons that begin in the ventral root of the spinal cord and innervate striated skeletal muscle; the preganglionic sympathetic and parasympathetic nerves that begin in the intermediolateral column of the spinal cord or brain stem and activate all the autonomic ganglia; the postganglionic parasympathetic nerves that innervate the viscera (i.e., heart, pulmonary bronchi, gastrointestinal tract, bladder, eye, and exocrine glands); and the postganglionic sympathetic cholinergic nerves to the major sweat glands of the skin. Because of their simplicity and the relative ease of studying them, much has been learned from these peripheral cholinergic nerves about nerve transmission.

Although all of these pathways respond to the same neurotransmitter, acetylcholine, the postsynaptic receptors that trigger the cellular event are different. In the neuromuscular junction and the autonomic ganglia, these receptors are called *nicotinic* because they specifically respond to the drug nicotine. In the case of the viscera (e.g., gut, heart, lungs) that are innervated by the parasympathetic nerves and the sweat glands, the receptors are called *muscarinic* because they are activated by the compound muscarine. Cholinergic receptors are discussed further in Chapter 22. Although the neurotransmitter is the same, the receptors are different. These different cholinergic receptors are also found in the brain, where the muscarinic types outnumber the nicotinic receptors by 10-fold to 100-fold.

After acetylcholine is release from the peripheral autonomic nerves, several events may occur. Acetylcholine triggers contraction of skeletal muscle. In the autonomic ganglia, acetylcholine stimulates the postganglionic nerves of sympathetic or parasympathetic nerves. Acetylcholine from parasympathetic nerves slows down the heart rate; constricts smooth muscle of the bronchi, stomach, intestines, bladder, and eye; and stimulates secretion of various enzymes, hydrochloric acid, and mucus from exocrine glands. A cholinergic receptor on the endothelium of the vasculature induces a biochemical event that produces nitric oxide from the amino acid arginine, which causes smooth muscle relaxation. Recent studies have shown that nitric oxide synthetase is present all over the brain and may represent a very important transmitter or second messenger system.

In the central nervous system, cholinergic pathways are present throughout the brain. Many of these pathways are diagramed in Figure 21-2. In the ventral forebrain, cholinergic cell bodies from the septum, diagonal band of Broca, and basal nucleus send axons to innervate the hippocampus, interpeduncular nuclei, and neocortex, respectively. Cholinergic cell bodies from the tegmentum of the brain stem innervate the hypothalamus and thalamus. There are also short cholinergic interneurons within the striatum.

The central cholinergic pathways in the striatum are important in the central motor control of muscles. This is evidenced by the fact that atropine, a drug that blocks muscarinic acetylcholine receptors, can alleviate the tremors associated with Parkinson's disease. Acetylcholine may also be important in memory consolidation, because cholinergic changes are found in the cortex of patients with Alzheimer's disease. Pharmacologic evidence suggests that blockade (e.g., scopolamine) of the central muscarinic cholinergic synapses results in sedation and amnesia, but activation of nicotinic synapses increases alertness and can be rewarding, as is evidenced by cigarette smoking.

Neurochemistry of cholinergic neurons

The cholinergic axon has the capacity to synthesize, store, and secrete acetylcholine in a manner relatively independent from the nerve cell body. The uptake of choline appears to be the rate-limiting step in acetylcholine synthesis.

Acetylcholine is synthesized from choline and acetyl-coenzyme A in the nerve terminal by the enzyme choline acetyltransferase (Fig. 21-3). This enzyme, which has a molecular weight of about 67,000, is ribosomally synthesized in the cell body and transported to the axon. The

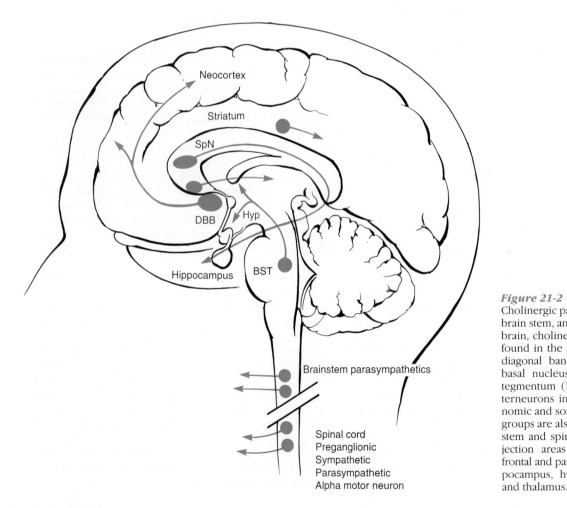

Figure 21-2
Cholinergic pathways in the brain, brain stem, and spinal cord. In the brain, cholinergic cell groups are found in the septal nuclei (SpN), diagonal band of Broca (DBB), basal nucleus (BN), brain stem tegmentum (BST), and small interneurons in the striatum. Autonomic and somatic motor nuclear groups are also found in the brain stem and spinal cord. Major projection areas are found in the frontal and parietal neocortex, hippocampus, hypothalamus (Hyp), and thalamus.

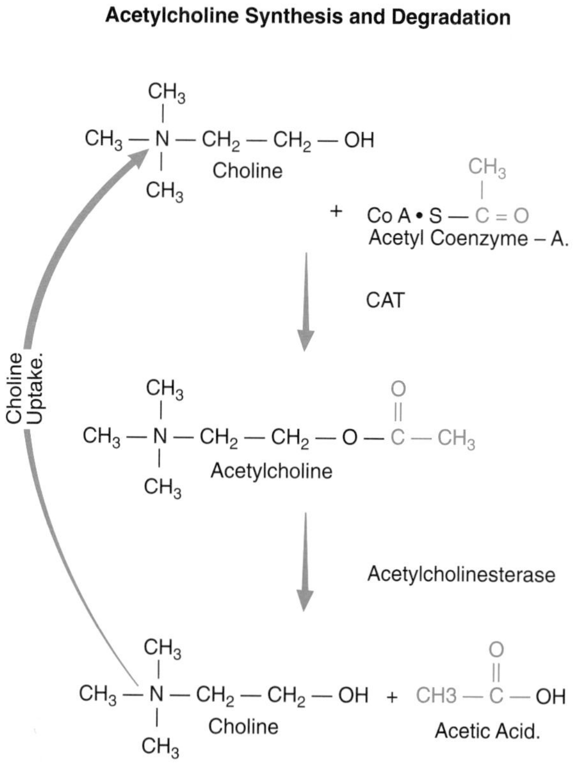

Acetylcholine Synthesis and Degradation

Figure 21-3
Acetylcholine synthesis is catalyzed in the cytoplasm of cholinergic cells by the enzyme choline acetyltransferase (CAT). The transmitter is degraded by the enzyme acetylcholinesterase into choline and acetic acid. The choline is taken back up in the nerve by a high-affinity pump for reuse. Shaded parts of the molecule denote additions and deletions by enzymatic activity.

enzyme is a cytoplasmic enzyme, but because it is positively charged, it is often associated with intracellular, mitochondrial, or vesicular membranes. Choline is a component of the complex lipids of all cell membranes (i.e., occurring as lecithin and sphingomyelin), and all cells have a choline uptake pump. However, uptake pumps in noncholinergic cells are of low affinity (K_m of 40 to 100 μM). Acetylcoenzyme A is produced within the nerve terminal itself using normal energy sources. Acetylcholine is synthesized in the cytoplasm and subsequently sequestered in small electron-opaque granules that are most prevalent in the active zone of the nerve terminal. Acetylcholine is stored within these granules with ATP at a concentration of about 1 M and with a charged anion protein called vesiculin.

When acetylcholine is released, it is rapidly degraded by the enzyme acetylcholinesterase, which is found in high concentrations within the synaptic cleft and in other areas. One form of this enzyme, called true acetylcholinesterase, is found in the synapse and is associated with the neuromuscular junction and cholinergic areas of the brain. This enzyme degrades acetylcholine to choline and acetic acid. True acetylcholinesterase is one of the fastest enzymes known. The maximal velocity of acetylcholine degradation has been measured at 75 g of substrate per hour using 1 mg of purified enzyme. It has been calcu-

lated that one molecule of cholinesterase can hydrolyze as many as 5000 molecules of acetylcholine per second. These biochemical events are diagramed in Figure 21-3.

The importance of acetylcholinesterase for the termination of the cholinergic transmission can be demonstrated if this process is pharmacologically inhibited. Many drugs, insecticides, and even war gases inhibit acetylcholinesterase in a reversible or irreversible fashion. The pharmacologic actions of these compounds produce overstimulation of the cholinergic receptor, which can lead to tetanic paralysis and death.

The choline that is formed from the degradation of acetylcholine is actively pumped back into the nerve terminal and reused to synthesize more acetylcholine. To accomplish this re-uptake, the cholinergic neuron has developed a high-affinity choline pump (K_m of 0.4 to 4.0 μM.) that rapidly sequesters any choline from the synaptic cleft. This process for supplying choline for transmitter synthesis appears to be the *rate-limiting step* for the production of this transmitter. Drugs that inhibit the uptake of choline (i.e., hemicholinium) produce rapid paralysis of muscular activity. The muscles that are used most are those that become paralyzed first, indicating that the rate of cholinergic nerve activity in these muscles leads to exhaustion and depletion of acetylcholine.

☐ Biogenic amine transmitters are decarboxylated amino acids

The aminergic transmitter systems have been given several names, which can cause confusion. The biogenic amines that are synthesized from tyrosine are called *catecholamines*, because their structure contains a catechol moiety (i.e., a phenol ring with two hydroxyls). They are also referred to as adrenergic after the English term, adrenalin. Serotonin or 5-hydroxytryptamine (5-HT) is also called a biogenic amine. Structurally, however, it is an indolamine because of its ring structure, which arises from tryptophan. Histamine is a biogenic amine that is present in the brain and the enteric nervous system of the gastrointestinal tract. Chemically, it is the decarboxylated amino acid histidine, and it is present in high concentrations in mast cells and in some neuronal pathways.

Catecholamine transmitters

The catecholamine transmitters, norepinephrine, epinephrine, and dopamine, serve a variety of functions in the peripheral and central nervous system.

Around the turn of the century, when acetylcholine was being established as a neurotransmitter, the sympathetic nerves and the adrenal medulla were being extracted for substances that raised the blood pressure and accelerated the heart rate. Otto Lowe, one of the discoverers of acetylcholine, which he called vagus-stuff, also worked with the sympathetic transmitter, and called it *acceleranse-stuff* because it increased the heart rate. The adrenal medulla is an autonomic ganglia where the postganglionic neurons (designated chromaffin cells) do not possess long axons. They have the capacity, as do a variety of neurons in the brain, to synthesize the catecholamine transmitter and neurohormone epinephrine.

Catecholamines, when exposed to light or alkaline pH, oxidize to colored substances called quinones. This helped to establish the chemical identity of these amines and helped identify some of their pathways in the brain. The substantia nigra of the midbrain, for example, is the Latin term for black substance. This area of the brain, which contains dopamine, turns dark when exposed to light or air. These neurons are unable to make dopamine in patients with Parkinson's disease, leading to the clinical manifestation of this disease.

The catecholamines serve as sympathetic nervous system transmitters and help to control most visceral activity. The major amine is norepinephrine, and epinephrine and norepinephrine are secreted in a ratio of 4 : 1 from the adrenal medulla. Dopamine may also be an unrealized sympathetic neurotransmitter, particularly because specific dopamine receptors have potent actions on kidney blood flow.

The peripheral actions of the catecholamines are well understood, because this is such an isolated system. The actions include constriction or dilation of the arteries controlling blood pressure, the increase of heart rate and strength, dilation of the bronchioles of the lungs, a decrease in gastrointestinal activity, dilation of the iris of the eye, and a variety of metabolic effects, including glycogenolysis, lipolysis, and the release of renin.

The central actions of catecholamines can best be described as arousal. Because of the complexity of the brain, their functions are not fully elucidated, but they can be somewhat understood from experiments using specific lesions of catecholamine pathways or the actions of selective drugs.

Tyrosine precursor. The amino acid tyrosine is the common precursor for all three catecholamine transmitters, dopamine norepinephrine, and epinephrine. The biosynthesis of the catecholamine neurotransmitters occurs primarily in the nerve terminal, as for other small-molecular-weight transmitters. The enzymes that catalyze this synthesis are made and packaged in the nerve cell body, where they are synthesized by ribosomes before transport down the axon to the terminal.

Within the terminal, specific membrane pumps supply tyrosine or phenylalanine as the precursors for amine biosynthesis. The details of this pathway are shown in Figure 21-4. Some of the biochemical parameters of the enzymes that synthesize the transmitters are also shown in Table 21-3. Tyrosine hydroxylase is the first and rate-limiting enzyme for the synthesis of all three catecholamines, and it is found in the cytoplasm. This enzyme requires molecular oxygen, iron, and a cofactor called tetrahydrobiopterin. This cofactor helps to maintain tyrosine hydroxylase in a reduced, active state. When catecholamines, such as dopamine or norepinephrine, build up in the cytoplasm, they inhibit the ability of the pteridine to activate tyrosine hydroxylase. This provides one of the major regulatory steps in catecholamine synthesis,

Catecholamine Biosynthesis

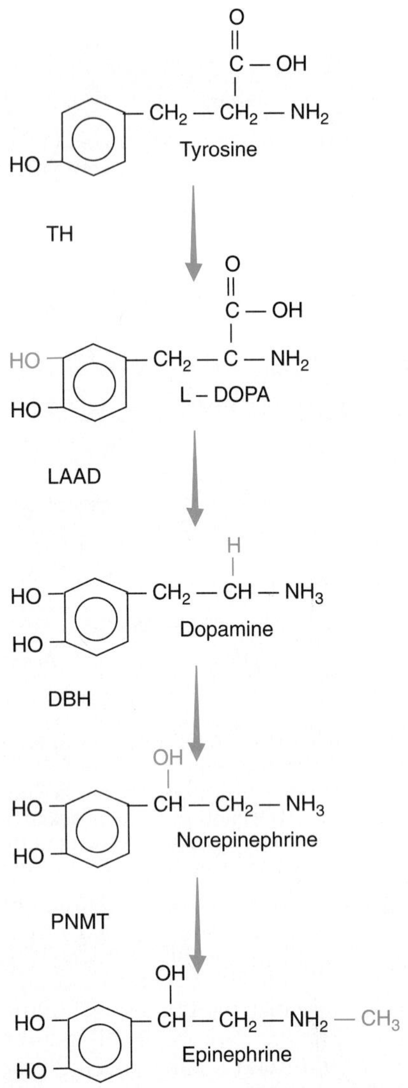

Figure 21-4
The biosynthesis of the catecholamines norepinephrine, epinephrine, and dopamine is illustrated from the common precursor, tyrosine. The enzymes involved are tyrosine hydroxylase (TH), L-amino acid decarboxylase (LAAD), dopamine-β-hydroxylase (DBH), and phenylethanolamine-N-methyltransferase (PNMT). The catalyzed additions are shaded.

which is essentially an end-product inhibition of the rate-limiting enzyme through a cofactor. Other important regulatory steps occur. Tyrosine hydroxylase can be phosphorylated, which increases its affinity for the pteridine cofactor. This allosteric activation seems to be mediated by a variety of presynaptic receptors that use cyclic $3',5'$-AMP or intracellular calcium concentration as second messengers. In this way, catecholamine synthesis is controlled by the rate-limiting enzyme tyrosine hydroxylase, which can be regulated by end-product feedback inhibition or by allosteric activation through second messenger systems.

Table 21-3
BIOCHEMICAL PROPERTIES OF ENZYMES

Enzyme	Molecular Weight	Cofactors	Affinity (K_m)
Tyrosine hydroxylase	60,000	Tetrahydrobiopterine, molecular O_2, Fe, NADPH	$0.4–2 \times 10^{-4}$ M (tyrosine)
L-amino acid decarboxylase	85,000–90,000	Pyridoxal phosphate (vitamin B_6)	4×10^{-4} M (L-dopa)
Dopamine-β-hydroxylase	75,000	Ascorbate (vitamin C), molecular O_2, Cu^{2+}	5×10^{-3} M (dopamine)
Phenylethanol-amine-N-methyl transferase		S-Adenosylmethionine	
Tryptophan hydroxylase	60,000	Molecular O_2, tetrahydrobiopteridine	5×10^{-5} M (tryptophan)
Choline acetyltransferase	67,000	Acetylcoenzyme A	7.5×10^{-4} M (choline) 1×10^{-5} M (coA)
Glutamic acid decarboxylase	85,000	Pyridoxal phosphate, vitamin B_6	7×10^{-4} M (glu) 5×10^{-5} M (B_6)

Hydroxylation of tyrosine by tyrosine hydroxylase yields the amino acid L-DOPA (L-3,4-dihydroxyphenylalanine). This intermediate never reaches high concentrations in the cytoplasm because it is immediately decarboxylated by the enzyme, L-amino acid decarboxylase. L-Amino acid decarboxylase and dopamine-β-hydroxylase have an activity that is 10 to 1000 times higher than tyrosine hydroxylase. The enzyme activity is not directly proportional to the affinity (K_m); it also depends on the amount of enzyme present and on cofactor regulation (see Table 21-3). L-Amino acid decarboxylase is very fast, uses the cofactor pyridoxal phosphate (i.e., vitamin B_6), and yields the first of the biogenic amine transmitters, dopamine. Although dopamine was originally thought to be only an intermediate in norepinephrine synthesis, it is the major catecholamine transmitter in the mammalian brain, is found in the autonomic ganglia, and is a neurohormone in the hypothalamus that controls pituitary prolactin secretion. In dopaminergic nerves, the transmitter is stored within secretory granules.

In noradrenergic nerves, dopamine is sequestered into a secretory granule that contains the enzyme dopamine-β-hydroxylase. This enzyme hydroxylates dopamine on the β-carbon atom using the cofactors ascorbate (i.e., vitamin C), molecular oxygen (O_2), and copper. The inside of these granules are relatively acidic, which is ideal for the pH maximum of this enzyme. Many of the peptidergic transmitters are amidated on their carboxyl-terminal ends by another vesicular enzyme that has identical pH and cofactor requirements. Noradrenergic nerves release the contents of these vesicles on stimulation. This type of nerve makes up the majority of sympathetic nerves in the periphery and many pathways in the brain. It has been estimated that 10,000 to 15,000 transmitter molecules are stored in a single granule as a salt with calcium, ATP, and a protein called chromogranin.

In adrenergic nerves in the brain or in the adrenal medulla, norepinephrine diffuses out of the granules that contain dopamine-β-hydroxylase, where it is activated on by the last synthetic enzyme, called phenylethanolamine-N-methyltransferase (PNMT). This enzyme uses S-adenosylmethionine as a cofactor to add a methyl group to the amino side of norepinephrine. The product, epinephrine (called adrenaline by the British) is repackaged into another set of granules that contain ATP, the protein chromogranin, and the catecholamines. The epinephrine to norepinephrine ratio is about 4 : 1 in the adrenal medulla.

The activity of PNMT can be regulated by the adrenal steroid, cortisone. A specific portal vasculature system between the adrenal cortex and medulla facilitates this activation. Early morning, diurnal increases in cortisol secretion activate PNMT to facilitate epinephrine synthesis. The increased epinephrine secretion stimulates liver glycogenolysis, gluconeogenesis, and prepares mammals for the anticipated daily activity.

Termination of catecholaminergic transmission. Termination of catecholaminergic transmission is accomplished by active uptake of the released amines into the nerve terminal and recycling or by enzymatic degradation. Although the termination of cholinergic neurotransmission is accomplished by the enzymatic degradation of acetylcholine, there are several routes for termination of catecholaminergic transmission. One of the most important routes is by active re-uptake of the amines into the nerve terminal, which essentially recycles the transmitter.

Most cells, including platelets, appear to have some ability to pump catecholamines across their membranes. Catecholaminergic neurons have a very active or high-affinity pump that sequesters the amines back into the axon before they can diffuse away from the synapse. This uptake, like that of choline, is sodium dependent and has a K_m of 1 to 5 μM. The importance of this system for terminating aminergic transmission is demonstrated by the action of the stimulant drug cocaine, which inhibits

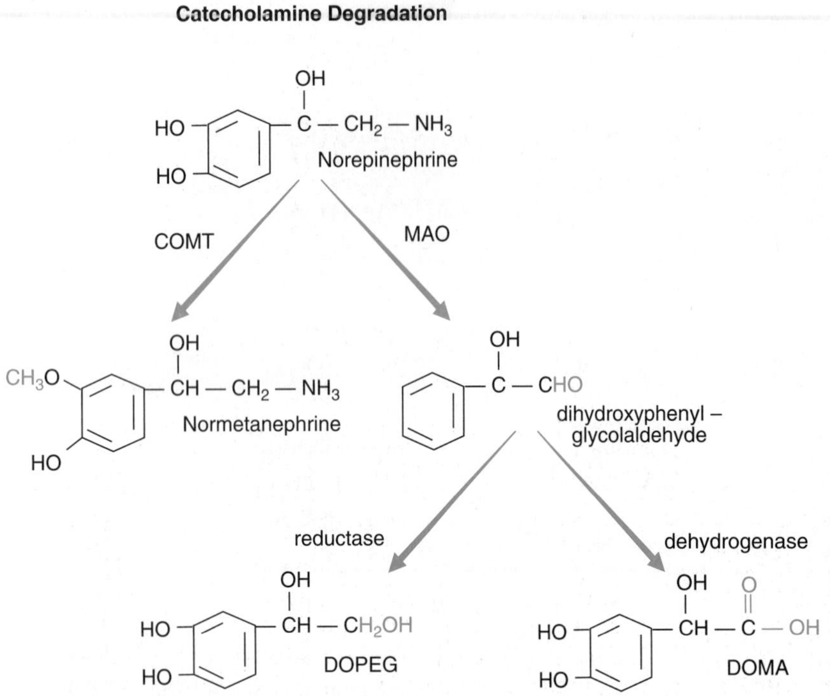

Catecholamine Degradation

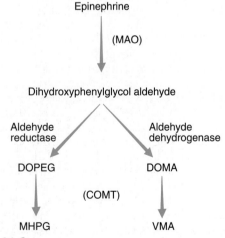

Figure 21-5

Structural changes induced by the degradation enzymes catechol-*O*-methyl transferase (COMT) and monoamine oxidase (MAO). The intermediate product of MAO is an aldehyde that can further be acted on by aldehyde reductase or dehydrogenase to 3,4-dihydroxyphenyl-ethylglycol (DOPEG) or 3,4-dihydroxymandelic acid (DOMA).

uptake of the amines and increases their availability for the postsynaptic receptors. The norepinephrine transporter protein has been cloned, and as expected, it is a sodium-dependent membrane pump and is similar to the GABA transporter. It has 617 amino acids and contains 12 or 13 hydrophobic (or lipophilic) amino acids sequences that span the lipid membrane.

The two enzymes that degrade the catecholamines are catechol-*O*-methyltransferase (COMT) and MAO. The biochemical action of these enzymes on the amine structure is shown in Figure 21-5. Although COMT attaches a methyl group to the ring, MAO oxidizes the amine part of the molecule. The catecholamines can be acted on by both enzymes, forming compounds that are methylated and oxidized. These metabolites are often measured in the urine to evaluate sympathetic activity or the presence of an adrenal tumor called pheochromocytoma.

MAO is an intracellular enzyme found in the mitochondrial membranes of neurons and glial cells of the brain and in the liver, kidney, glandular tissues, and intestines. Its wide distribution suggests its importance in metabolizing other toxic compounds. For example, when it is pharmacologically inhibited for depressive illnesses, tyramine, a common substance from aged cheeses and wines, can be absorbed intact from the gastrointestinal tract. When this occurs, the tyramine (normally oxidized by MAO) can cause the release of the sympathetic transmitter, which can lead to hypertensive crisis and death.

Monoamine oxidase has a molecular weight of about 102,000 and occurs in two forms, MAO-A and MAO-B, a subdivision based on pharmacologic specificity and inhibition. MAO inhibitors are useful in the treatment of depressive illness, perhaps because of their ability to make more biogenic amines available for the receptors.

This enzyme oxidizes the amines to their corresponding aldehydes. As shown in Figures 21-6 and 21-7, the aldehyde product of norepinephrine and epinephrine can then be acted on by aldehyde dehydrogenase or aldehyde reductase to produce dihydroxyphenylethylglycol (DHPG) or dihydroxymandelic acid, respectively. These

Catecholamine Degradation

Norepinephrine
or
Epinephrine

↓ (MAO)

Dihydroxyphenylglycol aldehyde

Aldehyde reductase ← → Aldehyde dehydrogenase

DOPEG   (COMT)   DOMA

↓   ↓

MHPG   VMA

Figure 21-6

The degradation of norepinephrine and epinephrine is catalyzed by monoamine oxidase (MAO), aldehyde reductase, and aldehyde dehydrogenase, and catechol-*O*-methyl transferase (COMT). The products include 3,4-dihydroxyphenylethylglycol (DOPEG), 3,4-dihydroxymandelic acid (DOMA), 3-methoxy-4-hydroxyphenylethylglycol (MHPG), and vanillylmandelic acid (VMA).

Catecholamine Degradation

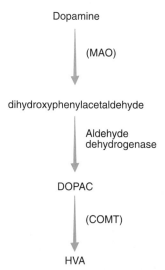

Figure 21-7

The enzymatic major degradation products of dopamine are catalyzed by the enzymes monoamine oxidase (MAO) and catechol-*O*-methyl transferase. The products are 3,4-dihydroxyphenyl acetic acid (DOPAC) and homovanillic acid (HVA).

can then be acted on by COMT to produce their 3-methoxy derivatives, which have been named 3-methoxy-4-hydroxy-phenylethylglycol (MHPG) and vanillylmandelic acid (VMA), respectively. The urinary excretion of these metabolic products, particularly VMA, are often used in the diagnosis of autonomic nervous system disorders. It has been suggested that the aldehyde reductase pathway is most active in the central nervous system, and MHPG or DHPG would be the products of increased central noradrenergic and adrenergic activity.

The primary MAO metabolite of central dopamine is dihydroxyphenylacetic acid (see Fig. 21-7). It can be further metabolized by COMT to produce homovanillic acid. Many of these metabolites are also used in research as a measure of noradrenergic or dopaminergic activity in the brain, and they may soon become clinically important.

COMT is a cytoplasmic enzyme that is distributed even more diffusely than MAO. It is found in high concentrations in the kidney and liver. The enzyme has a molecular weight of about 24,000 and requires *S*-adenosylmethionine and a divalent cation such as Mg^{2+}. The enzyme transfers a methyl group from *S*-adenosylmethionine to the 3-hydroxy group of the catecholamine. It can also methylate catechol-containing drugs such as isoproterenol, a β-adrenergic receptor–stimulating drug that is useful in asthma. Epinephrine, norepinephrine, and dopamine are acted on by COMT to produce metanephrine, normetanephrine, and 3-methoxytyramine, respectively. Urinary excretion of the these compounds is useful for the diagnosis of the adrenal tumor, pheochromocytoma.

Catecholamine pathways. Specific catecholamine pathways have been mapped throughout the brain and ap-

pear to function in a variety of behavioral, cognitive, and physiologic processes. The three catecholaminergic pathways can be visualized by techniques called fluorescent microscopy and immunocytochemistry.

Fluorescent histochemistry uses the properties of the catechol molecules to fluoresce when exposed to chemicals such as formaldehyde. This method was initially used to map the catecholamine and indolamine (e.g., serotonin) pathways within the central nervous system. The drawback of this technique was that it was difficult to differentiate different catecholaminergic or indolaminergic pathways because of the similar wavelengths of emitted light.

Immunocytochemistry is a method that uses specific antibodies that are generated against the purified enzymes involved in catecholamine biosynthesis. The antibodies bind to the enzymes (or specific transmitters) that are fixed on histologic sections of the brain. This technique can be used to label the cell bodies, axons, and terminals of a specific neuronal pathway. The bound antibodies are detected by using chemical reactions that produce a color change or electron-opaque product in the neurons containing the antigen. In practice, if a pathway contained the enzyme PNMT that methylates norepinephrine to epinephrine, it can be assumed that that pathway is adrenergic, although it would also contain dopamine-β-hydroxylase and tyrosine hydroxylase. If the pathway only contained dopamine-β-hydroxylase and tyrosine hydroxylase, it would probably be *noradrenergic*, but one that contains only tyrosine hydroxylase would be *dopaminergic*.

Dopamine is the major catecholamine neurotransmitter in the mammalian central nervous system. Dopamine was originally thought to be only a precursor in the synthesis of norepinephrine and epinephrine, but it was discovered that dopamine comprised as much as 50% of the total catecholamines in the brain of mammals. Moreover, the localization of dopamine and norepinephrine within the brain did not always coincide. The dopaminergic neuronal systems are heterogenous and have been classified by the Swedish investigators who developed the fluorescent technique into several specific nuclear groups designated A8 through A17.

Dopamine is found as an interneuron in the peripheral autonomic ganglia. Similar dopaminergic interneurons with very short axons are also found in the retina and in the olfactory bulb, where they appear to modify sensory input through inhibition of their target neurons. In the retina, this is called lateral inhibition, which is important for processing visual information.

Dopaminergic nerves with intermediate-length axons include the tuberoinfundibular and hypophysial, incertohypothalamic, and the medullary periventricular groups. The tuberoinfundibular nerves have a neurohumoral function; they secrete dopamine into a portal vascular system that supplies the anterior pituitary. This dopamine is responsible for inhibiting secretion of the anterior pituitary hormone, prolactin. Many of the dopaminergic pathways in the brain are diagrammed in Figure 21-8.

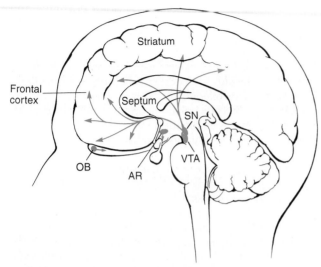

Figure 21-8
The major dopaminergic pathways in the central nervous system. Dopaminergic cell groups from the substantia nigra (SN) project to the caudate and putamen. Cell bodies near the substantia nigra in the ventral tegmental area (VTA) project axons to the septum, limbic cortex (including the frontal and cingulate cortex), amygdala, nucleus accumbens (NA), and olfactory tubercle (OB). Other discrete dopaminergic systems exist in the hypothalamic arcuate (AR) and periventricular nuclei in the olfactory bulb, and in the retina.

The final subdivision of dopaminergic neurons includes the midbrain group from the substantia nigra and the ventral tegmental area. These systems have long axons that innervate the basal ganglia, parts of the limbic system, and the frontal cortex. The neostriatal system, which has cell bodies in the substantia nigra, innervates the caudate and putamen. This suggests that dopamine released from neostriatal areas has motor functions. The motor problems associated with Parkinson's disease are caused by a decrease in dopamine in these areas. Administration of the dopamine precursor L-DOPA bypasses tyrosine hydroxylase and alleviates some of the motor disturbances of Parkinson's disease.

The specificity and complexity of the dopaminergic systems is further attested to by the mesolimbic system. These neurons originate in the ventral tegmental area of the midbrain, next to the substantia nigra. Long axons from these neurons project to many parts of the limbic system, including the nucleus accumbens, olfactory tubercle, septum, amygdala, and limbic cortex (i.e., frontal and cingulate cortex). These areas are associated with mood alterations and cognitive function, indicating another important role of central dopamine. The nucleus accumbens is involved with reward, and the release of dopamine in this area generates positive reinforcing feelings. It is in this area that the stimulant properties of cocaine and amphetamine (which releases axonal dopamine) are thought to act. The actions of many antidepressant drugs may also lie partially within these brain areas. These agents, which inhibit MAO or the amine uptake pump, result in more dopamine available for the do-

pamine receptor in these areas. The noradrenergic and serotonergic systems may also be involved in mood disorders.

The role of dopamine in cognition can be demonstrated by the action of a group of drugs used in treating schizophrenia. These agents, called the neuroleptics, block the dopamine receptor and alleviate many of the hallucinations and ideations associated with this disease. Prolonged therapy with the dopamine-blocking drugs affects the nigrostriatal system and leads to a Parkinson-like syndrome called tardive dyskinesia.

The *noradrenergic pathways* of the brain stem are quite diffuse, project all over the brain, and are probably involved in activation of cognitive function and mood. The noradrenergic system in the brain is less specific than the dopaminergic system and comprises far fewer cells. Although there are 30,000 to 40,000 dopamine cells in the midbrain system alone, there are only about 10,000 noradrenergic neurons in the whole brain. These are localized in the brain stem. Noradrenergic cell bodies are found in the locus ceruleus and lateral tegmental nuclei of the brain stem (Fig. 21-9). Ceruleus means blue and refers to the pigment associated with this area. Although there are very few cells in these nuclei, the axons are highly branched and project to all parts of the central nervous system. Noradrenergic axons can be found throughout the cortex, limbic system, hypothalamus, olfactory bulb, cerebellum, medulla, and spinal cord.

Rather than having discrete synapses, the noradrenergic "terminals" are located all along the axon as boutons. These are called *diffuse synapses* and resemble a string of beads or pearls when viewed by fluorescent microscopy.

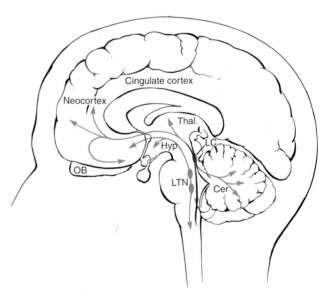

Figure 21-9
The noradrenergic pathways in the central nervous system are restricted to several cell groups in the brain stem. The limited number of noradrenergic cell bodies in the locus ceruleus (LC) and the lateral tegmental nuclei (LTN) send highly branching axons to all areas of the brain, including the neocortex, cingulate cortex, thalamus (Thal), hypothalamus (hyp), olfactory bulb (OB), and cerebellum (Cer).

Presumably, at this type of synapse, the transmitter bathes the postsynaptic target area in transmitter. This is also the case of noradrenergic terminals in the sympathetic nervous system, and this system can be differentiated from *contact synapses* in which specific synaptic processes are formed at the end of the axons. Contact synapses are found in many other transmitter systems, as exemplified by the neuromuscular junction.

Noradrenergic nerves, like dopaminergic neurons, can inhibit and excite the postsynaptic cells. Many of the receptor subtypes have been found to be β-adrenergic receptors, which use cyclic AMP as a second messenger. Noradrenergic activity inhibits spontaneous firing of large portions of the central nervous system and may enhance signal to noise levels in the brain.

The purpose of the noradrenergic system appears to be arousal or motivation. The stimulant drugs such as cocaine and amphetamine increase the amount of norepinephrine available for the receptor, and the antidepressants potentiate or augment the noradrenergic system. Most of the information suggests that increased noradrenergic activity is associated with increased vigilance and behavioral awareness, but decreased noradrenergic activity results in more vegetative processes. Central norepinephrine has also been implicated in pain pathways, memory, and the control of autonomic and endocrine function. The distribution of the noradrenergic neurons in the brain is diagramed in Figure 21-9.

Epinephrine-containing neurons in the brain are restricted to the brain stem and may be important in regulating autonomic and endocrine function. Adrenergic neurons in the brain stem have been visualized by staining for their unique enzyme, PNMT. Figure 21-10 shows the distribution of known adrenergic neurons in the

brain. These cells are restricted to nuclei in the lateral and dorsal tegmentum (designated C1–3 by the Swedish group that initially described them). They ascend to innervate the hypothalamus and descend the spinal cord to innervate the intermediolateral cell column that gives rise to the preganglionic nerve cell bodies of the sympathetic nervous system.

Little is known about the function of epinephrine in the brain, although it does inhibit firing of neurons in the locus ceruleus. Based on epinephrine's anatomic distribution, it seems to be important in the regulation of autonomic and neuroendocrine hypothalamic function. Many studies attest to this supposition.

Serotonin

Distribution. Serotonin is an indolamine present in neural pathways that parallel the distribution of norepinephrine in the brain. Serotonin is found in many cells of the body, such as blood platelets, mast cells, the enterochromaffin cells of the gastrointestinal tract, and in brain cells. Its original discovery was in blood, in which the isolated substance had a vasoconstrictor effect on the arteries. For this reason, it was thought to be the cause of high blood pressure. So much serotonin exists in peripheral platelets, mast cells, and the gastrointestinal tract that the amount in the brain only represents 1% to 2% of the total amount found in the body. Serotonin is also found in the pineal body of the brain, where it functions as the precursor for melatonin, an indolamine considered important in circadian cycles and reproductive function.

Serotonin is synthesized from the amino acid tryptophan, an essential amino acid. Tryptophan actively crosses the blood-brain barrier by a mechanism that also supplies aromatic and branched-chain amino acids. The concentration of tryptophan influences the concentration of serotonin in the brain, and this can be modified by dietary amounts of tryptophan or competition for uptake by other amino acids. In the serotonergic neuron, tryptophan is hydroxylated by an enzyme called tryptophan hydroxylase to 5-HT (Fig. 21-11). This enzyme is very much like tyrosine hydroxylase in that it uses molecular oxygen and a tetrahydropteridine cofactor. Like tyrosine hydroxylase, it is also regulated by phosphorylation, calcium, phospholipids, and partial proteolysis.

The genes for many of the hydroxylase enzymes have been cloned and characterized, making it possible to know the entire amino acid sequence. These studies show that tyrosine, tryptophan, and phenylalanine hydroxylases are different enzymes that probably originated or evolved from a common ancestral gene. The catalytic and regulatory elements of all three enzymes show marked similarity, but there are also differences.

Unlike tyrosine hydroxylase, tryptophan hydroxylase is not regulated by end-product inhibition. Although dopamine or norepinephrine can feed back to inhibit their own synthesis, increased cytoplasmic serotonin does not inhibit tryptophan hydroxylase. For example, if a MAO (i.e., enzyme that degrades catecholamines and serotonin) inhibitor is administered, the concentration of se-

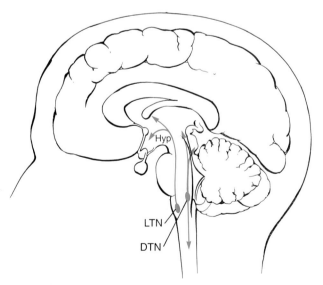

Figure 21-10
The adrenergic pathways in the brain originate from a few cell groups in the lateral (LTN) and dorsal tegmental nuclei (DTN). These cell bodies send axons to the hypothalamus (HYP), locus ceruleus (LC), and spinal cord to the intermediolateral cell columns.

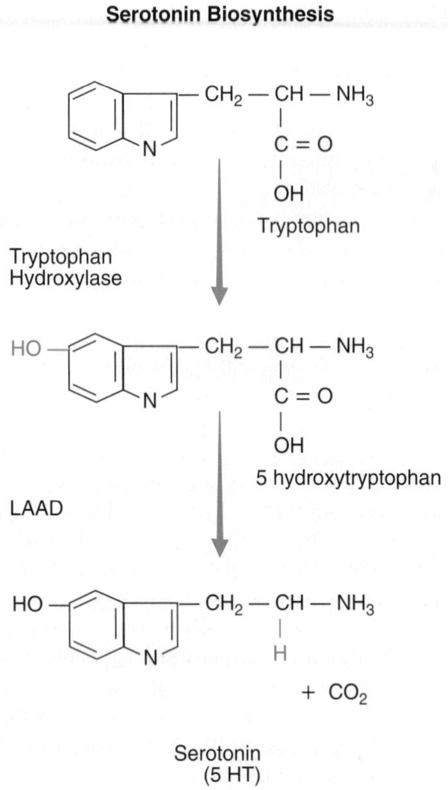

Serotonin Biosynthesis

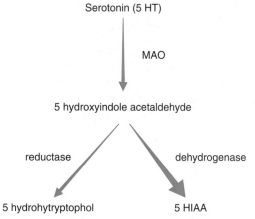

Figure 21-11
Serotonin biosynthesis begins with the hydroxylation of the amino acid tryptophan to 5-hydroxytryptophan by tryptophan hydroxylase. This amino acid is decarboxylated by L-amino acid decarboxylase (LAAD) to serotonin (5-HT). The shaded groups represent enzymatic additions or deletions.

rotonin in the brain increases 300%. This is not the case with the catecholamines.

What regulates serotonin turnover? Probably several factors are involved. Presynaptic neuromodulation can regulate tryptophan hydroxylase activity by activation of second messenger systems that phosphorylate and activate the enzyme. The K_m for tryptophan hydroxylase (see Table 21-3) is higher than the amount of tryptophan in the blood. This means that the enzyme is not saturated and that the availability of substrate may also regulate activity. This suggests that, in some circumstances, nutrition and diet can alter brain activity and mood.

The second enzymatic step in the formation of serotonin is the decarboxylation of 5-HT. This is accomplished by an L-amino acid decarboxylase enzyme that mimics the decarboxylase that produces dopamine from L-DOPA. As with the catecholamines, the activity of this enzyme system is very high. Because this is probably not a rate-limiting step, little 5-HT is expected to be present.

As in the case of the catecholamines, serotonin is actively taken up and stored in a specific set of secretory granules. The pumping mechanism responsible for the uptake of serotonin (and the catecholamines) is inhibited by the drug reserpine. If the storage of the amines is inhibited by this drug, all four of the biogenic amines are

depleted. Because of this action, this plant alkaloid, has been used for centuries as a sedative, in psychosis, and for hypertension.

In the pineal gland, two additional biochemical steps are involved to produce the pineal "hormone" melatonin. These are N-acetylation of serotonin to make N-acetyl serotonin and methylation of the hydroxy group to melatonin through the enzyme 5-hydroxyindole-O-methyltransferase. The first enzyme in this pathway, N-acetyltransferase, is altered during the day-night cycle by a sympathetic nervous system–mediated β-adrenergic activation of cyclic AMP formation.

Serotonin catabolism. The major pathways for the degradation of serotonin are re-uptake into the nerve and degradation by the enzyme MAO. There are many similarities between the catecholaminergic and indolaminergic nerve pathways. Unlike the catecholamines, indolamines can not be degraded by methylation by COMT because of the difference in the ring structure and the presence of only one hydroxyl on the phenolic ring. The pathway for serotonin degradation is shown on Figure 21-12. MAO oxidizes the indolamine to 5-hydroxyindoleacetic acid through an indoleacetaldehyde intermediate. 5-Hydroxytryptophol can also be formed, depending on the amount of NAD present in the brain.

The similarity between the uptake and degradation processes are such that many of the same drugs that inhibit catecholamine uptake and MAO degradation do the same for serotonin. For example, many of the tricyclic antidepressants also inhibit serotonin neuronal uptake, and all of the MAO inhibitors inhibit serotonin degradation. Their are, however, selective serotonin uptake inhibitors that are useful in the treatment of affective disorders (e.g., depression), which is further discussed in

Indolamine Degradation

Serotonin (5 HT)

↓ MAO

5 hydroxyindole acetaldehyde

reductase ↙   ↘ dehydrogenase

5 hydrohytryptophol   5 HIAA

Figure 21-12
Serotonin enzymatic degradation is catalyzed in a manner similar to that for the catecholamines. Serotonin or 5-hydroxytryptamine (5-HT) is acted on by monoamine oxidase to yield 5-hydroxyindole acetaldehyde. This intermediate is acted on by an aldehyde reductase or dehydrogenase to yield 5-hydroxytryptophol or the major product of 5-hydroxyindole acetic acid (5-HIAA).

the section on the aminergic theory of the affective disorders.

The *serotonergic pathways* originate in the raphe nucleus and parallel the noradrenergic system. The mapping of the serotonergic systems of the brain was initially hindered by some of the limitations of the histofluorescent microscopy. These system are well delineated anatomically. The cell bodies of the serotonergic nerves in the brain arise from several groups of cells in a midline area of the pons and upper brain stem called the *raphe nuclei* (Fig. 21-13). They have been classified as the B1 through B9 serotonergic nerve groups. The more rostral of the nuclei innervate the cortex, thalamus, and limbic systems, and the posterior nuclei (i.e., B1, B2, B3) innervate the medulla and spinal cord. Unlike the organized nuclear distribution of the dopaminergic system, the serotonergic nuclei innervate much of the telencephalon and diencephalon in an overlapping manner.

The major function of the serotonergic system appears to be relatively ill defined. More than 90% of the brain serotonin can be depleted with *p*-chlorophenylalanine, a drug that inhibits tryptophan hydroxylase, with few gross effects on animal behavior. The serotonergic cells of the raphe possess a pacemaker-like activity that is modified by 5-HT autoreceptors and noradrenergic receptors. The rate of activity of these cells is high during wakefulness, low during sleep, and absent during *rapid eye movements* (REM) sleep. The raphe nucleus displays pacemaker-type activity similar to the sinoatrial node of the heart that regulates the rate of the heart. Perhaps the raphe serotonergic neurons regulate the "rate of the brain," similar to the alteration of computer information

processing by 25- vs 33-MHz time clocks. Although the serotonergic systems have been implicated in such phenomena as sleep, vigilance, arousal, sensory perception, and emotion, they do not appear to cause these phenomena.

Pharmacologic observations may implicate the serotonergic system in higher cognitive function, schizophrenia, and hallucinations. Many of the hallucinogenic drugs resemble serotonin and interact with the 5-HT receptors. Dimethyltryptamine (DMT) and diethyltryptamine are hallucinogenic drugs of abuse and differ from serotonin only in having two methyl or ethyl groups on the amine terminal. The hallucinogenic drug lysergic acid diethylamide (LSD) also is chemically similar to serotonin and has been shown to interfere with autoreceptor function at the raphe nucleus and increase serotonergic firing rates. It was once thought that schizophrenia was caused by the pathologic production of an abnormal serotonin, such as DMT.

Aminergic theory of affective disorders. The *affective disorders* are a group of psychiatric diseases that include major and manic depression. The disease is characterized by exaggerated swings in mood that may cycle over months or years. It is a serious, debilitating disorder that can lead to suicide. About 80% of these patients respond to medication.

Although the medications fall into different groups, they all tend to make more biogenic amines available for receptor stimulation. The two major groups are the *tricyclic antidepressants*, which inhibit the uptake of the biogenic amines, and the *MAO inhibitors*, which inhibit their degradation. This information prompted the *amine theory of affective disorders*.

There are still many unanswered questions. The specific amine that is responsible for the depressed mood is unknown. Many of the tricyclic agents inhibit the uptake of norepinephrine, epinephrine, dopamine, and serotonin, and MAO is responsible for degrading all of the amine transmitters. There are selective uptake inhibitors for serotonin and dopamine, which are both effective in treating depression. The serotonin uptake inhibitors are widely prescribed drugs and have been shown to be effective. All of the antidepressants take several weeks to elevate mood. Although they are immediately effective in inhibiting amine uptake or MAO (in vivo and in vitro), they all take 2 to 4 weeks to alleviate depression. One explanation for the delay is that they lead to slow changes in receptor populations that result in stabilization of mood. Depression is still an ill-defined disease that may be the result of several biochemical imbalances.

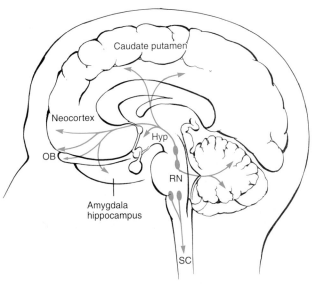

Figure 21-13
The major serotonergic pathways in the brain originate in the midline pontine and upper brain stem area called the raphe nuclei (RN). The posterior cell groups project to the spinal cord (SC), and the anterior serotonergic axons parallel the noradrenergic nerves to diffusely innervate the cortex, striatum, hypothalamus (HYP), olfactory bulb (OB), amygdala, hippocampus, and cerebellum.

Histamine

Histamine is found in mast cells and central neurons. The imidazole-containing substance, histamine, has been studied since the early 1900s, when Henry Dale described its many actions. It was isolated and purified around 1930, and it has often been considered as a neurotransmitter candidate. This role for histamine, however, has been hampered by several factors. It is found in high concentrations in mast cells (also present in the brain), where it is important for immune responses. The histochemical techniques to visualize histamine-containing neurons were difficult to develop and are only now being applied. Because the enzyme responsible for synthesizing this compound, histidine decarboxylase, is very much like the amino acid decarboxylase that produces catecholamines and serotonin, the specificity of its localization was questionable.

Histamine is produced by the decarboxylation of the essential amino acid histidine, as shown in Figure 21-14. The enzyme that catalyzes this reaction resembles L-amino acid decarboxylase, although it has been difficult to isolate from adult mammalian tissue. A fetal liver histidine decarboxylase has been purified and is thought to be similar to the enzyme in central neurons. Like the other decarboxylases, it requires vitamin B_6 (i.e., pyridoxal phosphate) and can be inhibited by drugs that interact with the same step in catecholamine and indolamine transmission. Similar to tryptophan hydroxylase, the affinity (K_m) for histidine decarboxylase is higher than the amount of substrate (i.e., histidine) available for decarboxylation. The enzyme is not usually saturated, and dietary intake of this essential amino acid may affect the activity of this system. Histidine loading increases the amount of histamine in the brain.

The major catabolic pathway for the elimination of histamine in the mammal is through N-methylation to methylhistamine (see Fig. 21-14) by the enzyme histamine methyltransferase. This enzyme uses S-adenosylmethionine as the methyl donor. Methylhistamine can be further catabolized by MAO.

The concentration of histamine in the brain is about 50 ng/g, some of which is undoubtedly in mast cells. Administration of mast cell–degranulating agents (i.e., histamine-depleting agents) results in a 50% reduction of brain concentrations of histamine. During postnatal development, when the blood-brain barrier matures, the elevated histamine levels and the number of mast cells concurrently decrease.

Using antibodies against histidine decarboxylase and histamine itself, nerve pathways have been observed that originate in posterior basal hypothalamus and premammillary area. These histaminergic neurons appear to ascend through the medial forebrain bundle to innervate the forebrain, including cortical, thalamic, and limbic structures. Lesioning of the medial forebrain bundle causes a complete depletion of the norepinephrine and serotonin, which has a time course similar to the degeneration of axons after lesioning. Studies have shown that similar lesions also cause a 70% decrease in histidine decarboxylase levels in the forebrain.

Many studies have indicated that central histamine pathways may be involved in the central control of autonomic and endocrine activity, food and water intake, and temperature regulation. Because blockers of the H_1 histamine receptor causes sedation, a role in arousal, like those of catecholamines and serotonin, is also suggested. This is easily demonstrated in the allergic patient, in

Histamine Biosynthesis

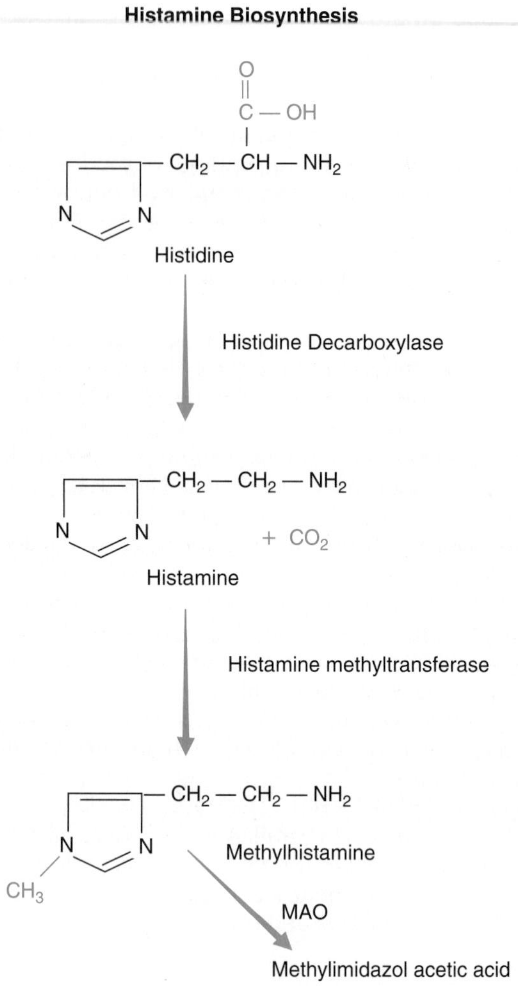

Figure 21-14
Histamine biosynthesis and degradation. The amino acid histidine is decarboxylated by histidine decarboxylase to produce the neurotransmitter histamine and CO_2. Histamine is degraded by methylation to methylhistamine and can be further oxidized by monoamine oxidase (MAO) to methylimidazol acetic acid. The shaded groups indicate enzymatic additions or deletions.

whom antihistamines have the side effects of sleepiness and increased appetite.

This chapter does not include a complete list of all neurotransmitters. It describes those that are known, best understood, and studied. Other potential candidates include taurine, octopamine, ATP, and even nitric oxide. The field offers a tremendous potential for further investigation in the future.

☐ γ-Aminobutyric acid, glutamate, and glycine are major transmitter systems in the mammalian brain

Although the amino acid tyrosine is the common precursor for all three of the catecholamine neurotransmitters, glutamate is equally pivotal for production of the amino acid transmitters in the central nervous system. Glutamate is a transmitter in the excitatory amino acid pathways and is the precursor for the major inhibitory amino acid transmitter, γ-aminobutyric acid (GABA). The concentration of these substances are almost 1000 times higher than the conventional amine transmitters in the brain, which may be indicative of their relative importance.

The excitatory amino acids, glutamate and aspartate, depolarize their postsynaptic target neurons and can be compared with the accelerator pedal of an automobile. GABA in the brain and glycine in the spinal chord hyperpolarize their target neurons and can similarly be compared with the brake pedal of the automobile. The inhibitory importance of GABA and glycine can be easily demonstrated if their receptors are blocked by drugs such as picrotoxin or strychnine, which block the GABA and glycine receptors, respectively. Administration of these drugs immediately induce life threatening seizures and death.

Why was such an important transmitter system overlooked for so many years? The reason lies in the fact that amino acids are a common constituent of all cells. Although norepinephrine or serotonin could be specifically localized to certain neurons and brain areas, glutamate and glycine are universally present. These obstacles

are being overcome by new neurochemical, molecular biologic, and immunochemical techniques.

γ-Aminobutyric acid

Function and distribution. γ-Aminobutyric acid serves as the principal transmitter involved in internal circuits within specific brain areas. GABA is restricted to the brain, spinal cord and retina, and it is only found in trace quantities in other types of tissues or peripheral nerves. Quantitatively, the concentration of the biogenic amine norepinephrine is in the range of nM/g, but GABA can be measured within the brain in concentrations of μM/g. If the concentration of a transmitter is 1000-fold more, is it proportionally that much more important? The answer to this question is being sought in many laboratories.

Decarboxylation of glutamate to GABA is not very different from the decarboxylation of L-DOPA or tryptophan to dopamine and serotonin. The enzyme that does this is called glutamic acid decarboxylase (GAD), and it removes the α-carboxyl group to produce a γ-carboxyl amino acid. The biosynthesis and catabolism of GABA is shown in Figure 21-15. Like the three other decarboxylases discussed in this chapter, GAD requires the cofactor pyridoxal phosphate (vitamin B_6). The saturation of GAD with its cofactor may be one of the rate-limiting steps in

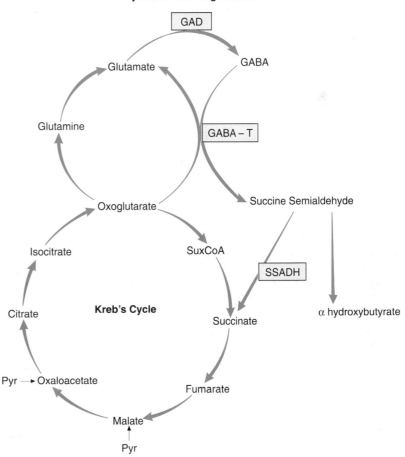

Synthesis and Degradation

Figure 21-15
γ-Aminobutyric acid (GABA) and glutamate synthesis and degradation are shown in relation to the Kreb's cycle and GABA shunt. The important enzymes in the pathway are shown in the shaded boxes. GABA is synthesized by the enzyme, glutamic acid decarboxylase (GAD) from glutamate. GABA can be degraded by GABA transaminase (GABA-T) to succinic semialdehyde in a reaction that regenerates glutamate from oxoglutarate. Succinic semialdehyde can reenter the Kreb's cycle by the action of succinic semialdehyde dehydrogenase (SSADH) or can be degraded to α-hydroxybutyrate. The entire system is fueled from glucose, which can enter the cycle as pyruvate (Pyr) as indicated.

GABA synthesis, controlled by stearic inhibition with ATP. This could explain why the concentrations of GABA in the brain increases rapidly from 30% to 45% after death when ATP levels drop. The distribution of GAD does appear to be selective to endogenous GABA neurons that generally have short axons. GAD is not present in glia or other types of nerves.

The GABA shunt, which is studied in conjunction with the Krebs cycle in biochemistry, is a unique feature of the brain and a viable form of energy production. The key aspects of this pathway are emphasized in Figure 21-15, and the Krebs cycle is indicated by abbreviations. GABA is produced from α-oxoglutarate, through glutamate, degraded by two enzymes present throughout the brain, and reenters the Krebs cycle as succinic acid.

What is the significance of this energy-producing shunt? First, it generates ATP (three), although a little less efficiently (by one GTP) than the Krebs cycle. Estimates have been made that the GABA shunt contributes to 10% to 40% of brain metabolism. Second, glial cells, which have high-affinity GABA pumps, are able to scavenge GABA and use it for energy. Third, glutamate is regenerated through the degradation of GABA in GABA-ergic nerves. Enzyme localization studies have shown that GAD is high in GABA-containing areas (e.g., GABA nerves), but the GABA-degrading enzymes, GABA transaminase (GABA-T) and succinic semialdhyde dehydrogenase (SSADH), are localized throughout the brain.

γ-Aminobutyric acid catabolism. GABA-ergic transmission is terminated by uptake into nerves and recycling, uptake into glia, and metabolism by enzymes that can regenerate glutamate and produce more ATP. The GABA uptake transporter has been characterized and cloned. It is a large protein that appears to have 12 membrane-spanning domains that are made up of hydrophobic amino acids that can insert through the lipid bilayer of the membrane. Like the catecholamine transporter, it is Na^+ dependent and has been shown to transport two Na^+ ions out of the cell in exchange for one GABA and one Cl^- ion. Because of the ionic redistribution, GABA uptake is electrogenic (i.e., changes the membrane potential) and can sequester GABA against a gradient of 10,000 to 1.

GABA is catabolized by several enzymes that are localized throughout the brain. GABA transaminase requires pyridoxal phosphate and may under certain circumstances compete with GAD for the cofactor. The K_m of GABA-T for vitamin B_6 is much lower that of GAD. The overall effect of vitamin B_6 deficiency is an increased susceptibility to seizures related to decreased GABA synthesis. These seizures are rapidly reversed after pyridoxine is administered. Transamination of GABA to succinic semialdehyde by GABA transaminase results in the regeneration of glutamate for decarboxylation to GABA. GABA-T catalyzes the formation of succinic semialdehyde, which is metabolized by succinic semialdehyde dehydrogenase to succinate. The K_m of SSADH is so low (10^{-6}) that very little of its substrate, succinic semialdehyde, ever accumulates in the brain. An alternate pathway results in the formation of the metabolite γ-hydroxybutyrate.

Localization studies have hinged on the immunocytochemical distribution of GAD. Many of these pathways are outlined in the brain map in Figure 21-16. Mapping studies have shown that GABA-ergic pathways make up the major endogenous circuits within specific brain areas. Most of these have short axonal pathways that interact locally. The best understood systems are those in the Purkinje cells of the cerebellar cortex that inhibit the deep nuclei and the GABA-ergic system in the striatum that project to the substantia nigra and globus pallidus. Other systems that are thought to be GABA-ergic are the granule cells of the olfactory bulb, some of the amacrine cells of the retina, basket cells of the hippocampus, and a long ascending projection from the hypothalamus to the cerebral cortex.

The major function of the GABA system is inhibition of the target pathways. This is accomplished by the opening of chloride channels and the resultant hyperpolarization. Several drugs interact with the GABA system that illustrate the inhibitory function of these nerves. GABA-blocking drugs, such as picrotoxin, or drugs that decrease the availability of GABA lead to convulsions similar to epileptic seizures. It is thought that this transmitter system may be important in the cause of epilepsy. Drugs that increase the amount of GABA or potentiate the action of GABA result in central inhibition and can lead to coma and death.

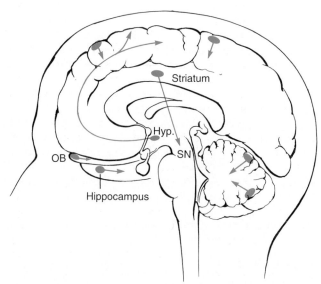

Figure 21-16
The γ-aminobutyric acid (GABA) pathways in the brain are present mostly as small-axon–possessing interneurons within the brain structure. GABA occurs in concentrations of about 100-fold higher than the catecholamines in most brain structures, including the cortex, limbic system, and cerebellum. A longer-axoned GABA neuron in the striatum projects to the substantia nigra (SN), and a group of cell bodies in the hypothalamus (Hyp) are thought to project to the forebrain. The olfactory bulb (OB) and hippocampus are also depicted.

There are many drugs that can potentiate the GABA system, including the barbiturates, benzodiazepines (e.g., Valium, Librium), anesthetic steroids, and alcohol. The antianxiety and sedative characteristics of these agents provide a pretty good idea of some of the actions and functions of central GABA systems. It is possible to speculate that death is not such an unpleasant experience because of the massive disinhibition and increase in GABA activity found postmortem.

This is an interesting, seemingly well-designed, inhibitory transmitter system that generates metabolic energy when it is active, regenerates its own substrate, and limits the rate of its own activity through production of ATP. It is well suited for the major inhibitory transmitter system in the brain.

Glycine

Glycine is an inhibitory transmitter that is principally localized in the brain stem and spinal chord. Like glutamate, the transmitter role of glycine is confused by the fact that it is localized in all cells as an important constituent of proteins, peptides and precursor for porphyrins and nucleic acids. The concentration of glycine is very high in the spinal cord as is glutamate and glutamine. Glycine is the simplest amino acid in the body and is not considered essential for diet.

Although glycine could originate from many metabolic sources, it appears to be predominantly generated from serine. This has been determined by administering radiolabeled substrates and after the incorporation of the radioactive carbon into glycine. In a similar manner, there are many possible degradation pathways. Active uptake, however, seems to be the preferred route of removal of this inhibitory amino acid from its receptor. Like the GABA and biogenic amine transporters, the glycine uptake mechanism is Na$^+$ dependent. Radiolabeled preloaded glycine can be release from spinal cord slices by depolarization in a Ca^{2+}-dependent manner. Glycinergic neurons are thought to be involved in motor and sensory functions in the spinal cord, brain stem, and the retina.

Like GABA, glycine has important anticonvulsant properties. However, the antianxiety and sedative actions of glycine have not been demonstrated pharmacologically. Strychnine is a specific glycine receptor blocker. It is a complicated plant alkaloid found in the herb *Nux vomica*. It has often been used as a poison and rat killer and causes convulsions by blockade of the glycine receptor. The Renshaw cell of the ventral root is an interneuron that has been characterized in terms of its ability to alter alpha motor neuron activity. This glycinergic interneuron receives its input from a collateral axon of the alpha motor neuron and is stimulated by the release of acetylcholine when the motor neuron fires. The stimulated Renshaw cell then releases glycine onto dendrites or the cell body of the motor neuron to inhibit further motor activity. When this classic feedback mechanism is blocked by strychnine, the motor neuron fires without control, and convulsions result.

Sensory actions of glycine in the spinal cord are also suggested by the use of small amounts of an extract of *Nux vomica* to increase tactile sensations. This has been an ingredient in herbal, homeopathic, and proprietary drugs.

Glycine may also be an important component of glutamatergic transmission. One of the glutamate receptors (i.e., NMDA receptor) has a glycine binding site that, when occupied, increases the frequency of Ca^{2+} channel opening. This neuromodulatory response to glycine is not blocked by strychnine.

Glutamate

Glutamate is the major excitatory transmitter in the brain. Amino acids such as glutamate and aspartate stimulate (i.e., depolarize) many different types of neurons. This action was often considered to be nonspecific, which seemed to disqualify glutamate as a transmitter candidate. Glutamate and its precursor glutamine are found in very high concentrations in the brain and spinal cord, but they do not appear to evolve from the brain uptake. The influx of glutamate from the blood seems to be slower than its efflux from the brain to the periphery. The massive amounts of glutamate and glutamine in the brain arise from glucose produced by means of the Krebs cycle. Like glycine, these amino acids occur in all brain cells, including glia, and this has made it difficult to map precise glutamate and aspartate pathways throughout the central nervous system. Much of the momentum for including it as an important excitatory transmitter has come from the isolation of several glutamate receptors and the cloning of one receptor.

There are three possible major pathways for the synthesis of glutamate in the central nervous system, which were discussed in the previous section on GABA and diagrammed in Figure 21-15. Synthesis proceeds from 2-oxoglutarate or aspartate by the actions of ornithine or aspartate aminotransferase or from glutamine by means of glutaminase. Termination of the response may be achieved through a variety of metabolic pathways or through the actions of a high-affinity glutamate pump that actively sequesters glutamate inside neurons and glia. Because of these active transporter mechanisms, little or none of the amino acid transmitters are measured extracellularly. In the dentate gyrus, for example, the active uptake of glutamine is required for continuing transmission. The system may have many similarities to the acetylcholine system, in which uptake of a precursor is a rate-limiting consideration in transmission.

Excitatory amino acid nerves serve the function of the major projection neurons between brain areas. Essentially every outgoing (i.e., efferent) system of the cortex appears to use glutamate, including the corticostriatal, thalamic, bulbar, and pontine pathways. The mossy fiber afferents of the cerebellum and pathways in the olfactory lobe may also use glutamate. Some of the glutamatergic pathways in the central nervous system are diagrammed in Figure 21-17.

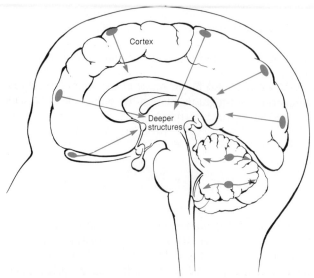

Figure 21-17
Central excitatory amino acid or glutamate pathways are the major efferent pathways from one brain area to another. These long-axon–possessing neurons project from the cortical and cerebellar areas to almost all brain areas. Concentrations of this transmitter are 1000-fold higher than the catecholamines. The glutamate neuronal pathways are shown by the black ovals and arrows.

As can be seen from the distribution of fibers, the glutamatergic pathways appear to be the major stimulatory or driving pathway within the central nervous system. Pharmacologically, the effects of underactivity of this system are unknown, although many of the antagonists such as ketamine and phencyclidine (both hallucinogenic) are anesthetic. Overactivity of the glutamate system and the action of some highly potent agonists such as quisqualic acid, ibotenic acid, and kainic acid lead to overstimulation and neurotoxicity.

GABA and glutamate appear to be the major inhibitory and excitatory transmitter systems in the central nervous system and provide braking and acceleration systems, respectively, for neural activity. It is fascinating that these major systems use substances that are so intimately involved in brain metabolism. Through these systems, central energy metabolism, transmission, and nerve activity are inseparably linked.

Neuroactive Peptides

There are a variety of neuroactive peptides in the brain that are synthesized in the cell body and axonally transported to the nerve terminal. Peptidergic transmission in the central nervous system is different from that of the small-molecular-weight transmitters. More than 50 peptide transmitters have been described in the central nervous system and more are being discovered every year. This may not be so surprising considering the fact that the human brain contains approximately 10^{12} neurons.

The history of neuropeptides can be broken down into two periods: a 40-year period from about 1930 to 1970, when only a few peptides were chemically characterized, and the 20-year period from 1970 to the present, during which many neurally active peptides have been discovered and sequenced, mostly because of newer techniques and advances in molecular biology.

The first peptide transmitter, *substance P*, was discovered serendipitously around 1930, when U.S. Von Euler and John Gaddum were screening various tissues for concentrations of acetylcholine. A substance was found in the gut and in the brain that lowered blood pressure and increased gastrointestinal activity. The substance was not acetylcholine and was called substance P because it was a powder that resembled a protein in several ways. The neurohypophysial peptides, *oxytocin* and *vasopressin*, were isolated and characterized in the 1950s, partially because of the immense concentrations that were stored in the neural lobe. In the common laboratory animal, the white rat, almost a half of a microgram of each of these peptides is stored and released into peripheral blood to effect physiologic functions such as water retention and lactation. Because of the very small size of the neural lobe and the amount of peptide present, it was relatively easy to extract and characterize these peptides, each of which is nine amino acids long. The amino acid sequence of substance P was confirmed around 1970, when the new method became available. At that time, myriad new brain peptides were isolated and characterized.

☐ Neuropeptides are synthesized in the cell body and processed from larger precursors in the endoplasmic reticulum, Golgi, and secretory granules

The brain peptides are present in the axon in larger secretory granules that are synthesized in the cell body and reach the nerve terminal by axoplasmic transport. These are classified as slow transmitters, because their actions on the target cells take some time to occur. The mechanism of their synthesis is also slow. This probably represents an important balance so that terminal depletion of these compounds does not usually occur.

The process of *peptide synthesis* begins with the nuclear translation of specific messenger RNAs and their binding to ribosomes. When the mRNA is translated into protein, a 20- to 30-amino acid sequence called the signal peptide is first formed. This hydrophobic sequence inserts itself into the endoplasmic reticulum (ER) and is followed by the gradually elongating protein chain. Some of the peptide processing occurs in the ER. The protein is glycosylated with a mannose-rich sugar that is attached to asparagine residues (flanked by other specific amino acids). These specific sequences of amino acids (or nucleotides) are called *consensus sequences*. In the endoplasmic reticulum the protein is folded and disulfide bonds are formed. The disulfide isomerase that accomplishes this is not entirely characterized, but these bonds are important for maintaining the folded structure of the neuroprotein.

The mature propeptide is transported to the Golgi apparatus where it can be further processed, sorted, and packaged into secretory granules. Much of the processing of the precursor to the active neuropeptide occurs within the secretory granule. These granules have a membrane-bound proton pump that maintains an acidic pH. An example of neuropeptide processing, using oxytocin, is shown in Figure 21-18. Peptides are represented (by con-

vention) with the amino-terminal on the left and carboxyl-terminal on the right.

A trypsin-like enzyme first cleaves the prohormone or proneuropeptide at a pair of basic amino acids made up of lysine and arginine. This enzyme may be a serine protease and has a acidic pH maximum that allows it to work within the secretory granule. In the case of oxytocin, this produces two products: an extended oxytocin containing Gly-Lys-Arg on the carboxyl-terminal end and a 93-amino acid protein called neurophysin. Some proneuropeptides have many more than a single neuropeptide sequence

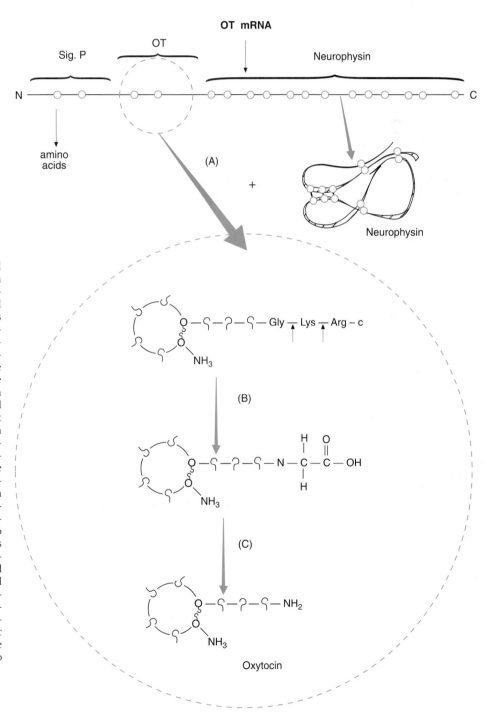

Figure 21-18
Peptide biosynthesis in the central nervous system. Neuropeptides in the central nervous system are synthesized on ribosomes in the cell body of peptidergic nerves. This diagram demonstrates the synthesis of oxytocin in the hypothalamus. The proneuropeptide contains the natural neuropeptide (OT) flanked by a signal peptide (Sig. P) and a binding protein called neurophysin. The shaded dots on the propeptide represent the 18 cysteine residues (drawn relatively to scale). In the endoplasmic reticulum and Golgi apparatus, the disulfide bonds are formed, and the peptide is packaged in secretory granules. Within the secretory granule the propeptide is cleaved by a serine protease (A) at a pair of basic amino acids. The extended oxytocin is then acted on by a carboxypeptidase (B) to remove the extended basic amino acids and peptidyl glycine α-amidating monooxygenase (C) to produce the biologically active neuropeptide. The activity of these three enzymatic steps occur in the synaptic vesicle while it is being transported to the axon terminal.

within their precursor. For example, proenkephalin-A has seven enkephalin sequences in the single precursor, and hypothalamic thyrotropin-releasing hormone (TRH) has five of these tripeptide sequences. The specificity of these proteases may differ in various tissues. In the anterior pituitary, the enzyme converts precursor proopiomelanocortin (POMC) into corticotropin (ACTH) and the opiate peptide β-endorphin. In the intermediate lobe, this same precursor POMC is cleaved to melanocyte-stimulating hormone (MSH), a peptide called CLIP, and β-endorphin.

The next enzymatic step eliminates the basic amino acids from the carboxyl-terminal of the neuropeptide. This enzyme is a carboxypeptidase, which has an acidic pH maximum and (in the case of oxytocin) yields the nonapeptide with an additional glycine. Amidation is the next step in peptide synthesis, accomplished by an enzyme called peptidylglycine α-amidating monooxygenase (PAM). Many peptides are amidated on the carboxyl-terminal end, including TRH, gonadotropin-releasing hormone, oxytocin, vasopressin, angiotensin II, substance P, and vasoactive intestinal peptide (VIP). The enzyme cleaves the additional glycine (of extended oxytocin or TRH) at the α-carbon atom, leaving the amide attached to the carboxyl residue (see Fig. 21-18). In terms of cofactor requirements, this enzyme is like dopamine-β-hydroxylase. As is the case for norepinephrine biosynthesis, copper and reduced ascorbic acid are required for the amidation of peptides to occur within the vesicles.

Much of this peptide processing occurs as the secretory granule is being transported down the axon to the nerve terminal. An investigator can follow the maturation of some peptides from the cell bodies down the length of the axons to the nerve terminal. Peptide-containing granules usually are larger than fast transmitter–containing granules. When they exist as co-transmitters, they can be differentially released, depending on the frequency of nerve action potentials and distribution of Ca^{2+} channels. This has been most completely worked out in the parasympathetic innervation of the salivary gland, where low-frequency stimulation releases acetylcholine, and high frequencies cause the secretion of VIP.

Termination of neuropeptide responses is achieved principally through enzymatic degradation. Uptake and reuse of peptides are not options for this classification of transmitters, and there are no known peptide transporters. The enzymes that degrade peptides are called *peptidases*, and they can be classified by amino acids or metals that are involved in their catalytic site (e.g., serine, metalloproteases) or by the location of the cleavage site of the peptide. Endopeptidases cleave internal peptide bonds at specific sequences. An example is the enzyme that degrades the opiate peptides, sometimes called enkephalinase. Exopeptidases remove one or two amino acids from the carboxyl or amino terminus of the peptide. The carboxypeptidases that remove the basic amino acids from peptide precursors have already been discussed. Angiotensin II–converting enzyme (ACE) also removes a carboxyl-terminal dipeptide from its substrate. Aminopeptidase A removes selectively acidic amino acids from the amino-terminal end of the peptide. A variety of peptidases are found in extracellular fluid or circulate in the blood. Aminopeptidase A and M, cathepsin D, and ACE all are found circulating in the blood.

Some of the characteristics of neuropeptides partially protect them from degradation by the peptidases. For example, the disulfide bonds and carboxyl-terminal amidation of oxytocin and vasopressin are essential for biologic activity and protect these peptides from immediate destruction by circulating peptidases.

□ The families of neuroactive peptides mediate endocrine, neural, and behavioral functions

The peptides can be subdivided into several families of neurotransmitters. Because much of this information is discussed elsewhere in the text, only their major groupings, distribution, and some aspects of their function are discussed.

Neurohypophysial peptides

Neurohypophysial peptides project to the posterior pituitary and higher brain centers. Oxytocin and vasopressin are synthesized in specific nuclei of the anterior hypothalamus and are transported by axons through the internal layer of the median eminence to the posterior pituitary (i.e., neural lobe). This small bundle of axon terminals and pituicytes stores large amounts of both peptides, which contributed to their early isolation and characterization.

Vasopressin is secreted into the blood, where its hormonal functions include constriction of vessels, as its name implies, and inhibition of water loss by the kidney (hence the name antidiuretic hormone). Oxytocin is usually associated with reproductive function and is released to induce contraction of the pregnant uterus during parturition and the letdown of milk during suckling and lactation.

The axons of these peptides project to other areas of the brain, brain stem, and spinal cord. Vasopressin has been associated with memory consolidation in the limbic system, and oxytocin has been implicated in memory, maternal and sexual behavior, and autonomic activity.

Hypothalamic neurohormones

The hypothalamus of the midbrain regulate autonomic functions, behavioral processes, and the endocrine system. Several small peptides reach the anterior pituitary through a portal blood system and regulate most humoral function. *Thyrotropin-releasing hormone* is a tripeptide that causes the release of thyroid-stimulating hormone from the pituitary. *Gonadotropin-releasing hormone* is a decapeptide that releases pituitary luteinizing hormone and follicle-stimulating hormone peripherally, and it has been implicated in some sexual behaviors (e.g., lordosis) in the rodent.

Somatostatin has 14 amino acids (i.e., a tetradecapeptide) and inhibits the release of growth hormone from the pituitary, but it also occurs in many higher brain areas, including the cortex and limbic system. Present in peripheral nerves, somatostatin also plays a role in sensory function.

Corticotropin-releasing hormone (CRH) is found in many higher brain areas. It is responsible for regulating the secretion of ACTH and modulating the immune system.

Many other neuroendocrine peptides and factors have been described.

Opiate peptides

Opiate peptides have similar peptapeptide sequences. In the mid-1970s, a peptide-like substance was isolated from brain that displaced morphine from a membrane-bound receptor present in many areas of the mammalian brain. Morphine that is derived from opium, the resinous juice of the poppy plant, has been used for centuries to alleviate pain and as a drug of abuse.

Opiate peptides exist in almost all areas of the brain. Three major opiate precursors give rise to the main opiate peptide groups: β-endorphin, the enkephalins, and the dynorphins. POMC, which was previously discussed, is the common precursor for ACTH and β-endorphin. It is present in the anterior and intermediate lobe of the pituitary and in an ascending group of neurons in the hypothalamus.

Proenkephalin-A gives rise to several pentapeptides (i.e., five methionine-enkephalins and one leu-enkephalin) that have potent opiate-like effects. Proenkephalin-B is the precursor for several leu-enkephalin pentapeptides, and the dynorphins. The common feature of all of these opiate peptides is the pentapeptide sequence Tyr-Gly-Gly-Phe-Met (or Leu), which seems to be important for binding to the several opiate receptors that have been described. This family of neuropeptides are important in analgesia and a variety of other important inhibitory functions. The enkephalins and dynorphins are present throughout the central nervous system, particularly in the striatum, the limbic system, raphe nuclei, and the hypothalamus.

Gut-brain peptides

Since the discovery in 1930 of substance P, several peptides that are found in high concentrations in the gastrointestinal tract have been isolated and mapped in the brain. It has been proposed that the gastrointestinal tract has its own endogenous brain, called the *enteric nervous system*, which is separate from the autonomic nervous system. The gut peptides also found in the brain include VIP, cholecystokinin, gastrin, secretin, neurotensin, pancreatic polypeptide Y, NPY, and many others. These have

been implicated in several physiologic processes, including pain, temperature regulation, satiety and hunger, and nausea and as major co-transmitters in the autonomic nervous system (e.g., VIP, NPY).

Other peptides

The list of peptides that are being isolated and characterized in the brain and periphery is growing every day. Several others should be mentioned. *Angiotensin II* and *bradykinin* are vasoactive peptides that have been localized within the brain. Angiotensin II has potent effects on water and fluid intake when injected directly into the hypothalamus. Various peptides, collectively called *sleep peptides*, are potent inducers of sleep. *Calcitonin gene–related peptide* was accidentally discovered when the gene for calcitonin in the thyroid was sequenced and translated into protein. Antibodies against a peptide coded for by the gene reacted strongly with peptidergic pathways in the trigeminal nerve and hypothalamus. Although it is a vasodilator, its central function is still unknown. *Galanin* is another peptide that is present in the brain stem and appears to regulate neuroendocrine activity.

Conclusion

In this chapter, about 100 neurotransmitters that are distributed among 1 to 10 billion nerves have been directly or indirectly discussed. If there were no other transmitters in the brain, each transmitter should occupy 10 million to 100 million neurons, but this is not the case. For example, the noradrenergic nerves in the brain only occupy about 10,000 nerve cells, and the dopaminergic neurons in the midbrain number only 30,000 to 40,000. These calculations are obviously misleading, because the 100 transmitters are not evenly distributed and there are probably many more transmitters that are currently unknown. Our knowledge of "how the brain works" is undoubtedly very primitive. In the future, many new transmitter substances will be discovered, and new concepts will be offered to explain the complex neural interactions involved in thought, memory, mood, sensation, and maintenance of somatic and visceral function.

SELECTED READINGS

Bjorklund A, Hökfelt T. GABA and neuropeptides in the CNS, part I. In Bjorklund A, Hökfelt T, Tohyama M (eds). Handbook of chemical neuroanatomy, vol 4. Amsterdam: Elsevier, 1985.

Bjorklund A, Hökfelt T. Classical transmitters and receptors in the CNS, parts I and II. In Bjorklund A, Hökfelt T, Kuhar MJ (eds). Handbook of chemical neuroanatomy, vols 2 and 3. Amsterdam: Elsevier, 1984.

Bjorklund A, Hökfelt T, Tohyama M. Ontogeny of transmitters and peptides in the CNS. In Bjorklund A, Hökfelt T, Tohyama M (eds). Handbook of chemical neuroanatomy, vol 10. Amsterdam: Elsevier, 1992.

Cooper JR, Bloom FE, Roth RH. The biochemical basis of neuropharmacology, 6th ed. New York: Oxford University Press, 1991.

Erecinska M, Silver IA. Metabolism and role of glutamate in mammalian brain. Prog Neurobiol 1990;35:245.

Gainer H, Brownstein MJ. Neuropeptides. In Seiger GJ, Albers RW, Agranoff BW, Katzman R (eds). Basic neurochemistry. Boston: Little, Brown, 1981:269.

Gilman AG, Rall TW, Nies AS, Taylor P (eds). Goodman and Gilman's the pharmacological basis of therapeutics, 8th ed. New York: Pergamon Press, 1990.

Partridge WB. Biological diversity of peptides. In Peptide drug delivery to the brain. New York: Raven Press, 1991.

Clinical Correlation
PARKINSON'S DISEASE

ROBERT L. RODNITZKY

In 1817, the English physician James Parkinson published a monograph entitled *An Essay on the Shaking Palsy*. He described six persons afflicted by a condition characterized by a flexed posture, slowness of movement, a tendency to walk rapidly with small steps, and an associated tremor of one or more extremities. Not all of the six cases were Parkinson's personal patients. Some were observed on the streets of London. To Parkinson's description of tremor and slowness of movement, the modern neurologist would add rigidity and loss of postural reflexes to complete the typical symptom complex of Parkinson's disease.

Parkinson's Disease Symptoms

☐ Parkinson's tremor is most prominent at rest

Tremor is prominent in a variety of neurologic conditions, many of which are confused with Parkinson's disease. However, unlike most tremors, the *parkinsonian tremor* is most prominent at rest and is typically suppressed, at least temporarily, during action. In essential tremor, the condition most commonly confused with Parkinson's disease, tremor is typically absent at rest and only appears when the affected body part assumes a sustained, active posture. For instance, in essential tremor involving the hands, the tremor may only appear while holding the outstretched hands in front of the body or when attempting to bring a coffee cup to the mouth. The distribution also helps identify Parkinson's tremor. In Parkinson's disease, the tremor most commonly affects the extremities, especially the hands. It occasionally affects the mandible, but unlike essential tremor, it almost never affects the head.

The term *pill-rolling tremor* is often used in Parkinson's disease. It refers to the tendency of the thumb and index finger to approximate one another while trembling as though an object were being rolled between the two fingers. This term is a reference to the turn-of-the-century technique used by pharmacists to fashion a pill by rolling a soft substance between the thumb and index finger. Like many of the symptoms of Parkinson's disease, tremor is often asymmetric, and in the early stages of the illness, it may be exclusively on one side. The typical frequency of parkinsonian tremor is four to six cycles per second. The exact mechanism by which tremor arises in this condition is not clear, but most evidence strongly favors a central rather than a peripheral origin.

☐ Parkinsonian rigidity differs from the rigidity of upper motor neuron lesions or other basal ganglia disorders

In Parkinson's disease, limb rigidity can be demonstrated throughout the entire range of a large-amplitude passive movement. In contrast, the stiffness associated with an upper motor neuron lesions such as a cerebral infarction is referred to as clasp-knife rigidity, because there is considerable resistance at the beginning of the examiner's passive movement, which suddenly gives away as the passive movement is continued. When the examiner attempts to demonstrate parkinsonian rigidity, a rachet-like sensation known as cogwheeling can often be felt as the limb is moved. In Parkinson's disease, rigidity involves the extremities more than the axial musculature. In other degenerative conditions mimicking Parkinson's disease, such as progressive supranuclear palsy, axial rigidity predominates.

☐ Bradykinesia is one of the most disabling symptoms of Parkinson's disease

Bradykinesia is slowness of movement or inability to initiate movement. As a result of this dysfunction, automatic movements such as swinging the arms while walking become diminished or are totally lost. Similarly, eye blinking is decreased, and the normal range of facial expression is lost, resulting in a typical fixed stare. When severe slowness of movement evolves into a lack of movement, it is referred to as *akinesia*.

Patients with severe bradykinesia or akinesia find it difficult to perform motor tasks that require repetitive motions, such as finger tapping or combing the hair, and tasks that require manipulation of small objects, such as placing a button through a buttonhole. Difficulty in initiating movement results in inability to arise from a chair,

427

exit a car, or roll over in bed. Of the four cardinal clinical features of Parkinson's disease—tremor, rigidity, brady-kinesia, and loss of postural reflexes—bradykinesia is the most severely disabling.

Another common clinical manifestation of bradykinesia is inability to properly manipulate a pen or pencil while writing. The resultant handwriting, referred to as micrographia, classically becomes progressively smaller as the affected person writes, reflecting an inability to carry out sustained repetitive movements in a normal fashion.

Like many of the clinical signs of Parkinson's disease, the exact physiologic basis of bradykinesia is not under-stood. The phenomenon of *kinesia paradoxia*, wherein an ordinarily bradykinetic patient demonstrates an un-expectedly rapid motor response to a sudden stimulus such as a thrown ball, suggests that alternate motor path-ways may exist in these persons that can support move-ments of normal range and speed.

☐ Impairment of postural reflexes is typically a late development in Parkinson's disease

Postural reflexes are tested by applying a backward-di-rected perturbation to both shoulders in an attempt to pull a person off his base. The normal person takes one step at most to compensate for the perturbation. The Par-kinson's patient with impaired postural reflexes cannot right himself and continues to move backwards with a series of small and uncontrollable steps until stopped by the examiner. This inability to stop oneself from propel-ling backwards is referred to as *retropulsion*. The ten-dency to take uncontrollable steps in increasingly rapid fashion in the forward direction is referred to as a *festin-ation* or *propulsion*. Patients with moderate to severe im-pairment of postural reflexes are at great risk for falls, because they cannot recover from the slightest naturally occurring perturbation, such as tripping over the edge of a rug.

☐ The cardinal symptoms of Parkinson's disease can appear in different combinations

Not all patients have all four of the cardinal symptoms of Parkinson's disease. Isolated resting tremor is a common presenting symptom of Parkinson's disease, and in some patients, it remains the predominant symptom for the en-tire course of the patient's illness. In a small percentage of patients, tremor never develops, and the parkinsonian syndrome consists solely of rigidity and bradykinesia with or without loss of postural reflexes. The absence of resting tremor makes establishing a diagnosis of Parkin-son's disease slightly more difficult. If bradykinesia ap-pears in isolation, it is commonly manifested as a change in handwriting or difficulty in performing fine motor tasks.

Age-Related Incidence of Parkinson's Disease

In the United States, the estimated prevalence of Parkin-son's disease is approximately 187 per 100,000 persons in the general population, with an annual incidence of 20 per 100,000. Although occasionally beginning in persons as young as 30 years of age, it classically presents in the sixth and seventh decades. Epidemiologic surveys sug-gest that the prevalence per 100,000 is between 30 and 50 in the fifth decade and between 300 and 700 in the seventh decade. The increasing occurrence with age has led to the speculation that normal age-related abiotrophy of dopaminergic cells is a contributory factor in the etio-pathogenesis of the illness.

The role of genetic factors in Parkinson's disease has not been fully clarified. There are rare instances of a fam-ily with an apparent autosomal dominant pattern of in-heritance, but for most patients, no familial pattern is ap-parent. Supporting the relatively minor role of heredity is the fact that twin studies have not consistently sug-gested a significant contribution of genetic factors.

Etiology and Treatment

☐ A toxin can cause a syndrome similar to that of Parkinson's disease

Parkinsonism developed in a group of narcotic addicts who mistakenly injected themselves with the compound MPTP (1-methyl-4-phenyl-1,2,3,6-tetrahydropyridine). MPTP is metabolized by brain MAO to MPP+. MPP+ is con-veyed into dopamine nerve terminals by the dopamine transporter, which normally acts to terminate dopami-nergic neurotransmission by reaccumulating dopamine into presynaptic nerve terminals. The transporter also ac-cumulates neurotoxins such as MPP+ that share struc-tural features with dopamine.

Once inside dopaminergic cells, MPP+ accumulates within mitochondria, where it inhibits complex I of the mitochondrial respiratory chain, resulting in cell death. Decreased complex I activity has also been reported in naturally occurring Parkinson's disease, suggesting that the underlying mechanism of cell death may be similar to that documented in MPTP parkinsonism. Based on these processes, it can be appreciated why this toxin pref-erentially affects dopaminergic cells and why it causes a syndrome almost identical to that of ordinary Parkinson's disease.

In the laboratory, injection of MPTP into primates reli-ably produces parkinsonism. The remarkable resem-blance of MPTP parkinsonism to naturally occurring Par-kinson's disease raises the possibility that Parkinson's disease itself may be due to a similar environmental toxin. Although MPTP itself has not been found in the environment, other potential toxins have been impli-cated. Epidemiologic studies have suggested that Parkin-son's disease may be related to rural living, raising the possibility that agrichemicals may play a role in its causation.

Monoamine oxidase inhibition may benefit patients with Parkinson's disease

Drugs that inhibit the enzyme MAO block MPTP from inducing experimental parkinsonism by preventing the conversion of this protoxin to its active toxic form, MPP+. The fact that MAO inhibitors prevent the development of MPTP-induced parkinsonism is one factor that has suggested their use to prevent or retard the development of naturally occurring Parkinson's disease, especially if that condition is caused by a neurotoxin biochemically related to MPTP.

There is another mechanism through which MAO inhibitors may benefit the Parkinson patient. MAO is normally involved in the catabolism of dopamine. MAO-inhibiting drugs slow this process, reducing the associated generation of potentially toxic free radicals, which may produce further nigral damage. Whether this effect of MAO-inhibiting drugs actually retards dopaminergic cell death and slows the progression of Parkinson's disease is not firmly established.

The Lewy body is the pathologic hallmark of Parkinson's disease

In Parkinson's disease there is severe neuronal loss in the pars compacta of the *substantia nigra*. Many of the remaining neurons contain a *Lewy body*, an eosinophilic cytoplasmic inclusion surrounded by a lighter halo. Lewy bodies can also be found, although in lesser numbers, in some other central nervous system degenerative conditions and occasionally in the brains of nonparkinsonian elderly persons. In Parkinson's disease, Lewy bodies are most prominent in the substantia nigra, but they are also found in the locus ceruleus, nucleus basalis of Meynert, raphe nuclei, thalamus, and cerebral cortex. To the extent the cerebral cortex and the nucleus basalis of Meynert contain Lewy bodies or other forms of pathology, dementia may appear in some Parkinson's patients.

Dopaminergic preparations are the most effective therapy for Parkinson's disease

In the 1950s, it was discovered that dopamine is profoundly depleted in the striatum of patients with Parkinson's disease. This led to initial attempts to treat parkinsonian patients with small, orally administered dosages of levodopa, a dopamine precursor that, unlike dopamine itself, can cross the blood-brain barrier. Once inside the brain, it was reasoned, levodopa could be transformed to dopamine by the enzyme dopa-decarboxylase. These initial attempts were unsuccessful because the small dosages of levodopa employed were almost entirely metabolized to dopamine peripherally by dopa-decarboxylase in the liver and gut and therefore lost to the brain. Subsequent attempts in the late 1960s using much larger dosages were successful in allowing some levodopa to escape peripheral decarboxylation and enter the brain. It was soon apparent that this strategy was remarkably effective in reversing the symptoms of Parkinson's disease. Levodopa remains the mainstay of therapy for Parkinson's disease. It is now administered with a peripheral dopa-decarboxylase inhibitor, and almost no conversion to dopamine occurs outside the brain, allowing smaller amounts of levodopa to be administered.

Similar but less reliable improvement of parkinsonian symptoms can be achieved by the administration of dopamine agonists, such as pergolide or bromocriptine. These substances readily pass the blood-brain barrier and directly stimulate dopamine receptors, simulating the effect of dopamine. They tend to be slightly less effective, have a slightly higher side-effect profile, and accordingly are more often used as adjunctive than primary therapy for Parkinson's disease.

Severe complications can occur after years of levodopa therapy

After five years of levodopa therapy, approximately 50% of Parkinson's disease patients develop complications consisting of involuntary writhing, twisting movements of the extremities, trunk, and face, or episodes of sudden, transient, near-total loss of dopamine effect on their symptoms. The sudden loss of antiparkinson effect has been referred to as the *on-off effect*. Patients experiencing this complication find that they may suddenly revert to total immobility as though someone had flipped a switch, followed by an equally sudden return to normal mobility. The unpredictability of these fluctuations in mobility can be extremely disabling. Both the on-off effect and the involuntary writhing movements that occur in patients with advanced Parkinson's disease are postulated to be related to the development of altered dopamine receptor sensitivity. Whether these putative receptor changes are a function of the duration of the underlying disease or result from long-term nonphysiologic receptor stimulation by dopaminergic drugs is unknown.

Tissue transplantation may improve the symptoms of Parkinson's disease

In the 1980s, there was considerable interest in transplantation of tissue into the central nervous system as a means of treating Parkinson's disease. Initial reports suggested improvement after transplantation of cells derived from the patients' own adrenal glands into the caudate nucleus. Cells from the adrenal medulla are metabolically capable of synthesizing dopamine, and it was reasoned that they could survive in the brain and elaborate this neurotransmitter. Although there were initial reports of improvement in patients who underwent this procedure, subsequent patients did not fare as well. The postmortem studies suggested that most transplanted cells did not survive.

A second wave of enthusiasm for cell grafting in treating Parkinson's disease has been generated by the early experience with the transplantation of tissue derived from human fetal mesencephalon and implanted into the caudate or putamen of patients. Careful study has shown that these cells do survive in the recipient's brain and effectively increase dopamine production, as demonstrated by positron emission tomography. Clinically, the procedure results in a moderate reduction in neurologic symptoms and a reduced requirement for levodopa therapy.

Future therapies using cell or tissue grafting may depend on cells other than fetal tissue that have been genetically modified to produce dopamine.

Section V

SENSORY PERCEPTION AND VESTIBULAR SYSTEMS

Neuroscience in Medicine, edited by P. Michael Conn.
J. B. Lippincott Company, Philadelphia © 1995

Chapter 22

Somatovisceral Sensation

G. F. GEBHART

Consideration of somatic and visceral sensations begins with an appreciation that the sense organs of the body each typically respond only to particular stimuli, which are forms of physical energy. Information about such stimuli is conveyed from specific receptors by afferent nerves to the central nervous system, where the stimulus is transformed into a sensation. This chapter considers the stimuli and sensory channels that lead to the conscious appreciation of sensations arising from skin, muscle, and joints (i.e., somatic structures) and the viscera.

Sensory Modalities

☐ Activation of specific receptors define a modality and its qualities

Classically, there were thought to be five basic senses or sensory modalities: hearing, sight, smell, taste, and touch. Other senses, such as proprioception, have since been recognized, but what constitutes a modality or sensation is not always a matter of unanimous agreement. For example, some consider pain to be a sensation, but pain is unique among the somatovisceral sensations.

The somatovisceral sensations considered here include the senses of touch (i.e., *mechanoreception*), heat and cold (i.e., *thermoreception*), position and movement (i.e., *proprioception*), and the sensory component of pain (i.e., *nociception*). Most sensations arise from stimuli in our external environment, and these sensations are called *exteroceptive*. Numerous sensations also arise internally, and these senses are referred to as *interoceptive*

because they provide information about the internal state of the body. We are typically unaware of most interoceptive stimuli (e.g., baroreceptor activation).

Within each modality, there are *qualities* (e.g., the qualities of vision include brightness and colors) produced by specific sensory stimuli. These stimuli acquire their qualities by action at specific receptors. Sensory receptors are specialized neurons that respond to a specific quality of the stimulus, called the *adequate stimulus* for the receptor. Different receptors have specialized structures and unique properties, but the end result of stimulus-receptor interaction is ultimately the same. Adequate stimuli produce changes in the membrane potential of the receptor. At low intensities of stimulation, small changes in membrane potential, called *generator potentials*, are produced that encode the stimulus intensity. Action potentials in the receptor are not generated until stimulus intensity is sufficient to produce a suprathreshold generator potential (Fig. 22-1). At suprathreshold intensities of stimulation, the action potentials generated are conducted by *afferent nerve fibers* to the central nervous system.

The process of converting the energy of a stimulus into a change in permeability of the sensory receptor membrane to ions and ultimately to the generation of an action potential is called *transduction*. The mechanisms of transduction of physical energies into action potentials are not understood for all stimuli and all senses, but vision is a notable exception. Action potentials are essentially the same for all sensory qualities; the information the action potentials convey is determined by the type of receptor from which the afferent fiber arises. The kind of sensation perceived or experienced arises from activation of a specific receptor, which defines modality and quality.

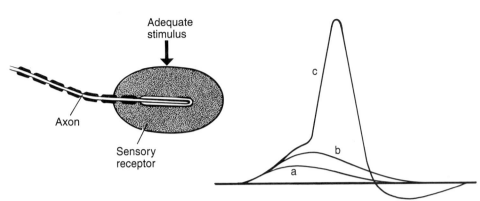

Figure 22-1
Generator potentials (*a* and *b*) and action potential (*c*) to graded intensities of an adequate stimulus applied to a sensory receptor.

☐ Stimulus strength and the intensity of sensations are related

The subjective interpretation of a sensation and its intensity is derived from the strength or magnitude of the stimulus. Within an appropriate range of stimulus intensity, which is different for different receptors, the stronger the stimulus, the greater is the response that can be objectively measured (e.g., action potentials in an afferent nerve fiber). As stimulus strength increases, the magnitude of the response also increases (Fig. 22-2). Each receptor has an absolute *stimulus threshold* (S_0), which is the intensity of the stimulus at threshold, below which no action potential or sensation is produced and above which the number of action potentials and the intensity of sensation increases.

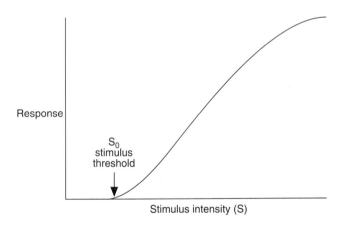

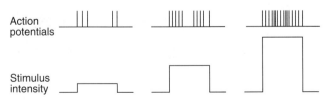

Figure 22-2
Relation between stimulus intensity (S) and response (e.g., action potentials).

The quantitative study of the intensity of sensations began more than 150 years ago by Weber and Fechner. Weber's work focused on the ability to discriminate between two weights when both were relatively light and when both were relatively heavy. He advanced a law to define the "just noticeable difference," which is the amount one stimulus must differ from another to be sensed as different. Weber established a means of measuring relative sensitivity. Fechner experimentally examined the relation of stimulus magnitude to the intensity of sensation, extending Weber's work and establishing the first psychophysical relation. *Sensory psychophysics* is study of the relation or comparison of objectively measured, intensity-dependent responses to a stimulus with subjective impression of the same stimulus. Most of the important questions in sensory physiology that are addressed today by sophisticated, quantitative methods were formulated initially on the basis of advances made first in subjective sensory physiology.

In psychophysical experiments, volunteers are asked to estimate the intensity of the sensation while stimulus strength is experimentally varied. There are several ways in which these experiments can be conducted, and psychophysicists have studied virtually all sensory modalities. The relation between stimulus strength and intensity of sensation in psychophysical experiments is best described by a power function experimentally validated by Stevens in the 1950s:

$$I = k \cdot (S - S_0)^n$$

in which I is the intensity of sensation, k is a constant, S is stimulus intensity, and S_0 is stimulus intensity at threshold; the exponent n depends on the modality and experimental conditions.

This same power function describes the dependence of a neuron's response to a stimulus and stimulus strength or intensity. There is remarkable agreement between the subjectively estimated intensity of sensation and the objectively measured response of sensory neurons recorded in humans (Fig. 22-3). For most sensory receptors, it has been determined experimentally that the stimulus-response function is nonlinear and that the *encoding* of stimulus intensity by receptors is best described by the power function. When the experimental

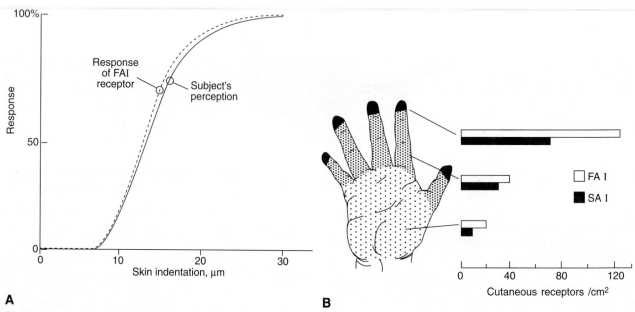

Figure 22-3
Microneurography of human nerves. (**A**) Correspondence between the response of FA I receptors and the person's perception in response to stimulation of the finger tip. (**B**) Density of innervation of FA I and SA I receptors in the human hand. (Adapted from Vallbo AB, Johansson RS. Hum Neurobiol 1984;3:3.)

results are plotted in a double logarithmic plot of stimulus intensity versus response, the exponent n is determined as the slope of the line:

$$\log I = \log k + n \cdot \log (S - S_0)$$

If the slope (n) is less than 1, the sensory receptor transforms a broad range of stimulus intensities into a narrow range of responses that are limited at the upper range of stimulus intensity (e.g., vision). If n is greater than 1, the receptor transforms a relatively narrow range of stimulus intensities into large magnitude increases in response (e.g., pain).

Somatovisceral sensation can be as selective as sensations arising from specialized sense organs

Unlike vision or audition, for which the sensory receptors form a specialized sense organ, sensory receptors for somatovisceral sensation are widely and inhomogeneously distributed throughout the skin and in deep tissue such as joints, muscle, and the viscera. The skin contains mechanoreceptors, thermoreceptors, and nociceptors, with properties suitable for giving accurate information about stimulus location, onset, intensity and duration.

Somatovisceral sensory receptors include *mechanoreceptors* and *nociceptors* in skin, muscle, joints, and the viscera; *thermoreceptors* (i.e., warm and cold) in skin and the viscera; and *proprioceptors* in joints and muscle. The chemosensitive nerve endings (i.e., *chemoreceptors*) in skin, muscle, joints, and the viscera are not well characterized. There is also a significant population of afferent sensory axons that are normally silent. These afferent fibers are called *silent nociceptors*.

Sensory nerve axons have different conduction velocities

Sensory (i.e., afferent) nerve axons differ with respect to diameter and myelination and therefore conduction velocity (Table 22-1). The most rapidly conducting axons have large diameters and are heavily myelinated. The largest-diameter and most rapidly conducting cutaneous axons are called Aβ, which conduct at 25 to 70 meters per second. Small-diameter, thinly myelinated axons belong to the Aδ group, which conduct at 10 to 30 meters per second. Unmyelinated, small-diameter axons are called *C fibers* and are the slowest-conducting sensory neurons (<2.5 m/second).

Muscle and joint sensory nerve axons are similarly myelinated and unmyelinated, but they are described with different terminology. Myelinated axons from muscles and joints are called group I, II, or III; the unmyelinated axons from muscles and joints are referred to as group IV (see Table 22-1). The conduction velocities of the myelinated group I through III axons range from 10 to 120

Table 22-1
CLASSIFICATION OF SENSORY NERVE FIBERS

Fibers					
Skin	*Muscle and Joint*	*Viscera*	**Examples**	**Myelination**	**Conduction velocity (m/sec)**
	Group I		Muscle spindle afferents	Yes	70–120
Aβ	Group II		Cutaneous mechanoreceptors	Yes	25–70
Aδ	Group III	Aδ	Myelinated nociceptors	Yes	10–30
C	Group IV	C	Unmyelinated nociceptors	No	<2.5

meters per second; group IV axons are like C fibers, conducting slowly at less than 2.5 meters per second.

There are no or few large-diameter sensory axons that innervate the viscera; thinly myelinated Aδ and unmyelinated C fibers convey information from the viscera to the central nervous system. Because visceral sensory nerves are often found in nerve bundles that also contain sympathetic efferent axons, visceral afferent nerve fibers are sometimes referred to as sympathetic afferents. This confusing terminology suggests that visceral sensory neurons have their cell bodies in sympathetic ganglia of the autonomic nervous system, which is not the case. Visceral sensory nerves, like sensory nerves from skin, muscle, and joints, have their cell bodies in spinal and cranial sensory ganglia (e.g., dorsal root ganglia, nodose ganglia, trigeminal ganglia).

Cutaneous Receptors

☐ Spatial resolution of cutaneous sensation depends on receptor density in skin

Because of the ease of access, sensory receptors in skin have been most completely characterized. The area of skin invested with a particular sensory axon and its branches defines the *receptive field* for a particular sensory neuron. The receptive field is the area of skin within which the adequate stimulus for a sensory receptor excites the sensory neuron. Because the density of particular sensory receptors in skin differs over the surface of the body, the ability to sense stimuli applied to different areas of skin also differs. For example, the density of cutaneous mechanoreceptors is much greater in the fingertips than in the palm or back of the hand (see Fig. 22-3).

A simple test for estimating the relative density of innervation of mechanoreceptors in the skin is to determine the shortest distance between two equivalent mechanical stimuli that can be discriminated when applied to the skin. This is called the *two-point discrimination threshold* and is only several millimeters for the lips and fingertips, but it is more than 60 mm for skin on the upper arm or thigh. The spatial resolution of tactile sensation (i.e., nonpainful mechanical sensation) depends on the innervation density of mechanoreceptors in the skin.

☐ Activation of fast-adapting, slowly adapting, and hair follicle receptors give rise to nonpainful mechanical sensations from the skin

Cutaneous mechanoreceptors, also called low-threshold mechanoreceptors, respond best to mechanical forces applied to the skin. Mechanoreceptors are neurons with specialized end-organs and their peripheral terminals are considered to be corpuscular or encapsulated. Activation of low-threshold mechanoreceptors leads to nonpainful sensations common in everyday experience.

The response of low-threshold mechanoreceptors to a maintained stimulus has been used to classify them as *slowly adapting (SA)* or *fast adapting (FA)*, as shown in Figure 22-4. SA I mechanoreceptors respond to skin indentation, signaling displacement of the skin and the velocity of skin displacement, and they are associated with Merkel's disks and tactile domes in the epidermis. SA II receptors respond to indentation and stretch of the skin. SA II receptors are associated with Ruffini endings in the dermis. It is thought that the ability of SA II receptors to provide information about the direction of lateral tension in the skin contributes to the sense of position (i.e., proprioception). SA I and SA II mechanoreceptors continue to discharge as long as the stimulus is maintained.

FA mechanoreceptors, like SA mechanoreceptors, respond to skin indentation, but they signal only the velocity of the applied stimulus (see Fig. 22-4). The FA receptors respond to tapping and vibratory stimuli but not during sustained pressure on the skin. FA mechanoreceptors are associated with Meissner's corpuscles in glabrous (i.e., hairless) skin and with pacinian corpuscles in subdermal glabrous and hairy skin.

The receptive fields of low-threshold mechanoreceptors differ significantly in size. SA I and FA I receptors have small, punctate receptive fields (10 to 15 mm²) and sharp borders. SA II and FA II receptors have much larger receptive fields (50 to 100 mm²) and indistinct borders.

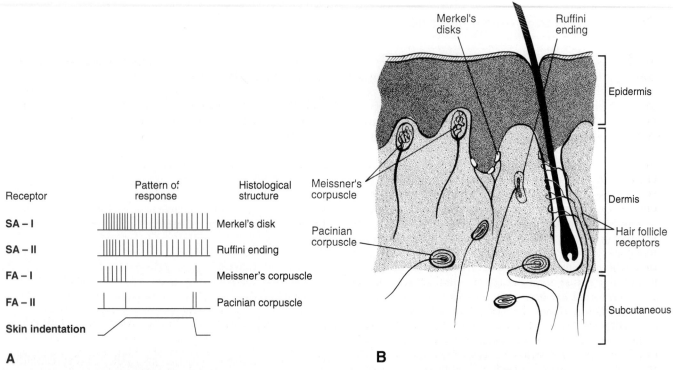

A

Receptor	Pattern of response	Histological structure
SA – I		Merkel's disk
SA – II		Ruffini ending
FA – I		Meissner's corpuscle
FA – II		Pacinian corpuscle
Skin indentation		

B

Figure 22-4
Stimulus-response characteristics of mechanoreceptors and their associated histologic structures. Slow-adapting (SA) receptors respond during sustained skin indentation; fast-adapting (FA) receptors signal the velocity of skin indentation and respond only during the dynamic phase of skin stimulation.

The movement of hairs in hairy skin gives rise to sensations. For example, movement of a single hair on the back of a hand is easily sensed. *Hair follicle receptors*, of which there are several kinds, respond to movement of down (i.e., short and soft) or guard (i.e., long) hairs and signal the velocity of movement or transient movement of hairs. Information from low-threshold mechanoreceptors and hair follicle receptors are conveyed rapidly to the central nervous system by large-diameter myelinated Aβ axons.

☐ Activation of nociceptors signals injury to the skin

Another major class of cutaneous receptors are the *nociceptors*, which Sherrington defined as sensory receptors that signal damage to the skin or the threat of damage. Stimuli that activate nociceptors are called *noxious*. The two principal nociceptors in skin are named for the stimuli to which they respond best (i.e., their adequate stimuli) and the afferent axons that innervate them: *Aδ mechanical nociceptors* are small-diameter, thinly myelinated fibers, and *C polymodal nociceptors* are small-diameter, unmyelinated fibers.

Aδ mechanical nociceptors respond to pinch, squeeze, and other mechanical stimuli that can damage the skin. They do not respond to chemical or thermal stimuli, except when they become sensitized in certain conditions.

C polymodal nociceptors are common in skin and respond to noxious mechanical, noxious thermal (i.e., heating of skin >45°C), and noxious chemical stimuli (e.g., acids, capsaicin). There are other cutaneous nociceptors, such as cold nociceptors that respond to intense cold, but they are not as common in skin as Aδ mechanical and C polymodal nociceptors.

Unlike low-threshold mechanoreceptors in the skin, which are identified with specific, different encapsulated endings (see Fig. 22-4), nociceptors innervate the epidermis, and their endings are not encapsulated. Consequently, nociceptors have been referred to as having free nerve endings, but ultrastructural studies suggest that nociceptors may have unique, nonencapsulated end-bulbs with characteristic axonal beads that contain unusually high concentrations of mitochondria.

☐ Tissue injury changes the responses of nociceptors

It is a common experience after an injury that the area of skin at and surrounding the injury becomes more sensitive. Typically, pain is produced by previously nonpainful stimuli, which is a phenomenon called *allodynia*, and the magnitudes of responses to previously noxious stimuli are increased and of greater duration, a phenomenon called *hyperalgesia*.

Hyperalgesia may be primary or secondary. *Primary hyperalgesia* refers to an increase in sensitivity at the site of injury and is mediated by endogenous substances released or synthesized locally in response to tissue injury. Nociceptors become sensitized in this circumstance. As a consequence of sensitization, nociceptors that previously responded only to mechanical stimuli may also respond to thermal stimuli. Among cutaneous sensory receptors, nociceptors are considered unique in their ability to become sensitized. *Secondary hyperalgesia* refers to an increase in the sensitivity of undamaged skin surrounding the injury. Secondary hyperalgesia is caused by an increase in the excitability of neurons in the central nervous system as a consequence of input from the peripheral injury.

When tissue is damaged, a variety of substances are released at the site of injury, are synthesized at the site of injury, or are attracted to the site of injury. They include bradykinin, histamine, serotonin, leukotrienes, prostaglandins, peptides such as substance P, cytokines, leukocytes, and platelets. In the aggregate, these substances (and probably others) lead to local inflammation and sensitization of nociceptors (i.e., primary hyperalgesia). Tissue injury also leads to an increase in the excitability of neurons in the central nervous system, probably arising from a persistent or greater than normal release of peptides (e.g., substance P and calcitonin gene-related peptide) and excitatory amino acids (e.g., glutamate, aspartate) from the central terminals of nociceptors. Second-order neurons respond to previously subthreshold intensities of noxious stimulation and produce greater responses to previously threshold intensities of stimulation (i.e., secondary hyperalgesia). The change in excitability of central neurons also leads to an increase in the size of receptive fields.

Demonstration that the mechanisms of primary and secondary hyperalgesia were different was accomplished experimentally in 1950 by Hardy, Wolff, and Goodell. They and others repeated the experiments of Lewis, who earlier studied the mechanisms of hyperalgesia. Hardy and coworkers established that secondary hyperalgesia did not arise from what Lewis called "a local nervous mechanism" (i.e., an *axon reflex*). Secondary hyperalgesia developed because neurons in the central nervous system became more sensitive to peripheral stimuli after tissue damage. The researchers showed that secondary hyperalgesia after experimentally produced damage to the skin could be prevented if conduction of the nociceptive message to the spinal cord was prevented by a local anesthetic block of the nerve. Primary hyperalgesia was not influenced by this same manipulation. The significance of these developments and the new knowledge that excitatory amino acids are involved in mechanisms of secondary hyperalgesia has led to development of new strategies to control prolonged pain arising from tissue injury.

Although the work of Hardy and colleagues and others focused on secondary hyperalgesia associated with cutaneous injury, the notion has been advanced that a variety of inflammatory-like disorders of the gastrointestinal tract may represent a visceral analog of cutaneous hyperalgesia. Although experimental evidence is not available, altered sensations arise from the inflamed gastrointestinal tract, and it may be that similar local and central mechanisms are involved in these altered sensations, which often include pain.

Another development suggests a contribution of previously unknown nociceptors to pain arising from injured skin, joints, and viscera. It is known that local inflammation in these tissues activates previously silent nociceptors, which may then contribute to the central changes in excitability that characterizes secondary hyperalgesia.

☐ Warming and cooling skin sensations arise from specific sensory receptors

An important group of cutaneous receptors are the thermoreceptors, which signal changes in temperature. Like nociceptors, *cold and warm thermoreceptors* are innervated by thinly myelinated Aδ and unmyelinated C fibers. Unlike nociceptors that respond to noxious heat or intense cold, thermoreceptors respond only to nonnoxious warming and cooling stimuli. Warm receptors are typically unresponsive to noxious intensities of skin heating (>45°C), and cold receptors respond best to applied temperatures between 18°C and 30°C and become unresponsive to intense cold stimuli (Fig. 22-5).

The warm and cold receptors are active at normal body temperature, and their discharges increase when the skin is warmed or cooled, respectively. Receptive fields for thermoreceptors in skin were first described as "cold spots" and "warm spots." In humans, these receptive fields are about 1 mm in diameter, and each is typically innervated by a single sensory neuron.

☐ Skin sensations and sensory receptors are confirmed by recording from sensory axons in humans

The study of cutaneous receptors in humans has been advanced significantly by development of the technique of *microneurography*. A metal microelectrode is inserted percutaneously into a nerve (e.g., peroneal nerve in the leg, radial nerve at the wrist), and activity from single axons can be recorded. Psychophysical studies can be done because verbal reports about the intensity of sensation can be obtained when the axon is electrically stimulated through the microelectrode, a process called intraneural microstimulation, or when natural stimuli are applied in the neuron's receptive field. These verbal reports can be compared with the simultaneously obtained measurements of neuronal activity (see Fig. 22-3). Microneurographic studies using humans have provided broad confirmation of results of studies of cutaneous receptors in nonhuman animals. For example, qualities of sensa-

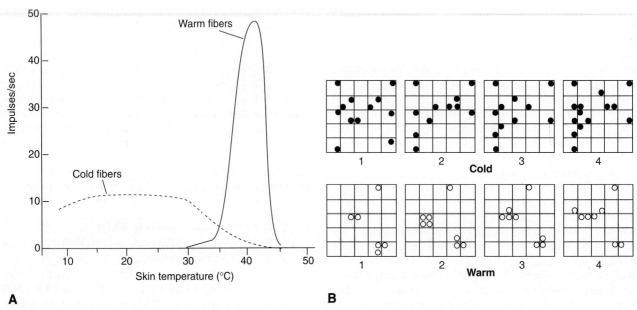

Figure 22-5
Response characteristics of thermoreceptors and distribution of cold and warm spots in a single person. (Adapted from Hensel, Kenshalo. J Physiol 1969;204:99 and from Dallenbach. Am J Psychol 1927;39:402.)

tion have been shown to be retained regardless of the intensity of intraneural microstimulation (i.e., pain is produced only when nociceptors are stimulated), and the intensity of the sensation increases as the frequency of intraneural microstimulation is increased experimentally (Table 22-2).

Muscle and Joint Receptors

☐ There are several types of muscle and joint receptors

Muscle and joint receptors include those responsible for proprioception and nociception. Proprioception includes the senses of position and posture, with or without eyes closed, movement (i.e., kinesthesia), and force (e.g., to maintain joint position against a load). *Stretch*

receptors in muscle spindles and Golgi tendon organs are associated with large-diameter myelinated axons (i.e., group I and II) and contribute to our conscious appreciation of motor activity. *Mechanoreceptors* in joints of the slowly adapting type (i.e., Golgi tendon organs and Ruffini endings that signal joint movement and torque, respectively) and fast-adapting type (i.e., pacinian corpuscles that respond to vibration) provide conscious appreciation of joint position and movement. They are also associated with rapidly conducting group I and II axons.

Although joint mechanoreceptors were once considered to be primarily responsible for proprioception, it is now known that muscle stretch receptors and, to a lesser extent, cutaneous SA II mechanoreceptors contribute to the total information that the brain integrates to derive the sense of proprioception. The same stretch receptors in muscle spindle and Golgi tendon organs constitute the afferent branch of important motor reflexes that occur

Table 22-2
SENSATIONS PRODUCED BY INTRANEURAL MICROSTIMULATION IN HUMANS

Receptor	Sensation
FA I (Meissner's corpuscle)	Tapping at 1 Hz; flutter at 10 Hz; vibration at 50 Hz
FA II (Pacinian corpuscle)	Tickling or vibration over 20–50 Hz
SA I (Merkel's disk)	Sustained pressure over 5–10 Hz
SA II (Ruffini ending)	No sensation
Aδ mechanical nociceptors	Sharp pain
C polymodal nociceptors	Dull pain, burning pain or itch

Modified from Willis WD, Coggeshall R. Sensory mechanisms of the spinal cord, 2nd ed. New York: Plenum Press, 1991:575.

"unconsciously." Similarly, postural adjustments that arise from joint mechanoreceptor and muscle spindle inputs are typically not sensed. When SA II cutaneous mechanoreceptors were electrically stimulated in humans, no sensation was reported (see Table 22-2). A significant component of proprioception is not conscious.

☐ The prominent conscious sensation from joints and muscles is pain

No conscious sensations regularly arise from joints. The principal sensation associated with joints is pain. *Joint nociceptors* are group III and group IV axons with unencapsulated endings located in the fibrous joint capsule, articular ligaments, menisci, and adjacent periosteum. Neither joint cartilage nor the synovial layer contain sensory endings, although the presence of sensory fibers in the synovial layer is argued. Cutaneous nociceptors respond only to stimuli that damage or threaten to damage the skin. Many group III and IV joint nociceptors, however, respond to normal joint movement, although they respond best when the joint is rotated or forcefully extended outside its normal range.

Joint afferent fibers have receptive fields in the joint defined by their sensitivity to mechanical stimulation. Some joint nociceptors also respond to chemicals introduced experimentally into the joint capsule. However, a proportion of group III and IV joint afferent fibers do not respond to noxious intensities of mechanical stimulation. They are normally silent, but they become active during injury or inflammation. These afferents may constitute a special class of nociceptors, particularly in deep tissue.

In striated muscle, group III and IV afferent fibers can be excited by pressure on the muscle or on the muscle-tendon junction and by stretch of the muscle. In addition to direct mechanical activation of muscle nociceptors, pain that accompanies muscle damage and the tenderness associated with sore muscles is probably a result of nociceptor sensitization. As in the skin, endogenous substances that can sensitize nociceptors are produced when muscle fibers are damaged or excessively exercised. Muscle pain, however, differs from cutaneous pain. Cutaneous pain is typically described as sharp, pricking, dull, or burning (see Table 22-2), but muscle pain is typically described as aching or cramping. In microneurographic studies of humans, electrical stimulation of muscle afferent axons classed as nociceptive gave rise to cramp-like sensations.

Pain from skin, joints, and muscle is usually localized to the appropriate area of skin, joint, or muscle that is the source of the pain. Conscious sensation from muscle shares with conscious sensation from the viscera the phenomenon of *referral*. Muscle pain is localized to the site of injury, and it is often localized to other deep, nonvisceral structures, such as nearby muscle, fascia, tendons, joints, and ligaments. This area of referral can exceed the borders of the dermatome of the muscles' innervation, but referral of muscle pain is not as problematic diagnostically as is referral of visceral pain.

Visceral Receptors

Activity in most visceral sensory neurons does not typically give rise to conscious sensation. We are aware of some interoceptive sensations such as hunger and thirst, but visceral sensory neurons also convey information that is important to regulation of the respiratory and circulatory systems, water and electrolyte balance, and digestion that is not sensed. We are usually not conscious of inputs from the heart, lungs, and other viscera, but irregular heart beats and sensations of stomach, bladder, or bowel fullness are commonly experienced. It was argued even as recently as the beginning of the 20th century that the viscera were insensate. There are, however, *mechanoreceptors* located in the mesentery, along blood vessels, and in the various tissue layers of the gastrointestinal tract that respond to distention or contraction of hollow organs and to movement of the viscera.

There are also *thermoreceptors* in the upper part of the gastrointestinal tract and at its end in the rectum and anus. Evidence for their existence is apparent when drinking a cold beverage on an empty stomach. *Chemoreceptors* that respond to fatty acids, glucose, amino acids, and other nutritional elements and receptors that respond to changes in pH or osmolarity of gastrointestinal contents are also present, but conscious sensation is not normally associated with their activation.

☐ Visceral nociception is unlike pain arising from other tissues

It had long been argued that pain did not arise directly from visceral organs and that there were no visceral nociceptors. Instead, pain from the viscera was thought to arise from excessively intense visceral stimulation or was encoded in the pattern of action potentials in visceral sensory nerves. Studies using animals have clearly established that there exist afferent fibers in the viscera that respond only at intensities of mechanical stimulation in excess of intensities that produce pain in humans. There are also probably chemical nociceptors in viscera, but our knowledge of adequate stimuli for the viscera is limited. It is probably an oversimplification to consider that visceral nociceptors in the heart, for example, are the same as those in the colon (i.e., have the same adequate stimulus).

Visceral sensory fibers, unlike sensory fibers from skin, muscle, or joints, traverse paravertebral and prevertebral ganglia on their way to the spinal cord (Fig. 22-6). Axon collaterals arising from visceral sensory fibers influence neurons in prevertebral autonomic ganglia. Activation of visceral sensory receptors can produce local reflexes by axon collaterals to autonomic ganglia or by visceral sensory neuron terminals in the spinal cord. Such reflexes are not usually sensed.

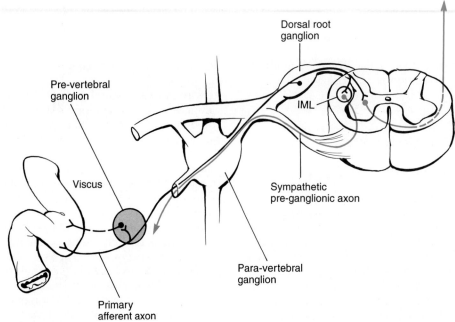

Figure 22-6
Anatomic organization of spinal visceral sensory innervation. Visceral sensory fibers have their cell bodies in the dorsal root ganglion, but they typically traverse prevertebral and paravertebral ganglia en route to the spinal dorsal horn. Collaterals of visceral sensory axons have been found in prevertebral ganglia, where they probably contribute to local visceral reflexes. In the spinal cord, visceral afferent input can activate second-order neurons, which project to supraspinal sites (e.g., thalamus through the spinothalamic tract) or can initiate sympathetic reflexes. *IML,* intermediolateral cell column.

□ Visceral pain is referred and difficult to localize

An important way in which visceral pain differs from most other pain is that it is typically referred to cutaneous and subcutaneous, deep sites. The pain is not localized to the site of its cause (e.g., a viscus) but instead is localized to a distant site (e.g., skin). The classic example is pain associated with angina pectoris, which is usually referred to the left chest, shoulder, and upper arm. Painful distention of the colon is referred to the periumbilical area; painful distention of the esophagus is referred to the chest in a pattern similar to pain arising from the heart. Visceral pain is not well localized, and the sensations arising from nearby viscera are usually confused. Cutaneous and joint pains are usually localized easily and are not confused with pain arising from other structures. Muscle pain is most typically referred to nearby muscle-related structures.

Visceral pain is referred because the central terminals of visceral sensory axons terminate in the spinal cord (principally lamina I or II, lamina V, and the area around the central canal, called lamina X) on the same neurons that receive input from the skin. It is also likely that some of these same spinal cord neurons receive input from muscle, because visceral pain is sometimes referred to deeper, subcutaneous muscle. This anatomic arrangement and referral of visceral pain to other structures has been called the *convergence-projection theory* of visceral pain. Visceral and other sensory neurons converge on the same spinal neurons, called *viscerosomatic neurons,* and that information is conveyed (i.e., projected) rostrally to supraspinal sites (Fig. 22-7). Consequently, localization of visceral sensations is difficult.

Most spinal cord neurons do not receive convergent visceral and cutaneous inputs. Because most second-order neurons in the central nervous system that receive input from cutaneous receptors receive no other convergent input, localization of cutaneous stimuli is precise.

When afferent inputs to the spinal cord are examined anatomically, visceral sensory afferent fibers constitute less than 10% of the total afferent input to the spinal cord. This partially explains why visceral pain is not well localized. The central terminals of visceral afferent fibers have a much greater rostrocaudal distribution within the spinal cord than afferent fibers arising from skin. This broader distribution in the spinal cord contributes to poor localization of visceral sensation, and the differential diagnosis of visceral pain is problematic. For example, visceral sensory afferent fibers from the esophagus, gallbladder, and heart have similar distributions to the thoracic spinal cord, where they converge on similar, if not the same second-order neurons, which also receive a cutaneous input.

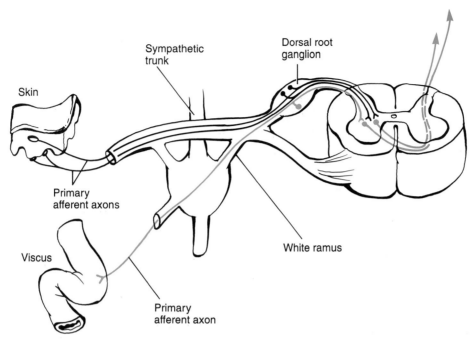

Figure 22-7
Convergence of cutaneous and visceral inputs onto the same second-order projection neuron in the dorsal horn of the spinal cord. Although virtually all visceral inputs converge with cutaneous inputs, most cutaneous inputs do not converge with other sensory inputs.

Silent Nociceptors

There exists in skin and deep tissues a sensory receptor not previously known, called the *silent nociceptor* or sleeping nociceptor. Silent nociceptors are normally inactive and unresponsive to high-intensity mechanical stimulation. They only become active and responsive to mechanical stimulation during and after injury and inflammation of tissue. Silent nociceptors are usually associated with group IV or C fibers and were discovered by electrically stimulating joint afferent nerves while recording from individual afferent axons. A group of axons were excited by electrical stimulation that could not be excited by mechanical stimulation (e.g., noxious squeezing or heating of the skin, overdistention of hollow organs, rotation of joints outside their normal range). After inducing inflammation experimentally, these fibers typically develop spontaneous activity and often begin to respond to mechanical stimuli, sometimes at intensities of stimulation that are not normally noxious. This previously unappreciated sensory receptor common to skin, joints, and viscera may be critically important to our understanding of pain processing and its modulation.

Silent nociceptors represent a new development in sensory physiology. In deep tissues such as muscle, joints, and particularly the viscera, adequate stimuli for sensory receptors are not well established or known at all. It has been suggested that silent nociceptors be called *mechanically insensitive nociceptors*, because typically only mechanical stimuli have been used to test their responsiveness. The discovery of silent nociceptors affords new possibilities for understanding pain and a new target to approach to control pain.

Sensory Cell Bodies

Cell bodies of sensory neurons in dorsal root ganglia have been subdivided on the basis of their size, which is related to myelination and conduction velocity of the cells' axon. Cytologic and histochemical differences also exist among these sensory cell bodies. The peptides in dorsal root ganglion cells, primarily in small cells and presumably representing neurotransmitters or modulators, have become a focus of study. In general, small-diameter cell bodies in dorsal root ganglia have thinly myelinated (i.e., Aδ or group III) or unmyelinated (i.e., C or group IV) axons and large-diameter cell bodies have myelinated axons with fast conduction velocities (i.e., Aβ or group I and II fibers). There are general associations among cell body size, axon diameter, conduction velocity, and function.

Because of the association of small cell size and axon diameter or conduction velocity with peptide content, nociceptors have been extensively studied. The chemical content of their cell bodies in dorsal root ganglia, and their putative central neurotransmitters, have been identified. Substance P was the first neuropeptide to become associated with pain and nociceptors, and it is often erroneously considered to be the central neurotransmitter for nociception. Cell bodies of nociceptors and the cell bodies of other sensory neurons in the spinal and cranial ganglia also synthesize calcitonin gene-related peptide and a variety of other putative peptide transmitters, such as vasoactive intestinal peptide, vasopressin, oxytocin, and somatostatin. These same cell bodies also synthesize nonpeptide transmitter substances such as the excitatory amino acids glutamate and aspartate. Typically, sensory neurons synthesize several putative neurotransmitters, and co-localization of peptides in the same sensory neuron is the rule rather than the exception. For example, most dorsal root ganglion cells that contain substance P also contain calcitonin gene-related peptide, although not all calcitonin gene-related peptide-containing cells contain substance P. As much as 80% of some neuropeptides synthesized in sensory cell bodies are transported peripherally to nerve terminals in skin, muscle, joints, and viscera. Peptides released peripherally may exert an *effector* role in the tissue, such as vasodilation and sensitization of the same or adjacent nerve terminals.

At the central terminals of nociceptors in the spinal cord, substance P, calcitonin gene-related peptide, other peptides, and excitatory amino acids are released, where they function as neurotransmitters or modulate the actions of other released substances. For example, calcitonin gene-related peptide seems to produce little effect when given experimentally directly onto the surface of the spinal cord, but calcitonin gene-related peptide enhances the actions of substance P by slowing the degradation of substance P. The co-release of peptides and amino acids from nociceptor terminals in the central nervous system function in concert to transmit information from skin, muscle, joints, and the viscera about noxious and probably other stimuli.

Sensory Channels

The central axons of sensory neurons enter the spinal cord through a dorsal root, where they eventually terminate on spinal neurons in different lamina of the dorsal horn. The spinal neurons give rise to several *sensory pathways* or *tracts* that ascend in the spinal cord to the brain, where the transmitted information ultimately leads to sensation. The sensory receptors, terminations of their central axons in the spinal cord, ascending spinal tracts, and supraspinal targets (e.g., thalamus, cortex) for a modality of somatovisceral sensation are serial components of what Willis called a *sensory channel* (Table 22-3).

□ The mechanoreceptive sensory channel includes touch-pressure and flutter-vibration sensations

Mechanoreceptors and hair follicle receptors in the skin give rise to the sensations of *touch-pressure*, derived primarily from activation of SA I mechanoreceptors, and *flutter-vibration*, derived from activation of FA I, FA II, and hair follicle receptors. The types of information that can be extracted from the touch-pressure channel include recognition of contact in hairy skin, two-point discrimination, detection of edges, and discrimination of shapes. The sensations that are evoked by activation of FA I and FA II receptors in the flutter-vibration channel include recognition of contact in glabrous skin, tapping, vibration, and texture or movement across a smooth surface with small imperfections.

SA I and SA II primary afferent fibers from the limbs and trunk project into lamina III and IV of the spinal dorsal horn, and *collaterals* of these axons ascend ipsilaterally in the dorsal columns to terminate on second-order neurons in the medullary dorsal column nuclei (Fig. 22-8). The input is somatotopically organized in that afferent fibers from the midthoracic and more caudal levels of the spinal cord are contained in the fasciculus gracilis and terminate in the more medially placed nucleus gracilis. Afferent fibers from the midthoracic to upper cervical levels of the spinal cord ascend in the more laterally displaced fasciculus cuneatus and terminate in the nucleus cuneatus. These ascending pathways are layered mediolaterally as input (i.e., axons) from legs, trunk, and arms successively enter the spinal cord. In a parallel fash-

Table 22-3
SENSORY CHANNELS

Sensation	Receptors	Spinal Termination	Major Ascending Spinal Pathway
Mechanoreception			
Touch-pressure	SA I + hair follicle	Lamina III–IV	Dorsal columns
Flutter-vibration	FA I, FA II + hair follicle	Lamina III–IV	Dorsal columns
Proprioception	Muscle spindle	Lamina IV–VII	Dorsal columns
	Golgi tendon	+ IX	
	SA II (distal joints)		
Temperature			
Cold	Aδ and C	Lamina I, II	Spinothalamic tract
Warm	C	+ V	Spinothalamic tract
Pain (nociception)	Aδ/group III +	Lamina I, II,	Spinothalamic tract
	C/group IV nociceptors	V + X	

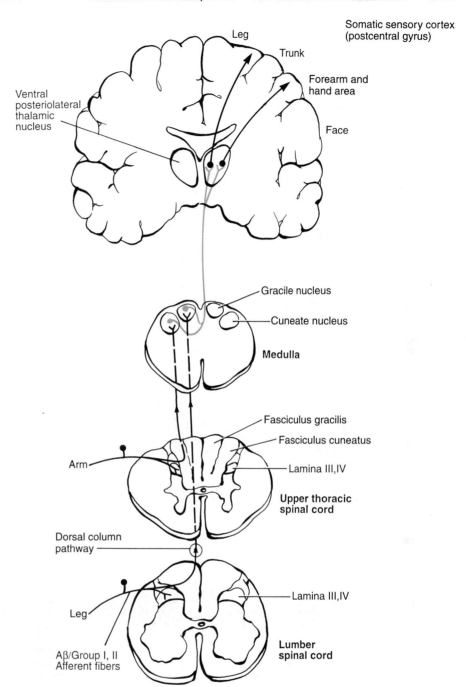

Somatic sensory cortex
(postcentral gyrus)

Leg

Trunk

Forearm and
hand area

Face

Ventral
posteriolateral
thalamic
nucleus

Gracile nucleus

Cuneate nucleus

Medulla

Fasciculus gracilis

Fasciculus cuneatus

Arm

Lamina III,IV

**Upper thoracic
spinal cord**

Dorsal column
pathway

Lamina III,IV

Leg

Aβ/Group I, II
Afferent fibers

**Lumber
spinal cord**

Figure 22-8
Cutaneous mechanosensitive and
conscious proprioception sensory
channels.

ion, flutter and vibratory sensations are also transmitted primarily in the dorsal columns.

☐ Touch-pressure and flutter-vibration sensations are somatotopically represented in the thalamus and cortex

Axons from the second-order neurons in the nuclei gracilis and cuneatus cross at the medullary level to project somatotopically to the contralateral ventrobasal thalamus through the medial lemniscus. Information in the touch-pressure and flutter-vibration sensory channels from the head and neck projects to the ventroposteriomedial thalamus, and information from the trunk and the limbs projects to the ventroposteriolateral thalamus; the upper extremities are more medial than the trunk or legs (see Table 22-3 and Fig. 22-8). These sensory channels are highly organized with respect to their terminations in the thalamus, reproducing a well-ordered map of the body that is subsequently projected to the somatosensory cortex. In general, the cutaneous sensory channels from the head and body reproduce a somatotopic map in the ventrobasal thalamus that is faithfully projected onto the somatosensory cortex. Central neurons in the thalamus and cortex have a receptive field that includes the receptive field of the peripheral sensory receptor that originally gave rise to the sensation. Although receptive fields of sensory receptors in the skin are relatively fixed, receptive fields of central neurons, particularly cortical neurons, can and do change under various circumstances.

Muscle stretch receptors, joint receptors, and SA II cutaneous mechanoreceptors in the skin of distal joints are components of the *proprioception sensory channel*. Proprioceptive input is rapidly conveyed by means of Aαβ or group I and II fibers to the spinal cord, where it is processed locally in spinal lamina IV through VII and IX. The information is transmitted rostrally in the dorsal columns, and conscious proprioception is represented in the thalamus and cortex, as is other cutaneous sensory input (see Table 22-3 and Fig. 22-8).

The *temperature sensory channel* is associated with separate cold and warm thermoreceptors in the skin. The receptive fields are spot-like and insensitive to mechanical or noxious thermal stimuli, but they are sensitive to small changes in temperature. Information in the temperature sensory channel is conveyed by Aδ or C fibers primarily to second-order neurons in the superficial dorsal horn. Axons of the second-order neurons cross at or near the segment of entry in the spinal cord and ascend to the contralateral thalamus in the spinothalamic tract, which is in the anterolateral funiculus of the spinal cord (Fig. 22-9 and see Table 22-3). Representation of thermoreception in the thalamus and cortex is also somatotopically organized. Input from the head and neck is represented medially in the ventrobasal thalamus and that from the upper extremities, trunk, and legs is represented progressively more laterally in the ventrobasal thalamus.

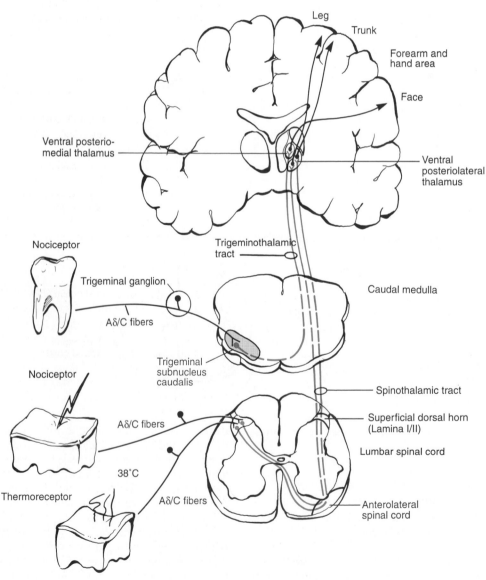

Figure 22-9
Thermoreceptive and nociceptive sensory channels.

Pain

Among the somatovisceral sensations, pain is arguably the most important. Nociceptors in tissue are the sensory receptors that give rise to the qualities of pain as we commonly experience them. Pain is a complex experience and can arise without activation of tissue nociceptors, such as after damage to a peripheral nerve or in the brain after a stroke. Typically, pain warns us of injury or potential injury and is a vital protective mechanism important to normal life. Individuals congenitally insensitive to pain are easily injured and become disfigured; most die at an early age. In circumstances of recurrent or chronic pain, some of which exist in the absence of obvious pathology, pain serves no useful purpose and can, if severe and intractable, contribute to depression and suicide.

Pain is commonly described in terms of its duration as *acute* or *chronic* (typically >6 months' duration). This distinction is not trivial. Different mechanisms come into play as pain becomes chronic, and different treatment strategies are required for acute and chronic types of pain. Consideration of pain as acute or chronic, however, ignores the common experience of prolonged (i.e., days to weeks), nonchronic pain.

Pain may be more appropriately described as acute, prolonged, or chronic. Acute pain arises from activation of nociceptors and is not associated with significant tissue injury (e.g., pin prick, paper cut). Prolonged pain lasting for days or weeks arises from tissue injury and inflammation (e.g., sunburn, sprain) and is associated with activation of nociceptors, sensitization of nociceptors, and allodynic and hyperalgesic phenomena. Acute and prolonged pain states are time limited, and neither are associated with long-term changes that could be described as abnormal. Chronic pain arises from nerve damage, permanent and abnormal responses of the central nervous system to tissue injury, tumor growth, or other circumstances that produce, usually by unknown mechanisms, disordered and abnormal pain sensations.

Pain can be described in terms of its location as superficial (i.e., skin) or deep (i.e., muscle, joints, viscera) or as somatic and visceral. Based on our understanding of mechanisms of pain, the qualities of pain are based on the site of origin. If somatic pain arises from skin, it is called *superficial pain*, and if it arises from muscle, joints, or connective tissue, it is called *deep pain*. *Visceral pain* is considered a separate quality.

□ Pain and nociception are not synonymous

A committee of the International Association for the Study of Pain has defined pain as "an unpleasant sensory and emotional experience associated with actual or potential tissue damage, or described in terms of such damage." This definition takes into account the range of unpleasant sensory experiences called pain and includes *sensory-discriminative* and *motivational-affective* components. Only the sensory-discriminative component of pain is directly linked to the stimulus and represents what is meant when pain is described as a sensation (e.g., as a sensory channel; see Table 22-3).

A noxious stimulus applied to skin activates nociceptors, second-order neurons in the central nervous system, neurons in the thalamus, and ultimately the cortex to allow determination of the *location* of the stimulus, the *intensity* of the stimulus, and its *onset* and *duration*. This sensory channel involves Aδ or group III and C or group IV fiber input to the superficial and deeper dorsal horn (i.e., lamina I, II, and V), where nociceptive information is transmitted to second-order neurons. Axons from these spinal (or medullary trigeminal) neurons cross at or near the spinal segment of entry to the contralateral side and ascend to the thalamus through the spinothalamic (or trigeminothalamic) tract in the anterolateral funiculus of the spinal cord. The sensory-discriminative nociceptive sensory channel is a somatotopically organized, direct pathway through the spinal cord to the thalamus to the somatosensory cortex.

There are also ascending pathways contained in the anterolateral funiculus that are thought to contribute to the motivational-affective component of pain. Such pathways are sometimes described as indirect because one or more synapses are interposed between the spinal cord and the thalamus or cortex, unlike the direct spinothalamic tract pathway.

The anatomic foundation of the motivational-affective component of pain is unknown, but it is thought to involve projections from the spinal cord to the brain stem reticular formation, called the spinoreticular pathway, or to the parabrachial area in the midbrain, called the spinomesencephalic pathway. Nociceptive information from these relays is then transmitted to other brain nuclei, including those in the medial thalamus and limbic system, where affective contributions to the noxious stimulus are thought to arise. The nociceptive sensory channel and other pain-related ascending pathways are illustrated in Figures 22-9 and 22-10.

The motivational-affective component of pain includes the nature and intensity of emotional responses that makes pain personal and unique for each person. Overlaying the sensory-discriminative and motivational-affective components of pain are learned cultural and *cognitive contributions* such as attention, anxiety, anticipation, and past experiences. These factors contribute to a person's interpretation and concerns about pain and allow modulation of the response to a painful stimulus.

Painful stimuli also produce autonomic and motor (e.g., withdrawal) reflexes. For example, painful stimuli often lead to increases in heart rate and blood pressure mediated by the autonomic nervous system. These autonomic reflexes are different from motor reflexes that are organized segmentally. *Autonomic reflex* responses to painful stimuli require supraspinal integration, and they are typically absent when the spinal cord has been severed. In general, autonomic reflexes are greater for deep somatic and particularly visceral pain than for superficial pain. Painful stimuli also produce *motor withdrawal reflexes*, as when a person unexpectedly touches a very hot surface. These are segmentally organized protective re-

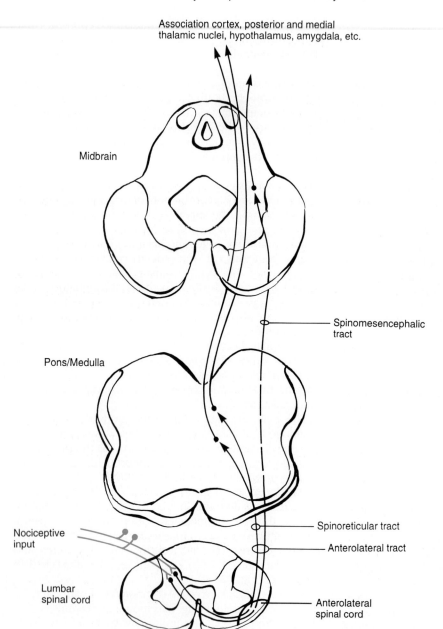

Figure 22-10
In addition to the spinothalamic tract, nociceptive information is conveyed in ascending spinoreticular and spinomesencephalic tracts.

flexes that help to prevent injury and are present in vertebrates and invertebrates alike.

☐ Early investigators hypothesized that pain was a specific sensation

An early theory of pain proposed that pain was a specific sensation with its own specialized sensory nerves, anatomically separate pain pathway in the spinal cord, and pain centers in the brain. This *specificity theory* of pain, derived from Müller's law of specific nerve energies, has often been denigrated as a concept inconsistent with the complex experience produced by a noxious stimulus.

Müller was a contemporary of Weber and Fechner and was similarly interested in human sensation. He proposed what became the foundation for the study of sensation, that the nature of sensation is determined by the nerve (sense organ) stimulated and not by the stimulus. Sensation or modality is a property of the sense organ stimulated. It was not until 1884, however, that Blix showed that distinct cutaneous sensations could be associated with separate, localized cooling, warming, pressure, or painful stimuli applied to the skin. Von Frey and colleagues made similar observations about pressure points and pain points on the back of the hand, and their work at the end of the 19th century suggested a causal relation between sensory receptors and specific sensations. In the mid-1930s, Zotterman provided early electrophysiologic evidence for the existence of nociceptors.

It was not until the 1960s that recordings were made from single Aδ and C fibers in peripheral nerves that had unique receptive fields and responded only to noxious stimuli. These were the nociceptors Sherrington defined almost 6 decades earlier.

Theories about the *intensity* of a stimulus (i.e., pain was produced by high-intensity stimulation of any sensory receptor) or the *pattern* of nerve impulses were advanced in competition with the specificity theory of pain, which had its detractors because the sensory receptors were not demonstrated until relatively recently. Intensity and pattern theories of pain were attractive because they were not based on morphologic specializations and were thought to explain pain arising from structures like the viscera, which were then and are today often considered to be lacking nociceptors. Visceral pain was thought to arise by intense activation of visceral mechanoreceptors or by the pattern of activity in sensory afferent fibers from the viscera.

☐ Modern theory includes the gate control of pain

In 1965, Melzack and Wall summarized arguments for and against the various theories of pain and advanced the hypothesis that there existed in the spinal cord a gate-like pain control mechanism. They recognized the contribution of specific, nociceptive inputs to pain perception and that patterns of input to the spinal cord were important.

They hypothesized that cells in the spinal cord that transmitted the nociceptive message rostrally to the brain constituted a functional gate that was opened by activity in small-diameter afferent fibers (i.e., group III or Aδ and group IV or C fibers) and closed by activity in large-diameter, myelinated nonnociceptive afferent fibers (i.e., groups I and II or Aβ fibers). The balance in activities between nociceptive and nonnociceptive afferent fiber inputs to the spinal cord could modulate the intensity of pain.

It was also appreciated that pain was subject to modulation *descending* from supraspinal sites. Consequently, modifications to the *gate-control theory* included the addition of components that modulated the spinal gate. The theory, although incorrect with respect to details of the neuronal circuitry, was extremely influential and served an important heuristic function.

Sensory-discriminative, motivational-affective, and cognitive components are now understood to contribute to the response to noxious inputs to the central nervous system, including modulation of the noxious input (Fig. 22-11).

☐ Pain control involves endogenous pathways

The principal approach to pain control is pharmacologic (e.g., opioids, nonsteroidal antiinflammatory drugs). Pharmacologic control of pain is generally effective for

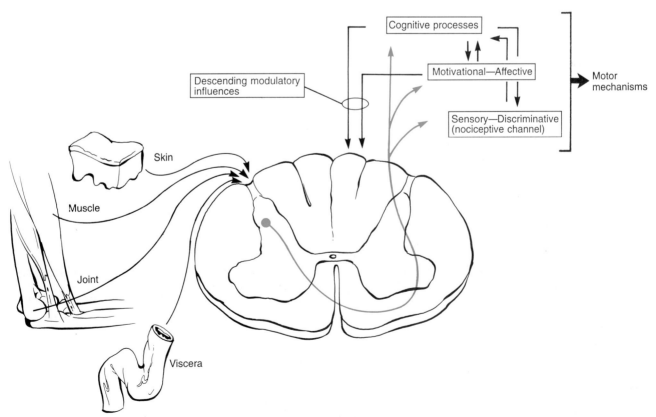

Figure 22-11
Representation of the sensory, motivational, and cognitive determinants of pain.

most acute and persistent pains and can be effective for most types of chronic pain if the drugs are used appropriately. There also exist *endogenous pain modulatory* systems, but their contribution to effective pain control is not completely understood.

Although descending modulation of spinal reflexes was a focus of study in the 1950s, the concept of *endogenous* pain control developed in the early 1970s when it was learned that electrical stimulation in the midbrain of experimental animals could produce analgesia, called *stimulation-produced analgesia*. Stimulation-produced analgesia was quickly established to be produced in humans and to be useful for relief of some chronic pain conditions. At about the same time, *opiate receptors* were discovered and *endogenous opioid peptides* were found in the central nervous system. It was later documented that the central terminals of nociceptive afferent fibers constituted one of many places in the central nervous system where opiate receptors were located and where exogenous (e.g., morphine) and endogenous (e.g., *enkephalins*) opioids could act to attenuate the release of their peptide content and their ability to transmit information about nociception.

Stimulation in periaqueductal and periventricular gray areas in the rostral midbrain and diencephalon is employed to control pain intractable to other treatment strategies. Endogenous opioid peptide–containing neurons play a role in stimulation-produced analgesia because opioid receptor antagonists attenuate the analgesia produced by electrical brain stimulation. The activation of *nonopioid* systems descending from the brain stem influences endogenous opioid peptide–containing interneurons in the spinal cord and acts directly to block the pain message at the first central synapse. These *descending monoamine systems*, activated by electrical stimulation in the brain and by opioids given systemically, have been shown to mediate stimulation-produced analgesia at the level of the spinal cord (Fig. 22-12).

In addition to inhibitory modulation (i.e., analgesia), activation of descending systems may also facilitate or enhance the nociceptive message arriving in the spinal cord. The functional significance of putative pain-producing systems is not understood but conceivably could mediate unusual pain syndromes that exist in the absence of tissue pathology and contribute to some chronic pain conditions.

Modulation of central nervous system neurons that receive nociceptive input occurs at the segmental level (e.g., by activity in nonnociceptive afferent fibers) and by inhibitory and facilitatory influences descending from supraspinal sites. There are intersegmental, propriospinal modulatory circuits within the spinal cord that probably contribute to the modulation of pain. The events that contribute to the overall balance of activity in nociceptive pathways include contributions from sensory-discriminative, motivational-affective, and cognitive components; disease states such as hypertension; and environmental factors such as exercise, stress, or pain itself. It is thought that internal and external environmental circumstances contribute to the tone of endogenous pain modulatory circuitry.

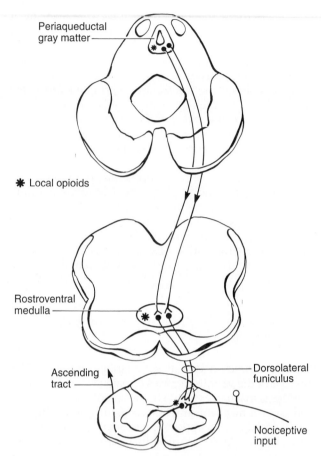

Figure 22-12
Descending monoaminergic and local opioid modulation of spinal nociceptive transmission. Inhibitory and facilitatory modulation can be produced by activation of cells in the midbrain periaqueductal gray matter or rostroventral medulla. Endogenous opioids and opioid receptors are present in the midbrain, medulla, and spinal dorsal horn, where endogenous opioid peptides or exogenously administered opiates, such as morphine, act to attenuate pain.

SELECTED READINGS

Kandel ER, Schwartz JH, Jessell TM. Principles of neural science, 3rd ed. Amsterdam: Elsevier Biomedical Publishers, 1991.

Lewis T. Pain. London: Macmillan Press, 1942:192.

Melzack R, Wall PD (eds). Textbook of pain, 3rd ed. London: Churchill Livingstone, 1994.

Mense S. Nociception from skeletal muscle in relation to clinical muscle pain. Pain 1993;54:241.

Ness TJ, Gebhart GF. Visceral pain: a review of experimental studies. Pain 1990;41:167.

Schaible HG, Grubb BD. Afferent and spinal mechanisms of joint pain. Pain 1993;55:5.

Schmidt RF (ed). Fundamentals of sensory physiology. Heidelberg: Springer-Verlag, 1978.

Schmidt RF, Thews G (eds). Human physiology, 2nd ed. Heidelberg: Springer-Verlag, 1989.

Willis WD, Coggeshall R. Sensory mechanisms of the spinal cord, 2nd ed. New York: Plenum Press, 1991.

Willis WD. Hyperalgesia and allodynia. New York: Raven Press, 1992.

Clinical Correlation
PHYSICAL TRAUMA TO NERVES

ROBERT L. RODNITZKY

Nerve injury can occur as a result of major acute trauma, such as a gunshot wound, or develop more slowly as a result of milder, but continuous trauma. The degree of neurologic deficit and the prospect for recovery depend on several factors. Typically, acute injuries resulting in anatomic disruption of the nerve cause more severe and lasting disability than slowly developing injuries that leave the nerve in anatomic continuity.

Nerve Severance Results in a Predictable Sequence of Changes in the Nerve and Muscle

If nerve continuity is disrupted as a result of trauma, the nerve segment distal to the injury undergoes *wallerian degeneration*. This process involves disintegration of axoplasm and axolemma and ultimately results in the breakdown and phagocytosis of myelin. The process takes several weeks to complete. Severe degeneration of the nerve implies that any recovery of function will require regeneration of nerve fibers from the intact nerve stump to the appropriate muscle or sensory organ.

After about 2 weeks, significant changes appear in the denervated muscle. Acetylcholine receptors begin to proliferate at locations on the muscle fibers outside the end-plate region, where they are usually concentrated. At the same time, spontaneous contractions of muscle fibers, known as *fibrillations*, appear. Fibrillations can be detected using electromyography, a common clinical electrodiagnostic technique in which the electrical potentials generated by muscle fibers can be measured through a small intramuscular needle electrode. Detection of fibrillation potentials suggests that the muscle is denervated. This finding can be considered indirect evidence that the nerve fiber innervating the muscle has undergone axonal disruption.

Fibrillations, which can only be detected by electromyographic techniques, should not be confused with *fasciculations*, which are gross muscle twitchings under the skin that are clearly visible to the naked eye. Unlike fibrillations, which strongly suggest muscle denervations, fasciculations can be seen in normal and denervated muscles.

Nerve Conduction Testing Is Useful in Determining the Extent and Location of Nerve Damage

Stimulation of a motor nerve segment distal to the point of injury can help determine whether the injury has disrupted the nerve's anatomic continuity. Within 8 days after severance of a nerve, stimulation of the distal segment no longer produces a response in the muscle it innervates. However, if the damage to the nerve has only resulted in a physiologic block of impulse conduction (i.e., *neuropraxia*) but not anatomic discontinuity, a muscle response occurs, implying intact conduction in the nerve segment beyond the point of injury.

Nerve stimulation can also help to determine whether an injury to a sensory nerve is preganglionic or postganglionic. In preganglionic injury, the distal nerve fibers are still in continuity with ganglion cells. Stimulation of the nerve produces propagated impulses that can be recorded over a distant portion of the nerve, proving its viability. In a postganglionic lesion, the distal segment of the nerve, after being disconnected from the nerve cell body, degenerates and looses its ability to be stimulated.

Chronic Compression and Entrapment Are Among the Most Common Forms of Nerve Injury

Compression, constriction, or stretching of nerves is common at certain anatomic sites that are vulnerable to these mechanical forces. A nerve may be susceptible to compression because of its superficial location. An example is the peroneal nerve below the knee, where it crosses over the lateral border of the top of the fibula just under the skin. Because of this superficial location, there is no protective layer of muscle or fat. The nerve is commonly damaged when one leg is crossed over the other and the fibular head comes to rest on the opposite knee. The ensuing peroneal nerve palsy results in a pattern of muscle weakness, causing footdrop.

Nerve compression resulting from passage through a confining space is exemplified by carpal tunnel syndrome. At the wrist, the median nerve passes under a thick, fibrous ligament and can be chronically compressed, especially during repetitive flexion of the wrist. This syndrome, unlike peroneal nerve palsy, typically develops over months as a result of less severe but more prolonged trauma. In severe forms of carpal tunnel syndrome, neurologic symptoms develop, including weakness of the muscle controlling the thumb and tingling of the thumb and the next two digits.

Stretching is a form of trauma that often affects the ulnar nerve. This nerve courses over the elbow and can be stretched when the elbow is flexed. With repeated stretching, an ulnar nerve palsy develops, producing symptoms that include weakness of the intrinsic hand muscles and numbness of the fourth and fifth digits of the hand. The susceptibility of the ulnar nerve to compression at the elbow is known to anyone who has struck his or her "funny bone."

Recovery After Nerve Trauma May Depend on Nerve Regeneration

After wallerian degeneration, nerve regeneration may occur, especially if surrounding connective tissue elements of the nerve sheath are intact and guide the regenerating fibers in the proper direction. Even under these favorable circumstances, recovery may be delayed, because nerve regeneration proceeds at the rate of approximately 1 mm per day. The forward progress of the regenerating nerve tip can be monitored by using Tinel's sign. Because the advancing edge of a regenerating nerve fiber is unmyelinated, it is unusually sensitive to minimal mechanical stimuli. By lightly tapping through the skin along the course of a regenerating nerve, the physician can identify the location of the advancing nerve tip. The tap produces a tingling sensation in the part of the body toward which the regenerating sensory nerve is growing.

One of the pitfalls of peripheral nerve regeneration is that fibers, especially those with motor and autonomic functions, may become misdirected to a different muscle or gland than was originally innervated. This is known as *aberrant regeneration*. This misdirected growth commonly occurs in the face when regenerating facial nerve fibers intended for eyelid muscles grow instead to perioral muscles. An attempted eye blink then produces a simultaneous twitch of the side of the mouth. This dual action, caused by aberrant regeneration is called *synkinetic movement*.

Clinical Correlation
PERIPHERAL NEUROPATHY

ROBERT L. RODNITZKY

The peripheral nerves are susceptible to toxic, metabolic, traumatic, and neoplastic damage. Regardless of the cause, the damage is known as *neuropathy*.

Circumstances can alter the nature of the damage to the nerve. In some neuropathies, the axon is the primary focus of involvement. In some, the myelin sheath is largely involved, and in others, the axon and sheath are equally affected. The anatomic distribution of nerve involvement varies as well. In some neuropathies, the most proximal portion of the nerve is involved, but in others, the most distal segment is primarily affected. Involvement of a single nerve is referred to as *mononeuropathy*, and the term *polyneuropathy* is used when nerves throughout the body are diffusely affected.

There are too many causes of peripheral neuropathy to list here, but among the most common are diabetes, kidney failure, chronic alcohol use and its associated nutritional deficiencies, autoimmune diseases, and trauma.

Axonal Degeneration Can Take Several Forms

Several processes can result in axonal degeneration. *Wallerian degeneration* is common after severe nerve trauma. After any insult that interrupts the axon, the segment distal to the lesion undergoes progressive degeneration over the next several days. The myelin components of the distal nerve segment degenerate, and fragments of the axon and myelin sheath are ultimately cleared by macrophages.

Many neuropathies that primarily affect the axon, especially those of toxic origin, are characterized by degeneration of the distal portion of the axon through a process known as the *dying back phenomenon*. This process most commonly involves the longest nerves in the body and is typically manifested first in the nerves innervating the feet and hands. This explains why sensory symptoms in many axonal neuropathies initially appear in the toes and feet or in the fingertips and hands. The dying back phenomenon is probably related to the failure of axonal transport to support sufficiently the portions of the nerve that are most distant from the cell body.

Neuronopathy refers to a destructive process primarily involving the nerve cell body, producing degeneration and loss of neurofilaments within the axon, which re-duces the caliber of the axon. The autoimmune sensory neuropathy associated with lung cancer is a neuronopathy of sensory neurons in the dorsal root ganglia.

Demyelination of Peripheral Nerves May Be Primary or Secondary

Segmental demyelination refers to the breakdown of the myelin sheath in the nerve segment between two nodes of Ranvier. These abnormal segments may be scattered along the length of the nerve. Because of the importance of the myelin sheath in facilitating rapid impulse conduction, segmental demyelination markedly slows nerve conduction and, in extreme forms, produces total conduction block. These conduction abnormalities can be easily demonstrated in the clinical electrophysiology laboratory using the technique of nerve conduction velocity testing. The term *secondary demyelination* is used to describe the myelin breakdown that occurs as the result of an antecedent primary axonal insult.

The Typical Symptoms of Peripheral Neuropathy Include Positive and Negative Phenomena

The most common negative signs of peripheral nerve dysfunction are losses of strength and sensation. In axonal neuropathies, these signs begin in the feet and progress proximally, correlating with the predominance of pathology in the distal portion of long peripheral nerves. Even in demyelinating neuropathies, a similar distal pattern of sensory and motor loss is often observed, because longer nerve fibers are more likely than shorter ones to contain randomly distributed demyelinated foci. In addition to muscle weakness, involved muscles may demonstrate atrophy and fasciculations, which are signs of denervation.

Any sensory modality can be lost in peripheral neuropathy. In conditions primarily affecting small fibers, pain and temperature sensation are preferentially lost, and in neuropathies involving large, myelinated fibers, there may a greater loss of vibratory and proprioceptive senses.

453

One of the most prominent negative signs of peripheral nerve disease is the loss of muscle stretch reflexes. Reduction or loss of stretch reflexes is one of the most sensitive signs of peripheral nerve disease. There are several mechanisms by which these reflexes can be affected. In neuropathies involving large, rapidly conducting fibers, Ia afferent fibers from the muscle spindle may be directly involved, interrupting the afferent arc of the reflex. In neuropathies involving smaller fibers, the gamma efferent fibers to the spindle may be the locus at which the reflex is affected. Although large and small fiber neuropathies can produce areflexia, it is more common in cases of large fiber involvement.

Involvement of autonomic fibers can produce symptoms such as a loss of sweating, abnormal heart rhythms, or impaired control of blood pressure. In most neuropathies, autonomic dysfunction is not prominent, but there can be significant autonomic symptoms in conditions resulting in acute demyelination or those primarily affecting small myelinated and unmyelinated fibers. An example of the former is Guillain-Barré syndrome, and an example of the latter is the neuropathy of diabetes.

Sensory ataxia is another negative feature of peripheral neuropathy. It results from the severe loss of position sense in the feet. Involvement of large, myelinated fibers conveying proprioceptive sense prevents the patient from appreciating the exact position of the feet, resulting in an unsteady gait.

The most prominent positive symptoms of neuropathies are in the sensory realm. The term *paresthesia* is used to describe uncomfortable sensory perceptions occurring spontaneously without an apparent stimulus. These sensations are variously described as burning, tightness, tingling, or pins and needles. The term *dysesthesia* refers to an unusual or distorted sensation evoked by a stimulus such as a simple touch. *Hyperalgesia* is exaggerated pain perception in response to a stimulus that would ordinarily be painless. Spontaneously painful sensations are slightly more common in neuropathies involving small-diameter fibers.

Positive phenomena related to motor nerve involvement include fasciculations and muscle cramps. A fasciculation is the spontaneous contraction of a denervated motor unit, which is a group of muscle fibers previously innervated by a single motor nerve fiber. Fasciculations appear as flickering movements of muscles that can be seen through the skin. Normal muscles may occasionally exhibit fasciculations under conditions such as extreme fatigue.

Nerve Conduction Testing Is Useful in Analyzing Peripheral Neuropathies

Nerve conduction velocity can be calculated by stimulating a nerve trunk at one point along its course and determining the time of arrival of the impulse at a measured distance along the nerve. This is a relatively painless procedure that requires a small electrical stimulus to be applied to the nerve through the skin. The technique can differentiate axonal from demyelinating neuropathies. Because an intact myelin sheath is critical to rapid nerve conduction, nerve conduction velocity is markedly diminished in demyelinating neuropathies. Conduction block, another electrophysiologic feature typical of demyelinating neuropathies, can be demonstrated by this technique. In axonal neuropathies, conduction velocity is only minimally slowed, but the the amplitude of the evoked response in the nerve or in a muscle innervated by the nerve is clearly diminished.

In demyelinating neuropathies, recovery can occur through remyelination. However, even after recovery, nerve conduction velocity may remain slow, because the nodes of Ranvier in remyelinated segments may be closer together, resulting in less efficient saltatory conduction.

Peripheral Neuropathy May Develop Acutely or Chronically

Most neuropathies due to metabolic, degenerative, or heritable abnormalities develop over several months or years. Chronic evolution is seen in the neuropathies associated with diabetes, kidney failure, or lead exposure, and this pattern is typical of most familial neuropathies. A smaller subset of neuropathies develop more rapidly.

The Guillain-Barré syndrome is an example of a neuropathy with a subacute onset. In this condition, autoimmune inflammatory demyelination of peripheral nerves occurs over several days to 2 weeks. Within this period, the affected person may revert from total normality to a profound state of weakness, immobility, and respiratory insufficiency.

The onset of traumatic nerve disorders can be acute or chronic. An example of acute traumatic neuropathy is the sudden development of footdrop caused by acute compression of the peroneal nerve as it crosses the outside of the knee. This condition is often caused by crossing one leg over the other. A more slowly developing traumatic nerve disorder is carpal tunnel syndrome. It is caused by chronic or repetitive compression of the median nerve as it courses under a tight ligament at the wrist. Dysfunction of the nerve results in sensory loss and paresthesia involving the first three or four digits of the hand and weakness of the muscles controlling the thumb. Footdrop due to peroneal nerve compression often recovers spontaneously if further trauma is avoided, but in carpal tunnel syndrome, trauma frequently continues, and the condition often requires treatment consisting of surgical release of the compressing ligament.

Therapy for Neuropathies Can Be Directed at the Symptoms and the Cause

For most metabolic or toxic neuropathies, medical treatment is aimed at correcting the underlying metabolic abnormality or reducing the amount of the toxic substance within the body. For example, in patients with kidney failure, renal dialysis is an extremely effective means of improving the associated neuropathy. In a toxic neuropathy such as that caused by lead intoxication, chelating agents, which promote the excretion of lead, are useful. Neuropathies that have an autoimmune basis are treated with immunosuppressive therapies. For patients with Guillain-Barré syndrome, plasma exchange has proven to be beneficial.

The treatment of the symptoms, as opposed to the cause of neuropathy is largely confined to attempts at reducing pain and uncomfortable paresthesias. Anticonvulsant medications and tricyclic antidepressant drugs have been used successfully to reduce neuropathic pain. In neuropathies in which pain is localized to a discrete area such as the foot, capsaicin can be applied to the skin. This agent works by locally depleting substance P, a neuropeptide involved in the transmission of pain impulses.

Neuroscience in Medicine, edited by P. Michael Conn.
J. B. Lippincott Company, Philadelphia © 1995

Chapter 23

Vision

J. FIELDING HEJTMANCIK

Components of the visual system include the cornea, aqueous humor, lens, vitreous humor, retina, optic nerves, optic tracts, optic radiations, visual cortex, and a variety of nuclei. Each of these structures plays an important role in receiving and interpreting visual signals.

The anterior structures focus light on the retina, which transduces the light signal into electrical signals and performs some initial processing before they are passed through the optic nerves and tracts to central structures that perform more elaborate processing, integrating their information with that of the other senses. This chapter considers the components of the phenomenon known as *vision*.

Refraction

Initially, light passes through the cornea, aqueous humor, lens, and vitreous humor (Fig. 23-1). Light travels through each of these relatively dense materials with a velocity inversely proportional to the material's density.

The *refractive index* of the material is defined as the ratio of the velocity of light in a vacuum to the velocity in that substance. When a light wave strikes the curved surface of the cornea at an angle, the wave front that enters the cornea first is slowed relative to that which travels a

longer distance through the air. By this process, the path of a light ray is bent, a phenomenon called *refraction*. If the components of the anterior chamber, especially the cornea and lens, are shaped correctly, the light rays that emanate from a single point are focussed onto a single point on the retina. The image of an object is projected onto the retina in an inverted fashion. The inferior part of the visual field is projected onto the superior part of the retina.

The refractive power of the eye resides in four surfaces: the anterior and posterior surfaces of the cornea and the anterior and posterior surfaces of the lens. Because the amount of refraction depends on the change in the refractive index between two substances, most refraction occurs at the anterior surface of the cornea, which is adjacent to air. Approximately one third of the refractive power of the eye results from the lens, whose curvature depends on the contraction of the circular ciliary muscle. Contraction of the ciliary muscle decreases its diameter, allowing the lens to assume a more spherical shape. This change in the shape of the lens allows the eye to shift its focus from a distant point to a near point and is called *accommodation*.

A variety of refractive errors can occur. *Hyperopia* (i.e., farsightedness) is usually caused by a globe that is shortened anteroposteriorly or a lens system that is too weak

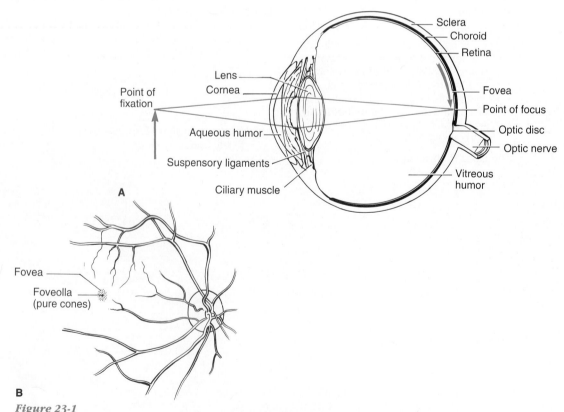

Figure 23-1

(**A**) Overview of the eye and refraction. The refraction of light rays from the fixation point at the tip of the arrow to the focal point on the retina represent the summed effects of the anterior and posterior surfaces of the cornea and lens. The image of the arrow is projected in an inverted orientation on the retina. (**B**) View of the central retina, including the macular region.

and focuses light on a point behind the retina. *Myopia* (i.e., nearsightedness) usually results from an elongated anteroposterior diameter, which causes light to be focussed anteriorly to a point in the vitreous humor.

Astigmatism, a more complex refractive error, results from an irregularly shaped lens or more commonly from an asymmetric cornea with stronger curvature in one pole than another, which focuses light into a line rather than a point. These refractive errors can usually be corrected with appropriate lenses. However, severe lens opacities (i.e., cataracts) may require surgery for correction.

After the light has been focussed on the retina, this structure effects the transformation from light into electricity. The retina and the higher neuronal pathways interpret the information in the visual signal. Despite intensive receptor research, the molecular events underlying the transformation of light energy into electrical information remain largely a mystery.

The Retina

☐ **The retina consists of the retinal pigment epithelium and the neural retina containing photoreceptors and neuronal processing cells**

The retina and its component cells previously have been described in great detail for vertebrate and invertebrate species. This discussion focuses on vertebrate vision, al-

though much of the progress in understanding visual processes has come from studies of invertebrate models, which continue to be a valuable resource.

The structure of the vertebrate retina is shown schematically in Figure 23-2. Many cell types contribute to the adult structure. The basic processes underlying retinal development are just beginning to be understood but are outside the purview of this discussion. It is sufficient for our purposes to mention that the retina consists of two structural and functional components: the retinal pigment epithelium (i.e., nonneural component) and the closely associated but physically distinct neural (i.e., sensory component) retina.

The *retinal pigment epithelium* (RPE) contains melanin granules, which prevent light that is passed through the retina from being internally reflected by the sclera and obscuring visual pattern reception, which is a problem, for example, in ocular albinism. The RPE cells also assist the photoreceptors with resynthesis of visual pigments and phagocytosis of shed outer segment tips. Because this requires the outer segments that contain these pigments to be closely approximated to the retinal pigment epithelial layer, the neural processing networks of the retina are the anterior-most structures, and light must pass through them before stimulating the photoreceptor cells. The neural retina can be further divided into two plexiform and three somatic layers, composed of six neuronal cell types and the nonneuronal glial (Müller) cell. The interactions of these cell types are described later in the chapter.

Phototransduction, the biochemical process of transforming light to electrical energy, occurs in the *photoreceptor cells*, the first of the retinal neural cell types. The

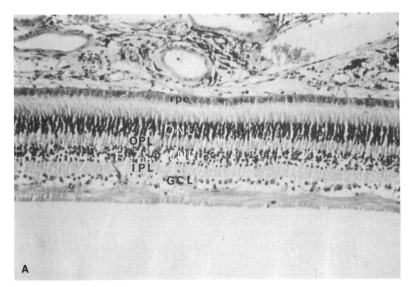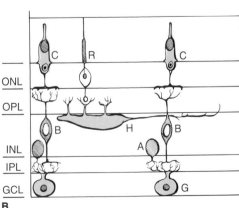

Figure 23-2
The vertebrate retina. (**A**) Light microscopy section of a mouse retina. (**B**) Schematic of the cells in a retina. *ONL*, outer nuclear layer; *OPL*, outer plexiform layer; *INL*, inner nuclear layer; *IPL*, inner plexiform layer; *GCL*, ganglion cell layer; *R*, rod cell; *C*, cone cell; *B*, bipolar cell; *H*, horizontal cell; *A*, amacrine cell; *G*, ganglion cell; *RPE*, retinal pigmented epithelium. (Photograph of hematoxylin-eosin stain; mouse retina at original magnification ×200; courtesy of Dr. Chi Chao Chan, National Eye Institute.)

highly specialized photoreceptor cells can be further divided into *rods* or *cones*, depending on their morphology. The most notable feature of these cells is the outer segment, which consists of a series of stacked disks, from which they derive their names. *Rhodopsin*, which absorbs light energy and initiates the transduction cascade, is located in the disks of the rod cells (Fig. 23-3). Rod cells, which are found at greater densities in the periphery of the retina and contain rhodopsin molecules able to detect light under dim illumination, are important in night vision. Cone cells are found primarily in the macular regions and contain opsins specialized for reception of specific colors and for strong (e.g., daylight) illumination.

The photoreceptor cells, whose cell bodies make up the outer *nuclear layer*, synapse in the *outer plexiform layer* of the retina with horizontal cells and bipolar cells. The bipolar cells synapse in the inner plexiform layer with amacrine and ganglion cells. The cell bodies of the amacrine, bipolar, horizontal, and interplexiform cells comprise the inner nuclear layer, and the cell bodies of the ganglion cells constitute the ganglion cell layer. The ganglion cell axons traverse the nerve fiber layer of the retina and collect in the optic nerve, which leads to the brain. What is known of the specific nature of these neural interactions is considered in the following sections.

☐ Continuously graded signals are generated in the photoreceptors by activation of opsin-chromophore complexes

Phototransduction

In the neural retina, the information contained in light absorption by the retinal surface is converted into neural signals in a process called *phototransduction*. Phototransduction is a model for understanding more general signal transduction processes. For example, opsins show homology to the family of hormone receptors, including adrenergic receptors that activate adenylyl cyclase through G-protein—coupled receptors.

Figure 23-4 shows a representation of the phototransduction cycle. A photon of light is absorbed by the opsin-chromophore complex (i.e., rhodopsin in rods, color pigments in cones) in the outer segment of the cone cell or rod cell, changing the chromophore retinal from the 11-*cis* to the all-*trans* conformation. This conformational change results in the activation of *transducin* (i.e., photoreceptor-specific G protein) by the exchange of a bound GDP for GTP. The precise nature of the rhodopsin-transducin interaction has not been elucidated. The activated transducin stimulates a cGMP phosphodiesterase (PDE), which then cleaves cyclic guanosine monophosphate (cGMP). Reduction of cGMP levels closes the cGMP-gated ion channels, causing plasma membrane hyperpolarization and signal generation.

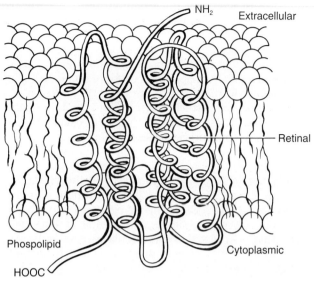

Figure 23-3

The structure of human rhodopsin and its putative position in the membrane. A lysine in the seventh transmembrane domain is the site of the chromophore attachment (retinal).

Rod cells have a low threshold of excitation and react readily to the low intensities of light required for black and white vision at twilight and at nighttime. The cones require a much higher intensity of stimulatory light and are needed for fine vision and color discrimination. This is especially true in the posterior pole, where the the high concentration of cones in the macula and fovea centralis are responsible for central vision. Because the eye must provide sensitivity over a range of light intensities and wavelengths, the basic transduction process is modulated by several mechanisms, many of which are just beginning to be understood. For example, it is known that the phosphorylation of rhodopsin by rhodopsin kinase after light stimulation is required for the effective quenching of the signal through the interaction of phosphoopsin and arrestin (also called S antigen). The phosphorylation or dephosphorylation of a PDE subunit is also important in modulating the light-induced response. Other mechanisms involved directly or indirectly in the *transduction cascade* include modulation by Ca^{2+} levels, which are implicated in dark and light adaptation, and the dissociation and reassociation of the opsin protein moiety with its chromophore after light activation. Phosphorylation and dephosphorylation of several different components of the cascade probably play a major regulatory role. Although the nuances of the transduction process are far from understood, the basic outline has begun to emerge.

Other biochemical cascades appear to be initiated by photoactivation of rhodopsin. For example, activation of retinal phospholipase C, which cleaves phosphatidylinositol bisphosphate to diacylglycerol and inositol triphosphate, and phospholipase A_2, which releases arachidonic acid, have been shown to be light dependent. The role of these pathways and probably others in the normal

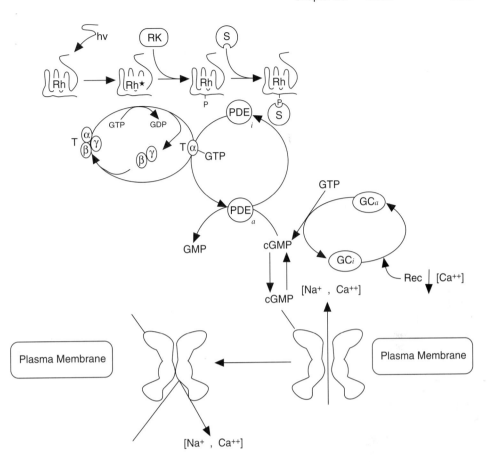

Figure 23-4

Phototransduction. The light activation of rhodopsin (Rh → Rh*) activates transducin (T), the photoreceptor-specific G protein, by the exchange of GTP for GDP. The T_α subunit activates the cGMP phosphodiesterase (PDE), cleaving cGMP and closing the ion channels, resulting in hyperpolarization and subsequent propagation of the electrical impulse. Two pathways involved in modulating the photoresponse are also shown. The first involves phosphorylation of the rhodopsin by rhodopsin kinase (RK), with subsequent binding by the *S*-antigen, also known as arrestin or the 48-kd protein (S). The second pathway involves the regeneration of cGMP through stimulation of guanylate cyclase by recoverin (rec) under reduced calcium level. Recoverin is also known as the 26-kd protein.

physiology of the retina is unclear. It is likely that they are involved in some type of second-order modulation of the photoresponse or are involved in housekeeping-type functions, such as signaling the turnover (i.e., shedding) of disks. Identification of molecules involved in these processes is important for understanding the molecular basis of vision and for identifying the possible causes of retinal lesions.

Neural transmission and processing in the retina

Photoreceptor cells transduce visual images to other cells of the neural retina as continuously graded changes in membrane potential. The retinal neurons pass these images to the optic pathways and transform them into a series of on and off signals (i.e., action potentials) with enhanced color, motion, and contrast detection. Interaction of the neural cells of the retina was described earlier and is shown in Figure 23-2. The photoreceptor cells send their input to the bipolar and horizontal cells. The horizontal cells appear to function as inhibitory neurons, using γ-aminobutyric acid as a neurotransmitter to perform negative-feedback control in the distal retina. This enhances contrast, emphasizing sharp edges and compensating for the blurring of the image caused by scattering of light by the various optical media of the eye.

Most bipolar cells are stimulatory (i.e., on), using glutamate to synapse to ganglion cells with their dendrites

in the inner portions of the inner plexiform layer. Some are inhibitory (i.e., off), synapsing to ganglion cells with dendrites in the outer portion of the inner plexiform layer. The bipolar cells convert the graded light signal that they have received from the photoreceptors into on and off signals, typical of information handling in the brain. Some ganglion cells receive stimuli from both types of bipolar cells, as do some amacrine cells. Although the interactions of amacrine cells are limited to the inner plexiform layer, they synapse with all the cell types found there.

Reflecting the various interactions, there appear to be different subsets of amacrine cells: inhibitory cells using glycine as a neurotransmitter and excitatory cells using acetylcholine. The ganglion cell layer consists of ganglion cell bodies and a few amacrine cells. The ganglion cells send the final relay, turning the input they receive from bipolar and amacrine cells into action potentials that are then sent to the brain for further processing. It is impressive that more than half of all ganglion cells receive input solely from the foveal area, which makes up only about 5% of the retinal surface. There is extreme bias toward central (i.e., fine) vision rather than generalized visual reception.

The number of bipolar cells is much smaller than the number of photoreceptor cells, and there is an even further reduction in the relative number of ganglion cells. In humans, there may be as many as 140 million rods and cones, but only 1 million ganglion cells. Each photoreceptor cell contacts a number of different neurons, and each neuron has several different synapses. The integrative abilities of the retinal neurons is probably indispensable for sorting out the signals.

Contrary to what may be expected, photoreceptor cells are actually depolarized in the dark and hyperpolarized in the light. In the on or dark state, glutamate is released by the photoreceptors at bipolar and horizontal cell synapses. The horizontal cells are also on in the dark as a result. The on bipolars are inhibited by the photoreceptors' release of glutamate and are excited in the light, and the off bipolars are excited by glutamate and therefore excited in the dark.

The importance of retinal processing is seen in the response of an individual ganglion cell to its receptive field. The receptive field of a cell is the part of the retina in which stimulation of photoreceptors with light causes activation of that cell, as demonstrated by an increase or decrease in its firing rate. For ganglion cells, the receptive field is a roughly circular area of retina corresponding to less than 1° of visual field at the fovea, which is the retinal area with tightly packed cones providing the finest visual discrimination, to 3° to 5° at the retinal periphery. Ganglion cells are maximally stimulated by large contrasts in the intensity of light striking the center and periphery of their receptive fields rather than an evenly spread illumination. It is a general principle that the visual system responds primarily to changes in visual input, time, location, or intensity.

The electrical activity of the various retinal neurons can be measured by a noninvasive procedure known as an electroretinogram (ERG), similar to the electrocardiogram and electroencephalogram in that it measures the activity of large numbers of cells. Figure 23-5 depicts a highly idealized ERG tracing. Because different components of the ERG can be attributed to different cell types in the retina, it is clinically useful to identify possible retinal dysfunction and to narrow the diagnostic focus to specific cell types when looking for the source of visual difficulties.

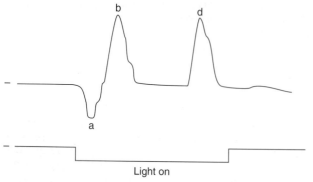

Figure 23-5
Idealized electroretinogram tracing. The a-wave indicates photoreceptor activity, the b-wave indicates activity in the inner nuclear layer, especially bipolar and Müller cells, and the d-wave indicates the off response of the inner nuclear layer.

The Optic Pathways

Axons of the ganglion cells collect in fine bundles that converge in a radiating pattern at the optic disc to form the optic nerve, connecting the eye to the brain. Because the optic disc contains no photoreceptors, it is insensitive to light, causing a small blind spot noticeable only on monocular vision. The spatial relations of the visual field are maintained in the fibers making up the optic nerves. After passing through the optic foramina, the right and left optic nerves meet at the optic chiasm. At this point, nerve fibers originating in the nasal halves of the retina cross to the opposite side, and those from the temporal retinal halves continue uncrossed. As a result of this pattern, the whole right visual field (received by the left temporal and right nasal halves of the retinas) projects to the left hemisphere, and the left visual field (received by the right temporal and left nasal halves of the retina) projects on the right hemisphere (Fig. 23-6). Beyond the optic chiasm, the optic fibers continue on as the optic tract.

Although most of the optic tract fibers extend to the lateral geniculate nucleus, portions continue to the superior colliculus and the pretectal area of the midbrain, which actuate reflex responses of the eyes and body to visual stimuli and the pupillary light reflex, respectively. The spatial relations of the retina and visual field continue to be maintained in the lateral geniculate body, consistent with its role as the most important subcortical region for further visual processing. Axons of the upper quadrants (receiving the lower visual fields) project to the medial laminae, and these cells project as the geniculostriate pathway or optic radiation to the superior edge of the calcarine sulcus in the striate cortex, also known as the primary visual cortex or Brodmann's area 17. Axons of the lower quadrants (receiving the upper visual fields) project to the lateral half of the lateral geniculate body, with these cells projecting to the inferior lip of the calcarine sulcus of the striate cortex. The layered arrangement of these projection terminals forms ocular dominance and orientation columns.

The visual field is mapped in a precise and orderly way on the visual cortex, just as in the lateral geniculate nucleus, although the relative spacing differs. Within this arrangement can be found distinct columns in which the cells are stimulated by input from the right or left visual fields (i.e., ocular dominance). This columnar arrangement allows cortical cells to convert input from multiple cells of the lateral geniculate nucleus into linear receptive fields. Further refinement of this pattern is based on the orientation preference of cells positioned vertically within a column and their relation to a specific retinal field. Cells of this region continue the visual processing begun in the retina, analyzing visual images in terms of linear contours and boundaries, to begin the process of abstracting information from visual signals.

From the striate cortex, there are additional projections to a surprisingly large number of other (extrastriate) visual areas of the cortex. For example, it is known that areas V_1 and V_2 integrate most of the visual information regarding color, movement, and form and that other

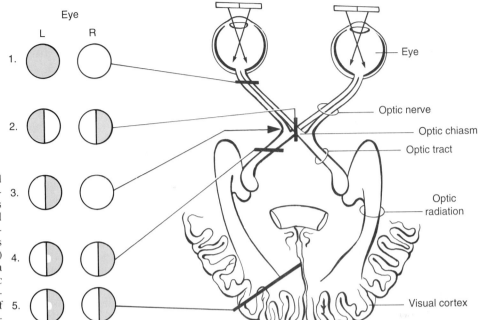

Figure 23-6
Optic pathways in the brain and visual field defects that their lesions produce: 1) total visual loss of the left eye; 2) bitemporal hemianopsia; 3) left nasal hemianopsia; 4) right homonymous (contralateral) hemianopsia; 5) right homonymous hemianopsia with macular sparing. At the optic chiasm, nerve fibers from the nasal retina cross to the other side of the brain, joining the opposite optic nerve to form the optic tract.

areas play some roles in fine tuning that information. The lateral geniculate pathway is the most important for the conscious experience of vision in mammals. People having lesions of the striate cortex claim to be completely or almost completely blind. However, they are also partially able to identify or locate some objects, indicating that the retinotectal pathway is able to supply some kind of vision, albeit subconsciously. These pathways are not independent of each other, because there is also a corticotectal path, providing input from the cortex to the superior colliculi. In addition to these two major pathways, there are several minor pathways, which remain poorly understood. These project to the ventral lateral geniculate nuclei, tegmentum, and hypothalamus.

Basis of Disease States

☐ Characteristic patterns of visual field loss are produced by anatomic lesions at various points in the visual pathways

The complexity of the visual system provides a vast array of things that can go wrong. Lesions at different locations along the optic pathways result in characteristic visual defects. For example, lesions on one side of the optic tract behind the chiasm (i.e., lateral geniculate body, optic radiation, or visual cortex) result in a loss of the corresponding opposite field of vision, ranging from specific field defects to complete opposite field loss, called *homonymous hemianopsia*. Cortical lesions often preserve central (i.e., macular) vision while ablating the rest of the visual field, which is called *macular sparing*. Rarely, one

or both lateral angles of the chiasm may be compressed, causing destruction of the noncrossing fibers and resultant *nasal or binasal hemianopsia*. A complete lesion of one optic nerve results in *blindness* of the corresponding eye. The pupil of the blind eye constricts consensually when light stimulates the opposite retina, because each pretectum receives input and return fibers bilaterally. Direct stimulation of the blind eye in a darkened room, however, reveals a dilated pupil. This is referred to as an *afferent pupillary defect* or Marcus-Gunn pupil. Lesions occurring beyond the pretectum and superior colliculus do not affect the pupillary light reflex.

Diseases of the visual system are not limited to gross morphologic damage. Many genetic disorders have been described, but it is only with the advent of molecular biology that the molecular basis of these has been described. Two are considered here: blue cone monochromasy, which is a form of color blindness, and retinitis pigmentosa.

☐ Loss of color vision can result from genetic lesions causing defects in the color pigments

An interesting ophthalmologic disease that has been elucidated at a molecular level is *blue cone monochromasy* (BCM). It is an X-linked form of color blindness in which the cones respond only to blue light. The highly similar red and green color pigment genes normally occur as a head-to-tail tandem array with a red pigment gene followed by one or more green pigment genes. Most instances of BCM arise from unequal homologous recombination events occurring in this array during meiosis, resulting in the deletion of all but a single remaining gene (in some cases, a red-green hybrid gene), or leaving two genes, one of which is inactivated by a point mutation. Deletions in a region approximately 4 kilobases upstream from the red pigment gene can result in inactivation of both genes. These molecular lesions leave the patient able to detect only black and white by means of rhodopsin in the rods and blue, because the blue pigment gene was unaffected.

The elegant delineation of the pathophysiology of BCM is a classic example of the power of combining the new techniques of molecular genetics with clinical acumen.

□ Retinitis pigmentosa can result from mutations in highly expressed retinal specific genes

Retinitis pigmentosa (RP) is the clinical term used to describe a large, heterogeneous group of visual disorders in which the retinal photoreceptors degenerate, resulting in partial or total blindness. Affecting approximately 1 of every 4000 persons worldwide, RP has a variable age of onset and rate of progression.

For the purposes of clinical diagnosis, the classic symptoms of RP include night blindness (i.e., nyctalopia), alterations in the electroretinogram, and the deposit of "bone spicules" of pigment on the retinal surface, accounting for the name of the disease.

RP shows genetic heterogeneity as well. Autosomal dominant (adRP), autosomal recessive, and X-linked forms of RP have been described. The inheritance of RP in these different fashions implies that multiple genes can cause RP.

Lesions in two different genes have been identified as causes of adRP. In 1989, Peter Humphries and colleagues established the linkage between one form of adRP and a marker on the long arm of chromosome 3 in a large Irish pedigree. The gene for human opsin had been previously shown to map to the long arm of chromosome 3. Using this knowledge, Dryja and his colleagues quickly identified a point mutation in the opsin gene that could be identified as the defect in 17 of 148 unrelated adRP families. The specific change resulted in the substitution of a histidine for a proline at amino acid 23, in the amino-terminal region of the mature opsin moiety.

Since then, approximately 30 different mutations in opsin have been identified. Studies are underway to elucidate the precise molecular mechanisms whereby the mutant opsin gives rise to the dominant RP phenotype. Preliminary studies of some alleles indicate that the defective opsins are poorly transported to and incorporated into the membrane, although the detailed pathophysiology leading to the relatively late onset of symptomatic RP has yet to be delineated.

Another gene was found to be defective in some cases of adRP. The gene encodes a protein called peripherin, which is essential to the correct assembly of photoreceptor rod outer segment disks. The peripherin gene was first identified as the site of the lesion in the retinal degeneration slow (rds) mouse and has subsequently been shown to cause adRP in some pedigrees. The specific biochemical requirement of peripherin in rds cases is even less understood than for opsin, and experiments to clarify its role in normal and impaired vision are underway.

Mutations in opsin and peripherin still account for a minority of RP cases, indicating that many more genes may be involved in other forms of the disorder. Likely candidates include genes encoding proteins involved in phototransduction specifically and genes that code for retinal housekeeping-type proteins. Several laboratories are actively looking for the genes and for lesions that may be linked to RP.

The clinical heterogeneity of RP even among siblings carrying the same RP-causing mutation is still not explained. Eventually, it is possible that such studies will lead to cures for selected cases of RP-induced blindness.

SELECTED READINGS

Albert D, Jacobiec F (eds). Principles and practice of ophthalmology: basic sciences. Philadelphia: WB Saunders, 1994.

Hubel DH. Eye, brain, and vision. Scientific American library series, number 22. New York: WH Freeman, 1988.

Kandle ER, Schwartz JH, Jessel TM (eds). Principles of neural science, 3rd ed. New York: Elsevier, 1991.

Nathans J. Rhodopsin: structure, function, and genetics. Biochemistry 1992;31:4923.

Palczewski K, Benovic J. G-protein–-coupled receptor kinases. Trends in Biochemical Sciences 1991;16:387.

Ryan SJ. Ogden TE (eds). Retina: basic science and inherited retinal disease, vol 1. St. Louis: CV Mosby, 1989.

Stryer L. The molecules of visual excitation. Sci Am 1987;257:42.

Neuroscience in Medicine, edited by P. Michael Conn.
J. B. Lippincott Company, Philadelphia © 1995

Chapter *24*

Critical Periods in Visual System Development

MARK F. BEAR

A *critical period of development* may be defined as a period of time in which intercellular communication alters a cell's fate. The concept is usually credited to the experimental embryologist Hans Spemann who, working around the turn of the century, showed that transplantation of a piece of early embryo from one location to another would often cause the "donor" tissue to take on the characteristics of the "host," but only if transplantation had taken place in a well-defined period. After the transplanted tissue had been induced to change its developmental fate, the outcome could not be reversed. The intercellular communication that altered the phenotype of the transplanted cells was shown to be mediated by contact and by chemical signals.

The term took on new significance with respect to brain development as a consequence of the work of Konrad Lorenz in the mid-1930s. Lorenz was interested in the process by which graylag goslings come to be socially attached to their mother. He discovered that, in the absence of the mother, the social attachment could occur instead to a wide variety of moving objects, including Lorenz himself. Once imprinted on an object, the goslings followed it and behaved toward it as they normally would their mother. The term *imprinting* was used by Lorenz to suggest that this first visual image was somehow permanently etched in the young bird's nervous system. Imprinting was also found to be limited to a finite time (i.e., the first 2 days after hatching), which Lorenz called the critical period for social attachment. Lorenz himself drew the analogy between this process of imprinting the external environment on the nervous system and the induction of tissue to change its developmental fate during critical periods of embryonic development.

This work had a tremendous impact in the field of developmental psychology. The terms imprinting and critical period conjured images (and generated heated debate) that changes in "behavioral phenotype" caused by early sensory experience were permanent and irreversible later in life, much like the determination of tissue phenotype during embryogenesis. Numerous studies extended the critical period concept to aspects of mammalian psychosocial development. The implication was that the fate of neurons and neural circuits in the brain depended on the experience of the animal during early postnatal life. It is not difficult to appreciate why research in this area took on political as well as scientific significance.

By necessity, the effects of experience on neuronal fate must be exercised by neural activity generated at the sensory epithelia and communicated by chemical synaptic transmission. The idea that synaptic activity can alter the fate of neuronal connectivity during central nervous system (CNS) development eventually received solid neurobiologic support from the study of mammalian visual system development, beginning with the experiments of Hubel and Wiesel that partly earned them the 1981 Nobel Prize in medicine. They found, using anatomic and neurophysiologic methods, that visual experience or lack thereof was an important determinant of the state of connectivity in the central visual pathways and that this en-

vironmental influence was restricted to a finite period of early postnatal life. Because much work has been devoted to the analysis of experience-dependent plasticity of connections in this system, this is an excellent model system to illustrate the principles of critical periods in nervous system development.

The focus of this chapter is on the critical periods during which intercellular communication by synaptic transmission alters the fate of connections in the primary visual pathway from retina to visual cortex. In keeping with the philosophy of this volume, general principles are emphasized instead of experimental results. Students who wish to learn more about the experimental bases of these principles are directed to the bibliography provided at the end of this chapter. Although the effects of synaptic transmission in the development of other systems (e.g., neuromuscular connections) and the effects of other types of intercellular communication (e.g., long-range hormonal signals) are not covered explicitly here, the general principles of other critical periods are thought to be quite similar to those illustrated by visual system development.

Adult Organization of the Visual Pathway From Retina to Cortex

☐ The Anatomical Organization of the Visual Pathway Is Precise

The detailed structure and function of the visual system is covered in Chapter 23. An overview of the adult organization is provided here to give perspective on what must be accomplished during development to ensure proper wiring and function of this sensory system.

The central visual pathway begins with the projection of ganglion cell axons from the retina into the optic nerve (Fig. 24-1). A useful rule of thumb is that the ganglion cells that "view" the right visual hemifield project to the left hemisphere of the brain. Conversely, the ganglion cells that are responsive to visual stimuli in the left visual hemifield project to the right hemisphere. Because human eyes are situated frontally in the head, a significant fraction of each visual hemifield is viewed by both retinas. This region of space is the *binocular visual field*. Each retina must project to both hemispheres; ganglion cells in the nasal retina (i.e., the half closer to the nose) project across the midline to the contralateral hemisphere, and ganglion cells in the temporal retina project axons into the hemisphere of the ipsilateral side. This partial decussation of the visual pathways occurs at the *optic chiasm*.

From the chiasm, the ganglion cell axons project into the thalamus by means of the optic tract. The target of this projection is the *lateral geniculate nucleus* (LGN). The ganglion cell axons make chemical synapses on LGN neurons, which project to the cerebral cortex. There are two remarkable features of the anatomy of this connection from the retina to the LGN. First, retinotopy is preserved. Neighboring ganglion cells synapse onto neigh-

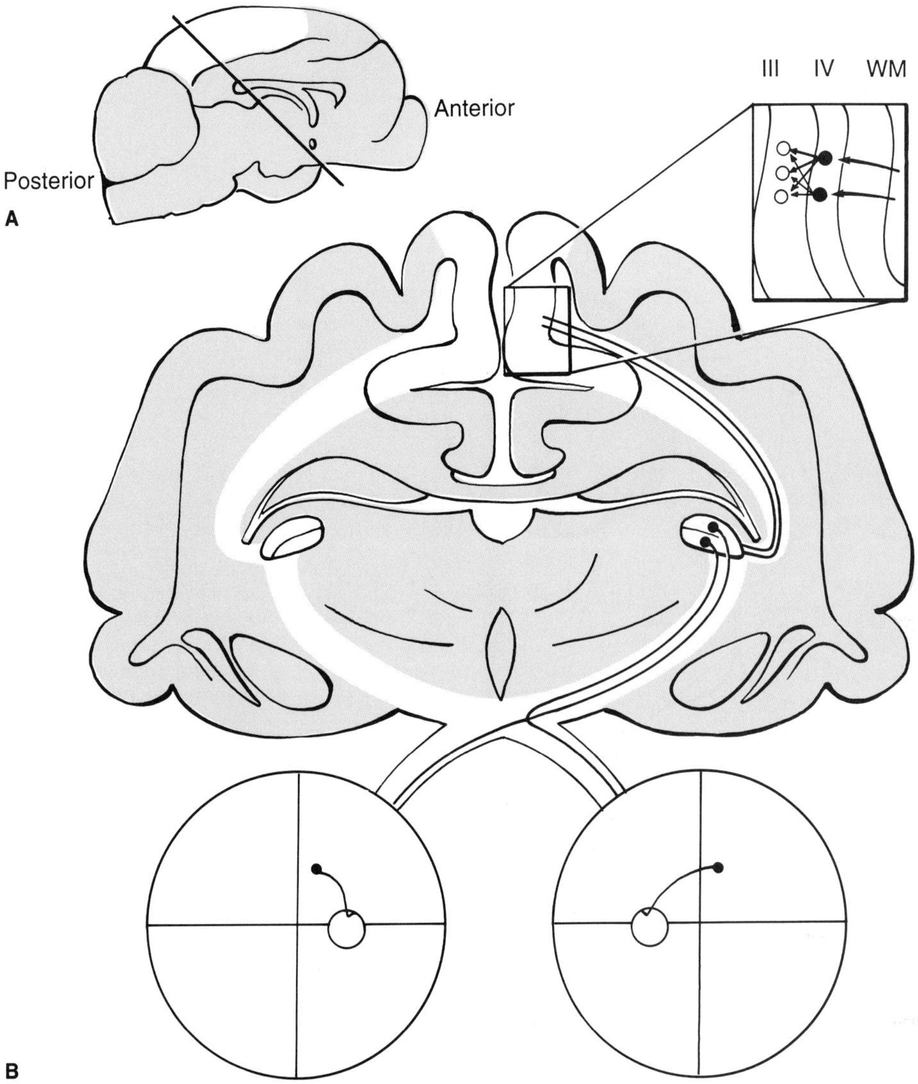

Figure 24-1

Organization of the mature mammalian retinogeniculocortical projection. (**A**) Midsagittal view of a cat brain, showing the location of the primary visual cortex (i.e., striate cortex or area 17). The line indicates a plane of section, illustrated in **B,** that reveals all the components of the ascending visual pathway. (**B**) The temporal retina of the left eye and the nasal retina of the right eye project axons through the optic nerve and optic tract to the lateral geniculate nucleus of the left dorsal thalamus. Inputs from the two eyes remain segregated in separate lamina at the level of this synaptic relay. The lateral geniculate cells project on to striate cortex through the optic radiations. These axons terminate mainly in layer IV, where inputs subserving the two eyes continue to be segregated. (**Inset**) The first site of major convergence of inputs from the two eyes is in the projection of layer IV cells onto cells in layer III. *WM,* white matter. (Modified from Bear MF, Cooper LN. Molecular mechanisms of synaptic modification in the visual cortex: interaction of theory and experiment. In Gluck M, Rumelhart D [eds]. Neuroscience and connectionist theory. Hillsdale, NJ: Lawrence Erlbaum Associates, 1990:65.)

boring LGN neurons such that there is a mapping of retinal position onto LGN position. A *retinotopic map* of visual space is said to exist in the LGN. The second structural feature of interest is that the inputs from the two eyes (i.e., the nasal retina of the eye on the opposite side and the temporal retina of the eye on the same side) remain anatomically segregated in the LGN. A separate layer of LGN neurons receives the inputs from each eye.

These LGN lamina can be thought of as a stack of pancakes, with each one containing a retinotopic map from one eye. These maps are in perfect register, except for a small, *monocular segment* that represents the far periphery of the visual field that is viewed only by the contralateral eye (Fig. 24-2).

In the human brain, each eye projects to three different layers, giving a total of six cell layers in each LGN. These

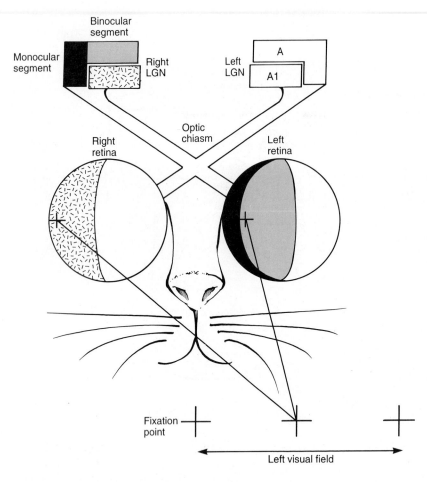

Figure 24-2

The general organization of the lateral geniculate nucleus. Each eye projects to a separate
cell layer in the lateral geniculate. In the cat,
the principal layers are called A and A1. Layer
A receives input from the nasal half of the contralateral retina, and layer A1 receives input
from the temporal half of the ipsilateral retina.
The retinotopic maps in the two layers of the
lateral geniculate occur in perfect register, except for a lateral region of layer A called the
monocular segment. Cells in the monocular
segment are activated by visual stimuli in the
far periphery that fall outside the binocular visual field and are viewed only by the contralateral eye.

different layers appear to be specialized to relay information from different classes of ganglion cells in each
eye. In general, at least two cell layers exist in the LGN of
mammalian species with binocular vision, and each of
these layers serves as a relay of information from one
retina to the cerebral cortex.

LGN neurons project to a specific region of the cerebral cortex in the occipital lobe called the primary visual
cortex, striate cortex, or area 17. The main target of LGN
neurons is a band of cells in the middle of the cortex,
called *layer IV*. Layer IV cells feed information to cells
in the superficial layers (closer to the pia), particularly
layer III. Layer III neurons communicate with other
cortical areas for the further processing of visual information.

The inputs from the different LGN layers that subserve
the two eyes are not intermingled in layer IV. This was
demonstrated by Hubel and Wiesel using an ingenious
anatomic method that takes advantage of the fact that proteins are synthesized in the neuronal cell body and are
subsequently transported in an anterograde direction
down the axon. When the tritium-labeled amino acid proline was injected into the eye of a monkey, it was incorporated into protein that was transported down the optic
nerve, through the chiasm, and into the appropriate layers of the LGN. Hubel and Wiesel found that if sufficient
radioactive proline was injected intraocularly, some of

the radioactive label spilled out of the retinal terminals
and was taken up by the postsynaptic LGN neurons and
was transported on to the cortex. By preparing histologic
sections of cortical tissue and exposing them to a tritium-
sensitive film (i.e., autoradiography), they could detect
where in layer IV the inputs from one eye terminated.
When the sheet of layer IV cells was viewed from above,
the distribution of LGN axons subserving the injected eye
had the appearance of zebra stripes (Fig. 24-3). The radioactive stripes were about 0.5 mm across and separated
by nonradioactive regions of the same width that corre-
sponded to the distribution of axons from the LGN lamina subserving the other eye. In cross-sectional histologic
sections, these stripes had the appearance of a series of
columns that spanned the full thickness of layer IV. Hubel
and Wiesel called these *ocular dominance columns*. As
in LGN, the inputs from the two eyes remain segregated
at the level of the first synaptic relay in layer IV. As in LGN,
the cortex is organized retinotopically, with the nearest
neighbor LGN neurons in each layer projecting to the
nearest neighbor layer IV cells. In the cortex, instead of
these segregated maps lying like pancakes on top of
each other, they are laid out as interdigitated fingers in
layer IV.

Fusion of the images formed in the two eyes requires
that eventually the information converges onto the same
cells. This is accomplished by the projections of layer IV

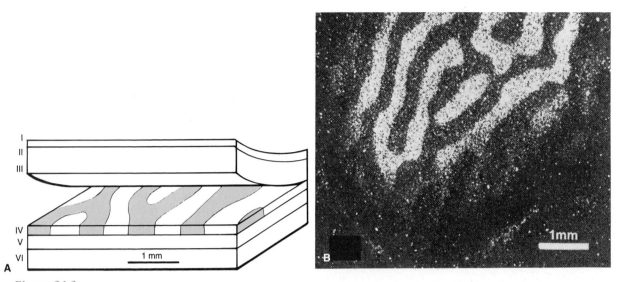

Figure 24-3

(**A**) The organization of ocular dominance columns in layer IV of the striate cortex of macaque monkeys. The distribution of geniculate afferents subserving one eye is darkly shaded. In cross section, these eye-specific zones appear as columns of approximately 0.5-mm width in layer IV. When the superficial layers are peeled back, allowing a view the ocular dominance columns in layer IV from above, these zones take on the appearance of zebra stripes. (**B**) Dark-field autoradiograph of a histologic section of layer IV viewed from above. Two weeks before sacrifice, this monkey had received an injection of ^{3}H-proline into one eye. In the autoradiograph, the radioactive lateral geniculate terminals appear bright on a dark background. (From Wiesel TN. Postnatal development of the visual cortex and the influence of the environment. Nature 1982;299:583.)

axons from different but adjacent ocular dominance columns onto the same cells in layer III. This convergence in layer III is the primary substrate for binocular vision in mammals.

The LGN is not the only source of extrinsic input to the visual cortex. The striate cortex receives input from the same cortical areas to which it projects, including the cortex of the opposite hemisphere through the corpus callosum. There are brain stem inputs originating in the *locus ceruleus*, using *norepinephrine* as a neurotransmitter, and in the *Raphe nuclei*, using *serotonin* as a neurotransmitter. Unlike the LGN inputs, the axons from the brain stem do not show much respect for the borders between cortical layers, and the fibers tend to ramify widely with little apparent order. A similar organization is seen in the projection from the nucleus *basalis of Meynert* in the forebrain, which uses *acetylcholine* as a neurotransmitter (Fig. 24-4). From the anatomy of these systems and from the effects of applying these neurotransmitters to cortical neurons, it is thought that these inputs function to modulate visual processing according to behavioral state.

☐ The Physiological Organization of the Visual Pathway Reflects Its Anatomical Precision

Because the physiology of the visual system is the topic of another chapter, this section focuses on the details that are required for a discussion of the critical period.

At each stage of processing in the visual system, cells are said to possess *receptive fields*. The receptive field of a visual cell is that area of visual space that, when illuminated with an appropriate stimulus, causes a change in the "resting" discharge rate of that cell. For a ganglion cell in the retina, it is possible to map with spots of light a region of visual space that influenced the cell's activity. This region would be the cell's receptive field and would reflect the convergence of photoreceptors by way of other retinal cells onto the ganglion cell. The receptive fields of ganglion cells usually are circular regions that range in size from less than a tenth of a degree across in the central retina to several degrees in the peripheral retina (for comparison, the moon as viewed from earth subtends an angle of about one half of a degree).

The roughly 1 million ganglion cells in each eye feed information to roughly a similar number of lateral geniculate neurons, and there is little transformation in the receptive fields by this synaptic relay. As expected from the anatomy, each LGN neuron responds to stimulation of only one eye. Similarly, the layer IV neurons that receive synaptic input from LGN are characterized by small, circular, monocular receptive fields (this is true for the macaque monkey and presumably for humans, but not for other species, such as cats). The first significant elaboration of receptive fields occurs at the projection from layer IV to layer III. First, layer III neurons have elongated receptive fields and are responsive particularly to elongated, high-contrast bars or edges with the same orientation as the long axis of the receptive field. This property

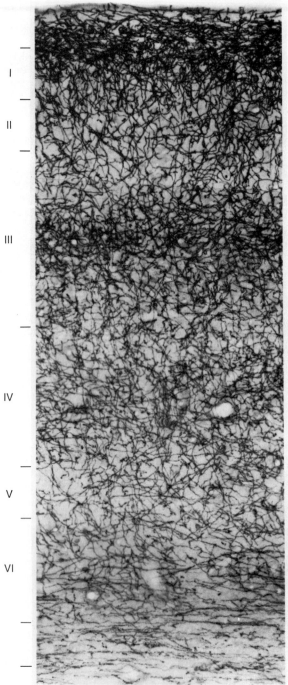

Figure 24-4
Distribution of fibers in cat striate cortex that use the neurotransmitter acetylcholine. Unlike the projection from the thalamus, this projection from the basal forebrain innervates all cortical layers. These cortical inputs modulate the activity patterns that arise from retinal stimulation and are thought to influence experience-dependent aspects of cortical development. (From Bear MF, Carnes KM, Ebner FF. An investigation of cholinergic circuitry in cat striate cortex using acetylcholinesterase histochemistry. J Comp Neurol 1985;234:411.)

is called *orientation selectivity*. Normally, receptive fields of all orientations are represented equally in the population of cortical neurons, and cells with like orientation preferences tend to be clustered together in stripes, not unlike the ocular dominance stripes. The second important elaboration of cortical receptive fields is that most neurons in layer III are responsive to stimulation of both eyes, and the receptive fields mapped through the two eyes are matched to the same position in space. The cortex is the most peripheral station in the ascending visual pathway where information from the two eyes is combined to form a single percept of visual space.

It is also important to recognize that most cortical neurons respond poorly or not at all to changes in diffuse illumination of the retina. Cortical neurons respond to contrast borders. This elaboration of cortical receptive fields reflects the wiring of intracortical connections, which are excitatory and inhibitory.

Activity-Independent Development of Order in the Visual System

The preceding discussion indicates that there is considerable precision in the connections of the mature visual pathway. There are several examples:

Only ganglion cell axons from nasal retinas cross at the chiasm.

The mixed population of axons in the optic tract is sorted out at the lateral geniculate by eye and by retinotopic position.

The LGN axons project to a specific layer of cells in the cortex, and within this layer, they segregate again according to retinotopic position and by eye.

Layer IV cells make connections with cells in other layers that are appropriate for binocular vision and are specialized to enable detection of contrast borders.

To what extent does the establishment of these highly specific connections depend on activity during critical periods of development? Before addressing this question, it is important to recognize that activity often precedes experience during CNS development. Activity, defined as the occurrence of action potentials and chemical synaptic transmission, occurs in the visual pathway in utero, even before the development of the photoreceptors. To say that a developmental process occurs before sensory experience, in this case vision, is not the same as saying that it is activity independent. Activity dependence can be assessed experimentally by using the drug *tetrodotoxin*, which binds tightly to voltage-sensitive sodium channels and blocks action potentials. Aspects of development that are tetrodotoxin insensitive are considered to occur independently of intercellular communication by synaptic transmission.

It is useful to think of the development of long-range connections in the CNS as occurring in three phases (Goodman and Shatz, 1993): *pathway selection, target selection,* and *address selection*. Examples of pathway selection are the decisions made by growing axons originating in the nasal retina to cross the midline at the optic chiasm and the decisions made by LGN axons to project to the cortex through the internal capsule rather than to the spinal cord via the cerebral peduncle. Examples of target selection are the decision made by optic tract axons to form connections with the LGN and not another part of the thalamus and the decision made by LGN axons to innervate cortical layer IV and not layer V. Examples of address selection are the sorting of inputs by retinotopic location in the LGN and cortex, the segregation of axons subserving the two eyes in the LGN and layer IV of cortex, and the convergence of retinotopically matched inputs from the two eyes onto layer III neurons.

Experimental evidence has now shown that pathway and target selection occur entirely in the absence of neural activity. Many aspects of address selection, such as the establishment of crude retinotopy, are also activity independent. This is not to say that these aspects of development occur independently of intercellular communication, because virtually all phases of visual pathway development depend critically on communication by cell-cell contact and by gradients of diffusible chemicals. The idea that specific connections are established by dif-ferential chemical attraction or repulsion was the basis of the *chemoaffinity hypothesis* of Roger Sperry, who shared the 1981 Nobel Prize with Hubel and Wiesel.

The axons from the retina grow along the substrate provided by the extracellular matrix of the ventral wall of the optic stalk. One glycoprotein in this matrix is *laminin*. The growing tips of the axons, called *growth cones*, express surface molecules called *integrins* that bind laminin, and this interaction promotes axonal elongation (Fig. 24-5). The extracellular matrix along the optic stalk forms a molecular highway on which retinal axons grow. Travel down this highway is aided by another mechanism that causes axons that are growing together to stick together, a process called *fasciculation*. This stickiness is due to the expression of specific *cell-adhesion molecules* on the surface of axonal membranes.

Extracellular matrix can be repulsive as well as attractive to growing axons, depending on the cell-surface receptors the axons express. Axons from the temporal retina grow toward the midline at the chiasm, but they encounter a signal there that causes them to veer sharply away. In contrast, nasal axons continue right across the midline and into the contralateral optic tract. These dif-

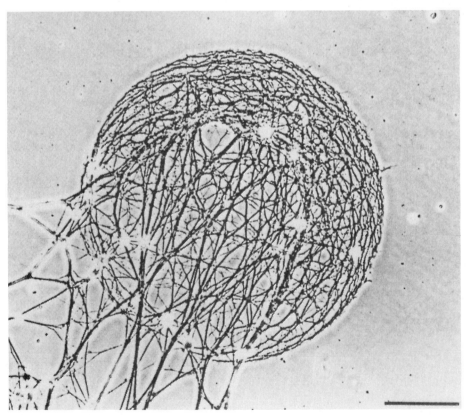

Figure 24-5
Neurites growing in a collagen-coated culture dish enter a region containing the extracellular matrix protein laminin. On this preferred substrate, the neurites branch profusely but stay within the border of the circular laminin dot. The interplay of extracellular matrix and specific axon-surface molecules is thought to be crucial for pathway and target selection during visual system development. (From Gunderson RW. Response of sensory neurites and growth cones to patterned substrata of laminin and fibronectin in vitro. Dev Biol 1987;121:423.)

ferences must be explained by differential expression of cell-surface molecules based on cell position in the retina. Such a nasal-temporal gradient in axon surface markers is thought to be matched to complementary gradients on the surfaces of cells in target structures, and this match gives rise to retinotopy.

When axons reach their target, they often encounter a new extracellular environment that retards further growth. These environmental signals can be the absence of specific glycoproteins in the extracellular matrix, such as the absence of laminin. Axonal growth can be inhibited by diffusible signals released from target structures. In some systems, application of neurotransmitters can inhibit axonal growth, probably by raising calcium concentrations in the growth cone. There is evidence that diffusible factors can promote axon growth (i.e., chemotrophism) and guide that growth to distant targets (i.e., chemotrophism). The number and nature of the chemical signals that guide activity-independent pathway formation are being investigated.

Considerable order can develop in the visual system solely under the influence of these molecular mechanisms. The role of activity appears to be reserved for the final refinement of the patterns of connectivity according to functional criteria.

Activity-Dependent Development of Order in the Visual System

Much of the organization of the visual pathway is specified without any contribution from neural activity and synaptic transmission. The retinal and lateral geniculate axons navigate down the appropriate paths and terminate in retinotopic order in the appropriate target structure.

The main features of visual system organization that depend on retinal activity are the segregation of axons in LGN and cortex according to the eye that drives activity in them and the establishment and maintenance of connections in the visual cortex that generate binocular and stimulus-selective receptive fields. The eye-specific segregation in LGN occurs entirely before birth, but many refinements of cortical circuitry occur postnatally and are under the influence of the visual environment during infancy.

☐ Segregation of axons in the lateral geniculate nucleus is influenced by activity

The first axons to reach the LGN are usually those from the contralateral retina, and they spread out to occupy the entire nucleus. Somewhat later, the ipsilateral projection arrives and intermingles with the axons of the contralateral eye. Over the next several weeks, the axons from the two eyes segregate into the eye-specific domains that are characteristic of the adult nucleus. Intraocular injection of tetrodotoxin prevents this process of segregation, showing that it depends on activity generated in the retina. What is the source of the activity, and how does it orchestrate segregation?

Because segregation occurs in utero, before the development of photoreceptors, the activity cannot be driven by photic stimulation. It appears that ganglion cells are spontaneously active during this period of fetal development. This activity is not random. Ganglion cells fire in quasisynchronous waves that spread across the retina (Fig. 24-6). The origin of the wave and its direction of propagation may be random, but during each wave, the

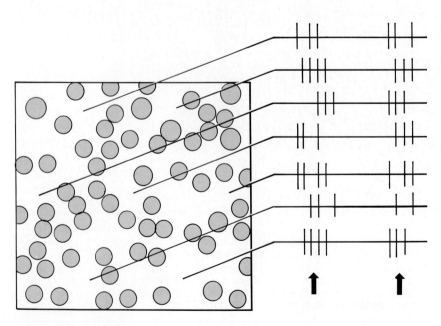

Figure 24-6
Illustration of an experiment in which action potential recordings were made from a whole-mount specimen of the fetal retina in vitro. Circles represent ganglion cell outlines. Action potentials occur in spontaneous bursts (*arrows*) that are almost synchronous in widely separated regions of retina. These local correlations in activity are thought to play a critical role in the sorting of retinal axons in the lateral geniculate nucleus. (Modified from Shatz CJ. The developing brain. Sci Am 1992;267:61.)

activity in a ganglion cell is highly correlated with the activity in its nearest neighbors. Because these waves are generated independently in the two retinas, the activity patterns arising in the two eyes are not correlated with one another.

Segregation is thought to depend on a process of synaptic stabilization in which only retinal terminals that are active at the same time as their postsynaptic LGN target neuron are retained. This hypothetical mechanism of synaptic plasticity was first articulated by Donald Hebb in the 1940s and can be remembered by a mnemonic, attributed to Sigrid Löwel and Wolf Singer: *neurons that fire together wire together*. Connections that are modified according to this rule are said to employ *Hebb synapses*. According to this hypothesis, when a wave of retinal activity drives a postsynaptic LGN neuron, the active retinal inputs onto this neuron are consolidated. Because the activity from the two eyes does not occur in register, the inputs compete on a winner-takes-all basis until one input is retained and the other is eliminated, leading to complete segregation of the two inputs.

☐ Segregation of axons in layer IV is influenced by activity

Similar to the situation in the LGN, the afferents subserving the two eyes are initially intermingled in cortical layer IV and then segregate under the influence of activity. In the macaque monkey, this segregation of ocular dominance columns begins before birth and is normally completed by the third postnatal week. In cats, the process occurs entirely after birth (Fig. 24-7).

To the extent that segregation occurs postnatally, the formation of ocular dominance columns can be affected by deprivation of normal pattern vision. This is most dramatically demonstrated by an experimental manipulation called *monocular deprivation*, in which one eyelid is sutured closed. This situation approximates a unilateral cataract in which the amount of light reaching the affected retina is only slightly decreased but image formation is greatly impaired. If lid suture is performed shortly after birth and continued for the duration of the period of

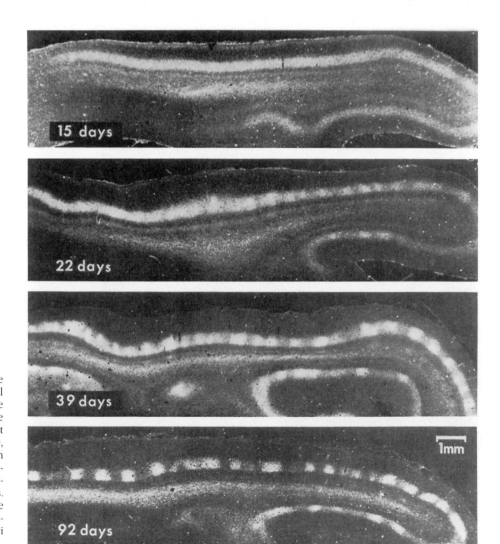

Figure 24-7
Dark-field autoradiographs of the cat visual cortex cut in horizontal sections. In each case, one eye had been injected with ³H-proline and the animals later sacrificed at the ages shown. At 15 days of age, the radioactivity is uniform in layer IV (*arrow*). Ocular dominance columns segregate progressively over the next several weeks. (From LeVay S, Stryker MP. The development of ocular dominance in the cat. Soc Neurosci Symp 1979;4:83.)

natural segregation, the striking result is that the open eye columns develop to be much wider than the closed eye columns (Fig. 24-8). If the deprivation is begun later, after the period of natural segregation, anatomic effects on LGN axon arbors are not observed in layer IV. A *critical period* exists for this type of plasticity, which in the macaque monkey lasts until about 6 weeks of age.

Within this critical period, the anatomic effects of monocular deprivation can be reversed by closing the previously open eye and opening the previously closed eye. The result of this reverse suture manipulation is that the shrunken ocular dominance columns of the formerly closed eye expand and the expanded columns of the formerly open eye shrink. These and other data suggest that within this 6-week critical period, even after ocular dominance column segregation appears to be anatomically complete (after 3 weeks), the afferents subserving the two eyes exist in a dynamic equilibrium that can still be disrupted by deprivation. At the end of the critical period, the afferents apparently lose their capacity for growth and retraction.

One correlate of the change in cortical ocular dominance columns during the critical period is a change in the size of the neurons in the LGN that relay information to visual cortex. For example, LGN cells deprived of normal vision are visibly shrunken. This change in soma size is thought to reflect the decreased axonal arbors of these cells in layer IV. Curiously, shrinkage is observed only in the segment of the nucleus where both retinas are represented. Cells in the monocular segment are largely unaffected by deprivation. This observation suggests that the loss of territory in layer IV by afferents deprived of normal visual input is not caused by simple disuse. The evidence suggests that it is an active process of *binocular competition* that requires patterned activity in the open eye. The classic test of this hypothesis was performed by Guillery in 1972. By destroying part of the central retina of the nondeprived eye, he created a region of LGN that was free of the effects of binocular competition. In this "critical segment" monocular deprivation caused far less shrinkage of LGN neurons (Fig. 24-9). Evidently, the activity in the open eye actively promotes the synaptic disconnection of afferents that subserve the closed eye.

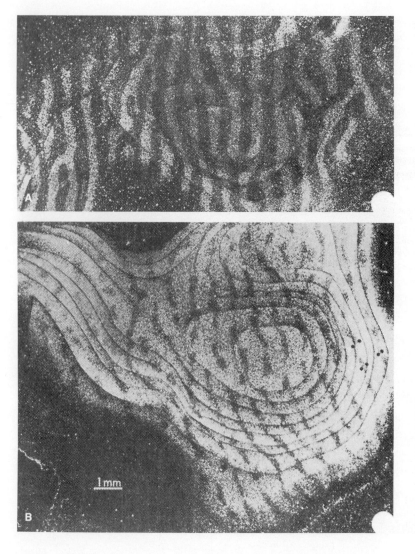

Figure 24-8
Dark-field autoradiographs of tangential sections through layer IV of monkey striate cortex after injection of one eye with ³H-proline. (**A**) Normal monkey. (**B**) Monkey that had been monocularly deprived for 18 months starting at 2 weeks of age. The nondeprived eye had been injected, revealing expanded ocular dominance columns in layer IV. (From Wiesel TN. Postnatal development of the visual cortex and the influence of the environment. Nature 1982;299:583.)

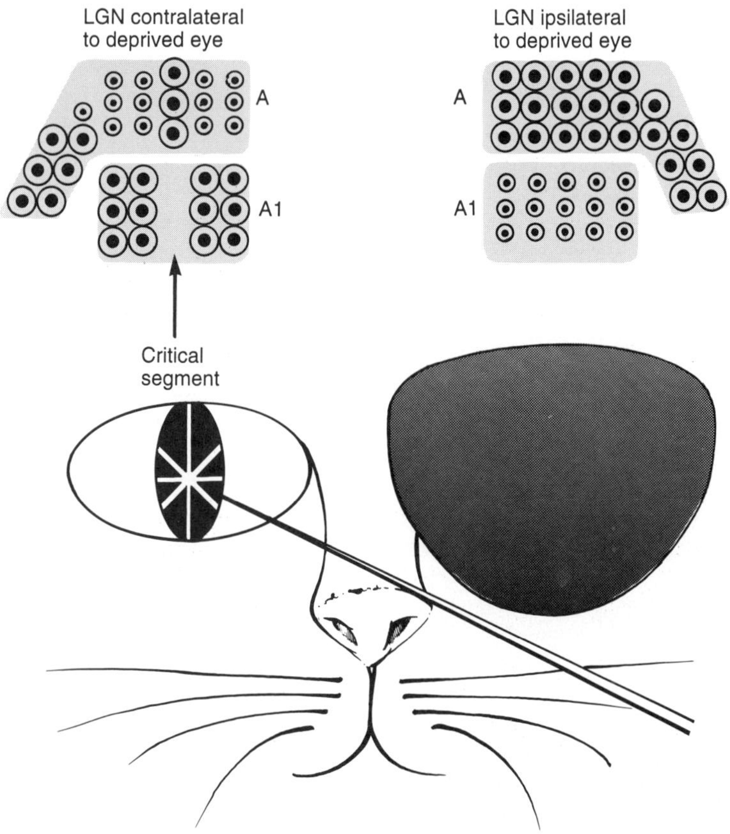

Figure 24-9
Illustration of the critical segment the experiment by R.W. Guillery, who found that cells in the monocular segment of the lateral geniculate layer A did not shrink like those in the binocular segment after monocular deprivation of the contralateral eye. To test the hypothesis that the shrinkage resulted from some competitive interaction of inputs arising from homotypic points in the two retinas, Guillery produced a lesion in the central region of the nondeprived retina. This removed the competition from the open eye in a critical segment of the lateral geniculate, and in this region, the cells in layer A did not shrink after monocular deprivation. Lateral geniculate cell size is thought to reflect accurately the extent of the axon arbor in cortical layer IV. These results supported the concept of binocular competition in the regulation of ocular dominance columns in the visual cortex. (From Guillery RW. Binocular competition in the control of geniculate cell growth. Comp Neurol 1972;144:117.)

☐ The establishment and maintenance of binocular connections is influenced by visual experience

The last connections to be specified during the development of the retinal-geniculate-striate pathway are those that subserve binocular vision. These are formed and modified under the influence of sensory experience during early postnatal life. Unlike the segregation of eye-specific domains, which evidently depends on asynchronous patterns of activity spontaneously generated by the two retinas, the establishment of binocular receptive fields depends instead on correlated patterns of activity that arise from the two eyes as a consequence of vision. This

has been demonstrated by experiments that bring the patterns of activity from the two eyes out of register. For example, monocular deprivation, which replaces patterned activity in one eye with uncorrelated visual "white noise," profoundly disrupts the binocular connections in striate cortex. Neurons outside of layer IV, which normally have binocular receptive fields, respond only to stimulation of the nondeprived eye after even a brief period of monocular deprivation. This change in the binocular organization of the cortex is called an *ocular dominance shift* (Fig. 24-10).

These effects of monocular deprivation are not merely a passive reflection of the anatomic changes in layer IV. The ocular dominance shift occurs in response to monocular deprivation initiated in the macaque at 1 year of age, well beyond the period of susceptibility of LGN axon

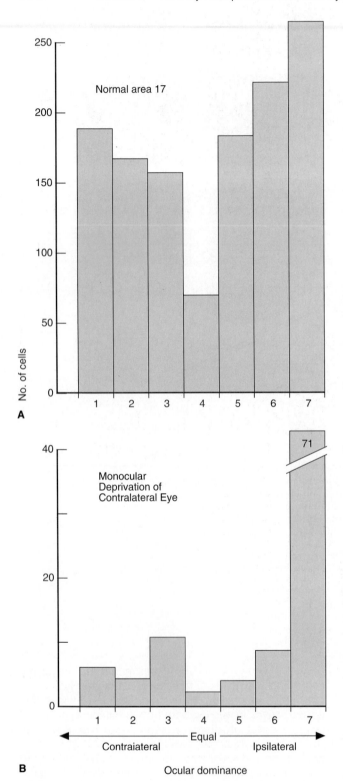

A

No. of cells

Normal area 17

B

Monocular
Deprivation of
Contralateral Eye

71

Contraiateral — Equal — Ipsilateral

Ocular dominance

Figure 24-10

The ocular dominance shift after monocular deprivation. Illustrated are histograms of ocular dominance data obtained from the striate cortex of (**A**) normal monkeys and (**B**) a monkey that had been monocularly deprived early in life. The bars show the number of neurons outside of layer IVc in each of the seven ocular dominance categories. Cells in groups 1 and 7 are activated by stimulation of the left or right eye, respectively, but not both. Cells in group 4 are activated equally well by either eye. Cells in groups 2 and 3 and in groups 5 and 6 are binocularly activated, but they show a preference for the left or right eye, respectively. The histogram in **A** reveals that most neurons in the visual cortex of a normal animal are driven binocularly. The histogram in **B** shows that a period of monocular deprivation leaves few neurons responsive to the deprived eye. (Modified from Wiesel TN. Postnatal development of the visual cortex and the influence of the environment. Nature 1982;299:583.)

arbors. However, this form of plasticity is also limited to a critical period of postnatal life. For example, in the cat, the ocular dominance plasticity peaks at about 1 month of age and then progressively declines to a very low level by 3 to 4 months (Fig. 24-11). The critical period for modification of binocular connections in the macaque extends to about 2 years, and estimates are that this plasticity ends in human children by about 10 years of age. These critical periods coincide with the times of greatest growth of the head and optical axes in these species. Plasticity of binocular connections probably is required to maintain good binocular vision throughout this period of

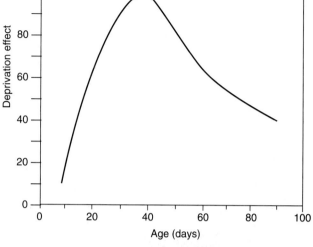

Figure 24-11
Sensitivity of binocular connections in cat striate cortex to monocular deprivation at different postnatal ages. The deprivation effect is the percentage of neurons in area 17 whose responses are dominated by stimulation of the nondeprived eye. This critical period begins at about 3 weeks of age and declines to low levels after 3 months of age. (Modified from Dudek SM, Bear MF. A biochemical correlate of the critical period for synaptic modification in kitten visual cortex. Science 1989;246:673.)

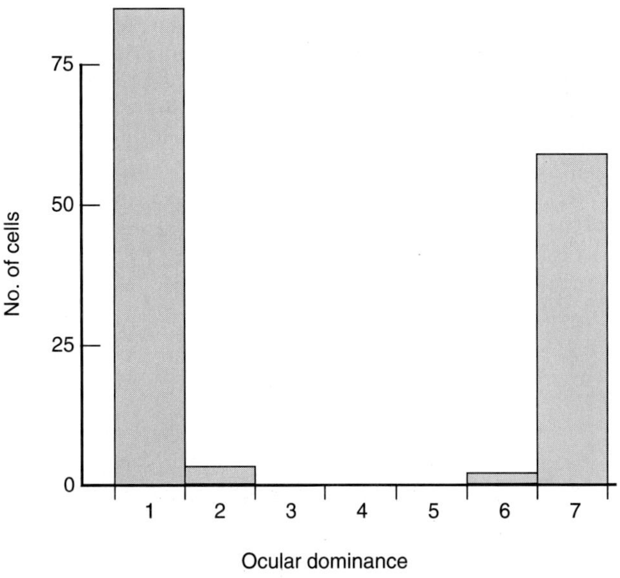

Figure 24-12
Ocular dominance histogram of cells recorded in the striate cortex of a 3-year-old strabismic monkey, in which the lateral rectus muscle of the right eye was sectioned at 3 weeks of age. Binocular cells are almost completely absent; the cells are driven exclusively by the right or the left eye. (Modified from Wiesel TN. Postnatal development of the visual cortex and the influence of the environment. Nature 1982;299:583.)

rapid growth. The hazard associated with this activity-dependent fine tuning is that these connections are also highly susceptible to deprivation.

Another experimental manipulation that disrupts cortical binocularity is *strabismus*. The misalignment of the two eyes causes visually evoked patterns of activity to arrive out of register in the cortex. As a result, there is a total loss of binocular receptive fields even though the two eyes retain equal representation in the cortex (Fig. 24-12). This is a clear demonstration that the disconnection of inputs from one eye occurs as the result of competition rather than disuse; the two eyes are equally active, but for each cell, a winner takes all. Strabismus, if produced early enough, can also sharpen the segregation of ocular dominance columns in layer IV.

The changes in ocular dominance and binocularity after deprivation have clear behavioral consequences. An ocular dominance shift after monocular deprivation leaves the animal visually impaired in the deprived eye, and the loss of binocularity associated with strabismus completely eliminates stereoscopic depth perception. However, neither of these effects are irreversible if they are corrected early enough in the critical period. The clinical lesson is that congenital cataracts or ocular misalignment must be corrected in early childhood, as soon as is surgically feasible, to avoid permanent visual disability.

☐ The development and modification of binocularity is influenced by extraretinal factors

With increasing age, there appear to be additional constraints on what forms of activity can modify cortical circuits. Before birth, spontaneously occurring bursts of retinal activity are sufficient to orchestrate aspects of address selection in the LGN and cortex. After birth, an interaction with the visual environment is of critical importance. However, even visually driven retinal activity may be insufficient for modifications of binocularity during this critical period. Such modifications seem to require that the animals attend to visual stimuli and use vision to guide behavior. For example, modifications of binocularity after monocular stimulation do not occur in anesthetized animals, even though it is known that cortical neurons respond briskly to visual stimulation under this condition. These and related observations have led to the proposal that synaptic plasticity in the cortex requires the release of extraretinal "enabling factors" that are linked to behavioral state. There is some evidence to suggest that this release may occur in response to eye movement.

Some progress has been made in identifying the physical basis of these enabling factors. Several modulatory systems converge at the level of striate cortex, including the noradrenergic inputs from the locus ceruleus and the cholinergic inputs from the basal forebrain. These axonal projections have a trajectory that is distinct from the optic radiation from the LGN to the cortex. It has been possible to study the effects of monocular deprivation in animals in which these modulatory inputs were surgically transected. This manipulation was found to cause a substantial impairment of ocular dominance plasticity outside of cortical layer IV, even though transmission in the retinal-geniculate-cortical pathway was apparently normal (Fig.

24-13). By making chemical lesions subcortically, it was confirmed that this impairment was caused specifically by the loss of the cholinergic and noradrenergic innervation of cortex.

The mechanism for modulation of synaptic plasticity by acetylcholine and norepinephrine remains to be determined. However, it is known that these neurotransmitters increase the excitability of neurons, making it more likely that cortical neurons will respond with action potentials to visual stimulation.

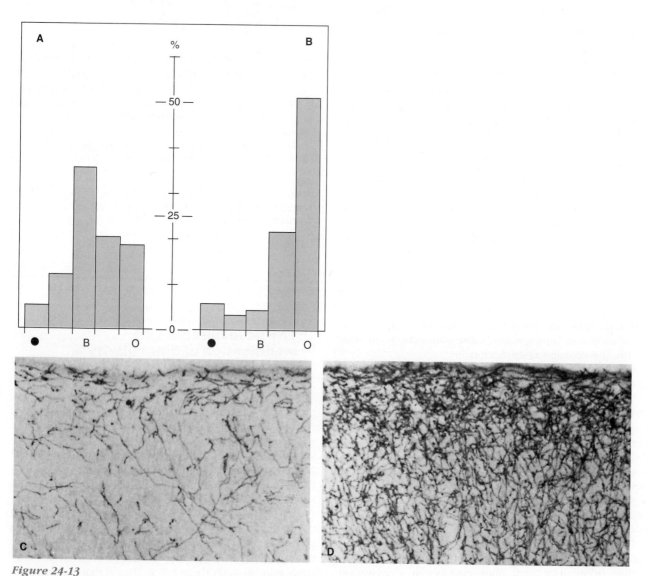

Figure 24-13

Destruction of the modulatory noradrenergic and cholinergic inputs to striate cortex interferes with the plasticity of binocular connections. (**A, B**) Percentages of cells in each of five ocular dominance categories in striate cortex of monocularly deprived kittens. Open circles indicate the monocular open-eye group; filled circles indicate the monocular closed-eye group, and the x-axis B indicates the strictly binocular group. Data in **A** were obtained from the striate cortex ipsilateral to a unilateral transection of the cingulate bundle, a fiber tract that brings the modulatory inputs to the visual cortex. Data in **B** were obtained from the contralateral hemisphere where the modulatory inputs were intact. The ocular dominance shift occurred only in the intact hemisphere. (**C, D**) Distribution of cholinergic axons in the striate cortex of a hemisphere with a lesion of the ascending modulatory fibers and of the contralateral hemisphere with these fibers intact. (From Bear MF, Singer W. Nature 1986;320:172.)

☐ The elementary mechanisms of synaptic plasticity during the critical period may involve NMDA receptors

Much of the activity-dependent development of the visual system can be explained using Hebb synapses. This is particularly transparent for the development of binocular connections, in which afferents that converge onto the same cell are consolidated only if they carry synchronous patterns of activity.

Why is it that neurons that fire together wire together and those that do not become unplugged? The answer to this question requires some knowledge of the mechanisms of excitatory synaptic transmission in the visual system. The transmitter at all of the modifiable synapses (e.g., retinogeniculate, geniculocortical, corticocortical) is likely to be an amino acid, glutamate or aspartate, and is known to recruit a family of postsynaptic receptors called *excitatory amino acid receptors*. These receptors may be divided into two broad categories called *metabotropic* and *ionotropic*. The metabotropic receptors are linked by G proteins to intracellular second messenger systems. The ionotropic receptors are ion channels that allow passage of positively charged ions into the postsynaptic cell. These receptors may be further divided into two categories named for compounds that act as selective agonists at these sites: the AMPA (a-amino-3-hydroxy-5-methyl-4-isoxazole propionic acid) receptor and the NMDA (N-methyl-D-aspartate) receptor. AMPA and NMDA receptors are co-localized at many synapses.

Synaptic activation of the AMPA receptor activates a monovalent cation conductance that shows a linear current-voltage association with a reversal potential of about 0 mV. Activation of this receptor by glutamate stimulates an inward (i.e., depolarizing) current whose amplitude diminishes as the postsynaptic membrane is depolarized because of decreased driving force.

The NMDA receptor has two unusual features that differentiate it from the AMPA receptor. First, the NMDA receptor conductance is voltage-dependent because of the action of Mg^{2+} at the channel. At the resting membrane potential, the inward current through the NMDA receptor is interrupted by the movement of Mg^{2+} ions into the channel, where they become lodged. However, as the membrane is depolarized, the Mg^{2+} block is displaced from the channel, and current is free to pass into the cell. Substantial current through the NMDA receptor channel requires concurrent release of glutamate by the presynaptic terminal and the depolarization of the postsynaptic membrane. The other distinguishing feature of this receptor is that the NMDA receptor channel conducts Ca^{2+} ions. The magnitude of the Ca^{2+} flux passing through the NMDA receptor channel specifically signals the level of presynaptic and postsynaptic coactivation. It is thought that NMDA receptors in the cortex and LGN serve as Hebbian detectors of coincident presynaptic and postsynaptic activity and that Ca^{2+} entry through the NMDA receptor channel triggers the biochemical mechanisms that modify synaptic effectiveness. Hebbian enhancements of synaptic effectiveness have been shown experimentally in the connections from layer IV onto layer III neurons in visual cortex.

Pairing low-frequency stimulation of layer IV with intracellular depolarization of a cell in layer III results in a *long-term potentiation* of the conditioned synapses (Fig. 24-14). This long-term potentiation is prevented by application of the drug 2-amino-5-phosphonovaleric acid (AP5), an antagonist of the NMDA receptor. That similar mechanisms contribute to naturally occurring synaptic remodeling is suggested by observations that application of AP5 in vivo can disrupt the natural segregation of eye-specific inputs in the LGN, binocular competition in layer IV ocular dominance domains after monocular deprivation, and the modification of binocular connections in the superficial layers after a number of experimental manipulations of visual experience (Fig. 24-15).

The strong activation of NMDA receptors that occurs when presynaptic and postsynaptic neurons fire together is thought to account partly for why they wire together during visual system development. However, the NMDA receptor does not function as a switch that is only "on" when input activity coincides with strong postsynaptic depolarization. Weak coincidences are signaled by lower levels of NMDA receptor activation and less Ca^{2+} influx. Experiments suggest that the lower level of Ca^{2+} admitted under these conditions triggers an opposite form of synaptic plasticity, *long-term depression*, by which the effectiveness of the active synapses is decreased. The maintenance of a connection formed during development may depend on its success in evoking an NMDA receptor-mediated response beyond some threshold level. Failure to achieve this threshold leads to disconnection. Both processes depend on activity in the retinofugal pathway and on postsynaptic Ca^{2+} entry.

☐ The visual cortex is plastic beyond the critical period

In visual system development, there are multiple critical periods. For example, the critical period for activity-dependent anatomic rearrangements of geniculate axonal arbors in layer IV ends much earlier than the critical period for the experience-dependent modification of binocular connections. In these discussions, it is important to avoid contending that the primary visual cortex of the adult brain is aged beyond the critical period and therefore is immutable. Critical periods must be defined according to which type of intercellular interaction is being considered.

An example of plasticity in the adult brain is the modification of striate cortical circuits that follows a circumscribed lesion in the retina. Not surprisingly, the cortical neurons at the corresponding point in the retinotopic map initially fall silent and are unresponsive to any type of visual stimulation. However, in a matter of a few weeks, these same neurons again become visually responsive but to the region of retina surrounding the lesion. The retinal lesion causes a "filling in" of the cortical retino-

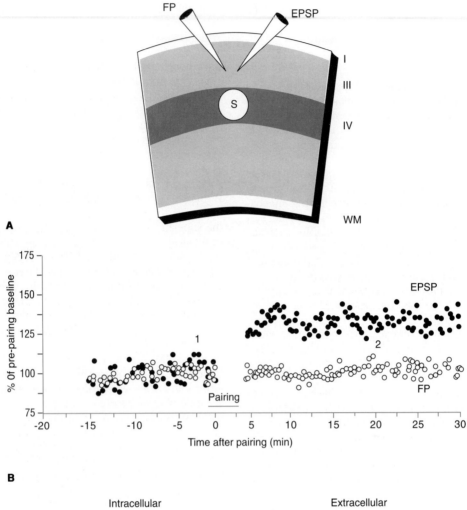

A

B

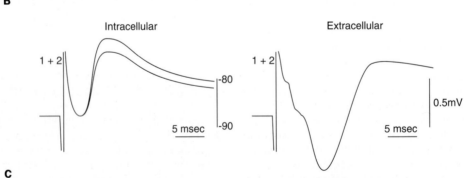

C

Figure 24-14

Record of an experiment demonstrating Hebbian synaptic modification of the connections between layer IV and layer III in the rat visual cortex. (**A**) Layer IV was electrically stimulated (at
the site marked Ⓢ) every 30 seconds, and (**B, C**) the intracellular (EPSP) and extracellular synaptic responses in layer III were monitored. At the time indicated, the electrical stimuli to layer
IV were paired with strong depolarization of the intracellularly recorded layer III neuron. As a
consequence of this pairing, the intracellular response of the layer III neuron to the test stimulation of layer IV was potentiated. The extracellular response, which reflects the summed activity
of a large population of synapses on many layer III neurons, is unchanged. *FP,* field potential.

topic map (Fig. 24-16). The anatomic substrate of this
map plasticity is thought to be an adjustment in the effectiveness of horizontal projections that interconnect the
layer III neurons in different parts of the retinotopic map.
This demonstrates that cortical circuits can be modified

by peripheral lesions well after the end of the critical
period that is defined by the effects of monocular deprivation.

It is assumed that synaptic adjustments to more subtle
changes in the sensory environment give the cortex the

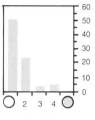
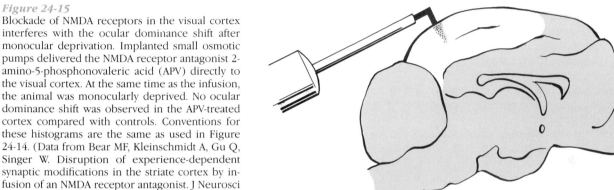

APV (50 nMol per hour for 48 hours)   Control

Figure 24-15
Blockade of NMDA receptors in the visual cortex interferes with the ocular dominance shift after monocular deprivation. Implanted small osmotic pumps delivered the NMDA receptor antagonist 2-amino-5-phosphonovaleric acid (APV) directly to the visual cortex. At the same time as the infusion, the animal was monocularly deprived. No ocular dominance shift was observed in the APV-treated cortex compared with controls. Conventions for these histograms are the same as used in Figure 24-14. (Data from Bear MF, Kleinschmidt A, Gu Q, Singer W. Disruption of experience-dependent synaptic modifications in the striate cortex by infusion of an NMDA receptor antagonist. J Neurosci 1990;10:909.)

life-long capability to reorganize itself. The elementary mechanisms of synaptic plasticity, long-term potentiation and long-term depression, persist in the superficial layers of the adult striate cortex. It is likely that experience-dependent synaptic remodeling in the cortex is one mechanism that contributes to environmental adaptation and possibly to learning and memory in the adult brain.

Why do critical periods end?

Although plasticity of visual connections persists in the adult brain, the dynamic range over which this plasticity occurs constricts with increasing age. Early in development, gross rearrangements of axonal arbors are possible, but in the adult, the plasticity appears to be restricted to local changes in synaptic efficacy. The adequate stimulus for evoking a change also appears to be increasingly constrained as the brain matures. An obvious example is the fact that patching one eye causes a profound alteration in the binocular connections of the superficial layers during infancy, but by adolescence, this type of experience fails to cause a lasting alteration in cortical circuitry.

There still is no satisfactory single explanation of why critical periods end. As more is learned about the elementary mechanisms of axonal path finding and synaptic plasticity, insights will be gained about how these processes are regulated. However, it is already possible to identify some of the rate-limiting factors that govern activity-dependent plasticity in the developing visual pathway.

One common feature in the establishment of connectivity in the LGN and cortex is the initial activity-indepen-

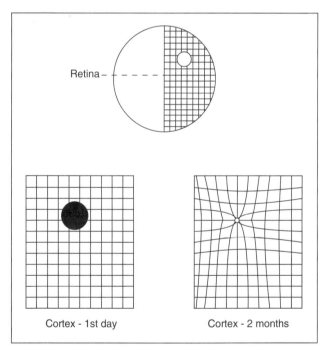

Figure 24-16
Plasticity in the adult visual cortex. A lesion in the retina initially causes a patch of cortex to fall silent. However, these silent neurons later become responsive to stimulation of the surrounding areas of retina, causing the map to "fill in." (Modified from Baringa M. The brain remaps its own contours. Science 1992;258:216.

dent establishment of widespread "exuberant" connections that are sculpted by activity to achieve their final form during a critical period. One explanation for the end of a critical period of, for example, ocular dominance column formation in layer IV is that, once segregation is complete and the afferents no longer contact the same postsynaptic cells, the substrate for the winner-takes-all competition is lost (Fig. 24-17). According to this view, monocular deprivation, by removing a competing input, only "saves" the initially widespread open eye axonal arbors from being retracted. If it is assumed that axons lose the ability to grow after they have invaded their target structure and established their initial pattern of connectivity, the final state of connectivity is permanently "imprinted" after segregation (by loss of synapses and axon retraction) is complete.

This is unlikely to be the full explanation in the visual cortex, because reverse suture experiments have shown that ocular dominance columns retain some capacity for reexpansion after segregation is anatomically complete. In the macaque monkey, there is a period of several weeks when the geniculate arbors can exhibit some limited growth within layer IV. Another factor that limits the critical period in layer IV appears to be a loss of the capability for axonal elongation, which may be caused by changes in the extracellular matrix or surface glycoproteins expressed by the geniculate axons.

A third possible reason the critical period ends reflects changes in the elementary mechanisms of synaptic plasticity. There is evidence that some EAA receptors change during postnatal development. For example, it has been shown that EAA-stimulated phosphoinositide turnover, mediated by one of the metabotropic receptors, peaks in visual cortex when binocular connections are most susceptible to monocular deprivation and then virtually disappears at the end of this critical period (Fig. 24-18). The effectiveness of NMDA receptors appears to be downregulated in layer IV at the same time as the end of the critical period for ocular dominance column segregation. This change in effectiveness can be detected at the single

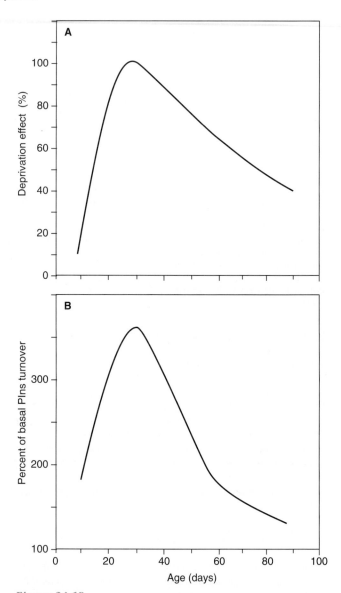

Figure 24-18
(**A**) Sensitivity of binocular connections in striate cortex to eyelid suture at different postnatal ages. (**B**) Phosphoinositide turnover stimulated by an excitatory amino acid in synaptoneurosomes prepared from kitten striate cortex at different postnatal ages. These data suggest that excitatory synaptic transmission during this critical period is characterized by unique patterns of second messenger activity. (Modified from Dudek SM, Bear MF. A biochemical correlate of the critical period for synaptic modification in the kitten visual cortex. Science 1989;246:673.)

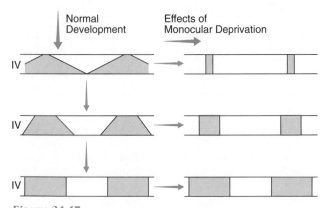

Figure 24-17
Segregation of ocular dominance columns in layer IV at three times during postnatal development and the effects of monocular deprivation initiated at these same times.

channel level and may result from a developmental change in the NMDA receptor subunit composition.

As development proceeds, certain types of activity may be filtered by successive synaptic relays to the point where they no longer activate NMDA receptors or other elementary mechanisms sufficiently to trigger plasticity. The neuromodulators acetylcholine and norepinephrine facilitate synaptic plasticity in the superficial cortical layers and may do this by enhancing polysynaptic intracortical transmission. A decline in the effectiveness of these

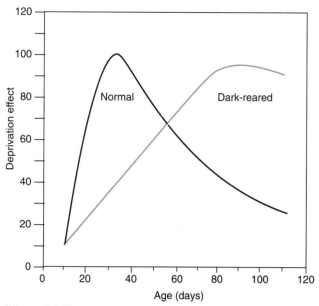

Figure 24-19
Raising cats in complete darkness before visual experience with one eye open slows the critical period for the modification of binocular connections. (Data from Mower GD. The effect of dark rearing on the time course of the critical period in cat visual cortex. Dev Brain Res 1991;58:151.)

neuromodulators or a change in the conditions under which they are released may contribute to the decline in plasticity. There is some evidence that supplementing the adult cortex with norepinephrine can restore some degree of modifiability. There is also evidence that intrinsic inhibitory circuitry is late to mature in the visual cortex.

Consequently, patterns of activity that may have gained access to modifiable synapses in superficial layers early in postnatal development may be damped by inhibition in the adult.

Whatever the reason for the decline in synaptic plasticity at the end of the critical periods, it is clear that the duration of the critical periods is not always locked to a certain postnatal age. For example, rearing kittens in complete darkness appears to slow the critical period for modification of binocular connections (Fig. 24-19). Dark rearing also slows segregation of ocular dominance columns and some of the changes in EAA receptor properties. It seems that the duration of this critical period may be measured by the history of stimulation rather than by age.

SELECTED READINGS

Bornstein MH (ed). Sensitive periods in development: interdisciplinary perspectives. Hillsdale, NJ: Lawrence Erlbaum Associates, 1987:290.

Goodman CS, Shatz CJ. Developmental mechanisms that generate precise patterns of neuronal connectivity. Cell 1993;72:77.

Greenough WT, Juraska JM (eds). Developmental neuropsychobiology. Orlando: Academic Press, 1986:485.

Landmesser LT (ed). The assembly of the nervous system. New York: Alan R Liss, 1989:313.

Löwel S, Singer W. Selection of intrinsic horizontal connections in the visual cortex by correlated neuronal activity. Science 1992;255(5041):209.

Rauschecker JP, Marler P (eds). Imprinting and cortical plasticity. Comparative aspects of sensitive periods. New York: John Wiley & Sons, 1987:377.

Shatz CJ. The developing brain. Sci Am 1992;267:61.

Wiesel TN. Postnatal development of the visual cortex and the influence of the environment. Nature 1982;299:583.

Neuroscience in Medicine, edited by P. Michael Conn.
J. B. Lippincott Company, Philadelphia © 1995

Chapter 25

Audition

TOM C. T. YIN

This chapter provides an introduction to the physiologic mechanisms underlying audition. We first consider the physical properties of sound, then examine the processing of acoustic input by the peripheral auditory system, and then explore the central auditory system.

The Physical Nature of Sound

A few basic properties of sound must be understood before analyzing the auditory system. Physically, sound is a mechanical disturbance that is propagated through an elastic medium. This medium usually is air, but sound can also propagate through solids or liquids. Air molecules are in constant, random motion such that the large number of air molecules striking any given point in space produces a static pressure, which depends on conditions of the system, such as the density of gas and air temperature. With a sudden change in the position of some large object in the air, such as clapping hands or vibrating the cone of a loudspeaker, the mechanical disturbance causes a temporary increase in pressure locally with a corresponding decrease elsewhere (e.g., increase on one side of the speaker cone and decrease on the other). The disturbance is propagated through the medium as the air molecules collide with each other and transfer energy to neighboring molecules. This variation in pressure as a function of time is called a *sound wave*. It represents the characteristics of the population of molecules as a whole rather than any single air molecule, which moves back and forth randomly without necessarily propagating energy.

The velocity at which the wave travels depends on the density and elasticity of the medium. Sound travels about four times faster in water than in air. The strength of the sound wave is usually measured as the deviation in sound pressure from atmospheric pressure.

Most sound waves result from the vibrations of an object, and the simplest form of vibration is a sine wave, which produces a pure tone. Figure 25-1 shows a diagram of the instantaneous pressure and the spatial distribution of air molecules as a function of distance in the medium for a pure tone at one instant in time. It can be considered to be a snapshot of the sound wave at some moment. Because sound propagates through air at a constant velocity (340 m/second), the abscissa in Figure 25-1 could also be time; it would then depict the variation in pressure at a given point in the medium as a function of time. The wavelength would then be the *period* of the tone.

The relation between the wavelength λ, conduction velocity c, and frequency f expressed in cycles/second or hertz (Hz) and the period T of the sine wave is given by: $\lambda = c/f$, in which $f = 1/T$. For a 100-Hz pure tone, which is near the lower limit of human hearing, λ is 3.3 m; at 10,000 Hz, it is 3 cm. The musical note middle C has a frequency of 256 Hz. These variations in wavelength with frequency are important in discussing the cues that are available for sound localization.

The sensitivity of the auditory system is quite remarkable. On the low end, the faintest sound that can be de-

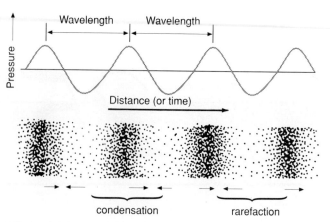

Figure 25-1
Diagram of air molecules in response to sinusoidal sound vibrations. The rarefactions and condensations are shown.

tected by humans is generated by movements of the eardrum of approximately 10^{-10} meters, which is in the range of diameters of air molecules. On the high end, sound pressures that are 10^6 larger begin to result in painful sensations. Because of this large dynamic range, the intensities of sounds are usually expressed in decibels (dB), which is a logarithmic unit of ratios. The arithmetic definition of a decibel is $dB = 20 \log_{10} P_1/P_2$, in which P_1 is the pressure of interest and P_2 is a reference pressure, which can be arbitrarily chosen. Usually, P_2 is chosen to be near the average normal threshold of hearing, and, when so chosen the decibels are denoted as dB sound pressure level (SPL). The dynamic range in humans is about $20 \log_{10} 10^6 = 120$ dB.

The physical parameters of frequency and SPL correspond to the perceptual qualities of *pitch* and *loudness*, respectively. In audiology and speech analysis, any particular sound can be represented by a spectrogram that plots the SPL as the darkness of spots on a time-frequency axis, as shown in Figure 25-2. A pure tone is represented by a single frequency that persists over time, a musical note generally has many harmonics, and speech consists of a complex mixture of frequencies varying with time. A topic of considerable clinical interest is how the auditory system can encode a complex sound, such as speech, and distinguish the subtle differences between similar sounds or the same sound uttered by different people that constitute daily human experience. We know very little about these questions.

The Peripheral Auditory System

It is convenient to divide the ear into an outer, middle, and inner ear, as shown in Figure 25-3, which depicts a cross section of the human ear. A standard sequence of events leads to activation of auditory nerve fibers in response to an acoustic stimulus. Sound waves enter the external ear and travel down the external auditory meatus to strike the tympanic membrane or eardrum. Move-

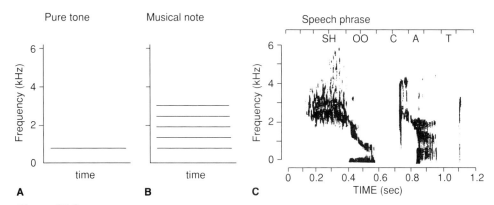

Figure 25-2

Spectrograms of (**A**) a pure tone of 2000 Hz, (**B**) a musical note from a flute, and (**C**) a speech phrase (i.e., shoo cat). Darkness represents acoustic energy in the signal at the corresponding frequency. (Adapted from Kiang NYS. Stimulus representation in the discharge patterns of auditory neurons. In Tower DB (ed). The nervous system, vol 3: human communication and its disorders. New York: Raven Press, 1975:81.)

ments of the tympanic membrane are transferred by means of the ossicular chain to the oval window of the cochlea, and back and forth movements of the oval window cause a traveling wave to be set up in the fluid-filled cochlear ducts. The traveling wave causes deflections of the basilar membrane, which vibrates and activates the receptor cells, the hair cells in the organ of Corti (Figs. 25-3 and 25-4).

☐ The outer ear is important for collecting sound waves

The outer ear consists of the pinna or external ear and the external auditory meatus or ear canal. This apparatus has important acoustic properties, because it accentuates, or attenuates sounds of certain frequencies and those coming from particular directions. The degree to which the pinna can influence a sound wave depends on the wavelength of the sound and hence its frequency. If the wavelength is much longer than the dimensions of the pinna, as it is for low-frequency sounds, it appears transparent to the sound and produces no diffractive or reflective effects. Only high-frequency sounds of at least 5 kHz are greatly affected by the pinna. Because of the directional properties of the pinna, these transformations are important for localizing sounds, particularly those in the vertical plane. In humans, unlike most other mammals, there is virtually no control over the direction in which the pinna is pointing other than movements of the head.

☐ The middle ear acts as an impedance matching device

The middle ear consists of the *tympanic membrane* and the three bony *ossicles* that conduct vibrations to the oval window: the *malleus, incus,* and *stapes* (i.e., hammer, anvil, and stirrup). The middle ear is important in providing an *impedance matching*, also called an interface, for the airborne sound waves to transmit their energy to the fluid in the cochlea. If there were no middle ear, only 0.1% of the energy of a sound wave would be transmitted to the fluid in the cochlea. The middle ear reduces this energy loss, chiefly by the mechanical advantage resulting from the reduction in surface area from the tympanic membrane to the footplate of the stapes. The principle is the same as that used in hydraulic jacks, in which force is applied through a combination of a large and a small cylinder connected by a pipe and filled with a fluid. A small fluid pressure applied to the large cylinder results in a large pressure on the small piston. The force on the tympanic membrane is equal to the pressure times the total area of the eardrum and, assuming negligible frictional losses in the transmission through the ossicles, it is equal to the force exerted at the stapes footplate on the oval window. Because the area of the stapes is much smaller than that of the tympanic membrane, there is a *pressure amplification* given by the ratio of these areas (approximately 20:1 or 30:1).

Two other factors amplify the pressure at the stapes: a lever arm action through the ossicular chain and a buckling factor that results from the conical shape of the tympanic membrane. The potential loss in energy transmission at the air-fluid boundary is reduced considerably, as is evident in the 20 to 30 dB hearing loss suffered by patients in which the ossicular chain has been broken.

There are two important muscles in the middle ear: the tensor tympani and stapedius. Both muscles are attached to the ossicles. The stapedius muscle is innervated by motor neurons in the facial nucleus that run with cranial nerve VII, and the tensor tympani is innervated by motor neurons in the motor division of the trigeminal nucleus that travel in the V nerve. Contraction of these muscles increases the stiffness of the ossicular chain and can decrease the sound transmission by as much as 15 to 20 dB. Loud sounds cause the muscles to contract reflexly, a response that protects the ear from very loud sounds by reducing energy transmission.

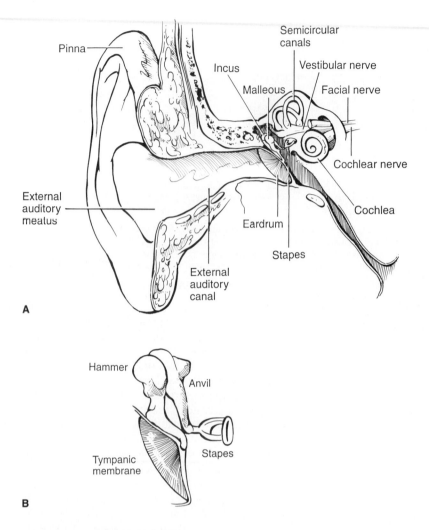

A

B

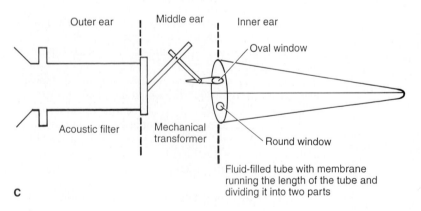

C

Figure 25-3
Major divisions of the peripheral auditory system. (**A**) The outer, middle, and inner ear. (**B**) An enlarged version of the middle ear. (**C**) The mechanical analogs of the outer and middle ear.

☐ The inner ear contains the mechanisms for sensory transduction

The inner ear is divided into three interconnected parts: the semicircular canals, the vestibule, and the cochlea, all of which are located in the temporal bone. Only the *cochlea* is considered here. The cochlea is the small, shell-shaped part of the bony labyrinth that contains the receptor organ of hearing. It resembles a tube that is coiled increasingly tightly on itself. One end of the tube is called the apex, and the opposite end, closest to the middle ear, is called the base. The fluid-filled spiral canal of the cochlea is divided along its length by a partition, the *basilar membrane*, which is attached to the bony walls of the cochlea. Cross sections of the cochlea show

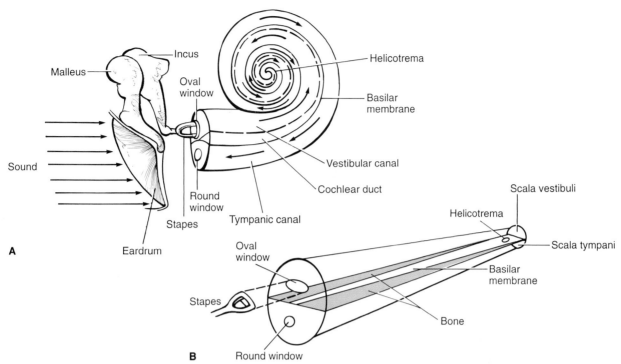

Figure 25-4
The major structural features of the cochlea. (**A**) Coupling of the middle ear to the coiled cochlea through the oval and round windows. (**B**) The cochlea is shown uncoiled. The basilar membrane is narrow near the round window and wider near the helicotrema, a taper opposite the cross-sectional area of the cochlea.

that the canal is divided into three ducts: scala vestibuli, scala tympani, and scala media (see Fig. 25-5).

The input to the cochlea is supplied by movements of the oval window by means of the stapes footplate. Because the fluid in the cochlea is incompressible, there must be some point where the pressure applied at the oval window which is in the scala vestibuli, is transferred. This occurs at the *round window*, which is in scala tympani, through the helicotrema at the apex of the cochlea, where the scala vestibuli joins the scala tympani (see Fig. 25-4). As the pressure wave travels the length of the cochlea, it creates a pressure differential across the basilar membrane (between scala vestibuli and scala tympani) that sets the membrane in motion in the form of a traveling wave.

Within the scala media and attached to the basilar membrane is the receptor organ of hearing, the *organ of Corti* (Fig. 25-5). The organ of Corti is composed of numerous structures, most of which are not considered here. Of special importance are the receptor cells, which are similar to the hair cells in the semicircular canals. The hairs are in contact with an auxiliary structure, the *tectorial membrane*. Sound waves reaching the inner ear set the basilar membrane and the organ of Corti into motion. This motion results in a shearing force between the tectorial membrane and the hairs of the hair cells. Displacement of the hairs results in transmitter release at the base of the hair cell, where fibers of the cranial nerve VIII

make synaptic contact. All information about the acoustic environment is carried to the central nervous system (CNS) by trains of all-or-none action potentials generated in primary afferent fibers.

The coding of acoustic information depends on the vibratory pattern of the basilar membrane. The basilar membrane is a relatively flaccid structure that increases in width and decreases in stiffness from its base to apex. The change in the width of the basilar membrane is the opposite of the change in width of the cochlear duct as it coils from base to apex (see Fig. 25-4). The vibratory undulations generated in the cochlea by sound waves contain all the information about the acoustic environment that must be coded into neural information. A key factor in this process is the mechanical response to sound waves of the basilar membrane and organ of Corti. Most of the pioneering work on the vibratory patterns of the basilar membrane was done by Georg von Bekesy who was recognized with the Nobel Prize in 1960.

To understand the motion of the basilar membrane and how the frequency of a sound affects this movement, we first consider the responses to a sinusoidal stimulus (i.e., tone). Each point along the basilar membrane that is set in motion vibrates at the same frequency as the acoustic stimulus. However, the amplitude of membrane vibration is different in different locations, depending on the frequency of the sound. A wave motion is set up along the membrane as the fluids of the inner ear are

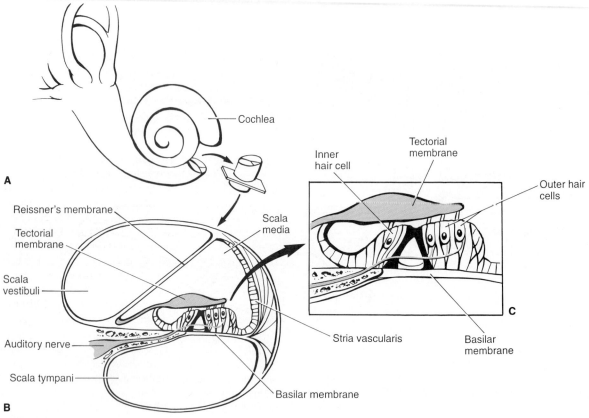

Figure 25-5
Anatomic features of the cochlea. (**A**) The cochlea in relation to the middle ear, the vestibular channels (partially illustrated), and the auditory nerve. (**B**) Cross section of the cochlea. (**C**) The structures within the scala media.

driven by motion of the stapes. This wave motion on the basilar membrane is referred to as a *traveling wave* (Fig. 25-6).

Because the basilar membrane becomes wider and more flaccid as the distance from the base increases, the natural frequency of vibration (i.e., resonance) of the basilar membrane decreases toward the helicotrema. Because of this variation in stiffness, different frequencies cause maximal vibration amplitudes at different points along the membrane (Fig. 25-7A). If two different frequencies are received by the cochlea simultaneously, they each create a maximal displacement at different points along the basilar membrane. This separation of a complex signal into different points of maximal displacement along the basilar membrane, corresponding to the frequency of the sinusoids of which the complex signal is composed, means that the basilar membrane performs like a series of filters. The basal part of the basilar membrane responds maximally to high-frequency sounds, and the region of maximal displacement of low-frequency tones is in the apical portion (Fig. 25-7B); the basilar membrane is tonotopically organized.

The receptor cells are hair cells that line the length of the cochlea. In most mammals, there are three rows of *outer hair cells* and a single row of *inner hair cells*. The tops of the outer hair cells are embedded in the overlying

tectorial membrane (see Fig. 25-5), but it is unknown whether the same is true of the inner hair cells. When the basilar membrane is set into motion by an acoustic stimulus, the basilar membrane and tectorial membranes do not move in unison, because they pivot about different points. The result is that a shearing force is applied to the cilia of the hair cells, which bends the cilia. Just as with the hair cells in the semicircular canals, this mechanical action results in a depolarization or hyperpolar-

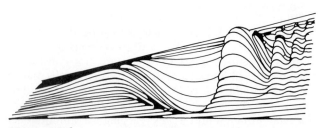

Figure 25-6
Representation of a traveling wave at one instant in time. Cochlear base is at left and the apex at right. Notice the rapid decline in amplitude just to the right of the region of maximal displacement amplitude. (Redrawn from Tonndorf J. Shearing motion in scala media of cochlear models. J Acoust Soc Am 1960:32;238.)

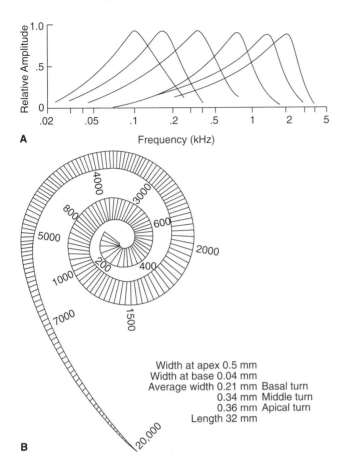

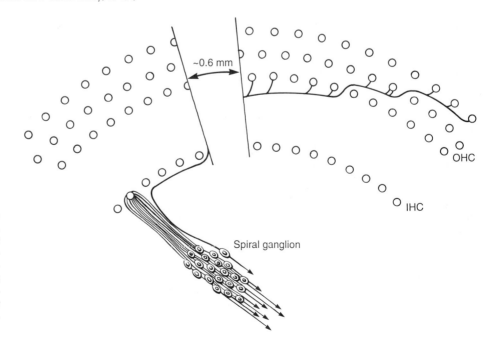

Width at apex 0.5 mm
Width at base 0.04 mm
Average width 0.21 mm Basal turn
          0.34 mm Middle turn
          0.36 mm Apical turn
Length 32 mm

B

Figure 25-7
(**A**) Envelopes of traveling waves at seven different frequencies. (Redrawn from Bekesy G von. Experiments in hearing. New York: McGraw-Hill, 1960.) (**B**) Diagram of the human basilar membrane, showing the approximate positions of maximal displacement to tones of different frequencies and changes in width going from the base near the stapes and oval window to the apex near the helicotrema. The ratio of width to length is exaggerated to show the variation in width more clearly. (Redrawn from Stuhlman O. An introduction to biophysics. New York: Wiley, 1943.)

Figure 25-8
Schematic view of cochlear innervation for afferent fibers in cat. *IHC,* inner hair cells; *OHC,* outer hair cells. (Redrawn from Spoendlin H. Structural basis of peripheral frequency analysis. In Plomp R, Smoorenburg GF (eds). Frequency analysis and periodicity detection in hearing. The Netherlands: AW Sitjhoff, 1970:2.)

ization of the membrane potential of the cell, depending on the direction of movement. The afferent nerve fibers of the cranial nerve VIII make chemical synapses onto the base of the hair cells, and depolarization of the hair cell increases the resting impulse discharge of the fiber. Hyperpolarization from movement in the opposite direction decreases the resting discharge level.

We do not understand the exact role of the two types of hair cells. A clue can be obtained from considering their innervation by auditory nerve fibers, because the pattern is quite different for the inner and outer hair cells. In the cat, there are about 50,000 auditory nerve fibers. Not all of these are afferent fibers; about 5% are efferent fibers that have cell bodies in and around the superior olivary complex and convey information back to the cochlea. The precise nature of this efferent innervation of the cochlea is not well understood. Of the remaining afferent fibers, about 95% terminate only on inner hair cells (Fig. 25-8) so that each inner hair cell is contacted by about 20 afferent fibers. The remaining afferent fibers innervate the outer hair cells in a much more diffuse manner. Typically, the afferent fiber crosses over to the rows of outer hair cells and travels toward the base before innervating a number of outer hair cells in all three rows. It is reasonable to conclude that the inner hair cells are most important for conveying information about the vibrations in the cochlea to the CNS.

Exactly what role the outer hair cells play and why there should be three times as many is not understood. Current speculations center on their role in efferent innervation and evidence for an active mechanism in the cochlea. Evidence for an active process comes from the surprising observation that the cochlea can emit sound spontaneously or in response to acoustic stimulation. The outer hair cells are motile and can be made to contract when stimulated electrically. One hypothetical scenario is that the efferent system enables the CNS to contract the cilia of the outer hair cells, changing the micromechanical sensitivity of the basilar membrane to low-level stimuli. This remains an active area of research.

Because the eighth cranial nerve afferent fibers are distributed along the whole length of the basilar membrane, a tone of one frequency can excite the afferents connected to the region undergoing suprathreshold vibration. The fact that a discrete population of fibers is activated by a pure tone and that this population changes when the stimulus frequency changes is the basis for the *place principle* of hearing. This principle states that the perceived pitch of a sound depends only on the particular population of nervous elements activated. The tonotopic organization of the array of auditory nerve terminals along the basilar membrane is maintained in all major areas of the central auditory system, but this tonotopic organization is probably not precise enough to explain a listener's ability to discriminate one frequency from another.

☐ Frequency tuning and temporal information is transmitted by auditory nerve fibers

There are several ways in which information about an acoustic stimulus is coded in the discharges of fibers of the auditory nerve. These results come from studies in which fine micropipette electrodes are used to record from single auditory nerve fibers.

Tuning properties

Auditory nerve fibers are responsive only within a restricted range of frequencies and intensities. This frequency-intensity domain is called the response area, which may be considered analogous to the receptive field in the visual and somatic sensory systems. A plot of the threshold sound intensity versus stimulus frequency is called the *tuning curve* of the fiber (Fig. 25-9). The

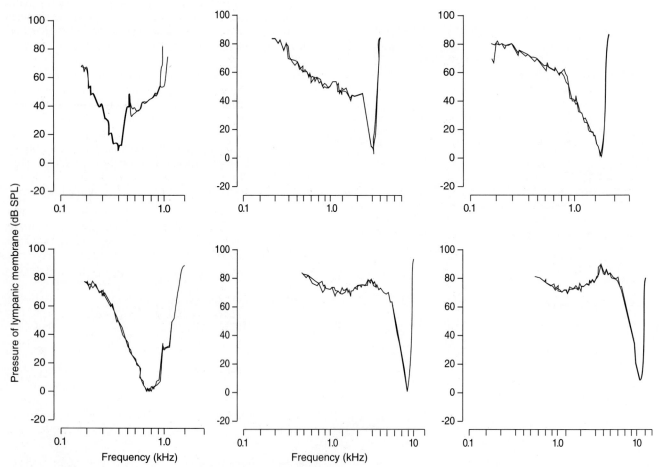

Figure 25-9

Representative tuning curves (i.e., frequency threshold curves) of cat auditory nerve fibers are shown for six different frequency regions. In each panel, two fibers from the same animal of similar characteristic frequency and threshold are shown, indicating the constancy of tuning under such circumstances. (Redrawn from Liberman MC, Kiang NYS. Acoustic trauma in cats: cochlear pathology and auditory-nerve activity. Acta Otolaryngol (suppl) 1978;358:1.)

frequency of lowest threshold is designated as the best or characteristic frequency. Different fibers have different best frequencies, reflecting their connections along the basilar membrane. Fibers with high best frequencies are connected to basal regions of the cochlea, and fibers with low best frequencies are connected more toward the apical regions.

The CNS receives information about stimulus frequency in terms of which fibers are activated in accord with the place principle of pitch perception. However, at even moderate intensity, the tuning curve is relatively broad, especially on the low-frequency side. If the place principle was the only mechanism that allowed pitch identification by the CNS, we could assume that all of the activity is ignored except at the peak or boundary of the array of active fibers or that the place principle operates only near the threshold of hearing. Because neither possibility seems reasonable, there must be other mechanisms for coding acoustic information. The *time principle* states that the acoustic waveform is encoded in the temporal discharge pattern of auditory nerve fibers.

Temporal properties

Information about the frequency of low-frequency tones is conveyed to the CNS by another method. When the fiber discharges, it tends to do so around the same phase of the stimulus waveform, corresponding to movements of the basilar membrane in only one direction (Fig. 25-10). This phenomenon is called *phase-locking* and is observed only at frequencies below about 4000 Hz. This temporal coding of low-frequency sounds is important because the energy in speech signals is predominantly below 4000 Hz (see Fig. 25-2), and the changes in sound pressure resulting from complex sounds consisting of many low-frequency spectral components are encoded in the temporal discharge in a similar fashion. With phase-locking, the time interval between spikes tends to be a multiple of the period of the stimulating tone. No single fiber fires on every cycle of a tone, and if we consider an array of many fibers, the phenomenon known as *volleying* is revealed. In the total array of responding fibers, there are discharges in some fibers on each of the stimulus cycles. Information about the period (and hence the frequency) of a low-frequency tone is represented in the temporal rhythm of nerve impulses. The CNS uses these rhythms to encode speech sounds and to localize a sound source in space. Mechanisms by which the timing of neural volleys from the two ears is used in sound localization are described in the following sections.

The Central Auditory System

The central auditory pathways are discussed in Chapter 26. We will briefly review the general organization here, but please refer to Chapters 15 and 26 for details.

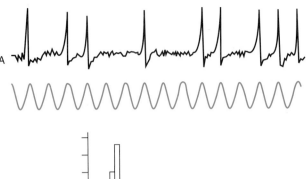
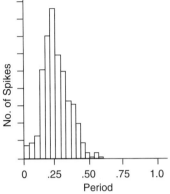

Figure 25-10
Representation of the phase-locked response of a single auditory nerve fiber to a low-frequency tone. The period histogram below it shows the distribution of phase angles at which the nerve preferentially discharges.

☐ The central auditory pathway consists of many relay nuclei

The central auditory pathway comprises a number of nuclear groups within the medulla, pons, midbrain, thalamus, and cerebral cortex. These cell groups are interconnected by fiber tracts that ascend from the cochlea to the auditory cortex (Fig. 25-11). These pathways carry information about the acoustic environment that reaches consciousness. A descending pathway carries information back to the cochlea, primarily to the outer hair cells, but we do not understand the function of these descending pathways. Other pathways that communicate acoustic reflexes that are activated by sound stimulation and involve various motor systems to move the head, eyes, and ears are not described here.

Cochlear nucleus

The cochlear nucleus is composed of a complex of cell groups on the lateral surfaces of the medulla within which all auditory nerve fibers terminate. Entering auditory nerve fibers bifurcate in an orderly way, sending an ascending branch to innervate cells in the *anteroventral cochlear nucleus* (AVCN) and a descending branch to innervate neurons in the *posteroventral* (PVCN) *and dorsal cochlear nuclei* (DCN). The ascending auditory pathways coming out of the cochlear nucleus to the lower brain stem show a bewildering combination of bilateral and unilateral connections. For example, the AVCN projects

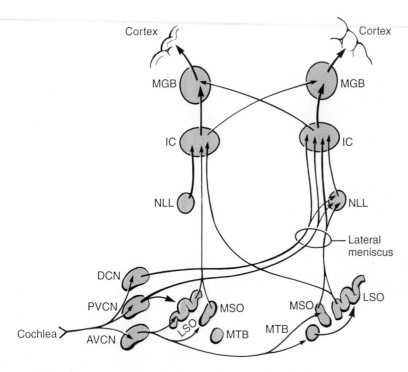

Figure 25-11
The ascending (afferent) neuronal chain from the cochlea to the cortex: spiral ganglion cell, cochlear nucleus neuron, superior olivary complex neuron, inferior colliculus neuron, and medial geniculate neuron. Notice the many crossed pathways that allow interactions between the outputs of the two ears. *AVCN,* anteroventral cochlear nucleus; *DCN,* dorsal cochlear nucleus; *IC,* inferior colliculus; *LSO,* lateral superior olive; *MGB,* medial geniculate body; *MSO,* medial superior olive; *MTB,* medial nucleus of trapezoid body; *NLL,* nucleus of lateral lemniscus; *PVCN,* posteroventral cochlear nucleus.

bilaterally to the medial superior olive but only unilaterally to the ipsilateral lateral superior olive. From medial superior olive, there is an ipsilateral projection to the inferior colliculus, but the lateral superior olive projects to the inferior colliculi of both sides. These particular connections do make sense functionally.

Superior olivary complex

The superior olivary complex lies in the tegmentum of the pons and consists of several subdivisions. One of the most significant aspects of the superior olivary complex is that it is the first point at which outputs from the two ears converge. Evidence shows that the different subdivisions of the superior olivary complex are involved in different aspects of the processing of the acoustic signal, especially for binaural processing. The superior olivary complex projects to the inferior colliculi of both sides by way of the lateral lemniscus and to the nuclei of the lateral lemniscus. The superior olivary complex also is the source of the efferent projection back to the cochlear hair cells.

Inferior colliculus

The inferior colliculus comprises the caudal pair of protuberances that make up the roof of the midbrain. Although the superior colliculus is primarily visual in function, the inferior colliculus is an important auditory relay station. It receives input from the cochlear nuclear complex, superior olivary complex, nuclei of the lateral lemniscus, and inferior colliculus of the opposite side. All fibers ascending in the *lateral lemniscus* synapse in the inferior colliculus.

Many neurons in the inferior colliculus, like those in the superior olivary complex, are sensitive to extraordinarily small changes in the time of arrival of the stimulus at the two ears or small differences in interaural intensity. The inferior colliculus is also tonotopically organized. Although the inferior colliculus sends its axons centrally only as far as the medial geniculate body, it receives an impressive descending projection from auditory cortex. The output of the inferior colliculus travels by way of the brachium of the inferior colliculus to innervate the medial geniculate body of the thalamus on the same side, and there is a smaller contralateral projection.

Medial geniculate body

The medial geniculate body represents the thalamic relay for auditory information. Its neurons project in an orderly fashion to the auditory areas of the cerebral cortex by way of the sublenticular portion of the internal capsule. The auditory cortex projects back on the medial geniculate body in a highly organized way. The medial geniculate body is tonotopically organized.

Auditory cortex

The cortical auditory receiving areas in humans and other primates is located on the dorsal surface of the superior temporal lobe. In humans, it occupies one or more of the transverse gyri of Heschl. The auditory cortex is not uniform in its cellular architecture. It consists of a primary receiving area which is tonotopically organized and which has a relatively uniform cellular structure throughout. Several other auditory cortical fields can

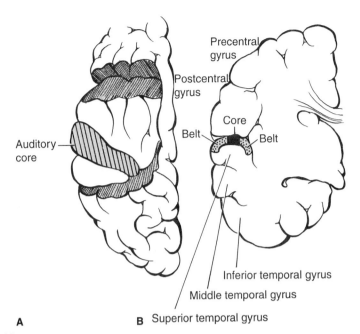

Figure 25-12
The human auditory cortex in the first transverse gyrus of the temporal lobe. (**A**) The parietal lobe has been removed to reveal the superior temporal plane as seen from above. (**B**) A frontal section passes through the auditory cortex, showing the core area (AI) buried in the sylvian fissure, surrounded by a belt area of auditory association cortex. (Redrawn from Neff WD, Diamond IT, Casseday JH. Behavioral studies of auditory discrimination: central nervous system. In Keidel WD, Neff WD (eds). Handbook of sensory physiology: auditory system: physiology (CNS): behavioral studies: psychoacoustics. Berlin: Springer-Verlag, 1975:307.)

be identified surrounding the auditory core (Fig. 25-12). Each auditory cortical area has complex afferent and efferent connections with the thalamus, with nearby cortical fields, and with auditory areas in the opposite hemisphere. The functional significance of multiple cortical areas is unknown, although it has been suggested that each processes a different aspect of the acoustic stimulus.

☐ Central auditory neurons show some common general properties

A microelectrode inserted into the auditory nerve or a central auditory nucleus records trains of action potentials that are evoked by sounds reaching the ears. This method has enabled study of the way in which acoustic information is encoded in trains of nerve spikes and in the transformations that take place in these spike trains at successive levels of the auditory system. Many of the response properties of auditory nerve fibers are also observed in neurons throughout the auditory pathway. However, numerous transformations in the sound-evoked discharge take place at each successive synaptic station as the result of convergence of excitatory and inhibitory activity from various sources.

One obvious transformation is illustrated in Figure 25-11. Notice that the projection from the cochlear nucleus to the superior olivary complex is bilateral, as is the pro-

jection from the superior olivary complex to the inferior colliculus. This means that, starting from the superior olivary complex, most auditory neurons respond to stimulation of either ear. The nature of binaural interactions is discussed later.

In general, most central auditory neurons have a best or characteristic frequency. Their response areas and tuning curves share similarities with those of auditory nerve fibers, but many differences are also apparent. One general property found in many central auditory neurons is the presence of *inhibitory sidebands* in the response area. Sidebands are presumably the result of inhibitory circuits that allow cells responding to a particular frequency to inhibit the responses of neighboring cells that are tuned to neighboring frequencies. These sidebands limit the response areas of single cells to narrower frequency ranges and are reminiscent of the center-surround receptive fields of visual neurons.

Another common characteristic of all the major auditory nuclei is a tonotopic organization. The distribution of best frequencies of single neurons can be studied systematically by microelectrode recording techniques. The resultant three-dimensional map is an orderly arrangement of best frequencies that is referred to as *tonotopic organization*. This organization is a reflection of the preservation of the orderly projection of eighth cranial nerve fibers from the cochlea to the CNS. Because frequency representation is related to the spatial distribution of points of maximal displacement along the basilar membrane, this organization also can be referred to as cochleotopic. It can be viewed in much the same way as retinotopic organization in the visual system and somatotopic organization in the somatic sensory system.

The information is transformed in the cochlear nucleus

The fibers of the cochlear nerve terminate in an orderly way within the three cochlear nucleus subdivisions, preserving the innervation pattern along the basilar membrane. The result is that each subdivision is tonotopically organized.

The morphology of the cells and auditory nerve terminals within each subdivision determine the kind of information that is relayed to higher auditory centers. Cells in the PVCN and DCN receive button endings from auditory nerve fibers and from interneurons and higher centers. This neuronal network results in interactions of excitation and inhibition that are reflected in the complex discharge patterns of single neurons, which differ considerably from the incoming volleys in eighth cranial nerve fibers. Some bushy cells in AVCN receive very specialized endings (e.g., *end bulbs of Held*) from just a few auditory nerve fibers. Figure 25-13 shows one of these end bulbs making contact with a "bushy cell" in the AVCN. Notice the unique morphology of the synaptic ending, which ensures that an action potential on the auditory nerve fiber is transmitted with great temporal fidelity to the AVCN. As a result, the activity of these bushy cells has been found to be similar to the activity in incoming auditory nerve fibers. In this case, the bushy cells serve as relay cells, with little transformation of the input. The end bulb on bushy cells is an example of a unique specialization in the auditory system, which is thought to preserve timing information.

An important function of the bushy cells in the AVCN is to relay the precise low-frequency time information carried by auditory nerve fibers. This time information is relayed primarily to the superior olivary complex, which is the first region in the brain where outputs from the two ears converge. On the other extreme are cells in the DCN, which only respond to acoustic inputs with a single spike or with some complex pattern of discharges that bear no simple relation to the acoustic stimulus. Some intrinsic circuitry in the DCN or the membrane properties of these cells are responsible for a different kind of transformation. We do not understand the reasons for all of these different transformations, and the only one addressed in this chapter is that of the bushy cell in the AVCN. Already at the first relay station, the auditory and visual systems manifest different processing schemes. Most of the lateral geniculate neurons cells relay the signals from the ganglion cells, but only a subset of the cochlear nucleus cells can be considered to be relay cells.

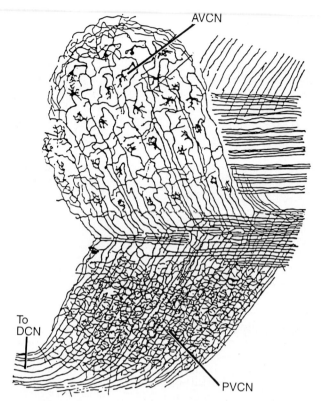

Figure 25-13
Axons of the cochlear nerve entering ventral cochlear nuclei. Each axon branches to send an ascending branch to the anteroventral (AVCN) nucleus, which is characterized by specialized end bulbs and limited convergence from the cochlea, and a descending branch to the dorsal (DCN) and posteroventral (PVCN) nuclei. Notice the difference in the morphology of the terminals and the orderly, cochleotopic arrangement of fibers. (Adapted from Histologie du systeme nerveux, vol 1. Madrid: Instituto Ramon y Cajal, 1909.)

Binaural interactions are important for sound localization

In analyzing the anatomic connections of the central auditory system, it has been emphasized that some cells receive input from both ears. What are the advantages of having two ears, instead of one? If one ear is plugged, it is clear that many acoustic tasks are not affected. It is still possible to understand speech, perceive music, and discriminate sounds of different pitch. However, there are certain tasks that are much more difficult to perform with only one ear. Two tasks especially depend on binaural hearing: the ability to localize the source of sound and the ability to differentiate one sound from background noise. Psychophysical tests have demonstrated that localizing a sound source along the horizontal plane depends largely on the differences in the sound reaching the two ears. Related to this is the ability to attend to a single sound source in noisy environment, which is commonly referred to as the "cocktail party" phenomenon. The latter ability is a common complaint of patients suffering from presbycusis (i.e., progressive, bilateral, symmetric hearing loss in the elderly).

There are two important cues for sound localization: *interaural level differences* (ILDs) and *interaural time differences* (ITDs), as shown in Figure 25-14). A listener employs one or both cues in localizing a sound source. For low-frequency tones (below about 1000 Hz), a listener relies on ITD cues for sound localization. The ability to localize tones becomes progressively poorer up to about 3000 Hz, at which point it begins to improve. For higher-frequency tones, the listener relies on ILDs. This dichotomy between the cues used for localizing high- and low-frequency tones is known as the *duplex theory of sound localization*, which is valid only for pure tones. The localization of complex sounds, which contain both high and low frequencies, uses both mechanisms. Because fundamentally different processes are necessary for encoding ITDs and ILDs, this processing could be expected to use different anatomic structures and physiologic mechanisms. This is the case in the superior olivary complex: some cells are important for encoding ITDs and others encode ILDs.

Acoustically, the ears can be regarded as a pair of holes separated by a spherical obstacle. Sound on one side of the head reaches the farther ear some 30 microseconds later for each additional centimeter it must travel. If we assign a radius of about 20 cm to the spherical head, the maximal ITD is about 600 microseconds. This maximal

ITD occurs if the stimulus is located directly to one side; it is as far as possible from the opposite ear.

The ability of the auditory system to detect ITDs is quite remarkable. Under optimal conditions, the just noticeable difference in ITD in humans is 6 microseconds. This is accomplished in a nervous system that uses action potentials whose width and synaptic delays are over 100 times longer! Animals specialized for acoustic communication can do even better; bats can detect interaural arrival times of echoes of less than 1 microsecond. Preserving timing information is clearly an important task for the central auditory system. How does the auditory system achieve this discrimination? The end bulbs of Held are one anatomic specialization for preserving timing information.

ILDs are not so simply behaved. The far ear lies in a sound shadow whose depth depends on the direction and wavelength of the sound. The head acts as an effective acoustic shadow only if the wavelength of the sound is smaller than the size of the head. The head can act as an effective acoustic shadow only at higher frequencies. For the size of human heads, interaural intensity differences are negligible at frequencies below about 1000 Hz and may be as great as 20 dB at 10 kHz.

For several of the auditory nuclei, spatial maps of ITDs or ILDs have been described, creating neural maps of the acoustic environment. Such maps are fundamentally different from the topographic maps of the body and retinotopic maps of the visual world seen at all levels of the somatosensory and visual systems. Auditory space does not map onto the cochlea, which differentiates it from the sensory maps of somesthesia and vision that are direct projections of the sensory surface. The auditory map is said to be a *computational map*, because it must be derived from neuronal processing of the cues, the ITDs and ILDs, that produce the map.

Studies of the encoding of sound localization cues in the central auditory system have shown that, above the level of the superior olivary complex, the representation of auditory space is primarily of the contralateral sound field. Patients with lesions of the auditory cortex, for example, are unable to localize sounds originating from the contralateral sound field, without deficits in detection or discrimination of the sound. The auditory system derives a map of the contralateral field just as the visual and somatosensory maps in the CNS are of the contralateral visual field and body. The contralateral representation in the auditory system is derived from the particular way in which some fibers cross and others remain uncrossed in the ascending projection from the cochlear nucleus to the inferior colliculus. There is order behind the seemingly haphazard interconnections of the various auditory nuclei.

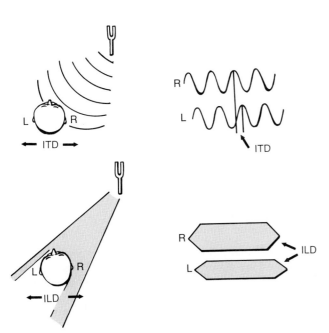

Figure 25-14
The drawings illustrate interaural time differences (ΔT) for a low–frequency tone (*top*) and intensity differences (ΔI) for a high–frequency tone (*bottom*) when a sound source is off the midline. Those on the left are schematic diagrams of the physical situation; those on the right are the signals received at the right (*R*) and left (*L*) ears. For the high–frequency tone, only the envelope of the sine curve is shown.

☐ The physiology of the auditory system is reflected clinically

Brain stem auditory evoked response

One of the clinically important consequences of the preservation of temporal information by the peripheral and central auditory system is that it allows a record of the *brain stem auditory evoked potential* (BAEP). This record consists of complex, time-locked, electrical slow waves, which can be recorded from the scalp (i.e., noninvasively, like electroencephalographic patterns) of human subjects in response to a brief, abrupt sound, such as a click or tone burst (Fig. 25-15). These electrical potentials are extremely low in amplitude (some <1 μV) and are usually lost within the higher-amplitude spontaneous activity recorded on the electroencephalogram (EEG). However, by using computer signal-averaging techniques, it is possible to extract from the EEG a well-defined complex waveform whose peaks and troughs can be associated with activity in various parts of the auditory system.

Many recognized components of an auditory evoked potential have been somewhat arbitrarily divided into three epochs: early-, middle-, and long-latency components. The early components, defined as those occurring during the first 8 milliseconds after stimulus onset, are thought to represent activation of the cochlea and auditory nuclei of the brain stem, and it is these early components that are of interest clinically. The middle and late components represent a mixture of activity in the brain stem and higher auditory centers and in the association cortex.

During the past few years, the early components have been employed in evaluating hearing loss and in helping to diagnose trauma or disease involving the brain stem. This technique is proving to be important in assessing the hearing of persons unwilling (e.g., malingerers) or unable (e.g., newborns, mentally retarded) to cooperate in routine audiometric testing. The ability to identify the activity of specific auditory nuclei by the presence of characteristic waves of the BAEP is of great clinical value. Many of the children who had previously been thought to be "slow learners" or developmentally disabled were instead found to be suffering from hearing loss. The absence or degradation of acoustic input during the first few years, when the recognition and expression of speech is so critical, can cause permanent developmental damage, which can be prevented by restoring the acoustic experience.

The reason that such analysis can be done in the auditory system and not in, for example, the visual system, is that the specialization within the auditory system to preserve timing information of the stimulus results in a synchronous activation of many cells at each of the early brain stem nuclei. This synchrony is able to be detected by computer averaging even when recording from the scalp of a subject.

Conductive and sensorineural deafness

Anything that interferes with the transmission of sound to the hearing apparatus or of the neural signal to the auditory cortex can produce deafness. There are two major categories of deafness: conductive and sensorineural deafness. *Conduction deafness* is caused by a malfunction in the transmission of sound through the middle ear. A common example, particularly in young children, is the buildup of fluid in the middle ear because of infection or otitis media. Treatment with antibiotics has greatly reduced the incidence of conductive hearing loss from otitis media. In adults, otosclerosis is the most frequent cause of conduction deafness, in which there is overgrowth of the labyrinthine bone around the oval window,

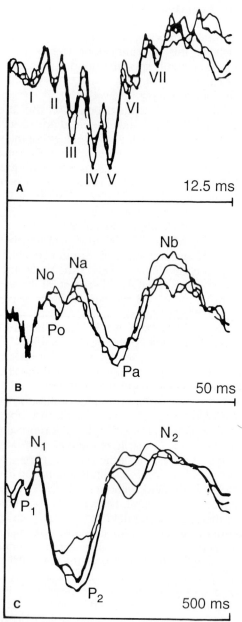

Figure 25-15
Scalp-evoked potentials shown on three different time scales. (**A**) Auditory brain stem responses. (**B**) Middle components. (**C**) Late components. Electrodes are vertex to mastoid (positive up). Each trace shows the average of 1024 clicks (60 dB SL) delivered at 1/sec to the right ear. Labels shown are commonly used for individual wave components. (Redrawn from Picton TW, Hillyard SA, Krausz HI, Galambos R. Human auditory evoked potentials, I: evaluation of components. Electroenceph Clin Neurophysiol 1974:36;179.)

leading to fixation of the stapes. This condition can often be treated by surgical intervention to loosen or replace the stapes. Boosting the input to the ear with a hearing aid alleviates many cases of conduction deafness.

Sensorineural or nerve deafness is caused by damage to the cochlea, to the auditory nerve, or to the central auditory system. There are many causes of sensorineural deafness. The hair cells of the inner ear are especially susceptible to damage from prolonged exposure to loud sounds, antibiotics (e.g., streptomycin, kanamycin, gentamicin), or rubella infection in utero. The most common type of hearing loss in the aged, presbycusis, is characterized by a progressive loss of high frequencies, which is probably caused by progressive changes in the cochlea. In the central auditory system, any damage to the pathways leading from the cochlear nucleus to the auditory cortex can result in deafness. Acoustic neuroma, a tumor that develops in the auditory nerve in the internal auditory meatus or at the the cerebellopontine angle, produces an ipsilateral deafness, tinnitus (i.e., sensation of ringing in the ears), or vestibular problems. Not much can be done about most cases of sensorineural deafness.

Because many cases of sensorineural deafness involve problems in the cochlea without involvement of the auditory nerve fibers or central connections, it may be possible to stimulate the auditory nerve fibers electrically. Much attention has been focussed in recent years on the development of a cochlear prosthesis in which an electrode is implanted directly into the cochlea. Electrical stimulation of the cochlear implant can activate the still-viable ends of auditory nerve fibers, and some sense of hearing can be restored.

It is important to differentiate conductive from sensorineural deafness, because many types of conductive deafness can be treated. This is particularly important in young children during the critical time during which they are learning to speak. Two tests with tuning forks are helpful. They both rely on the fact that the middle ear can be bypassed when the sound is delivered directly to the bone. Air conduction usually is much more efficient than bone conduction, but in cases of conductive deafness, bone conduction becomes more efficient. *Weber's test* can be used when a patient complains of deafness in one ear. A vibrating tuning fork (usually 512 Hz) is applied to the forehead at the midline, and the resulting vibrations are conducted through the bone to each ear. Patients with conduction deafness report that the sound is heard louder in the deaf ear, presumably because bone conduction is equal in the two ears but the deaf ear also hears no background noise from the air. Patients with sensorineural deafness cannot hear the sound even through bone conduction and report that it is louder in the normal ear. In *Rinne's test*, the tuning fork is applied to the mastoid process. As soon as the sound ceases, the fork is moved to just outside the auditory meatus. In a normal ear, the fork is heard again when the sound waves travel by air conduction, because it is more efficient than bone conduction. In conduction deafness, the fork's vibration remains imperceptible when held next to the ear. In sensorineural deafness, the reverse is true, although the thresholds for air and bone conduction are increased.

Effects of lesions in auditory cortex

Lesions of the auditory cortex make up a special case of sensorineural deafness. Many behavioral experiments have been conducted using cats, and extensive damage to these areas has little effect on the animal's ability to discriminate between tones of different frequencies or intensities. The major effect appears to be on the animal's ability to localize the source of a sound in space and to discriminate between complex sound patterns.

There have been numerous reports in the literature on the effects of temporal lobe damage in humans. Unfortunately, in only a few cases have complete audiometric tests been made and postmortem examination of the brain carried out. It is not surprising that reports have conflicted on the effects of temporal lobe damage on hearing in humans. However, several effects usually are detected in humans after auditory cortical damage. After a period following bilateral lesions of the auditory cortex, the pure tone hearing threshold appears to return toward normal, and there is usually little or no change in threshold. There is also little or no change in discrimination of alterations in intensity or frequency. Deficiencies are detected more commonly for the discrimination of changes in the temporal order or sequence of sounds, their duration, or in the ability to localize sounds in space.

SELECTED READINGS

Altschuler R, Hoffman D, Bobbin D, Clopton B. Neurobiology of hearing, vol III. The central auditory system. New York: Raven Press, 1991.

Edelman GM, Gall WE, Cowan WM (eds). Auditory function: neurobiological bases of hearing. New York: John Wiley & Sons, 1988.

Irvine DRF. Progress in sensory physiology, vol 7. The auditory brain stem. Berlin: Springer-Verlag, 1986.

Neuroscience in Medicine, edited by P. Michael Conn.
J. B. Lippincott Company, Philadelphia © 1995

Chapter 26

The Vestibular System

ROBERT F. SPENCER

Vestibular Pathways
   Vestibular Receptors in the
      Semicircular Canals and Otolith
      Organs are Innervated by Bipolar
      Neurons in Scarpa's Ganglion

Central Vestibular Connections and
   Pathways are Related to the Control
   of Gaze, Posture, and Balance

Nystagmus

Vestibular Pathways

☐ Vestibular receptors in the semicircular canals and otolith organs are innervated by bipolar neurons in Scarpa's ganglion

The vestibular portion of the membranous labyrinth consists of the three *semicircular canals*, the lateral, anterior, and posterior canals, and the *otolith organs*, the *utricle* and the *saccule* (Fig. 26-1). Like the cochlea, these structures are filled with endolymph, which communicates with that of the cochlea through the ductus reuniens. The lateral, anterior, and posterior semicircular canals are oriented almost orthogonal to one another and represent the three dimensions of space. Rotational movements about the x-, y-, and z-axes are classified as roll, pitch, and yaw, respectively (Fig. 26-2). When the head is tilted approximately 25° forward, the horizontal canal is in the horizontal plane, and the anterior and posterior canals are in the vertical plane. The utricle is oriented almost horizontally, and the saccule is oriented almost vertically in the sagittal plane. The receptors for the vestibular system, which are hair cells similar to those in the organ of Corti, are located in specialized regions of these structures: the crista ampullares of the semicircular canals and the maculae of the utricle and the saccule.

A *crista ampullaris* is in the widened (i.e., ampullated) end of each of the three semicircular canals. The neurosensory epithelium of each crista ampullaris is composed of supporting cells and two types of hair cells. Projecting from each hair cell are *stereocilia* (40 to 110 per cell) and a single *kinocilium* (Fig. 26-3). The cilia are embedded in a gelatinous *cupula* that extends to the roof of the ampulla. The stereocilia for each hair cell have graded lengths and are polarized toward the kinocilium. All hair cells in the crista ampullaris of a single canal are arranged in the same fashion, with all kinocilia in the same ampulla polarized to the same side. In the horizontal canals, the polarization of kinocilia is toward the utricle (i.e., utriculopetal), and in the vertical canals, the polarization of the kinocilia is away from the utricle (i.e., utriculofugal). The hair cells in the cristae of the semicircular canals are stimulated by an appropriate *angular acceleration* that causes movement of the endolymph, with a resulting deflection of the cilia.

The utricle and the saccule each contain a *macula* (Fig. 26-4). The neurosensory epithelium of each of the maculae consists of hair cells and supporting cells, the hair cells having bundles of stereocilia and a single kinocilium. These cells are embedded in an otolithic membrane, which is a gelatinous matrix that contains *statoconia* (calcium carbonate crystals). The hair cells in the maculae of the utricle and the saccule are stimulated by an appropriate *linear acceleration* that produces deflection of the cilia by the overlying otoliths.

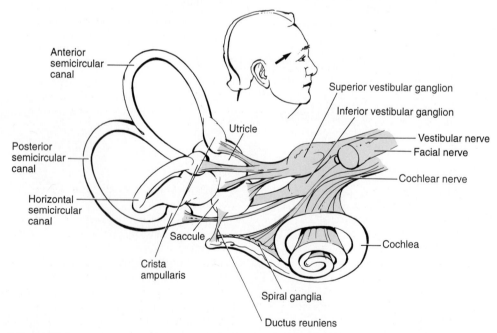

Figure 26-1
Components of the membranous labyrinth: the three semicircular canals (i.e., horizontal, anterior, and posterior), the otolith organs (i.e., saccule and utricle), and the cochlea. The spiral-shaped cochlea winds around a central bony modiolus. Bipolar neurons in the spiral ganglion innervate the hair cells in the organ of Corti, and their central processes form the cochlear nerve. Peripheral processes of bipolar neurons in the superior and inferior vestibular (Scarpa's) ganglia innervate hair cells in the cristae ampullaris of the semicircular canals and the maculae of the saccule and utricle, and their central processes form the vestibular nerve. The arrow indicates the plane of the horizontal semicircular canal.

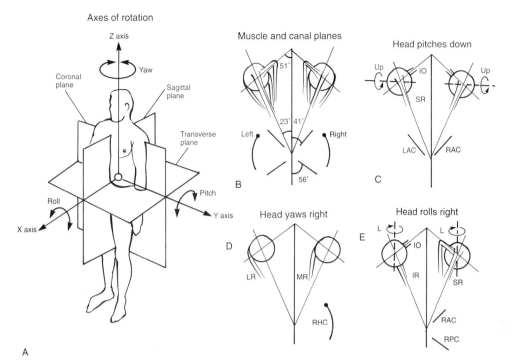

Figure 26-2

Compensatory eye movements produced by movements of the head in the vestibulo-ocular reflex. (**A**) The planes of reference (i.e., coronal, transverse, sagittal) are associated with the principal axes of rotation (i.e., roll in the x-axis, pitch in the y-axis, yaw in the z-axis) with respect to the center of gravity (c.g.). (**B**) Angular relation between the orientation of the semicircular canals and the pulling actions of the vertical extraocular muscles. The orientation of the anterior canal approximates the pulling action of the superior rectus muscle, and the orientation of the posterior canal approximates the pulling action of the superior oblique muscle. The orientation of the horizontal canal approximates the pulling actions of the lateral rectus and medial rectus muscles. (**C, D, E**) Activation of different pairs of extraocular muscles resulting from rotation in different axes. (**C**) When the head is pitched down, the anterior canals are excited bilaterally, producing a compensatory vertical upward eye movement that results from activation of inferior oblique and superior rectus motor neurons. (**D**) A yaw movement to the right excites the right horizontal canal and inhibits the left horizontal canal, which produces a compensatory eye movement to the left resulting from activation to the left lateral rectus and right medial rectus muscles. (**E**) A roll movement to the right co-activates the anterior and posterior canals on the right side, producing a compensatory torsional eye movement achieved by activating the inferior oblique and inferior rectus muscles in the left eye and the superior oblique and superior rectus muscles in the right eye.

The macula of the utricle is located on the floor of the utricle. When the head is upright, the macula is oriented almost parallel with the ground. The posterior end is horizontal, while the anterior end is elevated about 45° from horizontal. As in the crista ampullares of the semicircular canals, the hair cells in the utricular macula are polarized, but they have a different arrangement (see Fig. 26-4). Half of the macula is stimulated by displacement of the otoliths in one direction, but the other half is inhibited, and the opposite is true for the opposite direction. The utricular macula is the organ stimulated by vertically directed linear forces, particularly by the forces of gravity during tilting of the head.

The macula of the saccule is an elongated structure that lies vertical to the ground when the head is upright (see Fig. 26-4). The vertical orientation of the macula and the roof-like arrangement with which the hair cells are polarized suggest that the statoconia are displaced optimally by horizontally directed linear forces of the head in the dorsal-ventral direction.

The *hair cells* in the cristae ampullares of the semicircular canals and the maculae of the saccule and utricle are similar to those in the cochlea. There are two morphologic types of hair cells (see Fig. 26-3). *Type I hair cells* are spherical or goblet-shaped cells and are innervated by a large chalice-like afferent ending that encompasses most of the cell. These cells are innervated by large-diameter afferent axons and typically respond to stimulation with a phasic discharge. *Type II hair cells* are smaller and cylindrical, and afferent and efferent endings are distributed at the base of the cell. These cells are innervated by small-diameter afferent axons and exhibit tonic activity. Like in the cochlea, vestibular hair cells are activated or inhibited by deflection of the cilia. In the crista ampullaris, the cilia are displaced by movement of the cupula caused by the flow of the endolymph through

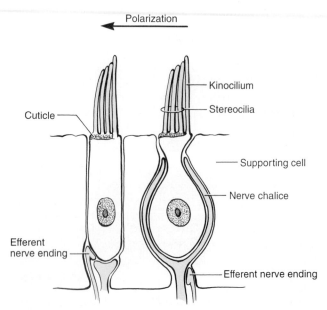

Figure 26-3

Two principal types of hair cells in the cristae ampullaris of the semicircular canals and the maculae of the saccule and utricle. Type I hair cells are goblet shaped and are innervated by chalice-type nerve endings that arise from large-diameter axons of bipolar neurons in the vestibular (Scarpa's) ganglion. Type II hair cells are columnar and have simple nerve endings associated with small-diameter axons at the base of the cells. Vestibular efferent nerve endings establish synaptic connections with the primary afferent ending on type I hair cells and directly on the base of type II hair cells. Both types of hair cells have an array of stereocilia that are polarized according to increasing length in the direction of the kinocilium. In the crista ampullaris, the stereocilia and kinocilia are embedded in the gelatinous cupula, and in the macula, they contact the crystalline statoconia matrix. The structural and innervational differences between the two types of hair cells are correlated with differences in their physiologic activity. Type I hair cells are phasic, and type II hair cells are tonic.

the canals when the head moves. In the maculae, changes in the position of the head produce a shearing effect of the otoliths on the cilia. In both cases, the hair cells are activated when the cilia are displaced toward the direction of the polarization of the kinocilium and are inhibited with displacement in the opposite direction.

The mechanism of activation of the vestibular hair cells is the same as that described for the cochlear hair cells. With deflection of the cilia toward the kinocilium, the hair cells are depolarized by an influx of K^+ from the surrounding endolymph, which activates voltage-sensitive Ca^{2+} channels. The influx of Ca^{2+} produces the release of neurotransmitter, which causes activation of the primary vestibular afferent terminal. Because the hair cells normally have a small influx of K^+ at rest, deflection of the cilia in the opposite direction closes the cation channels, resulting in a hyperpolarization of the cell and decreased neurotransmitter release.

The neural innervation of hair cells in the semicircular canals and the otolith organs is provided by *bipolar neurons* in the vestibular (Scarpa's) ganglion, which lies at the base of the internal auditory meatus. Peripheral processes from neurons in the superior division of the ganglion synapse in relation to hair cells in the anterior vertical and horizontal semicircular canals and the maculae of the utricle and part of the saccule. Bipolar neurons in the inferior division of the ganglion innervate hair cells in the posterior vertical canal and most of the macula of the saccule. The central processes of the bipolar neurons form the superior and inferior branches of the vestibular portion of the vestibulocochlear or eighth cranial nerve.

The information that is transmitted by vestibular nerve axons is related to head velocity and head position. This information is encoded in the frequency of discharge of primary vestibular axons in response to different velocities of movement. Although a topographic organization,

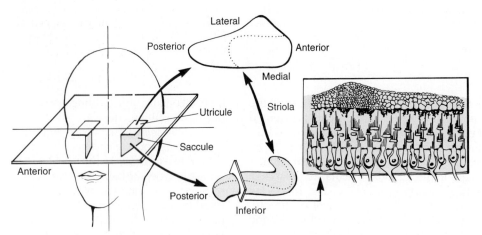

Figure 26-4

Spatial orientation of the utricle and saccule. The maculae of the utricle and saccule are curved, plate-like structures on which the hair cells are polarized in opposite directions in relation to the striola, which corresponds to the maximal thickness of the overlying otoconial layer. In the macula of the utricle, which is oriented approximately parallel to the ground, the hair cells are polarized toward the striola on each side. In the macula of the saccule, which is oriented approximately perpendicular to the ground in the sagittal plane, the hair cells are polarized away from the striola. The shape of the maculae of the utricle and saccule and the polarizations of their respective hair cells effectively provide a means of responding to linear accelerations in three dimensions.

like that in other sensory systems (e.g., somatotopic, retinotopic, tonotopic), is not apparent in the vestibular system, there is evidence that information is processed centrally through frequency-tuned channels that ultimately may be important for the recruitment of motor neurons in the control of eye and head movements.

□ Central vestibular connections and pathways are related to the control of gaze, posture, and balance

Vestibular nuclei

The vestibular nerve courses through the cerebellopontine angle medial to the cochlear nerve and enters the brain stem dorsally between the inferior cerebellar peduncle and the spinal trigeminal tract. Most vestibular fibers terminate in one or more of the vestibular nuclei (Fig. 26-5). These nuclei lie in the floor of the fourth ventricle and extend from a level rostral to the hypoglossal nucleus to slightly beyond the level of the abducens nu-

cleus. The nuclei of the vestibular complex are arranged in two longitudinal columns: the lateral column consists of the inferior (i.e., spinal or descending) vestibular nucleus, the lateral vestibular nucleus of Deiters, and the superior vestibular nucleus of Bechterew. The medial vestibular (i.e., triangular) nucleus of Schwalbe constitutes the medial cell column.

On entering the vestibular complex, most primary fibers bifurcate into ascending and descending rami. In general, primary vestibular fibers that innervate the cristae ampullaris of the semicircular canals project primarily to rostral portions of the vestibular complex, including the superior vestibular nucleus, ventral portions of the lateral vestibular nucleus, and rostral portions of the medial vestibular nucleus. The primary vestibular fibers that innervate the maculae of the saccule and the utricle terminate in caudal portions of the medial and inferior vestibular nuclei.

Some primary vestibular fibers bypass the vestibular nuclei and ascend to the cerebellum by means of the *juxtarestiform body* (Fig. 26-6). Cells in all parts of the vestibular ganglion send mossy fiber projections to the ipsilateral *nodulus* and *uvula* of the vermis and to the ipsilateral *flocculus*.

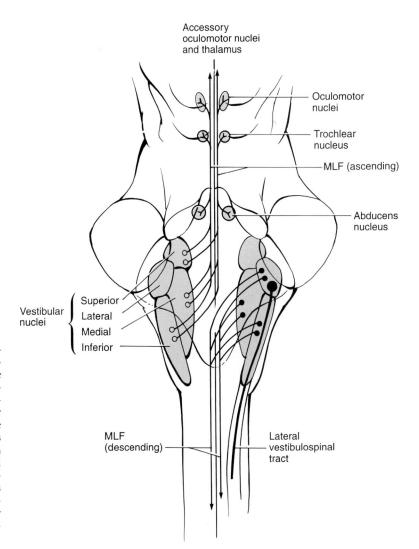

Accessory oculomotor nuclei and thalamus

Oculomotor nuclei

Trochlear nucleus

MLF (ascending)

Abducens nucleus

Vestibular nuclei
{
Superior
Lateral
Medial
Inferior
}

MLF (descending)

Lateral vestibulospinal tract

Figure 26-5
The ascending (*left*) and descending (*right*) vestibular pathways. Ipsilateral inhibitory and contralateral excitatory ascending projections from the vestibular nuclei course through the medial longitudinal fasciculus (MLF) and target motor neurons in the abducens, trochlear, and oculomotor nuclei. The ascending connections mediate the vestibulo-ocular reflex. Bilateral descending fibers in the MLF project primarily to motor neurons in the cervical spinal cord that innervate the dorsal neck muscles, forming the basis for the vestibulocollic reflex. The lateral vestibulospinal tract arises from the lateral vestibular nuclei and descends ipsilaterally in the spinal cord, targeting primarily motor neurons that innervate axial extensor muscles that maintain posture.

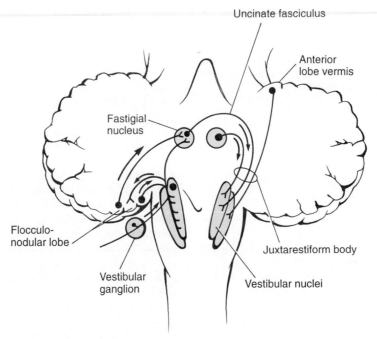

Figure 26-6

Vestibulocerebellar and cerebellovestibular connections. First-order fibers from bipolar neurons in the vestibular ganglion and second-order fibers from neurons in the vestibular nuclei project to the flocculus, nodulus, and uvula of the cerebellar cortex (i.e., vestibulocerebellum) and to the fastigial nucleus by way of the juxtarestiform body. Purkinje cells in the cerebellar cortex project directly and indirectly by the fastigial nucleus to the vestibular nuclei through the juxtarestiform body. Purkinje cells in the vermis of the anterior lobe of the cerebellum (i.e., spinocerebellum) project directly through the juxtarestiform body to neurons in the lateral vestibular nucleus that are the cells of origin of the lateral vestibulospinal tract.

Second-order vestibular connections

Unlike the auditory system and most other sensory systems, the central connections of the vestibular system are related primarily to motor behaviors involving the maintenance of balance and compensatory movements in response to changes in head and body position. This function is achieved by connections of the vestibular nuclei with motor neurons in the extraocular motor nuclei, the spinal cord, and the cerebellum.

Vestibulocerebellar fibers. The cerebellum has a major role in the integration of vestibular and visual information. In addition to first-order vestibular ganglion neurons, some neurons in the superior, medial, and inferior vestibular nuclei project their axons through the juxtarestiform body bilaterally to the nodulus, uvula, and flocculus of the cerebellar cortex and to the *fastigial nucleus* (see Fig. 26-6). Reciprocal connections with the cerebellum are achieved by cerebellovestibular fibers from the fastigial nuclei of both sides and from Purkinje cells in the ipsilateral flocculus that course through the juxtarestiform body to terminate in all four vestibular nuclei. In this manner, the cerebellum exerts an overall inhibitory influence on the entire vestibular complex.

Medial longitudinal fasciculus. The superior, medial, and inferior vestibular nuclei are the main sources of fibers that comprise the medial longitudinal fasciculus (MLF). Ascending fibers in the MLF from the superior, medial, and inferior vestibular nuclei project primarily to the oculomotor, trochlear, and abducens nuclei, which contain the motor neurons that innervate the extraocular muscles, and to accessory oculomotor nuclei (see Fig. 26-5).

Second-order vestibular fibers that cross the midline and ascend contralateral are excitatory, and those that are uncrossed and ascend on the ipsilateral side are inhibitory. The connections of second-order vestibular neurons with motor neurons in the extraocular motor nuclei are canal specific and are related to the spatial orientation of the semicircular canals and their alignment with the pulling actions of the extraocular muscles (see Figs. 15-1 and 26-2B). Excitatory second-order vestibular neurons that receive input from the posterior vertical semicircular canal project to inferior rectus motor neurons in the oculomotor nucleus and superior oblique motor neurons in the trochlear nucleus, controlling vertical, downward eye movements. Excitatory second-order vestibular neurons that receive input from the anterior vertical semicircular canal project to superior rectus and inferior oblique motor neurons in the oculomotor nucleus, controlling vertical, upward eye movements (see Fig. 26-2C). Excitatory second-order vestibular neurons that receive input from the horizontal canal project directly to lateral rectus motor neurons in the abducens nucleus and indirectly by a relay neuron in the abducens nucleus to medial rectus motor neurons in the oculomotor nucleus, controlling horizontal eye movements (see Fig. 26-2*D*). Torsional eye movements are produced by co-activation of the anterior and posterior canals on opposite sides of the head, such as during roll movements (see Fig. 26-2*E*).

In almost all cases, the motor neurons that innervate the antagonistic muscles are inhibited in the same canal-specific manner. Although transection of the MLF rostral to the abducens nucleus abolishes primary oculomotor responses, labyrinthine stimulation still produces nystagmus. This finding is attributable to some fibers from the ventral lateral vestibular nucleus that course uncrossed to the oculomotor nucleus through the ascending tract of Deiters, which lies dorsolateral to the MLF in the pontine tegmentum. Some second-order vestibular fibers also may course rostrally through a third pathway in the ventral tegmentum.

Descending fibers in the MLF originate from the medial and inferior vestibular nuclei and establish synaptic connections predominantly with motor neurons in the cervical spinal cord that innervate muscles of the neck and control movements of the head (see Fig. 26-5). The descending limb of the MLF also is called the *medial vestibulospinal tract*, particularly with reference to fibers that project caudal to cervical levels. Some second-order vestibular neurons have axons that bifurcate and project to the extraocular motor nuclei and the cervical spinal cord.

Second-order vestibular fibers in the MLF are concerned with two general functions: maintained postural deviation of the eyes, which is under the control of the maculae of the saccule and utricle, and postural adjustments of the eyes and the head and neck. The postural adjustments are made in response to stimulation of the cristae ampullares of the semicircular canals brought about by angular acceleration (rotation) of the head and proprioceptive impulses arising in neck muscles and coursing through the spinovestibular tract to terminate in the inferior vestibular nucleus from which information is relayed to the motor nuclei of the extraocular muscles via the MLF.

The *vestibulo-ocular reflex* (VOR) is one example of vestibular function. The VOR stabilizes images on the retina by making a compensatory eye movement that is an accurate reproduction of a head movement. These compensatory eye movements are always *conjugate*; both eyes move simultaneously in the same direction (see Fig. 26-2*C–E*). *Linear acceleration* stimulates the otolith organs, producing compensatory conjugate vertical and torsional eye movements.

Vestibular commissural connections. The vestibular nuclei on one side of the brain stem are connected homotopically to the same nuclei on the contralateral side by direct and indirect excitatory and inhibitory commissural pathways. The direct commissural pathway courses through the dorsal tegmentum of the brain stem. Indirect connections use the vestibulocerebellar pathways and the *prepositus hypoglossi nucleus*, which is intimately related to the vestibular nuclei by extensive reciprocal connections. Vestibular commissural connections play a significant role in the phenomenon of *vestibular compensation*, which is a recovery of function from the deficits that are associated with peripheral lesions of the vestibular apparatus or nerve.

Vestibular efferent connections. Cholinergic neurons in the vicinity of the abducens nucleus project their axons into the vestibular nerve and establish efferent connections with the hair cells in the cristae ampullaris of the semicircular canals and the maculae of the saccule and the utricle. Synaptic connections are made with the primary afferent nerve endings on type I hair cells and directly with the cell bodies of type II hair cells (see Fig. 26-3). Little is known about the function of the vestibular efferent pathway. Because the synaptic connections of vestibular efferent neurons with vestibular hair cells is reminiscent of those established by efferent olivocochlear neurons with hair cells in the cochlea, it seems likely that the vestibular efferent pathway is capable of modulating the activity of the hair cells during movements of the head.

Vestibular-thalamic-cortical pathway. Some second-order fibers that project to the oculomotor and trochlear nuclei and the accessory oculomotor nuclei continue rostrally to terminate in the *ventral posterior inferior nucleus* of the thalamus, which is located in the vicinity of the ventral posterior somatosensory relay nuclei. The ventral posterior inferior nucleus projects to a distinct region of the postcentral gyrus, coextensive with Brodmann's area 3a. This region appears to be functionally distinct and is thought to be responsible for complex position and movement perception, possibly by integrating vestibular information with proprioceptive information from joint afferents and group I muscle afferents.

Lateral vestibulospinal tract. The lateral vestibulospinal tract arises from neurons that are located in the lateral vestibular nucleus (see Fig. 26-5). Only the ventral portion of the lateral vestibular nucleus receives direct primary vestibular input from the semicircular canals. The dorsal region of the nucleus has connections with the vermis of the anterior lobe of the cerebellum, which is the site of termination of the dorsal spinocerebellar tract. The projection from the anterior cerebellar vermis to the lateral vestibular nucleus is somatotopically organized such that cervical is represented rostral and lumbar is represented caudal in the nucleus. The lateral vestibulospinal tract descends *uncrossed* in the anterior funiculus of the spinal cord and establishes *excitatory* synaptic connections with motor neurons in lamina IX, primarily at cervical and lumbar levels of the spinal cord. The lateral vestibulospinal tract is involved in the regulation of posture by influencing motor neurons that innervate primarily axial *extensor muscles*.

The vestibulospinal fibers are necessary to the development of *decerebrate rigidity* induced by rostral midbrain (i.e., intercollicular) transection, which interrupts descending rubrospinal fibers that control flexor muscles. The rigidity is abolished by destruction of the lateral vestibular nucleus or by interruption of the lateral vestibulospinal tract. Decerebrate rigidity is strictly an antigravity reaction that is characterized by tremendously increased tone in the antigravity muscles, apparently caused by an increased rate of firing of muscle spindles by gamma motor neurons. The increased firing rate of muscle spindle afferents activates alpha motor neurons, which increase muscle tonus.

Nystagmus

Nystagmus is a rhythmic, oscillatory, involuntary movement occurring in one or both eyes in any or all fields of gaze. Nystagmus is always pathologic if it occurs spontaneously or is sustained. It may be induced in normal persons by two different methods: rotation or caloric stimulation.

Postrotatory nystagmus is the result of angular acceleration produced by rotating a person in a Barany chair at a rate of 10 revolutions in 20 seconds, followed by a sudden stop. The fast phase of nystagmus (i.e., the direction for which the nystagmus is named) is toward the side opposite the direction of rotation. The slow component (i.e., the active phase) is in the same direction as rotation. When the head is rotated, the initial inertia of the endolymph in the semicircular canals deflects the cupula in one direction (Fig. 26-7A). If the rotation is continued in the same direction, the fluid eventually obtains the same velocity of rotation as the canal and stimulation ceases (Fig. 26-7B). When the movement suddenly stops, the inertia of the endolymph continues its flow, with a renewed stimulation of the vestibular cupula, but in an opposite direction (Fig. 26-7C). The stimulation of the vestibular receptors gives rise to potentials transmitted to the vestibular nuclei, which are then relayed through the MLF to the appropriate motor nuclei of the extraocular muscles, initiating a conjugate eye movement. When the person after rotation is asked to touch a certain point with the

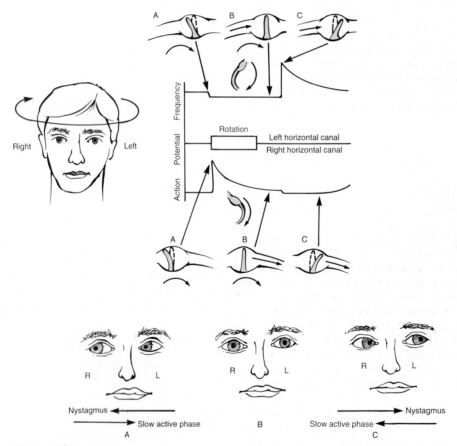

Figure 26-7
Principles of rotatory nystagmus at the start and during rotation and postrotatory nystagmus after cessation of rotation. Rotation of the head and body is to the right. (**A**) At the start of rotation, as the head and horizontal canals begin to move, the endolymph remains stationary and effectively causes excitation of the right horizontal canal, inhibition of the left horizontal canal, and nystagmus to the right (i.e., in the direction of rotation). (**B**) During rotation, the head, semicircular canals, and endolymph are moving at the same velocity, and the eyes remain stationary. (**C**) Immediately after stopping rotation, the head and the canals are stationary, but the endolymph continues to flow in the direction of rotation to the right, resulting in excitation of the left horizontal canal, inhibition of the right horizontal canal, nystagmus to the left (i.e., in the opposite direction of rotation).

eyes closed, the arm deviates in the opposite direction to the nystagmus (i.e., past pointing). There is also a tendency to fall in this direction. The direction of past pointing and the tendency to fall are in accord with the fact that the slow component of the nystagmus is the active phase. All phenomena tend to direct the rotated person to the same side.

Caloric nystagmus depends on the production of convection currents in the endolymph of the semicircular canals by extreme temperature stimuli. Because the horizontal canal responds best to caloric stimulation, the person sits erect with the head tilted backward at an angle of 60°. This brings the horizontal canal into a vertical position, where it is maximally influenced by the convection currents set up in the endolymph. Irrigating the ear with cool water (30°C) lowers the temperature of the prominence of the horizontal canal, condensing the endolymph and inducing a convection current in the direction of the cooling. This results in inhibition of the hair cells. The change in direction of the endolymph current causes the eyes to be turned slowly toward the side of the irrigated ear, the fast component of nystagmus directed toward the opposite side. If the ear is irrigated with warm water (40°C), the endolymph expands, and the resulting ampullopetal convection current, which causes stimulation of the hair cells, produces the reverse;

the slow component of nystagmus is away from the side of the irrigated ear, and the fast component toward the irrigated side. During caloric nystagmus, the endolymph is set in circulation by differences in temperature, with the warmer endolymph migrating upward, in the same direction as that in which the hair cells in the crista ampullaris are polarized.

SELECTED READINGS

Anniko M. Functional morphology of the vestibular system. In Jahn AF, Santos-Sacchi J (eds). Physiology of the ear. New York: Raven Press, 1988:457.

Gresty MA, Bronstein AM, Brandt T, Dieterich M. Neurology of otolith function. Brain 1992;115:647.

Markham CH. Vestibular control of muscular tone and posture. Can J Neurol Sci 1987;14:493.

Peterson BW, Richmond FJ. The control of head movements. New York: Oxford University Press, 1988:322.

Uemura T, Cohen B. Effects of vestibular nuclei lesions on vestibulo-ocular reflexes and posture in monkeys. Acta Otolaryngol Suppl 1973;315:5.

Waespe W. The physiology and pathophysiology of the vestibulo-ocular system. In Hofferberth B, Brune GG, Sitzer G, Weger HD (eds). Vascular brain stem diseases. Basel: Karger, 1990:37.

Wilson VJ, Jones MG. Mammalian vestibular physiology. New York: Plenum Press, 1979:365.

Neuroscience in Medicine, edited by P. Michael Conn.
J. B. Lippincott Company, Philadelphia © 1995

Chapter 27

*T*he *Gustatory System*

DAVID V. SMITH
MICHAEL T. SHIPLEY

The term *taste* is used to refer to the complex of sensations known as *flavor perception*, which includes afferent information from the olfactory, gustatory, and trigeminal systems. More strictly defined, taste refers to the sensations arising from stimulation of the gustatory receptors located within the oropharyngeal mucosa. Throughout this chapter, the terms taste and *gustation* are used interchangeably to refer to the gustatory system.

Taste information arises from stimulation by chemicals dissolved in saliva, which initiate the activation of receptor mechanisms located on specially modified epithelial cells distributed throughout the oral mucosa. Taste transduction initiates depolarization of these receptor cells, which make synaptic contact with first-order fibers of one of several cranial nerves. These fibers project to the medulla, into the nucleus of the solitary tract (NST), where second-order projections arise to connect to the pons or thalamus, depending on the species studied. Pontine neurons project to the thalamus and to various areas of the limbic forebrain involved in food and fluid regulation. Thalamic cells connect to the insular and orbitofrontal cortex.

The various populations of taste buds on the anterior and posterior tongue, palate, and laryngeal mucosa have somewhat different sensitivities and project in a topographic, overlapping manner throughout the taste pathway. Cells at all levels of the gustatory system are broadly tuned across taste quality. Different neuron types can be identified on the basis of their profiles of sensitivity and taste quality appears to be represented by the pattern of activity evoked across these neuron types. In addition to its role in the perception of salty, sweet, sour, and bitter sensations, taste input is important in regulating several visceral reflexes involved in ingestive and digestive functions.

Anatomy

□ Gustatory receptor cells are situated within taste buds

The sense of taste provides a gateway for monitoring and controlling the ingestion of food. It responds to chemical substances in the oral cavity and helps to regulate the interaction between ingestive behavior and the internal milieu. Taste is mediated through chemical stimulation of gustatory receptor cells, which are located within taste buds distributed throughout the oral, pharyngeal, and laryngeal mucosa.

Taste buds on the tongue are contained within distinct papillae; those in other areas are distributed across the surface of the epithelium. At the ultrastructural level, at least two kinds of cells can be discerned within the taste bud. They are called *dark cells* and *light cells* based on their ultrastructural appearance and the presence or lack of dense granules in their apical portion; *intermediate cells* have characteristics between these extremes. Cells

within a taste bud are arranged in a concentric columnar fashion, with their apical microvilli projecting toward a pore that opens through the epithelium into the oral cavity (Fig. 27-1); gustatory stimuli interact with receptors and ion channels on these apical microvilli. The base of the taste bud is penetrated by terminal branches of the afferent nerve, which make synaptic contact with the receptor cells. All three cell types exhibit synaptic specializations, suggesting that they all may be receptor cells. A single nerve fiber may innervate cells in more than one taste bud, each of which is innervated by several different afferent fibers.

□ The turnover and replacement of taste cells is a continuous process

Taste receptor cells arise continually from an underlying population of basal epithelial cells. It is not clear, primarily because the cells are in a constant state of turnover, whether the cell types identifiable on structural grounds are different cell types or a single type at different stages of maturation. In animals, the life span of a taste cell in a fungiform papilla is approximately 10 days. The afferent nerve maintains a trophic influence over the taste buds, which degenerate if the nerve supply is removed.

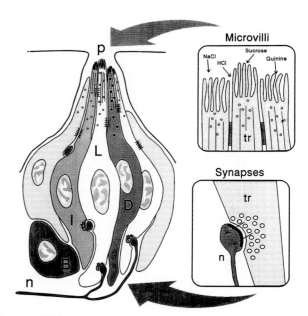

Figure 27-1

Mammalian taste bud. This barrel-shaped structure contains different cell types, including basal cells (B), dark cells (D), light cells (L), and intermediate cells (I). These epithelial receptor cells (tr) make synaptic contact with distal processes (n) of cranial nerves VII, IX, or X, whose cell bodies lie within the cranial nerve ganglia. The microvilli of the taste cells project into an opening in the epithelium, the taste pore (p), where they make contact with gustatory stimuli.

Although the innervation by gustatory nerve fibers is necessary to maintain the structural integrity of the taste bud, the gustatory sensitivities of the receptor cells appear to be determined by the epithelium itself. A classic experiment in which the glossopharyngeal nerve (i.e., cranial nerve IX) and the chorda tympani nerve were transected and crossanastomosed demonstrated that, when the ninth cranial nerve was rerouted to the anterior tongue, its fibers responded like those of the chorda tympani and vice versa, demonstrating that the receptor phenotype is a property of the target epithelium rather than the innervating nerve. Consistent with this observation, the several branches of a chorda tympani axon that innervate different fungiform papillae have shown similar profiles of sensitivity. Combined with the fact that the sensitivity of a given receptor field appears to be determined by the epithelium, this suggests that during cell turnover the nerve fibers are guided to make contact with particular types of receptor cells. Alternatively, the sensory code for taste could be maintained during taste cell turnover by reorganization of the central synaptic connections within the brain stem.

Although taste cells are modified epithelial cells, they possess many characteristics of neurons. Several investigations have demonstrated the presence of a variety of cell-surface molecules and other neural antigens on cells in mammalian taste buds. The neural cell adhesion molecule (NCAM) is expressed on a subset of vallate taste bud cells in rats and mice and on the innervating fibers of the glossopharyngeal nerve. Transection of the nerve results in a loss of NCAM expression as the taste buds degenerate. Reinnervation of the vallate papilla after bilateral nerve crush is accompanied by NCAM expression in the nerve, followed by differentiation of the epithelium and the subsequent expression of NCAM in the differentiated taste cells. A similar temporal sequence is seen during taste bud development in the mouse. These results are compatible with the idea that NCAM plays some role in triggering taste cell differentiation or in subsequent axon–taste cell recognition, which has to occur during cell turnover. Any of a number of cell-surface molecules could play a role in the structural integrity of the taste bud or in the mediation of axon–taste cell recognition.

□ Taste buds are distributed in distinct subpopulations and are innervated by several cranial nerves

Taste buds are found on the anterior portion of the tongue in fungiform papillae and in circumvallate and foliate papillae on the posterior tongue. There are also taste buds on the soft palate, pharynx, epiglottis, and upper third of the esophagus. The distribution of taste buds on the human tongue and within the oral cavity is shown schematically in Figure 27-2. In humans, the 200 to 300

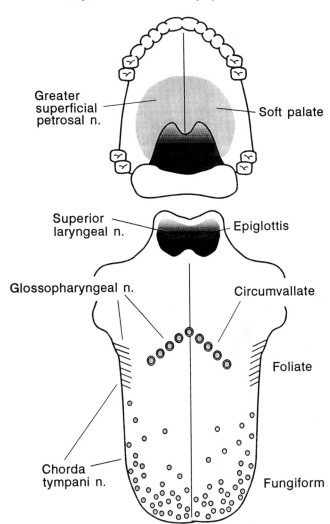

Figure 27-2
Diagram of the human oral cavity, showing the distribution of various taste bud populations, which are found on the anterior tongue (i.e., fungiform papillae), posterior tongue (i.e., circumvallate and foliate papillae), the soft palate, and the laryngeal surface of the epiglottis. These taste buds are innervated by several branches of cranial nerves VII, IX, and X.

fungiform papillae on the anterior portion of the tongue contain approximately 1600 taste buds, although there is considerable variation among persons. The 8 to 12 circumvallate papillae contain about 250 taste buds each, for a total of almost 3000 taste buds, and the foliate papillae have about 1300 taste buds. Although taste buds have been described on the soft palate of human adults only in biopsy material and only in very small numbers, a few studies have reported that human infants have about 2600 taste buds in the pharynx and larynx and on the soft palate. On the tongue of adult rhesus monkeys (i.e., fungiform, circumvallate, and foliate papillae), there are approximately 8000 to 10,000 taste buds, which are maintained well into old age. Most electrophysiologic studies of taste have employed stimulation of the fungiform papillae on the anterior portion of the tongue, although most taste buds in all mammalian species studied are located in other areas.

Taste receptors are innervated by branches of the seventh (i.e., facial), ninth (i.e., glossopharyngeal), and tenth (i.e., vagus) cranial nerves. These special visceral afferent fibers project centrally into the rostral pole of the NST, which is the rostral-most extension of the visceral afferent column in the medulla. Taste buds in the fungiform papillae on the anterior portion of the tongue are innervated by the chorda tympani branch of the facial nerve, and those on the soft palate are innervated by its greater superficial petrosal branch. The cell bodies of these fibers are located in the geniculate ganglion and project to the most rostral extension of the NST. Circumvallate and most foliate taste buds are supplied by the lingual-tonsillar branch of the glossopharyngeal nerve, although the most rostral foliate taste buds are innervated by the chorda tympani nerve. Afferent fibers of the glossopharyngeal nerve project through the inferior glossopharyngeal (i.e., petrosal) ganglion to the NST just caudal to, but overlapping with, the facial nerve termination. The pharyngeal branch of the glossopharyngeal nerve innervates taste buds in the nasopharynx. Taste buds on the epiglottis, aryepiglottal folds, and esophagus are innervated by the internal portion of the superior laryngeal nerve, which is a branch of the vagus nerve. Afferent fibers of the superior laryngeal nerve project by means of their cell bodies in the inferior vagal (i.e., nodose) ganglion to the NST caudal to, but overlapping with, the glossopharyngeal nerve termination.

□ There are both thalamocortical and limbic forebrain projections of gustatory afferent information

Secondary gustatory fibers arise from the NST to project rostrally more or less parallel to the projection of general visceral sensation, which arises from the more caudal aspects of the solitary nucleus. In most mammalian species, there is a second-order projection into the parabrachial nuclei (PbN) in the pons, from which third-order fibers arise to project ipsilaterally through a classic sensory path to the parvicellular division of the ventroposteromedial nucleus (VPMpc) of the thalamus and then to the agranular insular cortex (Fig. 27-3).

In addition to the thalamocortical projection, fibers travel from the pons, along with other visceral afferent fibers, into areas of the ventral forebrain involved in feeding and autonomic regulation, including the lateral hypothalamus, central amygdala, and the bed nucleus of the stria terminalis. In the monkey, taste fibers bypass the pontine relay and project ipsilaterally through the central tegmental tract directly to the VPMpc in the thalamus. The VPMpc projects in the primate to the insular and opercular cortex and to area 3b on the lateral convexity of the precentral gyrus. What may be a secondary cortical area, receiving input from the anterior insula, is found within the posterior orbitofrontal cortex. None of these cortical areas are purely gustatory because they all contain neurons responsive to other sensory modalities such as touch and temperature. There are few studies of the anatomy of these projections in humans, although a careful

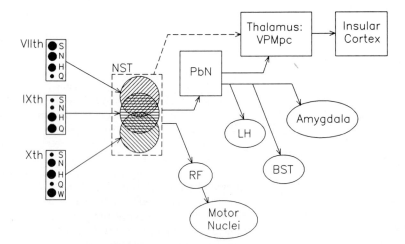

Figure 27-3
Mammalian gustatory afferent pathway. Peripheral fibers with different degrees of responsiveness to sucrose (S), NaCl (N), HCl (H), quinine hydrochloride (Q), or water (W) project into the nucleus of the solitary tract (NST). Fibers of cranial nerves VII, IX, and X project in an organized, overlapping termination within the rostral portion of the NST. Second-order cells project into the parabrachial nuclei (PbN) of the pons, where a classic lemniscal pathway proceeds to the parvicellular division of the ventroposteromedial nucleus (VPMpc) of the thalamus and then to the insular cortex. Another projection arises within PbN to connect to areas of the ventral forebrain, such as the lateral hypothalamus (LH), bed nucleus of the stria terminalis (BST), and amygdala, that are involved in the control of feeding and in autonomic regulation. Cells of the NST also make reflex connections through the reticular formation (RF) with cranial motor nuclei that control muscles involved in facial expression, licking, chewing, and swallowing. In primates, cells of the NST bypass the PbN and project directly to the thalamus (*dashed line*). (Smith DV, Shipley MT. *J Head Trauma Rehabil* 1992;7:1.)

review of the clinical literature suggests that the human gustatory system is quite similar to that of the Old World monkey. A schematic of the probable gustatory projections in humans is shown in Figure 27-4.

Taste Physiology

☐ Taste receptors are coupled to several transduction systems

The mechanisms of taste transduction are receiving much attention. Although the nature of gustatory transduction is not completely understood, a coherent story is beginning to emerge from studies of isolated amphibian and mammalian receptor cells (Fig. 27-5). Sour taste is produced by acids (H^+); voltage-dependent K^+ channels restricted to the apical membrane of the taste receptor cell are blocked by acidic stimuli, causing direct depolariza-

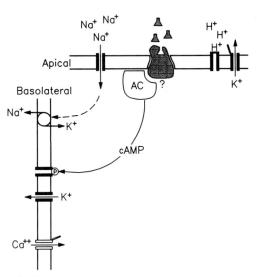

Figure 27-5
Taste transduction mechanisms. Transduction of the taste of acids involves H^+ blockage of voltage-dependent K^+ channels on the apical membrane of the taste receptor cell. Sodium salt (Na^+) transduction involves the passage of Na^+ into the receptor cell through passive, amiloride-blockable ion channels on the apical membrane. The sodium balance is then restored through a (Na^+, K^+)-ATPase on the basolateral membrane. The transduction of sweet and bitter tastes probably involves several different receptor proteins on the apical membrane that bind specifically to these substances. Sweet transduction involves receptor-mediated stimulation of adenylate cyclase (AC), which leads to closing of voltage-independent K^+ channels on the basolateral membrane by a cAMP-dependent phosphorylation. However, the nature of the mediation between receptor binding and adenylate cyclase activity has not been revealed. All of these mechanisms lead to depolarization and influx of Ca^{+2} through voltage-dependent Ca^{2+} channels. (Smith DV, Shipley MT. *J Head Trauma Rehabil* 1992;7:1.)

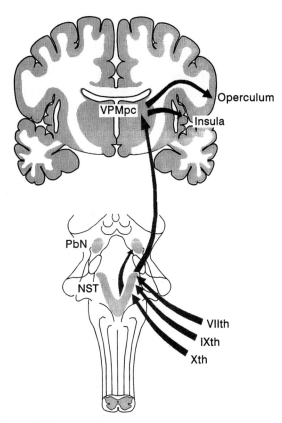

Figure 27-4
Drawing of the probable gustatory afferent pathway in humans. Fibers of cranial nerves VII, IX, and X project into the nucleus of the solitary tract (NST), from which there is a direct projection to the ipsilateral thalamus and to the parvicellular division of the ventroposteromedial nucleus (VPMpc). From the thalamus, fibers project into the insular cortex and frontal operculum. Visceral afferent fibers from the caudal NST project to the parabrachial nuclei (PbN), which also receives gustatory projections in nonprimates. In addition to this classic sensory projection, there are numerous subcortical projections of the gustatory nuclei in a variety of species (see Fig. 27-3).

tion of the cell and leading to excitation of the basolateral membrane. Sodium chloride (NaCl) produces a relatively pure salty taste in humans. Voltage-independent amiloride-blockable ion channels on the apical membranes of taste cells in mammals and frogs appear to partially mediate the transduction of the taste of Na^+ salts. Na^+ ions depolarize the cell by inward movement through these passive channels. Treatment of the tongue surface with amiloride, an epithelial sodium channel blocker, reduces the Na^+-evoked nerve activity in chorda tympani fibers and the short-circuit current across lingual epithelium and blocks whole-cell passive Na currents in isolated frog taste cells. Na^+ ions also appear to stimulate through a basolateral ion channel. The transduction of nonsodium salts, which are characterized by multiple taste qualities, are not understood, although their salty component may be mediated by means of cells that are also responsive to Na^+ ions; human studies show that the saltiness of a wide array of salts is eliminated by adaptation to NaCl, implicating a common mechanism.

The transduction events for acids and Na salts do not require the existence of specific membrane receptors but depend on the direct action of the stimulating ions on apical membrane channels. Transduction of sweet and bitter tastes is not as well understood, but it appears to involve specific membrane receptors linked to second messenger systems. There may be more than one type of sweet receptor and multiple bitter receptor mechanisms are also likely. Although there have not been extensive systematic studies of the profiles of sensitivity of gustatory receptor cells, the available evidence, mostly from micro-electrode recordings in situ, suggests that most taste cells in amphibians and mammals are broadly sensitive to stimuli representing different taste qualities, responding often to two or more of the four basic taste stimuli: NaCl, sucrose, quinine, and acid.

The neural representation of taste quality is a complex phenomenon

The perception of saltiness, sweetness, sourness, or bitterness emerges from neural activity within the central nervous system. The way in which this information is represented in neural activity is the subject of debate. Many textbooks of physiology show a diagram of the human tongue that suggests that saltiness and sweetness are appreciated at the tip, sour on the sides, and bitter on the back of the tongue. Although there are slight differences in the absolute threshold for different taste qualities in different regions of the human tongue and palate, all taste qualities (i.e., salty, sour, sweet, bitter) can be perceived by stimulation of each of the oral taste bud populations. Physiologic studies of rodents suggest some fairly striking differences between the various taste bud populations in their response to different tastants. Fibers of the seventh cranial nerve are much more responsive to sweet and salty stimuli, and those of the ninth cranial nerve are relatively more responsive to sour and bitter substances. These differences may relate to different functional roles for the separate taste bud populations.

Most of what is known about the neurophysiology of the mammalian gustatory pathway has been derived from studies of chorda tympani nerve fibers innervating taste buds on the anterior portion of the tongue. Like individual receptor cells, single fibers in the chorda tympani nerve typically respond to more than one taste quality. Individual second- and third-order gustatory neurons in the NST and parabrachial nuclei that respond to anterior tongue stimulation are similarly broadly tuned across taste qualities. Fibers of the glossopharyngeal nerve are also broadly tuned. Despite this broad tuning, individual gustatory fibers appear to fall into functional groups, which can be identified by their "best" stimulus. Among taste-responsive cells of the peripheral and central nervous system, there are distinct groups of "tuning curves," each responding best to one of the four taste qualities. Attempts to understand the neural processing of taste quality information have relied heavily on this best-stimulus classification (e.g., sucrose-best, NaCl-best), which assumes the existence of four basic taste qualities: salty, sour, sweet, and bitter.

Inputs from the separate peripheral taste fields project into the NST, where the cells receive converging input from separate peripheral fields and are even more broadly tuned than peripheral fibers. The response of a broadly tuned cell in the hamster NST is shown in Figure 27-6. Of the four prototypical stimuli, the cell illustrated responds best to hydrochloric acid (HCl), but it also responds to NaCl and quinine HCl (QHCl); it is inhibited by sucrose. Other salts and acids are excitatory for this cell, and DL-alanine, which tastes sweet to humans and is similar to sucrose in hamster behavioral studies, inhibits the cell. Further increases in the breadth of tuning have been shown in the third-order cells of the PbN.

Because taste fibers in peripheral gustatory nerves and cells in central taste nuclei are fairly broadly tuned across stimulus qualities, it is agreed that the representation of taste quality involves a comparison of activity across taste fibers. The across-fiber pattern theory of quality coding was first suggested when it was apparent from the earliest recordings from taste nerves that these afferent fibers lacked stimulus specificity. The across-neuron patterns of activity can be quantified by calculating the correlations among the responses evoked by a series of chemical stimuli across a sample of taste neurons. These correlations can then be subjected to a multivariate analysis to

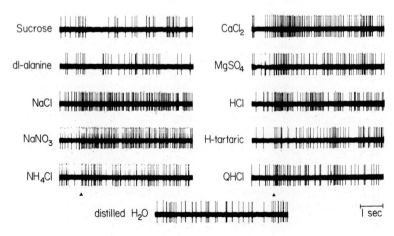

Figure 27-6
Responses of a neuron in the nucleus of the solitary tract (NST) of the hamster to several taste stimuli applied to the fungiform papillae. The triangles indicate the onset of the response, of which about 5 seconds is shown, preceded by about 1 second of response to distilled water. The concentrations of the stimuli are those that produce a half-maximal response to these chemicals in the hamster's chorda tympani nerve. This cell shows an excitatory response to all of these stimuli except sucrose and DL-alanine, which inhibit ongoing activity. (Smith DV, Travers JB, Van Buskirk RL. *Brain Res Bull* 1979;4:359.)

create a "taste space," which represents the neurophysiologic similarities and dissimilarities among the responses to the stimuli. The across-neuron taste space generated from the responses of cells in the hamster PbN to several stimuli is shown in Figure 27-7.

Stimuli with similar tastes (to hamsters) are grouped together, and those with different tastes are separated within this space. The similarities and differences in the patterns of activity evoked by these stimuli provide a basis for their discrimination. By relating the responses of the various fiber types (e.g., sucrose-best, NaCl-best) to these across-neuron patterns, it has been shown that a comparison of activity across fiber types appears to be necessary for the neural discrimination between stimuli of different taste quality. Experiments using the specific ion channel blocker, amiloride, have suggested that saltiness may be coded specifically by activity in a single class of NaCl-best cells. However, the discrimination of sodium salts from other stimuli requires a comparison of the activities in several fiber types.

Apart from the issue of quality-coding mechanisms is the interesting question of how any kind of code for quality is maintained in the face of the constant turnover of receptor cells. Electrophysiologic measures of cross-regenerated taste fibers in cases in which the chorda tympani and glossopharyngeal nerves were cut, crossed, and allowed to reinnervate the tongue demonstrated that the tongue epithelium determined the relative sensitivities of the two nerves. This was somewhat surprising, because the nerves themselves have a trophic influence over the taste buds. However, if the receptors are determined by the epithelium, the gustatory nerve fibers have to make contact with the appropriate receptor cells during turnover to maintain a constant sensory code. Another alternative is that the synaptic organization of the medullary relay could be continually altered to maintain a constant representation of a changing periphery. Whatever the mechanism, the representation of taste quality in gustatory nerve fibers and central neurons must somehow remain constant during the continual turnover of taste receptor cells and their reinnervation by fibers of the peripheral nerve. Because the taste system is composed of broadly tuned and redundant elements, it may be particularly suited to representing information during continual receptor replacement; the contribution of any small number of cells at any given point in time is much less important than the overall pattern of activity.

☐ Taste-mediated behavior requires the interplay of many neurons

Fibers of the various gustatory nerves project into the NST of the medulla. The second-order neurons give rise to ascending projections to the PbN of the pons (in nonprimates), which sends projections to the thalamus and insular cortex and to widespread areas of the limbic forebrain. There are also numerous connections of NST neurons to the oral motor nuclei through interneurons in the reticular formation. These anatomic associations, the differential sensitivities of the seventh, ninth, and tenth cranial nerves, and the contribution of gustatory afferent input to taste-mediated behaviors are summarized in the diagram of Figure 27-8. Although taste physiologists have

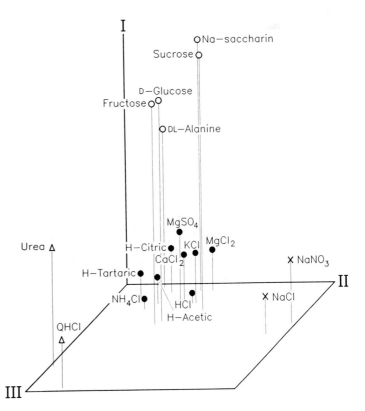

Figure 27-7
Three-dimensional "taste space," showing the similarities and differences in the across-neuron patterns evoked by 18 stimuli delivered to the anterior tongue of the hamster. This space was derived from multidimensional scaling (KYST, Bell Laboratories) of the across-neuron correlations among these stimuli recorded from neurons in the PbN of the hamster. The proximity of neurons within this space indicates a high degree of correlation between the across-neuron patterns elicited by these stimuli. Four groups of stimuli are indicated by different symbols: sweeteners, sodium salts, nonsodium salts, and acids, and bitter-tasting stimuli. (Modified from Smith DV, Van Buskirk RL, Travers JB, Bieber SL. *Neurophysiol* 1983;50:541.)

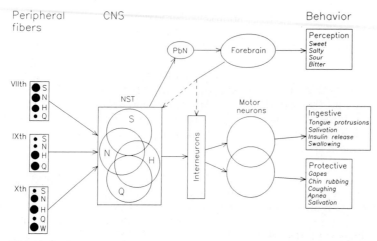

Figure 27-8

Chemosensory inputs of three cranial nerves to the taste-responsive portion of the nucleus of the solitary tract (NST) and their putative role in taste-mediated behaviors. The size of the filled circles for each of the peripheral nerves (i.e., VII, IX, X) depicts the relative responsiveness of these nerves to sucrose (S), NaCl (N), HCl (H), quinine hydrochloride (Q), and water (W). The sensitivities of NST cells are largely overlapping, with each cell type somewhat responsive to two or three of the basic stimuli, but sucrose and QHCl stimulate few of the same NST cells. Output from the NST ascends in the classic taste pathway to produce perceptions of sweetness, saltiness, sourness, and bitterness and to hedonic tone (not depicted). Local reflex circuits within the brain stem control the ingestive and protective responses evoked by taste stimulation. Behavioral data suggest that the ingestive and protective responses can be triggered in parallel, depending on the quality of the stimulus. (Smith DV, Frank ME. Sensory coding by peripheral taste fibers. In Simon SA, Roper SD (eds). Mechanisms of taste transduction. Boca Raton, FL: CRC Press, 1993:295.)

focused largely on the role of gustatory afferent fibers and central neurons in taste quality perception, there are several taste-mediated behaviors, ranging from tongue movements to salivation to preabsorptive insulin release (i.e., in response to gustatory stimulation), that have their neuronal substrate within the brain stem. Input from the various gustatory nerves contributes differentially to these diverse taste-mediated behaviors.

Taste may be viewed as the oral component of a visceral afferent system, which includes gustatory, respiratory, cardiovascular, and gastrointestinal functions. The functional contribution of the taste nerves seems to be organized in a rostral to caudal continuum, with ingestive substrates more rostral and protective components more caudal. For example, cranial nerve VII fibers are responsive primarily to preferred stimuli such as sucrose and NaCl; cranial nerve IX fibers are most sensitive to adverse stimuli such as HCl and QHCl; and cranial nerve X fibers respond to stimuli that deviate from the normal pH and ionic milieu of the larynx.

Sucrose predominantly stimulates fibers of the seventh cranial nerve, which project into the rostral-most NST. The output of these second-order neurons ascends to the forebrain to generate the perception of sweetness (see. Fig. 27-8). These cells also provide input to the motor system that drives the ingestive components of feeding behavior, including (in rodents) rhythmic mouth movements, tongue protrusions, lateral tongue protrusions, salivary secretion, preabsorptive insulin release, and

swallowing. Conversely, QHCl stimulates predominately the fibers of the ninth cranial nerve. These fibers project to more caudal parts of the NST; quinine-sensitive cells of th NST send ascending projections to the forebrain to give rise to sensations of bitterness (see Fig. 27-8), but they also provide input to motor systems that drive protective behaviors such as gaping, chin rubbing, forelimb flailing, locomotion, and fluid rejection. In rats, the number of gapes elicited by quinine stimulation is reduced by almost one half after bilateral transection of cranial nerve IX. Sucrose and quinine produce different patterns of ingestive and protective taste reactivity, and a combination of these behaviors can be triggered by mixtures of sucrose and quinine.

The superior laryngeal branch of the tenth cranial nerve is involved in swallowing, airway protection, and several other visceral reflexes. Respiratory apnea is produced by laryngeal stimulation with water, and chemosensory fibers of the rat superior laryngeal nerve mediate diuresis in response to stimulation of the laryngeal mucosa with water. Chemoreceptive fibers of the tenth cranial nerve are more involved in visceral functions than in taste quality perception. In addition to its role in controlling ingestive behavior, taste triggers several metabolic responses, including salivary, gastric, and pancreatic secretions, although the specific contributions of particular cranial nerves to these responses are not entirely understood. In addition to their mediation of gustatory sensation, taste buds may have a number of roles related to gustatory-visceral regulation, depending on their peripheral distribution and innervation.

SELECTED READINGS

Erickson RP. On the neural basis of behavior. Am Scientist 1984;72:233.

Erickson RP, Doetsch GS, Marshall DA. The gustatory neural response function. J Gen Physiol 1965;49:247.

Farbman AI. Renewal of taste bud cells in rat circumvallate papillae. Cell Tissue Kinetics 1980;13:349.

Frank M. An analysis of hamster afferent taste nerve response functions. J Gen Physiol 1973;61:588.

Frank ME, Bieber SL, Smith DV. The organization of taste sensibilities in hamster chorda tympani nerve fibers. J Gen Physiol 1988;91:861.

Hanamori T, Miller IJ Jr, Smith DV. Gustatory responsiveness of fibers in the hamster glossopharyngeal nerve. J Neurophysiol 1988; 60:478.

Heck GL, Mierson S, DeSimone JA. Salt taste transduction occurs through an amiloride-sensitive sodium transport pathway. Science 1984;223:403.

Kinnamon SC. Taste transduction: a diversity of mechanisms. Trends Neurosci 1988;11:491.

Miller IJ Jr, Bartoshuk LM. Taste perception, taste bud distribution, and spatial relationships. In Getchell TV, Bartoshuk LM, Doty RL, Snow JB Jr (eds). Smell and taste in health and disease. New York: Raven Press, 1991:205.

Norgren R. Gustatory system. In Paxinos G (ed). The human nervous system. San Diego: Academic Press, 1990:845.

Oakley B. Neuronal-epithelial interactions in mammalian gustatory epithelium. In Rubel E (ed). Regeneration of vertebrate sensory receptor cells. Chichester: John Wiley, 1991:277.

Pfaffmann C. Gustatory nerve impulses in rat, cat and rabbit. J Neurophysiol 1955;18:429.

Pritchard TC. The primate gustatory system. In Getchell TV, Bartoshuk LM, Doty RL, Snow JB Jr (eds). Smell and taste in health and disease. New York: Raven Press, 1991;109.

Smith DV, Frank ME. Sensory coding by peripheral taste fibers. In Simon SA, Roper SD (eds). Mechanisms of taste transduction. Boca Raton, FL: CRC Press, 1993:295.

Smith DV, Van Buskirk RL, Travers JB, Bieber SL. Coding of taste stimuli by hamster brain stem neurons. J Neurophysiol 1983;50:541.

Smith DV, Van Buskirk RL, Travers JB, Bieber SL. Gustatory neuron types in hamster brain stem. J Neurophysiol 1983;50:522.

Neuroscience in Medicine, edited by P. Michael Conn.
J. B. Lippincott Company, Philadelphia © 1995

Chapter 28

The Olfactory System

MICHAEL T. SHIPLEY
DAVID V. SMITH

Olfactory information is transduced when odorant molecules contact *olfactory receptor neurons* (ORNs), which are located in the *olfactory epithelium,* a specialized region of the nasal cavity. The axons of these neurons form the *olfactory nerve,* which transmits information to the glomerular layer of the *olfactory bulb*. In this area, the ORN axons synapse onto the dendrites of the *mitral and tufted cells,* which are the output neurons of the olfactory bulb. Axons from these cells project to the *primary olfactory cortex* by means of the *lateral olfactory tract*

The primary olfactory cortex is composed of several regions, including the anterior olfactory nucleus, the piriform cortex, parts of the amygdala, and the entorhinal cortex. These areas are interconnected with many areas of the brain, including the neocortex, hippocampus, mediodorsal thalamus, preoptic area, hypothalamus, and other parts of the limbic system. Through these connections, the olfactory system influences a wide range of behaviors and physiologic functions, including reproduction, social behavior, and communication (e.g., scent marking), food finding and selection, and maternal behavior in addition to regulating neuroendocrine functions.

Compared with vision, audition, touch, and even taste, our understanding of olfactory system function is rudimentary. A major reason for this is that we do not know how odors are "coded" by primary olfactory neurons in the nasal epithelium. Advances in molecular biology and membrane biophysics promise to close this critical gap. A second reason for slow progress in olfaction has been the lack of a critical mass of researchers in this field. This is also changing, and there are likely to be major advances in our understanding of olfactory system function in the next decade. This chapter concentrates on the anatomy and classic physiology of the olfactory system, because these have been relatively well characterized. We also summarize advances, point out gaps in our knowledge, and indicate where progress is expected.

Although there are many gaps in our knowledge, there are also many features of the olfactory system that make it an interesting object of study. It is the only sensory system that is essentially cortical from the first synaptic relay in the brain. The other sensory systems involve several layers of subcortical synaptic processing before information reaches the cortex. In the olfactory system, the primary sensory nerve endings synaptically terminate in the olfactory bulb, which is a cortical structure that derives embryologically from the same tissue as the neocortex. Mechanisms of cortical sensory processing are directly accessible in the olfactory system. Primary olfactory neurons are the only neurons in the mature mammalian nervous system that replace themselves after disease or injury, and the replacement neurons successfully reconnect with the brain to restore sensory function.

The olfactory system holds important clues to puzzles such as why the human brain cannot repair itself after damage and how growing axons find their appropriate postsynaptic targets. A better understanding of the olfactory system may provide important information about basic processes common to other sensory and integrative systems in the brain.

The Olfactory Epithelium

☐ The sense of smell is mediated through the stimulation of ORNs by volatile chemicals

Olfactory receptor neurons are contained in a neuroepithelium at the top of the nasal vault, along the upper portion of the nasal septum, the cribriform plate region, and the medial wall of the superior turbinate. Afferent information from these receptors is carried to the olfactory bulbs by the *olfactory nerve*, which is the first cranial nerve.

To stimulate the olfactory receptors, airborne molecules must enter the nasal cavity, where they are subject to relatively turbulent air currents. The duration, volume, and velocity of a sniff are important determinants of an odor's stimulating effectiveness. Although these parameters differ markedly among persons, they are quite constant for any one person. Once airborne volatiles reach the olfactory epithelium, they must pass through the layer of mucus that covers the olfactory epithelium. The stimulating effectiveness of an odor is also determined by the relative partitioning of the odor between air and mucus. Macromolecules in the mucus may bind odorants and present them to receptors; similarly macromolecules may be required to remove odorants from the receptors and or chemically inactivate odorants (Fig. 28-1).

The ORNs lie in a pseudostratified columnar epithelium, which is thicker than the surrounding respiratory epithelium of the nasal cavity. This epithelium rests on a vascular lamina propria. Within the epithelium are the bipolar ORNs, supporting cells (i.e., sustentacular cells), microvillar cells, and basal cells; Bowman's glands lie within the underlying lamina propria.

Unlike taste receptor cells, which are modified epithelial cells (see Chap. 27), the olfactory receptors are true neurons. Their cell bodies lie in the basal two thirds of the epithelium, and their apical dendrites extend to the surface. At its peripheral tip, the dendrite swells slightly to form the olfactory knob, from which several cilia extend into the mucous layer. Although human cilia do not appear to be motile, in some vertebrate species, ciliary length and motility have been related to receptor age and development. Much evidence suggests that the cilia are the sites of chemosensory transduction.

Basal to the ORN cell body, a nonmyelinated axon arises and joins a small bundle of other ORN axons. These axons penetrate the basal lamina, at which point the bundles become ensheathed by Schwann cells. These bundles join others to assemble the 15 to 20 fascicles (i.e., *fila olfactoria*) of the olfactory nerve, which pass through the cribriform plate to synapse in the olfactory bulb.

The supporting cells of the olfactory epithelium separate and partially wrap the ORNs. Their apical surface in

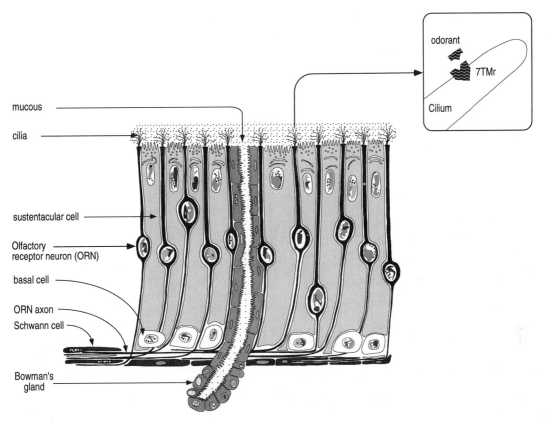

Figure 28-1
The olfactory epithelium, showing the major cell types. The inset shows the location of seven transmembrane (7TMr) odorant receptors on the cilia of the olfactory receptor neurons.

humans and some other vertebrates is covered with microvilli, which project along with the olfactory cilia into the mucous layer. A third cell type, the microvillar cells, present at about one-tenth the number of ORNs in humans, have microvilli on their apical surface, projecting into the mucous layer. Their basal end tapers into a cytoplasmic extension that appears to enter the lamina propria. It is not known if this projection is an axon, although the ultrastructural appearance of the microvillar cells appears to be neuronal. Deep to the ORNs, sustentacular, and microvillar cells are the basal cells, which sit on a basement membrane just above the lamina propria. The *basal cells* are *stem cells* for the replacement of the ORNs, which in the mouse have a life span of approximately 40 days. Within the lamina propria are the secretory Bowman's glands, which provide a serous component to the mucous layer covering the olfactory epithelium.

Olfactory receptor cells and axons contain *olfactory marker protein* (OMP), which is unique to olfactory neurons. OMP is found in a number of mammalian species, including humans. This protein, whose function is unknown, appears to be expressed in all ORNs and accounts for 1% of the total protein content of these cells.

In addition to input from the olfactory epithelium, some chemical stimuli give rise to activity in the trigeminal system through the stimulation of nerve endings in the nasal and oral cavities. The burning or irritation arising from stimuli like ammonia or hot peppers interacts with olfactory and gustatory input in what has been called the *common chemical sense*. Fibers in each of the three branches of the fifth cranial nerve carry information about intranasal or intraoral chemical irritation. These sensations usually remain intact in patients complaining of taste or smell dysfunction and can be useful in the assessment of the malingering patient.

☐ The function of the olfactory system has been studied by several electrophysiologic methods

In one of the earliest studies, a surface electrode placed on the olfactory epithelium was used to record a slow monophasic negative potential on an *electro-olfactogram* (EOG) in response to odorant stimulation. The EOG is thought to be the summed generator potentials from a large number of ORNs. It has also been possible to record spike discharge activity from individual ORNs. These measures suggest that ORNs are typically responsive to several odors.

Even though a given cell responds better to some odors than to others, attempts to classify the receptors into groups on the basis of their profiles of sensitivity have not been successful. Some experiments suggest that different regions of the olfactory epithelium give maximal responses to different odorants, as though the nose and mucosa were somehow operating like a chromatograph, separating odors on the basis of their physical and chemical properties. The measurement of radioactive odorants delivered to the nose suggests that different odors result in distinctive temporospatial patterns of absorption across the olfactory mucosa. Nevertheless, there is no general agreement about the way in which quality is coded by the olfactory system.

☐ There has been much progress in understanding the mechanisms of olfactory transduction

It is widely thought that electrical signals are generated in ORNs when odorants bind to specific membrane receptors located on their cilia. According to this hypothesis, these receptors are linked through G proteins to cyclic nucleotides and possibly other second messenger systems that directly control ionic channels to depolarize the ORN. This proposed transduction mechanism is analogous to the rhodopsin-transducin-cGMP cascade that mediates phototransduction in vertebrate rods. According to this scenario, odorants bind to specific receptor proteins in the membrane of the olfactory cilia (Fig. 28-2).

☐ Olfactory receptors may be encoded by a large multigene family

Until recently, the existence of olfactory receptors as discrete, specific molecules was largely a theoretical construct advanced to account for the initial step in the transduction process. However, the existence of olfactory

receptors is strongly supported by the discovery of a very large (at least several hundred distinct genes) multigene family expressed by ORNs. Members of this putative olfactory receptor gene family have a high degree of base sequence homology with other genes that encode *seven transmembrane (7TM) receptor molecules* that function as receptors for neurotransmitters. Putative olfactory receptor gene transcripts (mRNAs) appear to be preferentially expressed in ORNs and other chemosensory cells such as taste cells and vomeronasal neurons. Significantly, these genes do not appear to be expressed by any other neurons or any other tissues except sperm cells. Although it remains to be formally demonstrated, it is likely that these gene transcripts are translated to yield an equally large family of putative olfactory receptor proteins that are targeted to their cilia, where they they bind specific odorants and activate a signal transduction cascade.

Much important research remains to be done, but assuming that this outline is fleshed out by subsequent investigation, the discovery of this putative olfactory receptor multigene family should make it possible to solve many fundamental problems in olfactory stimulus transduction. For example, it should be possible to determine if receptors are broadly or narrowly "tuned" for different odors, and it can be determined whether individual ORNs express many or only a few different receptors. It may be possible to learn if the receptors expressed by different ORNs are under strict genetic control or whether epigenetic factors (e.g., cell-cell interactions, odor experience) can regulate the receptor types expressed by individual ORNs.

These 7TM receptors are thought to activate G proteins that are specific to ORNs. Activation of the G proteins stimulates adenylate cyclase, which increases cAMP production. Patch clamp studies on excised patches of olfactory cilia have demonstrated a cyclic nucleotide-gated, cation-permeable channel. This channel is opened by the binding of intracellular cyclic nucleotides, resulting in the transduction of cAMP levels into alterations in the membrane potential. Sufficient membrane depolarization leads to the generation of action potentials that are

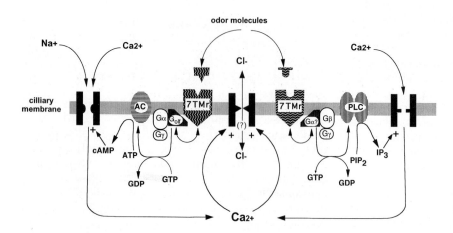

HYPOTHESIZED OLFACTORY RECEPTOR-TRANSDUCTION MECHANISMS

Figure 28-2
Hypothesized olfactory receptor transduction mechanisms. Evidence suggests that odor molecules bind to specific seven transmembrane receptor (7TMr) proteins located in the cilia of olfactory receptor neurons (ORN). These 7TMrs are thought to be coupled to G-proteins that activate either adenyl cyclase (AC) to generate cyclic AMP (cAMP) or phospholipase C (PLC) to generate phosphitidal inositol (IP3). These second messengers open channels that admit calcium (Ca^{2+}) or sodium (Na^+) into the cilium. These ions lead to membrane depolarization and may modulate intracellular free Ca^{2+} levels, both of which lead to the generation of action potentials that are conducted along ORN axons to the olfactory bulb.

conducted along ORN axons to the olfactory bulb. There is evidence that cAMP and inositol triphosphate may be involved as second messengers in response to different odorants.

☐ The olfactory system has the unique capacity to replace itself

ORNs undergo constant turnover. Cutting the olfactory nerve results in retrograde degeneration of the olfactory epithelium, which is subsequently reconstituted. New ORNs are generated from basal cells. The new ORNs mature and send axons to the olfactory bulb, where they form new synaptic connections. The olfactory epithelium may have this capacity to replace ORNs because these cells, which are directly exposed to airborne molecules, are vulnerable to injury from environmental toxins and viruses. The regenerative capacity of the olfactory epithelium allows recovery of function after damage to the olfactory nerve, toxic exposure, viral infection, and other conditions that injure the olfactory receptor sheet. The ability of the adult epithelium to generate new neurons and the ability of these neurons to grow axons that reestablish functional synaptic connections with target cells in the olfactory bulb are unique features among all mammalian neural structures. Understanding the mechanisms that allow this remarkable neural replacement to take place may eventually provide therapies for promoting repair in other parts of the brain.

☐ The olfactory nerve is a portal for the entry of foreign substances into the brain

Because ORNs are in the nasal cavity, they are exposed to airborne substances, including viruses, industrial pollutants, and toxins. There is evidence that some viruses and other substances can be incorporated into ORNs and transported along their axons to the olfactory bulb. In animal studies, it has been demonstrated that some of these substances escape from ORN synaptic terminals and are incorporated into neurons and axons within the olfactory bulb. These substances are then further disseminated to other parts of the brain by anterograde or retrograde axonal transport. The olfactory nerves are a potential conduit for the entry and spread of foreign substances into the brain. It has been suggested that some forms of Alzheimer's disease and other degenerative diseases could be the result of viruses, metals, or silicates entering the brain from the olfactory epithelium.

☐ There is a second, parallel olfactory system

In many species, there is a second olfactory organ in the nasal cavity called the *vomeronasal organ* (VNO), which is at the base of the midline nasal septum. The VNO contains receptor neurons that are morphologically similar to ORNs. These cells are thought to be preferentially sensitive to nonvolatile odorants, including relatively large proteins. In many species, VNO receptor neurons (VORNs) respond to *pheromones*, molecules emitted by members of the same species (i.e., conspecifics); the pheromones often signal the gender or reproductive status of the sender. In species with a VNO, the axons of the VORNs project to a specialized structure at the dorsocaudal end of the "main" olfactory bulb (MOB) called the *accessory olfactory bulb* (AOB). The structure and neuronal cell types of the AOB are remarkably similar to those of the MOB, but the axons of AOB mitral and tufted cells project to sites in the brain that are contiguous but nonoverlapping with the projections of mitral and tufted cells from the MOB. This anatomic organization has given rise to the concept of two parallel olfactory systems: the main and accessory olfactory systems. Lesions of the VNO or the AOB cause significant impairment of reproductive behaviors and gonadosteriod function.

The VNO appears to be present in most human fetuses but persists to adulthood in only a few persons. It is not known if humans have a functional VNO, because there is no histologically recognizable AOB. However, the central structures that receive AOB innervation in mammals with VNOs are present in humans, and these structures play important roles in human reproductive behavior and endocrine function. An important question is whether odors play any role in human sexual and endocrine functions, and if so, what are the anatomic substrates for olfactory modulation of these functions. It is possible that ORNs specialized to transduce VORN-specific odorants exist but are not anatomically segregated from ORNs in humans. Future research on 7TM receptor molecules may be able to resolve this question.

The Olfactory Bulb

☐ Second-order olfactory neurons are located in the olfactory bulb

The *olfactory bulb* is an oval structure that lies on the ventral surface of each frontal lobe and dorsal to the cribriform plate of the ethmoid bone. The olfactory bulb is arranged in layers, from superficial to deep: the olfactory nerve layer, the glomerular layer, the external plexiform layer, the mitral cell layer, the internal plexiform layer, and the granule cell layer.

There are two major neuronal groups in the olfactory bulb: second-order neurons, such as the mitral and tufted cells, and interneurons, called the juxtaglomerular and granule cells (Figs. 28-3 and 28-4). Axons of the mitral and tufted cells constitute the primary output pathway of the olfactory bulb, the *lateral olfactory tract*. Juxtaglomerular and granule cells are interneurons that modulate the activity of the mitral and tufted cells and regulate the transfer of information from olfactory nerve inputs to mitral and tufted outputs from the olfactory bulb. This reg-

Olfactory Bulb: Basic Circuit

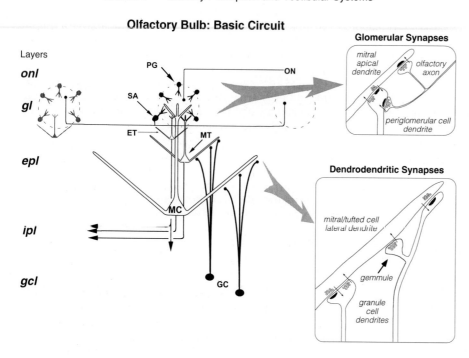

The basic circuitry of the olfactory bulb. The axons of the olfactory neurons form the olfactory nerve (ON). These axons terminate in the glomeruli into mitral (M) and tufted cells (external tufted cell [ET]; middle tufted cell [MT] and onto juxtaglomerular neurons, including periglomerular cells (PG), ET cells, and short-axon cells (SA). There are one-way and reciprocal synapses between the apical dendritic branches of mitral and tufted cells and the dendrites of juxtaglomerular neurons (*upper inset*). The lateral dendrites of mitral and tufted cells form one-way and reciprocal synapses with the apical dendrites of granule cells (*lower inset*).

ulation is organized into two distinct and largely separate levels—the glomeruli and the external plexiform layer—and is targeted at the two classes of dendrites of the mitral and tufted cells—the apical and the lateral dendrites.

Glomeruli

The axons of ORNs terminate exclusively in the *glomeruli*, the superficial-most cell layer in the bulb. The glomeruli are spheroid structures composed of a cellular shell surrounding a core that is rich in neuropil. The cellular shell contains glial cells and several classes of juxtaglomerular neurons, including the periglomerular cells, which receive synapses from olfactory nerve terminals. The glomerular core is richly invested by ramifications of the apical dendrites of mitral and tufted cells and the dendrites of juxtaglomerular neurons; these dendrites are heavily targeted by olfactory nerve synaptic endings. Mitral and tufted cells have a single *apical or 1°dendrite* that extends from the cell body through the external plexiform layer to ramify in a single glomerulus. Mitral and tufted cells' apical dendrites are synaptically contacted by olfactory nerve terminals and, through dendrodendritic synapses, with the dendrites of periglomer-

ular and possibly other (e.g., short-axon cells) juxtaglomerular neurons.

Most juxtaglomerular cells are periglomerular cells and external tufted cells. The predominant action of juxtaglomerular neurons is probably inhibitory to mitral and tufted cells or to olfactory nerve terminals. Axons of some juxtaglomerular neurons project to nearby glomeruli. These dendrites also may be influenced by the axons of neurons from other parts of the brain (i.e., centrifugal afferents). Cholinergic axons from the basal forebrain terminate in the glomeruli, as do serotonergic axons from the midbrain raphe. The apical dendrites of the mitral and tufted cells are influenced by olfactory nerve terminals, interneurons (e.g., juxtaglomerular cells), and centrifugal modulatory systems. The predominant synaptic input, however, is from olfactory nerve terminals. Accordingly, the apical dendrite of mitral and tufted cells may be considered the "sensory dendrite."

The glomeruli are the first site of synaptic integration in the olfactory pathway. Olfactory receptor axons synapse on the dendrites of the two classes of bulb output neurons—the mitral and tufted cells—but this message is modified by the actions of the juxtaglomerular neurons, including the periglomerular, external tufted, and

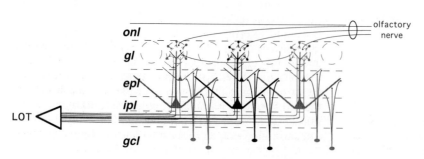

The lateral olfactory tract. The axons of mitral and tufted cells collect at the caudal end of the olfactory bulb to form the lateral olfactory tract (LOT). The LOT distributes terminals throughout the olfactory cortex.

various short-axon cells. The nature of this modulation is unknown.

External plexiform layer

Mitral and tufted cells also have four to six *lateral* or *2° dendrites*. These dendrites extend for considerable distances in the *external plexiform layer*, below the glomerular layer. These lateral or secondary dendrites are influenced chiefly by the granule cells. The granule cells lie deep to the mitral cell layer. The apical dendrites of granule cells form dendrodendritic inhibitory synapses onto the lateral dendrites of the mitral and tufted cells, and the mitral and tufted cells' dendrites form excitatory synapses onto the dendrites of the granule cells. These dendrodendritic synapses probably provide lateral inhibition or a temporal filter in the processing of olfactory information from the primary olfactory neuron to the output of the mitral and tufted cells.

Because the two classes of mitral and tufted cells' dendrites ramify in two distinct layers of the bulb, they are regulated by different interneurons. The apical dendrites are regulated by juxtaglomerular cells, and the lateral dendrites are regulated by granule cells. As a result, the input-output functions of the olfactory bulb are critically determined by neural signals that influence juxtaglomerular and granule cells and signals that directly influence mitral and tufted cells.

Granule cell layer

The granule cells are heavily targeted by excitatory synaptic inputs from ipsilateral and contralateral olfactory cortical structures, the very structures that are targeted by the mitral and tufted cells. These excitatory inputs cause the granule cells to release γ-aminobutyric acid (GABA), inhibiting mitral and tufted cells. The major excitatory input to the granule cells is from the areas targeted by the mitral and tufted cells. The result of this circuitry is that the output of the olfactory bulb causes a massive negative feedback to the bulb's output neurons. The functional significance of this circuitry is incompletely understood, but it would appear that the coding of olfactory information involves feedback mechanisms in which the output of the bulb at any moment is fed back to the bulb to modify its subsequent outputs. In addition to these olfactory cortical inputs, the granule cells appear to be heavily and selectively targeted by norepinephrine-containing terminals from the locus coeruleus. Granule cell somata are also coupled by gap junctions. These gap junctions could serve to electrically couple clusters of granule cells such that synaptic activation of one cell can activate other cells in the cluster. This arrangement could "amplify" granule cell inhibition of the mitral and tufted cells' lateral dendrites.

Convergence in the olfactory bulb

In the rabbit, there are 50,000,000 olfactory receptors and 2000 glomeruli, and each glomerulus receives 25,000 olfactory axons. Because there are approximately 25 mitral cells associated with each glomerulus, the average number of ORN axons terminating on each mitral cells is about 1000. The sensory input to mitral cells is highly convergent. There is no evidence that the mitral cells associated with a single glomerulus project to common targets in olfactory cortex, and there appears to be no additional convergence of the mitral cells from a single glomerulus onto a relatively small number of target neurons. The significance of the convergence of ORN inputs onto mitral cells may not be clarified until there is a better understanding of the organization of putative olfactory 7TM receptors on the ORN axons terminating to specific glomeruli.

☐ The neurotransmitters used by ORNs are unknown

ORNs are the only known neurons to synthesize the dipeptide carnosine; carnosine is transported to olfactory synaptic terminals and can be released from storage granules in a Ca^{2+}-dependent manner, but there is no convincing electrophysiologic evidence that carnosine is a transmitter in ORNs. Whole-cell patch clamp recordings from mitral cells show that the synaptic current caused by activation of ORN axon terminals has a rapid onset, which is a characteristic of ligand-gated ion channels, but has also a prolonged duration, which is a feature of second messenger–dependent transmitters. It is possible that ORNs release more than one transmitter. Mitral and tufted cells contain the neuropeptide corticotropin-releasing hormone and many tufted cells contain cholecystokinin. These bulb output neurons may release excitatory neuropeptides as transmitters.

There is pharmacologic evidence that mitral cells also release an excitatory amino acid, and bulb output neurons may release more than one transmitter. Juxtaglomerular neurons contain several different transmitters including dopamine, GABA, and in some species, different neuropeptides. Granule cells contain GABA. Olfactory bulb neurons also receive cholinergic, noradrenergic, and serotonergic synaptic inputs from subcortical and brain stem "modulatory" systems. The extensive excitatory inputs to the bulb from higher-order cortical areas are probably excitatory amino acids.

☐ Much of the basic electrophysiology of the olfactory bulb is well-characterized

Classic field potential analysis

The study of extracellular field potentials generated by antidromic activation of lateral olfactory tract or by orthodromic activation of the olfactory nerve has played an important role in the analysis of the basic synaptic events in the olfactory bulb. These waveforms, recorded after lateral olfactory tract or olfactory nerve stimulation, change in a characteristic manner as a recording electrode is passed through the layers of the bulb.

The initial event is a fast potential that reflects the invasion of the mitral cell soma and possibly of the primary and secondary dendrites. This fast potential is followed by a slower wave that is positive when the recording electrode is in the granule cell layer and negative when it is above the mitral cell layer. After a single shock to lateral olfactory tract or the olfactory nerve, there is a period of inhibition of the spontaneous activity of mitral cells. This inhibition is accompanied by a reduction in amplitude of the response to a test shock to olfactory nerve or lateral olfactory tract. Electrophysiologic analysis, combined with ultrastructural studies of the synaptic connections of the external plexiform layer, provided an explanation for these observations. The positive wave in the granule cell layer results from synaptic currents flowing into cells that send processes across the mitral cell layer. The granule cells are the only large population of neurons fitting this description. The positive potential is the result of excitatory currents flowing into granule cells from synapses on their dendrites in the external plexiform layer, above the mitral cell layer. These conclusions have been confirmed by intracellular recordings from mitral cells.

Cellular responses to electrical and odor stimulation

There have been few studies of the responses of olfactory bulb neurons to odors, partially because of technical problems. Many olfactory bulb neurons are small and difficult to record with microelectrodes, and the accurate presentation of controlled concentrations of specific odors requires sophisticated equipment. Part of the problem is our ignorance of the coding properties of primary olfactory neurons, and part is the result of the limited number of workers who have studied this system. The few studies published indicate that mitral cell responses to odors are complex. Mitral cells are excited by some odors, inhibited by others, and do not respond at all to some odors. As primary olfactory neurons adapt rapidly to continuous exposure to the same odor, it is difficult to obtain repeated measures of bulb neurons to the same odor. Most electrophysiologic studies of bulb neurons have been done in anesthetized animals and most anesthetics profoundly alter the firing characteristics of these cells.

Glomerular neurons. Extracellularly recorded units in the glomerular layer fire one or a few spikes after electrical stimulation of the olfactory nerve. Intracellular recordings from a small population of periglomerular neurons showed that olfactory nerve stimulation produces an excitatory postsynaptic potential (EPSP) lasting for about 30 milliseconds. Intracellular studies indicate that juxtaglomerular neurons respond to odor stimulation with a burst of spikes associated with a depolarization. Periglomerular neurons are facilitated after weak shocks to the olfactory nerve; at longer stimulus intervals or with stronger stimuli, these neurons are inhibited. Intracellular recordings from periglomerular cells revealed an inhibitory postsynaptic potential (IPSP) lasting for 200 milliseconds that was produced by olfactory nerve stimulation, indicating an intraglomerular slow inhibitory system.

A slow inhibitory system in the glomerular layer activated by olfactory nerve and regulating transmission from olfactory nerve to secondary neurons would be ideally situated to function as an "automatic volume control of input," which could maintain relatively constant excitation of mitral and tufted cells over a wide range of odor intensities. Increased activity in olfactory nerve terminals recruits additional inhibition, increasing the "dynamic range" of mitral and tufted cells. The processes of some juxtaglomerular interneurons appear to extend over a distance corresponding to about two to three glomeruli; an automatic volume control may respond to the input level over a region of four to nine glomeruli. Within a glomerulus, there is presumably strong electrotonic coupling of synaptic currents between parts of individual neurons; coupling between glomeruli is presumably achieved through axons and requires action potentials.

Mitral and tufted cells. Responses of mitral cells to odors are typically complex: during odor exposure, units may be initially excited and then inhibited, inhibited and then excited, or exhibit more complex responses. The character of these responses may alter with odor concentration. Individual units may more reliably discriminate between odors if unit activity is recorded in relation to an artificial sniff cycle. Some researchers have emphasized the necessity of testing the response of each cell over a range of odor concentrations. Testing with several odors, each at several different concentrations, showed that a significant number of cells respond differently to at least two odors at all odor concentrations. Similar results in the salamander led to the concept of *concentration tuning* of bulb units: individual cells appeared to respond best to a particular concentration of each odorant.

Our understanding of the neural correlates of odor perception lag far behind our knowledge of the stimulus-response characteristics of neurons in other sensory systems. It may be that odor quality and concentration are not uniquely represented by single neurons but are instead perceived by populations or ensembles of cells. Analysis of this possibility will require the use of experimental techniques that allow simultaneous measures of large numbers of cells during the presentation of odors. Recent advances in the use of imaging techniques and the use of vital dyes whose spectral emission or absorption properties change systematically with changes in cell membrane potential or ionic concentrations may allow workers to determine if populations of neurons "encode" specific aspects of odors in the olfactory bulb or higher olfactory structures.

Higher Olfactory Structures

☐ Anterior olfactory nucleus

The olfactory bulb is connected to the base of the temporal lobe by a stalk of tissue called the *olfactory peduncle* (Figs. 28-5 and 28-6). In most infraprimate species, the peduncle consists of a population of neurons, the *anterior olfactory nucleus* (AON), and two major tracts of fi-

MAIN OLFACTORY SYSTEM

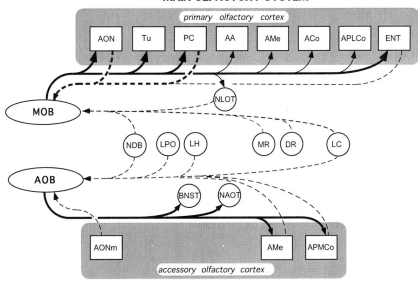

ACCESSORY OLFACTORY SYSTEM

Figure 28-5

Major connections of the main and accessory olfactory systems. The diagram shows the connections of the main (MOB) and accessory bulbs (AOB), with cortical (*gray panels*) and subcortical (*circles*) structures. The relative strengths of the connections are represented by line thickness. Output projections of MOB and AOB are shown by solid lines; reciprocal and centrifugal projections to MOB and AOB are shown by dashed lines. Cortical areas comprising the primary and accessory olfactory cortex are indicated by squares. *AA*, anterior amygdala; *AOB*, accessory olfactory bulb; *ACo*, anterior cortical amygdaloid nucleus; *AMe*, medial amygdaloid nucleus; *AON*, anterior olfactory nucleus (*m*, medial division); *APLCo*, posterolateral cortical amygdaloid nucleus; *APMCo*, posteromedial cortical amygdaloid nucleus; *BNST*, bed nucleus of the stria terminalis; *DR*, dorsal raphe nucleus; *ENT*, entorhinal cortex; *LH*, lateral hypothalamus; *LC*, locus coeruleus; *LPO*, lateral preoptic area; *MOB*, main olfactory bulb; *MR*, median raphe; *NAOT*, nucleus of the accessory olfactory tract; NLOT, nucleus of the lateral olfactory tract; *NDB*, nucleus of the diagonal band; *PC*, piriform cortex; *Tu*, olfactory tubercle.

Figure 28-6

Basic olfactory network. Schematic of the networks linking the olfactory bulb and primary olfactory cortex. Olfactory nerve axons (ON) terminate in the glomeruli (glom) onto mitral (m) and tufted (t) cells, which project through the lateral olfactory tract (LOT) to layer Ia of the primary olfactory cortex to terminate on the dendrites of layer II–III pyramidal (p) cells. Layer II–III pyramidal cells in the rostral olfactory cortex project to layer Ib in the caudal olfactory cortex and vice versa. Olfactory cortical pyramidal cells send reciprocal projections back to the olfactory bulb. The olfactory bulb output is continuously modified by feedback from the areas it targets. Inhibitory interneurons in the olfactory bulb and olfactory cortex (*gray*) modulate the network function. Neurons in the ipsilateral (AONi) and contralateral anterior olfactory nuclei (AON) link olfactory network function in the two hemispheres through the anterior commissure.

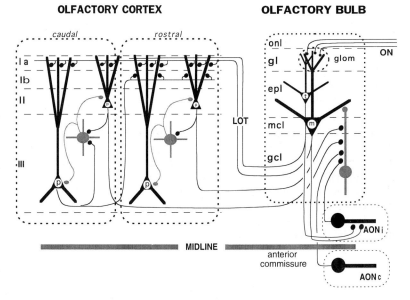

OLFACTORY CORTEX   **OLFACTORY BULB**

bers, the *lateral olfactory tract* and the *anterior limb of the anterior commissure*. Though historically referred to as a nucleus, the AON is now considered a cortical structure with several subdivisions differentiated on the basis of cellular architecture and connectional patterns. A substantial number of AON neurons are pyramidal cells whose apical dendrites extend toward the pial surface of the AON. The major afferents to the AON arise in the ipsilateral bulb and the contralateral AON. The AON also receives afferents from more caudal primary olfactory cortical structures and from a variety of cortical and subcortical structures associated with the limbic system.

The major efferents of the AON are to the *ipsilateral bulb* or olfactory cortex and the *contralateral bulb* and AON. Our understanding of the functional significance of the AON is still rudimentary. The major interbulbar connections of the AON implicate this structure in the interhemispheric processing of olfactory information. There is evidence that binasal mechanisms may function in spatial localization of odors and the AON system would be suspected to play a significant role in such mechanisms. There is also evidence from animal studies that the AON plays a key role in the interhemispheric transfer of olfactory memories.

□ Primary olfactory cortex

Beginning at the caudal limits of the AON, the cortex of the basal temporal lobe expands in the caudal direction, forming a pear-shaped structure, the *piriform lobe*. The rostral part of this structure contains the piriform cortex, the major cortical component of the primary olfactory cortex. Further caudally, the medial part of this cortex overlies the amygdaloid complex; this part of the piriform cortex is referred to as the *periamygdaloid cortex*. Further caudally and medially the piriform periamygdaloid cortex gives way gradually to the parahippocampal cortex and then to the hippocampus. There is a continuous expanse of gradually changing cortical architecture leading from the olfactory bulb, the AON, piriform, periamygdaloid, entorhinal, and subicular cortices directly into the fields of the hippocampus. This orderly anatomic arrangement is one reason that early anatomists referred to the olfactory cortex, parahippocampus, and hippocampus as the rhinencephalon, believing that the entire expanse of cortex constituted the "smell brain." This view was tempered by subsequent research showing that the projections of the olfactory bulb directly innervate only the AON, piriform, periamygdaloid, and lateral entorhinal cortices.

Most of the parahippocampal region and the hippocampus are now considered to be part of the limbic system. Notwithstanding this considerable loss of cortical real estate, the olfactory system can still claim to be the sensory system with the most direct access to the hippocampus, because there are direct projections of the olfactory bulb and piriform cortex to the entorhinal cortex, and the entorhinal cortex is the major source of afferent input to the hippocampus.

The piriform, periamygdaloid, and parts of the entorhinal cortex together are often referred to as the *primary olfactory cortex* (POC), because they are all directly targeted by synaptic inputs from the olfactory bulb. The rostral parts of the POC receive terminals from tufted and mitral cells. The caudal parts of the piriform, the periamygdaloid, and lateral entorhinal cortex receive terminals primarily or exclusively from mitral cells. In all parts of the POC, the olfactory bulb projection terminates in the superficial half of layer I, designated layer Ia. In the olfactory bulb and AON, there is no indication of point-to-point topography in the projections of the olfactory bulb to the POC. This further reinforces the idea that point-to-point topography does not play a significant role in the central representation of odor space and has prompted several workers in the field to suggest that processing of olfactory information in olfactory cortex can be best understood by considering olfactory cortex as a content addressable, distributed neural network. In this view, the functioning of a network is inherent in the organization of its microcircuits, the patterns of connection between input and output of the circuits do not matter so much as the ability of the circuitry to form associative (e.g., memory) linkages with no discernible spatial patterns of anatomic connectivity.

Connections of primary olfactory cortex

The connections of the POC can be discussed as four classes: *intrinsic* or *local*, which are short connections between neurons in different layers of POC; *associative* connections with different parts of POC; *extrinsic*, which are connections with other structures; and *modulatory inputs*, which are afferents that terminate in the POC as part of a broader innervation of other cortical and subcortical neural systems.

Intrinsic or local connections. The POC has two principal layers of pyramidal neurons, layers II and III, which comprise several morphologic classes, and has several classes of nonpyramidal neurons. There are extensive translaminar or local connections among the POC neurons. Layer II neurons give off axon collaterals to the deeper layer III pyramidal cells, and there are local inhibitory interneurons in layers I and II that are contacted by olfactory bulb terminals and by local collaterals of pyramidal cells. Deeper pyramidal cells also give rise to extensive local collaterals that may synapse with local interneurons or with more superficial pyramidal cells. There are extensive translaminar connections from the superficial to deeper layers and vice versa. There also are several classes of GABA-ergic and neuropeptide-containing neurons in the POC, and although the connection of these neurons is not known, many of them have the appearance of local interneurons.

Association connections. Corticocortical projections within the POC are extensive and exhibit some degree of laminar and regional organization. Axons from the pyramidal cells of layer II are primarily directed at more caudal sites in the POC; cells in layer III project predominantly to rostral parts of the POC. Commissural fibers to the contralateral POC arise from layer II of the anterior parts of the POC. The ipsilateral and commissural association projections of the POC terminate in a highly laminar fashion in layer Ib, immediately below the zone that contains the inputs from the olfactory bulb; a lighter projection terminates in layer III. The POC projections back to the AON also terminate in layer Ib, below the bulb recipient zone. Neurons in layers II and III send a dense projection back to the olfactory bulb; this feedback pathway terminates primarily in the granule cell layer.

Extrinsic connections. The major extrinsic connections of the POC are its reciprocal connections with the olfactory bulb and AON and its efferent projections to various nonolfactory cortical-subcortical targets.

Modulatory inputs. The POC also receives subcortical modulatory inputs from the locus coeruleus (i.e., norepinephrine), midbrain raphe nuclei (i.e., serotonin), and magnocellular basal forebrain, from the nucleus of the diagonal band (i.e., cholinergic, GABA-ergic) and from the ventral tegmental area (i.e., dopamine).

Beyond the olfactory cortex

Two classes of the POC outputs were discussed previously: the feedback projection to the olfactory bulb and the association connections between rostral and caudal olfactory cortex. A third class of outputs is treated separately, because it represents the projections of the POC to brain regions not generally included in the olfactory system, although their receipt of inputs from the POC implicates these POC targets in olfactory function. The extrinsic outputs of piriform cortex are to cortical and subcortical structures.

Neocortical projections. The olfactory bulb projection to the POC extends dorsally beyond the cytoarchitectural limits of the POC into the ventral parts of the granular insular and perirhinal cortices. There are also direct projections from the POC to the insular and orbital cortices. The insular and orbital cortices are the primary cortical targets of ascending pathways arising in the nucleus of the solitary tract in the medulla and appear to contain the primary cortical representations for gustatory and visceral sensation. Olfactory projections to the insular and orbital cortices may be sites that integrate olfactory and gustatory signals to generate the integrated perception of flavor. These same cortical areas have descending projections to the hypothalamus and back to the nucleus of the solitary tract; these corticofugal projections influence visceral-autonomic and possibly gustatory functions. This circuitry may allow olfactory modulation of autonomic function. Neurons in these cortical areas in primates respond to odors with a higher degree of selectivity than neurons in the olfactory bulb or the POC. These neocortical sites may play a role in the discrimination of different odors.

Subcortical Projections

Hypothalamus. The heaviest and most direct projections to the hypothalamus arise from neurons in the deepest layers of piriform cortex and the anterior olfactory nucleus. These projections terminate most heavily in the lateral hypothalamic area. Olfactory-recipient parts of the cortical and medial amygdaloid nuclei also project to medial and anterior parts of the hypothalamus.

Thalamus. There is a strong projection from the POC to the magnocellular, medial part of the mediodorsal thalamic nucleus and the submedial nucleus (i.e., nucleus gelatinosa). These thalamic nuclei project to the orbital cortex and frontal lobes.

Impact of Olfaction on Behavior

Complete removal of the olfactory bulbs completely eliminates the ability to detect or discriminate odors. This is not particularly surprising, because the bulb is the sole target of ORNs, and after bulbectomy, the ORNs degenerate. What is somewhat surprising is that, after removal of all but 10% of the olfactory bulbs, animals can detect and discriminate odors. This suggests that odors are not represented in discrete sites in the olfactory bulb, because animals with 90% of the bulb removed should otherwise be unable to smell some odors. Contrasting with this conclusion are experiments showing that discrete sites in the bulb have increased metabolic activity after exposure of the animal to different odors. The question of whether different odors are selectively represented by spatially discrete populations of neurons in the olfactory bulb remains unresolved.

Olfaction is important to many behaviors. For many animals, odors play an essential role in reproductive and maternal behaviors. Bulbectomized males of some mammalian species do not mate with receptive females. In other species, olfactory cues are not essential, but mating behavior is reduced by damage to the olfactory system. In females, olfactory bulb lesions severely impair gonadal steroid function.

In humans, a developmental disorder called *Kallmann's syndrome* is caused by impaired migration of certain neurons from the developing olfactory epithelium into the brain (see Clinical Correlation in Chap. 8). One of the neuron types that fails to enter the brain contains the neuropeptide gonadotropin-releasing hormone (GnRH). Normally, GnRH cells migrate from the epithelium to the hypothalamus and preoptic area during fetal development; many of these cells generate axons that terminate in the median eminence, where they release GnRH. The GnRH acts on cells in the pituitary, causing the release of luteinizing hormone (LH) into circulation. LH is necessary for the proper development of reproductive organs at puberty. In Kallmann's syndrome, the GnRH cells do not enter the brain, and these patients have gonadal atrophy. The genetic defect that prevents the normal migration of GnRH cells into the brain often also prevents ORN axons from reaching the olfactory bulb. As a result, the olfactory system fails to develop, and these persons are unable to smell (i.e., anosmia).

In some species, odors also signal identity and social status. For example, if a pregnant female mouse is exposed to the odor of the urine of a strange male, she aborts. There is also evidence that the hormonal status of some animals influences their ability to detect certain odors. Female sheep become unusually sensitive to odors of their own lambs at parturition.

In humans, odors do not appear to have such profound effects on behavior and endocrine function, although there have been relatively few studies of this subject. It should not be concluded, however, that olfaction is unimportant in humans. A casual glance at the ledger sheets of the food, beverage (i.e., flavor is more olfaction than taste), and fragrance industries leaves little doubt about the importance of olfaction and taste to the quality of our lives!

□ Why don't we understand olfaction?

Our inability to understand how odors are coded by olfactory receptor neurons in the olfactory epithelium has been a formidable obstacle to understanding how olfactory signals are processed by the olfactory bulb and subsequent stages in the olfactory pathway. However, this situation may change, with new insights provided by advances in understanding the molecular and biophysical bases of odor transduction in olfactory receptor neurons.

There is considerable information about the anatomy and classic electrophysiology of the olfactory system. Unfortunately, this body of knowledge gives little indication how the system works to allow us to detect and recognize odors or how odors are able to influence behavior and endocrine function. Compared with other sensory systems, our knowledge of the integrative properties of the olfactory system is rudimentary. Because it is the primary output target of the olfactory bulb, it has been tempting to think that the olfactory cortex is involved in some kind of hierarchic or higher-order stimulus feature extraction. This expectation is based on analogy with the organization of the other major sensory systems, for which it has been possible to infer how neurons, at successive levels from the periphery through subcortical relays to the primary sensory cortex, transform inputs to extract different features of the sensory signal.

In the case of the olfactory system, the prevailing *sensory systems paradigm* has provided little insight into the neural operations that lead to the perception of odor qualities or the features of odors that modify behavior. It is reasonable to wonder why the paradigm that has worked so well for understanding the associations among sensory stimuli and the neural mechanisms that encode stimulus features in other sensory systems has fallen so short in the olfactory system.

There are many possible answers to this puzzle, but none seems to be uniquely compelling. One problem is that we have a poor understanding of olfactory stimuli and the relation between olfactory stimuli and sensory transduction in olfactory receptor neurons. In the case of vision, audition, and somatic sensations, there is a fairly intuitive, Newtonian appreciation of the relevant dimensions of light, sound, and pressure that is not immediately obvious for the relation between a molecule and an odor. Complex patterns of light can be thought of in terms of contrast, hue, reflection, and absorption of light, but such dimensions do not leap readily to mind for odors.

Similarly, there is an almost intuitive grasp of the utility of a "place code" in other sensory systems. Light patterns falling on different parts of the retina correspond to different sites of origin of light rays. An itch on the tip of the nose defines a spatial locus that invites a directed behavioral response. The place of the stimulus is used to guide the motor system to a correct location, but foveation and scratching have no obvious equivalents in the olfactory system. It is difficult to overemphasize how critical the simple place coding principle has been to the analysis of neural mechanisms in other sensory systems. The poverty of our intuitive sense of the odor world, combined with the lack of any obvious meaning to a place-code relation between the olfactory receptor sheet and subsequent levels of the olfactory system, have greatly impeded our attempts to understand the olfactory system by analogy with other sensory systems.

Despite these conceptual problems, there have been repeated efforts to understand olfactory anatomy and physiology by comparison with other sensory systems. Almost every decade, someone tries to make the olfactory bulb like the retina. The similarity between the alternating layers of neuron types with intervening plexuses of synaptic integration in the retina and the olfactory bulb make it irresistibly tempting to try to understand the circuitry of the bulb in terms of the retina. This analogy is maintained only by ignoring equally compelling and fundamental differences between the retina and the bulb. Chief among these is the existence of massive feedback inputs to the olfactory bulb and the total absence of centrifugal inputs to the retina in mammals. In some non-mammalian vertebrates, there is a midbrain nucleus that projects to the retina, but this feature is lacking in mammals. The retina transduces light, processes the neural responses by means of networks of intrinsic neurons, and then conducts the output to the geniculocortical or collicular systems. In the olfactory bulb system, the outputs of the bulb are paralleled at every step by massive feedback pathways that modulate populations of bulbar interneurons, which directly regulate the output neurons. In this sense, the bulb is not so much a relay to higher olfactory structures as it is the first stage in a circuit that feeds back on itself at every stage of synaptic transfer. From this perspective, attempts to analogize the olfactory bulb with the retina may be misleading, because the functional status of bulb neurons at any moment reflects activity in sensory afferents, interneurons, and feedback pathways from the AON and olfactory cortex.

Because the AON and POC are the predominant output targets of the bulb, it follows that olfactory bulb neurons are continuously modulated by signals that represent transformations of their own output. The retina does not work this way. The output of ganglion cells is modulated by intraretinal processing but not by operations performed in the output targets of the retina. Progress in understanding the functions of the olfactory system may depend in large measure on our ability to develop new paradigms of neural network operation, because traditional paradigms lack the power necessary to encompass the complexity of olfactory circuits. This may be a useful and important endeavor, because the classic sensory paradigm has probably outlived its utility and is not adequate to model the neural operations that underlie cognitive functions. The olfactory system is an inherently cortical sensory system. The eventual understanding of olfactory network operations may provide new insights for analyzing higher cortical functions.

Sleep, dreams, and states of consciousness

Neuroscience in Medicine, edited by P. Michael Conn.
J. B. Lippincott Company, Philadelphia © 1995

Chapter 29

Sleep, Dreams, and States of Consciousness

ROBERT W. MCCARLEY

Organization of Sleep and Wakefulness

Understanding the brain basis of consciousness is one of the oldest dreams of neuroscientists and physicians. We are beginning to understand some of the basic mechanisms controlling the changes of consciousness associated with sleep and wakefulness.

☐ The electroencephalogram is an important but limited tool for the study of states of consciousness

We begin with a brief review of the biologic basis of the electroencephalogram (EEG) and evoked potentials. When activity is synchronous, neurons in the cerebral cortex generate electrical signals strong enough to be detected through the skull by sensors (i.e., electrodes) placed on the scalp. These small (i.e., microvolt) electrical signals are amplified and filtered to produce EEG recordings. Although the EEG is a crude way to determine brain activity—similar to figuring out what is happening in a football game by putting microphones outside the stadium—it has proved to be a remarkably useful tool for studying the basic structure of sleep in humans.

Figure 29-1 illustrates, for one cortical neuron, the source of currents underlying the EEG. The influx of positive ions into the soma of the neuron after a depolarizing postsynaptic potential generates a current "sink," be-

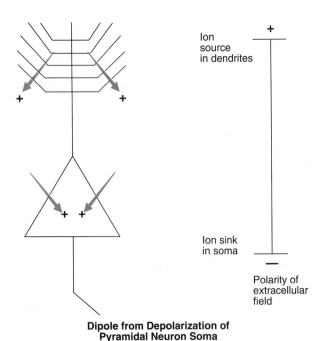

Dipole from Depolarization of Pyramidal Neuron Soma

Figure 29-1
Example of a cortical neuronal dipole occurring with a depolarizing postsynaptic potential on the soma. This pattern of current flow, repeated over many thousands of cortical neurons, is probably the main generator source for electroencephalographic and evoked potential waves.

cause, by convention, current is composed of positive ion flow. The apical dendrites of this cortical neuron act as a "source" of positive ions and current flow. This current flow pattern of a source in the dendrites and a sink in the soma creates a *dipole* (i.e., a "two pole") with the positive pole or source in the dendrites and the negative pole or sink in the soma. In cerebral structures with a regular laminar structure such as the cerebral cortex, this simple dipole model repeated over many constituent neurons provides a reasonable first approximation to how positive EEG waves are generated. In the case of a hyperpolarizing postsynaptic potential in the soma, the dipole polarity is reversed, with the source in the soma and sink in the dendrites. Membrane hyperpolarization arises as the consequence of an net efflux of positive ions.

Most of the components of the EEG arise from the currents generated by postsynaptic potentials, as described in our example, and not from the currents generated by action potentials. Action potentials usually are too brief and too asynchronous to sum and produce a signal detectable from the scalp. Evoked potentials may be thought of as EEG waves that occur after a sensory stimulus. Because they are time locked to the stimulus, they may be averaged to improve the signal to noise ratio.

Many investigators are exploring the utility of modeling "equivalent dipoles" as a representation of the average amplitude and polarity of EEG and evoked potentials arising within a cerebral region, such as the sensory receiving areas of the cortex. Practical constraints to localizing the source of evoked potentials and EEG include the use of scalp recordings and the consequent "smearing" of current flow as the boundaries between zones of different conductivities are traversed. For example, the brain and its extracellular fluid is a much better conductor than the skull. Cohen and Cuffin (1990) applied an experimental approach to the question of the accuracy of source localization possible with electrical signals. Using patients who had deep electrodes implanted to locate the seizure source before surgery, these investigators passed a low level signal through two deep electrodes (a true dipole source!) and examined how closely the signal source within the brain could be localized from scalp electrode recordings. For this single source, brain localization was found to be accurate to within about 1 cm. However, as the number of sources increases, as is usually the case in brain processing, the ability to localize them diminishes.

The EEG is perhaps most useful in detecting the presence of seizure activity and in pinpointing changes in alertness and sleep stages. The EEG is described in terms of the amplitude of its waves and their frequency. As shown in Table 29-1, EEG frequencies are grouped into bands, which range from the very low frequencies (i.e., delta, 0.5 to 4 Hz) to the very fast (i.e., beta, 14 to 32 Hz). As a general rule, the delta EEG frequencies are associated with states of consciousness with little complex processing, such as non-rapid eye movement (non-REM) sleep, and those with higher frequencies are associated with more complex processing, as in wakefulness and REM sleep (i.e., dreaming).

Table 29-1
ELECTROENCEPHALOGRAPHIC FREQUENCY BANDS

Name	Frequency Range (Hz)
Delta	0.5–4
Theta	4–8
Alpha	8–14
Beta	14–32

The *alpha rhythm* has a frequency range of 8 to 13 Hz and is best recorded over the occipital scalp region. It occurs during wakefulness, often appearing on eye closure and disappearing with eye opening. Depth recordings in animals indicate alpha rhythm frequencies may also be present in the visual thalamus (i.e., lateral geniculate body, pulvinar), and the cortical component appears to be generated in relatively small cortical areas that act as epicenters. There are no definitive studies of the genesis of this rhythm, although interaction of corticocortical and thalamocortical neurons has been postulated. Origins of the delta waves are discussed later.

☐ Sleep is organized into a definite structure

One third of our lives is spent in sleep. No other single behavior occupies so much of our time, but few other behaviors have been so mysterious. We know that there are two main states of sleep, REM sleep, typically associated with a high level of brain neuronal activity and the distinctive conscious state of dreaming, and non-REM sleep, typically associated with a low level of neuronal activity and nonvisual, ruminative thinking.

A typical study of sleep includes records of the EEG, of eye movements by the electro-oculogram (EOG), and of muscle tone by the electromyogram (EMG). This ensemble of records is known as a polysomnogram and the recording process is called polysomnography. These key records enable us to describe the main stages of sleep. As sleep onset approaches, the low amplitude, fast frequency EEG of alert wakefulness, often with alpha present (Fig. 29-2A), yields to stage 1 sleep, a brief transitional phase between wakefulness and "true" sleep. This stage is often called *descending stage 1* because it is a prelude to deeper sleep stages and because it is characterized by low-voltage (amplitude), relatively fast-frequency EEG patterns and slow, rolling eye movements. During stage 2 sleep, there are episodic bursts of rhythmic, 14- to 16-Hz waveforms in the EEG, known as *sleep spindles*, interspersed with occasional short-duration, high-amplitude *K complexes*, so named because of their morphologic resemblance to this letter. During stage 2, the EEG slows still further. Stages 3 and 4 are

defined, respectively, by lesser and greater occurrence of high-amplitude, slow (0.5 to 4 Hz) waveforms, called *delta waves*. The low-voltage, fast-frequency EEG pattern of REM sleep is in marked contrast to delta sleep and resembles the non-alpha EEG pattern of active wakefulness and stage 1 descending (see Fig. 29-1A).

REM sleep is further characterized by the presence of bursts of rapid eye movements and by loss of muscle tone in certain major muscle groups of the limbs, trunk, and neck. The other sleep stages are often lumped together and called non-REM sleep. (Researchers working with animals often use the term "slow wave sleep" for non-REM sleep, and this term sometimes occurs in the human literature, although, properly speaking, stages 1 and 2 do not have slow waves.) Table 29-2 summarizes the chief differences between waking, non-REM, and REM sleep in a polysomnographic recording.

There is a rather predictable pattern of shifting between one sleep state and another pattern during a typical night's sleep (Fig. 29-2B). As the night begins, there is a stepwise descent from wakefulness to stage 1 through to stage 4 sleep, followed by a more abrupt ascent back toward stage 1. However, in place of stage 1, the first REM sleep episode usually occurs at this transition point, about 70 to 90 minutes after sleep onset. The first REM sleep episode in humans is short. After the first REM sleep episode, the sleep cycle repeats itself with the appearance of non-REM sleep and then, about 90 minutes after the start of the first REM period, another REM sleep episode occurs. This rhythmic cycling persists throughout the night. The REM sleep cycle length is 90 minutes in humans, and the duration of each REM sleep episode after the first is approximately 30 minutes. Over the course of the night, delta wave activity tends to diminish, and non-REM sleep has waves of higher frequencies and lower amplitude. As Figure 29-2B makes clear, body movements during sleep tend to cluster just before and during REM sleep. In general, the ease of arousal from sleep parallels the ordering of the sleep stages, with REM and stage 1 being the easiest for arousal and stage 4 the most difficult.

Ontogenetic and Phylogenetic Features of Sleep

Periods of immobility and rest are experienced by many lower animals, including insects and lizards. Because of the absence of a cortical brain structure like that of humans, it is difficult to say whether the absence of slow waves in these animals means they don't have the equivalent of human non-REM sleep or this occurs but is expressed in different forms not detectable with EEG recordings. REM sleep is experienced by all mammals, except egg-laying mammals (i.e., monotremes) such as the echidna (i.e., spiny anteater). Birds have very brief bouts of REM sleep. REM sleep cycles vary in duration according to the size of the animal; elephants have the longest cycle, and smaller animals have shorter cycles.

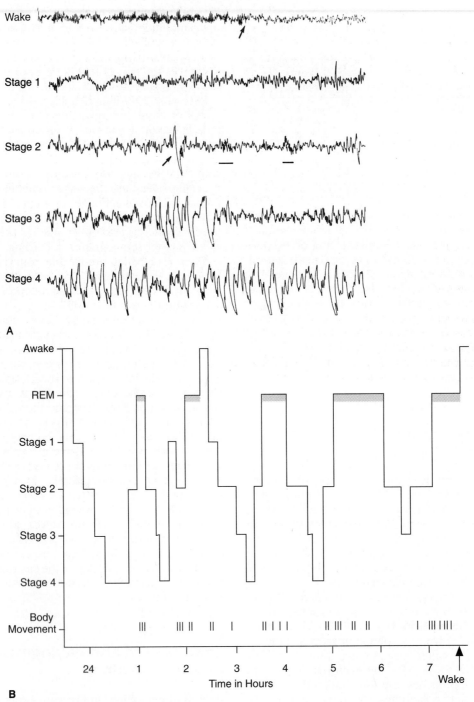

Figure 29-2
(**A**) The electroencephalographic (EEG) patterns associated with wakefulness and the stages of sleep. (**B**) The time course of sleep stages during a night's sleep in a healthy young man. During wakefulness, there is a low-voltage fast EEG pattern, often with alpha waves, as shown here. At the arrow, there is a transition to stage 1 sleep, with the loss of the alpha rhythm and the presence of a low-voltage fast EEG. As sleep deepens, the EEG frequency slows more and more. Stage 2 is characterized by the presence of K complexes (*arrow*) and sleep spindles (*underlined*). During stage 3, delta waves (0.5–4 Hz) appear, and in stage 4, they occur more than 50% of the time. During rapid eye movement (REM) sleep (*black bars*), the EEG pattern returns to a low-voltage fast pattern. The percentage of time spent in REM sleep increases with successive sleep cycles, and the percentage of stages 3 and 4 decreases. The EEG segments were recorded over parietal lobe (C3) except in waking, where occipital recording (O2) was used to show the alpha rhythm most clearly. (Adapted from Carskadon MA, Dement WA. Normal human sleep: an overview. In Kryger MH, Roth T, Dement WC (eds). Principles and practices of sleep medicine. New York: WB Saunders, 1989:3.)

Table 29-2
POLYSOMNOGRAPHIC DEFINITION OF WAKEFULNESS, NON-REM AND REM SLEEP

State	EEG Amplitude and Main Frequencies	Rapid Eye Movement (EOG)	Muscle Tone (EMG)
Waking	Low voltage, fast	+	+
Non-REM sleep	High voltage, slow	−	−
REM sleep	Low voltage, fast	+	−

EEG, electroencephalogram; *EOG*, electro-oculogram; *EMG*, electromyogram; *non-REM*, non–rapid eye movement sleep; *REM*, rapid eye movement sleep.

For example, the cat has a sleep cycle of approximately 22 minutes, and the rat cycle is about 12 minutes.

In utero, mammals spend a large percentage of time in REM sleep, ranging from 50% to 80% of a 24-hour day. Animals born with immature nervous systems have a much higher percentage of REM sleep at birth than the adults of the same species. For example, sleep in the human newborn occupies two thirds of the time, with REM sleep occupying one half of the total sleep time or about one third of the entire 24-hour period (Fig. 29-3A). In infants born 10 weeks prematurely, the percentage of REM sleep in the total sleep time reaches 80%. The percentage of REM sleep declines rapidly in early childhood, and by approximately 10 years of age, the adult percentage of 20% to 25% of total sleep time is reached. The predominance of REM sleep in the young suggests an important function in promoting nervous system growth and development. The absolute amount of REM sleep is greater at birth in animals that are born immature (i.e., altricial) than those that are born more mature (i.e., precocious).

Stage 4 sleep, defined by the presence of delta EEG frequencies, is minimally present in the newborn but increases over the first years of life, reaching a maximum at about 10 years of age and declining thereafter (Fig. 29-3B). Feinberg and coworkers (1990) have examined the time course of intensity of delta waves over the first three decades of life. They find that this time course of delta wave intensity is fit by a particular probability distribution (i.e., gamma distribution) and that approximately the same time course obtains for synaptic density and PET measurements of metabolic rate in human frontal cortex. They speculate the correlated reduction in these three variables may reflect a pruning of redundant cortical synapses that is a key factor in cognitive maturation, allowing greater specialization and sustained problem solving.

A frequent question asked of physicians is how much sleep is needed. The answer partly depends on the age of the person. One guideline is that enough sleep is needed to prevent daytime drowsiness. Each person seems have a particular set-point requirement. In adults, the modal value of sleep need is close to the traditional 8 hours, but there is considerable individual variation. If someone functions and feels well on less sleep, there is little need for concern.

Determination of Sleep Onset and Sleepiness

☐ Circadian factors partially determine sleepiness

In adult humans, the period of maximal sleepiness occurs at the time of the *circadian* low point of the temperature rhythm (Fig. 29-4). Circadian means about a day (circa = about), and the circadian temperature rhythm of humans can be thought of as a sine wave function with a minimum that occurs between 4 and 7 AM in subjects with a normal daytime activity schedule. Accidents occur most frequently at times near this circadian temperature minimum, because this is the time of maximal sleepiness. Per vehicle mile, the risk for truck accidents is greatest at this time. The nuclear reactor incidents at Chernobyl and Three Mile Island also occurred in the early morning hours. There is a secondary peak of sleepiness that occurs about 3 PM (see Fig. 29-4), corresponding to a favored time for naps. The main functional consequence of deprivation of sleep seems to be the presence of *microsleeps*, which are very brief episodes of sleep during which sensory input from the outside is diminished and cognitive function is markedly altered. As every parent knows, human newborns do not have a strong circadian modulation of sleep, and some species, such as the cat, do not have much circadian modulation even as adults.

☐ Prior wakefulness partially determines sleepiness

Mathematical models of sleep propensity have been developed by Kronauer and coworkers (1982), who emphasized circadian control, and by Borbély and coworkers (1982), who emphasized the extent of prior wakefulness. Borbély and coworkers' model postulates that the intensity and amplitude of delta wave activity, as measured by power spectral analysis, indexes the level of sleep factors and slow wave sleep drive. In this model, the time course of delta activity over the night, a declining exponential, reflects the dissipation of the sleep factors. The nature of the postulated underlying sleep factors has not been specified.

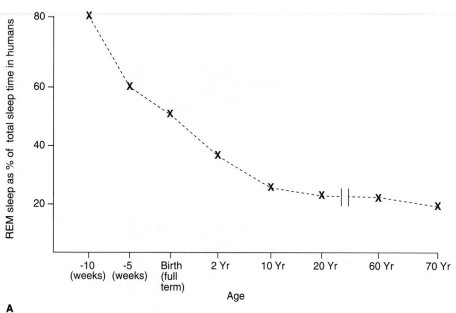

A

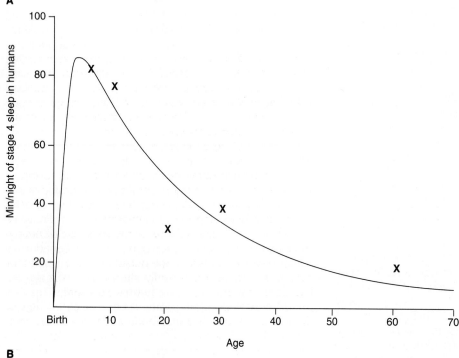

B

Figure 29-3
(**A**) REM sleep as a percentage of total sleep time in infants born 10 and 5 weeks prematurely, at full-term birth, and in children and adults at the indicated years of age. Notice the dramatic decline of REM sleep during early life and the long plateau during maturity, with a decline observed only in the seventh decade. (**B**) Stage 4 (i.e., delta) sleep is shown in minutes as a function of age. There is little delta wave activity at birth, presumably reflecting cortical immaturity. Delta wave activity peaks between 3 and 5 years of age and declines exponentially thereafter. (Adapted from Feinberg I. Effects of age on human sleep patterns. In Kales A (ed). Sleep, physiology and psychology. Philadelphia: JB Lippincott, 1969:39.)

Slow Wave Sleep Factors and the Neurophysiologic Basis of Slow Waves

☐ Several humoral factors may cause non-REM sleep

The status of sleep factors is one of uncertainty. It is unknown whether they participate in the natural regulation of sleep, although they are certainly active as pharmacologic agents that decrease wakefulness and increase non-REM sleep.

Pappenheimer, Karnovsky, and Krueger (reviewed in Krueger, 1990) demonstrated that *muramyl peptides* were concentrated in the cerebrospinal fluid and urine of sleep deprived animals. These muramyl peptides reliably induce non-REM sleep when injected into the lateral ventricles or into the basal part of the forebrain. The compounds also induce hyperthermia. They are derived from bacterial cell walls, and it has been theorized they may be absorbed from the gut and act like "vitamins" for the production of sleep. Another sleep factor is interleukin-1. Interleukin-1 is a cytokine that is produced in response to infections and by injections of components of

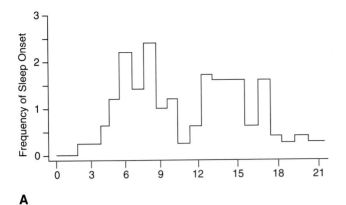

A

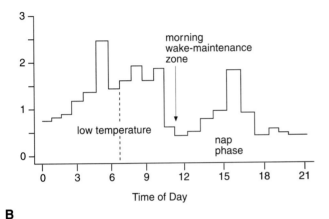

B

Figure 29-4

Circadian control of sleepiness and sleep onset. (**A**) Sleepiness at various clock times for persons on a constant routine. Sleepiness was measured by Carskadon as frequency of unintended microsleeps in patients instructed to stay awake, with a frequency of 1 indicating the average across all measurements. Notice the major peak about 6 AM, at the presumptive time of circadian temperature minimum, and a secondary peak about 3 PM, a favored time for a nap. (Carskadon MA, Dement WA. Normal human sleep: An overview. In Kryger MH, Roth T, Dement WC (eds). Principles and practices of sleep medicine. New York: WB Saunders, 1989:3.) (**B**) Sleep propensity measured as the number of self-selected bedtimes or sleep onsets in subjects in whom temperature was continuously monitored. As in **A**, the maximal number of sleep onsets occur near the temperature minimum, and a secondary peak occurs at a circadian phase corresponding to about 3 PM. These patients were maintained without circadian cues and showed decoupling of the activity and the temperature rhythms (i.e., internal desynchronization) that are otherwise synchronized by external circadian cues, such as dawn and dusk. The sleep onsets were converted to approximate times of day by assuming a temperature minimum at 6:30 AM. (Adapted from Strogatz SH. The mathematical structure of the human sleep-wake cycle. New York: Springer-Verlag, 1986.)

bacterial cell walls, such as muramyl peptides. It increases non-REM sleep and produces hyperthermia. Hyperthermia itself may increase non-REM sleep, but blocking the hyperthermic effects of interleukin-1 does not block the non-REM sleep-inducing effects (Krueger, 1990). The argument that interleukin-1 and muramyl

peptides are important in the hypersomnia associated with infections is strong, although the determination that they are natural sleep factors awaits further research.

In another line of research, Hayaishi and coworkers (1988) found that injections of prostaglandin D_2 into the third ventricle reliably produced non-REM sleep. They proposed that it is a natural sleep regulatory factor. However, Krueger and coworkers were unable to confirm these data. Delta sleep-inducing peptide has been proposed as a sleep factor by Monnier and coworkers; however, the literature does not suggest a robust effect. Rainnie and coworkers' recent finding of adenosine inhibition of cholinergic neurons important for arousal (see below) suggests that a buildup of extracellular adenosine concentration as a result of brain metabolism might act to promote sleep after prolonged wakefulness.

☐ Neurophysiologic basis for electroencephalographic synchronization and desynchronization

The high-voltage slow-wave activity in the cortex during most non-REM sleep, called *EEG synchronization*, contrasts sharply with the low-voltage, fast-frequency pattern, called *desynchronized* or *activated*, characteristic of waking and REM sleep and consisting of frequencies in the beta range (approximately 14 to 32 Hz).

One of the major advances of the past few years has been the establishment of the importance of a cholinergic activating system in EEG desynchronization. This is probably the major component of the "ascending reticular activating system," a concept that arose from the work of Moruzzi and Magoun (1949) before methods were available for labeling of neurons using specific neurotransmitters. We now know that a group of the neurons in the brain stem cholinergic nuclei near the pons-midbrain junction have high discharge rates in waking and REM sleep and low discharge rates in non-REM sleep. Figure 29-5 indicates the location of the cholinergic laterodorsal (LDT) and pedunculopontine tegmental nuclei (PPT). There is also extensive anatomic evidence that these cholinergic neurons project to thalamic nuclei important in EEG desynchronization and synchronization. In vivo and in vitro neurophysiologic studies have indicated that the target neurons in the thalamus respond to cholinergic agonists in a way consistent with EEG activation.

Cholinergic systems are not the exclusive substrate of EEG desynchronization; brain stem reticular neuronal projections to thalamus, probably using excitatory amino acid (EAA) neurotransmission, and noradrenergic projections from locus ceruleus (for waking, because locus ceruleus neurons are silent in REM sleep) may also play important roles. In addition to brain stem cholinergic systems, cholinergic input to thalamus and cortex from the basal forebrain cholinergic nucleus basalis of Meynert also plays a role in EEG desynchronization.

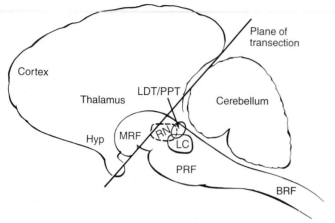

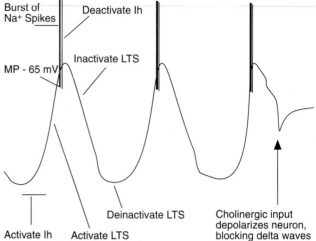

Depolarizing Intrinsic Currents:
Ih = hyperpolarization-activated current
LTS = low -threshold calcium spike

Figure 29-5

Schematic of a sagittal section of a mammalian brain (cat), showing the location of the nuclei especially important for sleep. *BRF, PRF,* and *MRF,* bulbar, pontine, and mesencephalic reticular formation; *LDT/PPT,* laterodorsal and pedunculopontine tegmental nuclei, the principal site of cholinergic (acetylcholine-containing) neurons important for rapid eye movement (REM) sleep and electroencephalographic desynchronization. *LC,* locus ceruleus, where most norepinephrine-containing neurons are located; *RN,* dorsal raphe nucleus, the site of many serotonin-containing neurons; *HYP,* hypothalamus. The oblique line is the plane of transection that preserves REM sleep signs caudal to the transection but abolishes them rostral to the transection. (Adapted from McCarley RW. The biology of dreaming sleep. In Kryger MH, Roth T, Dement WC (eds). Principles and practice of sleep medicine. Philadelphia: WB Saunders, 1989:173.)

Sleep spindles

Spindles occur during stage 2 of human sleep and in the light slow-wave sleep phase of animals. They are composed of waves of approximately 10 to 12 Hz frequency; the wave amplitudes waxes and then wanes over the spindle duration of 1 to 2 seconds. Wave frequency varies between species and is higher in primates. Spindles are relatively well understood at the cellular level. Studies by Steriade and coworkers (reviewed in Steriade and McCarley, 1990) indicate spindle waves arise as the result of interactions between spindle pacemaker GABA-ergic thalamic nucleus reticularis (RE) neurons and thalamocortical neurons. Spindle waves are blocked by cholinergic brain stem–thalamus projections, which hyperpolarize the RE neurons. The forebrain nucleus basalis also provides cholinergic and hyperpolarizing GABA-ergic input to RE that assists brain stem input in disrupting the spindles.

Delta electroencephalographic activity

The cellular basis of delta waves (0.5 to 4 Hz) is an area of intense investigation. Figure 29-6 is a schematic of the mechanisms proposed for the generation of delta waves by thalamocortical neurons. This sketch portrays intracellularly recorded events in a thalamocortical neuron during delta wave generation and is based on in vivo and in vitro recordings by McCormick and by Steriade and

Figure 29-6

The mechanisms proposed for the generation of delta waves by thalamocortical neurons. Without exogenous input, these neurons show a spontaneous oscillation of membrane potential and action potential production that is in the delta frequency range and probably drives the cortical delta rhythm. Oscillation occurs because of an interplay of intrinsic membrane currents. When the membrane potential (MP) is hyperpolarized, (− 80 mV), a particular cation current, called Ih (I, current, h, hyperpolarized), is activated. This inward flow of positive ions from Ih depolarizes the membrane and thereby activates a calcium current called the low-threshold spike (LTS). The inrush of calcium ions further depolarizes the neuron to the firing threshold of the sodium action potential, and a burst of action potentials is produced. Ih is turned off or deactivated at depolarized potentials. The LTS current is automatically turned off by another process called inactivation. The membrane then returns to its previous hyperpolarized level, which removes the LTS current inactivation and renders it ready for activation. The cycle then repeats. Delta oscillations are halted by exogenous, depolarizing input, such as the illustrated brainstem cholinergic input. (Based on data from McCormick DA, Pape H-C. Properties of a hyperpolarization-activated cation current and its role in rhythmic oscillation in thalamic relay neurons. J Physiol 1990;431:291; McCormick DA, Pape H-C. Noradrenergic and serotonergic modulation of a hyperpolarization-activated cation current in thalamic relay neurons. J Physiol 1990;431:319; and Steriade M, Curró Dossi R, Nuñez A. Network modulation of a slow intrinsic oscillation of cat thalamocortical neurons implicated in sleep delta waves: cortically induced synchronization and brain stem cholinergic suppression. J Neurosci 1991;11:3200.

their coworkers. The basic concept is that a hyperpolarized membrane potential permits the occurrence of delta waves in thalamocortical circuits. Any factors depolarizing the membrane block delta waves. During waking, input from the cholinergic forebrain nucleus basalis is important for suppression of slow-wave activity, as shown by lesion studies (Buzsaki et al., 1988). Brain stem norepinephrinergic and serotonergic projections may disrupt delta activity in waking, although they are inactive during REM sleep. During REM sleep, cholinergic input from brain stem is a major factor, producing membrane depolarization, with reticular formation input, probably using EAA neurotransmission, also playing an important role. This membrane depolarization leads to suppression of delta wave activity. The delta waves during sleep may be seen to represent thalamocortical oscillations occurring in the absence of activating inputs.

From the standpoint of the cellular physiologist, the relative intensity of cortical desynchronization correlates well with the intensity of cholinergic input to thalamus; conversely, the relative intensity of cortical synchronization, including delta waves, correlates well with the relative absence of cholinergic activity. The identification of desynchronizing processes in sleep with ascending brain stem cholinergic and reticular activation means that the increasing intensity of EEG desynchronization preceding REM sleep is related to the increasing level of activity of REM-related cholinergic and reticular activity that precedes this state.

□ Slow wave sleep at the cellular level in the thalamus: the burst mode of relay cell discharge is responsible for the failure of information transmission

Extracellular recordings by the Benoit and the author demonstrated that dorsal lateral geniculate relay neurons discharged in stereotyped bursts during non-REM sleep but not during waking or REM. Subsequent in vivo (by Steriade and coworkers) and in vitro investigations (by McCormick and coworkers) indicate the bursting in thal-amocortical neurons occurs when the membrane is hyperpolarized, as illustrated in Figure 29-6, in association with the delta EEG rhythm. This hyperpolarization removes the inactivation of particular Ca^{2+} channels and enables the production of a "calcium spike" (e.g., an inrush of depolarizing calcium ions) when a small depolarization occurs. This depolarizing calcium spike is called a *low threshold spike* (LTS) to differentiate it from other calcium currents with different triggering thresholds. The LTS depolarizes the neuron sufficiently to reach the threshold for fast sodium action potentials, and a burst of these action potentials rides on the LTS. However, the production of a LTS limits the discharge frequency of relay neurons and hence blocks rapid information transmission, as illustrated in Figure 29-7.

Rapid Eye Movement Sleep Physiology and Relevant Brain Anatomy

Figure 29-8 illustrates the time course of discharge activity of neurons thought to be important in the generation of REM sleep and characterized as *REM-on neurons* because of their selective activity during this state. These neurons are located in the LDT/PPT nuclei and use acetylcholine as a neurotransmitter. Activity in this group of neurons recruits activity in effector neurons for REM sleep phenomena that are located in the brain stem reticular formation. Neurons in the locus ceruleus using norepinephrine and neurons in the dorsal raphe using

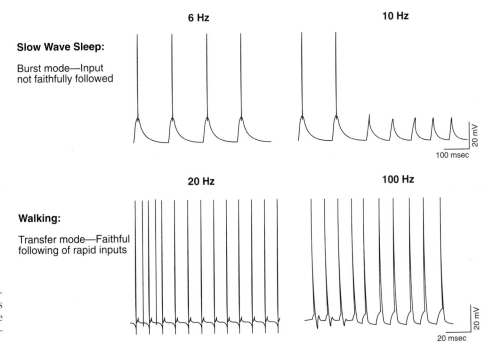

Figure 29-7
Visual system transmission in lateral geniculate relay neurons is blocked during non–rapid eye movement sleep when the neurons are in the burst mode.

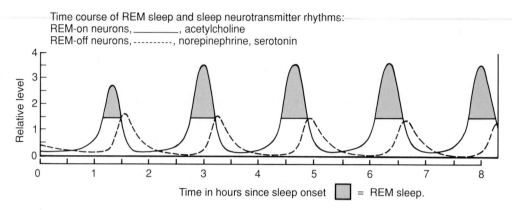

Time course of REM sleep and sleep neurotransmitter rhythms:
REM-on neurons, _____, acetylcholine
REM-off neurons, ----------, norepinephrine, serotonin

Time in hours since sleep onset   ▮ = REM sleep.

Figure 29-8

Schematic of a night's course of rapid eye movement (REM) sleep in humans, showing the occurrence and intensity of REM sleep as depending on the activity of populations of "REM-on" (i.e., REM-promoting) neurons (*solid line*). As the REM-promoting neuronal activity reaches a certain threshold, the full set of REM signs (*black areas under curve*) occurs. However, unlike the step-like electroencephalographic diagnosis of stage in Figure 29-2, the underlying neuronal activity is a continuous function. The neurotransmitter acetylcholine is thought to be important in REM sleep production, exciting populations of brain stem reticular formation neurons to produce the set of REM signs. Other neuronal populations using the monoamine neurotransmitters serotonin and norepinephrine probably suppress REM. The time course of their activity is sketched by the dotted line. These curves mimic actual time courses of neuronal activity recorded in animals and were generated by a mathematical model of REM sleep, the limit cycle reciprocal interaction model of McCarley and Massaquoi. (McCarley RW, Massaquoi SG. A limit cycle mathematical model of the REM sleep oscillator system. Am J Physiol 1986;251:R1011.)

serotonin have an opposite time course, becoming selectively inactive during REM. They act to suppress REM sleep-promoting activity.

☐ The brain stem contains the neural machinery of the rapid eye movement sleep rhythm

As illustrated in Figure 29-5, a transection just above the junction of the pons and midbrain produces a state in which periodic occurrence of REM sleep can be found in recordings made in the isolated brain stem while, in contrast, recordings in the isolated forebrain show no sign of REM sleep. These lesion studies by Jouvet and coworkers in France established the importance of the brain stem in REM sleep.

The cardinal signs of REM sleep in lower animals, as in humans, are muscle atonia, EEG desynchronization (i.e., low-voltage, fast-frequency pattern), and rapid eye movements. EEG depth recordings in animals show another important component of REM sleep, the *PGO waves*, so called because they are recorded from the pons, the lateral geniculate nucleus, and the occipital cortex. They are visible in the recording from the cat lateral geniculate nucleus in Figure 29-9. (The depth recordings necessary to establish their presence in humans have not been done.) PGO waves arise in the pons and are then transmitted to the thalamic lateral geniculate nucleus and to the visual occipital cortex. PGO waves represent an im-

portant mode of brain stem activation of the forebrain during REM sleep and are also present in nonvisual thalamic nuclei.

Most of the physiologic events of REM sleep have effector neurons located in the brain stem reticular formation, with many important neurons concentrated in the *pontine reticular formation* (PRF). PRF neuronal recordings are of special interest for information on the mechanisms of production of these events. Intracellular recordings of PRF neurons (see Fig. 29-9) show that these neurons have relatively hyperpolarized membrane potentials and generate almost no action potentials during non-REM sleep. As illustrated in Figure 29-9, PRF neurons begin to depolarize even before the occurrence of the first EEG sign of the approach of REM sleep, the PGO waves that begin 30 to 60 seconds before the onset of the rest of the polysomnographic signs of REM sleep. As PRF neuronal depolarization proceeds and the threshold for action potential production is reached, these neurons begin to discharge (i.e., generate action potentials). Their discharge rate increases as REM sleep is approached, and the high level of discharge is maintained throughout REM sleep because of the maintenance of a membrane depolarization.

Throughout the entire REM sleep episode, almost the entire population of PRF neurons remains depolarized. The resultant increased action potential activity leads to the production of the REM sleep physiologic signs that have their physiologic bases in PRF neurons. Figure 29-10 provides a schematic overview of REM sleep as arising from increases in excitability and discharge activity of the

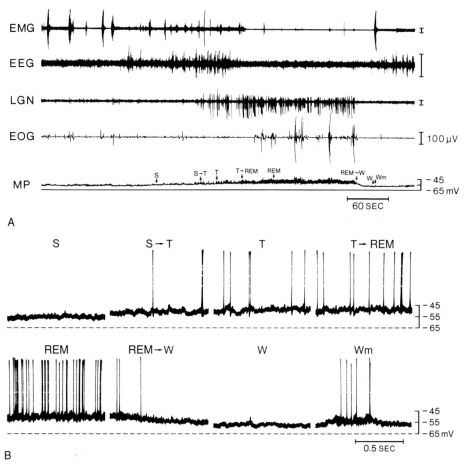

Figure 29-9

Rapid eye movement (REM) sleep is marked by a membrane potential depolarization in pontine reticular formation neurons. (**A**) Continuous polysomnographic record of waking, non-REM sleep, REM sleep, and return to waking in the cat. Waking is indicated by electromyographic (EMG) activity, low-voltage fast-electroencephalographic (EEG) eye movements (EOG); and non-REM sleep shows high-voltage slow-EEG waves. The transition to REM sleep is heralded by the onset of spiky waves in the lateral geniculate nucleus EEG recording (PGO waves), and with the occurrence of REM sleep, there is muscle atonia, low-voltage fast-EEG PGO waves, and rapid eye movements. The bottom trace is of the membrane potential (MP) of an intracellularly recorded pontine reticular formation neuron with the action potentials filtered out; notice the membrane potential depolarization that begins before and remains throughout REM sleep. (**B**) Samples of the oscilloscope record of the intracellular recording and the occurrence of action potentials and postsynaptic potentials at the times indicated on the MP tracing. *S*, slow wave or non-REM sleep; *S→T*, beginning of transition to REM sleep, as indicated by PGO waves; *T*, transition; *T→REM*, onset of REM sleep; *REM*, REM sleep; *REM→W*, transition to waking; *Wm*, waking with body movement and action potentials in the neuron. (Data from Ito K, McCarley RW. In Steriade M, McCarley RW. Brain stem control of wakefulness and sleep. New York: Plenum Press, 1990.)

various populations of reticular formation neurons that are important as effectors of REM sleep phenomena. PRF neurons are important for the rapid eye movements (i.e., generator for saccades is in PRF) and the PGO waves (i.e., different group of neurons); a group of dorsolateral PRF neurons controls the muscle atonia of REM sleep, and these neurons become active just before the onset of muscle atonia.

Neurons in midbrain reticular formation (MRF, see Fig. 29-5) are especially important for EEG desynchroniza-

tion, for the low-voltage, fast-frequency EEG pattern. These neurons were originally described Moruzzi and Magoun as making up the *ascending reticular activating system* (ARAS), the set of neurons responsible for EEG desynchronization. Subsequent work has enlarged this original ARAS concept to include cholinergic neurons. Neurons in the bulbar reticular formation are important for muscle atonia. Figure 29-11 illustrates the location of reticular formation zones important in the production of the muscle atonia of REM sleep.

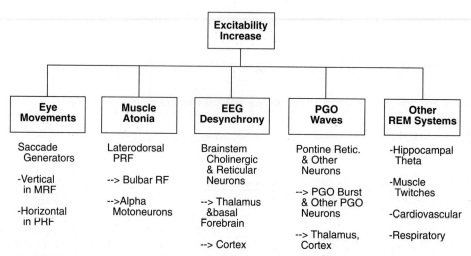

Figure 29-10

Schematic of rapid eye movement (REM) sleep control and the components of the mechanisms of REM sleep. Increasing the excitability or activity of brain stem neuronal pools that serve each of the major components of the state causes the occurrence of this component. For example, the neuronal pool important for the REMs is thought to be the brain stem saccade generating system, whose main machinery is in the paramedian pontine reticular formation. Although vertical saccades are fewer in REM, their presence suggests similar involvement of the mesencephalic reticular formation. The information under the other system components sketches the major features of the anatomy and the projections of neuronal pools important for muscle atonia, electroencephalographic desynchronization and PGO waves. The last part of the diagram lists other components of REM sleep. *MRF*, midbrain reticular formation; *PRF*, pontine reticular formation; *PGO*, waves recorded from the pons, lateral genticulate nucleus, and occipital cortex.

☐ Cholinergic mechanisms are important for rapid eye movement sleep

Extensive work by many investigators has led to an appreciation of the importance of the neurotransmitter acetylcholine for REM sleep and to a reasonably detailed knowledge of the nature of the anatomy and physiology of the cholinergic influences on REM sleep. The essential points are outlined below; detailed references can be found in many reviews, including that of Steriade and McCarley (1990).

Injection of compounds that are acetylcholine agonists into the pontine reticular formation produces a REM-like state that closely mimics natural REM sleep. The latency to onset and duration are dose dependent. Muscarinic receptors appear to be especially critical, with nicotinic receptors of lesser importance.

There are naturally occurring cholinergic projections to effector neurons in brain stem reticular formation. These arise in the two nuclei at the pons-midbrain junction (Fig. 29-12): the LDT and the PPT.

In vitro studies of PRF slice preparations by Greene and coworkers showed that 80% of reticular formation neurons are excited by cholinergic agonists, with muscarinic effects especially pronounced. In vitro studies of PRF slice preparations showed that the increased excitability and membrane depolarization produced by cholinergic agonists is a direct effect, because it persists after

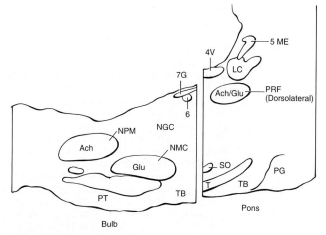

Figure 29-11

Schematic sagittal section of cat pons and bulb, showing reticular formation zones important in producing the muscle atonia of rapid eye movement (REM) sleep and where microinjections of acetylcholine agonists (ACh) or glutamate (Glu) induce muscle atonia. The dorsolateral pontine reticular formation (PRF) zone is ACh and Glu sensitive, receives laterodorsal tegmental nucleus (LDT) and pedunculopontine tegmental nucleus (PPT) cholinergic input, and projects to the bulbar nucleus magnocellularis (NMC), which is Glu sensitive. The nucleus paramedianus (NPM) is ACh sensitive and receives LDT and PPT cholinergic input. *IO*, inferior olive; *LC*, locus ceruleus; *7G & 6*, genu of 7th nerve and 6th nucleus; *NGC*, bulbar nucleus gigantocellularis; *PG*, pontine gray; *PT*, pyramidal tract; *SO*, superior olivary nucleus; *TB*, trapezoid body. (Adapted from Lai YY, Siegel JM. Medullary regions mediating atonia. J Neurosci 1988;8:4790.)

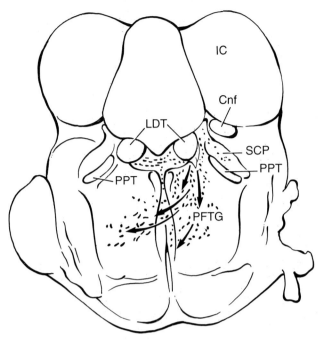

Figure 29-12

Frontal section of the brain stem at the pons-midbrain junction, showing the location of the acetylcholine-containing neurons most important for rapid eye movement sleep in the laterodorsal tegmental nucleus (LDT) and pedunculopontine tegmental nucleus (PPT) and a schematic of projections of the LDT to the pontine reticular formation. *IC*, inferior colliculus; *Cnf*, cuneiform nucleus; *PFTG*, one component of pontine reticular formation; *scp*, superior cerebellar peduncle. (Adapted from Mitani A, Ito K, Hallanger AH, et al. Cholinergic projections from the laterodorsal and pedunculopontine tegmental nuclei to the pontine gigantocellular tegmental field in the cat. Brain Res 1988;451:397.)

synaptic input has been abolished by addition of tetrodotoxin, which blocks sodium-dependent action potentials.

Experiments lesioning the LDT/PPT nuclei confirm their importance in producing REM sleep phenomena. Jones and coworkers have shown that destruction of the cell bodies of LDT/PPT neurons by local injections of excitatory amino acids leads to a marked reduction of REM sleep.

Work by McCarley, Sakai, Steriade, and their coworkers have shown that one subgroup of LDT/PPT neurons discharges selectively in REM sleep and that the onset of activity begins before the onset of REM sleep. This LDT/PPT discharge pattern and the presence of excitatory projections to the PRF suggest that these cholinergic neurons may be important in producing the depolarization of reticular effector neurons for REM sleep events. The LDT/PPT and reticular formation neurons that become active in REM sleep are often referred to as *REM-on neurons*.

Cholinergic neurons are important in the production of the low-voltage, fast-frequency EEG pattern (representing cortical activation) in REM sleep and in waking. As shown by Steriade and coworkers, a different subgroup of cholinergic neurons in LDT/PPT is active during this low-voltage, fast-frequency pattern in both REM and

waking. This cholinergic system is especially important in generating the low-voltage, fast-frequency EEG pattern, often called the activated EEG. Also playing a role in forebrain activation are projections from midbrain reticular neurons and aminergic neurons, especially those in locus ceruleus. Together these neuronal groups form the ARAS. Evidence that multiple systems are involved in EEG desynchronization comes from the inability of lesions of any single one of these systems to disrupt EEG desynchronization on a permanent basis.

Many *peptides* are co-localized with the neurotransmitter acetylcholine in LDT/PPT neurons; this co-localization probably means they are co-released with acetylcholine, and they may modify responsiveness to acetylcholine and have independent actions. The peptide substance P is found in about 40% of LDT/PPT neurons, and overall, more than 15 different co-localized peptides have been described. The role of these peptides in modulating acetylcholine activity relevant to wakefulness and sleep remains to be elucidated, but the co-localized vasoactive intestinal peptide has been reported by several different investigators to enhance REM sleep when it is injected intraventricularly.

☐ REM-off neurons may suppress rapid eye movement sleep phenomena

As illustrated in Figure 29-8, *REM-on neurons* are those neurons that become active in REM sleep compared with slow wave sleep and waking, and they presumably have a protagonist role in the production of REM sleep phenomena. Neurons with an opposite discharge time course that become inactive in REM sleep are called *REM-off neurons*. REM-off neurons are most active in waking, have discharge activity that declines in slow-wave sleep, and are virtually silent in REM until they resume discharge near the end of the REM sleep episode. This inverse pattern of activity to REM-on neurons to REM sleep phenomena such as PGO waves has led to the hypothesis that these neurons may be REM-suppressive and interact with REM-on neurons in control of the REM sleep cycle. This concept is indirectly supported by production of REM sleep from cooling (i.e., inactivating) the nuclei where REM-off neurons are found and by some in vivo pharmacologic studies. In vitro data from Luebke and coworkers have provided direct support for the inhibition of cholinergic LDT neurons by serotonin (Fig. 29-13).

Several classes of neurons are REM-off (see Fig. 29-5 for anatomy). *Norepinephrine-containing neurons* are principally located in the locus ceruleus, called the "blue spot" because of its appearance in unstained brain. *Serotonin-containing neurons* are located the *raphe system* of the brain stem, the midline collection of neurons that extends from the bulb to the midbrain, with higher concentrations of serotonin-containing neurons in the more rostral neurons. Before it was known that serotonin-containing neurons in the dorsal raphe had maximal discharge activity in waking, slowed discharge during non-

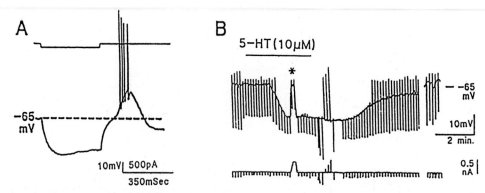

Figure 29-13

Serotonin directly inhibits cholinergic neurons in the laterodorsal tegmental nucleus, as shown in the intracellular recordings from rat in vitro preparation. (**A**) With the membrane potential held at − 65 mV, a hyperpolarizing current step results in a rebound low threshold spike. This is the likely mechanism for the production of action potential bursts seen in neurons transmitting PGO wave information from the brain stem to the lateral geniculate nucleus in REM sleep. (**B**) Application of 10 μM serotonin (5-HT) to a cholinergic bursting neuron results in a hyperpolarization of the membrane potential (*top trace*), a response that other experiments showed persisted in the presence of tetrodotoxin, indicating the inhibition was a direct serotonergic effect. The voltage deflections in the top trace are responses to constant current pulses used to measure input resistance. Repolarization of the membrane to resting levels (*asterisk*) demonstrated that there was a decrease in the input resistance that was independent of the change in the membrane potential and the result of a direct drug action. The same current that was effective in producing a burst in **A** did not result in a burst during the application of 5-HT (not shown); other experiments indicated a 5-HT$_{1A}$ receptor mediated these effects. (Modified from Luebke JI, Greene RW, Semba K, Kamondi A, McCarley RW, Reiner PB. Serotonin hyperpolarizes cholinergic low threshold burst neurons in the rat laterodorsal tegmental nucleus in vitro. Proc Natl Acad Sci U S A 1992;89:743.)

REM sleep, and virtually ceased discharge in REM sleep, it was proposed that serotonin neurotransmission might induce REM sleep. This theory, proposed by Jouvet and others, was based on indirect evidence. With the current knowledge of the REM-off nature of dorsal raphe neurons, there is now no solid evidence for this theory. Unfortunately, this theory is still stated as fact in some textbooks, although it is now nearly 20 years out of date! *Histamine-containing neurons* are located in the posterior hypothalamus and are REM-off. This system has been conceptualized as one of the wakefulness-promoting systems, in agreement with drowsiness as a common side effect of antihistamines, but transection studies indicate that the histaminergic neurons are not essential for the REM sleep oscillation.

☐ A mathematical and structural model of rapid eye movement sleep has been proposed

A version of the reciprocal interaction model, originally proposed by McCarley and Hobson and revised by McCarley and Massaquoi, rather accurately predicts the timing and percentage of REM sleep over a night of human sleep and its variation with the circadian temperature rhythm, and it is the basis for Figure 29-14.

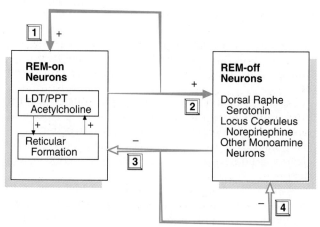

Figure 29-14

Reciprocal interaction model of rapid eye movement (REM) sleep control. (**Arrow 1**) Cholinergic neurons activate reticular formation neurons through positive feedback to produce the onset of REM sleep. (**Arrow 3**) REM sleep is terminated by the inhibitory activity of REM-off aminergic neurons (*right box*), which become active at the end of a REM sleep period (**Arrow 2**) because of their recruitment by REM-on activity. (**Arrow 4**) REM-off neuronal activity decreases in slow-wave sleep and becomes minimal at the onset of REM sleep because of self-inhibitory feedback. This decreased REM-off activity disinhibits REM-on neurons and allows the onset of a REM sleep episode. The cycle than repeats. (Adapted from McCarley RW, Massaquoi SG. Neurobiological structure of the revised limit cycle reciprocal interaction model of REM cycle control. J Sleep Res 1992;1:132.)

Neuronal Input
(specific sensory- or motor-task-related)

||||||||| || || || | ||

Waking Output
(Faithful following of inputs: Graded, precise responses.)

||||||||| || || || | ||

Slow Wave Sleep Output
(Poor response to rapid inputs, details of information lost.)

|| | || || ||

REM Sleep Output
(Responsivity high, but specificity lost due to other inputs, and altered firing patterns from low norepinephrine/serotonin.)

|| |||||||| ||||||||| ||||||| || || || | |||||

Figure 29-15
Schematic of cortical neuronal processing during waking, slow-wave sleep, and rapid eye movement (REM) sleep. The spikes represent neuronal discharge.

The Form of Dreams Parallels the Biology of Rapid Eye Movement Sleep

REM sleep is strongly associated with dreaming. In experiments involving awakenings at random intervals throughout the night, 80% of all such randomly elicited dream reports have been found to occur in REM sleep. The dreams that occur in non-REM sleep have been found to be less vivid and intense than REM sleep dreams, suggesting they may represent a pre-REM state in which brain stem neuronal activity is approximating that of REM sleep but the EEG has not yet changed.

Dreams have a long history of interest in popular culture and in psychiatry. Sigmund Freud, writing before the biologic state of REM sleep was known, suggested that dreams represented a symbolic disguise of an unacceptable unconscious wish (e.g., sexual or aggressive wishes); the purpose of the disguise was to prevent the disruption of sleep that would occur with consciousness of the undisguised wish. The activation of the neural systems responsible for REM sleep appears to be a more accurate and simple explanation for the instigation of the dream state that is linked to the cyclic appearance of REM sleep.

There remains the question of why dreams have their own distinctive characteristics and are different from waking consciousness. One obvious hypothesis is that the conscious states are different because the brain states differ. Figure 29-15 shows the input-output associations in a cortical neuron during the various states of consciousness; this schematic is based on inferences from in vivo recordings in animal and in vitro experiments examining firing patterns as a function of membrane potential. This schematic suggests obvious differences in processing during the different behavioral states.

The *activation-synthesis hypothesis* proposed by Hobson and McCarley (1977) suggests that many of the characteristic formal features of dreams are isomorphic or parallel with distinctive features of the physiology of REM sleep. Formal features refer to the universal aspects of dreams, distinct from the dream content particular to a person. As an example of a formal feature of a dream, consider the presence of motor activity in dreams. At the physiologic level, it is known that motor systems at the motor cortex and at the brain stem level are activated during REM sleep episodes. Paralleling motor system activation at the physiologic level is the finding that movement in dreams is extremely common, with almost one third of all verbs in dreams indicating movement and 80% of dreams having some occurrence of leg movement (a movement that was easily and reliably scored in dream reports by McCarley and Hoffman).

Similarly, there is activation of sensory systems during REM sleep (Fig. 29-16). The visual system is intensely ac-

Figure 29-16
Percentage of rapid eye movement sleep dreams assessed by various sensory modalities in a sample of 100 dreams from 15 healthy young men. (Adapted from McCarley RW, Hoffman EA. REM sleep dreams and the activation-synthesis hypothesis. Am J Psychiatry 1981;138:904.)

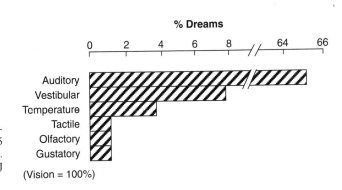

% Dreams

0   2   4   6   8   //   64   66

Auditory
Vestibular
Temperature
Tactile
Olfactory
Gustatory

(Vision = 100%)

tivated in REM, and all dreams have visual experiences, considered to be one of the defining features of a dream. An important source of visual system activation during REM sleep is from the PGO waves. The activation synthesis theory suggests that this intense activation of visual and other sensory systems are the substrate for dream sensory experiences. Supporting this theory is the rather frequent occurrence—about 9% of all REM sleep dreams—of dreams with intense "vestibular sensations," such as dreams of flying, floating, falling, soaring, or tumbling that are easily related to the vestibular system functions of sensing position of the body in space and changes of position. The presence of dreams with vestibular sensations was highly atypical of the daytime sensory experience of the subjects whose dream reports were examined and is incompatible with any dream theory linked to a simple recall of previous experiences. The dream experience may instead reflect the intense REM sleep vestibular system activation, followed by its elaboration and synthesis into dream content. The final product, the dream, represents the synthesis of the brain stem–induced motor and sensory activation with the particular memories and personality characteristics of the dreamer.

Lesion-induced release of REM sleep motor activity supports the presence of neural commands for patterned motor activity in REM sleep and a direct correspondence of the motor system commands and the subjective content of the dream. Activation of motor systems in REM can be observed in cats with a lesion of the muscle atonia zone of the PRF and a consequent *REM sleep without atonia*, a state in which motor activity is released but all of the other signs of REM sleep are present. The failure of muscle atonia is also observed in a human disorder reported by Schenck and Mahowald and called *REM sleep behavior disorder*. In persons with this disorder, the muscle activity observed has been found always to parallel the dreamed activity. This close linkage between the physiology and psychology of REM sleep and dreams supports the activation-synthesis hypothesis.

When the activation-synthesis hypothesis was first proposed, it aroused considerable controversy, perhaps because it seemed to threaten psychologic interpretation of dreams. Although this theory clearly places instigation of the dream state as a concomitant of a basic biologic rhythm, there appears to be more than ample room for addition of personal characteristics in the process of synthesis of brain stem–instigated activation. For example, interpretations of Rorschach cards are rich sources of information on personality, although the images on the cards themselves were not generated by psychologically meaningful mechanisms.

As more is learned about forebrain processing during REM sleep, a more complete and complex theory will emerge. Crick and Mitchison (1983) have proposed that REM sleep is a state in which unwanted processing modes are eliminated; chief among these are cortical "parasitic oscillations," a kind of neural analog to obsessive thoughts. In this theory, the PGO wave activity acts to reset and redirect the unwanted neural loops. The authors themselves state that a full test of this theory will have to be postponed until our ability to monitor complex neural processing has greatly increased.

Neuroimaging studies of humans and cats have indicated activation of the limbic system, suggesting a biologic basis of activation of memories and emotions in REM sleep and perhaps a mechanism of their linkage. Winson (1990) built on his own work in hippocampal systems to suggest a basis of dreams and a function of REM sleep. Extracellular recordings in animals of CA1 neurons indicated that those neurons that had been active during the day in encoding "place fields" (i.e., spatial position) fired at a significantly higher rate in REM sleep than in CA1 neurons that had not mapped place fields during the day. Winson suggests that a general function of REM sleep is "off line" processing of information acquired during the day by comparison with other information acquired throughout the person's lifetime. In humans, dreams may come by their unusual character as a result of the complex associations that are culled from memory.

Functions of Rapid Eye Movement and Other States of Sleep

There are many theories but relatively little solid data on the function of sleep. In this section we summarize some of the most plausible functional theories for each sleep phase.

☐ Is non-REM sleep a time for rest and recovery?

Rest theory

Neuronal recordings and brain metabolic studies indicate the presence of rest on the neural and the behavioral levels during non-REM sleep.

Behavioral immobilization

Another theory suggests that sleep evolved as a way of arresting behavior at a time when it might not be advantageous, such as night activity in animals with poor night vision and vulnerability to predators.

☐ Does REM sleep promote growth and development?

The following theories are not mutually exclusive. It seems a cogent argument that a complex behavioral state such as REM sleep may have multiple functions. Just as for wakefulness, it may not be meaningful to speak of "the" function of REM sleep.

Promotion of growth and development of the nervous system

The abundance of this metabolically and neurally active state in the young argues for a growth and development hypothesis. The discovery of NMDA receptors in the PRF offers a potential mechanism for regulation of neuronal growth and development in this region with high activation during REM sleep (Stevens et al., 1992). Work has also indicated that REM-like activity in the brain stem may alter the activity of "immediate early gene" systems, such as *FOS*, suggesting a mechanism by which REM activity may affect DNA transcription and influence developmental and structural changes. Along this line of reasoning, the French scientist Jouvet suggested that the stereotyped motor command patterns of REM sleep are useful in promoting epigenetic development of these circuits.

Circuit exercise and maintenance function in adults

It is postulated that maintenance of neural circuits requires use and that with increasing diversity of behaviors possible in more advanced animals, REM sleep serves as a "fail safe" mode for ensuring activation and consequent maintenance of sensorimotor circuits. Crick and Mitchison (1983) suggested that the REM sleep activity involves removal of unwanted, "parasitic" modes of neural circuit processing.

Memory processing

Memories may be consolidated or processed during sleep. Hippocampal neurons that encode spatial location and that are activated during wakefulness are preferentially activated in subsequent REM periods compared with the non-wake-activated neurons; the inference is that "memories" are being related to other brain information (summarized in Winson, 1990).

Acknowledgments

Work supported by awards from the Department of Veterans Affairs, Medical Research Service and NIMH (R37 MH39,683 and R01 MH40,799).

SELECTED READINGS

Borbély AA. A two process model of sleep regulation. Human Neurobiol 1982;1:195.

Borbély AA, Tobler I. Endogenous sleep-promoting substances and sleep regulation. Physiol Rev 1989;69:605.

Buzsaki G, Bickford RG, Ponomareff G, Thal LJ, Mandel R, Gage FH. Nucleus basalis and thalamic control of neocortical activity in the freely moving rat. J Neurosci 1988;8:4007.

Carskadon MA, Dement WA. Normal human sleep: an overview. In Kryger MH, Roth T, Dement WC (eds). Principles and practices of sleep medicine. New York: WB Saunders, 1989:3.

Cohen D, Cuffin BN, Yunokuchi K, et al. MEG versus EEG localization test using implanted sources in the human brain. Annal Neurol 1990;28:811.

Crick F, Mitchison G. The function of dream sleep. Nature 1983;304:111.

Feinberg I. Effects of age on human sleep patterns. In Kales A (ed). Sleep, physiology and psychology. Philadelphia: JB Lippincott, 1969:39.

Feinberg I, Thode HC, Chugani HT, March JD. Gamma function describes maturational curves for delta wave amplitude, cortical metabolic rate and synaptic density. J Theor Biol 1990;142:149.

Greene RW, Gerber U, McCarley RW. Cholinergic activation of medial pontine reticular formation neurons in vitro. Brain Res 1989;476:154.

Hayaishi O. Sleep-wake regulation by prostaglandins D_2 and E_2. J Biol Chem 198;263:14593.

Hobson JA, McCarley RW. The brain as a dream state generator: an activation-synthesis hypothesis of the dream process. Am J Psychiatry 1977;134:l335.

Jouvet M. What does a cat dream about? Trends Neurosci 1979;2:15.

Kronauer RE, Czeisler CA, Pilato SF, Moore-Ede MC, Weitzman ED. Mathematical model of the human circadian system with two interacting oscillators. Am J Physiol 1982;242:R3.

Krueger JM. Somnogenic activity of immune response modifiers. Trends Pharmacol Sci 1990;11:122.

Lai YY, Siegel JM. Medullary regions mediating atonia. J Neurosci 1988;8:4790.

Luebke JI, Greene RW, Semba K, Kamondi A, McCarley RW, Reiner PB. Serotonin hyperpolarizes cholinergic low threshold burst neurons in the rat laterodorsal tegmental nucleus in vitro. Proc Natl Acad Sci U S A 1992;89:743.

Mahowald MW, Schenck CH. REM sleep behavior disorder. In Kryger MH, Roth T, Dement WC (eds). Principles and practices of sleep medicine. New York: WB Saunders, 1989:389.

McCarley RW. The biology of dreaming sleep. In Kryger MH, Roth T, Dement WC (eds). Principles and practice of sleep medicine. Philadelphia: WB Saunders, 1989:173.

McCarley RW, Hoffman EA. REM sleep dreams and the activation-synthesis hypothesis. Am J Psychiatry 1981;138:904.

McCarley RW, Massaquoi SG. A limit cycle mathematical model of the REM sleep oscillator system. Am J Physiol 1986;251:R1011.

McCarley RW, Massaquoi SG. Neurobiological structure of the revised limit cycle reciprocal interaction model of REM cycle control. J Sleep Res 1992;1:132.

McCormick DA, Pape H-C. Properties of a hyperpolarization-activated cation current and its role in rhythmic oscillation in thalamic relay neurons. J Physiol 1990;431:291.

McCormick DA, Pape H-C. Noradrenergic and serotonergic modulation of a hyperpolarization-activated cation current in thalamic relay neurons. J Physiol 1990;431:319.

Mitani A, Ito K, Hallanger AH, et al. Cholinergic projections from the laterodorsal and pedunculopontine tegmental nuclei to the pontine gigantocellular tegmental field in the cat. Brain Res 1988;451:397.

Moruzzi G, Magoun HW. Brain stem reticular formation and activation of the EEG. Electroencephalogr Clin Neurophysiol 1949;1:455.

Rainnie DG, Grunze HCR, McCarley RW, Greene RW. Adenosine inhibits mesopontine cholinergic neurons: implicatons for EEG arousal. Science 1994;263:689.

Steriade M, McCarley RW. Brain stem control of wakefulness and sleep. New York: Plenum Press, 1990.

Steriade M, Curró Dossi R, Nuñez A. Network modulation of a slow intrinsic oscillation of cat thalamocortical neurons implicated in sleep delta waves: cortically induced synchronization and brain stem cholinergic suppression. J Neurosci 1991;11:3200.

Stevens DR, McCarley RW, Greene RW. Excitatory amino acid–mediated responses and synaptic potentials in medial pontine reticular formation neurons of the rat in vitro. J Neurosci 1992;12:4188.

Strogatz SH. The mathematical structure of the human sleep-wake cycle. New York: Springer-Verlag, 1986.

Winson J. The meaning of dreams. Sci Am 1990;263:86.

Neuroscience in Medicine, edited by P. Michael Conn.
J. B. Lippincott Company, Philadelphia © 1995

Chapter *30*

Higher Brain Functions

DANIEL TRANEL

Clinical Correlation: Aphasias and Other
Disorders of Language
ROBERT L. RODNITZKY

Aphasia

Aphasic Disorders Are Usually Caused by
Lesions of the Left Hemisphere
Several Forms of Brain Pathology Can
Cause Aphasia

Several Facets of Language Must Be
Assessed in Evaluating an Aphasic
Disorder
Aphasias Can Be Divided Into Fluent
and Nonfluent Types
Repetition is Usually Abnormal in
Aphasias Caused by Lesions in the
Perisylvian Region

Disorders of Written Language

Higher brain functions are the operations of the brain that stand at the pinnacle of neural evolution. Using language, manipulating ideas in the future, and remembering vast amounts of information are examples of higher mental functions linked in unique fashion to the human brain. Capacities such as intellectual function, memory, speech and language, complex perception, orientation, attention, judgment, planning, and decision-making comprise what is known as *cognition*. Another domain of higher brain function is captured by the term *personality*, which describes the traits and response styles that typify a person's behaviors across a range of situations and circumstances. Cognition and personality cover the range of human behaviors that represent the higher-order functions of the human brain.

Most neuroscientists have come to view *brain functions* as synonymous with *mind functions*, even if we do not yet know all the details of how mental operations are served by neural machinery. It remains to be seen whether we will ever be able to account for all human behavior in neural terms. Nonetheless, breakthroughs in the last two decades in the field of neuroscience have provided evidence to support the notion that brain facts and mind facts are one and the same. This chapter uses as a fundamental premise the fact that there are orderly associations among certain neural operations and certain cognitive capacities.

Brain and Behavior Associations

More than a century ago, scientists began to notice that damage to certain brain regions could produce highly selective defects in behavior. In the 1860s, the surgeon and physical anthropologist Paul Broca made the observation that damage to the *anterior* part of the left side of the brain led to an inability to produce speech, while sparing the comprehension of speech. A complementary observation was reported some 10 years later by the neuropsychiatrist Carl Wernicke, who observed that damage to the *posterior* part of the left hemisphere led to a disturbed capacity to comprehend speech, but the production of speech was spared. The combined observations of Broca and Wernicke led to the notion that humans speak and process language with the left side of our brains. These early writings became the fundamental underpinnings on which the fields of neuropsychology and cognitive neuroscience have been established. Since then, several associations between brain and behavior have been established, and various cognitive and behavioral capacities that tend to be strongly associated with particular brain regions can be highlighted.

In this chapter, three perspectives will be used to describe higher brain functions. First, fundamental neuroanatomic subdivisions of the brain are outlined, and for each, the principal cognitive and behavioral correlates of the region are described. Second, specific disorders of higher brain function are enumerated. For each disorder, the primary neuropsychologic manifestations are described, along with the main neuroanatomic correlates, which provides another perspective on the way in which particular brain regions are dedicated to particular cognitive functions. A brief review of the principal methods of measuring higher brain functions is presented to provide a hint for the complexity involved in trying to quantify capacities such as memory, language, and problem-solving.

☐ The human brain has gross left–right and anterior–posterior subdivisions

Lateral specialization: left versus right

As suggested by the pioneering discoveries of Broca and Wernicke, there are several fundamental differences between the left and right hemispheres of the human brain, one of which constitutes perhaps the most robust principle of neuropsychology. In most adults, the left side of the brain is specialized for language, for the processing and mediation of verbally coded information. This principle applies regardless of the mode of input. Verbal information apprehended through the auditory (e.g., spoken speech sounds) or visual (e.g., reading material) channel is processed by the left hemisphere. The principle also applies to the input and output aspects of language; we understand language with our left hemisphere, and we produce spoken and written language with our left hemisphere. Figure 30-1 illustrates the typical arrangement of language in the left hemisphere.

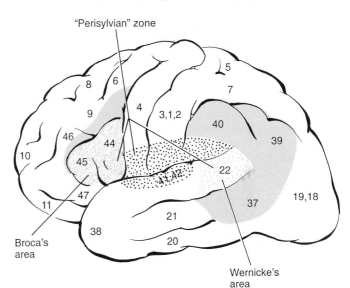

Figure 30-1
Lateral view of the left hemisphere. The principal language-related regions are highlighted, including Broca's area and Wernicke's area. The "perisylvian" zone includes Broca's and Wernicke's areas and the dotted zone. Broca's area is dedicated to language *expression,* and Wernicke's area is responsible for language *comprehension.* Other language-related regions include the supramarginal gyrus (i.e., area 40), the angular gyrus (i.e., area 39), part of area 37, and the region immediately above and anterior to Broca's area (*blue*). Not pictured are the left-sided subcortical structures (i.e., basal ganglia, thalamus) that also participate in speech and language functions.

For many years in the early history of neuropsychology, the right hemisphere was thought to be the "silent" or minor hemisphere, because it did not participate to any great extent in language. We know now that the right hemisphere has an entirely different type of specialization, and it has cognitive and behavioral capacities that are as important as those of the left hemisphere. The right side is specialized for nonverbal processing, and it handles information such as complex visual patterns (e.g., faces) or auditory signals (e.g., music) that are not coded in verbal form. The right side of the brain is also dedicated to the mapping of feeling states, which are patterns of bodily sensations that are linked to certain emotions such as happiness, anger, and fear. A related right-hemisphere capacity concerns the perception of our bod-

ies in space, in intrapersonal and extrapersonal terms. For example, an understanding of where our limbs are in relation to our trunk and where our body is in relation to the space around us is under the purview of the right hemisphere. Figure 30-2 depicts some of the fundamental capacities of the right hemisphere.

In early conceptualizations of the differences between the left and right hemispheres, the prevailing notion was that the left hemisphere was the major or *dominant side* and the right hemisphere was the minor or *nondominant side.* This attitude reflected the emphasis on language; because language is a highly observable and uniquely human capacity, it received the most attention from neurologists and neuropsychologists and was considered the quintessential human faculty. For many de-

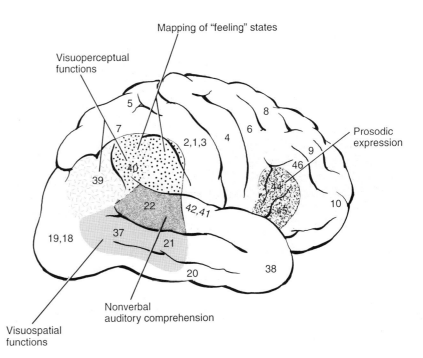

Figure 30-2
Lateral view of right hemisphere, depicting several primary regions with a label corresponding to their functional correlates. Many of these functions overlap psychologically and anatomically, and the areas depicted should be considered approximate.

cades, the right hemisphere was thought to contribute little to higher-level cognitive functioning. Lesions to the right side did not produce language disturbances, and it was often concluded that the patient had lost little in the way of higher-order functions after right-sided brain injury.

As the field evolved and it became clear that each hemisphere was dedicated to certain cognitive capacities, the notion of dominance gave way to the idea of specialization. Each hemisphere was dominant for certain types of cognitive functions. Early work had already established the role of the left hemisphere in language function, and subsequent investigations confirmed this conclusion. Other studies documented the role of the right hemisphere in various visuospatial capacities. Many of the breakthroughs in this area of investigation came from studies of *split-brain patients*, a line of work led by the late psychologist Roger Sperry (Sperry, 1968). These patients had received a special operation for the control of seizures, in which the *corpus callosum*, the large bundle of fibers that connects the left and right hemispheres, was surgically cut, and the left and right sides were no longer in communication. Careful investigations of these patients revealed that each side of the brain had its own "consciousness," with the left side operating in a verbal mode and the right in a nonverbal mode. Sperry's work and that of others led to several fundamental distinctions between the types of operations for which the left and right hemispheres are specialized. Several of the most important dichotomies are listed in Table 30-1.

Longitudinal specialization: anterior versus posterior

Another useful organizational principle for understanding brain-behavior associations is an anterior and posterior distinction. The major demarcation points are the *Rolandic sulcus*, which is the major fissure separating the frontal lobes (i.e., anterior) from the parietal lobes (i.e., posterior), and the *Sylvian fissure*, which forms a boundary between the temporal lobes (i.e., inferior) and the frontal and parietal lobes (i.e., superior). Figure 30-3 depicts these areas.

Table 30-1
FUNCTIONAL DICHOTOMIES OF LEFT AND RIGHT HEMISPHERIC DOMINANCE

Left Side	Right Side
Verbal	Nonverbal
Serial	Parallel
Analytic	Holistic
Controlled	Creative
Logical	Pictorial
Propositional	Appositional
Rational	Intuitive
Social	Physical

Adapted from Benton A. The Hecaen-Zangwill legacy: hemispheric dominance examined. Neuropsychol Rev 1991;2:267.

As a general principle, the posterior regions of the brain are dedicated to perception. The primary sensory cortices for vision, audition, and somatosensory perception are in the posterior sectors of the brain, in the occipital, temporal, and parietal regions, respectively. Apprehension of sensory data from the world outside is mediated by posterior brain structures. The world outside refers to two domains: the world that is outside our bodies and brains and the world that is outside our brains but inside our bodies. The latter, the *soma*, comprises the smooth muscle, the viscera, and other bodily structures innervated by the central nervous system.

Anterior brain regions generally comprise effector systems, which are systems specialized for the execution of behavior. For example, the primary motor cortices are located in the strip of cortex immediately anterior to the Rolandic sulcus. The motor area for speech, known as *Broca's area*, is in the left frontal operculum. The right-hemisphere counterpart of Broca's area, in the *right frontal operculum*, is important for executing stresses and intonations that infuse speech with emotional meaning (e.g., prosody). A variety of executive functions, such as judgment and decision-making, and the capacity to construct and implement various plans of action are associated with structures in the frontal lobes.

☐ The posterior part of the brain is comprised of the occipital lobes, which are important for vision

Primary visual cortex

The occipital lobes comprise the posterior part of the hemispheres, and they can be subdivided into primary visual cortex and the visual association cortices (Fig. 30-4). The numbering system of Brodmann (1909) is used to refer to particular brain regions (Damasio and Damasio, 1989). The primary cortices are composed of Brodmann's area 17, located in the region directly above and below the calcarine fissure on the mesial aspect of the hemispheres and extending some extent onto the lateral surface (see Fig. 30-4). This region is dedicated to *form vision*, and damage to these cortices produced blindness in the corresponding visual field. The system is wired in a crossed fashion in the vertical and horizontal dimensions. Visual information from the space to the right of the vertical meridian is perceived with the left visual cortex, and information from the left hemispace is perceived with the right visual cortex. Similarly, information above the horizontal meridian reaches visual cortex below the calcarine fissure, and information below the midline reaches cortex above the calcarine fissure.

Other component features of the visual world are processed in or near the primary visual cortices. The processing of color, for example, is strongly linked to the lingual gyrus, immediately below primary visual cortex on the inferior bank of the calcarine sulcus. *Depth perception* (i.e., stereopsis) and motion are associated with cortices in and near the superior component of the pri-

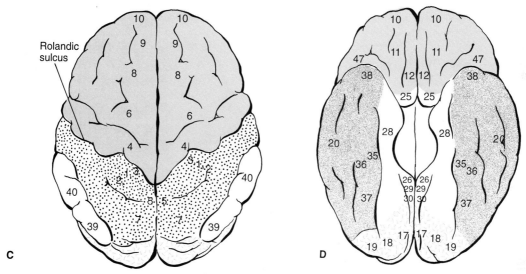

A

Rolandic sulcus

Sylvian fissure

B

C

Rolandic sulcus

D

Figure 30-3
Line drawings of (**A**) lateral, (**B**) mesial, (**C**) superior, and (**D**) inferior views of the brain, depicting major demarcation points, including the Rolandic sulcus and the Sylvian fissure. The four main lobes are shown in different shades: *blue,* frontal; *dots,* parietal; *dashes,* occipital; *shading,* temporal. Only the left hemisphere is depicted in the lateral and mesial views, but the mapping would be the same on the right hemisphere. The unmarked zone, including the cingulate gyrus (i.e., areas 24 and 23) and areas 25, 26, 27, and 28, corresponds to a region commonly referred to as the *limbic lobe.* In the superior perspective, the left hemisphere is on the left, and the right hemisphere is on the right; the sides are reversed in the inferior perspective.

mary visual region, in an area known as the cuneus (see Fig. 30-4).

Visual association cortices

The visual association cortices include areas 18 and 19 on the lateral and mesial aspects of the hemispheres (see Fig. 30-4). Areas 37, 20, and 21 in the inferior and ventral banks of the temporal lobes are also dedicated primarily to the processing of visual information. The visual association cortices, which communicate with primary visual cortex posteriorly and with more anterior regions in the temporal and parietal lobes through a series of extensive feedforward and feedback connections, are specialized for progressively higher-order aspects of visual processing. For example, the ability to perceive and assign meaning to orthographic symbols (i.e., reading) is associated with the region of cortex and white matter in the lower part of visual association cortex in the left hemisphere. In the right hemisphere, the ventral visual association cortices are specialized for the registration and decoding of nonverbal patterns, such as the holistic perception of faces.

Superior parts of the visual association cortices are important for deriving meaningful information from motion and depth. The anterior part of this region, which

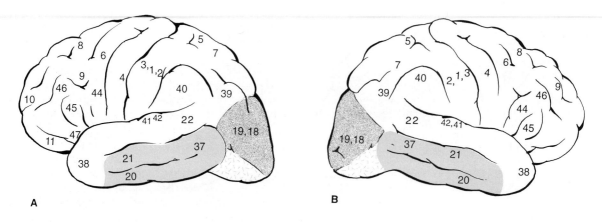

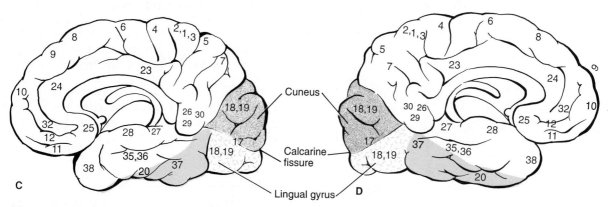

Figure 30-4
The (**A**) left hemisphere and (**B**) right hemisphere views show the major subdivisions of the occipital lobes and visually related systems in the temporal lobe, mapped on the lateral aspect of the brain. The (**C**) right hemisphere and (**D**) left hemisphere views show the same subdivisions, mapped on the mesial aspect of the brain. The primary visual cortex (i.e., area 17) is formed along the upper and lower banks of the calcarine fissure. On the mesial aspect of the hemispheres, the *lingual gyrus* is immediately below the primary visual cortex, and the *cuneus* is immediately above. Visual association cortices are composed of the superior (*shading*) and inferior (*dashes*) regions. Higher-order visual systems are marked in blue.

overlaps with the posterior part of the superior parietal region, is specialized for visuospatial capacities that relate to the placement of external stimuli in space and to the tracking those stimuli when they are in motion. This region is also important for the accurate mental assemblage of extrapersonal space and the subsequent placement and location of oneself in that space. For example, the ability to know where your body is in relation to the chair on which you are sitting and to guide your arm down to a book to turn the page depends on association cortices in the upper parts of the occipital and posterior parietal regions. The upper part of the visual system is important for attending to multiple visual elements concurrently. When watching a politician delivering a speech from center stage, flanked on both sides by other dignitaries, we may be aware of the speech giver and the person just to the left of the podium who is having trouble staying awake. The capacity to do this is known as *simultaneous*

visual perception. Table 30-2 lists some of the main brain-behavior associations of the occipital region.

The ventral and dorsal systems

The visual system characteristics described earlier can be conceptualized along anatomic and functional lines as comprising two distinct subsystems: a ventral system and a dorsal system. The ventral system, located below the calcarine fissure, is specialized for decoding what things are. The dorsal system, located above the calcarine fissure, is specialized for decoding where things are. These have been called the "what" and "where" systems (Ungerleider and Mishkin, 1982). Consistent with its dedication to the what aspect of stimuli, the ventral system processes primarily the features of shape and color. The dorsal where system processes primarily the features of motion, depth, and the position of stimuli in space.

Table 30-2
BRAIN-BEHAVIOR ASSOCIATIONS IN THE OCCIPITAL REGION

Neural Region	Cognitive or Behavioral Function
Ventral occipital	
Left	Perception of shapes and contours (features)
	Color perception and color naming
	Reading
	Face recognition (features)
	The "what" recognition system
Right	Perception of shapes and contours (global)
	Color perception
	Nonverbal pattern recognition
	Face recognition (holistic)
	The "what" recognition system
Dorsal occipital	
Left	Depth perception; stereopsis; motion perception
	Visual attention
	Visually guided reaching
	Recognition of identity from movement
	The "where" recognition system
Right	Depth perception; stereopsis; motion perception
	Visual attention
	Visually guided reaching
	Recognition of identity from movement
	Mental rotation
	The "where" recognition system

□ The temporal lobes contain regions that are important for auditory processing and memory

The temporal lobes can be subdivided into three basic systems according to the type of specialization the regions have for different cognitive operations. Figure 30-5 illustrates the arrangement of the regions, and the cognitive or behavioral correlates of each are reviewed in the sections that follow.

Superior region: primary auditory cortex and auditory association cortices

Heschl's gyrus (i.e., areas 41 and 42) in the horizontal plane in the depths of the Sylvian fissure in the posterior aspect of the temporal lobes constitutes primary auditory cortex that is crucial for basic perception of auditory information. The system is relatively crossed, and that auditory information from the right ear is sent primarily to left Heschl's gyrus, and information from the left ear is sent primarily to right Heschl's gyrus. However, this decussation is incomplete, and a significant amount of information is perceived ipsilaterally (i.e., by primary auditory cortex on the same side as the ear).

Information can be "forced" into more complete crossing using a special paradigm known as *dichotic listening*. In this task, auditory information is presented to each ear simultaneously. If the information varies sufficiently between the left and right ears (e.g., if the word rabbit is

input to the left ear and the word teacher is input to the right ear), the auditory system will be driven in such a way that the subject will perceive the word rabbit with the right auditory cortex and the word teacher with the left auditory cortex. In keeping with the basic verbal-left, nonverbal-right arrangements of the brain, when verbal material is presented in dichotic listening paradigms, the right ear (i.e., left brain) shows a relative advantage (i.e., perceives information first and more strongly) over the left ear; if nonverbal material is presented (e.g., music), the left ear (right brain) shows a relative advantage.

The posterior third of the superior temporal gyrus (i.e., posterior area 22) contains important auditory association cortices. On the left, this region, which comprises *Wernicke's area*, is specialized for decoding aural verbal information, such as deciphering the meaning of speech input. The right side is specialized for nonspeech auditory information, such as environmental sounds, musical melodies, timbre, and prosody. In general, the left auditory association cortices are specialized for the perception and decoding of *temporal components* of auditory information: information pertaining to timing, the sequential aspects of auditory signals, the pace of information delivery (i.e., cadence), and the duration of signals and intervals between sounds. The right side is specialized for *spectral information*: the pitch (i.e., fundamental frequency) of signals and their spectral complexity (i.e., harmonic structures).

Inferolateral sector

The inferolateral part of the temporal lobe is comprised by areas 37, 20, 21, and 38 on the lateral and inferior aspect of the hemisphere. The left and right sides of the posterior part of this region, an area known as the inferotemporal (IT) area, have important roles in higher-order aspects of visual processing. In this context, "inferolateral" includes both the inferotemporal region and the polar temporal region (area 38). The posterior parts of the second, third, and fourth temporal gyri formed by areas 37, 20, and 21 are strongly linked to the higher-order decoding of visual information. This region is critical for face recognition and for visual recognition of other entities such as animals and fruits and vegetables. On the left side, the more anterior parts of the region are specialized for the retrieval of lexical items that denote various entities, including common nouns such as the names of animals, fruits and vegetables, and tools and utensils. The system remains strongly linked to the visual modality, and recognition and lexical retrieval defects resulting from damage in this region are specific to information presented through the visual modality. More forward parts of the system on the left are concerned with progressively more unique levels of information, and in the temporal polar region (i.e., area 38), there is specialization for the retrieval of proper nouns, which are highly unique lexical entries that denote items that constitute a class of one. The functional correlates of the anterolateral system on the right are less well understood. The region is probably important for the comprehension of the emo-

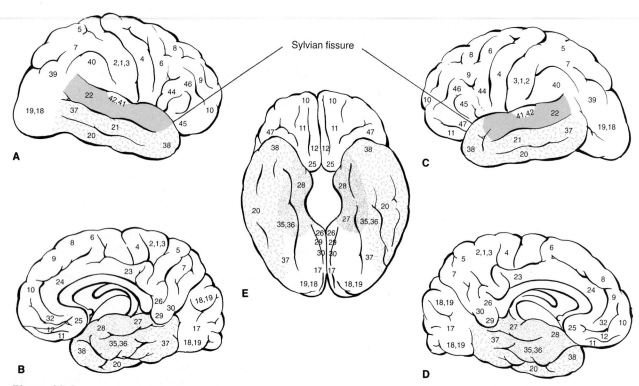

Figure 30-5
The major subdivisions of the temporal lobes are depicted on the lateral and mesial aspects of the (**A,B**) right and (**C,D**) left hemispheres, respectively. (**E**) An inferior view of these regions is depicted. The primary auditory cortices (i.e., areas 41, 42; Heschl's gyri) are buried in the depths of the Sylvian fissure. Blue denotes auditory association cortices. The inferolateral sector is marked with dashes, and the mesial temporal region, which includes the amygdala, entorhinal cortex, and hippocampus, is marked with shading.

tional meanings of nonverbal stimuli (e.g., the capacity to assign positive or negative valence to a visual nonverbal stimulus such as a face or complex scene) and may play a role in the retrieval of nonverbal information from retrograde memory.

Mesial temporal sector

The mesial sector of the temporal lobes includes the hippocampal system and the parahippocampal gyrus. The hippocampal system, composed of the amygdala, entorhinal cortex, and hippocampus proper, is specialized for memory. It mediates the acquisition of new material of a type that is known as *declarative* (i.e., information such as facts, words, and data that can be brought to mind and consciously inspected). The hippocampal system follows the general principles of hemispheric specialization; the left hippocampal system is dedicated to verbal material and the right to nonverbal material. The system is most important for the *acquisition of new information* (i.e., anterograde memory) and plays a lesser or negligible role in retrieval of previously learned material (i.e., retrograde memory).

The relative contributions of the amygdala and hippocampus to memory remain controversial. Scientists agree that the hippocampus plays a critical role in anterograde memory, but the amygdala may or may not figure as prominently. One possibility is that the amygdala is important for complex cross-modal learning, such as learning complex relations between stimuli perceived through different sensory modalities (e.g., visual, tactile). The amygdala may also have an important role in learning of relations between specific entities or events and the emotional or affective overtones associated with the entities or events. The principal brain-behavior associations associated with the three subregions of the temporal lobes are enumerated in Table 30-3.

☐ The parietal lobes are comprised of primary somatosensory cortices and important association areas

The parietal lobes, located behind the Rolandic sulcus and above the Sylvian fissure (Fig. 30-6), comprise a heterogeneous collection of primary sensory and association cortices. Basic somatosensory perception, including perception of touch, vibration, and temperature, takes place in the strip of cortex formed by the post-central

Table 30-3
BRAIN-BEHAVIOR ASSOCIATIONS IN THE TEMPORAL REGION

Neural Region	Cognitive or Behavioral Function
Superior temporal	
Primary cortices (Heschl's gyrus)	Basic auditory perception
Association cortices	
Left	Speech perception and comprehension
	Processing of temporal aspects of aural signals
Right	Perception and comprehension of environmental nonverbal sounds, music, timbre, prosody
	Perception of spectral aspects of aural signals
Inferolateral temporal	
Posterior sector	Visual object recognition; face recognition
Anterior sector	
Left	Common and proper noun retrieval
	Retrieval of verbal information from retrograde compartment
Right	Comprehension of emotional meanings of nonverbal stimuli
	Retrieval of nonverbal information from retrograde compartment
Mesial temporal	
Left	Anterograde verbal memory; acquisition of new verbal information
Right	Anterograde nonverbal memory; acquisition of new nonverbal information

gyrus (i.e., areas 3, 1, and 2). A secondary somatosensory area is in the inferior parietal operculum, and this region may play an important role in higher-order tactile perception (e.g., recognition of objects from touch). In general, the inferior parietal lobule, composed of Brodmann's areas 40 and 39, is linked to the auditory modality and is strongly connected to nearby auditory association cortices in the temporal lobe, but the superior parietal lobule, composed of Brodmann's areas 7 and 5, is linked to the visual modality and is strongly connected to visual association cortices in the occipital lobe. However, the region formed by the transition zone between the occipital, temporal, and parietal cortices on the lateral aspect of the hemisphere is highly heteromodal and is specialized for polymodal sensory integration, such as integration of visual, auditory, and somatosensory signals.

On the left, the inferior parietal lobule is involved in language functions. The ability to repeat verbatim words, digits, or sentences, for example, depends on intact parietal opercular cortices and underlying white matter. This function requires accurate perception of auditory information, the retrieval of identical information from the person's own store of acoustic records, and the triggering of anterior motor cortices to produce the information. Inferior parietal cortices on the right side play a significant role in self-perception, the placement of a person's body in space, and in the mapping of physical and emotional states. The ability to direct attention to the external and internal milieu, for example, depends critically on right inferior parietal cortices. Brain-behavior associations of the parietal region are summarized in Table 30-4.

☐ The anterior part of the brain is comprised of the frontal lobes, which are important for social conduct and decision-making

The frontal lobes comprise a vast expanse of cortex and white matter anterior to the Rolandic sulcus (Fig. 30-7). They represent the highest level of neural evolution, and in the prefrontal region in particular, humans show a tremendous increase in brain size compared with other higher primates. The cognitive operations mediated by the frontal lobes, such as foresight, complex decision-making, and social conduct, also stand at the zenith of evolution of mental processes. The frontal lobes can be subdivided into several functional units that have distinctive behavioral correlates. Four major subdivisions are reviewed in the sections that follow.

Premotor region

Immediately anterior to the Rolandic sulcus is the strip of motor cortex (i.e., area 4) that mediates basic motor activity of all parts of the body. Anterior to this is area 6, which with area 44 comprises the premotor region. The cortex in area 6 also participates integrally in motor behavior, but it is not dedicated to movement but to a sort of motor association cortex, and it is involved with the planning and initiation of motor behaviors. The premotor region has access to complex information in all major sensory modalities and participates in the formation of a plan for motor activity in response to a particular stimu-

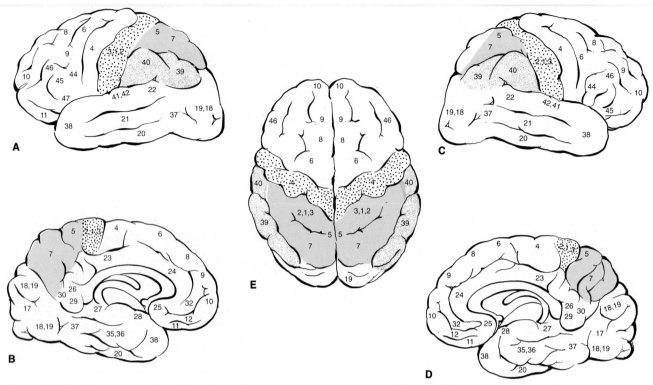

Figure 30-6
The major subdivisions of the parietal lobes are depicted on the lateral and mesial aspects of the (**A,B**) left and (**C,D**) right hemispheres, respectively. (**E**) A superior view of these regions is depicted. The primary somatosensory cortex (i.e., areas 3, 1, 2) is marked with dots. The superior parietal lobule, formed by areas 5 and 7, is denoted by blue. Areas 40 (i.e., supramarginal gyrus) and 39 (i.e., angular gyrus) form the inferior parietal lobule, marked with shading.

lus configuration and provides the impetus to set the plan into action.

The frontal operculum includes areas 44 and 45. On the left side, this region constitutes *Broca's area*, and it is the main speech output center; it is the region responsible for motor aspects of linguistic expression. On the right, the frontal opercular region plays a significant role in expressive prosody, the infusion of speech with various intonations that add emotional coloring to the content.

Dorsolateral prefrontal region

The dorsolateral prefrontal region is composed of the cortices and white matter formed by the lateral expanses of areas 8, 9, 46, and 10 (see Fig. 30-7). The precise neuropsychological correlates of this region are not well established, although it is likely that the dorsolateral frontal cortices play an important role in many types of higher-order intellectual behavior. Investigations taking advantage of fine-grained neuroimaging techniques, especially

Table 30-4
BRAIN-BEHAVIOR ASSOCIATIONS IN THE PARIETAL REGION

Neural Region	Cognitive or Behavioral Function
Primary somatosensory cortex	Basic perception of touch, vibration, temperature, pain
Inferior parietal lobule	
Left	Tactile object recognition
	Verbatim repetition
Right	Tactile object recognition
	Self-perception; mapping of physical and emotional states
	Placement of oneself in space

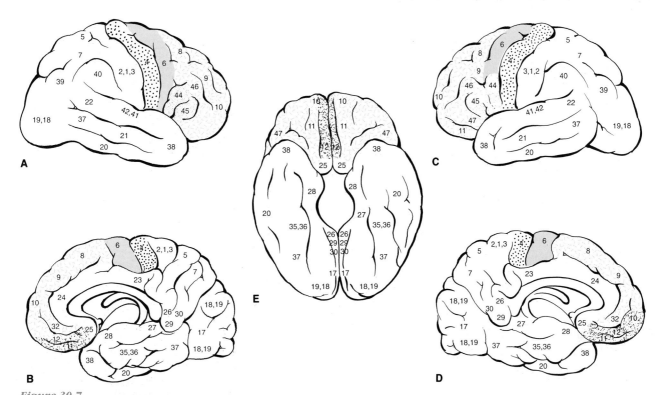

Figure 30-7

The major subdivisions of the frontal lobes are depicted on the lateral and mesial aspects of the (**A,B**) right and (**C,D**) left hemispheres, respectively. (**E**) An inferior view of these regions is depicted. The primary motor cortex (i.e., area 4) is marked with dots. The premotor region includes area 6 (*blue*) and area 44. The latter, together with area 45, (*gray shading*) comprises the functional unit known as Broca's area. The large expanse of prefrontal cortex on the dorsolateral, mesial, and inferior surfaces of the hemispheres is marked with dashes. The ventromedial frontal region, which includes the orbitofrontal and lower mesial frontal cortex, is marked with marbling. The posterior-most component of the orbitofrontal region, formed by area 25, is not included in the ventromedial designation, because area 25 is usually considered to be part of the limbic lobe formed by the cingulate and other structures around the corpus callosum. The region corresponding to the *basal forebrain* is not colored, but it is immediately posterior to area 25.

computed tomography (CT) and magnetic resonance imaging (MRI), and sophisticated neuropsychological assessment procedures have provided more conclusive information, and several conclusions can be advanced.

Various types of high-level intellectual abilities appear to depend in part on the dorsolateral frontal region. Examples include mental operations that require the retrieval and manipulation of information, especially when there is a demand for *unique, creative, and original* production. To the extent that such operations involve verbal material and the use of language symbols, the left dorsolateral region has the dominant role. When operations depend on and require the use of nonverbal information, the right dorsolateral region takes the lead. Consider the following two tasks, which are analogous in their demand for creative, flexible mental production but which vary in the use of verbal versus nonverbal mediation. In one, known as *verbal fluency*, the patient is required to generate as many words as possible that begin with a particular letter of the alphabet (e.g., f) within a specified time limit. In the other task, known as *design fluency,* the pa-

tient is asked to generate as many original and nonredundant designs as possible within a particular time constraint. Both tasks require retrieval, manipulation, and execution in a manner that depends critically on flexible, creative thought processes. The dorsolateral frontal region is an important neural substrate for performance of tasks such as these.

Superior mesial region

On the mesial surface of the cerebral hemispheres, there is a region of the frontal lobes composed of the supplementary motor area (SMA; mesial area 6) and the mesial parts of areas 8, 9, and 32. Together with the anterior part of the cingulate gyrus (i.e., area 24), this region is closely linked to emotional behavior and to basic drive states associated with motivation and arousal. The SMA plays a crucial role in motivating a person to execute motor behaviors. Without the input of the SMA, the person may never get past the stage of intention to act. The same principle applies to speech production, because the SMA plays an important role in the basic drive to produce speech.

The superior mesial region also is important for a variety of emotional, motivational, and other features that can be grouped as *personality characteristics*. This region, for example, participates in behaviors that include elements of obsessiveness, compulsiveness, anxiety, and introversion or extroversion. The maintenance of an adaptive and optimal state of arousal and alertness also is linked to this region. Having one's surveillance systems "set" in such a way to as to maximize the relation between energy expenditure and the detection of important external and internal events is a function that is linked to the superior mesial frontal region. Having a sense of how long to persevere on a particular line of goal-directed behavior or when to desist and seek an alternate track is another example. The spontaneous, automatic selection and implementation of an appropriate emotional response to a particular stimulus configuration is linked to the superior mesial frontal region and anterior cingulate. Examples include having a mirthful response to a good joke, having a feeling of sorrow on being told of a sad event, or having a response of empathy while watching a child endure a painful social learning experience.

Ventromedial region

The ventromedial region includes the orbitofrontal cortex and nearby regions, formed by areas 12 and 11 on the ventral and lower mesial aspects of the frontal lobes. This region forms a functional unit that is highly specialized for the mediation of internal and external stimuli in terms of their psychological significance. Detecting and evaluating correspondences between a particular stimulus configuration (e.g., being given the "what-for" by a boss) and a particular emotional state (e.g., feeling guilty and shameful) is an example of this capacity. The ventromedial frontal region is critical for the permanent registration of these correspondences, so that when particular stimulus configurations recur, the person can avoid punishment or gain reward by pursuing the most propitious courses of action. Left-right differences have not been firmly established, although it appears likely that the left side will prove to be relatively specialized for verbal information and the right for nonverbal data. The region is important for judgment, planning, and decision-making, and for a wide range of behaviors that can be grouped under the heading of social conduct.

Social conduct, when executed in such a way as to maximize a person's personal gain in the short and long term and the overall well-being of his or her species, depends on the rapid selection of appropriate courses of action, often in highly ambiguous situations. Such selection requires several component processes, including bringing on-line a range of viable potential courses of action, the assignment to each of these a degree of likelihood in terms of its short-term and long-term consequences for reward or punishment, and consideration of the implications of such courses in the context of the per-

son's current position in the world. It is precisely this type of decision-making process that is intimately linked to the ventromedial frontal region. As discussed in a later section, damage to the ventromedial region produces severe disturbances in social conduct, while sparing most basic components of cognition. Table 30-5 outlines basic brain-behavior associations associated with the main subdivisions of the frontal lobes.

☐ Subcortical structures such as the basal ganglia and thalamus are important for many basic and higher-level functions

Basal ganglia

The basal ganglia are a collection of gray matter nuclei deep in the brain, comprised mostly by the *striatum*, including the caudate nucleus and the putamen, and *pallidum* (i.e., globus pallidus). The striatum and pallidum play basic roles in motor behavior, particularly in the automatic execution of highly learned motor patterns. For example, the kinds of motor behaviors that are called into play when a person drives a car, rides a bicycle, or ice skates depend on the basal ganglia.

In addition to their role in automatic motor behaviors, evidence has pointed consistently to the importance of the striatum and pallidum in higher-order aspects of cognition and behavior. The left caudate nucleus (especially the head) plays an important role in language function, paralleling to some extent the roles played by interconnected cortical regions in the perisylvian region of the left hemisphere. The left putamen has a role in speech articulation: the implementation of strings of phonemes in the motor system in a smooth, seamless manner that allows fluent, canonical articulation. On the right side, the striatum plays an important role in prosody. It is also involved in various visuospatial functions, including visual perception and visual construction, in a manner generally analogous to the cortical regions of the right hemisphere.

The striatum and pallidum are important for the acquisition of new information that falls under the designation of procedural or nondeclarative. Such information is characterized by requiring a motor output for its instantiation and by its lack of amenability to being "brought to mind" for conscious inspection, unlike declarative memorial information. For example, learning new perceptuomotor skills such as rotor pursuit or mirror tracing depends on the basal ganglia. In a pursuit rotor task, the subject leans to follow a moving target with a hand-held stylus. The mirror-tracing task requires the subject to trace the shape of a geometric figure (e.g., triangle) using a mirror image to guide the hand. Learning new habits (i.e., a tendency to respond automatically and consistently to a particular stimulus configuration) is also probably linked to the basal ganglia, and it has been suggested that the learning of "gut-level" likes or dislikes of certain persons may depend on certain parts of the stria-

Table 30-5
BRAIN-BEHAVIOR ASSOCIATIONS IN THE FRONTAL REGION

Neural Region	Cognitive or Behavioral Function
Premotor cortex	
Area 6	Planning and initiation of motor behavior
Area 44 (Broca's area)	
Left	Speech output
Right	Prosodic output
Dorsolateral prefrontal	
Left	Verbal intellectual capacities
	Creative, flexible verbal thinking
	Verbal fluency
Right	Nonverbal intellectual capacities
	Creative, flexible nonverbal thinking
	Design fluency
Superior mesial	Emotional behavior
	Motivation; basic drive states
	Maintenance of adaptive state of arousal and alertness
	Personality characteristics
Ventromedial	Social conduct; judgment; planning; decision making
	Triggering of bodily states associated with emotions

tum. Brain-behavior associations of the basal ganglia are listed in Table 30-6.

Thalamus

The thalamus has a critical role in the transmission of motor and sensory information from lower subcortical systems to various regions of the cortex. Evidence also supports a role for the thalamus in higher-order cognition and behavior, especially in attention, memory, and language. The thalamus contains the reticular nuclei, which have an important role in the network of neural structures responsible for arousal and attention (Mesulam, 1985). The capacity to direct attention to a particular stimulus and to attend to it sufficiently to allow the selection of an appropriate response are linked to basic arousal and attentional mechanisms mediated in part by the reticular nuclei.

The thalamus has a left-right bias that mirrors that of the cortex; structures in the left thalamus are geared more toward verbal material, and structures in the right thalamus are geared more for nonverbal material. The left anterior thalamus plays a role in language that is related to comprehension and production aspects of linguistic function. The anterior part of the thalamus, which contains important pathways linking the thalamus to the hippocampus and amygdala (especially the mammillothalamic tract and the ventroamygdalofugal pathway), has been linked to material-specific learning capacities; the left is important for verbal material, and the right is important for nonverbal information. There are midline thalamic nuclei, including the medial dorsal nucleus, which have an important role in memory. In general, the thalamus is more important for anterograde memory than retrograde and in learning declarative types of information than in learning procedural knowledge. Brain-behavior linkages associated with the thalamus are enumerated in Table 30-6.

Acquired Disorders of Higher Brain Function

Much of what has been learned regarding the cognitive and behavioral correlates of neural structures has come from investigations using the lesion method as the basic paradigm of inquiry (Damasio and Damasio, 1989). A review of the most frequent disorders of higher brain function, with emphasis on the the neural correlates of such disorders, provides considerable insight into several brain-behavior associations. In the lesion method, focal

Table 30-6
BRAIN-BEHAVIOR ASSOCIATIONS IN THE BASAL GANGLIA AND THALAMUS

Neural Region	Cognitive or Behavioral Function
Basal ganglia	
Left	Motor behavior; automatic execution of well-learned motor actions
	Procedural memory; learning of new motor skills; learning of new habits
	Language; speech articulation
Right	Motor behavior; automatic execution of well-learned motor actions
	Procedural memory; learning of new motor skills; learning of new habits
	Prosody
	Visuospatial and visuoconstructional functions
Thalamus	
Left	Arousal and attention
	Language
	Anterograde verbal memory (declarative)
Right	Arousal and attention
	Visuospatial capacities
	Anterograde nonverbal memory (declarative)

damage to the brain is carefully specified in neuroanatomic terms and is related to a detailed neuropsychological profile. Patients with similar regions of damage can be studied to determine similarities in the pattern of neuropsychological deficit, lending support to conclusions about the relation between a specific brain region and a particular cognitive capacity. From another perspective, patients with a common neuropsychological defect can be studied to determine similarities in their neuroanatomic status. The resultant commonalties lend support to the notion that a particular cognitive capacity is linked to a particular neural region. The lesion method has enabled many important advances in our understanding of brain-behavior associations. However, the lesion method and the results it produces are not tantamount to the idea of *localization of function*. The method does not assume nor imply that particular cognitive functions are located in particular brain regions. Brain regions instead comprise parts of *neural systems* that are specialized or dedicated to particular cognitive operations.

There are several major causes of neuropsychological disorders, which differ in their demographic associates. Five of the most common forms of brain injury are *cerebrovascular disease, degenerative diseases* (including Alzheimer's disease and Pick's disease), *traumatic brain injury, cerebral tumors,* and *viral infections of the central nervous system,* especially herpes simplex encephalitis. These pathologies occur with different base rates in different demographic groups and tend to affect different brain regions, and these correlations can furnish important diagnostic clues when a patient first presents with a condition. Damage to the ventromedial frontal region of the brain, for example, which often produces a disturbance of personality, is usually caused by traumatic brain injury or cerebral tumor, and most of the patients are relatively young. Disorders of language occur most frequently in patients with stroke, who tend to be relatively older. Memory disturbances are overwhelmingly caused by degenerative conditions, especially Alzheimer's disease, which occur almost exclusively in elderly populations; much smaller percentages of memory disorders are produced by herpes encephalitis, traumatic brain injury, and surgical ablation (e.g., temporal lobectomy for intractable seizure disorder).

□ Disorders of perception affect the processing of sensory information in primary and early association cortices

Disorders of perception can affect any sensory modality. Damage to regions that serve primary perception in a particular modality (e.g., primary visual cortex, primary auditory cortex) produces complete or partial abolition of perception in that modality. This chapter focuses on disorders caused by damage to association cortices, which involve higher-order processes beyond the level of primary perception. The review is concentrated on visually related perceptual disturbances, which have been studied extensively and are particularly revealing of some of the most fascinating links between brain and mind. The disorders explicated in this section and their neural correlates are summarized in Table 30-7.

Impairment of color perception: acquired achromatopsia

Acquired or central *achromatopsia* is a disorder of color perception that affects part or all of the visual field and is associated with damage to the ventral part of the visual association cortices. In a striking dissociation, the patient loses the perception of color but preserves the ability to perceive form, depth, and motion. The color loss commonly affects a quadrant or hemifield, and it may be partial, in which case the patient complains that colors look "washed out" or dull, or complete, in which case everything is reduced to shades of black and white. The disorder has a sudden onset in a person with previously normal color vision, and it is distinct from hereditary (i.e., retinal) defects in color perception that are not attributable to central causes. Achromatopsia is a disturbance of visual perception and not of language; the problem is that the patient's brain cannot see colors normally, not that colors cannot be named correctly. The latter condition, known as *color anomia*, can occur in patients who have perfectly normal color perception, and it is associated with a different region of neural destruction.

Table 30-7
DISORDERS OF PERCEPTION

Disorder	Examples of Neural Correlates
Acquired achromatopsia	Lingual gyrus (left and right)
Apperceptive visual agnosia	Right inferior and superior visual association cortices
	Early visual association cortices (right and left)
Pure alexia	Periventricular white matter in lower left visual association cortices
	Left lower occipital *plus* splenium of corpus callosum
Balint syndrome	Superior visual association cortices; posterior superior parietal lobule (right and left)
Neglect	Right parietal and occipitoparietal

Achromatopsia is frequently seen in patients who also manifest disturbances of reading (if the lesion is on the left) or of facial recognition (if the lesion is on the right or bilateral). This is because the neural regions that are most critical for color perception are located in the ventral visual association cortices in regions that are also important for reading and face recognition. The lingual gyrus, immediately below the inferior bank of primary visual cortex along the calcarine fissure, is a primary brain region responsible for color perception.

There are three common presentations of achromatopsia: *left hemiachromatopsia* (i.e., color loss in the left field only), associated with a unilateral right occipitotemporal lesion, which may be unaccompanied by other neuropsychological defects; *right hemiachromatopsia* associated with a unilateral left occipitotemporal lesion, usually associated with a reading disturbance (i.e., alexia); and *full-field achromatopsia* associated with bilateral occipitotemporal lesions, commonly occurring in conjunction with face agnosia (i.e., prosopagnosia).

Findings from the lesion method have been supported by data from studies using positron emission tomography (PET), which indicate that there is specific activation (e.g., an increase in blood flow and metabolic activity) of cortices in the region of the lingual gyrus when subjects are concentrating on looking at different colors.

Impairment of form perception: apperceptive visual agnosia

Lesions in the right visual association cortices that involve ventral and dorsal sectors of the occipital region can produce a condition called *apperceptive visual agnosia*, in which patients lose the ability to recognize familiar visual stimuli because of a disturbance in the capacity to integrate various features of visual perception. A defect for familiar faces, called *face agnosia* or *prosopagnosia*, is the most common form of this disorder. When confronted with faces of those well known to them, patients fail to recognize who those faces belong to or even that they are familiar. The recognition defect may be partial, in which case the patient has trouble with many faces but can still perform accurately with a few highly known persons such as the self and family members, or it can be quite pervasive and affect virtually all faces the patient should know, including the self.

Two intriguing associated neuropsychological defects commonly accompany prosopagnosia of the apperceptive variety. First, patients have a profound inability to assemble puzzle parts to make a whole. Even a simple 6- or 9-piece puzzle can prove almost impossible. The main problem is that the patient cannot formulate a mental hypothesis about what the pieces may comprise; without this blueprint, the patient is reduced to merely fitting together a few individual pieces according to local features and borders, without any overarching framework to guide the construction. A second and related defect is that the patients cannot imagine faces; they cannot bring to their "mind's eye" the image of a face. As a consequence, a patient may construct a face puzzle in such a way that there are two eyes on one side or the mouth at the top—featural arrangements that violate basic rules about how features must relate and conjoin in a human face.

Apperceptive visual agnosia may affect other classes of objects. For instance, a patient may lose the ability to recognize cars (e.g., different makes of cars) or geographic entities (e.g., landmarks or buildings, such as Devil's Tower or The White House). Patients with apperceptive visual agnosia commonly have severe impairments in their ability to learn new geographic routes, and they may have considerable difficulty executing routes that should be well-known from previous experience. These defects are related to the patient's inability to use various landmarks to help guide the way. Another common neuropsychological defect in such patients is in the recognition of entities presented in noncanonical views. For example, when patients are shown pictures of common entities in which the object has been photographed from a highly unusual viewing perspective (e.g., an elephant as seen from behind, a spoon as viewed from the end of the handle), they have tremendous difficulty figuring out what the objects are.

Apperceptive visual agnosia can result from a right-sided lesion that involves superior and inferior parts of the visual association cortices. Another common cause of this type of visual agnosia is a bilateral lesion in relatively early visual association cortices (i.e., regions immediately adjacent to primary visual cortex), which can produce severe visual recognition problems while sparing enough basic visual processing that the patient can still "see" things. Apperceptive visual agnosia in the setting of a single right-sided lesion tends to be more severe and more persistent if the lesion also encroaches into the parietal region, especially area 39.

Impairment of reading: pure alexia

Acquired impairments of reading (i.e., alexia) are common in patients with language disturbances (i.e., aphasia) associated with left-hemisphere lesions. There is one special type of reading dysfunction, known as *pure alexia*, in which patients lose the ability to read (partially or completely) while having no other disturbance of language. Even writing is spared, producing the bizarre condition in which the patient cannot read what he or she wrote only moments before. Pure alexia, also known as alexia without agraphia, is associated with lesions to the periventricular white matter in the lower part of the left visual association cortices in a region behind and beneath the occipital horn. The lesion disconnects both visual association cortices from the dominant, language-related temporoparietal cortices on the left, and orthographic visual information (i.e., printed letters and words) cannot be decoded and made meaningful. Another lesion that can produce pure alexia is a combination of damage in the left occipital lobe with damage to the posterior corpus callosum; this effectively disconnects both visual association cortices from left-sided language-related regions.

Impairments of visual attention, motion detection, and visual guidance of movement

Balint syndrome and its components. Three components comprise the Balint syndrome: *visual disorientation* (i.e., simultanagnosia); *ocular apraxia* (i.e., psychic gaze paralysis), and *optic ataxia*. Development of the Balint syndrome is related to bilateral damage to occipitoparietal cortices in the superior sector of the visual association region. A lesion confined to the right occipitoparietal region may produce a less severe form of the condition, and lesions confined to the superior occipital region (without extension into the parietal region) usually cause only visual disorientation, without ocular apraxia and optic ataxia. The Balint syndrome is not caused by damage to the inferior sector of visual association cortices, illustrating the clear specialization of the inferior and superior systems for different types of visual processing.

Visual disorientation is the key component of the Balint syndrome. It entails an inability to attend to more than one limited sector of the visual panorama at any given time. The patient can hone in on only one small portion of the visual field and is unable to monitor simultaneously what is happening in other parts of the field. The sector of clear vision may be unstable, and it can shift unpredictably so that the patient suddenly loses a particular stimulus from view. Patients notice that their visual focus appears to jump from one part of the field to another. This phenomenon precludes the ability to form a spatially coherent visual field, and the patients are unable to follow the trajectories of moving objects or place stimuli in their proper locations in space. Defective perception of motion is a frequent manifestation in patients with Balint syndrome. Patients may fail to notice that objects have moved about in their visual field, or they may fail to recognize the meaning of motion. For example, they may fail to recognize the meaning of pantomime (e.g., wiggling a finger to signal "come here") or to recognize the identity of a familiar person based on gait.

Ocular apraxia is a deficit of visual scanning. It consists of the inability to direct the gaze voluntarily toward some particular target in the visual field. For example, patients may detect that something has entered the field on the left, but whey they try to shift their gaze so that the target will fall in central vision, they miss, making incorrectly calibrated saccades that fail to hit the target accurately. *Optic ataxia* is a disturbance of visually guided reaching behavior, manifested by the patient's inability to point accurately at a target or to reach out and deftly pick up an object. The reaching disturbance is confined to movements that are visually guided, and if the patients try to point toward sound sources, they do so accurately; they can also point accurately to targets on their own body, using somatosensory information (i.e., proprioception).

Neglect. Neglect refers to a phenomenon in which a patient fails to attend normally to a portion of extrapersonal or intrapersonal space, despite an adequate capacity for basic perception of information. In the visual modality, in which neglect is most commonly manifested, the patient may fail to attend to the left hemispace, ignoring objects, persons, and movements that occur on the left side. Such a presentation is most commonly caused by lesions in the right parietal and occipitoparietal region, including cortices composed of areas 39 and 40. Neglect can occur in other sensory modalities, albeit less frequently. For example, a patient may ignore auditory signals delivered from one side or may fail to acknowledge somatosensory information presented on one side. In all modalities, there is a highly prominent tendency for the phenomenon to develop in relation to the left hemispace of the patient, and right-sided neglect is rare. This tendency illustrates the dominant role that the right parietal region plays in the mapping of extrapersonal and intrapersonal space.

Severe neglect most often occurs in the acute stages of brain injury, often in the first few days or weeks after the onset of brain damage. After several months after the onset of the lesion, most patients have recovered substantially, although subtle attentional defects for left-sided stimuli may persist. However, even though more blatant signs of neglect may have recovered, lingering attentional problems can cause considerable cognitive disability. Several effective rehabilitation programs have been developed to treat these problems.

☐ Disorders of memory involve impairments in the ability to acquire and retrieve knowledge

Disorders of memory are the most common types of cognitive impairments in brain-injured persons, and in persons with degenerative disease or head injury—two conditions that are among the most frequent causes of brain damage—memory defects are virtually ubiquitous. This review focuses on several types of memory problems that have shed light on brain-behavior associations. A listing of these according to their neuropsychological manifestations and neural correlates is provided in Table 30-8.

Agnosia

In the condition of *apperceptive agnosia*, patients fail to recognize familiar stimuli because of difficulty integrating various components of perception. *Associative agnosia* refers to recognition failure that occurs in the setting of normal or near-normal perception and which is more a *disturbance of memory* than of perception. When confronted with a stimulus that should be known, a patient with associative agnosia fails to activate and retrieve pertinent associated memories that would give the stimulus meaning and make it familiar, recognizable, or identifiable. The recognition failure is confined to one sensory modality, and stimuli that go unrecognized in the affected modality can be identified normally through other channels. If recognition failure cuts across more

Table 30-8
DISORDERS OF MEMORY

Disorder	Examples of Neural Correlates
Agnosia	
Visual-associative	Bilateral occipitotemporal junction
Auditory	Bilateral posterior superior temporal region (posterior area 22)
Tactile	Left or right inferior parietal lobule; superior mesial parietal lobe (areas 7 and 5)
Amnesia	
Anterograde (declarative)	
Verbal	Left mesial temporal region; left thalamus; basal forebrain
Nonverbal	Right mesial temporal region; right thalamus; basal forebrain
Retrograde	
Verbal	Left nonmesial anterior temporal
Nonverbal	Right nonmesial anterior temporal
Procedural (nondeclarative)	Basal ganglia; cerebellum

than one sensory modality (e.g., if a patient fails to recognize a stimulus presented through the visual and auditory modalities), the term *amnesia* rather than agnosia should be applied. *Visual agnosia* is by far the most common variety; *auditory and tactile agnosia* are rare. Visual and auditory agnosia are strongly linked to bilateral cerebral disease of the visual association and auditory association cortices, respectively, and the presence of one of these conditions should prompt a close inspection for possible bilateral lesions.

Visual agnosia. Visual agnosia refers to an impairment of recognition of stimuli presented through the visual channel, despite normal visual perception. The patient fails to arrive at the meaning of a visual stimulus, such as a particular animal or a familiar face, and cannot describe its characteristic features, functions, or operations. The patient may remain capable of assigning the stimulus to a general class (e.g., "That's an animal" or "That's a face"), but cannot produce specific identification (e.g., "groundhog" or "John F. Kennedy"). It is important to differentiate visual agnosia from anomia; in the latter condition, the patient fails to name stimuli but still recognizes their identities and can produce accurate and specific descriptions.

Agnosia for faces, or *prosopagnosia*, is the most common and best studied form of visual agnosia. In this condition, patients fail to recognize familiar faces such as family members, close friends, and even their own faces, and there is usually a defect in the ability to learn new faces. Prosopagnosia usually has a sudden onset, and it is most typically produced by bilateral disease in the lower visual association cortices in and near the region of the occipitotemporal junction (i.e., lower areas 18 and 19 and area 37). The recognition failure is confined to the visual modality, and if the patient is allowed to hear the voices of persons who cannot be recognized visually, the patient instantly uncovers their identities.

There are several curious and relatively common dissociations in patients with prosopagnosia, which reveal important clues regarding the manner in which the brain processes visual learning and recognition. For instance, prosopagnosic patients fail to recognize identity (i.e., who people are), but they remain capable of recognizing other types of facial information, such as emotional expressions, age, and gender (Tranel, Damasio, and Damasio, 1988). A patient may look at a well-known face (e.g., his mother) and fail to recognize it, and yet be able to state correctly that the face is showing happiness and that it is the face of an elderly woman. Such a dissociation indicates that the cognitive demands of recognizing identity and something like facial expressions are quite different. The former task requires disambiguating one specific face from among many hundreds or thousands that a person may know, and the latter task requires disambiguation of one type of expression from among a mere handful that are most frequent.

Prosopagnosic patients usually remain capable of recognizing identity from motion. For example, they can identify a particular person based on that person's gait or characteristic manner of moving, even if they cannot recognize the face. This dissociation reflects the functional segregation of visual processing in the posterior visual association cortices, whereby information related to form and shape is processed by the ventral system, which is damaged in prosopagnosia, and information related to movement is processed by the dorsal system, which is often spared in prosopagnosia.

Another intriguing discovery regarding prosopagnosia is the recent demonstration that patients often have preserved *nonconscious recognition* of familiar faces, despite their failure to recognize at conscious level. For example, patients with severe inability to recognize familiar faces still produce discriminatory autonomic responses to familiar faces, in the same way as normal persons (Tranel and Damasio, 1985). This dissociation between overt and covert recognition has been demonstrated in a number of different experimental paradigms, and it indicates that the brain processes information on several different levels, not all of which are made available to consciousness.

Prosopagnosia can occur in remarkably pure form, in which the patient has a severe face recognition impairment without other neuropsychological defects, but there are several neuropsychological defects that are common accompaniments. One of these is *achromatopsia*. The presence of a color perception defect in prosopagnosia is attributable to the fact that the brain regions that process color perception and face recognition are in proximity in the ventral visual association cortices, and lesions in this region probably encroach on both territories. Another common manifestation in prosopagnosic patients is *topographical agnosia*, a condition in which the patient fails to navigate previously familiar geographic routes and cannot learn new ones. Most patients with prosopagnosia have recognition impairments in other classes of visual stimuli which, like faces, require specific identification among many members that are visually similar. For example, prosopagnosics may have difficulty recognizing particular buildings, cars, or articles of clothing.

Auditory agnosia. Auditory agnosia is a disorder of recognition confined to the auditory realm, in which a patient with normal or near-normal auditory perception fails to identify and detect the meaning of previously known auditory stimuli. In the full-blown version of the condition, the patient fails to recognize speech (i.e., words) and nonspeech (i.e., environmental sounds) auditory signals. The patient is not capable of understanding what is being said and does not respond appropriately to sounds such as a telephone ringing or a knock at the door. The disorder is caused by bilateral damage to auditory association cortices in the posterior third of the superior temporal gyrus. Unilateral damage on the right can produce a partial, more selective form of auditory agnosia. For example, the patient may lose the capacity to recognize familiar music or familiar voices. Unilateral damage on the left causes *Wernicke's aphasia*, which if severe can entail a complete inability to understand speech sounds.

Tactile agnosia. Tactile agnosia involves loss of the ability to recognize objects presented through the tactile modality, even when basic aspects of somatosensory function are normal or near normal. Patients fail to recognize objects such as keys, pencils, or eating utensils. In keeping with the definition of agnosia, the disorder is confined to the tactile modality, and if the patient is allowed to see the object or hear its characteristic sound, a correct identification can be made. Unlike prosopagnosia, which affects recognition of unique stimuli (e.g., specific identification of faces), tactile agnosia affects recognition of basic objects, such as utensils and tools.

Tactile agnosia has been associated with lesions to the left or right inferior parietal lobule in and near the region of the parietal operculum. It has been reported also in connection with a lesion in the superomesial part of the parietal lobe (i.e., mesial aspect of areas 7 and 5). The condition is far less disabling than visual agnosia. Patients may not complain about it, and it may not even be detected except through specialized laboratory assessment.

Amnesia

Memory is a complex and broad topic, and the disorder has been subjected to more dichotomies and subdivisions than perhaps any other domain of cognition. This discussion focuses on some of the best-known distinctions and on neural correlates that have been most firmly established.

Amnesia after temporal lobe lesions. The mesial temporal lobe, formed by the hippocampal system (i.e., amygdala, entorhinal cortex, and hippocampus proper) and adjoining parahippocampal gyrus, has been linked unequivocally to memory function for several decades. The connection was first established in the landmark report by Scoville and Milner (1957) on patient H.M., who became severely amnesic after bilateral mesial temporal resection for control of intractable seizures. Extensive investigations of H.M. and several other revealing cases over the past three decades have established that the hippocampus in particular is critical for the acquisition of new information.

Several types of memory impairment have been associated with hippocampal damage. First, there is a consistent correspondence between the side of damage and the type of learning impairment. Left-sided lesions produce defects in the learning of verbal information (e.g., words, written material), but learning of nonverbal information is spared; damage to the right hippocampus produces a defect in the learning of nonverbal material (e.g., complex visual and auditory patterns), but verbal information is spared. A second consistent finding is that, although the hippocampus is critical for the acquisition of new information (i.e., *anterograde memory*), it is much less important for the retrieval of previously learning knowledge (i.e., *retrograde memory*). Even if there is extensive bilateral hippocampal damage, the ability to retrieve information that was acquired before the onset of the lesion is largely spared. A third conclusion is that the hippocampus plays a role in the acquisition of *declarative information*, which is knowledge such as facts and words that can be brought to mind for conscious inspection. However, the hippocampus does not appear to be needed for the acquisition of *nondeclarative or procedural information*, which is motor skill learning and other types of knowledge that depend on a motor output and that cannot be brought to mind and consciously inspected (Tranel and Damasio, 1993).

Damage confined to the mesial temporal lobes produces anterograde amnesia. If the damage extends into nonmesial structures in the anterior region, including various higher-order association cortices (i.e., areas 38, 35, 36, 20, 21), the memory defect will invade the retrograde compartment as well. The patient becomes incapable of learning new information and unable to retrieve information that was acquired before the onset of brain injury. More specific and unique levels of knowledge tend to be most affected, and more nonspecific, "generic" types of information are less affected. For instance, the patient has difficulty retrieving detailed information about episodes such as the birth of children, weddings, and the purchase of a home, but remains capable of retrieving knowledge about the meanings of words, the locations of different countries, and the names of major cities in various states. The right nonmesial temporal region may play a predominant role in retrograde memory, compared with a lesser role for the left nonmesial temporal cortices.

One of the most common causes of damage to the hippocampal system is Alzheimer's disease, in which the first onset of neuropathology (i.e., neurofibrillary tangles) is typically in the entorhinal cortex and hippocampus. This correlates well with the finding that anterograde memory impairment is the hallmark neuropsychological feature of Alzheimer's disease. Another frequent cause of mesial temporal damage is herpes simplex encephalitis, which produces various patterns of unilateral and bilateral damage, with or without extension into nonmesial temporal structures and with a number of distinctive neuropsychological profiles (Damasio, Tranel, and Damasio, 1989). Head injury is another common cause of mesial temporal damage. Another cause is surgical resection of anterior and mesial temporal structures for control of intractable seizure disorders; such operations not infrequently extend into the entorhinal cortex and even the anterior hippocampus, and material-specific anterograde memory defects may ensue.

Amnesia after basal forebrain damage. The basal forebrain, situated immediately posterior to the orbitofrontal cortices, comprises a set of bilateral paramidline gray nuclei including the septal nuclei, the diagonal band of Broca, the nucleus accumbens, and the substantia innominata. Damage to this region is common in the setting of ruptured aneurysms located in the anterior communicating artery or in the anterior cerebral artery. A characteristic amnesia develops in connection with basal forebrain lesions. Although patients are able to acquire some new information and can remember most information from the past, they have difficulty associating specific components of memorial episodes in correct connection with one another. They may recall a particular family event but place the event entirely out of context with respect to other events that occurred about the same time. This problem affects new learning as well, and patients may learn the name, face, and occupation of a person but then be unable to correlate these separate pieces of information in recall. They put the wrong name with a face or the wrong occupation with a name.

A related phenomenon has been called source amnesia. A patient may learn a particular fact or piece of information accurately but be unable to recall the circumstances under which the learning took place. The patient recalls the content but not the source of the learning experience. Basal forebrain amnesics also have a tendency to produce wild confabulations, such as making up fantasy stories about various bizarre adventures that have no basis in fact.

The amnesia associated with basal forebrain damage often involves a defect in what is known as *prospective memory* (i.e., remembering to remember). This refers to the capacity of creating mentally certain markers for remembering in the future. For example, when a certain hour arrives or when a certain event occurs, we remember to execute some task (e.g., "It's five o'clock; I will call my mother." or "My daughter is home; I will get the package that came for her in the mail.").

Amnesia after diencephalic lesions. Severe alcoholism and stroke are frequent causes of damage to portions of the diencephalon, especially the mammillary bodies and the dorsomedial nucleus of the thalamus. Diencephalic lesions generally produce a severe anterograde amnesia that resembles the amnesia associated with mesial temporal damage; it is a major defect in the acquisition of declarative knowledge, with sparing of learning of procedural information. However, there is usually some impairment in the retrograde compartment as well. The retrograde amnesia associated with diencephalic lesions tends to have a temporal gradient, and retrieval of recently acquired memories is more defective, but retrieval of remote memories is much better preserved.

☐ Disorders of language involve difficulties in the comprehension and formulation of verbal messages

The study of language impairments associated with focal brain damage is one of the most extensively developed areas of research in neuropsychology, and much has been learned about the neural substrates that mediate processes such as comprehension, speech production, and reading and writing (Damasio, 1992). Our understanding of brain-language associations has been facilitated by multidisciplinary approaches to language disorders, and scientists from fields such as cognitive psychology and linguistics, together with neuropsychologists and neurologists, are making important contributions to the investigation of speech and language.

Aphasia

Aphasia is an acquired disturbance of the comprehension and formulation of verbal messages, caused by dysfunction in a set of language-related cortical and subcortical structures in the left hemisphere (see Fig. 30-1). Aphasia can be further specified as a defect in the two-way translation mechanism between thought processes and language, between the nonverbal mental representations whose organized manipulation constitutes thought and the verbal symbols and grammatical rules whose organized processing constitutes sentences. Aphasia can compromise the formulation or comprehension of language, or both, and it can affect *syntax* (i.e., grammatical structure of sentences), the *lexicon* (i.e., dictionary of words that denote meanings), or *word morphology* (i.e., combination of phonemes that results in word structure). Deficits in various subcomponents of language occur with different severity and in different patterns, producing a number of distinctive syndromes of aphasia. Each syndrome has a fairly regular set of neuropsychological manifestations and a typical site of neural dysfunction. Stroke is the most common cause. Several of the main subtypes of aphasia are reviewed in the sections that follow, and an overall summary is provided in Table 30-9.

Broca's aphasia. Patients with Broca's aphasia are *nonfluent*. They are unable to speak fluently, and their speech production is effortful, sparse, and agrammatic. Paraphasias (i.e., word substitutions) are common, usually involving omission of phonemes or substitution of incorrect phonemes (e.g., "hostal" for "hospital" or "cephalot" for "elephant"). Repetition is defective, and naming and writing are also impaired. However, Broca's aphasics remain relatively normal in their capacity to comprehend language in its aural and written forms. The syndrome of Broca's aphasia is related to damage in the frontal operculum, especially areas 44 and 45, in the region known as *Broca's area*.

Wernicke's aphasia. The counterpart of Broca's aphasia is Wernicke's aphasia. The affected patient, unlike a Broca's aphasic, produces fluent, well-articulated speech, but the speech is punctuated by paraphasic errors that are semantic (e.g., "superintendent" for "president") and phonemic. Repetition and naming are defective. A distinctive feature of Wernicke's aphasia is a severe impairment of comprehension of aural and written forms of language. Wernicke's aphasia is related to damage in the left posterior superior temporal gyrus (i.e., *Wernicke's area*) and nearby regions in the temporal and lower parietal cortices.

Conduction aphasia. This aphasia subtype is hallmarked by a severe defect in verbatim repetition. Other aspects of language, including fluency and comprehension, are relatively less affected. Phonemic paraphasic errors are common, and naming is usually impaired. The patient is unable to write in response to dictation (a defect related to the repetition impairment) but can write much better when copying another script or when producing spontaneous compositions. Conduction aphasia is related to damage to the supramarginal gyrus (i.e., area 40) in the lower parietal region or to the primary auditory cortices with extension into the insular cortex and underlying white matter. The primary auditory association cortices in posterior area 22 are spared, preserving comprehension.

Global aphasia. As the term implies, global aphasia involves a near-complete dilapidation of linguistic capacities, and the patient is rendered incapable of comprehending and producing verbal messages. Repetition, naming, reading, and writing are all severely impaired. The syndrome is related to two main patterns of lesion: extensive destruction of the perisylvian region, including Broca's area, the inferior parietal cortices, Wernicke's area, and underlying white matter and subcortical structures, or to two noncontiguous lesions, one involving Broca's area and the other involving Wernicke's area. In the latter presentation, the patient does not have a right hemiparesis, because of sparing of motor cortex between the two lesions, and the prognosis for recovery of some linguistic abilities is considerably better.

Transcortical aphasias. In the previous four aphasia subtypes reviewed, all shared the feature that *verbatim repetition* is impaired, and all are associated with damage to the some part of the perisylvian region on the left. When the lesion is situated anterior (in front of or above Broca's area) or posterior (behind and below Wernicke's area) to the perisylvian region, the aphasias that develop, known as the *transcortical aphasias*, lack the feature of impaired repetition. The anterior lesion, for example, causes a Broca-like aphasia, except that the patient can repeat—this syndrome is known as *transcortical motor*

Table 30-9
DISORDERS OF LANGUAGE

Disorder	Examples of Neural Correlates
Aphasia	
Broca's aphasia	Left frontal operculum (area 44; Broca's area)
Wernicke's aphasia	Left posterior superior temporal gyrus (posterior area 22; Wernicke's area)
Conduction aphasia	Left inferior parietal operculum; supramarginal gyrus; insula
Global aphasia	Entire perisylvian region on left, including frontal operculum, parietal operculum, and posterior temporal region
Transcortical aphasias	Left prefrontal (motor version)
	Left posterior temporal inferior to Wernicke's area (sensory version)
Subcortical aphasia	Left basal ganglia
Aprosodia	Right perisylvian region, including posterior superior temporal, inferior parietal, frontal operculum

aphasia. The posterior lesion causes a Wernicke-like aphasia; with the exception that repetition is spared; this is known as *transcortical sensory aphasia*. Verbatim repetition is a useful marker in diagnosing aphasia subtypes and in the formulation of an hypothesis regarding probable site of neural damage.

Subcortical aphasia. Lesions to the dominant basal ganglia, especially to the head of the caudate nucleus, cause a characteristic disturbance of speech and language that is marked by variably fluent, paraphasic, and highly dysarthric speech, accompanied by poor auditory comprehension, right hemiparesis, and in some cases, repetition impairment. This profile, which does not conform to any of the typical cortical-related aphasia subtypes, has been called *atypical or basal ganglia aphasia*. Disorders of speech and language may also follow lesions to the left thalamus, especially if the damage involves the ventrolateral and anteroventral nuclei. The aphasia associated with left anterior thalamic lesions shares several of the characteristics of the transcortical aphasias.

Sign language aphasia. Deaf persons who normally communicate with a sign language become aphasic for signing as a result of focal brain injury (Bellugi et al, 1983). The lesions that produce sign language aphasia are located in the left hemisphere, in precisely the same regions as those that are associated with the traditional aphasic syndromes in hearing persons. This finding supports the conclusion that even a visually based language, such as American Sign Language, is mediated by structures in the left hemisphere. This discovery provides considerable support for the contention that the left hemisphere is genetically endowed for language.

Anomic aphasia. Most classification systems for aphasia subtypes include a type known as *anomic aphasia*. However, anomia (i.e., lack of ability to name) is an extremely frequent manifestation in most types of aphasia, and the localizing value of this sign with regard to specific neuroanatomic regions is minimal. Some findings have shed light on different types of naming disorders associated with different regions of damage. It has been shown, for example, that damage to the IT region on the left, particularly in areas 20 and 21 and the anterior part of area 37, produces defects in the retrieval of common nouns; the patient cannot name entities such as various animals, fruits and vegetables, and tools and utensils (Damasio et al, 1990). Some of these patients have a remarkably pure disorder, in which the naming defect is the only real language impairment, and in these cases, the label of anomic aphasia is not inappropriate. Care is taken to differentiate this from the sign of anomia, which is the manifestation of naming impairment as part of an aphasia syndrome. If damage is in the left prefrontal region just anterior to Broca's area, patients may have naming defects only for actions and relations; the defect is in retrieval of verbs. Noun retrieval is not impaired. Such variations in the patterns of naming defects that develop with different lesion loci indicate why the neural correlates of anomic aphasia have been elusive.

Impairment of supralinguistic and paralinguistic processes

Normal language communication depends on a linguistic skills such as grammar, syntax, and morphology and on *supralinguistic* and *paralinguistic processes*. For example, the suprasentential assembly of speech, the architecture that links multiple utterances and that generates discourse, is a vital supralinguistic process. Prosody, a term that refers collectively to the melody, pauses, intonations, stresses, and accents that are applied to propositional speech to convey information about mood or urgency, is an important paralinguistic process. Although these aspects of communication can be disrupted by lesions in the left hemisphere, it has recently been shown that right-hemisphere lesions can produce selective impairments in discourse and prosody, even when basic propositional language is intact. Severe aprosodia, in which the patient loses the ability to comprehend the emotional overtones of speech and to produce such intonations, is associated with damage to structures in the perisylvian region in the right hemisphere.

☐ Disorders of emotion and affect involve changes in mood, social conduct, and motivation

Disturbances of emotion and affect are common in brain-injured persons, and the development of depression and related manifestations such as anxiety and withdrawal is extremely common in the brain-damaged population. Many of these disturbances are clearly a direct reaction on the part of the patient to the experience of having newly acquired and often quite debilitating defects in movement, sensation, and cognition. Other disorders of emotion and affect are linked directly to damage in particular neural regions. These disorders develop reliably in connection with damage to specific brain areas, and they are not readily attributable to nonspecific factors such as psychological reaction to illness. Table 30-10 enumerates some of these disorders and their neural correlates.

Akinetic mutism

Lesions in the cingulate gyrus or the SMA often cause a combination of akinesia and mutism. The patient pro-

Table 30-10
DISORDERS OF EMOTION AND AFFECT

Disorder	Examples of Neural Correlates
Akinetic mutism	Cingulate gyrus; supplemental motor area (mesial area 6)
Anosognosia	Right inferior parietal region; ventromedial frontal region
Depression	Left prefrontal area

duces no speech, even when spoken to, and makes few facial expressions. Movements are also lacking, except for a few automatic types of behaviors such as getting up to go to the bathroom. Unlike patients with aphasia, who invariably struggle to produce language even when they are having severe difficulty, patients with akinetic mutism make no attempt to communicate. They act as though they have lost the will to interact verbally and motorically with their environment. Akinetic mutism can be conceptualized as a disorder of motivation.

Akinetic mutism tends to be more severe and long-lasting when there is bilateral damage to the anterior cingulate or SMA region than for unilateral lesions in this region. The side of damage does not seem to be critical, because the same basic syndrome develops in relation to lesions on the right or left, supporting the notion that the cingulate and SMA structures are dedicated primarily to motivation and affective control, rather than to language processes. Most patients recover well from this condition, and with unilateral lesions, recovery often takes place within the first few weeks.

Anosognosia

Anosognosia is a term used to denote the condition in which a patient denies or otherwise fails to acknowledge an acquired impairment in movement, sensation, or cognition. A patient may deny that a paretic limb is in fact paretic or that appreciation of touch on one side of the body is lacking. Failure to acknowledge cognitive impairments is perhaps one of the most common manifestations of anosognosia. Patients may deny that their memory is disturbed or that they are having trouble keeping track of their surroundings and the passage of time. *Anosodiaphoria* refers to a related condition in which a patient minimizes, underestimates, or is pathologically unconcerned with acquired defects in movement, sensation, or cognition.

In patients with focal lesions, anosognosia is most commonly seen in association with right-hemisphere lesions, especially damage to the right inferior parietal cortices. This finding is in keeping with the specialization of this region for the mapping of intrapersonal and extrapersonal space, as elaborated earlier. Common neurologic and neuropsychological concomitants of anosognosia are left hemiparesis, left neglect than can affect the visual, auditory, and somatosensory modalities, and defects in visuospatial and visuoconstructional abilities.

Anosognosia for cognitive and behavioral impairments is also common after damage to the ventromedial frontal region. Patients may demonstrate a remarkable lack of understanding of their behavioral abnormalities, and they may fail to appreciate the extent to which their problematic behavior is having an adverse affect on those around them. They may adamantly deny the existence of cognitive limitations, even when there are severe defects in judgment and problem-solving and insist on returning to normal activities and occupational pursuits.

Stroke is a common cause of damage in the right parietal region leading to anosognosia. The ventromedial region is susceptible to damage caused by rupture of aneurysms of the anterior communicating and anterior cerebral arteries. Surgical resection of meningiomas is a frequent cause of damage in this region. Another common cause of anosognosia is head injury, because head trauma frequently causes damage in the orbitofrontal region. Anosognosia is frequent in patients with degenerative demential conditions, especially Alzheimer's and Pick's disease. Especially when the condition advances to the intermediate and severe stages, these patients show remarkable defects in their appreciation of cognitive and behavioral limitations. This development may actually provide a protective psychological effect for the patients, because they suffer less than they would if they were more fully cognizant of their decline; however, the lack of awareness can pose substantial challenges to caregivers and professionals who have to manage the patients daily.

Depression after focal brain damage

Depression occurs frequently after brain injury, exceeding by several times the base rate of depression in the general population. It is likely that there are direct and indirect factors related to the development of depression in brain-injured persons. Brain damage can produce direct neurophysiologic and neurochemical changes that may induce depressive symptomatology. The secondary reaction of the patient to the implications of newly acquired disabilities certainly is a frequent and understandable source of depression. There is evidence that the site of brain damage is an important factor in the development of depression in stroke victims. It has been reported that lesions in and near the left frontal region are more likely to produce depression than left-sided lesions located more posteriorly and compared with right hemisphere lesions.

Disorders of conduct

The ventromedial frontal region, composed of orbital and lower mesial frontal cortices, plays a critical role in the domain of behaviors that can be grouped under the rubric of social conduct. Lesions in this region frequently lead to a severe disruption of social conduct; remarkably, the patients are free of impairments in most basic domains of cognition, such as intellectual capacity, memory, speech and language, perception, and attention. Affected patients have drastic changes in personality, and they show profound defects in planning, judgment, and decision-making. These changes occur against the background of an entirely normal premorbid personality and few or no defects in basic neuropsychological function.

The disorder of social conduct associated with ventromedial frontal lobe lesions has been called *acquired sociopathy* (Damasio, Tranel, and Damasio, 1990). The term captures well the proclivity of the patients to engage in decisions and behaviors that have repeatedly negative consequences for their well-being. The patients tend not to be destructive and harmful to others in society, a feature that differentiates the acquired form of the disorder from the standard developmental form, but they repeatedly select courses of action that are not in their best interest in the long run. They make poor decisions about personal relationships, occupational endeavors, and finances. In short, they act as though they have lost the ability to ponder different courses of action and then select the option that promises the best blend of short- and long-term benefit.

Measurement of Higher Brain Functions

The measurement of higher brain functions falls under the purview of clinical neuropsychology, which has developed a variety of techniques for measuring and quantifying capacities such as intellect, memory, problem-solving, and attention. This endeavor is critical to the understanding of brain-behavior associations. Just as modern neuroimaging techniques such as CT and MRI have allowed dramatic insights into neuroanatomy (e.g., the precise location of a brain lesion), the development of well-standardized, reliable methods for quantifying cognition and behavior has facilitated the establishment of robust principles for diagnosing and managing neuropsychological disorders.

☐ Neuropsychological assessment provides standardized measurement of higher brain functions

Neuropsychological assessment refers to a process in which standardized tests are used to measure intellect, orientation and memory, speech and language, perception, and judgment and decision-making. There are several different approaches and styles, but they have several features in common, which are at the core of the endeavor. First, the tests are standardized. They have empirically derived normative standards that take into account variables such as age, gender, level of education, and occupational background, allowing precise quantification of cognition and behavior in individual patients. Second, the neuropsychological examination is comprehensive. It provides measurement of all major facets of higher brain function, and in each principal domain of cognition, the examination is sufficiently detailed to allow quantification of the patient's level of functioning in that area. Third, the assessment process is objective. The tests are administered under standard, controlled conditions, with materials, instructions, and response formats that are invariant across different patients and different examiners, permitting the minimization of many common sources of unreliability and bias. Fourth, the assessment procedure is repeatable; it can be administered to a given patient on multiple occasions. For many neuropsychological tests, multiple forms have been developed. The different versions are equivalent except that the actual items vary. The examiner can employ alternate versions of tests when conducting reassessments of patients to reduce "practice effects." This feature allows serial examination of patients over time, permitting the monitoring of the course of disease and the effects of treatments and interventions.

The interpretation of neuropsychological data depends on knowledge of several key background variables. Foremost among these is the estimated level of premorbid functioning (i.e., what was the patient like before the onset of brain disease?) A standard, objective method for calculating this has been developed. A formula that incorporates various demographic data including level of education and occupation is used to calculate the premorbid caliber of the patient's cognitive capacities. Another important variable is handedness. Which hand a person prefers for activities such as writing, eating, and throwing is an important indicator of hemispheric specialization. Full right-handedness—preference of the right hand for virtually all skilled manual operations (typically accompanied by right-footedness and right-eyedness)—is highly predictive of standard cerebral laterality, wherein the left hemisphere is dominant for language, and the right for visuospatial functions (about 99% of right-handed persons show this pattern). In fully left-handed persons, the chance of standard cerebral laterality is reduced to about two in three; almost one third of left-handed persons have reversed dominance (i.e., language-right, visuospatial-left) or a mixed pattern (i.e., bilateral hemispheric representation of many functions). The factors of age and gender are also important background variables in the interpretation of neuropsychological data.

☐ Intellectual abilities include many higher-order cognitive functions, measured with IQ tests

Intellect refers to a heterogeneous collection of capacities such as fund of information, concentration and attention, verbal and nonverbal problem-solving, and vocabulary. Several well-standardized instruments have been developed for measuring intellect, most of which yield a summary score in the form of an *intelligence quotient* (IQ). The IQ score provides a broad indication of an person's general intellectual abilities. More specific information can be derived by examining various subcomponents of intellect, such as scores on particular subtests that relate to specific functions. The patient's ability to perform verbal abstractions is investigated with a subtest that calls for drawing comparisons between two entities or concepts, such as "How are an axe and a hammer alike?" or "How are fear and surprise alike?" The Wechsler Adult Intelligence Scale-Revised (WAIS-R) is the most widely used IQ test, and it has been studied extensively in the context of neuropsychological assessment.

The WAIS-R is divided into two separate sets of scales, one aimed at measuring verbal abilities and the other aimed at measuring nonverbal, performance abilities.

The relative position of a patient's scores on these scales can provide clues as to the integrity of various brain regions. A patient who performs better on the verbal subtests, compared with the performance tests, may be suspected of having right-sided brain dysfunction; the reverse would apply in the case of performance scores superior to verbal scores. These overall scales provide only a gross indication, and specific patterns of scoring on individual subtests provide more important clues regarding location and degree of brain dysfunction.

☐ Memory is measured with tests aimed at probing the ability to acquire, retain, and retrieve information

Measurement of memory is an important part of neuropsychologic assessment, because memory defects are early, often inaugural manifestations of commonly encountered neurologic disorders. Standard tests are available to assess several important subcomponents of memory, including short- and long-term memory, acquisition of verbal and nonverbal information (i.e., anterograde memory), and recall of previously learned material (i.e., retrograde memory). To measure anterograde memory, the patient is administered a list of words, a short story, or a series of geometric designs. Immediately thereafter or following a delay of some length, the patient is asked to reproduce the material from memory. The delay period interposed between the time at which the material is learned and the time at which it must be recalled, can be varied (e.g., several minutes, hours, days), and by measuring the amount of recall at various delay intervals, it is possible to determine the forgetting curve, i.e., the rate at which newly acquired information is lost over time.

The two most frequent formats for quantifying memory are recall and recognition. In a recall paradigm, the patient is required to recall as much information as possible and to reproduce it in some way (e.g., by reciting, writing, or drawing). In a recognition paradigm, the patient is provided several alternative answers and must choose from among these the one that is correct. This format is easier than a recall format and can provide important information about the degree of preserved memory in a patient who is too impaired to produce correct answers in a recall paradigm. Some of the most widely used anterograde memory tests include the Wechsler Memory Scale and its recent revision, the Rey Auditory-Verbal Learning Test, the California Verbal-Learning Test, and the Benton Visual Retention Test.

Retrograde memory is measured with tests that require recall and recognition of information from the past. The patient may be required to recall various famous events from the past and to recognize the persons, places, and other facts associated with those events. More "personalized" questionnaires have also been developed, in which the patient is asked to provide information about events that occurred during various life epochs, such as childhood, adolescence, and early and middle adulthood.

☐ Measurement of speech and language is used to determine the presence of aphasia and related disorders

In the earlier review of various aphasia syndromes, the importance of capacities such as fluency, naming, repetition, and comprehension was emphasized. Several standard measures can be used to assess these abilities. The Multilingual Aphasia Examination and Boston Diagnostic Aphasia Examination provide comprehensive assessment of the major aspects of speech and language and allow characterization of the patient's linguistic functions, permitting classification of the language disturbance according to standard aphasia subtyping.

To measure repetition, the patient is given a short sentence and asked to repeat it verbatim. The initial sentences are short, two or three words, and the length gradually increases up to more than a dozen. Most normal persons can repeat a 10- or 12-word sentence with little difficulty. In aphasia, however, this span may be sharply reduced, and the patient may not be able to repeat any sentences and may even have trouble repeating single words. To measure naming, the patient is shown a series of pictures or drawings and is asked to name them. The items vary in difficulty, ranging from simple ones such as bed and tree to more difficult ones such as trellis and seahorse. Comprehension is measured by having the patient select a response based on an aural or written command; the extent to which the response is pertinent to the request is used as a measure of comprehension.

☐ Visually-mediated tasks are used to measure visuoperceptual, visuospatial, and visuoconstructional functions

There are several standardized procedures for measuring these visually mediated capacities. The Facial Recognition Test requires the patient to make perceptual discriminations of unfamiliar faces presented with various angles and lighting conditions. The Judgment of Line Orientation Test requires the identification of the slopes and angles of visually presented lines. In the Hooper Visual Organization Test, the patient is required to perform a mental reassembly of various puzzle fragments to determine the whole object.

Visuoconstructional abilities are assessed with drawing tests in which the patient is asked to draw certain entities (e.g., a clock or bicycle) or to copy another drawing. There are several block construction tests that require the patient to build with blocks in a way that matches a model. Visuoconstructional tasks depend on several separate cognitive capacities, including perception, manual manipulation of materials, and planning, and such tasks tend to be quite sensitive to generalized and specific brain dysfunction. Such tasks can reveal neglect; for example, a patient's failure to attend to one part of space may be glaringly evident in a drawing of a clock that includes numbers only on the right side of the clock face.

☐ Executive functions are measured with tests of planning, decision-making, and judgment

The term *executive functions* is used to denote capacities such as concept formation, abstraction, planning, judgment, and decision-making, which are intimately linked to the frontal lobes. In the Wisconsin Card Sorting Test, the patient is required to sort cards according to a matching principle that is never explicitly stated and that changes unexpectedly several times during the test. The Wisconsin Card Sorting Test is a good indicator of how well the patient can formulate, maintain, and shift cognitive sets. Another useful procedure is the Controlled Oral Word Association (COWA) test, which requires the patient to produce as many words as possible that begin with a certain letter, under a time constraint. For example, the patient is given one minute to generate words that begin with "s." Because lexicons are not arranged in the human brain in alphabetical order, this task requires a flexible cognitive approach, a thinking pattern than is novel and must be tailored for the situation at hand.

There are several other widely used procedures to measure executive functions, including the Trail-Making Test and the Category Test. They share in common demands for novel, flexible problem-solving strategies and for the subject to manipulate information in an unusual manner. Executive functions overlap with other capacities such as intellect and memory; however, they are not fully co-extensive, and it is possible to observe fairly remarkable dissociations between executive functioning and capacities such as intellect, memory, and language.

☐ Personality tests are used to measure traits, states, and other psychological characteristics

Assessment of personality is another important feature of neuropsychologic assessment. An understanding of factors such as response style, affective status, and motivation is crucial for interpretation of other performances and for providing direct clues regarding the presence and severity of brain dysfunction. The most widely used instrument to assess personality and psychopathology is the Minnesota Multiphasic Personality Inventory (MMPI), which provides comprehensive investigation of the psychological state of a patient. Based on the patient's responses to an extensive series of true or false statements, the MMPI yields quantitative information about the patient's status on various dimensions of personality, such as hypochondriasis, depression, sociopathy, anxiety, schizophrenia, mania, and extroversion.

Although personality is difficult to measure and quantify and is not closely linked to discrete brain regions like other functions such as memory and language, an understanding of a patient's personality is nonetheless an indispensable part of neuropsychologic assessment. It can aid in diagnosis and in decisions regarding management and rehabilitation. The existence of depression is quite common in brain-injured persons, as a direct conse-

quence of the lesion and as a secondary reaction to disability. Timely detection of depression can facilitate the development of effective treatment programs.

Conclusion

Higher brain functions have regained a place of respect and prominence in modern neuroscience, after decades of eschewal prompted by the notion that such functions could never be localized in the human brain and by the influence of strict behavioristic psychology. The evolution of fine-grained neuroimaging techniques, including CT and MR, and the concurrent development of sophisticated neuropsychologic procedures have permitted a flurry of advances in the understanding of brain-behavior associations in the past couple of decades. It has become fashionable again to investigate the contents of human consciousness and the phenomenon of consciousness itself. The contemporary neuroscience student, unlike his or her counterpart of several decades ago, will not only find mastery of the fundamentals of higher brain functions a requisite learning experience, but she or he may discover that this arena constitutes one of the most exciting and fascinating domains of all neuroscience.

Acknowledgment

I have been fortunate to have Antonio and Hanna Damasio as my teachers. I thank them for their loyal mentorship and for their unwavering support of my acquisition of knowledge about mind facts and brain facts.

SELECTED READINGS

Bellugi U, Poizner H, Klima ES. Brain organization for language: clues from sign aphasia. Hum Neurobiol 1983;2:155.

Benton A. The Hecaen-Zangwill legacy: hemispheric dominance examined. Neuropsychol Rev 1991;2:267.

Benton A, Tranel D. Visuoperceptual, visuospatial, and visuoconstructional disorders. In Heilman KM, Valenstein E (eds). Clinical neuropsychology, 3rd ed. New York: Oxford University Press 1993:165.

Brodmann K. Vergleichende Lokalisationlehre der Grosshirnrinde in ihren Principien dargestellt auf Grund des Zellenbaues. Leipzig: JA Barth, 1909.

Damasio AR. Aphasia. N Engl J Med 1992;326:531.

Damasio AR, Anderson SW. The frontal lobes. In Heilman KM, Valenstein E (eds). Clinical neuropsychology, 3rd ed. New York: Oxford University Press 1993:409.

Damasio AR, Damasio H, Tranel D, Brandt JP. Neural regionalization of knowledge access: preliminary evidence. Symposia on quantitative biology, vol 55. Cold Spring Harbor, NY: Cold Spring Harbor Laboratory Press, 1990:1039.

Damasio AR, Tranel D. Verbs and nouns are retrieved with differently distributed neural systems. PNAS 1993;90:4957.

Damasio AR, Tranel D, Damasio H. Amnesia caused by herpes simplex encephalitis, infarctions in basal forebrain, Alzheimer's disease, and anoxia. In Boller F, Grafman J (eds). Handbook of neuropsychology, vol 3. Amsterdam: Elsevier, 1989:149.

Damasio AR, Tranel D, Damasio H. Individuals with sociopathic behavior caused by frontal damage fail to respond autonomically to social stimuli. Behav Brain Res 1990;41:81.

Damasio AR, Van Hoesen GW. Emotional disturbances associated with focal lesions of the limbic frontal lobe. In Heilman K, Satz P (eds). Neuropsychology of human emotion. New York: Guilford Press, 1983:85.

Damasio H, Damasio AR. Lesion analysis in neuropsychology. New York: Oxford University Press, 1989.

Lezak MD. Neuropsychological assessment, 2nd ed. New York: Oxford University Press, 1983.

Mesulam M-M. Attention, confusional states, and neglect. In Mesulam M-M (ed). Principles of behavioral neurology. Philadelphia: FA Davis, 1985:125.

Scoville WB, Milner B. Loss of recent memory after bilateral hippocampal lesions. J Neurol Neurosurg Psychiatry 1957;20:11.

Sperry RW. The great cerebral commissure. Sci Am 1968;210:42.

Spreen O, Strauss E. A compendium of neuropsychological tests: administration, norms, and commentary. New York: Oxford University Press, 1991.

Tranel D, Damasio AR. Knowledge without awareness: an autonomic index of facial recognition by prosopagnosics. Science 1985; 228:1453.

Tranel D, Damasio AR. The learning of affective valence does not require structures in hippocampal system. Journal of Cognitive Neuroscience 1993;5:79.

Tranel D, Damasio AR, Damasio H. Intact recognition of facial expression, gender, and age in patients with impaired recognition of face identity. Neurology 1988;38:690.

Ungerleider LG, Mishkin M. Two cortical visual systems. In Ingle DJ, Mansfield RJW, Goodale MA (eds). The analysis of visual behavior. Cambridge, MA: MIT Press, 1982:549.

Clinical Correlation

THE APHASIAS AND OTHER DISORDERS OF LANGUAGE

ROBERT L. RODNITZKY

Language is the mechanism by which communication is achieved through the use of specific sound or symbols. In the clinical realm, it is extremely important to differentiate this process from that of *speech*, the coordinated motor mechanism which allows the utterance of sound. Either language or speech can be abnormal with total preservation of the other, although in some clinical circumstances, both are involved simultaneously. Conceptual abnormalities of production or understanding of spoken language are referred to as *aphasia*, and mechanical abnormalities in the motor production of speech are known as *dysarthria* or *anarthria*.

Aphasia

☐ Aphasic disorders are usually caused by lesions of the left hemisphere

Language dominance is located in the left hemisphere in approximately 95% of right-handed persons and in approximately 66% of those who are left handed. Because most of the population is right handed, a useful clinical principle is that the left hemisphere must be considered dominant in all persons until proven otherwise. In some circumstances, it is critical to determine the hemisphere in which language dominance resides with absolute certainty. For example, in patients undergoing surgical removal of a brain tumor or surgical excision of a portion of the temporal lobe as a treatment for epilepsy, the surgeon must known whether cortical structures in the hemisphere being operated on can be ablated without fear of inducing aphasia. Several techniques can be used to determine hemisphere dominance for language. Positron emission tomography can be used to identify the cortical area(s) activated during speech, although its accuracy is still questioned. The most widely used and time honored technique involves the injection of an anesthetizing drug, first into the carotid artery that perfuses the right hemisphere and then into the carotid artery system that perfuses the left hemisphere of the brain. The drug produces transient aphasia when injected into the artery

perfusing the dominant hemisphere, but it has no effect when injected into the nondominant side.

☐ Several forms of brain pathology can cause aphasia

Damage to the speech area of the dominant hemisphere may be caused by stroke, tumor, infection, trauma, or neurodegenerative conditions. Generally, the more rapid the development of the damage (e.g., sudden stroke), the more severe is the aphasia. The extent of the damage is also related to the severity of language dysfunction. Regardless of the form of brain pathology, the age of the affected person may be a modifying factor. Dominant hemisphere pathology acquired in infancy or early childhood does not usually cause aphasia, reflecting the plasticity of the brain at that age.

☐ Several facets of language must be assessed in evaluating an aphasic disorder

Suspicion of an aphasic disorder usually develops as the examiner is listening to a patient's ordinary conversational or *propositional speech*. There may be abnormalities in the rhythm and inflection of speech (i.e., *dysprosody*), in the use of proper word sequence or connecting words (i.e., *agrammatism*), in naming objects (i.e., *anomia*) or in repetition. In addition, words or sounds may be incorrectly substituted (i.e., *paraphasias*). The substitution of an incorrect sound (e.g., "spoot" for spoon) is known as a *phonemic paraphasia*. The substitution of a word related in meaning (e.g., fork for spoon) is known as a *verbal paraphasia*. When the substitution in a word is so severe as to render it unrecognizable (i.e., "sporaker" for spoon) it is referred to as a *neologism*. A person whose language is characterized by agrammatic speech containing many neologisms rendering it is virtually incomprehensible is said to exhibit *jargon aphasia*.

☐ Aphasias can be divided into fluent and nonfluent types

Nonfluent aphasia typically results from dysfunction of the anterior portion of the speech area. Broca's aphasia, resulting from a lesion of the posterior portion of the inferior frontal gyrus, is the classic example of a non-fluent aphasia. In this condition, speech is hesitant and effortful and the affected person appears frustrated. Connecting words such as articles and conjunctions are omitted, resulting in agrammatic and telegraphic speech, but the retention of appropriate nouns and verbs preserves meaning. Instead of "I want to see the doctor," the non-fluent, Broca aphasic may utter "want, see, doctor." In this form of aphasia, comprehension of language is largely intact. Because of the proximity of Broca's area to the motor cortex, most Broca aphasics also suffer from significant weakness on the right side of the body.

Fluent aphasia is characterized by a normal or increased output of words. Prosody and construction are often normal, but speech contains many paraphasic errors. This form of aphasia is typically caused by a lesion in the posterior speech area. A lesion of the posterior third of the superior temporal gyrus gives rise to Wernicke's aphasia, the classic aphasia of the nonfluent type. In this aphasia, comprehension of language is distinctly impaired. Unlike Broca aphasics, who are frustrated by their language impairment, Wernicke aphasics shows little awareness and concern over their deficit but after a time can become paranoid and agitated, perhaps related in part to inability to communicate their needs and to verbal isolation stemming from impaired comprehension. Unlike Broca's aphasia, there is typically no weakness associated with Wernicke's aphasia, because the causative lesion is usually far removed from the motor cortex.

☐ Repetition is usually abnormal in aphasias caused by lesions in the perisylvian region

Broca and Wernicke aphasics manifest abnormalities of repetition. This tends to be true of any aphasia resulting from a lesion involving perisylvian structures. Aphasias resulting from lesions distant from this region are characterized by intact repetition and are known as the trans-cortical aphasias. A lesion just anterior or superior to Broca's area results in language dysfunction similar to Broca aphasia with the exception that repetition is intact. This is known as a *motor transcortical aphasia*. Similarly, a lesion just posterior to Wernicke's area results in a fluent aphasia with intact repetition known as a *sensory transcortical aphasia*. Persons for whom impaired repetition is the predominant abnormality are said to have *conduction aphasia*. The lesion is considered to be in one of two loci, the supramarginal gyrus or the area encompassing the primary auditory cortices, insula, and underlying white matter, especially the arcuate fasciculus.

Disorders of Written Language

Impaired production and comprehension of written words is another common form of language disturbance. Impairment in production of written language is called *agraphia*. Agraphia almost always accompanies aphasia of spoken language. The writing abnormality may take the form of misspelling, agrammatism, or imperfectly constituted letters and words. Agraphia is such a common accompaniment of aphasia that its absence in a patient apparently afflicted with aphasia raises serious doubt about the diagnosis. Isolated agraphia without other disorders of language sometimes occurs when the dominant angular gyrus is damaged.

Abnormal comprehension of written words is called *alexia*. Acquired alexia, typically associated with agraphia, may result from lesions in a variety of locations, but damage to the dominant parietal lobe has traditionally been thought to be the most common cause of this combination of deficits. *Alexia without agraphia* is a distinct and relatively common syndrome. It results from infarction of the dominant (typical the left) occipital lobe and the splenium of the corpus callosum. In this circumstance, visual language information is prevented from gaining access to the language areas of the left hemisphere from the left occipital lobe, which is infarcted, and from the intact right occipital lobe, which has been disconnected from the dominant hemisphere by destruction of the splenium of the corpus callosum. Because the language areas are spared, agraphia is conspicuously absent in this syndrome. Prominent loss of vision in the right half of the visual field is a usual accompaniment of this syndrome because of involvement of the primary visual cortex on the left.

Developmental dyslexia refers to impaired development of reading and writing skills relative to that expected, based on overall intelligence. Typically, it becomes apparent in the school-aged child. Intellectual functions often are not impaired, but affected children are sometimes mistakenly thought to be mentally dull until the specific isolated language dysfunction is discovered. Although there are no gross abnormalities of the brain in these persons, postmortem and MRI studies have shown that developmental dyslexics often exhibit subtle structural abnormalities. For example, they may lack the expected interhemispheric asymmetry in a portion of the temporal lobe known as the planum temporale. In normal persons, this structure is usually larger in the dominant hemisphere. Abnormal architecture and cell clusters of the cerebral cortex (i.e., cortical dysplasia and ectopias) have been identified in these persons, suggesting that there may have been failure of normal neuronal migration or excessive neuronal death during fetal development.

SELECTED READINGS

Damasio AR. Aphasia. N Engl J Med 1992;326:531.
Ramsey JM. The biology of developmental dyslexia. JAMA 1992; 268:912.

Section *VII*

Neuroimmunology

Neuroscience in Medicine, edited by P. Michael Conn.
J. B. Lippincott Company, Philadelphia © 1995

Chapter 31

Neuroimmunology

MICHAEL D. LUMPKIN

The term *neuroimmunology* means different things to different biomedical scientists and clinicians. This newly evolving and broad discipline includes the study of secreted immune cell products known as cytokines, lymphokines, and monokines, and their actions in the central and peripheral nervous systems. Another area of study considers the nerve supply and nervous system regulation of lymphoid organs such as the spleen and thymus. A third aproach investigates how neuroendocrine substances of the hypothalamic-pituitary unit regulate the proliferation and activities of monocytes, macrophages, lymphocytes, and glial cells, which are considered by some to be the "macrophages of the brain." A fourth area of neuroimmunology investigates immune cells that can produce neuroendocrine peptides, such adrenocorticotropic hormone (ACTH) and gonadotropin-releasing hormone (GnRH). A fifth realm of neuroimmunology examines the ability of the body to produce antibodies that can attack cholinergic receptors at the neuromuscular junction and the myelin components of the white matter of nervous tissue.

There is *bidirectional communication* between the cells of the nervous system and the immune system. In some instances, neurons communicate with the immune cells (and vice versa) by way of endocrine messengers interposed between these other two systems. For example, a stressful event can cause the cytokine interleukin-1 (IL-1) to be produced by activated monocytes, glial cells, or neurons residing in or traveling through the hypothalamus of the brain. The IL-1 causes certain hypothalamic neurons to release corticotropin-releasing factor (CRF) which stimulates ACTH secretion from the anterior pituitary gland. The ACTH stimulates the production of cortisol by the adrenal gland, and this glucocorticoid suppresses further proliferation and cytokine secretion by monocytes, lymphocytes, and glial cells. The cortisol also exerts *negative-feedback* control of the CRF neurons and the pituitary corticotropes that manufactures ACTH to reduce their levels of secretion. Communication occurs among these three great systems of the body (Fig. 31-1).

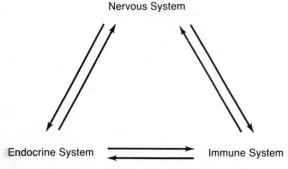

Nervous System

Endocrine System ⟶ Immune System

Figure 31-1
Bidirectional communication occurs among components of the nervous, endocrine, and immune systems. Neuroimmunomodulation occurs throughout the organism by means of these interactions and feedback loops, and a stimulus to any one of these systems inevitably involves the others in the resultant reactions.

Innervation of Lymphoid Organs

Some of the original concepts about neuroimmunology have come from the work of David Felten. He and others have shown that noradrenergic nerve fibers originating from the postganglionic sympathetic nervous system innervate the thymus, spleen, lymph nodes, and intestinal Peyer's patches. These norepinephrine-containing nerve fibers have been seen in contact with thymocytes, B lymphocytes, and macrophages. Both α- and β-adrenergic receptors have been found on the surfaces of these immune cells. In general, activation of the sympathetic nervous system or administration of epinephrine produces leukocytosis, lymphopenia, and suppression of natural killer cell activity. In these ways, activation of catecholaminergic β receptors on lymphocytes inhibits their proliferation and results in a suppressed immune response. The role of catecholamines in suppressing lymphocyte activity has caused some investigators to think this is one explanation for why prolonged stress, which activates the sympathetic nervous system, leads to greater susceptibility to disease, infections, and cancer in humans.

Felten's laboratory has also been instrumental in showing that several peptides found in autonomic nerves innervating the thymus, spleen, and lymph nodes are also regulators of cells of the immune system. Vasoactive intestinal polypeptide (VIP), neuropeptide Y, and substance P have been observed in nerve fibers that make contact with thymocytes, lymphocytes, and macrophages in the thymus gland and spleen.

The receptors for these peptides have been described on lymphoid cells. The most studied of these nerve-ending peptides is VIP and its lymphoid receptor. When VIP is released from these nerve endings, it may inhibit the proliferation of lymphocytes, inhibit natural killer cell activity, and alter antibody production.

Some peptidergic nerve fibers that innervate lymphoid organs produce the peptide *substance P*. In keeping with its known proinflammatory action and its role in mediating pain responses through peripheral nerves, activation of receptors by substance P stimulates lymphocyte proliferation and the release of pain mediators such as histamine and leukotrienes from mast cells. The peptide *somatostatin*, which is found in the brain and peripheral nerves and which is mostly inhibitory for all types of cellular activity, can block the release of substance P from peripheral nerve endings.

The peptide *neuropeptide Y* has been found in the same nerve fibers and terminals that contain norepinephrine and that innervate the thymus, spleen, and lymph nodes. It is likely that neuropeptide Y and norepinephrine are co-secreted and act in synergistic fashion to suppress lymphocyte proliferation and cytokine release in keeping with the general idea of sympathetic nervous system inhibition of immune function.

Other neuropeptides and their receptors are found in the brain and immune system. These include CRF, ACTH, the enkephalins, β-endorphin, vasopressin, oxytocin, and neurotensin. Most of these neuropeptides are released during episodes of physical or psychologic stress. The peptides mostly inhibit immune cell proliferation and function. They may mediate the suppression of the immune system and the subsequent susceptibility to disease and infection that often follows periods of prolonged stress.

Cytokines in the Nervous System

Immune system cells produce and secrete soluble mediators known as *cytokines*, whose role was originally defined as modifying the proliferation and activity of other cells of the immune system. However, activated immune cells, including monocytes, lymphocytes, and macrophages, can also breach the blood-brain barrier and establish themselves in the brain, where they may release their specific cytokines into the cerebral vasculature, cerebrospinal fluid, and parenchyma of the central nervous system. Microglial cells and, to a lesser extent, astroglial cells of the brain secrete the cytokines IL-1, IL-2, IL-4, and IL-6 and tumor necrosis factor-α (TNF-α). The receptors for IL-1, IL-2, and IL-6 have also been identified in the brain. Most provocative is the finding that IL-1 is also produced by neurons, particularly in the hypothalamic and hippocampal neuronal populations that are involved in stress and homeostatic responses. The IL-1 receptors have also been identified in the hypothalamus and hippocampus.

Physical stressors such as trauma, infection, and inflammation of brain tissue and psychologic stressors stimulate the production of cytokines such as IL-1 by glial cells and neurons. IL-1, IL-6, and TNF, when manufactured by brain cells for prolonged periods in pathologic states, can produce anorexia, fever, sleep induction, dementia, and neuronal death. During invasion of the brain by the human immunodeficiency virus (HIV) and in patients with Alzheimer's disease, the levels of IL-1 are increased and contribute to their behavioral disturbances. However, the brain has apparently developed a protective mechanism against the persistent presence of pathologically elevated IL-1 levels, because such situations also lead to the production of an *endogenous IL-1 receptor antagonist*, which is homologous to the structure of IL-1 and competes for the same receptor, dampening IL-1 activity. The IL-1 receptor antagonist has no known inherent activity of its own except to occupy the IL-1 receptor.

Although sustained high levels of cytokines in the brain produce deleterious effects on neuronal function, normal physiologic oscillations in brain cytokines in response to acute changes in homeostatic activity or intermittent stressors are beneficial to the organism. The presence of physiologic concentrations of IL-1, IL-2, and IL-6 in the hypothalamus appears to regulate certain neuroendocrine axes. For instance, IL-1 produced in or injected into the brains of experimental animals increases the expression CRF mRNA and synthesis and the release of its neuropeptide product. This causes the secretion of ACTH from the anterior pituitary gland. The ACTH then binds to its receptors on the cortex of the adrenal gland,

leading to the increased secretion of glucocorticoids such as cortisol. This corticosteroid provides a classic negative-feedback signal to the neurons, glia, and monocytes or macrophages that provided the elevated levels of CRF-stimulating cytokines in the first place. Elevated cortisol levels then decrease the secretion of cytokines from glial cells and neurons and reduce the proliferation and secretion of cytokines from monocytes and lymphocytes. In this fashion, the increased activities of immune cells and neuroendocrine cells in response to physical or emotional stressors are prevented from running amok. The cortisol down-regulates IL-1, IL-2, IL-6, and CRF secretion, thereby maintaining homeostatic concentrations of cytokines and hormones in the body.

Another beneficial effect of IL-1 may occur after nervous tissue injury. This positive influence is derived from the ability of IL-1 to stimulate the synthesis and secretion of nerve growth factor, which acts as a neutrophic growth factor that promotes the healing and regeneration of certain types of nerve cells.

IL-1 and its related cytokines influence other neuropeptides of the brain. IL-1 acts in the hypothalamus to increase somatostatin synthesis and secretion, and in a coordinated fashion, it inhibits the release of growth hormone–releasing hormone (GHRH). The overall consequence of this action is to reduce growth hormone secretion from the anterior pituitary. This mechanism may contribute to the protein wasting syndrome in adult patients with HIV infection of the brain and to the failure of somatic growth in children suffering from acquired immunodeficiency syndrome (AIDS). Growth hormone is anabolic for protein synthesis of muscle and connective tissue and stimulates the growth of long bones and soft tissues in children. An IL-1–induced growth hormone deficiency could account for the failure to thrive and negative nitrogen balance characteristic of AIDS patients.

IL-1 in the hypothalamus inhibits the secretion of GnRH and thyrotropin-releasing hormone (TRH). Sustained increases in IL-1 that may be encountered during prolonged stressful situations such as chronic viral infections can lead to amenorrhea in women and decreased spermatogenesis accompanied by lowered testosterone levels in men; the outcomes are decreased libido and reproductive failure. Hormonally, this deficit is seen as the cessation of the normal pulsatile secretion of luteinizing hormone (LH) and follicle-stimulating hormone (FSH) from the pituitary gland. Without sufficient stimulation of the gonads by LH and FSH, sex steroid synthesis and germ cell production fail.

Consistent with their many inhibitory activities on various brain functions, IL-1 and TNF-α suppress thyroid function directly at the level of the thyroid gland and through the inhibition of TRH production by the hypothalamus. Incidents of inflammation and sepsis decrease thyroid-stimulating hormone (TSH) secretion from the pituitary, because IL-1 from peripheral and central sources inhibits TRH secretion, and TNF-α lowers the hypothalamic mRNA content of TRH. These actions result in lowered plasma thyroid hormone levels, leading to a reduced metabolic rate and perhaps aggravating the fatigue and lethargy caused by the action of IL-1 and TNF directly on the brain.

Neuroendocrine Peptides in Immune Cells

Another development in the field of neuroimmunology was the discovery by Edwin Blalock and his colleagues that neuroendocrine peptides are produced by immunocompetent cells. For instance, B lymphocytes contain ACTH and enkephalins, the secretion of which can be stimulated by CRF and inhibited by glucocorticoids. The T cells synthesize growth hormone, TSH, LH, and FSH. Monocytes are a source of prolactin, VIP, and somatostatin. When a pituitary hormone is found in a cell of the immune system, the corresponding releasing factor peptide usually is found there and has the ability to stimulate the secretion of its "target" hormone from a particular type of immune cell. An example is the presence of GHRH in lymphocytes; it may be secreted locally among other lymphocytes and act on GHRH receptors on the lymphocyte cell membrane, eliciting the local secretion of growth hormone by like cell types. The same is true for TRH and TSH and for GnRH and LH, which have been located in immune cells. However, such mechanisms of local control are controversial and not extensively studied.

Although the secretion of neuropeptides by monocytes and lymphocytes may provide significant levels of *paracrine* and *autocrine* regulation among immune cells, there is little evidence to suggest that lymphocyte-derived releasing factor peptides or pituitary hormones exert actions on peripheral endocrine glands. The extremely small quantities of neuropeptides secreted by lymphocytes essentially preclude them from playing major roles as circulating hormones with peripheral endocrine glands as their targets.

Neuroimmunomodulation by Neuroendocrine Peptides

Having established the presence of various neuroendocrine peptides in and among the immune cell population, it is useful to understand how they affect the immunoregulatory activities of these cells. In the inhibitory category, ACTH suppresses macrophage activation and the synthesis of antibodies by B cells. Gonadotropins decrease the activity of T cells and natural killer cells. Somatostatin and VIP inhibit T-cell proliferation and the inflammatory cascade. These actions are somewhat opposed by growth hormone and prolactin (which are homologous in their structures), because these peptides can stimulate lymphocyte proliferation, antibody synthesis, and oppose the inhibitory effects of glucocorticoids on lymphocyte proliferation and cytokine production. Growth hormone and prolactin stimulate the growth and

activity of the thymus gland and lymphoid cells. Growth hormone also functions as a macrophage-activating factor, and prolactin enhances the tumoricidal activity of macrophages and the synthesis of interferon-γ. TSH enhances antibody synthesis, and β-endorphin stimulates the activity of T, B, and natural killer cells.

Autoimmunity and Neuroimmunology

Perhaps the best example of autoimmunity in human neurologic disorders is that of myasthenia gravis, in which autoantibodies attack acetylcholine receptors, rendering them unresponsive to the acetylcholine secreted into the synaptic cleft. Two mechanisms involved in this loss of receptor responsiveness are complement-mediated lysis and down-regulation of the acetylcholine receptor secondary to cross-linking of receptors. Although the precipitating event in myasthenia gravis is not fully understood, it is possible that T cells of the thymus gland may mediate this type of autoimmune response. Myasthenia gravis is explored in more detail at the end of this chapter.

Another example of deranged neuroimmunologic function is that represented by the chronic inflammatory neural disease of multiple sclerosis, in which demyelination of central nervous system neurons produces a loss of neuronal function. T-cell–mediated immunity is the principal mechanism by which autoimmunization to myelin antigens results from actual immunization with myelin components or infection with cross-reactive viruses or viral proteins. Antibodies to myelin basic protein and other myelin breakdown products appear in the cerebrospinal fluid and plasma. This condition also results in an elevation of other products in blood and cerebrospinal fluid. The activation of lymphocytes and macrophages in multiple sclerosis increases the levels of interferon-γ and prostaglandin E$_2$. The activated immune cells in multiple sclerosis patients secrete elevated levels of ACTH, prolactin, β-endorphin, and substance P, although the roles of these peptides and hormones on the clinical state of these patients is unknown.

SELECTED READINGS

Blalock JE. A molecular basis for bidirectional communication between the immune and neuroendocrine systems. Physiol Rev 1989;69:1.

Lumpkin MD. Cytokine regulation of hypothalamic and pituitary hormone secretion. In Foa PP (ed). Endocrinology and metabolism 5: Humoral factors in the regulation of tissue growth. New York: Springer-Verlag, 1993:139.

Raine CS (ed). Advances in neuroimmunology. Ann N Y Acad Sci 1988:540.

Reichlin S. Mechanisms of disease. Neuroendocrine-immune interaction. N Engl J Med 1993;329:1246.

Scarborough DE. Cytokine modulation of pituitary hormone secretion. Ann N Y Acad Sci 1990;594:169.

Clinical Correlation
MYASTHENIA GRAVIS

ROBERT L. RODNITZKY

Myasthenia gravis is the most common disorder of neuromuscular transmission. Patients experience weakness with abnormally rapid fatigue of the muscles that move the extremities, face, eyelids, and eyes and the muscles required for swallowing and respiration. Involvement of these last two groups of muscles can be fatal if severe and untreated. Myasthenia is an autoimmune condition in which there is antibody-mediated alteration of the postsynaptic membrane of the neuromuscular junction and destruction of acetylcholine receptors. Although its clinical manifestations are protean, there are several useful serologic, pharmacologic, and electrophysiologic means of confirming the diagnosis.

Acetylcholine Receptors in Myasthenia Gravis

☐ The discovery of α-bungarotoxin began the modern understanding of the acetylcholine receptor

The discovery that a Taiwanese poisonous snake's toxin, α-bungarotoxin, selectively binds to the nicotinic acetylcholine receptor at the neuromuscular junction provided a scientific tool that significantly enhanced efforts to understand the physioanatomic basis of myasthenia gravis. Using radioactive α-bungarotoxin to label acetylcholine receptors, it was confirmed that their numbers are significantly reduced in myasthenic patients. It was demonstrated that inoculation of laboratory animals with acetylcholine receptor protein resulted in the development of an experimental form of myasthenia gravis, strongly suggesting an autoimmune basis for the illness. Antibodies against acetylcholine receptors can be demonstrated in most human patients with myasthenia gravis, and their presence is a useful means of confirming the diagnosis.

☐ Antibodies can affect acetylcholine receptor functions in several ways

Several types of antibodies can be found in myasthenic patients. Approximately 90% of myasthenic patients with generalized weakness harbor one or more types of acetylcholine receptor antibody. The most commonly found antibody binds to the acetylcholine receptor but does not block its receptive sites. The acetylcholine receptor has five constituent subunits arrayed around a central ion channel. The acetylcholine and α-bungarotoxin binding sites are separate, but they are both located on the two α-subunits of the receptor. Antibodies that do block the acetylcholine receptive site are found in a much smaller percentage of patients but are often associated with more severe disease. Collectively, these abnormal antibodies accelerate the degradation of receptors by enhancing the normal process of endocytosis, and they impair the function of the remaining receptor by blocking their active sites and promoting a complement-mediated alteration of the architecture of the postjunctional membrane. The latter process includes widening of the synaptic cleft and smoothing of the normally infolded architecture of the postsynaptic membrane, with a resultant loss of receptor-containing surface area.

The specificity of circulating acetylcholine receptor antibodies for myasthenia gravis is high. False-positive results have been recorded for few persons without myasthenia gravis, such as those who received snake venom as a form of therapy, patients with tumors of the thymus, and those who received the drug penicillamine.

Epidemiology and Diagnosis

Myasthenia gravis occurs most commonly in young women and older men, peaking in the second and third decades in women and the sixth and seventh decades in men.

Neonatal myasthenia gravis is a transient condition resulting from transplacental transfer of acetylcholine receptor antibodies from a myasthenic mother to her infant. As the antibodies are cleared by the affected infant, the myasthenic symptoms gradually resolve. This phenomenon confirms the importance of circulating antibodies in the pathogenesis of myasthenia gravis.

☐ Weakness and easy fatigability are the clinical hallmarks of myasthenia gravis

Virtually any muscle in the body can become weakened in myasthenia gravis. Commonly, the first muscles to be-

589

come involved are those responsible for movement of the eye or eyelid. Affected patients complain of diplopia and eyelid ptosis. Ocular symptoms are the presenting sign of myasthenia gravis in more than 50% of patients and ultimately affect more than 90% of patients. In a few patients, the symptoms remain confined to the eyes, resulting in the syndrome of ocular myasthenia gravis.

Appendicular, bulbar, and respiratory muscles are also commonly involved. In the arms and legs, proximal muscles tend to be involved more severely, resulting in difficulty arising from a chair, climbing stairs, or raising the arms over the head. Bulbar muscle involvement results in symptoms such as difficulty in swallowing, nasal regurgitation of ingested fluids, and a characteristic nasal quality of the voice. Weakness of the facial muscles results in an inability to grimace, pucker the lips, or whistle. Involvement of respiratory muscles severely limits the volume of air that can be maximally moved in and out of the lungs.

One feature of myasthenic weakness that differentiates it from weakness caused by most other conditions is extreme and early muscle fatigability. Patients ascending stairs may notice that the first few steps can be accomplished without difficulty but that the next several steps become progressively more difficult.

☐ Serologic, pharmacologic, and electrophysiologic testing can help confirm the diagnosis

The usefulness of assaying acetylcholine receptor antibodies was previously discussed. Acetylcholine receptor antibodies can be demonstrated in only 90% of myasthenic patients with generalized weakness, and other tests may be needed to confirm the diagnosis.

Pharmacologic testing involves the use of cholinesterase-inhibiting drugs. These agents improve myasthenic symptoms by preventing the normal hydrolysis and inactivation of elaborated acetylcholine in the neuromuscular synaptic cleft. The duration of the effect of elaborated acetylcholine is prolonged, allowing an increased number of interactions with the remaining acetylcholine receptors and some repair of neuromuscular transmission. Edrophonium (Tensilon) is a rapid and short-acting cholinesterase-inhibiting agent that results in a brief but striking improvement in myasthenic symptoms. After injecting it intravenously, marked improvement in symptoms can be demonstrated within minutes, which is helpful in confirming the diagnosis.

Electrophysiologic testing for myasthenia gravis consists of measuring the amplitude of a muscle's response to repetitive stimulation of its nerve. In normal persons, the amplitude of the muscle response remains constant during repetitive nerve stimulation at three cycles per second. In myasthenia gravis, the muscle response to stimulation at this frequency becomes progressively smaller, mirroring the fatigue experienced after repetitive limb movements. Typically, this *decremental response* can only be demonstrated in clinically weak muscles.

Treatment

With therapy, myasthenia gravis is seldom a fatal disorder. Cholinesterase inhibitors are useful in testing and in treating myasthenia gravis. Administered in proper amounts, they produce a mild improvement in muscle strength. If given in excessive amounts, they can produce side effects related to overstimulation of muscarinic acetylcholine receptor, including excessive salivation and tearing, diarrhea, and slowing of the heart rate. A similar excessive effect on acetylcholine receptors in muscle may paradoxically lead to increased weakness, presumably resulting from continuous end-plate depolarization. In its most extreme form, a cholinergic crisis develops, in which there is marked weakness of the respiratory muscles.

Because cholinesterase inhibitors are only mildly effective, most severe myasthenic patients require immunotherapy. This approach, used with improved ventilatory support, has reduced the mortality rate for generalized myasthenia from the 40% to 70% reported before 1960 to less than 10% in 1994. One of the earliest immunologic therapies used for myasthenia gravis was removal of the thymus. In most myasthenia gravis patients, the thymus is microscopically hyperplastic, and in 20% of these patients, a tumor of the gland (i.e., thymoma) develops. Many myasthenic patients are improved after removal of the thymus, probably because of the gland's role in promoting and sustaining an autoimmune attack on acetylcholine receptors. The finding of acetylcholine receptor protein in the thymus supports this idea.

Immunosuppressive drugs, such as corticosteroids and azathioprine, are extremely effective in inhibiting the aberrant immune response underlying myasthenia gravis. Often, as the patient improves on these therapies, there is a corresponding drop in the titer of antibodies to the acetylcholine receptor. A more direct way to lower this antibody titer is through the use of plasmapheresis. In this process, the patient's plasma, which contains the abnormal antibody, is removed and replaced with a plasma substitute. This procedure effectively lowers the titer of abnormal antibodies and can result in remarkable improvement until more antibody is produced.

Section VIII

PLASTICITY AND REGENERATION

Neuroscience in Medicine, edited by P. Michael Conn.
J. B. Lippincott Company, Philadelphia © 1995

Chapter *32*

Degeneration and Regeneration in the Nervous System

PAUL J. REIER

Obtaining functional recovery after damage to the nervous system has been one of the more intriguing and perplexing issues of basic and clinical neuroscience. The challenging nature of this problem is underscored by the multitude of elaborate synaptic networks and axonal projections that contribute to the complex functional organization of the adult nervous system. These intricate associations arise during development as a product of dynamic cellular mechanisms involving neurogenesis, neuronal differentiation, axonal elongation, postsynaptic target recognition, and synaptogenesis. Ultimately, these circuitries represent precisely coordinated neurophysiologic pathways that serve a wide range of sensory, motor, cognitive, and autonomic modalities.

An insult to the nervous system causes widespread and devastating functional losses through the anatomic and physiologic disruption of neuronal networks. Whether such multidimensional consequences are reversible depends on the capacity of the nervous system to retain or reinvoke developmental mechanisms and of synaptic circuitries to be repaired spontaneously or with therapeutic strategies. In this context, certain intrinsic neuronal properties and extrinsic nonneuronal conditions must be available for optimal regeneration to be achieved.

The prospect for restoring a significant degree of function, especially after a brain or spinal cord injury, was once viewed with considerable pessimism. This negativism can largely be attributed to the perception that most neuronal populations lack the innate capacities needed to recapitulate the ontogenetic cellular dynamics needed to replace cells that have died or required to sustain the regrowth of cytoplasmic processes (i.e., axons) that have been interrupted. This view is rooted in the early stages of neurobiologic research and underscored by Santiago Ramon y Cajal's insightful and inspiring volumes on *Degeneration and Regeneration of the Nervous System*, first published in 1913 and 1914.

A more optimistic outlook began to evolve during the last 2 decades. Advanced anatomic and molecular approaches have provided greater insights into the neuronal and glial reactions to neural damage that appear to govern the success or failure of neural regeneration. Other research has been instrumental in defining new strategies by which functional repair of the damaged central nervous system can be more effectively encouraged by preventing progressive tissue deterioration or by replacing cells that had been lost in the aftermath of trauma or disease.

This chapter describes the fundamental principles underlying neural tissue reactions to trauma. These concepts provide a framework for a discussion of possible therapeutic approaches. The emphasis is restricted to neuronal changes after blunt mechanical damage to the central nervous system (CNS) or severance of nerve fibers in the peripheral nervous system (PNS), but many cellular features associated with trauma can be exhibited in other neuropathologic states. This is particularly true of neurodegenerative disorders, such as Parkinson's disease, Alzheimer's disease, and amyotrophic lateral sclerosis, and the diseases affecting peripheral nerves, such as neuropathies, for which the issues of degeneration and regeneration are equally pertinent.

Principles of Neural Regeneration

☐ Neural regeneration often implies regrowth of cytoplasmic processes

The term *regeneration*, when used in reference to the nervous system, has a meaning fundamentally distinct from that applied to most other tissues. Unlike the liver, in which cell proliferation can lead to substantial tissue replacement, the mammalian nervous system, with rare exception (e.g., primary olfactory neuronal regeneration), lacks the capacity to reconstitute itself by virtue of large-scale stem cell mitoses and subsequent axonal regrowth from newly formed cells. The worse prognosis for functional recovery is the consequence of direct trauma to gray matter leading to neuronal cell death. However, after axons are severed, neurons can survive. In the PNS, *axotomized cells* often support renewed growth of the severed axon. Neural regeneration implies the regrowth of damaged nerve fibers rather than cellular replacement.

☐ Regional differences exist in regenerative potential

Neurons with cell bodies, dendrites, and axonal processes totally confined to the CNS usually fail to regrow their damaged axons for distances more than 0.5 to 2.0 mm beyond the lesion. However, cells that usually reside completely in the periphery, such as sensory or autonomic ganglion cells, or that have their cell bodies in the CNS but project axons to the PNS, such as spinal motor neurons, have a much greater potential for regeneration and restoration of functional connections with targets such as autonomic ganglia and preganglionic cells, muscles, peripheral sensory receptors, or specialized end-organs.

Peripheral sensory neurons, such as spinal dorsal root ganglion cells, present a different situation. Although fully capable of regenerating their peripheral process, the cells' regrowth of the centrally projecting axons usually ceases at the PNS-CNS transition region of the dorsal root entry zone.

Neuronal Responses to Injury

☐ Direct damage to the cell body has the most immediate and severe consequences

Neuronal responses to physical insult can be diverse, and they depend on many variables, including the nature of the injury, the location of the lesion, and the maturational status of the nervous system. The histopathologic picture of injured neurons can range from hypertrophy or enlargement of the cell body due to increased metabolic activity to progressive atrophy or shrinkage of the cell soma and neuronal death. Lesions directly affecting gray matter can cause neuronal death directly by the physical insult or indirectly through the resultant vascular compromise. Such *primary neuronal damage* involves degeneration of the neuronal cell body and of all the cytoplasmic processes normally sustained by the biosynthesis occurring in the neuronal soma.

☐ Indirect neuronal damage can result from a host of pathophysiological reactions to trauma

Damage to the PNS or CNS can adversely affect the survival of neuronal populations that did not sustain direct insult. This phenomenon, called *secondary cell death*, can ensue over a period of hours or weeks after the trauma. For example, after spinal cord injury, considerable devastation of gray matter occurs at various distances rostral and caudal to the site of injury (i.e., lesion epicenter) because of the disruption of blood flow and the integrity of the blood-brain barrier, inflammation, ischemia, hypoxia, edema, and many biochemical, enzymatic, and ionic alterations.

Oxygen free radicals can be generated from a variety of sources in the injured CNS, such as the arachidonic acid cascade and the oxidation of catecholamines. Radical-initiated peroxidation of cellular membranes can alter membrane structure, fluidity, and permeability. Associated perturbations of membrane-bound Na^+, K^+-ATPase and the loss of cellular ionic homeostasis lead to K^+ ef-

flux from cells and elevated intracellular concentrations of Na^+ and Ca^{2+}. Swelling of the neuronal cell body, promoted by the influx of Na^+ and water, is one of the primary cytologic hallmarks of neuronal damage.

The augmentation of intracellular Ca^{2+} levels can be even more deleterious. The unchecked influx of Ca^{2+} can promote mitochondrial dysfunction, compromising cellular energy metabolism. An increased activation of Ca^{2+}-dependent proteases degrades the neuronal cytoskeleton and disrupts axoplasmic transport. Intracellular Ca^{2+} overload can stimulate lipolytic enzymes that affect cellular membrane integrity and its physiologic properties. Triggering other enzyme systems, such as the protein kinases, can result in altered gene expression, such as DNA damage and chromosomal breakage, and protein synthesis in ways that are detrimental to cell survival.

Trauma and certain disease states can stimulate an excessive release of excitatory amino acid neurotransmitters, such as glutamate or aspartate, that become concentrated in the extracellular compartment because of reduced cellular uptake. Excitatory amino acid–associated secondary neuronal death (i.e., *excitotoxicity*) follows because of overactivation of excitatory amino acid receptors. In the case of glutamate, three classes of postsynaptic receptors—NMDA (*N*-methyl-D-aspartate), AMPA (α-amino-3-hydroxy-5-methyl-4-isoxazole propionic acid), and kainate—have been implicated. Binding to the kainate and quisqualate receptors results in a massive influx of Na^+ and efflux of K^+, evoking repetitive or sustained depolarizations and cellular swelling due to altered influxes of electrolytes and water. Activation of the NMDA receptor complex (see Chap. 7) has a more profound consequence on cell viability, because this entails a receptor-mediated divalent cation channel that allows additional Ca^{2+} entry into the cell. Although it may seem maladaptive, excessive extracellular concentrations of excitatory amino acids contribute to the up-regulation of postsynaptic NMDA receptors. An abnormal inward flow of Ca^{2+} occurs through these NMDA and other voltage-sensitive channels as a result of cell depolarization. The resultant elevated intracellular concentrations of Ca^{2+} can have profoundly deleterious effects on neurons for reasons similar to those described previously.

☐ Therapeutic interventions to minimize secondary neuronal damage address secondary neuronal responses to trauma

Insights pertaining to these and other pathophysiologic mechanisms have stimulated the development of pharmacologic interventions to modify postinjury neurochemical changes. By attenuating secondary events, the rescue of cells could translate into less severe neurologic deficits and improved functional outcomes in many neurologic disorders. For example, administration of the steroid methylprednisolone within a few hours after spinal trauma has had significant neuroprotective effects. This is attributed to the antioxidant action of this compound, which minimizes lipid peroxidation and membrane lysis that otherwise lead to neuronal and glial cell death.

Additional pharmacologic interventions are focused at the excitotoxic component of the secondary neuronal responses occurring as part of certain neurodegenerative diseases (e.g., Huntington's disease) or as the sequelae to spinal cord or brain injury, stroke, hypoglycemia, and hypoxia. Certain competitive and noncompetitive NMDA-receptor antagonists have shown therapeutic promise in blocking the binding of excitatory amino acids, reducing cellular toxicity after CNS injury. One example is the noncompetitive antagonist, MK-801, which has been reported to promote anatomic and functional sparing. Although, many compounds such as MK-801 have significant contraindications that limit their clinical applicability, experimental findings have suggested important avenues for the future development of safer pharmacologic therapies aimed at the reduction of excitatory amino acid–associated tissue deterioration.

☐ Neuronal cell body alterations are exhibited after axotomy

Primary and secondary neuronal damage is not exclusively limited to global lesions affecting gray matter and constituent nerve cell bodies; many secondary injury mechanisms can promote axonal damage. Axotomy often stimulates a variety of retrograde cytologic and biosynthetic changes at the level of the neuronal cell body, collectively referred to as the *axon reaction*. The axon reaction can be characterized by anabolic responses that are characteristic of cells having the capacity to regenerate their axonal processes. Alternatively, this phenomenon may be largely regressive in nature and more typical of neurons that are unable to reconstitute their severed nerve fibers. The axon reaction is usually highly unpredictable and depends on a number of variables, including the age of the subject at the time of injury, distance of the lesion from the cell body, species differences, nature and severity of the lesion, extent of axonal collateralization between the site of axotomy and the cell body, the functional type of cells involved, and peripheral or central location of the axotomized neuronal population.

Chromatolysis

Some basic neuronal responses to axotomy represent a range of axonal reactions that can be exhibited at the level of the neuronal cell body. The prototypical neuronal response to axotomy, occurring in some neurons is a phenomenon referred to as *chromatolysis*. Within as little as 24 hours after axotomy, chromatolytic neurons begin to exhibit by light microscopy vacuolation, enlargement of the nucleolus and formation of multiple nucleoli, displacement of the nucleus from a central to an eccentric or peripheral somal location beneath the cell membrane, and swelling of the cell body. The hallmark feature of chromatolysis is the dissolution of Nissl substance, as demonstrated by decreased cytoplasmic staining and granularity when tissue sections are stained with basic dyes. The ultrastructural correlate of this reaction is a breakdown of rough endoplasmic reticulum, with a concomitant increase in the density of free polyribosomes.

The cytologic changes are consistent with the heightened RNA metabolism and protein synthesis required for the sustained regrowth of damaged axons. Neurons sustaining axonal damage usually exhibit a shift toward the biosynthesis of structural proteins, materials associated with axonal transport, and membrane lipids, all of which are necessary for axonal growth. Reduction in the synthesis of constituents that are more typically involved in neurotransmitter metabolism is consistent with the loss of neuronal excitability after axotomy.

Although displaying correlative histologic, histochemical, and biochemical features of anabolic reactions that support axonal regeneration, chromatolysis is not a faithful index of an injured neuron's ultimate fate, because it can be exhibited by cells whether they possess or lack inherent regenerative capacities. Certain neuron types do not exhibit chromatolysis at all or only display some chromatolytic features. The persistence or disappearance of a chromatolytic response does not consistently correlate with axonal growth dynamics; chromatolytic features eventually subside whether axonal regrowth is successful or not. Although extensively investigated, the signal that triggers chromatolysis is still unknown. It appears that retrograde transport may play a role in delivering some form of molecular message from the site of axonal injury to the cell body.

Figure 32-1   ▶

The series of diagrams illustrates neuronal responses to axonal damage. Three neurons are shown, with the middle cell as the axotomized one (*outlined region*). (**A**) The progressive demise of a neuron after axonal interruption is referred to as retrograde cell death. (**B**) The postsynaptic neuron is deafferented or denervated as a result of reduced presynaptic inputs, sometimes leading to the eventual degeneration of the target cell, a process called anterograde or orthograde transneuronal or transsynaptic degeneration. (**C**) Neurons that normally make connections with axotomized cells can also die by virtue of retrograde transneuronal or transsynaptic degeneration. (**D**) Transsynaptic phenomena may also entail the survival of shrunken or atrophic cells, a condition called transsynaptic neuronal atrophy. (**E**) The axotomized cell may have numerous collaterals emerging from the axon segment that remains connected to the cell body. Projections to other targets, representing alternative sources of neurotrophic substances, would be preserved, lending a survival advantage to the damaged cell, even though it may exhibit some degree of atrophy, depending on the extent of collaterization and amount of target-derived trophic support.

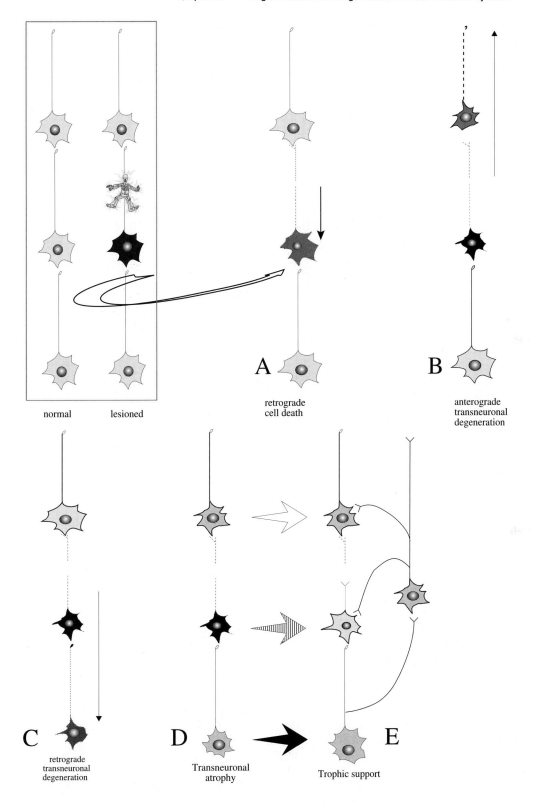

normal   lesioned

A   retrograde
cell death

B   anterograde
transneuronal
degeneration

C   retrograde
transneuronal
degeneration

D   Transneuronal
atrophy

E   Trophic support

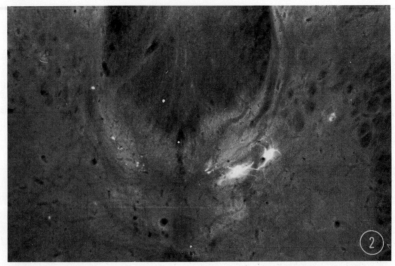

Figure 32-2
A fluorescence micrograph shows an example of retrograde cell death in the dorsal nucleus of Clarke after hemisection of the spinal cord (rat). A fluorescent substance was injected into the cerebellum before the spinal cord lesion. This neuroanatomic tracer was taken up at the terminal endings of the ascending spinocerebellar fibers and then transported back to the cell bodies (i.e., retrograde transport) that are located on both sides of the spinal cord above the central canal. Contralateral to the site of injury, two retrogradely labeled neurons are present, but no cells are evident on the lesioned side. The absence of cells is caused by retrograde cell death as a consequence of axotomy.

Other axon reactions

If the injured neuron successfully regenerates an its axon and restores viable connections with a postsynaptic target, the cell body will revert to a normal profile in terms of its size, general cytologic appearance, and biosynthetic profile. Cells that are unable to establish synaptic relations with suitable targets undergo atrophy and persist in that state or die. The progressive demise of a neuron after axonal interruption is referred to as *retrograde cell death* (Fig. 32-1A). One experimental illustration of this phenomenon is the retrograde degeneration of neurons in the dorsal nucleus of Clarke of the rat spinal cord. After lesions that interrupt the ascending axons originating from these cells (see Chap. 11), neurons in Clarke's nucleus undergo initial hypertrophy, followed by progressive degeneration during the subsequent 5 to 9 weeks (Fig. 32-2).

As in the case of secondary cell death, axotomy can cause neurons to die that were not originally injured. Axonal damage leads to the degeneration of the part of the axon that had become disconnected from the cell body. The postsynaptic neuron is *deafferented* or *denervated* as a result of reduced presynaptic inputs (Fig. 32-1B), and in certain cases, this can lead to the eventual degeneration of the target cell, a condition called *anterograde or orthograde transneuronal or transsynaptic degeneration*. A frequently cited example is the lateral geniculate nucleus of the thalamus, in which neurons receive retinal projections through the optic nerves and tracts, respectively. After section of the optic nerves, the postsynaptic cells of the lateral geniculate nucleus become deafferented and die.

Alternatively, neurons that normally make connections with axotomized cells can die by virtue of *retrograde transneuronal or transsynaptic degeneration* (Fig. 32-1C). These transsynaptic phenomena are not limited to degeneration but may also entail survival of shrunken or atrophic cells, a condition called *transsynaptic neuronal atrophy* (Fig. 32-1D).

An interesting corollary of these secondary degenerative responses to axotomy is that similar principles may apply to certain neurodegenerative diseases. For instance, the primary disease-related degeneration of certain neuronal populations could lead to more widespread cell loss elsewhere in the nervous system as a consequence of deafferentation (i.e., orthograde transneuronal degeneration) or the loss of targets (i.e., retrograde transneuronal degeneration).

Trophic influences

The illustrations of primary and secondary neuronal degeneration induced by axotomy underscore the important fundamental principle of *neurotropism*, which depends on and is responsible for the maintenance of certain sustaining cell-cell interactions. An example is denervation atrophy of muscles associated with peripheral nerve injury. The deterioration of muscle can be caused by a loss of muscle activity or absence of neuronal "nutritive" or trophic support. In analogous ways, diminished synaptic drive or neurotrophic support, which are not mutually exclusive, can account for orthograde transneuronal degeneration or atrophy. The death of a previously axotomized postsynaptic target cell can contribute to a neurotrophic imbalance, leading to retrograde transneuronal degeneration or atrophy (see Fig. 32-1D).

Trophic substances, derived from the target cell, are normally conveyed to the cell body of the presynaptic element by retrograde axonal transport. This principle also applies to retrograde degeneration of axotomized neurons, which are no longer in contact with their source of trophic support because of axonal interruption and failure to reestablish such connections. Retrograde degeneration may be curtailed by trophically effective inputs from other cells onto the injured neuron, and sustaining inputs may prevent anterograde and retrograde transneuronal degeneration.

The fact that retrograde neuronal degeneration and transneuronal loss are not more widespread after axotomy may be explained, at least in part, by the presence of *sustaining axonal collaterals* or *accessory sustaining inputs* (Fig. 32-1E). The axotomized cell, for example, may have numerous collaterals emerging from the axon segment still connected to the cell body. Projections to other targets, representing alternative sources of neurotrophic substances, may be preserved, supporting the survival of a damaged cell, even though it may still exhibit some degree of atrophy, depending on the extent of collateralization and amount of target-derived trophic support.

Several neurotrophic factors play a variety of important roles in the developing and adult CNS and PNS. The more extensively investigated of these is nerve growth factor (NGF) and brain-derived neurotrophic factor. Both neurotrophins can exert an effect on cell survival in the adult CNS. The hippocampus, to which cholinergic neurons in the septum project, contain relatively high levels of NGF and brain-derived neurotrophic factor. Transection of the fimbria or fornix area results in an interruption of the septohippocampal projection, after which approximately one half of the cholinergic neurons in the septum die. Many of these vulnerable cells can be rescued by infusion of NGF or brain-derived neurotrophic factor beginning immediately after the fimbria or fornix lesion. Because these basal forebrain cells also degenerate in Alzheimer's disease, the findings of such studies have raised provocative possibilities for at least one mode of therapeutic intervention for that condition. A disruption of neurotrophic support could be involved in some of the secondary pathology discussed earlier in this chapter, and replacement of target-derived neurotrophic support represents an interesting therapeutic possibility that warrants more investigation.

How NGF or other neurotrophins promote survival of neurons after trauma is still largely unknown. It has been suggested that they suppress genes that program the biosynthesis of substances, such as proteases, that would be neurotoxic. Neurotrophic support is not limited to presynaptic or postsynaptic neuronal interactions; it is probably also derived from nonneuronal support cells, such as glia.

Damage to Fiber Pathways and Peripheral Nerves

Although injury responses have been emphasized, neuronal responses to injury are seen at the level of the neuronal cell body, primarily in terms of cell survival. The following sections describe axonal and associated presynaptic terminal changes and provide a framework for subsequent discussion of neural regenerative dynamics in the CNS and PNS.

☐ Degenerative axonal responses occur proximal and distal to the site of injury

When nerve fibers in the CNS or PNS are completely severed or simply compressed, the axons are essentially divided into two cytoplasmic compartments. The portion of the axon that is still in continuity with the cell body represents the *proximal segment*, and the region isolated from the neuronal soma is called the *distal segment* (Fig. 32-3A). Because of the general absence of inherent biosynthetic capabilities, the axon depends on the cell body for its metabolic maintenance and integrity. This is accomplished primarily through fast and slow axonal transport (see Chap. 4). After separation from the cell soma, the distal segment cannot persist independently and undergoes a progressive demise, called *wallerian or anterograde degeneration* (see Fig. 32-3A). The rate at which total axonal breakdown and cellular debris removal occurs can vary. These events generally proceed more quickly in the PNS. In the CNS, some remnants of distal axonal segments can be seen several months after injury to certain fiber systems.

For the same reason that anterograde degeneration occurs, there is a concomitant loss of the axonal terminals, called *terminal degeneration*, that had been originally maintained by neurons before axotomy. How quickly this occurs depends on the distance between the lesion site and terminals. The longer persistence of terminal buttons reflects the fact that, in some cases, anterograde transport capacity is not immediately lost. Constituents residing in the distal axon segment before damage can be conveyed orthogradely and provide residual metabolic support of presynaptic endings for a limited time. Axonal damage also results in a loss of synaptic transmission, which can occur even before the first visible anatomic signs of terminal disruption. In some cases, electrical stimulation of the distal nerve segment can elicit conduction of action potentials and transsynaptic activation of the postsynaptic targets for periods ranging from several hours to a few days.

Axonal degeneration can also progress in a retrograde direction. Such *proximal axonal die-back* may sometimes extend only to the level at which the first axonal collateral emerges, but other factors may be involved. If the lesion occurs close to the cell body, the retrograde axonal degeneration frequently advances to the soma and results in complete neuronal degeneration.

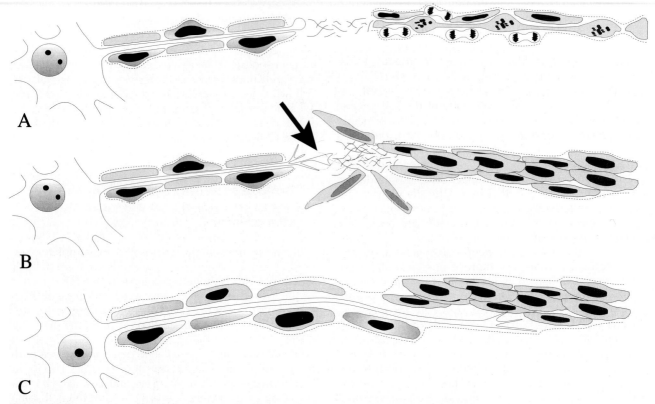

Figure 32-3
Features associated with axotomy and regeneration. (**A**) Wallerian degeneration distal to axotomy with the proliferation of Schwann cells. The distal axonal segment can assume a beaded appearance during the course of axonal breakdown. At the proximal cut end, the axon assumes a bulbous profile. An intervening gap of fibrotic scarring is indicated. (**B**) An early stage of axonal outgrowth with development of the growth cone profile (*arrow*). Distal to the lesion, Schwann cells are aligned in longitudinal columns within an investing basal lamina (*dashed line*). Extensive scarring within the lesion gap can contribute to the formation of a neuroma. (**C**) More advanced stage of regeneration, in which the growth cone has traversed the lesion gap and is growing close to the Schwann cell tubes (i.e., bands of Büngner) formed during the course of wallerian degeneration.

These degenerative changes involving the axon probably apply as well to dendritic damage (i.e., dendrotomy), but this has not been extensively studied because of technical difficulties. Such lesions must occur, especially if brain damage involves some of the superficial cortical layers to which pyramidal apical dendrites extend.

☐ Histopathologic features of axonal injury are seen in axonal segments, presynaptic terminals, and surrounding myelin sheaths

Axonal degeneration

After transection of a myelinated or nonmyelinated nerve fiber, axoplasm seeps from the proximal and distal segments until healing of the cut ends occurs by fusion of the axolemma. The cut ends retract after injury, leaving an intervening gap. As resealing of the damaged fibers takes place, the segments exhibit some swelling, characterized by an accumulation of organelles (e.g., mitochondria, lysosomes) that form a thick band immediately be-

low the axolemma and surrounding a dense core of cytoskeletal elements (Fig. 32-4A). The abundant aggregation of organelles contributing to axonal hypertrophy appears related to anterograde and retrograde axonal transport and blockage at the cut end. Fiber swelling is typically more pronounced at the proximal cut end, although such may occur at various distances back from the actual lesion site itself because of fiber retraction. In addition to the swelling seen at the level of nerve injury, distal axonal segments begin to exhibit sites of focal enlargement that are pronounced at the nodes of Ranvier within 1 to 4 cm from the site of injury.

Dramatic changes take place in the axoplasm with regard to cytoskeletal composition and distribution during the course of wallerian degeneration. Intact axons contain linearly arranged microtubules and intermediate filaments (i.e., neurofilaments). After nerve injury, amputated axons display a transient beaded conformation that is attributed to intermittent and focal varicose dilations of the fiber. At the ultrastructural level, the predominant feature of these enlarged axonal profiles is an accumulation of densely packed and highly tangled disarrays of intermediate filaments (Fig. 32-4B). Within 2 to 3 days

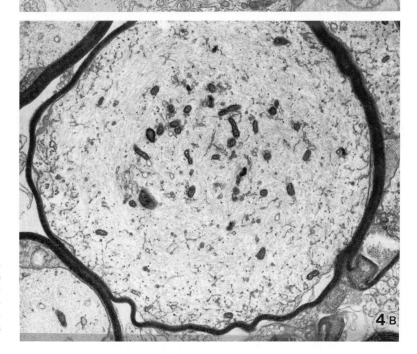

Figure 32-4
Two electron micrographs show profiles of transected axons distal to the site of the lesion. (**A**) Stage of degeneration in which abnormal organelles have accumulated within the axon. (**B**) Swollen axonal profile containing an abundance of cytoskeletal elements. The myelin sheath also has become attenuated.

after injury, the axoplasm begins to assume a more granular appearance due to a breakdown of cytoskeletal components. Soon thereafter, the axoplasm appears watery, and axonal morphology looks grossly distorted under light microscopy. The axolemma may become partially or completely fragmented. Disruption of microtubules and intermediate filaments after injury is caused by the unabated entry of Ca^{2+}, leading to the activation of Ca^{2+}-dependent neutral proteases or calpains. Degradation of the cytoskeleton probably contributes directly to axonal fragmentation and collapse.

Myelin breakdown

Changes in surrounding myelin sheaths, another aspect of axonal degeneration, progress in almost overlapping fashion. The first overt white matter alterations occur within hours after trauma, at which time the sheaths retract from axons and break apart, starting at the nodal-paranodal regions. The internodal myelin becomes segmented and wrinkled in appearance. Eventually, myelin debris is collected into linear groups of pleomorphic ovoid masses of membranous materials or amorphic lipid aggregates that become shorter as myelin breakdown continues.

Myelin disintegration implies an intimate dependency of the myelin-producing glial cell (i.e., Schwann cell in the PNS and oligodendrocyte in the CNS) on axonal structural and functional integrity. In some respects, this principle is an extension of the pattern of development, during which the axon dictates whether myelin is formed and the ultimate thickness of the sheath to be established.

Degeneration of terminals

As observed with the electron microscope, a generalized appearance of synaptic buttons after axonal interruption involves an increase in the electron density of the axon terminal. In some cases, this occurs immediately after the lesion, but in others, the process is more gradual. Organelles within presynaptic axoplasm become distorted in shape, synaptic vesicles are reduced in number, and whorls of intermediate filaments may appear. The terminals shrink, and in some cases, the buttons are detached from the postsynaptic target due to the insinuation of slender, glial cytoplasmic processes. At later stages, degenerating terminals are irregularly shaped masses of various size, and organelles within these profiles are barely discernible.

Damage to a motor neuron can result in the breakdown of the neuromuscular junction and phagocytosis of the debris by Schwann cells over the course of a few days. Transmission at the neuromuscular junction can cease within a few hours of nerve damage; the actual time correlates with the length of the distal stump preserved. It has been suggested that for every centimeter of distal nerve there is an additional delay of approximately 2 hours before excitability is lost at the neuromuscular junction.

Nonneuronal Cellular Responses

☐ **Peripheral nerve injury causes mesodermal and Schwann cell responses**

Trauma and organic neurologic disorders evoke other cellular reactions that, in principle, represent a type of healing response. The other cells involved are primarily neuroglia. Depending on the type of injury sustained, certain nonneural elements may be recruited to participate in functions such as the removal of cellular debris and scar formation. In the PNS, the most prominent cell types to consider are the intrinsic glia (i.e., Schwann cells), invading macrophages, and various connective tissue elements. These cells play significant roles during the immediate aftermath of nerve injury, and their functional repertoires influence the degree to which any subsequent regeneration is successful.

Mesodermal infiltration according to the type of nerve injury

Collagenous-fibroblastic scar formation at the site of nerve injury can have an adverse effect on axonal regrowth in the PNS. This applies particularly to transection-type lesions in which the continuity of a peripheral nerve is interrupted and the proximal and distal nerve stumps are separated by an intervening gap. This space eventually becomes occupied by proliferating cells that have migrated from nerve compartments (e.g., endoneurium, epineurium) and surrounding nonneural tissues. In the worst case, damage to the nerves and surrounding tissues can result in such extreme fibrotic reactivity that the nerve ends become blindly embedded in scar tissue or fixed by adhesions to adjacent structures.

Surgical repair of these injuries usually entails end-to-end suturing, called *coaptation*, of the proximal and distal ends if the gap is relatively short or grafting of an autologous peripheral nerve segment if the gap is much longer. Although it would seem that either procedure should limit the degree of mesodermal scarring, fibrosis frequently occurs at the suture interfaces. Scarring along the suture lines can interfere with regeneration through the site of injury.

Fibroblastic scarring is less of a concern after compression or crush injuries of a peripheral nerve. The continuity of the damaged nerve is preserved in terms of its intrinsic endoneurial fascicular organization and in relation to its investing connective tissue sheaths (e.g., perineurium). This condition yields the most favorable circumstances for axonal regrowth through the site of injury and the highest probability for functional return.

Debris removal and Schwann cell reactions associated with axonal degeneration

The removal of axonal and myelinic debris constitutes an important aspect of intrinsic nerve repair, and the rate at which this occurs probably influences the success or failure of neural regeneration. Myelin debris, for example, is taken away in the PNS over a period of 1 week to 3 months, with most removed during the second to fourth weeks. Schwann cells play an initial role in sequestering myelin ovoids; macrophages are subsequently recruited and appear more actively involved in the cleanup process. Most phagocytes are of hematogenous origin. The entry of these cells into the degenerating nerve stump is facilitated by an increase in capillary permeability generated by the liberation of histamine and serotonin from mast cells. Fibroblasts are also active during the period of axonal degeneration and contribute to endoneurial collagen deposition. The extent to which this process occurs depends on the duration that the nerve segment is denervated (i.e., free of regenerating fibers) after injury. After a nerve is severed, this interval can be lengthy, if not permanent, in which case collagen deposition is maximized to the extent that much of the internal anatomy of the nerve is radically altered.

The most significant nonneuronal regenerative response to peripheral nerve injury is that of the Schwann cell. Within the first 24 hours after injury, these cells begin to proliferate and reach a peak of mitotic activity by the end of the first week. Thereafter, their rate of division declines sharply. If a nerve is severed, Schwann cell proliferation at the cut ends can contribute to infiltration of the gap by these cells. The extent to which Schwann cells migrating from the proximal and distal ends meet depends on the length of the gap and degree of fibrotic scarring. If a nerve is compressed and its continuity is retained, the distribution of Schwann cells is virtually identical to that seen below the level of a transection or crush injury.

Within the distal nerve stump or at the site of compression injury, Schwann cell proliferation is confined within the basement membrane that originally invests each myelinated axon or group of unmyelinated fibers. Schwann cells are aligned in linear arrays parallel to the initial trajectory of the axons they had surrounded (Fig. 32-3*B*). The chains of Schwann cells contained within a common basal lamina are frequently referred to as *bands of Büngner*. This geometric organization, in conjunction with other Schwann cell properties, can yield a favorable setting for nerve regeneration.

☐ CNS injury activates a variety of cell types

Lesions to the brain, optic nerve, or spinal cord lead to the mobilization of a variety of neuroglial and non-neural cell types. Among the neuroglial cells, astrocytes have received considerable attention, and microglia have become the subject of greater interest in recent years.

Astrocyte and connective tissue element scars at lesioned sites

The type of pathology that evolves at the site of CNS trauma depends on the nature of the injury sustained. As in the case of peripheral nerve damage, the most fundamental lesions directly sever fiber tracts or interrupt white matter systems by bruising the brain or spinal cord. In a mechanical transection type of injury, as caused by laceration, the interior of the CNS is exposed to non-CNS tissue environment. Normally, the CNS is partitioned from non-CNS tissue by an external limiting membrane, called the *glia limitans*. At the ultrastructural level, this membrane is formed by the marginal end-feet of astrocytic processes that are blanketed on their subpial surfaces by a basal lamina. After trauma, astrocytes become reactive and are chiefly responsible for reinstating a glial limiting membrane along the externalized regions of the CNS neural parenchyma.

The process of glial encapsulation and repartitioning of CNS from non-CNS elements can result in the development of a region of dense astroglial scarring, called gliosis, characterized by a cobweb of interlacing cytoplasmic processes. In some laboratory experiments, astrocytes accumulated at the margin of a spinal cord transection by 1 week after injury, and a thick, tortuous astrocytic scar was evident by 2 weeks along the externalized spinal cord parenchyma immediately rostral and caudal to the wound site. Eventually, this glial capsule walls off the interior of the spinal cord from a dense, collagenous mesodermal matrix that fills the lesion (Fig. 32-5). That the invasion of connective tissue elements might induce the encapsulation response exhibited by astrocytes has been suggested by studies in which no glial scar formation was observed at lesion sites made in the immature CNS. In such instances, there is substantially less connective tissue infiltration of the lesion.

Astroglial scars along damaged white matter tracts and in gray matter

Astroglial scars are formed along damaged white matter tracts and in gray matter in the wake of wallerian degeneration, terminal degeneration, or neuronal death. Gliosis involving astroglial proliferation and hypertrophy ensues in response to the degeneration of axonal or dendritic processes, synaptic terminals, or neuronal cell bodies. A dense astrocytic meshwork is frequently established over time in regions of white matter that have undergone axonal die-back or frank wallerian degeneration (Fig. 32-6).

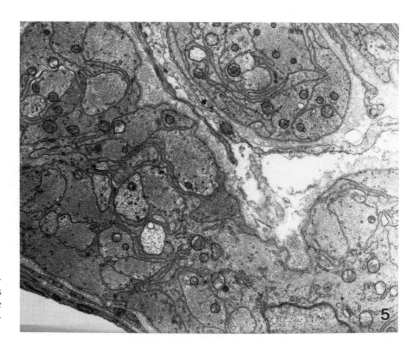

Figure 32-5
An electron micrograph illustrates a region of astroglial scarring at the site of a central nervous system injury. Notice the tight packing of astrocyte cytoplasmic processes associated with the reconstitution of the external glial limiting membrane.

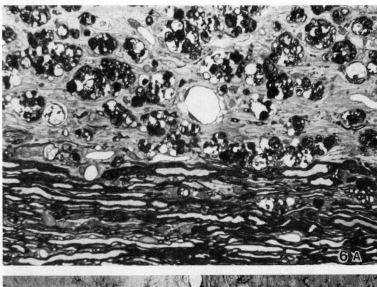

Figure 32-6

(**A**) A longitudinal section of the rat spinal cord after hemisection. Bundles of normal, myelinated axons are seen in white matter on the intact side of the spinal cord (*bottom half*), but on the lesioned side (*top half*), advanced degeneration is apparent and is indicated by macrophages and myelin debris. The section was obtained caudal to the level of injury. (**B**) Astroglial pathology resulting from the same type of surgical preparation is illustrated. An immunocytochemical stain (anti-glial fibrillary acidic protein), was used to demonstrate astrocytes caudal to the level of hemisection (*right half*). Notice the extensive astroglial reactivity on the lesioned side of the cord, which is particularly apparent in the white matter.

Astroglia also exhibit considerable reactivity in regions of gray matter after fiber and terminal degeneration and in response to primary or secondary neuronal death (Fig. 32-7). The intensity of gliosis in gray matter varies, and scar formation depends on the degree of degeneration that has occurred within a particular region. Glial reactivity in gray matter may be linked with a phenomenon referred to as *synaptic stripping*, in which presynaptic terminals are shed from the surfaces of axotomized cells. The response of astrocytes observed in gray matter reflects another integral function of these cells in addition to the encapsulation of injured surfaces. This second role is directed at filling spaces previously occupied by various neuronal elements.

Astocytic and microglial debris removal

A third function that has been ascribed to astrocytes after CNS injury is that of debris removal, but astrocytes and oligodendrocytes are minimally involved in this process. During wallerian degeneration in the PNS, macrophages are quickly recruited, and the removal of myelinic ovoid debris is completed within days. Myelin clearance is a more protracted process in the CNS, lasting several months or longer. The phagocytosis of degeneration debris in the CNS is primarily carried out by macrophages derived from the circulation and apparently even more so by resident microglia, which can transform into brain macrophages. Microglia can proliferate and become more ramified soon after axotomy distal to the site of injury as shown in Figure 32-8. Microglia can likewise become highly activated around the cell bodies of neurons sustaining axonal damage, and they may participate in synaptic stripping.

An important feature of microglial reactivity is this cell type's augmented expression of class I and class II antigens of the major histocompatibility complex. Such immunophenotypic changes in microglia occur under a variety of pathologic conditions and underscore the view that resting microglia in the normal CNS constitute a network of potential immunoeffector cells capable of responding to a variety of stimuli independent of any infil-

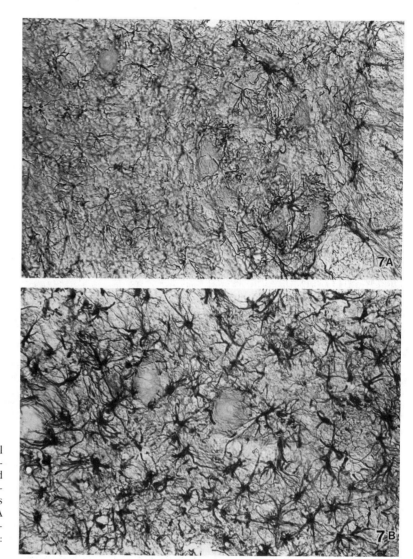

Figure 32-7
Comparison of astrocyte distributions in rat spinal
gray matter (**A**) above and (**B**) below a lesion. Caudal to the injury, astroglia appear more enlarged
and intensely stained. (**B** from Reier PJ. Gliosis following CNS injury: its microanatomy and effects
on axonal elongation. In Fedoroff S, Vernadakis A
[eds.]. Astrocytes: cell biology and pathology of astrocytes, vol 3. Orlando: Academic Press, 1986:
263.)

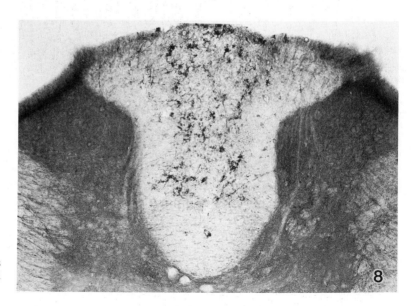

Figure 32-8
Microglial reactivity (*darkly stained profiles*) is
seen within the dorsal columns of the rat spinal
cord rostral to a dorsal transection lesion.

tration by hematogenous inflammatory cells. By acting as antigen-presenting cells, microglia can be regarded as the effector cell responsible for the recruitment of lymphocytes into the CNS. This appears to be a fundamental basis for the inflammatory reactions that influence neuronal survival and regeneration in some cases.

Neuroplasticity

The events after neuronal injury that have been described are essential to a basic understanding of what is functionally lost or retained and to an appreciation of the biologic setting affecting which motor, sensory, cognitive, or autonomic modalities can be restored through regeneration. Some functional gains may also be achieved by way of *neuroplasticity*, which can be considered as compensatory anatomic and functional changes in intrinsic neuronal circuits or end-organ connectivity.

Point-to-point regeneration represents the optimal format for functional recovery. Various features of neuronal degeneration help to define of the some basic ingredients of spontaneous or therapeutically induced repair. Broadly speaking, the extent to which any regenerative attempt can be successful depends on the neuron's intrinsic growth capacity and nonneuronal cellular responses at the lesion and those proximal and distal to it. In the event of neuronal death, the problem becomes more complex, because the inherent capacity is lacking for large-scale neurogenesis and neuronal replacement in the mature nervous system.

□ Dynamics of peripheral nerve regeneration ideally culminate in appropriately directed axonal regrowth

Neurons located exclusively within the PNS or that send axons from the CNS into peripheral nerves, such as motor neurons, have the innate capacity to initiate and maintain axonal regrowth, even for long distances, unless severely compromised by the injury. The overall functional outcome after PNS injury therefore is not a matter of the vigor of regeneration, which is often substantial. Instead, directed axonal regrowth to appropriate targets is much more of an issue. Robust regeneration is not synonymous with recovery, because failure to reach the proper target destination can lead to incomplete or aberrant functioning.

□ The initiation of axonal regrowth from the proximal stump entails growth cone formation and proliferation of neuritic processes

At the preserved end of the proximal nerve segment, the spherical axonal end-bulbs appearing after injury exhibit morphologic changes that often lead to the emergence of one or more fiber outgrowths. The ends of these newly formed branches can be slender, elongated profiles or more club-shaped endings (Fig. 32-3C). They represent the regenerating tips, called growth cones, that in most cases advance toward the lesion site. The onset of regeneration is not an immediate event and is governed by the rate at which the injured neuron is able to recover. The length of delay depends on the severity of the injury or the proximity of the lesion to the cell body.

As the newly formed axons approach the lesion, they may continue to branch, and a single parent fiber can give rise to as many as 50 or more sprouts. The resulting daughter fibers are usually of much smaller diameter than the parent axon and may persist as such throughout the course of regeneration. This can have functional impact, because the original conduction velocities may not be fully restored as a consequence of this reduced fiber diameter.

□ Traversal of the lesion site is influenced by the nature of the injury

The cellular composition and geometric orientation at the site of nerve damage influences the ultimate success of regeneration. A crush lesion in which the original alignment of endoneurial fascicles and the integrity of surrounding connective tissue sheaths are preserved yields the most favorable setting for unimpeded regeneration and end-organ reinnervation.

After a nerve is severed, cellular infiltration of the resulting gap between the retracted cut ends usually provides a less compatible condition for appropriately directed fiber outgrowth. Axons emerging from the proximal stump are given the opportunity to wander and can be diverted from the distal stump. The fibers reaching the distal stump often encounter foreign endoneurial fascicles and ultimately become directed toward inappropriate targets. This principle applies regardless of whether surgical repair of the severed nerve is attempted, because perfect realignment of the nerve fascicles is not always achievable.

Gaps between the cut ends that extend for 10 mm or more and gaps or proximal cut ends that are heavily scarred tend to be incompatible with axonal regrowth through the lesion site. In some instances, dense collagenous matrices in the wound can act as a physical barrier and prevent axons from advancing further, completely halting regeneration. In other cases, the progression of fibers through the lesion may be substantially delayed.

In a third situation, scarring promotes the development of highly disorganized bulbous enlargements of the proximal cut end of a nerve that contain a tangled meshwork of regenerating fibers, with many escaping into the surrounding connective tissue. This disorganized complex of profusely branched and often rebranched fibers, Schwann cells, fibroblasts, and collagenous matrices is commonly referred to as a *posttraumatic neuroma*. The neuroma can severely impede further regeneration, and it can be a source of painful stimuli. This nociceptive complication is related to the close approximation of fine-caliber fibers that are incompletely ensheathed by Schwann cells. In normal peripheral nerves, several unmyelinated fibers can be surrounded by a single Schwann cell; each fiber is isolated within a trough of Schwann cytoplasm. Unmyelinated axons in a neuroma can be bundled together, as seen during early stages of peripheral nerve development. This close approximation of fibers can permit the spread of electrical current from one fiber to another. Such *ephaptic transmission* to sensory fibers can evoke disordered sensory phenomena, including pain.

□ Schwann cells play key roles related to axonal elongation and remyelination in the distal stump

As axons emerge from the cut proximal end of a nerve, many group into small fascicles that reach the distal stump by approximately 3 to 28 days after injury, depending on the cellular makeup of the lesion gap. Schwann cells proliferate and co-migrate with these axons (see Fig. 32-3C). Within the distal stump, the regenerating axons encounter a more favorable terrain established by endoneurial tubes, consisting of linearly oriented Schwann cell arrays (i.e., bands of Büngner) that became orphaned as a result of wallerian degeneration. During their advance distally, growth cones of the regenerating axons adhere closely to the inner surface of the Schwann cell basal lamina. The Schwann cell microenvironment pre-

sents a type of cellular highway that permits relatively unimpeded elongation of axons toward peripheral targets.

Schwann cells are also responsible for remyelination, which progresses in a manner analogous to the deposition of peripheral white matter during development. During development, the myelin internodes (i.e., myelinated segments between successive nodes of Ranvier) increase in length. This continues until maturation is complete, at which time internodal myelin achieves a more static length. In regenerating nerves, the internodal distances are shorter than normal (Fig. 32-9) and undergo persistent remodeling for a year or more. This accounts for the altered conduction properties often exhibited by regenerated nerves.

Rate of axonal elongation

The total time for regeneration to be completed entails an initial proximal stump delay, lesion delay, distal stump growth phase, and functional recovery period. The *initial proximal stump delay* reflects the time for the cell body to respond to axotomy and commence regrowth and the interval involved in retrograde die-back. The time it takes axons to traverse the site of injury defines the period of *lesion delay*, which together with the initial delay establishes the total time for axons to reach endoneurial tubes of the distal stump after injury. The period of axonal growth from entry into the distal stump to reaching a site of peripheral termination constitutes the *distal stump growth delay*. The *period of functional recovery* is related to the time required for appropriate axonal conduction properties to be reinstated, the period necessary

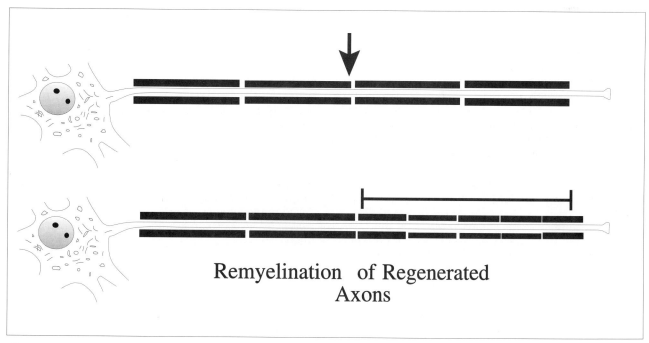

Figure 32-9
Myelin internodes are shorter after regeneration, as seen in the peripheral nervous system.

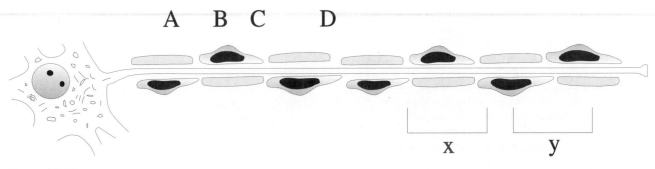

Figure 32-10
Intrinsic rates of axonal regeneration relative to the distance from the cell body. Axonal growth over segment x can be, for example, 1.5 mm per day whether the lesion occurred at point B or D. When axons reach segment y, a rate of 1 mm per day can be assumed which is intrinsic to that region of the nerve fiber. The rate at which elongation commences is defined by the segment at which the lesion had occurred.

for functional innervation to take place, and the time needed by the target tissue to recover from denervation and become responsive to new synaptic inputs.

The rate of axonal regeneration is thought to be influenced by a variety of factors, including the age of the subject, the location and type of injury, and the duration that the distal stump is denervated before the arrival of regenerating fibers. In humans, it has commonly been held that axonal regrowth proceeds at a relatively constant rate of 1 mm/day, but it has been shown that there is a diminishing rate over time and that the rate is a function of the growing tip's distance from the cell body. Rates of regeneration range from 1 to 4.5 mm/day, but measurements defining these rates can be inconsistent and depend on a number of considerations that cannot always be assessed rigorously, especially in humans.

According to Sir Sidney Sunderlund (1991), an eminent neurosurgeon who compiled extensive information about peripheral nerve injury and regeneration, the rate of axonal advance is always the same at a given distance from the cell body, regardless of where the original lesion occurred. As illustrated in Figure 32-10, axonal growth over segment x can be, for example, 1.5 mm/day, whether the lesion occurred at point B or D. When axons reach segment y, a rate of 1 mm/day can can be assumed, which is intrinsic to that region of the nerve fiber. The rate at which elongation commences is defined by the segment at which the lesion had occurred.

Targeting of regenerating axons and functional recovery

Although Schwann cell columns provide an amenable substratum for axon regrowth, this matrix does not necessarily guarantee that axons will reach appropriate or their original target sites. The best outcome potential is for crush injuries, which usually result in minimal disruption of the intrinsic funicular arrangement of endoneurium–Schwann cell tubes at or distal to the injury. A far more complex problem is posed by transections. In the case of a compression injury, in which nerve continuity is preserved, much of the internal anatomy is retained, and axons above the lesion can progress rather

faithfully through the injury and onward to their targets within the original funiculi. Disruption of this anatomy by most types of transection damage would require axons to traverse a complex lesion terrain between the nerve stumps. Within the lesion gap, they would be subjected to scarring and alternative directional cues that could easily guide axons into foreign funiculi. Many axons would be diverted to inappropriate targets, producing poor functional recovery under most circumstances.

The extent to which functional improvement occurs after peripheral nerve injury and regeneration depends on a variety of factors that require a discussion beyond the scope of this chapter. As a general rule, proximal muscles show greater functional recovery after nerve injury at high limb levels than do distal muscles. The transection of a nerve close to its target can also be of advantage by virtue of a reduction in the number of options for misdirection within the distal stump and the shorter time required for axons to reach the end-organ.

Despite the many variables that can affect the quality of peripheral nerve regeneration, under optimal conditions, axonal regrowth can occur with exquisite precision, frequently reinstating original target associations with considerable fidelity. One example is the reinnervation of skeletal muscle in which the new neuromuscular junctions appear and function much as their original counterparts. Many of the new terminals are formed in register with the postsynaptic junctional folds of the original synapses, even if no endoneurial or other cellular guidance is available. Evidence indicates that the proper address for the previous neuromuscular junction is within the basal lamina that is normally in the synaptic cleft between the axon terminal and junctional folds (see Chap. 20).

☐ Studies of peripheral nervous system regeneration have led to several basic theoretical considerations

The inherent capacity for regeneration exhibited by some neurons has stimulated efforts to identify changes in gene expression that could account for this ability. One molecule of particular interest has been a growth-associated protein with an apparent molecular weight of 43,000 (GAP-43). GAP-43 is produced by neurons during development and by mature neurons capable of sustaining renewed growth after axotomy. This molecule is conveyed from the cell body to the axon by fast axoplasmic transport and represents one of the more abundant constituents of the growth cone membrane. The deposition and stabilization of GAP-43 can be subject to axonal posttranslational mechanisms that may be influenced by different cues in the cellular microenvironment surrounding injured axons.

The PNS also affords a compatible cellular microenvironment for axonal elongation, which is best seen throughout the nerve after crush injury or within the distal nerve stump after transection. Schwann cells have long been considered to be a major player in this process. Santiago Ramon y Cajal proposed that these peripheral glial elements fostered axonal elongation by virtue of some form of trophic support. Subsequent experiments have demonstrated the preferential growth of axons toward distal nerve segments using a Y-tube model in which the proximal cut end is inserted into the base and the distal segment is inserted into one of the branches of the tube, with an alternative target placed into the other branch. Many of the regenerating fibers grew toward the distal segment even when it was not connected to any target structure. Such findings led to investigations directed at the definition of the Schwann cell molecular properties that play a role in the regenerative process.

A variety of molecules have been identified, including various neural cell adhesion molecules, constituents of the extracellular matrix secreted by Schwann cells, and neurotrophic factors such as NGF. Schwann cells express low-affinity, fast-dissociating NGF receptors on their surfaces after axotomy and the resulting loss of axonal contact. NGF synthesis and release is up-regulated in these cells, increasing the binding of NGF to Schwann cell receptors. Regenerating axons possessing high-affinity, slow-dissociating NGF receptors later bind and retain the NGF released from the Schwann cell receptors, engaging a mechanism of receptor-mediated intercellular transfer of NGF. From the axon, NGF is conveyed to the neuronal cell body by retrograde transport. As regeneration progresses, NGF production and receptor expression by Schwann cells are suppressed. Although it has been suggested that such an NGF-related mechanism could be important for regrowth of some NGF-responsive neuronal systems (e.g., sensory axons), there are some contradictory lines of evidence. Other neurotrophic factors may play a more pivotal role in the regenerative process, and the observations pertaining to NGF provide a potential model of the important underlying neurotrophic mechanisms during PNS regeneration.

Whether Schwann cells have the capacity for promoting selective axonal elongation remains unresolved. Some studies suggest that the initial random outgrowth of axons and their plethora of branches results in motor axons extending into predominantly sensory and motor

Schwann cell tubes of the distal nerve segment. Eventually, the collaterals in the sensory sectors are pruned, but those in the motor pathways persist. This result can be obtained regardless of whether a peripheral target is available. Such preferential motor axon outgrowth suggests some form of Schwann cell property is retained in the wake of wallerian degeneration that can lure fibers along appropriate endoneurial tubes. Research is focusing on a carbohydrate moiety, common to several cell adhesion molecules, that is differentially expressed by Schwann cells in motor and sensory nerves.

In addition to Schwann cells, macrophages may play significant roles in the regenerative process by their actions in wound healing, debris removal, and production of growth factors. Axonal elongation after peripheral nerve injury may be the result of degradation of inhibitory molecules or modification of the extracellular matrix. By rapidly removing myelin debris, macrophages may be integrally involved in regeneration. These cells also function as secretory elements and release apolipoprotein E, which may be important for regeneration and remyelination after nerve injury.

☐ Regeneration in the CNS is usually abortive

The relative success of regeneration in the PNS stands in sharp contrast with what has been repeatedly observed over many years after trauma to the CNS. Santiago Ramon y Cajal and other investigators of his era provided seminal observations leading to notion of *abortive regeneration*. The early investigations demonstrated that nerve fibers intrinsic to the CNS showed an initial growth response after axotomy that can extend for approximately 0.5 to 2 mm, but within a short time, regeneration ceased, and the newly formed sprouts were resorbed or exhibited sedentary terminal enlargements. Although there are some exceptions to this rule, the concept of abortive regeneration has endured.

Inhibitory factors contributing to regenerative failure

It has been inferred that most neurons confined to the brain, spinal cord, and retina lack the intrinsic metabolic ability to sustain regeneration. It was recognized that deficient neuronal growth dynamics were not the sole explanation for regenerative failure and that the cellular environment of the injured CNS also could be responsible. Influenced by observations made of peripheral nerve regeneration, early researchers speculated in a rather prophetic fashion that a major limitation in the CNS was the absence of elements equivalent to Schwann cells that could trophically nurture and physically support the elongation of newly formed neuritic sprouts. Scar tissue forming around lesions also was implicated as a major deterrent to effective axonal outgrowth.

Over the years, several other extrinsic conditions have been proposed to be adverse to CNS regeneration. These can be divided into two basic, although not necessarily

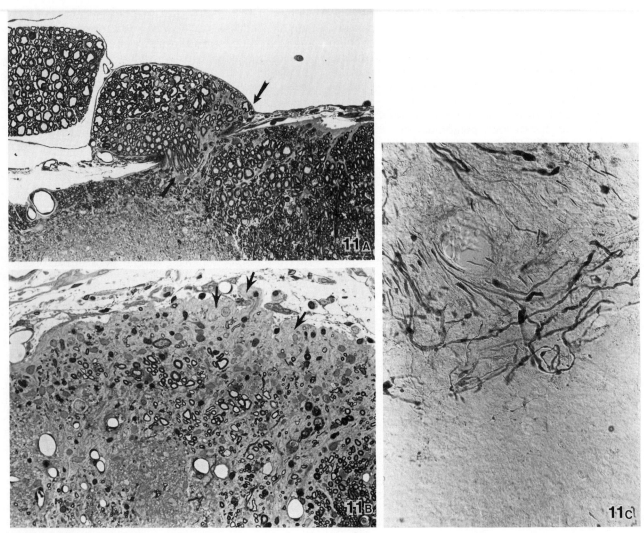

Figure 32-11
Dorsal root entry zone (DREZ) of the rat spinal cord. (**A**) The normal DREZ has myelinated axons entering the spinal cord at the transition zone (*arrow*) between the peripheral and central nervous systems. (**B**) After crush of the dorsal root and wallerian degeneration of axons central to the injury, a zone of glial scarring has formed at the DREZ (*arrows*). (**C**) Regenerated dorsal root axons have been filled with a neuroanatomic tracer. The micrograph is oriented such that the dorsal root portion is at the upper half of the field, and the unstained lower half is the zone of glial reactivity at the DREZ. On reaching the DREZ, axons begin to turn back toward the dorsal root rather than continuing on to reenter the spinal cord.

mutually exclusive, reasons for regenerative failure: the presence of inhibitory conditions or molecules or the absence of a permissive milieu provided by certain tissue terrains, growth-promoting substances, and neurotrophic-neurotropic factors.

In terms of axonal growth-inhibiting cellular conditions, virtually any aspect of trauma-related neuropathology can have significant bearing alone or in combination with one or more other cellular responses. The three hypotheses that have been most extensively explored center on the inhibitory effects of astroglial scarring, CNS white matter, and protracted debris removal.

Astroglial scarring. Astroglial scarring, a hallmark of CNS neuropathology, has frequently been viewed as a

physical barrier to regeneration. One line of evidence favoring the glial scar hypothesis derives from studies of dorsal root axon regeneration (Fig. 32-11). After crush damage, axonal outgrowth actively proceeds in the dorsal root by virtue of interactions with Schwann cells at and distal to the site of injury. After the advancing fibers reach the spinal cord-dorsal root entry zone (see Fig. 32-11), they terminate as large bulbous enlargements, assume random growth patterns, or are deflected back toward the spinal ganglion of origin. The dorsal root entry zone of the mature spinal cord is characterized by a transition from a PNS cellular environment to a CNS tissue terrain, consisting of a convoluted fringe of astroglial processes that interdigitate with Schwann cells and connective tissue elements of the dorsal root. Altered axonal growth patterns at this interface may be associated with astrocytic processes, which are associated with extracellular matrix molecules that can be inhibitory to axonal elongation. Such information underscores the fact that astroglial scarring may not simply represent a mechanical obstacle but could possess molecular attributes that are not conducive to regeneration.

If a relatively homogeneous population of astrocytes is grafted between the cut ends of a peripheral nerve, axons capable of substantial regeneration fail to enter that ectopic CNS tissue terrain. Migration of Schwann cells into the astrocyte-rich grafts likewise does not occur, suggesting that the apparent nonpermissive nature of the astrocytic scar is not exclusively limited to growth cone mobility.

White matter–associated inhibition. Glial scars are not always homogeneous in composition and may consist of other cellular elements, including persisting oligodendrocytes, intact myelin, and copious amounts of myelin debris. The potential significance of white matter in terms of limited CNS regeneration is a long-held view that has been bolstered by experimental evidence.

Under appropriate conditions, neurons grown in tissue culture extend neurites with prominent growth cones over the surface of the culture dish. If grown in the presence of white matter or in combination with populations of isolated, myelin-forming oligodendrocytes, axonal elongation ceases when contact is made between the growing fibers and these white matter constituents. The growth cones remain stationary until retraction of the fiber ensues. Two protein fractions detected in CNS myelin exert striking inhibitory effects on neuritic outgrowth in tissue culture. Such proteins have not been identified in PNS myelin sheaths. Monoclonal antibodies directed against these molecules have been reported to enhance axonal regeneration in the injured spinal cord.

Slow removal of cellular debris. Myelin-associated inhibitors of axonal outgrowth can complicate the slow rate of myelin and axonal debris removal in the injured CNS. Monocyte recruitment from the blood and macrophage activity in the CNS do not evolve as quickly as in the PNS after nerve injury, and the removal of potentially powerful inhibitors of axonal outgrowth is not as effective.

There are other significant implications related to macrophage activation that begin to address deficiencies in the CNS that can influence regeneration. Trauma is a powerful stimulus for interleukin-1 (IL-1) synthesis and release by macrophages and microglia. Neurons and glia express IL-1 receptors on their surfaces, and there is increasing evidence for cytokine-stimulated expression of neurotrophic factors with the induction of neurotransmitters and other neuropeptides. IL-1 increases NGF production in central and peripheral glial cells. Controlled enhancement of macrophage activation in the CNS could be beneficial in terms of neuronal survival and axonal growth. Macrophage-derived interleukins can also stimulate astroglial proliferation and potential glial scar formation.

Absence of growth-promoting conditions in injured tissue

Primarily reflecting the focus given to Schwann cells during peripheral nerve regeneration and the neurotrophic and neurite-promoting properties that have been assigned to them, the lack of CNS regeneration has been traditionally viewed as reflecting deficiencies in indigenous cell types capable of producing such molecules. Schwann cells produce the extracellular matrix molecule, *laminin*, which has been shown by many investigators to be highly conducive to neurite outgrowth. In the adult CNS, laminin is lacking.

The production of some neurotrophic factors has been attributed to Schwann cells distal to the site of nerve injury, and for many years, it appeared that the CNS lacked cells that could manufacture such substances. This view changed radically with the identification of a host of molecules in the CNS that have neurotrophic or growth-promoting actions. There is evidence for astroglial and neuronal synthesis of neurotrophins in different parts of the brain under certain lesion conditions. The apparent lack of growth factors in the damaged CNS reflects limited up-regulation of the expression of specific genes and possibly related signal transduction mechanisms. Provision of such substances in greater quantity may not resolve the problem. Experiments involving tissue culture models have shown that neuritic outgrowth can be impaired by the addition of axon growth inhibitors (e.g., CNS myelin) even when potent stimulators of neuritic elongation are readily accessible.

CNS neurons express intrinsic growth capacities under certain conditions

The fact that environmental factors can throttle regeneration in the CNS does not preclude the possibility that neurons are unable to muster sufficient metabolic energy to foster axonal regrowth over potentially long distances, as would be required in the case of a cervical spinal cord injury. Moreover, enhancement of regeneration in the CNS would not necessarily translate into beneficial functional outcomes. Two important sets of observations about neuroplasticity and neuronal growth capacities have prompted a more optimistic perspective of CNS repair.

Anatomic neuroplasticity

More than 40 years ago, the phenomenon of *collateral sprouting* was first described, and this has become a fundamental principle in neurobiology. Experiments were performed in which some nerve fibers to a muscle were transected. Neighboring intact fibers that had also projected to the same muscle began to extend preterminal outgrowths, which subsequently innervated the partially deafferented muscle (Fig. 32-12A). Later studies extended these findings to the CNS by showing that denervation of

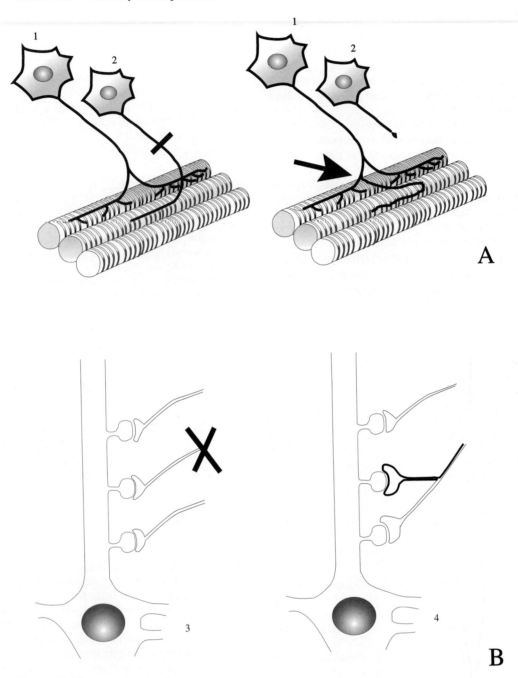

Figure 32-12
Collateral or paraterminal sprouting in the (**A**) peripheral and (**B**) central nervous system. Axotomy of a neuron (1) results in the emergence of a new collateral (*arrow*) from the uninjured cell (1). The new axonal outgrowth reinnervates the postsynaptic region originally occupied by the presynaptic terminal of neuron 2. In the central nervous system, a neuron has three separate sources of axonal input on neighboring dendritic spines (3). After one input is elimated by virtue of a lesion, reactive synaptogenesis occurs (4), whereby a new synapse is made with the denervated spine by short-range outgrowth of a collateral axonal process from an uninjured neuronal source.

a neuronal target can lead to the sprouting of adjacent fiber systems and formation of new synapses onto the denervated cell (Fig. 32-12*B*). Collateral sprouting and other morphologic variants of new synapse formation in the injured CNS have been collectively considered as expressions of *reactive synaptogenesis*. This growth represents the cellular basis for *neuroplasticity*, which can be associated with restored normal functions, the emergence of inappropriate functions, or the development of compensatory behaviors.

Reactive synaptogenesis does not take place over long distances, regardless of whether it occurs by collateral sprouting or some other mechanism. Such forms of axonal growth have little significance for regeneration, because the response is generated by previously uninjured cells. The importance of this phenomenon is that it pro-

vided the first clue that mature CNS neurons have a potential to support axonal growth.

Growth capacities of central nervous system neurons in the presence of peripheral nerve tissue

The inherent growth properties of CNS neurons have become even more profoundly recognized as a result of an elegant study performed by Sam David and Albert Aguayo at the Montreal General Hospital more than a decade ago. The experimental design involved grafting one end of an autologous peripheral nerve segment into the midthoracic spinal cord and inserting the other end into the medulla (Fig. 32-13). An axon-free bridge (i.e., wallerian degeneration is in progress), consisting of Schwann cells and other peripheral nerve tissue ele-

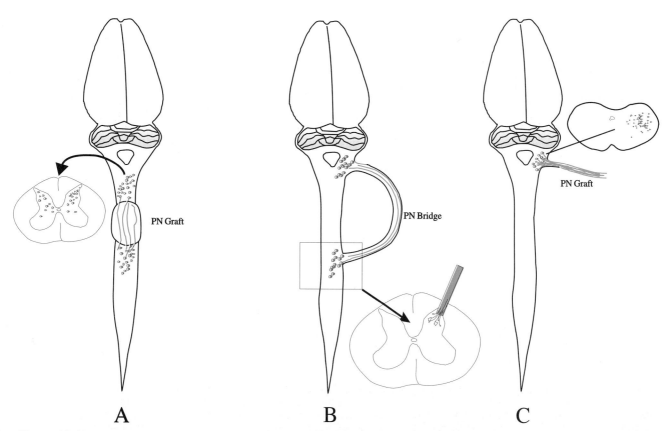

A **B** **C**

Figure 32-13
Three examples of peripheral (PNS) to central nervous system (CNS) grafting experiments that have illustrated the intrinsic capacity of CNS neurons to initiate and sustain axonal growth over substantial distances when the CNS cellular microenvironment is replaced by that of the PNS. (**A**) A PN graft placed into a complete transection lesion of the rat spinal cord. Neuroanatomic tracing subsequently showed axons extending through such grafts. Neurons from which the axons in PN grafts arose were demonstrated by retrograde labeling, and their distribution in spinal gray matter is illustrated by the accompanying transverse section of the spinal cord (*arrow*). (**B**) A PN bridge formed between the medulla and spinal cord. Neurons (*red filled circles*) at both levels near the graft insertion site extended axons through the bridge. As shown by the cross-sectional diagram (*arrow*), once axons reentered the spinal cord, their advance halted within a short distance of the graft insertion site. (**C**) A third situation in which an essentially blind-ended graft was able to support the growth of axons from intrinsic CNS cells.

ments, was formed and extended several millimeters. Subsequently, neuroanatomic tracing methods were used to determine whether axons had grown into the peripheral nerve bridge and, if so, to determine the cells from which these axons originated and the destination of fibers within the peripheral nerve graft.

The results demonstrated that neurons intrinsic to the CNS could extend fibers into the peripheral nerve bridges, that these axons could traverse the complete length of the peripheral nerve grafts, and that growth of these axons terminated shortly after reentering the CNS. Similar results have been obtained with peripheral nerve grafts to other regions of the brain and spinal cord.

PNS-CNS grafting experiments have been instrumental in reversing previous dogma by showing that neurons in the brain, spinal cord, or retina have an inherent capacity to sustain long-distance axonal elongation even after previous injury. In some cases, the elongation of axons exceeds the distance originally exhibited by comparable neuronal populations during development. The importance of the cellular environment is clearly emphasized, because growth only occurs if the CNS tissue milieu is replaced by peripheral cellular elements. Axonal regrowth supported by these neurons ceases soon after the growing tips of axons in peripheral nerve grafts encounter the CNS microenvironment.

Coupled with these intriguing developments are demonstrations showing that some functional synaptic connections can be reestablished in the mature CNS through PNS-CNS grafts. Studies of regenerating optic nerve axons in peripheral nerve bridges between the retina and superior colliculus have produced some interesting observations. On exiting the peripheral nerve grafts, optic nerve fibers formed extensive arborizations in the colliculus within a distance of 300 to 500 μm. Several of these axons established well-developed synaptic connections within the superficial layers of the tectum, where they can persist for the lifespan of the animal. Some of these synapses exhibited morphologic features characteristic of normal retinocollicular projections. In parallel experiments using similar retinocollicular peripheral nerve bridges, photic stimulation of the retina elicited transsynaptic extracellular evoked potentials in superficial regions of the superior colliculus where retinal axons normally terminate. Even in the mature CNS, some degree of appropriate target recognition and functional reinnervation can occur when regeneration is induced.

Pharmacologic and Transplantation Approaches to Central Nervous System Repair

☐ Neurotrophic and neurite-promoting factors may lead to neuronal survival and regrowth of damaged processes

Peripheral nerve grafting studies have played a more important role in underscoring essential neurobiologic principles than in identifying a specific therapeutic strategy. The insights gained inspire greater optimism than has previously existed about the possibility for someday engineering significant functional improvements in the CNS. As more is learned, the peripheral nerve grafting approach may yet provide regeneration after certain types of CNS trauma. These and other transplantation approaches and the many other experimental paradigms have served as valuable tools in defining some of the essential chemical requirements for neuronal survival and regeneration. The continually expanding list of factors may yield rational drug interventions in the future.

☐ Fetal neural tissue transplantation can foster anatomical and functional reconstruction

Based on experimental results primarily obtained from rodents and subhuman primates, transplants of fetal CNS tissue can be used to replace discrete populations of dying or dysfunctional neurons in the CNS of newborn, juvenile, and adult recipients. An extensive scientific literature exists showing that fetal grafts obtained from dif-

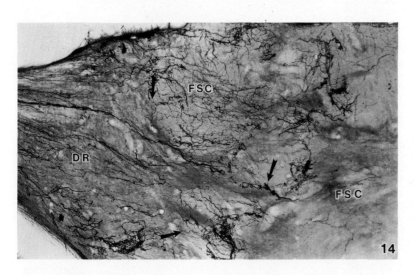

Figure 32-14
Innervation of a fetal spinal cord graft (*right half*) by host spinal primary afferent axons originating from a dorsal root that was sandwiched with the graft at the time of transplantation. (Tessler A, Himes BT, Houlé JD, Reier PJ. Regeneration of adult dorsal root axons into transplants of embryonic spinal cord. J Comp Neurol 1988;270:537). Axons enter the graft and then form extensive ramifications analogous to those characteristic of primary afferent trajectories in the normal spinal cord.

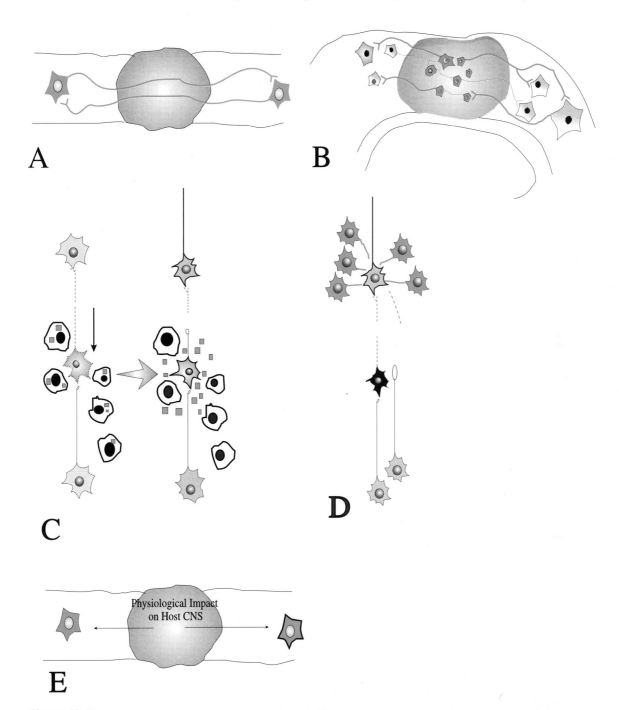

Figure 32-15

Some mechanisms by which fetal grafting can lead to anatomic and functional repair of the central nervous system (CNS). (**A**) Bridging of the lesion analogous to a peripheral nervous system (PNS) to CNS graft. (**B**) Formation of a functional relay allowing neurons in the transplant to receive axonal projections from host cells and form efferent projections to other host neurons. Theoretically, functional information may be transmitted across a lesion independent of long-distance regeneration of host axons or extensive axonal outgrowth from the donor neurons. (**C**) Grafts providing trophic factors or other substances that can lead to rescue, for example, of damaged neurons in the host that would ultimately die (Bregman BS, Bernstein-Goral H, Kunkel-Bagden E. CNS transplants promote anatomical plasticity and recovery of function after spinal cord injury. Restor Neurol Neurosci 1991;2:327). (**D**) Graft-mediated restoration of tonic or unregulated synaptic inputs or neurotransmitters depleted because of injury or disease. In this context, the grafted cells can serve as minipumps, producing substances necessary for functional improvement. This represents one way in which fetal neuron transplantation appears to be of some benefit in Parkinson's disease. (**E**) Fetal or other grafts may have inductive metabolic effects that could lead to activation of neuronal circuits in the host CNS. (Stokes BT, Reier PJ. Oxygen transport in intraspinal fetal grafts: graft-host relations. Exp Neurol 1991;111:312).

ferent levels of the embryonic neuraxis can become highly vascularized, survive for long posttransplantation intervals, and exhibit relatively normal cellular maturation. It has also been shown that during their maturation, fetal CNS grafts acquire many appropriate region-specific topographic features in terms of neuronal types and the laminar distributions of particular cell populations. Fetal CNS grafts do not become disorganized masses of neoplastic tissue; cell proliferation and differentiation appear to follow a schedule comparable to what would occur under normal developmental conditions in the intact fetus.

Various neuroanatomic methods have demonstrated the development of axonal projections between host and donor tissues (Fig. 32-14). Host neurons also are capable of forming some afferent and efferent synaptic interactions with grafted embryonic cells. That these connections can actually contribute to the development of functional neural circuitries has been demonstrated electrophysiologically in several regions of the mature CNS. Many examples of behavioral recovery have been reported for experimental models of neurologic disease and trauma. Several mechanisms of graft-mediated recovery have been proposed (Fig. 32-15) that illustrate the many opportunities fetal grafts can offer under different neuropathologic conditions.

Fetal grafts in models of neurodegenerative disease and trauma

The potential therapeutic value of transplants to attenuate clinically relevant deficits has been most extensively explored in relation to Parkinson's disease (see Chap. 21). This condition is caused by the death of dopamine-producing neurons in the substantia nigra and subsequent loss of dopaminergic projections to the neostriatum. A Parkinson-like disorder can be induced in rodents by injecting a neurotoxic chemical into the substantia nigra on one side of the brain stem, destroying dopaminergic neurons. The affected animal displays asymmetric posture and movement, often turning spontaneously toward the lesioned side. Placement of fetal dopaminergic neurons into or near the denervated striatum reduces

this deficit. The fetal cells probably act as minipumps that can restore appropriate levels of dopamine at the level of the target neurons, but the specific mechanisms underlying the functional changes seen have not been fully defined, and other possibilities have been suggested, including graft-induced (e.g., trophically stimulated) sprouting of fibers from other neighboring dopaminergic systems in the striatum.

Later studies demonstrated similar functional improvements in primates with parkinsonian deficits induced by the drug MPTP (1-methyl-4-phenyl-1,2,3,6-tetrahydropyridine), which destroys dopaminergic neurons in the nigrostriatal pathway in primates and humans. The promising results of laboratory studies have spawned clinical trials by groups at the University of Lund in Sweden and in the United States at Yale University and the University of Colorado. Although the efficacy of this approach requires confirmation, various degrees of benefit have been observed. With further technical improvements and better understanding of the biology of these grafts and of the disease process itself, neural grafting may someday offer more superior therapeutic benefits than are achievable with current pharmacologic interventions directed at Parkinson's disease.

The application of fetal neural tissue grafting to CNS trauma has been investigated, particularly in models of spinal cord injury in which grafting has been found to promote extensive anatomic reconstruction, even in contusive spinal cord lesions having a pathology closely resembling that frequently seen in humans (Fig. 32-16). Physiologic and behavioral indices indicate that intraspinal transplantation may ameliorate some functions involving motor, sensory, or autonomic modalities. Although spinal cord injuries can cause functionally complete lesions, some fiber tracts are preserved, except in the most severe injuries. The spared fiber systems are rendered dysfunctional. In addition to other mechanisms of graft-mediated function that have been proposed, the introduction of fetal cells could possibly contribute to a physiologic rejuvenation of persisting fibers and synaptic circuits above and below the site of injury by remyelination or the initiation of compensatory functions through activation of alternative circuitries.

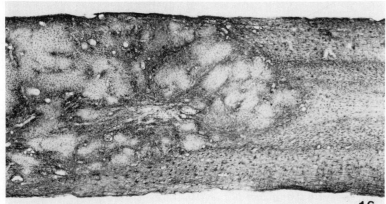

16

Figure 32-16
A longitudinal section of a rat spinal cord that had sustained a contusion injury several days before transplantation surgery. A large mass of graft tissue is seen in an area that otherwise would be represented by a region of massive host tissue erosion and cyst formation.

*Genetically engineered cells
and other alternatives to fetal tissue*

Fetal cell transplantation raises several legal and ethical issues, and the use of human fetal donor tissue presents a variety of logistic problems, such as immunohistocompatibility and an adequate amount of donor material for each procedure, that need to be circumvented before large-scale clinical trials can be realistically envisioned.

Other sources of donor tissue are being considered. Genetic engineering is being explored as a method for the delivery of trophic factors (e.g., NGF) and neurotransmitters or their related enzymes (e.g., tyrosine hydroxylase, glutamic acid decarboxylase, choline acetyltransferase), which could promote cell survival, regeneration, and functional improvement in the CNS. One approach entails the introduction of gene constructs, coding for specific molecules, into cell lines that are subsequently grafted into the CNS. Cells genetically engineered to express NGF have contributed to an enhanced survival and axonal regeneration by NGF-responsive septal neurons that would otherwise die after axotomy. In other experiments, L-dopa–secreting cells have induced behavioral improvements in a rat model of Parkinson's disease after being grafted into a denervated striatum. Conditions are being investigated that can maximize gene transfer, transgene expression, and greater survival of genetically altered cells in the brain and spinal cord.

☐ Mature and progenitor cell lines from the adult CNS offer other possibilities for CNS repair

Another intriguing alternative to fetal tissue grafts entails isolation of cells from the mature CNS. Mature oligodendrocytes have been obtained from the adult rat spinal cord. When grafted into the spinal cords of rats with inherited myelin deficiency, the cells formed myelin around host axons in the dorsal, ventral, and lateral white matter. The extent of myelin formation in such experiments is not widespread, because the transplanted cells tend to cluster. Additional studies are needed to enhance the migration and proliferation of these donor oligodendrocytes after they are in the recipient CNS.

The progenitor neural cell populations in the adult CNS offer another area of exciting possibilities. In certain nonmammalian species, such as fish, amphibians, and song birds, some neurogenesis occurs in the brain throughout life and may even be amplified by trauma. In mammals, the formation of new neurons in the adult CNS has been considered essentially nonexistent. Within the last few years, the generation of new neural elements has been observed in certain regions of the mature mammalian brain. This was previously documented only in the hippocampal dentate gyrus and in the olfactory bulb, but mitogen-induced proliferation has been reported for multipotent progenitor cells from explant cultures of postnatal rat retina and adult mouse brain. Glial and neu-

ronal progenitors apparently reside within the mature CNS in a dormant, nonproliferative state.

Investigators at the University of Calgary have demonstrated that cells isolated from the striatum of the adult mouse brain can be induced to proliferate in vitro by exposing the tissue explants to epidermal growth factor. Newly generated cells developed the morphology and chemical phenotypic properties of neurons and astrocytes. Cells expressing epidermal growth factor receptors and the receptors to other mitogenically active growth factors have been described even in the mature human CNS. Although a seemingly futuristic concept, the presence of these cells in the adult mammalian brain suggests another possible source of tissue for use in neural grafting to replace damaged or diseased cells.

Results of Regeneration Attempts and Developing Methods

The ultimate outcome after PNS or CNS nerve injury depends on a host of variables. This area of neurobiology is still evolving, and few absolute rules apply, but even with all of the complexities involved, the basic requirements for successful neural regeneration provide a useful guide for reviewing this topic and for in-depth examination of specific literature.

The neuron's innate ability to survive and reinitiate axonal growth represents is important, as is the influence of the cellular environment at and below the site of damage. In the PNS, achieving a relatively high degree of functional improvement after peripheral nerve injuries is feasible. In that context, the PNS satisfies many of the basic prerequisites for regeneration. The issue of ultimate functional recovery is not purely a matter of the possible vigor and robustness of the regenerative response but is more a question of guidance and targeting. Intrinsically, neurons projecting axons to the periphery can exhibit considerable metabolic ability to initiate and maintain axonal regrowth over long distances if axotomy does not precipitate a deleterious retrograde response. These neurons appear to retain an ability to recognize appropriate target sites when given the opportunity by providing optimal regeneration conditions.

Peripheral nerve regeneration under the most ideal conditions emphasizes the significance of supporting cells, growth-promoting extracellular matrix components, growth factors, and cell adhesion molecules. The predominant impression about abortive regeneration in the CNS is the presence of inhibitory molecules and nonpermissive cellular terrains or the lack of more compatible cellular terrains and associated growth-promoting molecules. However, as PNS-to-CNS grafting experiments have illustrated, a variety of neuronal types in the adult brain and spinal cord can regenerate axons for long distances when presented with an appropriate tissue environment. As in the best cases of regeneration in the PNS, neurons in the brain and spinal cord appear capable of making new connections with appropriate targets when presented with more compatible conditions for axonal growth. Because of neuroplasticity, opportunities are available for promoting recovery through uninjured neurons and the establishment of useful alternative synaptic associations.

The large body of evidence obtained from fetal cell grafting studies indicates that it may be possible to circumvent the problems associated with posttraumatic secondary cell loss or neurodegenerative disease by employing cellular replacement strategies and growth factors. Graft rejection is still a major concern, and the drugs used for immunosuppression may not be practical in all cases. Despite many challenging considerations, fetal CNS tissue transplantation provides an important experimental approach that can yield a better understanding of how amenable the nervous system can be to repair processes. The fact that some potential for neurogenesis exists even in the adult CNS underscores fundamental growth properties that may be enhanced with growth factors, mitogens, and other compounds.

Despite the many exciting possibilities for repairing damaged regions of the CNS, it may still be difficult for the student or experienced clinician to envision how restored function in complex neural circuits may be achieved. However, this conservative view assumes that restoration of normal function is the sole objective. Even relatively modest functional gains, obtained through a combination of surgical or pharmacologic interventions and rehabilitative therapy, could translate into a measurably improved quality of life and an added level of independence for someone afflicted with a neurologic disorder. The recovery of even a crude movement after a spinal cord injury can be meaningful to the patient's life, and seemingly trivial improvements can signal important scientific progress.

SELECTED READINGS

Aguayo AJ. Axonal regeneration from injured neurons in the adult mammalian central nervous system. In Cotman CW (ed). Synaptic plasticity. New York: Guilford Press, 1985:457.

Alvarez-Buylla A, Nottebohm F. Migration of young neurons in adult avian brain. Nature 1988;335:353.

Anderson DK, Hall ED. Pathophysiology of spinal cord trauma. Ann Emerg Med 1993;22:987.

Barron KD. Neuronal responses to axotomy: consequences and possibilities for rescue from permanent atrophy or cell death. In Seil FJ (ed). Neural regeneration and transplantation. New York: Alan R Liss, 1989:79.

Bayer SA, Yackel JW, Puri PS. Neurons in the rat dentate gyrus granular layer substantially increase during juvenile and adult life. Science 1982;216:890.

Benowitz LI, Routtenberg A. A membrane phosphoprotein associated with neural development, axonal regeneration, phospholipid metabolism, and synaptic plasticity. Trends Neurosci 1987;12:527.

Björklund A, Stenevi U. Neural grafting in the mammalian CNS. Fernström Foundation series, vol 5. Amsterdam: Elsevier, 1985.

Brown MC, Lunn ER, Perry VH. Consequences of slow wallerian degeneration for regenerating motor and sensory axons. J Neurobiol 1992;23:521.

Brown MC, Perry VH, Lunn ER, Gordon S, Heumann H. Macrophage dependence of peripheral sensory nerve regeneration—possible involvement of nerve growth factor. Neuron 1991;6:359.

Brushart TME. Motor axons preferentially reinnervate motor pathways. J Neurosci 1993;13:2730.

Caroni P, Schwab M. Two membrane protein fractions from rat central myelin with inhibitory properties for neurite growth and fibroblast spreading. J Cell Biol 1988;106:1281.

Carter DA, Bray GM, Aguayo AJ. Regenerated retinal ganglion cell axons can form well-differentiated synapses in the superior colliculus of adult hamsters. J Neurosci 1989;9:4042.

Cotman CW. Synaptic plasticity. New York: Guilford Press, 1985.

David S, Aguayo AJ. Axonal elongation into peripheral nervous system "bridges" after central nervous system injury in adult rats. Science 1981;214:931.

Defelipe J, Jones EG. Cajal's degeneration and regeneration of the nervous system (translated by RM May). New York: Oxford University Press, 1991.

Dunnett SB, Richards SJ. Neural transplantation: from molecular basis to clinical applications. Prog Brain Res 1990;82.

Fawcett JW. Intrinsic neuronal determinants of regeneration. Trends Neurosci 1992;15:5.

Freed CR, Breeze RE, Rosenberg NL, et al. Survival of implanted fetal dopamine cells and neurologic improvement 12 to 46 months after transplantation for Parkinson's disease. N Engl J Med 1992; 327:1549.

Gage FH, Buzsáki G. CNS grafting: potential mechanisms of action. In Seil FJ (ed). Neural regeneration and transplantation. New York: Alan R Liss, 1989:211.

Gage FH, Fisher LJ. Intracerebral grafting—a tool for the neurobiologist. Neuron 1991:1.

Gage FH, Kang UJ, Fisher LJ. Intracerebral grafting in the dopaminergic system: issues and controversy. Curr Opin Neurobiol 1991; 1:414.

Gage FH, Kawaja MD, Fisher LJ. Genetically modified cells: applications for intracerebral grafting. Trends Neurosci 1991;14:328.

Gash DM, Sladek JR Jr. Transplantation into the mammalian CNS. Prog Brain Res 1988;78.

Goldberger ME, Murray M. Patterns of sprouting and implications for recovery of function. Adv Neurol 1988;78:361.

Gorio A, Millesi H, Mingrino S. Posttraumatic peripheral nerve regeneration: experimental and clinical implications. New York: Raven Press, 1981.

Grafstein B, McQuarrie IG. Role of the nerve cell body in axonal regeneration. In Cotman CW (ed). Neuronal plasticity. New York: Raven Press, 1978:155.

Graziadei PPC, Graziadei GAM. The olfactory system: a model for the study of neurogenesis and axon regeneration in mammals. In Cotman CW (ed). Neuronal plasticity. New York: Raven Press, 1978:131.

Jewett DL, McCarroll JR. Nerve repair and regeneration: its clinical and experimental basis. St. Louis: CV Mosby, 1980.

Johnson EM, Taniuchi M, Distefano PS. Expression and possible function of nerve growth factor receptors on Schwann cells. Trends Neurosci 1988;11:299.

Kaplan MS, Bell DH. Mitotic neuroblasts in the 9 day old and 11 month old rodent hippocampus. J Neurosci 1984;4:1429.

Keirstead SA, Rasminsky M, Fukuda Y, et al. Electrophysiologic responses in hamster superior colliculus evoked by regenerating retinal axons. Science 1989;246:255.

Keirstead SA, Vidal-Sanz V, Rasminsky M, et al. Responses to light of retinal neurons regenerating axons into peripheral nerve grafts in the rat. Brain Res 1985;359:402.

Knüsel B, Beck KD, Winslow JW, et al. Brain-derived neurotrophic factor administration protects basal forebrain cholinergic but not nigral dopaminergic neurons from degenerative changes after axotomy in the adult rat brain. J Neurosci 1992;12:4391.

Koliatsos VE, Applegate MD, Knusel B, et al. Recombinant human nerve growth factor prevents retrograde degeneration of axotomized basal forebrain cholinergic neurons in the rat. Exp Neurol 1991;112:161.

Kromer LF. Nerve growth factor treatment after brain injury prevents neuronal death. Science 1987;235:214.

Lieberman AR. The axon reaction: a review of the principal features of perikaryal responses to axon injury. Int Rev Neurobiol 1971; 14:49.

Lundborg G, Dahlin LB, Danielsen N, et al. Nerve regeneration in silicone chambers: influence of gap length and of distal stump components. Exp Neurol 1982;76:361.

Maffei L, Carmignoto G, Perry VH, Candeo P, Ferrari G. Schwann cells promote the survival of rat retinal ganglion cells after optic nerve section. Proc Natl Acad Sci U S A 1990;87:1855.

Meiri KF, Pfenninger KH, Willard MB. Growth associated protein, GAP-43, a polypeptide that is induced when neurons extend axons, is a component of growth cones and corresponds to pp46, a major polypeptide of a subcellular fraction enriched in growth cones. Proc Natl Acad Sci U S A 1986;83:3537.

Morse JK, Wiegand SJ, Anderson K, et al. Brain-derived neurotrophic factor (BDNF) prevents the degeneration of medial septal cholinergic neurons following fimbria transection. J Neurosci 1993;13:4146.

Oorschot DE, Jones DG. Axonal regeneration in the mammalian central nervous system: a critique of hypotheses. Advances Anat Embryol Cell Biol 1990;119:1.

Paton JA, Nottebohm FN. Neurons generated in the adult brain are recruited into function circuits. Science 1984;225:1046.

Reier PJ, Anderson DK, Stokes BT. Neural tissue transplantation and CNS trauma: anatomic and functional repair of the injured spinal cord. J Neurotrauma 1992;9(supp 1):S223.

Reier PJ, Houlé JD. The glial scar: its bearing on axonal elongation and transplantation approaches to CNS repair. In Waxman SG (ed). Physiologic basis for functional recovery in neurological disease. New York: Raven Press, 1988:87.

Reynolds BA, Weiss S. Generation of neurons and astrocytes from isolated cells of the adult mammalian central nervous system. Science 1992;255:1707.

Schnell L, Schwab ME. Axonal regeneration in the rat spinal cord produced by an antibody against myelin-associated neurite growth inhibitors. Nature 1990;343:269.

Schwab ME. Myelin-associated inhibitors of neurite growth and regeneration in the CNS. Trends Neurosci 1990;13:452.

Siesjö BK. Basic mechanisms of traumatic brain damage. Ann Emerg Med 1993;22:959.

Sladek JR Jr, Gash DM. Neural transplants: development and function. New York: Plenum Press, 1984.

Spencer DD, Robbins RJ, Naftolin F, et al. Unilateral transplantation of human fetal mesencephalic tissue into the caudate nucleus of patients with Parkinson's disease. N Engl J Med 1992;327:1541.

Steward O. Principles of cellular, molecular, and developmental neuroscience. New York: Springer-Verlag, 1989.

Sunderlund S. Nerve injuries and their repair: a critical appraisal. Melbourne: Churchill Livingstone, 1991.

Thomas PK. Clinical aspects of PNS regeneration. Adv Neurol 1988;47:9.

Varon S, Hagg T, Manthorpe M. Neuronal growth factors. In Seil FJ (ed). Neural regeneration and transplantation. New York: Alan R Liss, 1989:101.

Widner H, Tetrud J, Rehncrona S, et al. Bilateral fetal mesencephalic grafting in two patients with parkinsonism induced by 1-methyl-4-phenyl-1,2,3,6-tetrahydropyridine (MPTP). N Engl J Med 1992;327:1556.

Williams LR, Varon S, Peterson GM, et al. Continuous infusion of nerve growth factor prevents basal forebrain neuronal death after fimbria fornix transection. Proc Natl Acad Sci U S A 1986;83:9231.

Yoshida K, Kakihana M, Chen LS, Ong M, Baird A, Gage FH. Cytokine regulation of nerve growth factor-mediated cholinergic neurotrophic activity synthesized by astrocytes and fibroblasts. J Neurochem 1992;59:919.

Neuroscience in Medicine, edited by P. Michael Conn.
J. B. Lippincott Company, Philadelphia © 1995

Chapter *33*

*C*linical Probes

NANCY C. ANDREASEN

Understanding of the perceptual, behavioral, cognitive, and emotional functions described in the previous chapters and the disease states described in this chapter has been facilitated by the development of clinical tools that are used for in vivo study of the human brain. These techniques include electroencephalography (EEG) and related neurophysiologic techniques, computerized tomography (CT), magnetic resonance (MR), single photon emission computed tomography (SPECT), and positron emission tomography (PET). These techniques have substantially expanded the capabilities of basic and clinical neuroscience during the past decade. Studies that were previously completed using animal models, human post-mortem tissue, or indirect measures such as peripheral metabolites or neuropsychological tests, are now complemented by an array of methods that permit examination of the human brain in its living and functioning state. Because of these techniques, it is possible to study directly human developmental anatomy, structure-function relationships, structure-disease relationships, physiologic and metabolic activity with experimentally controlled cognitive and pharmacologic manipulation, and the distribution and functional activity of neurotransmitter systems in the normal brain and during disease.

Electroencephalography and Related Electrophysiologic Techniques

A variety of electrophysiologic techniques are available for clinical and experimental use. These include EEG, polysomnography, and evoked potentials. The earliest of the electrophysiologic techniques, EEG, was first described by the psychiatrist Hans Berger in 1929. EEG, like other electrophysiologic techniques, reflects the summation of electrical activity throughout the brain. The EEG is recorded by placing multiple electrodes on the surface of the brain, using leads placed over frontal, temporal, parietal, and occipital regions. Because of their location on the scalp, these electrodes reflect a summation of electrical activity primarily derived from the electrical activity of the cortex, although the signal may also be influenced by action potentials deep within the brain that derive from the reticular activating system and the thalamus, particularly when the brain is at rest. During the waking state, the electrical activity measured by the EEG is desynchronous, with differential characteristic frequencies in some cortical electrodes.

The frequency range of the EEG is subdivided into four frequency bands: delta (<3 Hz), theta (4 to 7 Hz), alpha (8 to 13 Hz), and beta (>13 Hz). Alpha activity is typically seen in the occipital region when the subject has his eyes closed; anterior regions are characterized by faster beta activity. Theta and delta activity emerge during drowsiness and may reflect pathologic conditions such as damaged tissue (e.g., trauma, tumor) or delirium. Abnormalities in the frequency bands, such as the spike and wave discharges observed in epilepsy, may also be noted with an EEG. Special electrodes, such as nasopharyngeal leads, may be placed to detect possible focal abnormalities, such as those that occur in temporal lobe epilepsy, which may not be observed with traditional surface electrodes.

The standard EEG is usually obtained during waking and sleeping conditions. A sleeping EEG is desirable, because epileptiform activity may emerge during sleep. In addition, photic stimulation may also release epileptiform activity. The most characteristic epileptiform abnormality is the spike and slow wave pattern seen in patients who have grand mal seizures. Individual spikes, polyspikes, and sharp waves may also be observed. Although epileptiform abnormalities are not seen in all patients with seizure disorders, assessment for epilepsy is the commonest application of EEG. EEG may also be useful in assessing for tumors, dementia, delirium, encephalopathy, or encephalitis.

EEG patterns are usually read visually, but computer-based techniques have also been developed for the analysis of EEGs, including spectral analysis and brain mapping.

Polysomnography involves the application of EEG techniques to the study of sleep. Polysomnographic studies are conducted in sleep laboratories, where brain electrical activity is usually recorded during several all-night sessions. When the brain sleeps, a variety of synchronous patterns are observed that do not occur during the waking state. During sleep, the brain progresses through four stages, with periods of rapid eye movement (REM) activity interspersed. Polysomnography is used for evaluating a variety of sleep disorders, including chronic insomnia, hypersomnolence, and sleep-related respiratory impairments.

Evoked potentials involve measurement of stimulus-specific activities in individual brain regions that reflect the response of the brain to particular types of specific stimuli. Evoked potentials require the use of repeated stimuli and signal averaging to separate the response from the background EEG by improving the signal-to-noise ratio. Evoked potentials have been mapped for auditory, visual, and somatosensory pathways. Other specific types of evoked potentials have been observed during cognitive activity, such as the P300, a positive wave that occurs 300 milliseconds after the original stimulus, which involves identifying a rare target from among other nontarget stimuli.

Structural Neuroimaging Techniques

CT and MR are the two major structural imaging techniques. Both techniques permit visualization of the anatomic structure of the brain and its disruption in a variety of disease processes; MR also has potential applications for the study of brain physiology and biochemistry in addition to anatomy. CT, the oldest of the neuroimaging techniques, was developed by Hounsfield and Cormack, who were recognized with a Nobel Prize for this important clinical contribution. Its wide availability and relative ease of use make it the most accessible of the various neuroimaging techniques. Because it has many advantages over CT, MR has become an important clinical complement and will probably ultimately surpass CT. The relative strengths and weaknesses of these two structural imaging techniques are summarized in Table 33-1.

CT involves the use of ionizing radiation, while MR does not. MR is therefore considered to be a safer procedure, which is particularly desirable for the evaluation of children. CT has no other major risks and is relatively comfortable for the patient, but some patients experience claustrophobia during an MR scan because of the limited visual field produced by the magnetic coil that surrounds their heads and the extensive shielding that is placed around their bodies. Tissue resolution is far superior with MR, with the exception of bone, which cannot be seen with MR except for bone marrow. CT slices are limited primarily to transaxial views, but MR can acquire information in any plane. CT takes less time than MR imaging, making it more appropriate for patients who are likely to be restless, although MR scanning times are becoming steadily shorter. In addition to being a powerful anatomic probe, MR is also expanding its application to include the study of tissue physiology and metabolism, through the application of techniques such as magnetic resonance spectroscopy and echo planar imaging to permit measurement of cerebral blood flow. CT is less expensive than MR, and it is also widely available; MR usually is limited to university and urban centers.

Because the principles involved in producing CT images are common to all four neuroimaging modalities (i.e., CT, MR, SPECT, PET), a detailed description of the image acquisition and processing involved in a CT study provides a useful foundation for obtaining a general grasp of the basic principles of neuroimaging.

CT images are generated by passing an x-ray beam through the body and measuring its degree of attenuation. Cerebrospinal fluid (CSF), bone, and brain tissue vary in their ability to attenuate x-rays, with bone producing the greatest attenuation and CSF the least.

The degree of attenuation is picked up by detectors on the opposite side and mapped slice by slice. Each slice is subdivided into a grid of tiny cubes, referred to as volume elements or *voxels*; the extent of attenuation within each of these voxels can be measured numerically as a tissue density number. CT is called *computed* tomography because high-speed computers are required to collect and process the large volume of information generated and to turn it into a picture that visually displays the degree of attenuation within each tiny chunk of tissue; the term *tomography* refers to the fact that the brain is "cut" and mapped in a series of slices. The density numbers from each of the voxels are converted to gray scale values, with low numbers (e.g., CSF) coded as black and high numbers (e.g., bone) coded as white, and the whole is visually displayed in a dot matrix of picture elements or *pixels*. A variety of reconstruction and filtering techniques are applied to smooth the boundaries between pixels and make the picture more visually appealing.

All four of the existing neuroimaging techniques depend on this same set of principles: applying some stimulus to collect information about tissue characteristics, storing the information in the format of slices, subdividing the slices into a grid of voxels, and using high-speed computers and filtering techniques to turn this information into a visual image that is a dot matrix of pixels. The major difference between CT, MR, SPECT, and PET is the type of stimulus used to collect information about tissue. For CT, an x-ray beam is attenuated. For MR, the hydrogen protons in the brain are perturbated by being placed in a magnetic field and stimulated by a radiofrequency signal; subsequently, their rate of return to their original condition, called the *relaxation time*, is measured and expressed as a signal intensity. For SPECT and PET, radioactive tracers are injected, localized in particular brain regions on the basis of physiologic or biochemical activity, and the number of counts in the site of localization or uptake is measured. SPECT uses tracers that emit single photons, while PET uses positron emitters.

The various differences in CT, MR, SPECT, and PET turn on "what goes in," which affects the type of information that can be derived. Obtaining a CT scan is a relatively simple process, because it depends only on the attenuation of an x-ray beam to generate information about tissue characteristics. In general, the other three neuroim-

Table 33-1

COMPARISON OF COMPUTED TOMOGRAPHY AND MAGNETIC RESONANCE IMAGING

Characteristic	Computed Tomography	Magnetic Resonance Imaging
Risks	Ionizing radiation	No ionizing radiation
Resolution	1 cm	.5–1.5 mm
Tissue Discrimination	Fair	Superb
Flexibility	Single plane	Multiple planes
	Anatomic structure	Anatomic structure, spectroscopy, blood flow
Time	Seconds to minutes	1 to 40 min
Cost	$200–$500 in 1994	Approximately twice CT in 1994

aging techniques are far more complex than CT, because they measure tissue characteristics in a more complicated manner.

MR, the other major structural imaging technique, is a particularly powerful clinical probe. Purcell and Bloch, who discovered the basic principles of nuclear magnetic resonance, were also awarded a Nobel Prize for this achievement in 1952. Magnetic resonance imaging involves an extension of these principles to the clinical situation. Unlike the CT signal, which has a single component, the MR signal is produced by four components, which are summarized in Table 33-2. A clinical study is obtained by placing the patient's head (or body) in a relatively high-field-strength magnetic field (typically 1.5 Tesla), which has the effect of concentrating the magnetic moment produced by the individual protons in the tissue so that it is large enough to be measurable. The net magnetic moment is then deflected or tipped by sending radio-frequency signals that excite the tissue; after excitation, the protons relax and return to their original position. The MR signal is produced through the combination of flow (typically an unimportant component for brain imaging), hydrogen proton density, and T1 and T2 relaxation times. The various contributions of these components to the MR signal can be manipulated by varying the timing of the excitations, called pulse sequences.

Depending on the weighting of the various components of the MR signal, MR images are of three general types: proton density, T1 weighted, or T2 weighted. The pulse sequence variables that are manipulated to produce these different types of images are known as the echo time (TE) and the repetition time (TR). The associations among TE, TR, and the three components of the MR signal are summarized in Figure 33-1. A short TE and long TR produce a proton-density image; a short TE and a short TR produce a T1-weighted image; and a long TE and a long TR produce a T2-weighted image. The three types of images create pictures of the brain that are visually quite different from one another. The proton-density image has similar color shadings for CSF and white matter and makes gray matter relatively bright. The T1-weighted image shows the best tissue discrimination, with CSF shown as very dark, gray matter as lighter, and white matter as relatively bright. The T2-weighted image

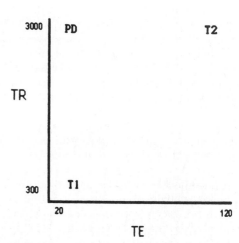

SPIN ECHO

Figure 33-1

Representation of the associations among echo time (TE) and repetition time (TR) and the three components of the MR signal (i.e., proton density, T1, and T2). A short TE and a long TR produce a proton-density image; a short TE and a short TR produce a T1-weighted image; and a long TE and a long TR produce a T2-weighted image.

makes CSF appear very bright and has the poorest discrimination between gray and white matter. Examples of proton-density, T1-weighted, and T2-weighted images appear in Figure 33-2.

The capacity to vary the components of the MR signal and the type of image generated makes MR a very complex and flexible clinical probe. Not only do these different types of scanning sequences produce a different gray scale for different tissue types, but they also have different values for identifying pathologic tissue. This difference is illustrated in Figure 33-3, which compares T1- and T2-weighted images of a child with a posterior fossa tumor. The tumor is so large that it is prominent on both images. On the T1-weighted image, the general brain anatomy is well visualized (e.g., corpus callosum, brain stem, cerebral hemispheres, cerebellum), but the anatomy is poorly visualized on the T2-weighted image. However, the T2-weighted image shows the tumor as a prominent bright signal area within the posterior fossa. T2-weighted sequences are particularly good for identifying areas of tissue pathology, such as multiple sclerosis plaques or areas of infarction. A T2-weighted image of multiple sclerosis plaques, which are poorly visualized with CT, is shown in Figure 33-4.

Clinicians ordering a CT scan should indicate whether they wish a contrast agent, which enhances tissue discrimination, to be used and the nature of the question to be answered, such as "Is there any evidence for a space-occupying lesion or other pathology in prefrontal regions?" Other options are limited, because CT can yield only limited information. Because of the greater complexity of MR, clinicians must be much clearer about the

Table 33-2
COMPONENTS OF THE MAGNETIC RESONANCE SIGNAL

Flow velocity: Flow measure, with a value of one in stationary tissue

Proton density: A measure of the number of hydrogen nuclei

T1 relaxation time: An exponential growth constant that reflects return of nuclei to their resting state in the z-axis

T2 relaxation time: An exponential decay constant that reflects loss of signal strength as dephasing of the spin occurs

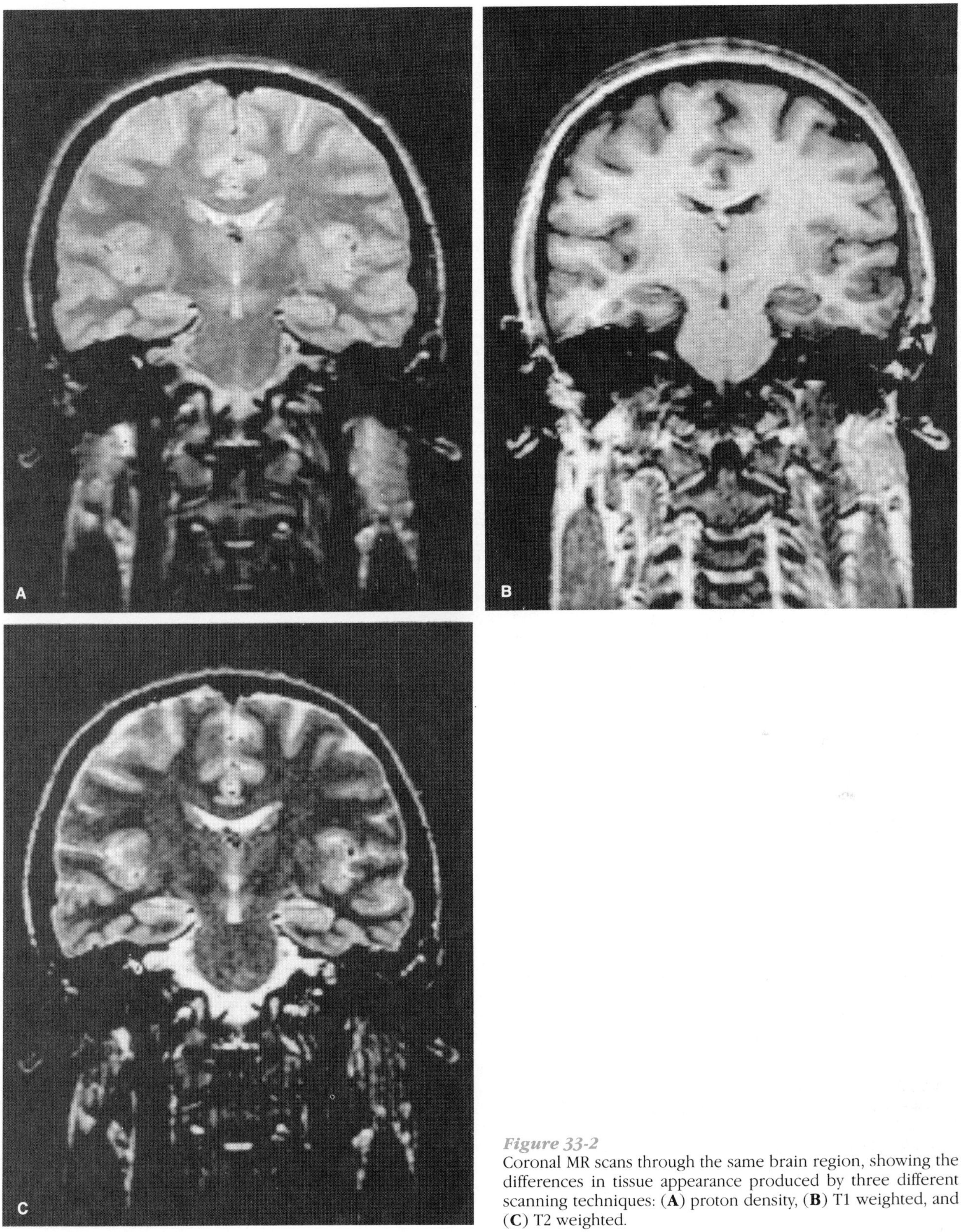

Figure 33-2
Coronal MR scans through the same brain region, showing the differences in tissue appearance produced by three different scanning techniques: (**A**) proton density, (**B**) T1 weighted, and (**C**) T2 weighted.

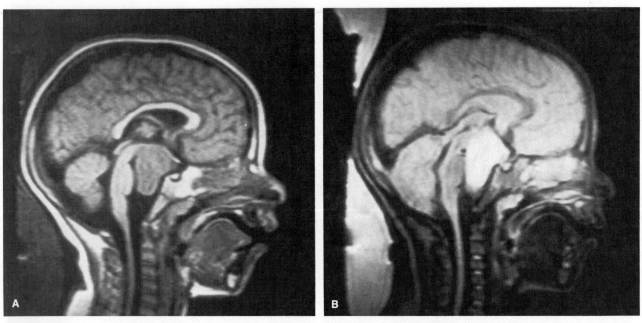

Figure 33-3

(**A**) A T1-weighted and (**B**) T2-weighted image of a child with a large tumor in the posterior fossa. The tumor appears as a centrally located gray mass in the T1-weighted image, and the remainder of the brain tissue is relatively clearly defined, especially the corpus callosum and the general outlines of the brain, cerebellum, brain stem, and spinal cord. In the T2-weighted image, the midline tumor appears quite bright, and specific aspects of brain anatomy are poorly delineated.

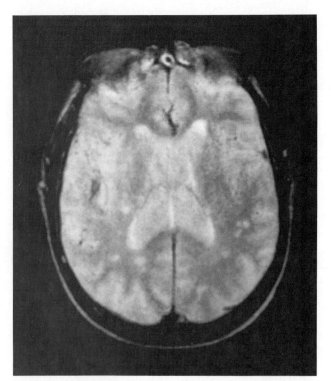

Figure 33-4

A transaxial slice, obtained using a T2-weighted sequence, shows ventricular enlargement and multiple small bright areas around the ventricles and in the white matter. This type of image is consistent with multiple sclerosis or multiple small vascular infarcts.

question that is being asked to ensure that the appropriate scanning sequence is obtained. In general, if the goal is to obtain excellent gray matter, white matter, and CSF discrimination and to visualize anatomy well, a T1-weighted sequence should be chosen. If the goal is to identify areas of tissue pathology, a T2-weighted sequence is chosen. The clinician may wish to specify a particular slice plane (e.g., coronal) or may request multiple views. Sometimes several different scanning sequences are chosen. Rapid scanning sequences permit the entire brain to be mapped in almost cubic 1-mm voxels in as little as 10 minutes using a T1-weighted sequence. Scanning sequences of this type permit three-dimensional reconstruction of sulcal and gyral patterns and allow internal resampling in any plane, aspects that permit MR to be used for precise lesion or structure localization and measurement, facilitating clinical and research purposes such as morphometric studies of the normal and diseased brain.

Clinical Applications of Computed Tomography and Magnetic Resonance Imaging

The structural imaging techniques have a broad range of clinical applications. In general, CT is used if a rapid and inexpensive screening procedure is needed to evaluate the possibility of gross lesions, and MR is used to answer questions that require finer detail in visualization or a more sensitive evaluation of tissue pathology. Applications include assessment for degenerative processes (e.g., sulcal enlargement, ventricular enlargement in dementias), demyelinating diseases (e.g., multiple sclerosis), ischemia and infarction, tumors, developmental anomalies and congenital malformations, hemorrhage, and infections.

MR technology is expanding and developing applications for the study of metabolism and blood flow. These rapidly developing areas have also made MR a tool that can be used to explore functional brain activity, a domain that has been primarily covered by PET and SPECT. MR spectroscopy has focused on the measurement of hydrogen and phosphorous spectra. For example, measurement of the ^{31}P spectrum permits determination of the concentration of substances such as adenosine triphosphate, phosphomonoesters, phosphodiesters, phosphocreatinine, or inorganic phosphate. Depending on the relative concentration of such tissue components, inferences can be made concerning tissue buildup versus breakdown in the lipid-rich membranes of nerve tissue. Because monoesters are building blocks and diesters are breakdown products, a different proportion of these substances may provide information concerning the presence of neurodegenerative disorders, such as Alzheimer's disease. MR spectroscopy is also being developed to measure drug concentrations in the brain, using spectra for atoms such as fluorine.

Another potential application of MR is the study of cerebral blood flow. Using echo planar imaging, regional blood flow can be measured quantitatively, much as is done through SPECT or PET, but without exposure to ionizing radiation or other invasive techniques.

Functional Imaging Techniques

SPECT and PET are the two major functional imaging techniques. Both are used to evaluate brain physiology, metabolism, and neurochemistry. Both techniques are relatively new, and they have been used primarily as research tools, although clinical applications also exist. SPECT is the newer and simpler of the two techniques, although its historic roots go back many years. The foundations of a method for measuring cerebral blood flow in vivo were laid by Kety and Schmidt, who measured the cellular uptake of nitrous oxide by monitoring the difference between arterial input and venous output during the 1940s. Their technique could be used only for the whole brain, but subsequent developments led to the use of small gamma detectors on the surface of the brain (i.e., *cortical probes*) to produce a surface map of cerebral blood flow using new nondiffusible tracers (e.g., xenon 133). The cortical probe technique, widely used during the 1970s, was surpassed through the development of tomographic equipment during the 1980s. With the development of these instruments, the era of SPECT arrived.

Two types of tracers are used in current SPECT studies: *dynamic tracers* (e.g., xenon 133) and *static tracers* (e.g.,

^{99}TC-HMPAO). Each of these tracers or techniques has inherent advantages and disadvantages, which are summarized in Table 33-3. A dynamic tracer diffuses into tissue rapidly, but it also washes out rapidly. Its washout curve can be measured, and a quantitative estimate of regional cerebral blood flow can be determined. Existing dynamic tracers such as xenon have a relatively low energy, a factor that reduces the quality of resolution to the range of 2 to 3 cm. Because dynamic tracers are rapidly washed out of the brain, they are suitable for repeated back to back studies, making them adaptable to mapping cognitive functions through the use of paired or multiple studies, much as is done in the ^{15}O-water PET technique. The higher-energy static tracers, such as ^{99}Tc-HMPAO, are thought to be removed from the blood in a single first-pass extraction, taken up into cells, and statically fixed there for as long as 24 hours. Their static characteristics make quantitative measurements of cerebral blood flow difficult, because input and output cannot be dynamically measured; however, their higher energy permits improved resolution (i.e., in the 8-mm range, compared with the cruder 2- to 3-cm range of the dynamic tracers). In general, imaging time is longer for the static tracers than for the dynamic tracers, approximately 30 vs. 5 minutes respectively. Although some innovative strategies have been developed recently to adapt static tracers to the study of effects of drugs on blood flow or of human cognition, these techniques have been used primarily to measure regional blood flow in a single study, typically the resting state.

Although most SPECT studies have focused on the use of tracers to detect blood flow, SPECT is also adaptable to the study of neuroreceptors. Several agents have been developed that label neuroreceptor systems in the brain. QNB, a muscarinic cholinergic agonist, is available for the study of the cholinergic system, and a D$_2$ antagonist has been developed for the study of the dopamine system.

PET is the other major functional imaging technique. PET is differentiated from SPECT in several ways. The tracers used in PET are positron emitters, while those used in SPECT emit single photons. In the case of PET, the positron-emitting tracer, attached to some informative label such as deoxyglucose, is taken up in a physiologically active brain area; the positron then collides with an electron, producing an "annihilation event" and the formation of two 5 11-keV photons, as predicted by the law of the conservation of energy. In both cases, the tracer is taken into brain tissue, and the imaging process generates a picture of its regional distribution, providing an index of blood flow, metabolic activity, and neurochemical activity. The picture is produced through the detection of the photons by crystal detectors that surround the brain; the simplest level of detector is the rotating *gamma camera* used in some types of SPECT studies, and the most elegant consists of multiple rings of small detectors used in the most advanced PET equipment.

Although the tracers used in SPECT have a long half-life (i.e., hours to days), those used in PET have a short half-life. The most widely used for brain imaging studies are fluorine 18 (110 minutes), carbon 11 (20 minutes),

Table 33-3

Table 33-3

COMPARISON OF CURRENT STATIC AND DYNAMIC BLOOD FLOW AGENTS USED IN SINGLE PHOTON EMISSION COMPUTED TOMOGRAPHY

Characteristic	Dynamic Tracers*	Static Tracers†
Risks	Ionizing radiation	Ionizing radiation
Resolution	2–3 cm	0.8–1.2 cm
Tissue discrimination	Poor	Fair to good
Flexibility	Transaxial plane (multiple back-to-back scans in a single study)	All planes (typically single scan)
Applications	Cognitive challenge studies	Single condition studies (cognitive challenge possible with dual-injection technique)
Quantification	Tissue blood flow (g/mL/min)	Tissue blood flow (ratio measures)
Time	5–10 min/scan	30–60 min/scan
Cost	Approximately $400 in 1992	Approximately $500 in 1992

*Xenon 133 is an example.
†^{199}Tc-HMPAO is an example.

and oxygen 15 (2 minutes). Tracers with a longer half-life permit a longer imaging time and improved resolution. Fluorodeoxyglucose (FDG), labeled with ^{18}F, is a widely used PET tracer that permits evaluation of glucose utilization and therefore serves as an indicator of tissue metabolism. At the opposite extreme, ^{15}O–H_2O, a tracer that can be used to measure cerebral blood flow, produces images with a relatively lower resolution.

Each of these tracers provides a direct or indirect index of tissue metabolism, and each has its own inherent advantages and disadvantages. Typically, an FDG study takes 1 to 2 hours, can be done only once, and provides a high-resolution image of tissue metabolism. The ^{15}O–H_2O "water" study takes only a few minutes, images a 40-second acquisition time window, is used for as many as 10 serial back to back studies, but has relatively poorer tissue resolution. The ^{15}O–H_2O technique is especially useful for taking multiple "snapshots" of the brain while it performs a series of carefully selected cognitive tasks that can be separated from one another through image subtraction techniques, permitting the mapping of cerebral cognitive functions.

Clinical Applications of Single Photon Emission Computed Tomography and Positron Emission Tomography

SPECT and PET have been used extensively as research tools to study normal brain physiology and metabolic function, but their clinical applications are still relatively modest. The primary uses of SPECT and PET have been to map cognitive and neuroreceptor systems within the brain and to study the effects of gender and aging on brain physiology.

Studies have consistently shown effects of gender, aging, and handedness. Cerebral blood flow and metabo-

lism steadily decreases with age, probably as a consequence of underlying loss in brain tissue, especially gray matter. Women appear to have higher rates of blood flow and metabolism than males, perhaps because of their smaller cranial cavities, causing gray matter to be more tightly packed within the brain. D_2 receptors have been reported to decrease as a consequence of the aging process. Several other factors influence cerebral blood flow or metabolism. Chronic nicotine use produces decreases in cerebral blood flow, probably as a consequence of vasoconstriction, as does caffeine. Anxiety produces a U-shaped curve, producing high levels of flow with low levels of anxiety, a steady reduction of flow as anxiety increases, and a return to high flow when anxiety reaches extreme levels.

The studies of cerebral blood flow and metabolism have been used to map functional areas of activity in the human brain. The application of PET to the study of cognitive neuroscience is slowly yielding the functional equivalent of Brodmann's maps. The earliest studies with FDG used simple stimuli such as visual input or language production. These early single studies using FDG produced a map suggesting the existence of regional processing centers and led for a time to models in which these were assumed to be linked serially to one another. The FDG approach has been complemented by the more elegant "water" paradigm, which permits the serial dissection of cognitive operations involved in related tasks. Early studies involved dissection of components of the visual system, showing that different areas of the visual cortex could be activated by different types of stimuli. Later studies using the water paradigm have focused on increasingly complex problems, such as the mechanisms of attention, language, volition, or pain perception. The results of most of these studies are consistent with extensive distributed networks of activated regions that interlinked with one another through complex parallel circuits. Water studies especially lend themselves to evaluation of the distributed parallel processing model of cognitive function.

SPECT and PET have been used to study a variety of disease conditions. Both techniques permit the visualization of areas of ischemia or infarction, although these techniques are rarely needed to make a diagnosis of stroke, limiting the clinical applications of functional imaging for this type of disease process. PET is used clinically to identify epileptic foci preoperatively in cases of intractable seizures. It may sometimes be useful for evaluating dementias, because reports have suggested that patients with Alzheimer's disease have a characteristic decrease in activity in posterior temporal or parietal regions. PET has also been useful in the study of Huntington's disease, consistently demonstrating patterns of hypometabolic activity in the basal ganglia in asymptomatic persons and often demonstrating such abnormalities in symptomatic persons at risk for Huntington's disease. Clinical applications may also be developing for the evaluation of Parkinson's disease, tumors, and mental illnesses such as schizophrenia or mood disorders. Metabolic abnormalities have been observed in alcohol and substance abuse and in patients with acquired immunodeficiency syndrome.

Conclusion

Electrophysiology and neuroimaging techniques serve as basic science tools and as clinical probes in clinical neuroscience. They provide the only methods for studying the human brain in vivo. They have provided substantial knowledge about the fundamental issues in normal brain structure and function, such as the effects of gender and aging, brain regions involved in various cognitive operations, and neuroreceptor distribution. Some techniques, such as MR, already have extensive clinical applications. As the science of neuroimaging matures, the clinical applications for the other techniques promise to expand as well.

SELECTED READINGS

Andreasen NC. Brain imaging: applications in psychiatry. Washington, DC: American Psychiatric Press, 1988.

Belliveau JW, Kennedy DN, McKinstry RC, et al. Functional mapping of the human visual cortex by magnetic resonance imaging. Science 1991;254:716.

Bradley WG, Bydder G. MRI atlas of the brain. New York: Raven Press, 1990.

Brant-Zawadzki M, Norman D. Magnetic resonance imaging of the central nervous system. New York: Raven Press, 1987.

Daniels DL, Houghton VM, Naidich TP. Cranial and spinal magnetic resonance imaging: an atlas and guide. New York: Raven Press, 1987.

Frost JJ, Wagner HN (eds). Quantitative imaging: neuroreceptors, neurotransmitters, and enzymes, New York: Raven Press, 1990.

Hounsfield GN. Computerized transverse axial scanning (tomography): Part I. Description of system. Br J Radiol 1973;46:1016.

Kety SS, Schmidt CF. The determination of cerebral blood flow in man by the use of nitrous oxide in low concentrations. Am J Physiol 1945;143:53.

Kooi KA: Fundamentals of electroencephalography. New York: Harper & Row, 1971.

Pettegrew JW. NMR: principles and applications to biomedical research. New York: Springer-Verlag, 1989.

Phelps ME, Mazziotta JC, Schelbert HR. Positron emission tomography and autoradiography principles and applications for the brain and heart. New York: Raven Press, 1986.

Reivich M, Abass A (eds). Positron emission tomography. New York: Alan R Liss, 1985.

Index

Page numbers followed by *f* indicate figures; page numbers followed by *t* indicate tables.